Family name	*Functional group structure*[a]	*Simple example*	*Name ending*
Carbonyl, $-\overset{:O:}{\overset{\|}{C}}-$			
Aldehyde	$-\overset{\vert}{\underset{\vert}{C}}-\overset{:O:}{\overset{\|}{C}}-H$	$H_3C-\overset{O}{\overset{\|}{C}}-H$	*-al* Ethanal (Acetaldehyde)
Ketone	$-\overset{\vert}{\underset{\vert}{C}}-\overset{:O:}{\overset{\|}{C}}-\overset{\vert}{\underset{\vert}{C}}-$	$H_3C-\overset{O}{\overset{\|}{C}}-CH_3$	*-one* Propanone (Acetone)
Carboxylic acid	$-\overset{\vert}{\underset{\vert}{C}}-\overset{:O:}{\overset{\|}{C}}-\ddot{\underset{\cdot\cdot}{O}}H$	$H_3C-\overset{O}{\overset{\|}{C}}-OH$	*-oic acid* Ethanoic acid (Acetic acid)
Ester	$-\overset{\vert}{\underset{\vert}{C}}-\overset{:O:}{\overset{\|}{C}}-\ddot{\underset{\cdot\cdot}{O}}-\overset{\vert}{\underset{\vert}{C}}-$	$H_3C-\overset{O}{\overset{\|}{C}}-O-CH_3$	*-oate* Methyl ethanoate (Methyl acetate)
Amide	$-\overset{\vert}{\underset{\vert}{C}}-\overset{:O:}{\overset{\|}{C}}-\ddot{N}H_2,$	$H_3C-\overset{O}{\overset{\|}{C}}-NH_2$	*-amide* Ethanamide (Acetamide)
	$-\overset{\vert}{\underset{\vert}{C}}-\overset{:O:}{\overset{\|}{C}}-\underset{\vert}{\ddot{N}}-H,$		
	$-\overset{\vert}{\underset{\vert}{C}}-\overset{:O:}{\overset{\|}{C}}-\underset{\vert}{\ddot{N}}-$		
Carboxylic acid chloride	$-\overset{\vert}{\underset{\vert}{C}}-\overset{:O:}{\overset{\|}{C}}-Cl$	$H_3C-\overset{O}{\overset{\|}{C}}-Cl$	*-oyl chloride* Ethanoyl chloride (Acetyl chloride)
Carboxylic acid anhydride	$-\overset{\vert}{\underset{\vert}{C}}-\overset{:O:}{\overset{\|}{C}}-\ddot{\underset{\cdot\cdot}{O}}-\overset{:O:}{\overset{\|}{C}}-\overset{\vert}{\underset{\vert}{C}}-$	$H_3C-\overset{O}{\overset{\|}{C}}-O-\overset{O}{\overset{\|}{C}}-CH_3$	*-oic anhydride* Ethanoic anhydride (Acetyl anhydride)

WARNING! If this envelope is opened, this book is not returnable.

Notice to users: Do not open the package containing the CD-ROM until you have read and agreed to this licensing agreement. You will be bound by the terms of this agreement if you open the sealed CD-ROM or otherwise signify acceptance of this agreement. If you do not agree to the terms contained in this agreement, return the entire product, along with any receipt to the store where the product was purchased.

Limited Warranty: The warranty for the media on which the software is provided is for ninety (90) days from the original purchase and valid only if the packaging for the Software was purchased unopened. If, during that time, you find defects in the workmanship or material, the Publisher will replace the defective media. The Publisher provides no other warranties, expressed or implied, including the implied warranties of merchantability or fitness for a particular purpose, and shall not be liable for any damages, including direct, special, indirect, incidental, consequential, or otherwise.

For Technical Support:
Email: support@kdc.com
Voice: 1-800-423-0563
Fax: 1-606-647-5045

at your first log-on.

For tech support, e-mail:
wp-support@infotrac-college.com

Fundamentals of Organic Chemistry

Fourth Edition

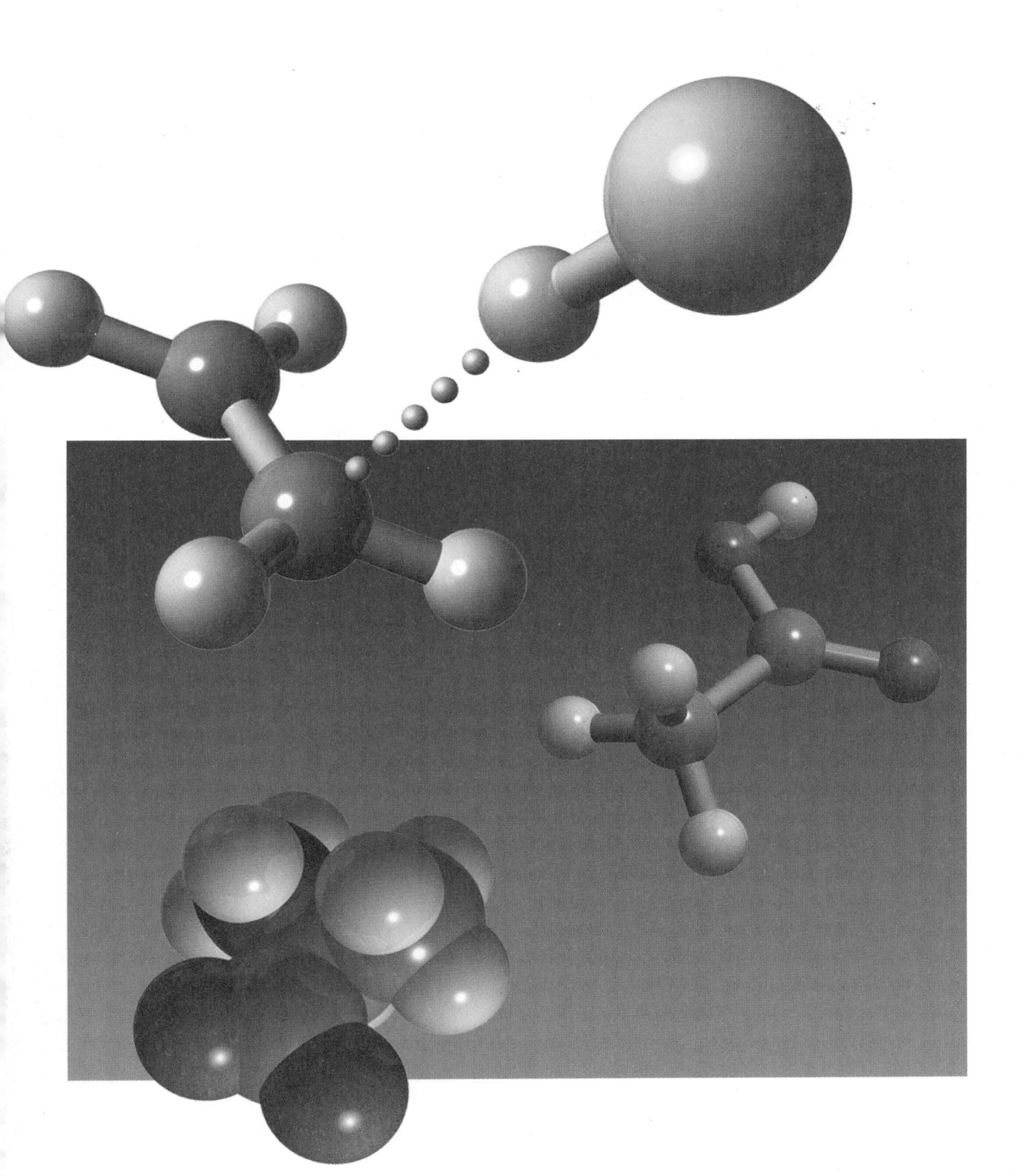

Fundamentals of Organic Chemistry

Fourth Edition

John McMurry
Cornell University

Brooks/Cole Publishing Company
I(T)P™ An International Thomson Publishing Company

Pacific Grove • Albany • Belmont • Bonn • Boston • Cincinnati • Detroit • Johannesburg • London • Madrid • Melbourne
Mexico City • New York • Paris • Singapore • Tokyo • Toronto • Washington

A division of International Thomson Publishing Inc.
I(T)P The ITP logo is a registered trademark under license.

For more information, contact:

BROOKS/COLE PUBLISHING COMPANY
511 Forest Lodge Road
Pacific Grove, CA 93950
USA

International Thomson Publishing Europe
Berkshire House 168-173
High Holborn
London WC1V 7AA
England

Thomas Nelson Australia
102 Dodds Street
South Melbourne, 3205
Victoria, Australia

Nelson Canada
1120 Birchmount Road
Scarborough, Ontario
Canada M1K 5G4

International Thomson Editores
Seneca 53
Col. Polanco
11560 México, D. F., México

International Thomson Publishing GmbH
Königswinterer Strasse 418
53227 Bonn
Germany

International Thomson Publishing Asia
221 Henderson Road
#05–10 Henderson Building
Singapore 0315

International Thomson Publishing Japan
Hirakawacho Kyowa Building, 3F
2-2-1 Hirakawacho
Chiyoda-ku, Tokyo 102
Japan

Printed in the United States of America.

10 9 8 7 6 5 4 3

Library of Congress Cataloging-in-Publication Data

McMurry, John.
Fundamentals of organic chemistry / John McMurry. — 4th ed.
p. cm.
Includes index.
ISBN 0-534-35215-4
1. Chemistry, Organic. I. Title.
QD251.2.M4 1998 97-27737
547—dc21 CIP

Brief Contents

Contents

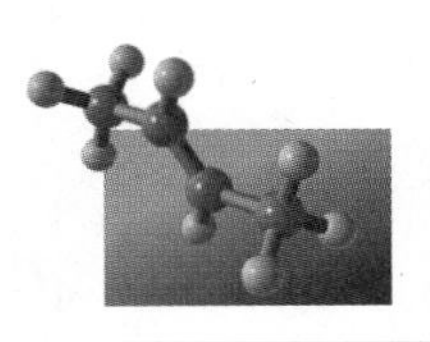

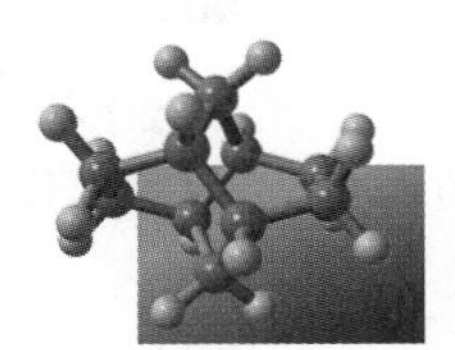

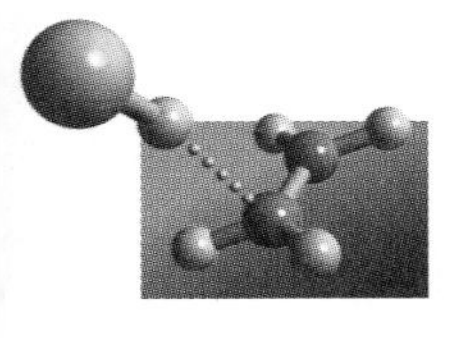

3 Alkenes: The Nature of Organic Reactions 76

4 Alkenes and Alkynes 108

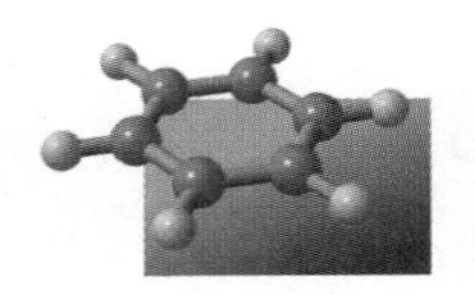

5 Aromatic Compounds 150

6 Stereochemistry 183

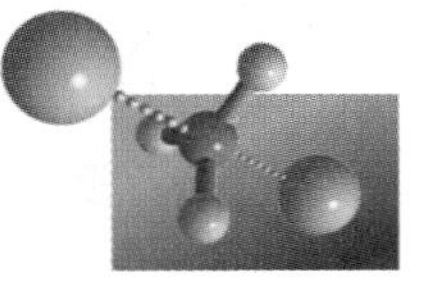

7 Alkyl Halides 217

8 Alcohols, Ethers, and Phenols 251

9 Aldehydes and Ketones: Nucleophilic Addition Reactions 281

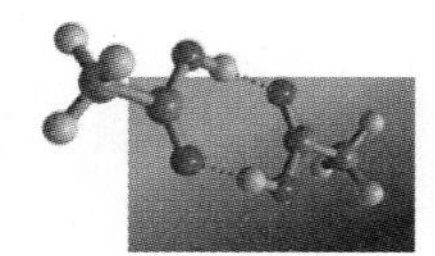

10 Carboxylic Acids and Derivatives 308

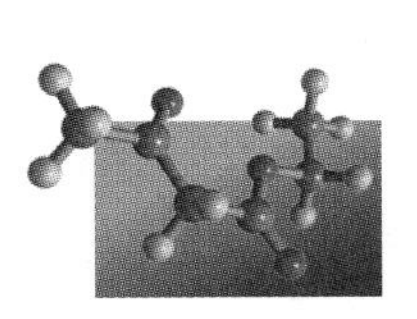

11 Carbonyl Alpha-Substitution Reactions and Condensation Reactions 352

12 Amines 380

13 Structure Determination 406

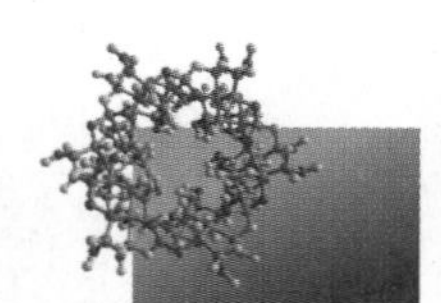

14 Biomolecules: Carbohydrates 440

15 Biomolecules: Amino Acids, Peptides, and Proteins 474

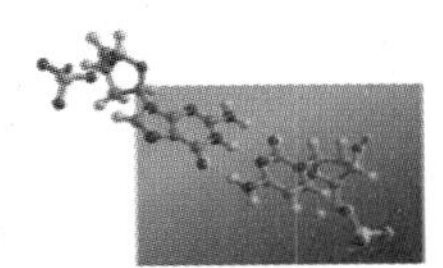

16 Biomolecules: Lipids and Nucleic Acids 507

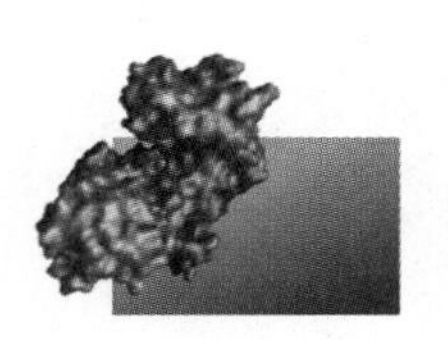

17 The Organic Chemistry of Metabolic Pathways 543

Reaction Mechanisms

Interludes

Preface

I wrote this book for a very simple reason: I love writing. I get great satisfaction from taking a complicated subject, turning it around until I see it from a new angle, and then explaining it in simple words. I write to explain chemistry to students today the way I wish it had been explained to me years ago.

The enthusiastic response of both students and faculty to the three previous editions has been very gratifying and suggests that this book has served students well. I hope you will find that this fourth edition of *Fundamentals of Organic Chemistry* builds on the strengths of the first three and serves students even better. I have made every effort to make this fourth edition as effective, clear, and readable as possible; to show the beauty and logic of organic chemistry; and to make the subject enjoyable to learn.

Organization and Teaching Strategies

The primary organization of this book is by functional group, beginning with the simple (alkanes) and progressing to the more complex. Within this primary organization, however, there is also an emphasis on explaining the fundamental mechanistic similarities of reactions. This emphasis is particularly evident in the chapters on carbonyl-group chemistry (Chapters 9–11), where mechanistically related reactions like the aldol and Claisen condensations are covered together. Memorization is minimized and understanding maximized with this approach.

Reaction Mechanisms

In the first edition, I introduced an innovative vertical format for explaining reaction mechanisms that met with an enthusiastic response. Now set off by color panels, mechanisms shown in this format have the reaction steps printed vertically while the changes taking place in each step are explained next to the reaction arrow. This format allows the reader to see what is occurring at each step in a reaction without having to jump back and forth between structures and text. Pages 115 and 240 show examples.

Basic Learning Aids

Clarity of explanation and smoothness of information flow are crucial requirements for any textbook. In writing and revising this text, I consistently aim for summary sentences at the beginning of paragraphs, lucid explanations, and smooth transitions between paragraphs and between topics. New concepts are introduced only when they are needed—not before—and are immediately illustrated with concrete examples. Frequent cross-references to earlier material are given, and numerous summaries are provided to draw information together, both within chapters and at the ends of chapters. In addition, the back of this book contains a wealth of material helpful for learning organic chemistry, including a large glossary, an explanation of how to name polyfunctional organic compounds, and answers to most in-text problems. For still further aid, an accompanying *Study Guide and Solutions Manual* gives summaries of reaction mechanisms and of methods for preparing functional groups, a list of named reactions with examples, and a list of the uses of important reagents.

Features of the Fourth Edition

The primary reason for preparing a new edition is to keep the book up-to-date, both in its scientific coverage and in its pedagogy. My overall aim has been to retain and refine the features that made earlier editions so successful, while adding new ones.

- **Full color** has now been added throughout the text, both for its visual appeal and for its pedagogical value in highlighting the reacting parts of molecules.
- **The writing** has again been revised at the sentence level, paying particular attention to such traditionally difficult subjects as stereochemistry and nucleophilic substitution reactions.
- **The artwork** has been redone, and many new computer-generated models have been added. Figures frequently present structures in several different formats, side by side, so that students learn structures thoroughly and become used to the various ways chemists graphically represent their work. Look at pages 14 and 63 to see some examples.
- **Stereo views** of computer-generated ball-and-stick molecular models have been added as an aid for three-dimensional perception. A stereo viewer is bound into the back of the book. Even the problems make use of stereo views. Some examples appear on pages 184 and 230.
- **The organic chemistry of metabolic pathways** is presented in Chapter 17. Several of the most important pathways—glycolysis and the citric acid cycle, for example—are dissected and analyzed according to the organic reaction mechanisms by which the various steps occur. This chapter will be of particular interest to the large number of students in health sciences who traditionally take this organic chemistry course.
- **Interlude boxes** at the end of each chapter present interesting applications of organic chemistry relevant to the main chapter subject. Including several topics from science, medicine, and day-to-day life, these

applications enliven and reinforce the material presented in each chapter. Some Interlude topics address environmental concerns and examine popular assumptions about chlorinated organic compounds or chemical toxins in our food and water. Other Interludes discuss such topics as antibiotics and antiinflammatory agents.

- **Biologically important organic reaction mechanisms** are specially identified by the use of a margin icon. Students often wonder about what topics are "important," and this icon helps students in the biological sciences answer that question. See pages 293 and 366, for example.
- **Spectra** are all new. They have been redrawn for clarity and accuracy. Some examples appear on pages 420 and 427.
- **New problems** have been added at the end of each chapter. Some are a new kind of problem called **Visualizing Chemistry,** in which students are challenged to make the connection between typical line-bond drawings and computer-generated molecular models. See pages 35 and 215, for example.

To facilitate the changes outlined above, some material from the previous edition has been compressed, several reactions (such as ozonolysis) that have little relevance to biological chemistry have been deleted, and other material has been rearranged.

Acknowledgments

I sincerely thank the many people whose help and suggestions were so valuable in preparing this fourth edition, particularly Phyllis Niklas, Jamie Sue Brooks, Beth Wilbur, and Harvey Pantzis. It is a real pleasure to work with such consummate professionals.

Among the reviewers providing valuable comments were Jean C. Beckman, University of Evansville; Claudia P. Cartaya, Appalachian State University—Rainking Science Center; Mildred V. Hall, Pennsylvania State University—Dubois Campus; Miroslav Krumpolc, University of Illinois; Keith F. McDaniel, Ohio University; John A. Miller, Western Washington University; David Minter, Texas Christian University; Roger K. Murray, University of Delaware; George V. Odell, Oklahoma State University; James Piper, Simmons College; Stanley Raucher, University of Washington; Gary Richmond, Grand Valley State University; David J. Rislove, Winona State University; Kevin Smith, University of California—Davis; Ronald Starkey, University of Wisconsin; Kathleen M. Trahanovsky, Iowa State University; and Carl C. Wamser, Portland State University.

A Complete Ancillary Package

For Instructors

Test Items
by Tammy Tiner, Texas A & M University

- Contains over 500 short-answer and multiple-choice test items.
- Computerized versions are also available.

Transparencies

- A complete new set of approximately 112 full-color transparencies covering most topics in the course.

Computerized Test Items

- Contains over 500 short-answer and multiple-choice test items.
- Available for DOS/Windows and Macintosh platforms.
- Print version also available.

For Students

Study Guide and Solutions Manual
by Susan McMurry

- Includes solutions to all end-of-chapter problems, detailed explanations of how answers are obtained, and skills to master for each chapter.
- Also contains a summary of general reaction mechanisms, a summary of methods for preparing functional groups, and a summary of the uses of important reagents.

Organic Chemistry Toolbox
by Norbert J. Pienta, University of North Carolina, Chapel Hill

- A text-specific program for constructing molecular models, drawing Lewis dot structures, creating animations of reactions, solving chemistry problems, and studying structures.
- Available on hybrid CD-ROM.

Beaker 2.1

- Software that is an expert system for the organic chemistry student. Available for DOS/Windows and Macintosh platforms.
- Sophisticated, yet easy to use program for exploring organic chemistry principles; for studying and solving, sketching, and analyzing molecular structures; for constructing NMR spectra; and for performing reactions.

A Note for Students

We have similar goals. Yours is to learn organic chemistry; mine is to do everything possible to help you learn. It's going to require work on your part, but the following suggestions should prove helpful:

Don't read the text immediately. As you begin each new chapter, look it over first. Read the introductory paragraphs, find out what topics will be covered, and then turn to the end of the chapter and read the summary. You'll be in a much better position to understand new material if you first have a general idea of where you're heading. Once you've begun a chapter, read it several times. First read the chapter rapidly, making checks or comments in the margin next to important or difficult points; then return for an in-depth study.

Keep up with the material. Who's likely to do a better job—the runner who trains five miles per day for weeks before a race, or the one who suddenly trains twenty miles the day before the race? Organic chemistry is a subject that builds on previous knowledge. You have to keep up with the material on a daily basis.

Work the problems. There are no shortcuts here. Working problems is the only way to learn organic chemistry. The practice problems show you how to approach the material, the in-text problems provide immediate practice, and the end-of-chapter problems provide additional drill and some real challenges. Answers and explanations for all problems are given in the accompanying *Study Guide and Solutions Manual.*

Ask questions. Faculty members and teaching assistants are there to help you. Most of them will turn out to be extremely helpful and genuinely interested in seeing you learn.

Use molecular models. Organic chemistry is a three-dimensional science. Although this book uses many careful drawings and stereo views to help you visualize molecules, there's no substitute for building a molecular model, turning it in your hands, and looking at it from different views.

Use the study guide. The *Study Guide and Solutions Manual* that accompanies this text gives complete solutions to all problems and provides a wealth of supplementary material. Included are a list of study goals for each chapter, outlines of each chapter, a summary of name reactions, a summary of methods for preparing functional groups, a summary of the uses of important reagents, and tables of spectroscopic information. Find out ahead of time what's there so that you'll know where to go when you need help.

Good luck. I sincerely hope you enjoy learning organic chemistry and that you come to see the logic and beauty of its structure. I would be glad to receive comments and suggestions from any who have learned from this book.

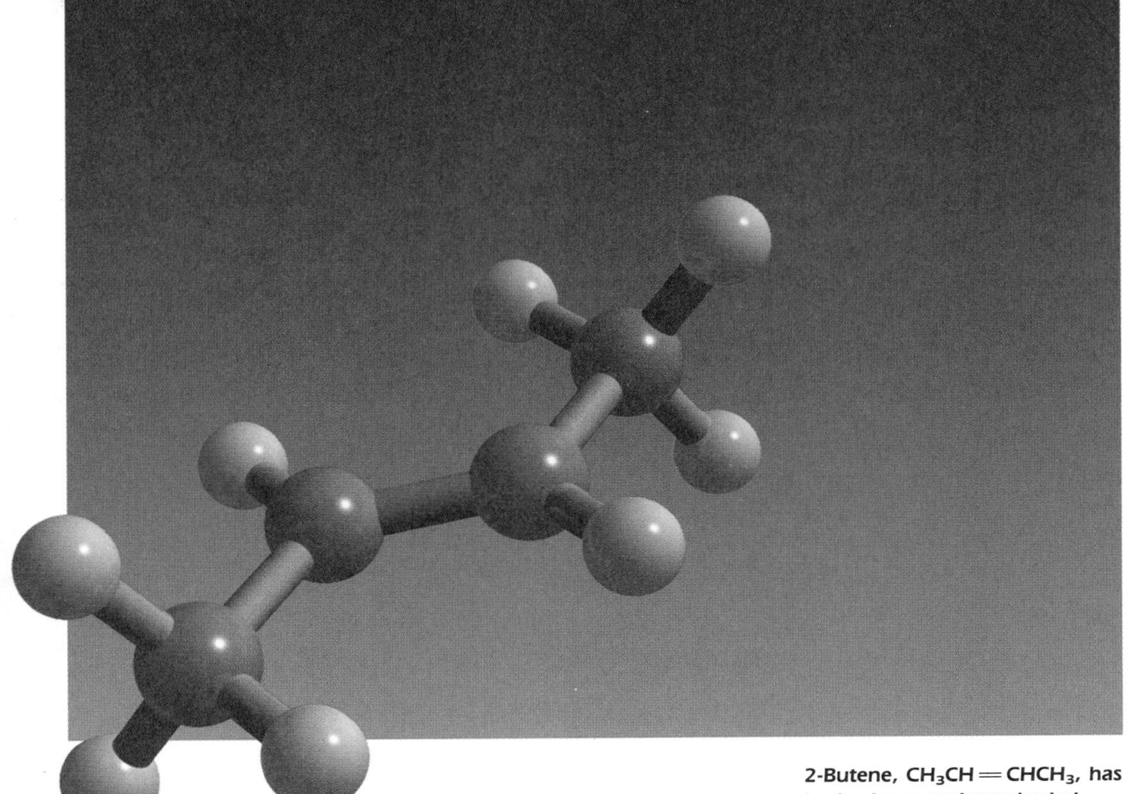

2-Butene, $CH_3CH{=}CHCH_3$, has both planar and tetrahedral carbon atoms.

1 Structure and Bonding; Acids and Bases

What is organic chemistry, and why should you study it? The answers are all around you. Every living organism is made of organic chemicals. The foods you eat; the medicines you take; the wood, paper, plastics, and fibers that make modern life possible, are all organic chemicals. Anyone with a curiosity about life and living things must have a fundamental understanding of organic chemistry.

The foundations of organic chemistry date from the mid-1700s when alchemists noticed unexplainable differences between compounds derived from living sources and those derived from minerals. Compounds from plants and animals were often difficult to isolate and purify. Even when pure, they were difficult to work with and tended to decompose more easily than compounds from minerals. The Swedish chemist Torbern Bergman was the first person to express this difference between "organic" and "inorganic" substances in 1770, and the term *organic chemistry* soon came to mean the chemistry of compounds from living organisms.

To many chemists of the time, the only explanation for the difference in behavior between organic and inorganic compounds was that organic compounds contained an undefinable "vital force" as a result of their origin in living sources. With time, however, it became clear that organic compounds could be manipulated in the laboratory just like inorganic compounds. Friedrich Wöhler discovered in 1828, for example, that it was possible to convert the "inorganic" salt ammonium cyanate into urea, an "organic" compound isolated from urine.

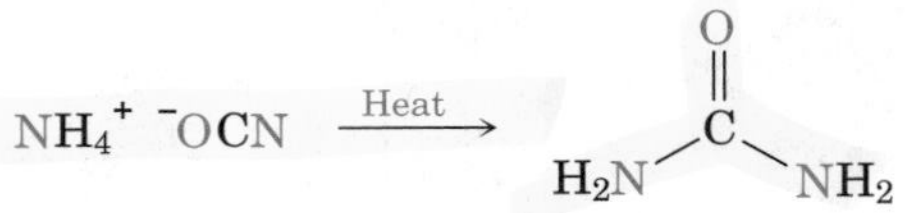

$$\underset{\textbf{Ammonium cyanate}}{NH_4^+ \ {}^-OCN} \xrightarrow{\text{Heat}} \underset{\textbf{Urea}}{H_2N{-}\overset{\overset{\Large O}{\|}}{C}{-}NH_2}$$

By the mid-1800s, the weight of evidence was against the vitalistic theory, and it had become clear that the same basic scientific principles are applicable to all compounds. The only distinguishing characteristic of organic compounds is that all contain the element carbon (Figure 1.1).

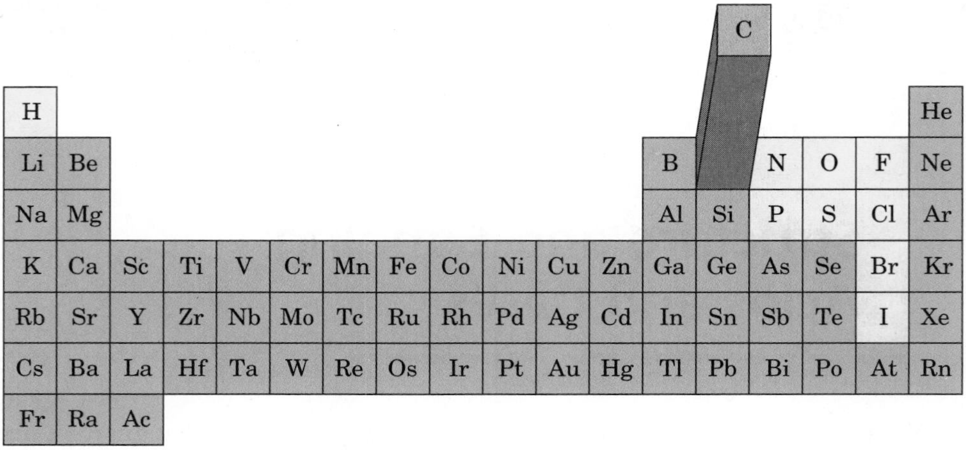

H																	He
Li	Be											B	C	N	O	F	Ne
Na	Mg											Al	Si	P	S	Cl	Ar
K	Ca	Sc	Ti	V	Cr	Mn	Fe	Co	Ni	Cu	Zn	Ga	Ge	As	Se	Br	Kr
Rb	Sr	Y	Zr	Nb	Mo	Tc	Ru	Rh	Pd	Ag	Cd	In	Sn	Sb	Te	I	Xe
Cs	Ba	La	Hf	Ta	W	Re	Os	Ir	Pt	Au	Hg	Tl	Pb	Bi	Po	At	Rn
Fr	Ra	Ac															

Figure 1.1 The position of carbon in the periodic table. Other elements commonly found in organic compounds are shown in yellow.

Organic chemistry, then, is the study of the compounds of carbon. But why is carbon special? What is it that sets carbon apart from all other elements in the periodic table? The answers to these questions derive from the unique ability of carbon atoms to bond together, forming rings and long chains. Carbon, alone of all elements, is able to form an immense diversity of compounds, from the simple to the staggeringly complex—from methane, containing one carbon atom, to DNA, which can contain tens of *billions.*

Not all organic compounds are derived from living organisms, of course. Modern chemists are extremely sophisticated in their ability to synthesize new organic compounds in the laboratory. Medicines, dyes, polymers, plastics, pesticides, and a host of other organic substances are all prepared in the laboratory. Organic chemistry is a subject that touches the lives of everyone. Its study is a fascinating undertaking.

1.1 Atomic Structure

Before beginning a study of organic chemistry, let's review some general ideas about atoms and bonds. Atoms consist of a dense, positively charged *nucleus* surrounded at a relatively large distance by negatively charged *electrons* (Figure 1.2). The nucleus consists of subatomic particles called *neutrons,* which are electrically neutral, and *protons,* which are positively charged. Though extremely small—about 10^{-14} to 10^{-15} meter (m) in diameter—the nucleus nevertheless contains essentially all the mass of the atom. Electrons have negligible mass and orbit the nucleus at a distance of approximately 10^{-10} m. Thus, the diameter of a typical atom is about 2×10^{-10} m, or 2 *angstroms* (Å), where 1 Å = 10^{-10} m. To give you an idea how small this is, a thin pencil line is about 3 *million* carbon atoms wide.

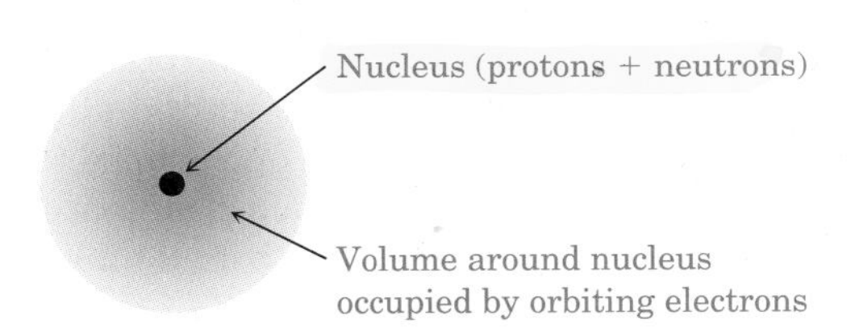

Figure 1.2 A schematic view of an atom. The dense, positively charged nucleus contains most of the atom's mass and is surrounded by negatively charged electrons.

An atom is described by its **atomic number (*Z*),** which gives the number of protons in the atom's nucleus, and its **mass number (*A*)**, which gives the total of protons plus neutrons. All the atoms of a given element have the same atomic number—1 for hydrogen, 6 for carbon, 17 for chlorine, and so on—but they can have different mass numbers depending on how many neutrons they contain. The *average* mass number of a great many atoms of an element is called the element's **atomic weight**—1.008 for hydrogen, 12.011 for carbon, 35.453 for chlorine, and so on.

How are the electrons distributed in an atom? It turns out that electrons aren't completely free to move about but are confined to different regions within the atom according to the amount of energy they have. Electrons can be thought of as belonging to different layers, or **shells,** around the nucleus. Within each shell, electrons are further grouped into **subshells,** denoted *s, p, d,* and *f,* and within each subshell, electrons are grouped by pairs into **orbitals.**

Different shells have different numbers and kinds of subshells, and different subshells have different numbers and kinds of orbitals. As indicated

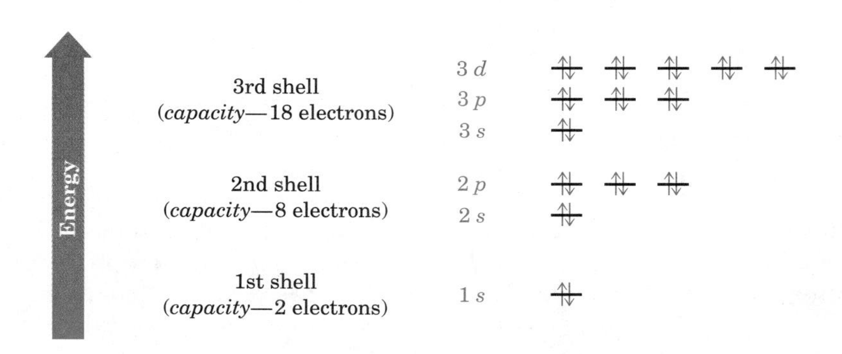

Figure 1.3 The distribution of electrons in an atom. The first shell holds a maximum of 2 electrons in one 1*s* orbital; the second shell holds a maximum of 8 electrons in one 2*s* and three 2*p* orbitals; the third shell holds a maximum of 18 electrons in one 3*s*, three 3*p*, and five 3*d* orbitals; and so on. The 2 electrons in each orbital are represented by up and down arrows, ⇅. Although not shown, the 4*s* orbital has an energy level between 3*p* and 3*d*.

in Figure 1.3, the first shell contains 2 electrons, which occupy an *s* subshell and are paired in an orbital designated 1*s*. The second shell contains 8 electrons, which occupy an *s* subshell that contains one orbital (designated 2*s*) and a *p* subshell that contains three orbitals (each designated 2*p*). The third shell contains 18 electrons, which occupy an *s* subshell (one 3*s* orbital), a *p* subshell (three 3*p* orbitals), and a *d* subshell (five 3*d* orbitals).

It's helpful to think of an orbital as a blurry photograph of an electron's movement, indicating the region of space surrounding the nucleus where the electron has been. This electron cloud doesn't have a sharp boundary, but for practical purposes we can set the limits by saying that an orbital represents the space where an electron spends most (95%) of its time.

What do orbitals look like? The shape of an orbital depends on whether it's in an *s, p, d,* or *f* subshell. Of the four, we'll be concerned only with *s* and *p* orbitals because most atoms found in living organisms use only these. The *s* orbitals have a spherical shape with the nucleus at the center, and the *p* orbitals have a dumbbell shape, as shown in Figure 1.4. Note that a given shell contains three different *p* orbitals, oriented in space so that each is perpendicular to the other two. They are arbitrarily denoted p_x, p_y, and p_z.

1.2 Electron Configuration of Atoms

The lowest-energy arrangement, or **ground-state electron configuration,** of an atom is a description of the orbitals that the atom's electrons occupy. We can predict this arrangement by following three rules:

Rule 1 The orbitals of lowest energy are filled first, according to the order 1*s* → 2*s* → 2*p* → 3*s* → 3*p* → 4*s* → 3*d,* as shown in Figure 1.3.

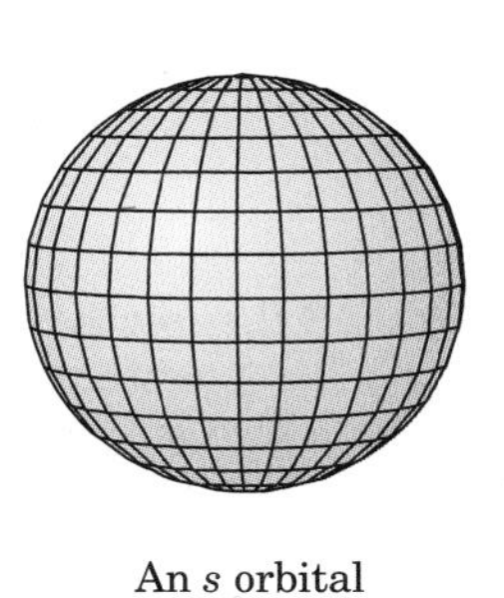

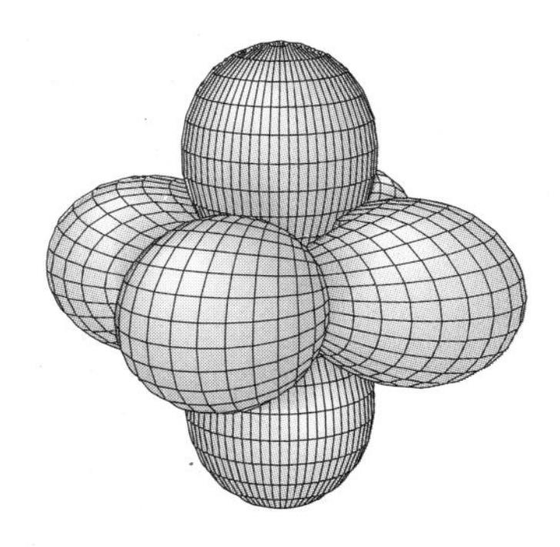

Figure 1.4 Computer-generated shapes of s and p orbitals. The s orbitals are spherical, and the p orbitals are dumbbell-shaped. There are three mutually perpendicular p orbitals, denoted p_x, p_y, and p_z.

Rule 2 Only two electrons can occupy an orbital, and they must be of opposite spin.[1]

Rule 3 If two or more empty orbitals of equal energy are available, one electron is placed in each with their spins parallel until all are half-full.

Some examples of how these rules apply are shown in Table 1.1. Hydrogen, for instance, has only one electron, which must occupy the lowest-

Table 1.1 Ground-State Electron Configuration of Some Elements

Element	*Atomic number*	*Configuration*	*Element*	*Atomic number*	*Configuration*
Hydrogen	1	$1s$ ↑	Lithium	3	$2s$ ↑ $1s$ ↑↓
Carbon	6	$2p$ ↑ ↑ — $2s$ ↑↓ $1s$ ↑↓	Neon	10	$2p$ ↑↓ ↑↓ ↑↓ $2s$ ↑↓ $1s$ ↑↓
Sodium	11	$3s$ ↑ $2p$ ↑↓ ↑↓ ↑↓ $2s$ ↑↓ $1s$ ↑↓	Argon	18	$3p$ ↑↓ ↑↓ ↑↓ $3s$ ↑↓ $2p$ ↑↓ ↑↓ ↑↓ $2s$ ↑↓ $1s$ ↑↓

[1]Electrons can be thought of as spinning on an axis in much the same way that the earth spins. This spin can have two orientations, denoted as up ↑ and down ↓.

energy orbital. Thus, hydrogen has a 1*s* ground-state electron configuration.[2] Carbon has six electrons and the ground-state electron configuration $1s^2 2s^2 2p^2$.

PRACTICE PROBLEM 1.1

Give the ground-state electron configuration of nitrogen.

Solution As shown by the periodic table on the rear inside cover, nitrogen has atomic number 7 and thus has seven electrons. Using Figure 1.3 to find the relative energy levels of orbitals, and applying the three rules to assign the seven electrons, the first two electrons go into the lowest-energy orbital ($1s^2$), the next two go into the second-lowest-energy orbital ($2s^2$), and the remaining three go into the three next-lowest-energy orbitals ($2p^3$). Thus, the configuration of nitrogen is $1s^2 2s^2 2p^3$.

PROBLEM

1.1 How many electrons does each of the following elements have in its outermost electron shell?
(a) Potassium (b) Calcium (c) Aluminum

PROBLEM

1.2 Give the ground-state electron configuration of the following elements:
(a) Boron (b) Phosphorus (c) Oxygen (d) Argon

1.3 Development of Chemical Bonding Theory

By the mid-1800s, the new science of chemistry was developing rapidly, and chemists had begun to probe the forces holding molecules together. In 1858, August Kekulé and Archibald Couper independently proposed that, in all organic compounds, carbon always has four "affinity units." That is, carbon is *tetravalent*; it always forms four bonds when it joins other elements to form chemical compounds. Furthermore, said Kekulé, carbon atoms can bond to one another to form extended chains, and chains can double back on themselves to form rings.

Although Kekulé was correct in describing the tetravalent nature of carbon, chemistry was still viewed in a two-dimensional way until 1874. In that year, Jacobus van't Hoff and Joseph Le Bel added a third dimension to our ideas about chemistry. They proposed that the four bonds of carbon are not randomly oriented but have a specific spatial orientation. Van't Hoff went even further and proposed that the four atoms to which a carbon atom is bonded sit at the corners of a regular tetrahedron, with carbon in the center.

[2]A superscript is used to represent the number of electrons at a particular energy level. For example, $1s^2$ indicates that there are two electrons in the 1*s* orbital. No superscript is used when there is only one electron in an orbital.

A representation of a tetrahedral carbon atom is shown in Figure 1.5. Note the conventions used to show three-dimensionality: Solid lines represent bonds in the plane of the paper, heavy wedged lines represent bonds coming out of the plane of the paper toward the viewer, and dashed lines represent bonds receding into the plane away from the viewer. Such representations will be used throughout this text.

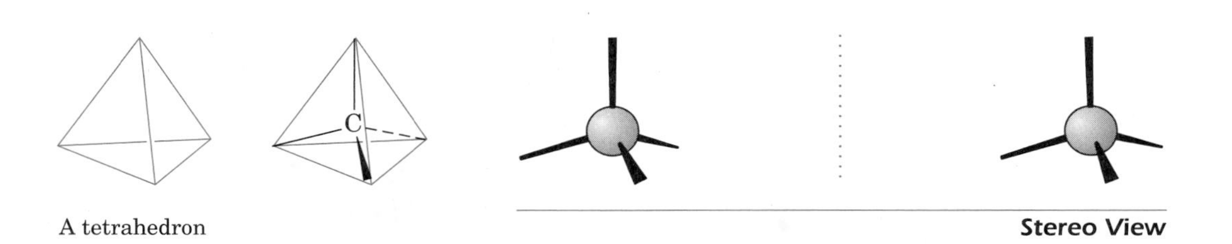

Figure 1.5 Van't Hoff's tetrahedral carbon atom. The solid lines are in the plane of the paper, the heavy wedged line comes out of the plane of the paper, and the dashed line goes back into the plane. The three-dimensional stereo view can be seen with the viewer that comes with this text.

The ability to visualize complex organic and biological molecules in three dimensions is a critical skill in organic chemistry. To help you develop this skill, a stereo viewer is bound inside the back cover, and more than 65 three-dimensional stereo views like that in Figure 1.5 are placed throughout this book. Don't overlook this valuable learning tool.

PROBLEM

1.3 Draw a molecule of chloromethane, CH_3Cl, using solid, wedged, and dashed lines to show its tetrahedral geometry.

1.4 The Nature of Chemical Bonds: Ionic Bonds

Why do atoms bond together, and how can bonds be described? The *why* question is relatively easy to answer: Atoms bond together because the compound that results is more stable (has less energy) than the separate atoms. Energy always flows *out of* the chemical system when a chemical bond is formed. Conversely, energy must be put *into* the system when a chemical bond is broken. Put another way, making bonds releases energy, and breaking bonds absorbs energy. The *how* question is more difficult. To answer it, we need to know more about the properties of atoms.

We know through observation that eight electrons (an electron octet) in the outermost shell (the **valence shell**) impart special stability to the noble-gas elements in group 8A: neon (2 + *8*), argon (2 + 8 + *8*), krypton (2 + 8 + 18 + *8*). We also know that the chemistry of many main-group elements is governed by a tendency for them to take on the stable noble-gas electronic makeup. The alkali metals in group 1A, for example, have a single *s* electron in their outer shells. By losing this valence electron, they can achieve a noble-gas configuration.

Just as the alkali metals and other reactive elements on the left side of the periodic table tend to form *positive* ions by *losing* one or more electrons, the halogens (group 7A elements) and other reactive nonmetals on the right side of the periodic table tend to form *negative* ions by *gaining* one or more electrons.

The simplest kind of chemical bonding is that between a reactive metal on the left of the periodic table and a reactive nonmetal on the right. For example, when sodium metal reacts with chlorine gas, each sodium atom gives an electron to a chlorine atom to form Na^+ ions and Cl^- ions. The NaCl product that results is said to have **ionic bonding,** meaning that the Na^+ and Cl^- ions are held together by an electrical attraction between their unlike charges. Note, though, that there is no such thing as an individual NaCl "molecule," and we can't speak of specific ionic bonds between specific pairs of ions. Rather, there are many ionic bonds between a given ion and its neighbors, and so we speak of the whole crystal as being an **ionic solid.**

PROBLEM

1.4 How many valence electrons does each of the following elements have?
(a) Be (b) S (c) Br

PROBLEM

1.5 Judging from their positions in the periodic table, which element in each of the following pairs gains an electron more easily?
(a) Potassium or oxygen (b) Calcium or bromine

1.5 The Nature of Chemical Bonds: Covalent Bonds

We've just seen that elements on the left (sodium) and right (chlorine) of the periodic table form ionic bonds by losing or gaining electrons. How, though, do the elements near the middle of the periodic table form bonds? Look at methane, CH_4, the main constituent of natural gas, for example. The bonding in methane is not ionic because it would be very difficult for carbon ($1s^22s^22p^2$) either to gain or lose *four* electrons to achieve a noble-gas configuration. In fact, carbon bonds to other atoms, not by losing or gaining electrons, but by *sharing* them. Such shared-electron bonds, first proposed in 1916 by G. N. Lewis, are called **covalent bonds.** The neutral collection of atoms held together by covalent bonds is called a **molecule.**

A simple shorthand way of indicating covalent bonds in molecules is to use **Lewis structures,** or *electron-dot structures,* in which an atom's valence electrons are represented by dots. Thus, hydrogen has one dot ($1s$), carbon has four dots ($2s^22p^2$), oxygen has six dots ($2s^22p^4$), and so on. A stable molecule results when a valence octet of electrons has been achieved for all atoms in the molecule, as in the following examples:

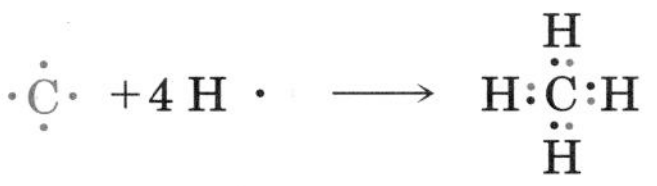

Methane (CH_4)

Ammonia (NH_3)

Water (H_2O)

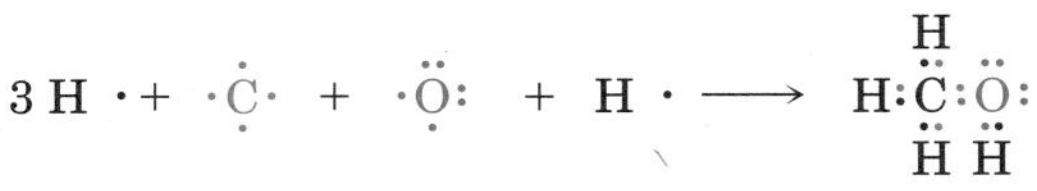

Methanol (CH_3OH)

The number of covalent bonds an atom forms depends on how many valence electrons it has and on how many additional valence electrons it needs to complete an octet. Atoms with one, two, or three valence electrons form one, two, or three bonds, but atoms with four or more valence electrons form as many bonds as they need electrons to fill the *s* and *p* levels of their valence shells and thereby reach a stable octet. Thus, boron has three valence electrons ($2s^22p^1$) and forms three bonds as in BH_3; carbon ($2s^22p^2$) fills its valence shell by forming four bonds as in CH_4; nitrogen ($2s^22p^3$) forms three bonds as in NH_3; and oxygen ($2s^22p^4$) forms two bonds, as in H_2O.

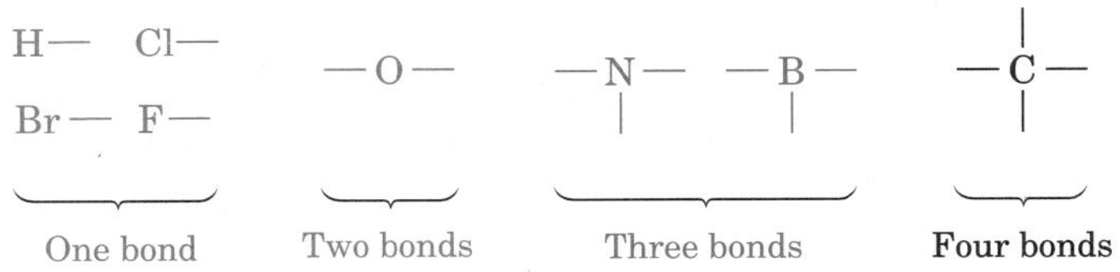

Valence electrons not used for bonding are called **nonbonding electrons, or lone-pair electrons.** The nitrogen atom in ammonia, for instance, shares six of its eight valence electrons in three covalent bonds with hydrogens, and has its remaining two valence electrons in a nonbonding lone pair.

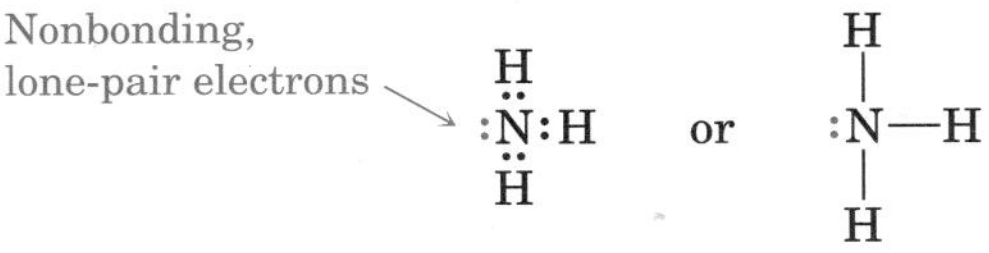

Ammonia

The Lewis structures we've been using are valuable because they make electron bookkeeping possible and act as reminders of the number of valence electrons present. Simpler still is the use of **Kekulé structures, or line-bond structures.** In a line-bond structure, the two electrons in a covalent bond are indicated simply by a line. Lone pairs of nonbonding valence electrons are often ignored when drawing line-bond structures, but it's still necessary to keep them in mind. Some examples are shown in Table 1.2.

Table 1.2 Lewis and Kekulé Structures of Some Simple Molecules

Name	*Lewis structure*	*Kekulé structure*
Water (H_2O)	H:Ö:H (with a pair of dots below O)	H—O—H
Ammonia (NH_3)	H H:N:H (with a pair of dots below N)	H | H—N—H
Methane (CH_4)	H H:C:H H	H | H—C—H | H
Methanol (CH_3OH)	H H:C:Ö:H (with a pair of dots below O) H	H | H—C—O—H | H

PRACTICE PROBLEM 1.2 ....

How many hydrogen atoms does phosphorus bond to in forming phosphine, $PH_?$?

Solution Because phosphorus is in group 5A of the periodic table, it has five valence electrons. It needs to share three more electrons to make an octet, and it therefore bonds to three hydrogen atoms, giving PH_3.

PRACTICE PROBLEM 1.3 ....

Draw a Lewis structure for chloromethane, CH_3Cl.

Solution Hydrogen has one valence electron, carbon has four valence electrons, and chlorine has seven valence electrons. Thus, chloromethane is represented as

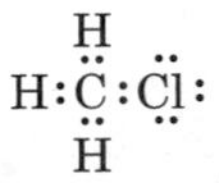

H
H:C:C̤l: **Chloromethane**
H

PROBLEM ....

1.6 What are likely formulas for the following molecules?
(a) $CCl_?$ (b) $AlH_?$ (c) $CH_?Cl_2$ (d) $SiF_?$

PROBLEM ....

1.7 Write both Lewis and line-bond structures for the following molecules, showing all nonbonded electrons:
(a) $CHCl_3$, chloroform (b) H_2S, hydrogen sulfide
(c) CH_3NH_2, methylamine

PROBLEM ....

1.8 Which of the following substances are likely to have covalent bonds and which ionic bonds?
(a) CH_4 (b) CH_2Cl_2 (c) LiI (d) KBr (e) $MgCl_2$ (f) Cl_2

PROBLEM ....

1.9 Write both a Lewis electron-dot structure and a line-bond structure for ethane, C_2H_6.

1.6 Formation of Covalent Bonds

The simplest way to picture the formation of a covalent bond is to imagine an *overlapping* of two atomic orbitals, each of which contains one electron. For example, we can picture the hydrogen molecule (H–H) by imagining what might happen if two hydrogen atoms, each with one electron in an atomic 1*s* orbital, come together. As the two spherical atomic orbitals approach and combine, a new, egg-shaped H_2 orbital results. The new orbital is filled by two electrons, one donated by each hydrogen:

$$\underset{1s}{H\uparrow} + \underset{1s}{\downarrow H} \longrightarrow \underset{\textbf{H}_2\textbf{ molecule}}{H \uparrow\downarrow H}$$

During the reaction 2 H· → H_2, 436 kJ/mol (104 kcal/mol) of energy is released.[3] Because the product H_2 molecule has 436 kJ/mol less energy than the starting 2 H· atoms, we say that the product is more stable than the reactant and that the new H–H bond has a **bond strength** of 436 kJ/mol. In other words, we would have to put 436 kJ/mol of energy *into* the H–H bond to break the H_2 molecule apart into two H atoms.

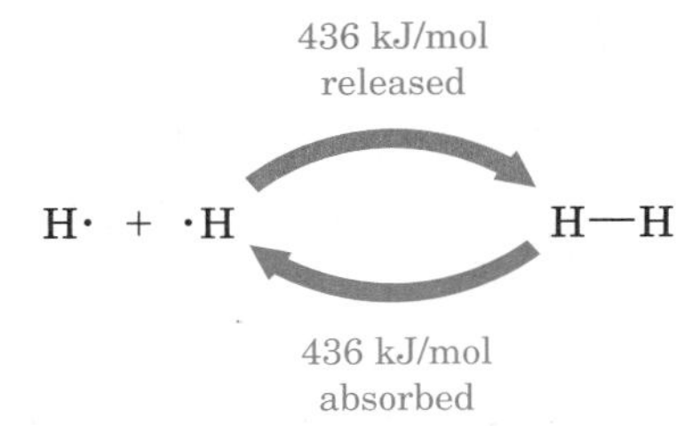

How close are the two nuclei in the hydrogen molecule? If they are too close, they will repel each other because both are positively charged, yet if they are too far apart, they won't be able to share the bonding electrons. Thus, there is an optimum distance between nuclei that leads to maximum bond stability, a distance called the **bond length** (Figure 1.6, p. 12). In the hydrogen molecule, the bond length is 0.74 Å. Every covalent bond has both a characteristic bond strength and bond length.

The orbital in the hydrogen molecule has the elongated egg shape that we might get by pressing two spheres together, and the intersection of a plane cutting through the middle of the H–H bond looks like a circle. In other words, the H–H bond is *cylindrically symmetrical,* as shown in Figure 1.7. Such bonds, in which the shared electrons are symmetrically centered around an imaginary line drawn between the two nuclei, are called **sigma (σ) bonds.**

---

[3]Organic chemists have been slow to adopt SI units, preferring to use kilocalories (kcal) rather than kilojoules (kJ) as a measure of energy. This book will generally show values in both units. The conversion factors are: 1 kcal = 4.184 kJ; 1 kJ = 0.239 kcal.

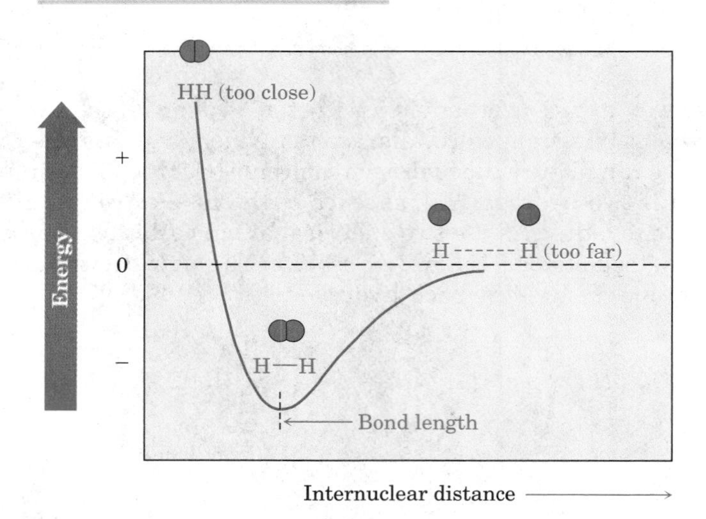

Figure 1.6 A plot of energy versus internuclear distance for two hydrogen atoms. The distance at the minimum energy point is called the bond length.

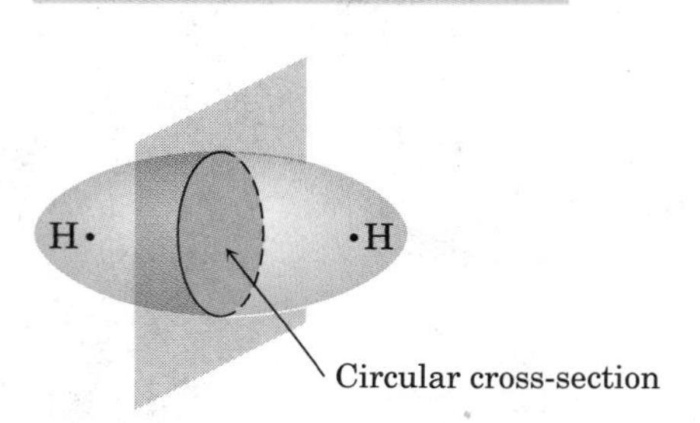

Figure 1.7 Cylindrical symmetry of the H–H sigma (σ) bond. The intersection of a plane cutting through the bond looks like a circle.

1.7 Hybridization: sp^3 Orbitals and the Structure of Methane

The bonding in the hydrogen molecule is fairly straightforward, but the situation is more complex in organic molecules with tetravalent carbon atoms. Let's start with the simplest case and consider methane, CH_4. Carbon has four electrons in its valence shell and can form four bonds to hydrogens. In Lewis structures:

$$\cdot\dot{\underset{\cdot}{C}}\cdot \xrightarrow{4\,H\cdot} H:\overset{H}{\underset{H}{\ddot{\underset{\cdot\cdot}{C}}}}:H$$

What are the four C–H bonds in methane like? Because carbon uses two kinds of orbitals (2*s* and 2*p*) to form bonds, we might expect methane to have two kinds of C–H bonds. In fact, though, all four C–H bonds in methane are identical and are spatially oriented toward the four corners of a tetrahedron. How can we explain this?

The answer was provided in 1931 by Linus Pauling, who showed that an *s* orbital and three *p* orbitals can combine, or *hybridize,* to form four equivalent atomic orbitals with tetrahedral orientation. Shown in Figure 1.8, these tetrahedrally oriented orbitals are called $\boldsymbol{sp^3}$ **hybrids**[4] because they arise from a combination of one *s* orbital with three *p* orbitals.

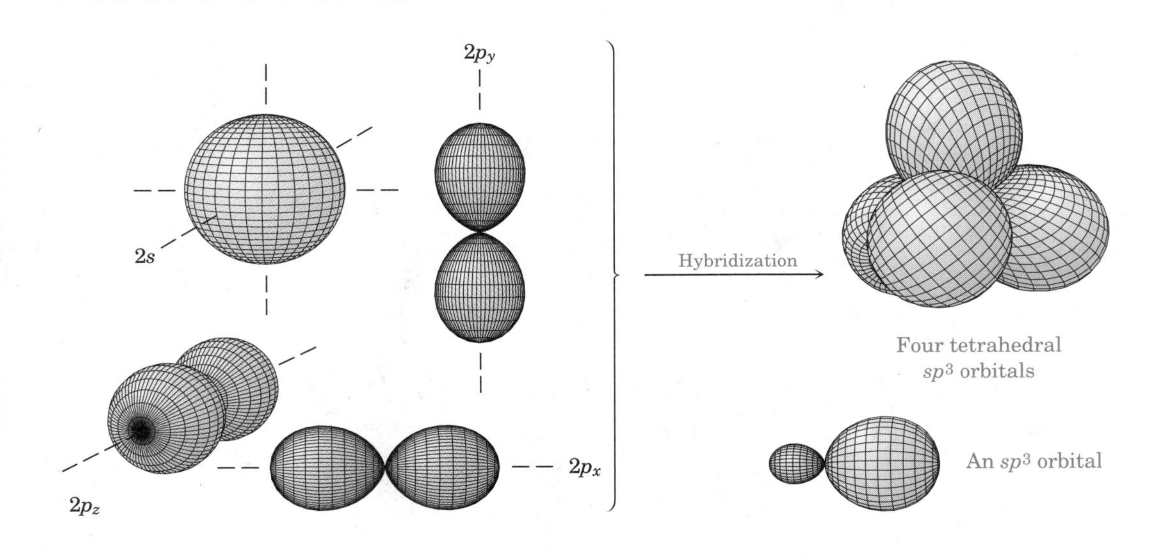

Figure 1.8 The formation of four sp^3 hybrid orbitals by combination of one *s* orbital with three *p* orbitals. The orbitals are unsymmetrical about the nucleus and have their large lobes oriented toward the corners of a regular tetrahedron.

The concept of hybridization explains *how* carbon forms four equivalent tetrahedral bonds but doesn't explain *why* it does so. Looking at an sp^3 hybrid orbital from the side suggests the answer. When an *s* orbital hybridizes with three *p* orbitals, the resultant hybrid orbitals are unsymmetrical about the nucleus. One of the two lobes of an sp^3 orbital is much larger than the other (Figure 1.8) and can therefore overlap better with another orbital when it forms a bond. As a result, sp^3 hybrid orbitals form stronger bonds than do unhybridized *s* or *p* orbitals.

When the four identical orbitals of an sp^3-hybridized carbon atom overlap with four hydrogen atoms, four identical C–H bonds are formed and methane results. Each C–H bond of methane has a strength of 438 kJ/mol

[4]Note that the superscript used to identify an sp^3 hybrid orbital tells how many of each type of atomic orbital combine in the hybrid, not how many electrons occupy that orbital.

(105 kcal/mol) and a length of 1.09 Å. Because the four bonds have a specific geometry, we can also define a property called a **bond angle.** The angle formed by each H–C–H is 109.5°, the so-called *tetrahedral angle.* Methane thus has the structure shown in Figure 1.9.

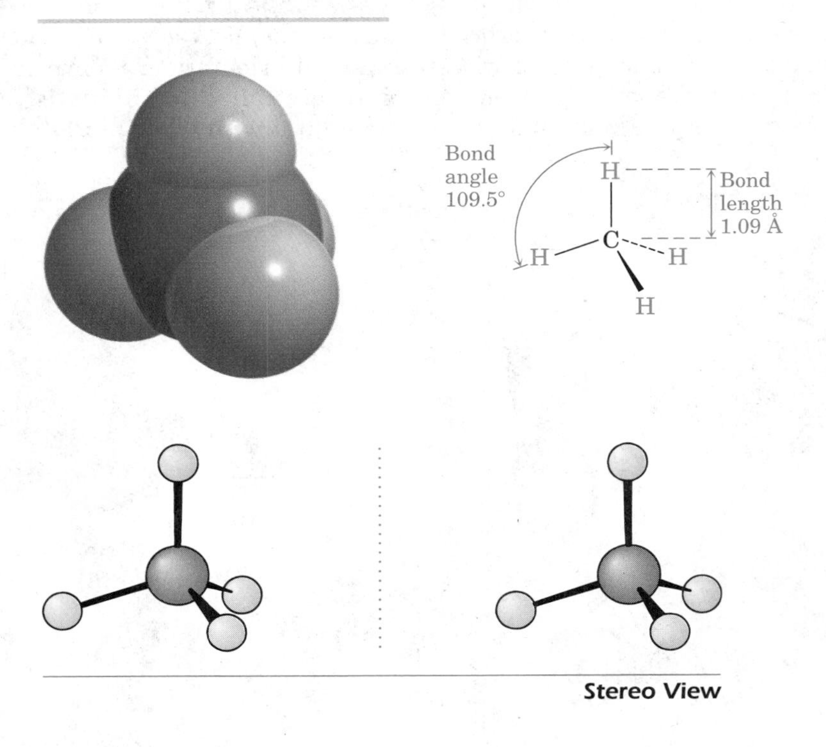

Figure 1.9 The structure of methane. The drawings are computer-generated for accuracy.

PROBLEM

1.10 Draw a tetrahedral representation of tetrachloromethane, CCl_4, using the standard convention of solid, dashed, and wedged lines.

PROBLEM

1.11 Why do you think a C–H bond (1.09 Å) is longer than an H–H bond (0.74 Å)?

1.8 The Structure of Ethane

The same kind of hybridization that explains the methane structure also explains how one carbon atom can bond to another to form a chain. Ethane, C_2H_6, is the simplest molecule containing a carbon–carbon bond:

```
   H H            H   H
   ..  ..         |   |
 H:C:C:H      H—C—C—H
   ..  ..         |   |
   H H            H   H
```

CH_3CH_3

Some representations of ethane

We can picture the ethane molecule by imagining that the two carbon atoms bond to each other by σ overlap of an sp^3 hybrid orbital from each. The remaining three sp^3 hybrid orbitals on each carbon form the six C–H bonds, as shown in Figure 1.10. The C–H bonds in ethane are similar to those in methane, though a bit weaker—420 kJ/mol (100 kcal/mol) for ethane versus 438 kJ/mol for methane. The C–C bond is 1.54 Å long and has a strength of 376 kJ/mol (90 kcal/mol). All the bond angles of ethane are near the tetrahedral value of 109.5°.

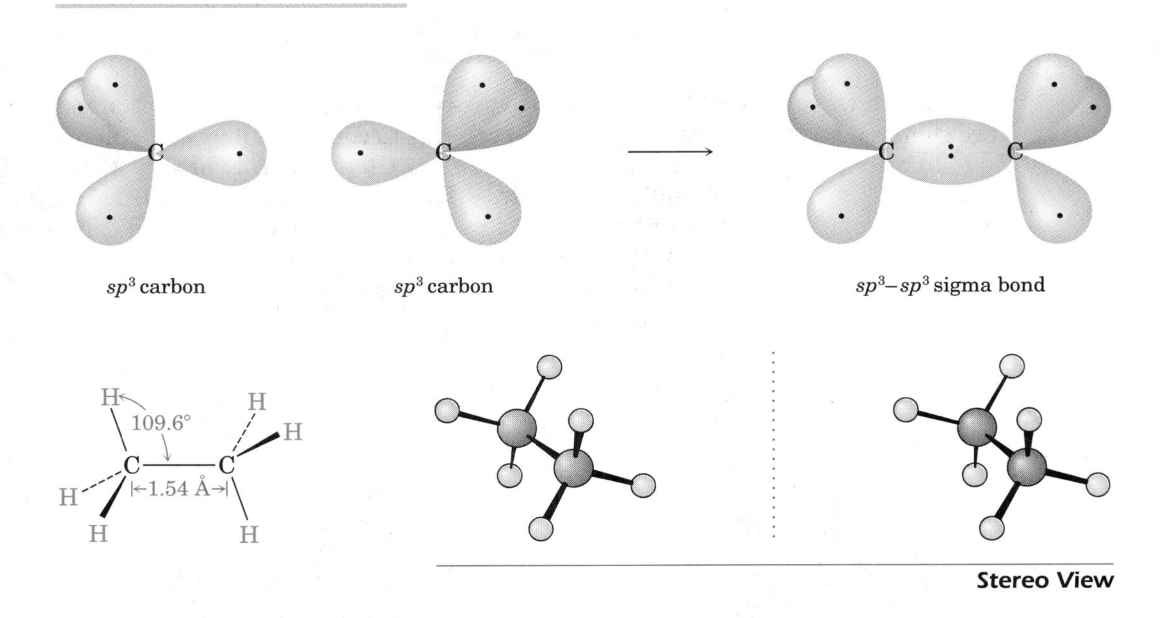

Figure 1.10 The structure of ethane. The carbon–carbon bond is formed by σ overlap of two carbon sp^3 hybrid orbitals. (For clarity, the smaller lobes of the hybrid orbitals are not shown.)

PROBLEM

1.12 Draw a line-bond structure for propane, $CH_3CH_2CH_3$. Predict the value of each bond angle and indicate the overall shape of the molecule.

PROBLEM

1.13 Why can't an organic molecule have the formula C_2H_7?

1.9 Hybridization: sp^2 Orbitals and the Structure of Ethylene

Although sp^3 hybridization is the most common electronic state of carbon, it's not the only possibility. Look at ethylene, C_2H_4, for example. It was recognized over 100 years ago that the carbon atoms in ethylene can be tetravalent only if they share *four* electrons and are linked by a *double* bond. Furthermore, ethylene is planar (flat) rather than tetrahedral and has bond angles of approximately 120° rather than 109.5°.

Top view   Side view

Ethylene

When we formed sp^3 hybrid orbitals to explain the bonding in methane, we combined all four of carbon's valence orbitals to construct four equivalent sp^3 hybrids. Imagine instead that we combine the carbon 2*s* orbital with only *two* of the three available 2*p* orbitals. Three hybrid orbitals called **sp^2 hybrids** result, and one unhybridized 2*p* orbital remains unchanged. The three sp^2 orbitals lie in a plane at angles of 120° to one another, with the remaining *p* orbital perpendicular to the sp^2 plane, as shown in Figure 1.11.

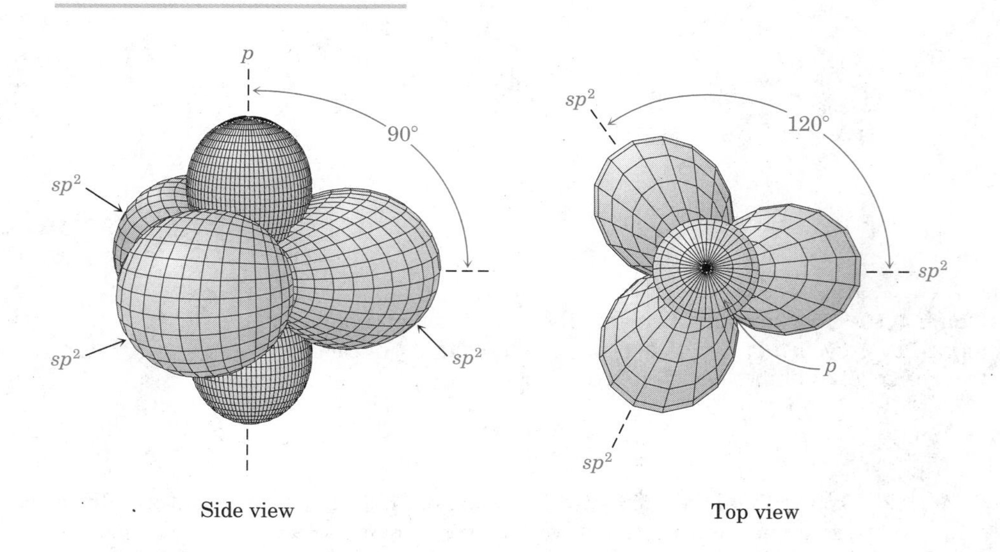

Figure 1.11 An sp^2-hybridized carbon atom. The three equivalent sp^2 hybrid orbitals (green) lie in a plane at angles of 120° to one another, with a single unhybridized *p* orbital (blue) perpendicular to the sp^2 plane.

When two sp^2-hybridized carbon atoms approach each other, they can form a strong σ bond by sp^2–sp^2 overlap. At the same time, the unhybridized *p* orbitals on each carbon can interact by *sideways* rather than head-on overlap, leading to the formation of what is called a **pi (π) bond.** The combination of sp^2–sp^2 σ overlap and 2*p*–2*p* π overlap results in the net sharing of four electrons and the formation of a C=C double bond (Figure 1.12). Note that in the π bond, the shared electrons occupy regions on either side of a line drawn between nuclei rather than the region centered directly on the internuclear line.

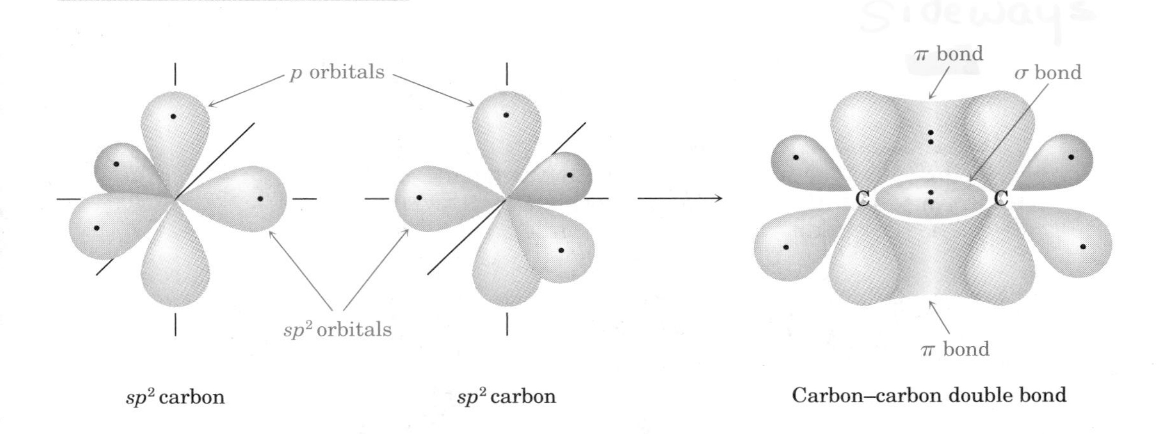

Figure 1.12 Orbital overlap of two sp^2-hybridized carbon atoms in a C=C double bond. One part of the double bond results from σ (head-on) overlap of sp^2 orbitals (green); the other part results from π (sideways) overlap of unhybridized *p* orbitals (blue). The π bond has regions of electron density on either side of a line drawn between nuclei.

To complete the structure of ethylene, four hydrogen atoms form σ bonds to the remaining four carbon sp^2 orbitals. The resultant ethylene molecule has a planar (flat) structure with H–C–H and H–C=C bond angles of approximately 120°. Each C–H bond has a length of 1.076 Å and a strength of 444 kJ/mol (106 kcal/mol).

As you might expect, the carbon–carbon double bond in ethylene is both shorter and stronger than the ethane single bond because it results from the sharing of four electrons rather than two. Ethylene has a C=C bond length of 1.33 Å and a bond strength of 611 kJ/mol (146 kcal/mol), whereas ethane has values of 1.54 Å and 376 kJ/mol. The structure of ethylene is shown in Figure 1.13 (p. 18).

PRACTICE PROBLEM 1.4

Formaldehyde, CH_2O, contains a carbon–*oxygen* double bond. Draw Lewis and line-bond structures of formaldehyde, and indicate the hybridization of the carbon atom.

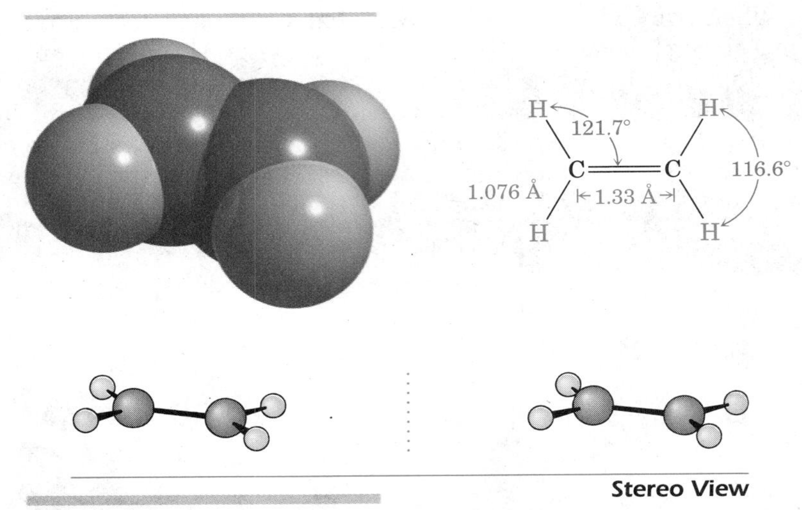

Figure 1.13 The structure of ethylene. Note that these computer-generated structures show only the connections between atoms and do not explicitly indicate the C=C double bond.

Solution There is only one way that two hydrogens, one carbon, and one oxygen can combine:

H:C::O: (with second H on C) — Lewis structure

$H_2C{=}O$ — Line-bond structure

Like the carbon atoms in ethylene, the carbon atom in formaldehyde is sp^2-hybridized.

PROBLEM

1.14 Draw both a Lewis structure and a line-bond structure for acetaldehyde, CH_3CHO.

PROBLEM

1.15 Draw a line-bond structure for propene, $CH_3CH{=}CH_2$, indicate the hybridization of each carbon, and predict the value of each bond angle.

PROBLEM

1.16 Draw a line-bond structure for 1,3-butadiene, $H_2C{=}CH{-}CH{=}CH_2$, indicate the hybridization of each carbon, and predict a value for each bond angle.

1.10 Hybridization: *sp* Orbitals and the Structure of Acetylene

In addition to forming single and double bonds by sharing two and four electrons, respectively, carbon can form *triple* bonds by sharing *six* electrons.

To account for triple bonds, such as that in acetylene, C_2H_2, we need a third kind of hybrid orbital, an ***sp* hybrid.**

H:C:::C:H H—C≡C—H

Acetylene

Imagine that, instead of combining with two or three 2*p* orbitals, the carbon 2*s* orbital hybridizes with only a single 2*p* orbital. Two *sp* hybrid orbitals result, and two *p* orbitals remain unchanged. The two *sp* orbitals are linear, or 180° apart on the *x*-axis, while the remaining two *p* orbitals are perpendicular on the *y*-axis and the *z*-axis, as shown in Figure 1.14.

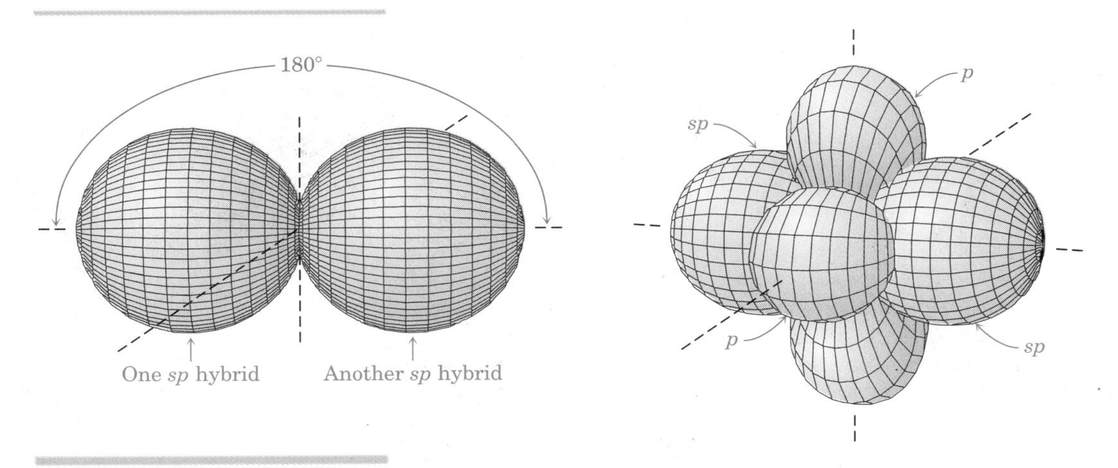

Figure 1.14 An *sp*-hybridized carbon atom. The two *sp* hybrid orbitals (green) are oriented 180° away from each other, perpendicular to the remaining two *p* orbitals (blue).

When two *sp*-hybridized carbon atoms approach each other, *sp* hybrid orbitals from each overlap head-on to form a strong *sp–sp* σ bond. In addition, the p_z orbitals from each carbon form a p_z–p_z π bond by sideways overlap, and the p_y orbitals from each carbon overlap similarly to form a p_y–p_y π bond. The net effect is the formation of one σ bond and two π bonds—a carbon–carbon triple bond. The remaining *sp* hybrid orbitals form σ bonds to hydrogen to complete the acetylene molecule (Figure 1.15, p. 20).

As suggested by *sp* hybridization, acetylene is a linear molecule with H–C≡C bond angles of 180°. The C–H bond has a length of 1.06 Å and a strength of 552 kJ/mol (132 kcal/mol). The C≡C bond length is 1.20 Å, and its strength is 835 kJ/mol (200 kcal/mol), making it the shortest and strongest of any carbon–carbon bond.

PROBLEM

1.17 Draw a line-bond structure for propyne, $CH_3C{\equiv}CH$, indicate the hybridization of each carbon, and predict a value for each bond angle.

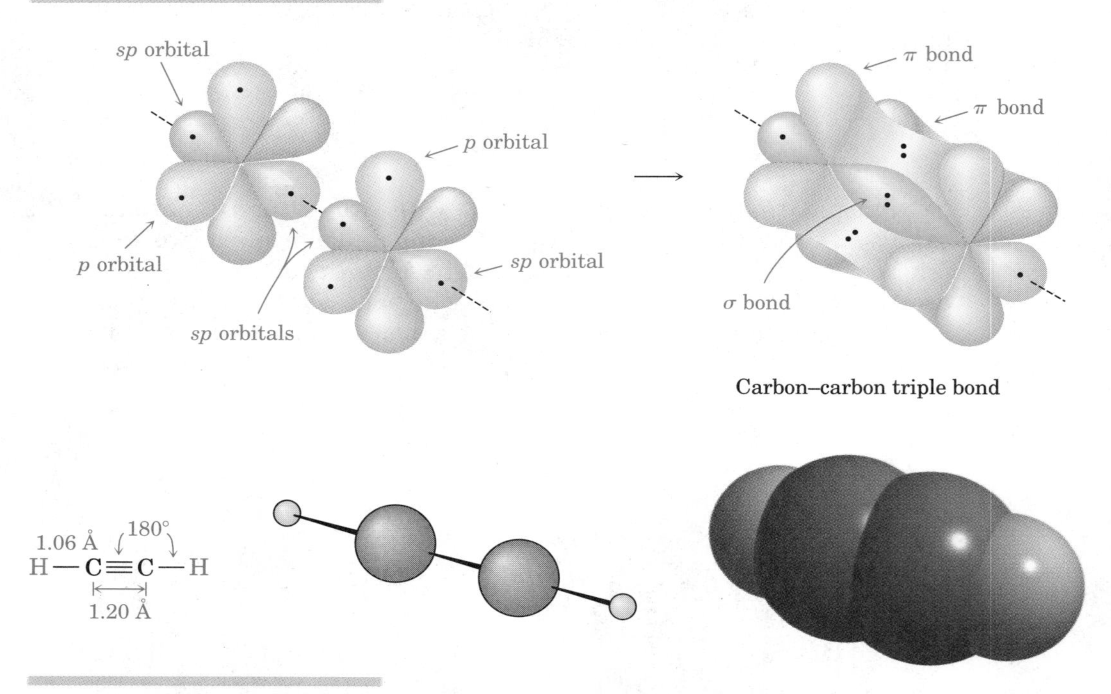

Figure 1.15 The carbon–carbon triple bond in acetylene. There is one σ bond, formed by head-on overlap of *sp* hybrid orbitals, and two mutually perpendicular π bonds, formed by sideways overlap of unhybridized *p* orbitals.

1.11 Bond Polarity and Electronegativity

Up to this point, we've viewed chemical bonding in an either/or manner: A given bond is either ionic or covalent. It's more accurate, though, to look at bonding as a continuum of possibilities, from a fully covalent bond with a symmetrical electron distribution on the one hand, to a fully ionic bond between anions and cations on the other (Figure 1.16).

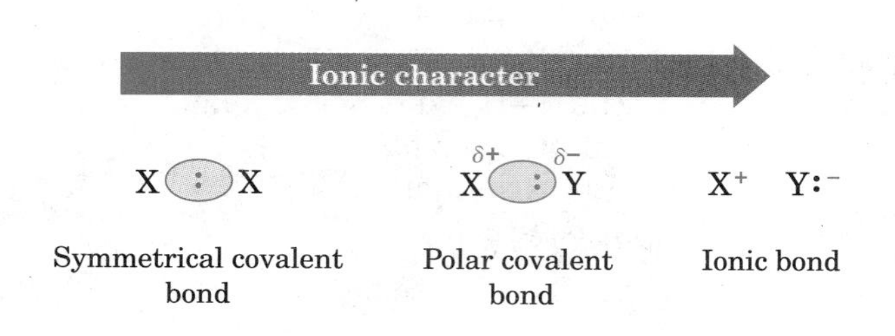

Figure 1.16 The bonding continuum from covalent to ionic is a result of unsymmetrical electron distribution. The symbol δ (lowercase Greek delta) means *partial* charge, either positive (δ+) or negative (δ−).

The carbon–carbon bond in ethane, for example, is electronically symmetrical and therefore fully covalent; the two bonding electrons are equally shared by the two equivalent carbon atoms. The bond in sodium chloride, by contrast, is largely ionic.[5] An electron has been transferred from sodium to chlorine to give Na^+ and Cl^- ions. Between these two extremes lie the majority of chemical bonds, in which the bonding electrons are attracted *somewhat* more strongly by one atom than by the other. We call such bonds with unsymmetrical electron distribution **polar covalent bonds.**

Bond polarity is due to differences in **electronegativity,** the intrinsic ability of an atom to attract electrons in a covalent bond. As shown in Figure 1.17, metallic elements on the left side of the periodic table attract electrons weakly, whereas the halogens and other reactive nonmetal elements on the right side of the periodic table attract electrons strongly.

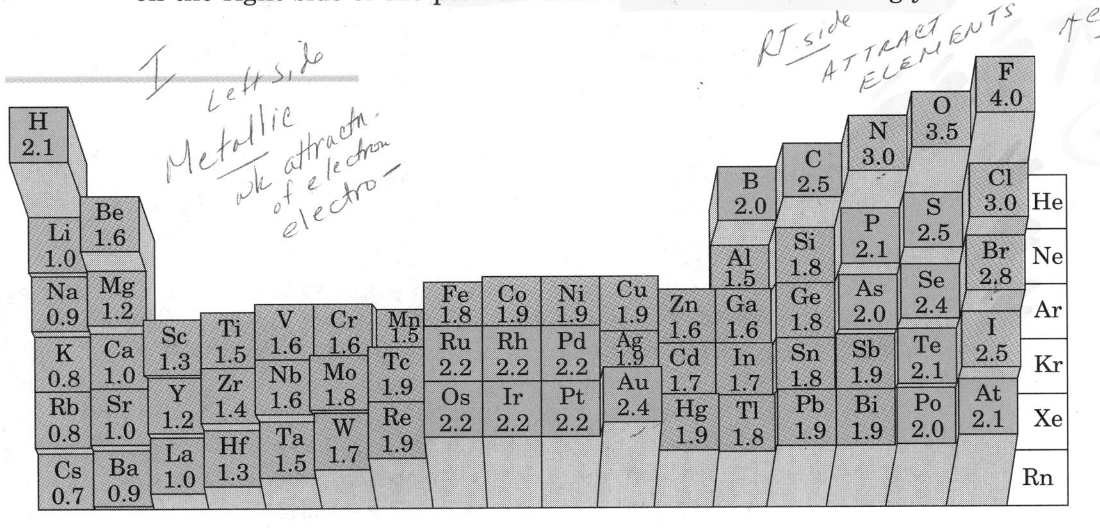

Figure 1.17 Electronegativity trends in the periodic table. Elements on the right side of the table are more electronegative than elements on the left side. The values are on an arbitrary scale, with H = 2.1 and F = 4.0. Carbon has an electronegativity value of 2.5. Any element more electronegative than carbon has a value greater than 2.5, and any element less electronegative than carbon has a value less than 2.5.

As a rule, bonds between atoms with similar electronegativities are covalent, bonds between atoms whose electronegativities (EN) differ by less than 2 units are polar covalent, and bonds between atoms whose electronegativities differ by more than 2 units are largely ionic. Carbon–hydrogen bonds, for example, are relatively nonpolar because carbon and hydrogen have similar electronegativities. The bonds between carbon and more electronegative elements such as oxygen, fluorine, and chlorine, however, are polar covalent. The electrons in such bonds are drawn away from carbon toward the more electronegative atom, leaving the carbon with a partial positive charge, denoted by $\delta+$, and the more electronegative atom with a

[5] Even in NaCl, the bond is only about 80% ionic rather than 100%.

partial negative charge, δ− (δ is the lowercase Greek letter delta). For example, the C–O bond in methanol (wood alcohol) is polar covalent:

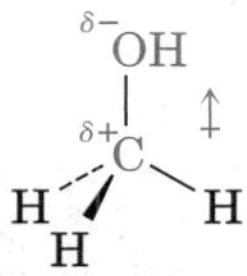

δ−OH, δ+C, H, H, H — **Methanol**

Oxygen:	EN = 3.5
Carbon:	EN = 2.5
Difference =	1.0

An arrow ↦ is sometimes used to indicate the direction of bond polarity. By convention, *electrons move in the direction of the arrow.* The tail of the arrow is electron-poor (δ+), and the head of the arrow is electron-rich (δ−).

Bonds between carbon and less electronegative elements are polarized so that carbon bears a partial negative charge and the other atom bears a partial positive charge. Organometallic compounds such as tetraethyllead, the "lead" in gasoline, provide good examples of this kind of polar covalent bond.

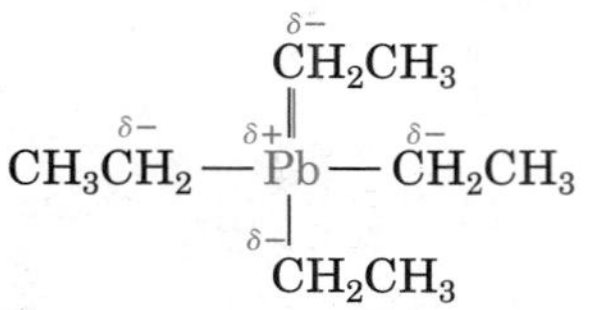

$$\overset{\delta-}{CH_3CH_2}-\overset{\delta+}{Pb}(\overset{\delta-}{CH_2CH_3})_2-\overset{\delta-}{CH_2CH_3}$$

Tetraethyllead

Carbon:	EN = 2.5
Lead:	EN = 1.9
Difference =	0.6

When speaking of an atom's ability to polarize a bond, we often use the term **inductive effect.** An inductive effect is simply the shifting of electrons in a bond in response to the electronegativity of nearby atoms. Metals, such as lithium and magnesium, inductively donate electrons, whereas electronegative elements, such as oxygen and chlorine, inductively withdraw electrons. Inductive effects play a major role in understanding chemical reactivity, and we'll use them many times throughout this text to explain a wide variety of chemical phenomena.

PRACTICE PROBLEM 1.5

Predict the amount and direction of polarization of an O–H bond in water.

Solution Oxygen (EN = 3.5) is more electronegative than hydrogen (EN = 2.1) according to Figure 1.17, and therefore attracts electrons more strongly. The difference in electronegativities (3.5 − 2.1 = 1.4) implies that an O–H bond is strongly polarized.

δ+H–Oδ−–Hδ+

PROBLEM

1.18 Which element in each of the following pairs is more electronegative?
(a) Li or H (b) Be or Br (c) Cl or I

PROBLEM

1.19 Use the δ+/δ− convention to indicate the direction of expected polarity for each of the bonds shown:
(a) Br–CH_3 (b) H_2N–CH_3 (c) Li–CH_3 (d) H–NH_2
(e) HO–CH_3 (f) BrMg–CH_3 (g) F–CH_3

PROBLEM

1.20 Order the bonds in the following compounds according to their increasing ionic character: CCl_4, $MgCl_2$, $TiCl_3$, Cl_2O.

1.12 Acids and Bases: The Brønsted–Lowry Definition

Another important concept related to electronegativity and bond polarity is that of *acidity* and *basicity*. We'll soon see that the acid–base behavior of organic molecules helps explain much of their chemistry. There are two frequently used definitions of acidity: the *Brønsted–Lowry definition* and the *Lewis definition*. Let's look at the Brønsted–Lowry definition first.

A **Brønsted–Lowry acid** is a substance that donates a proton (hydrogen ion, H^+), and a **Brønsted–Lowry base** is a substance that accepts a proton.[6] When HCl gas dissolves in water, for example, HCl donates H^+ and H_2O accepts it, yielding Cl^- and H_3O^+ ions. Chloride ion (Cl^-), the product that results when the acid HCl loses a proton, is called the **conjugate base** of the acid, and H_3O^+, the product that results when the base H_2O gains a proton, is called the **conjugate acid** of the base.

In a general sense:

$$\underset{\text{Acid}}{H{-}A} + \underset{\text{Base}}{:B} \rightleftharpoons \underset{\text{Conjugate base}}{A:^-} + \underset{\text{Conjugate acid}}{H{-}B^+}$$

For example:

$$\underset{\text{Acid}}{H{-}\ddot{\underset{..}{Cl}}:} + \underset{\text{Base}}{:\ddot{O}(H){-}H} \rightleftharpoons \underset{\text{Conjugate base}}{:\ddot{\underset{..}{Cl}}:^-} + \underset{\text{Conjugate acid}}{H{-}\ddot{O}^+(H){-}H}$$

$$\underset{\text{Acid}}{CH_3C(=O)\ddot{\underset{..}{O}}{-}H} + \underset{\text{Base}}{^-:\ddot{\underset{..}{O}}{-}H} \rightleftharpoons \underset{\text{Conjugate base}}{CH_3C(=O)\ddot{\underset{..}{O}}:^-} + \underset{\text{Conjugate acid}}{:\ddot{O}(H){-}H}$$

Acids differ in their proton-donating ability. Stronger acids such as HCl react almost completely with water, whereas weaker acids such as acetic

[6]The name *proton* is often used as a synonym for the hydrogen ion, H^+, because loss of the valence electron from a neutral hydrogen atom leaves only the hydrogen nucleus—a proton. Bare protons are far too reactive to exist alone in solution, however, and are always bonded to some other species. In the presence of water, for example, a proton bonds to the oxygen atom, forming H_3O^+.

acid (CH_3COOH) react only slightly. The exact strength of a given acid in water solution is expressed by its **acidity constant, K_a.**

For the reaction of any generalized acid HA in water, the acidity constant, K_a, is:[7]

$$HA + H_2O \rightleftharpoons A^- + H_3O^+$$

$$K_a = \frac{[H_3O^+][A^-]}{[HA]}$$

Stronger acids have their equilibria toward the right and thus have larger acidity constants; weaker acids have their equilibria toward the left and have smaller acidity constants. The range of K_a values for different acids is enormous, running from about 10^{15} for the strongest acids to about 10^{-60} for the weakest. The common inorganic acids such as H_2SO_4, HNO_3, and HCl have K_a's in the range 10^2–10^9, while many organic acids have K_a's in the range 10^{-5}–10^{-15}. As you gain more experience, you'll develop a rough feeling for which acids are "strong" and which are "weak" (always remembering that the terms are relative).

Acid strengths are normally expressed using pK_a values, where the pK_a is equal to the negative logarithm of the acidity constant:

$$pK_a = -\log K_a$$

A stronger acid (larger K_a) has a *smaller* pK_a, and a weaker acid (smaller K_a) has a *larger* pK_a. Table 1.3 lists the pK_a's of some common acids in order of their strength.

Table 1.3 Relative Strength of Some Common Acids

	Acid	*Name*	pK_a	*Conjugate base*	*Name*	
Weaker acid	CH_3CH_2OH	Ethanol	16.00	$CH_3CH_2O^-$	Ethoxide ion	Stronger base
	H_2O	Water	15.74	HO^-	Hydroxide ion	
↓	HCN	Hydrocyanic acid	9.31	CN^-	Cyanide ion	↑
	CH_3COOH	Acetic acid	4.76	CH_3COO^-	Acetate ion	
	HF	Hydrofluoric acid	3.45	F^-	Fluoride ion	
	HNO_3	Nitric acid	−1.3	NO_3^-	Nitrate ion	
Stronger acid	HCl	Hydrochloric acid	−7.0	Cl^-	Chloride ion	Weaker base

[7]Remember that brackets [ ] refer to the concentration of the enclosed species in moles per liter (mol/L), M. Note that the concentration of water [H_2O] is left out of the expression for K_a, since it remains effectively constant.

Notice in Table 1.3 that there is an inverse relationship between the acid strength of an acid and the base strength of its conjugate base. To understand this inverse relationship, think about what is happening to the acidic proton: A *strong acid* is one that loses its proton easily, meaning that its conjugate base does not hold the proton tightly and is therefore a *weak base.* A *weak acid* loses its proton with difficulty, meaning that its conjugate base holds the proton tightly and is therefore a *strong base.* The fact that HCl is a strong acid, for example, means that Cl^- does not hold the proton tightly and is thus a weak base. Water, on the other hand, is a weak acid, meaning that OH^- holds the proton tightly and is a strong base.

In general, an acid will donate a proton to the conjugate base of any acid with a larger pK_a. Conversely, the conjugate base of an acid will remove a proton from any acid with a smaller pK_a. For example, the data in Table 1.3 indicate that OH^- will react with acetic acid, CH_3COOH, to yield acetate ion, $CH_3CO_2^-$, and H_2O. Since water ($pK_a = 15.74$) is a weaker acid than acetic acid ($pK_a = 4.76$), hydroxide ion holds a proton more tightly than acetate ion does.

$$CH_3COOH + {}^-:\ddot{O}H \rightleftharpoons CH_3CO_2^- + H_2O$$

Acetic acid (pK_a= 4.76)	Hydroxide ion	Acetate ion	Water (pK_a= 15.74)

An easy way to predict acid–base reactivity is to remember that the products must be more stable than the reactants for reaction to occur. In other words, the product acid and base must be weaker and less reactive than the starting acid and base. In the reaction of acetic acid with hydroxide ion, for example, the product conjugate base (acetate ion) is weaker than the starting base (hydroxide ion), and the product conjugate acid (water) is weaker than the starting acid (acetic acid).

$$CH_3\overset{O}{\overset{\|}{C}}OH + HO^- \rightleftharpoons H_2O + CH_3\overset{O}{\overset{\|}{C}}O^-$$

Stronger acid	Stronger base	Weaker acid	Weaker base

PRACTICE PROBLEM 1.6

Water has $pK_a = 15.74$ and acetylene has $pK_a = 25$. Which of the two is more acidic? Will hydroxide ion react with acetylene?

$$H{-}C{\equiv}C{-}H + H{-}O^- \xrightarrow{?} H{-}C{\equiv}C:^- + H{-}O{-}H$$

Solution In comparing two acids, the one with the smaller pK_a is stronger. Thus, water is a stronger acid than acetylene. Since water loses a proton more easily than acetylene, the HO^- ion has less affinity for a proton than the $HC{\equiv}C:^-$ ion. In other words, the anion of acetylene is a stronger base than hydroxide ion, and the reaction will not proceed as written.

PRACTICE PROBLEM 1.7

Butanoic acid, the substance responsible for the odor of rancid butter, has pK_a = 4.82. What is its K_a?

Solution Since pK_a is the negative logarithm of K_a, it's necessary to use a calculator with an ANTILOG or INV LOG function. Enter the value of the pK_a (4.82), change the sign (−4.82), and then find the antilog (1.5×10^{-5}). Thus, $K_a = 1.5 \times 10^{-5}$.

PROBLEM

1.21 Formic acid, HCO_2H, has pK_a = 3.7, and picric acid, $C_6H_3N_3O_7$, has pK_a = 0.6. What is the K_a of each?

PROBLEM

1.22 Which is stronger, formic acid or picric acid? (See Problem 1.21.)

PROBLEM

1.23 Amide ion, H_2N^-, is a stronger base than hydroxide ion, HO^-. Which is the stronger acid, H_2N–H (ammonia) or HO–H (water)? Explain.

PROBLEM

1.24 Is either of the following reactions likely to take place according to the pK_a data in Table 1.3?
(a) $HCN + CH_3COO^-\ Na^+ \longrightarrow Na^+\ ^-CN + CH_3COOH$
(b) $CH_3CH_2OH + Na^+\ ^-CN \longrightarrow CH_3CH_2O^-\ Na^+ + HCN$

1.13 Acids and Bases: The Lewis Definition

The Lewis definition of acids and bases differs from the Brønsted–Lowry definition in that it's not limited to substances that donate or accept protons. A **Lewis acid** is a substance that accepts an electron pair, and a **Lewis base** is a substance that donates an electron pair. The donated pair of electrons is shared between Lewis acid and base in a newly formed covalent bond.

Lewis acids include not only proton donors but many other species as well. Thus, a proton (H^+) is a Lewis acid because it accepts a pair of electrons to fill its vacant 1*s* orbital when it reacts with a base. A compound such as $AlCl_3$ is a Lewis acid because it too accepts an electron pair from a Lewis base to fill a vacant valence orbital (Figure 1.18).

The Lewis definition of basicity—a compound with a pair of nonbonding electrons that it can use in forming a bond to a Lewis acid—is similar to the Brønsted–Lowry definition. Thus, H_2O, with its two pairs of non-

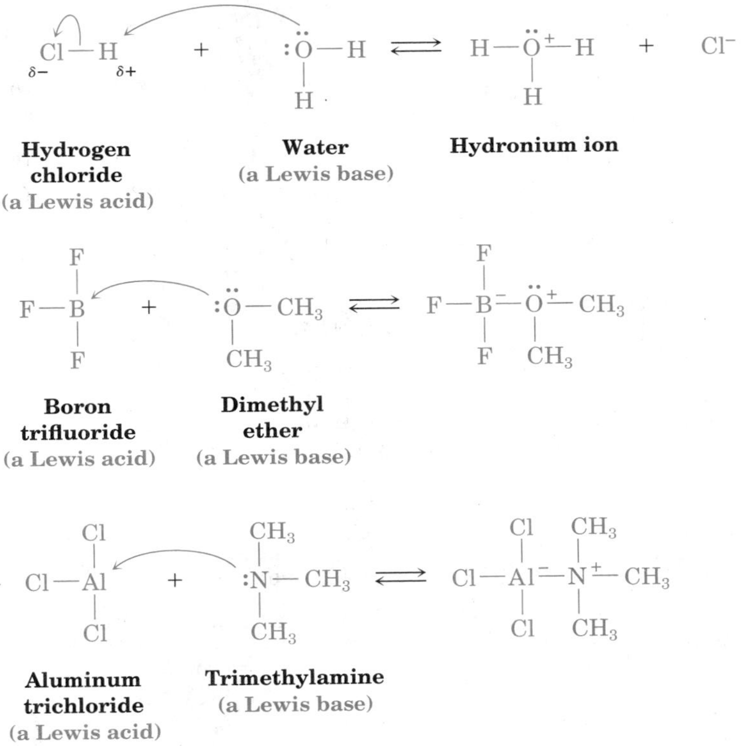

$$\overset{\delta-}{\mathrm{Cl}}-\overset{\delta+}{\mathrm{H}} \quad + \quad :\ddot{\mathrm{O}}(\mathrm{H})-\mathrm{H} \quad \rightleftharpoons \quad \mathrm{H}-\ddot{\mathrm{O}}^{+}(\mathrm{H})-\mathrm{H} \quad + \quad \mathrm{Cl}^-$$

Hydrogen chloride (a Lewis acid) | **Water** (a Lewis base) | **Hydronium ion**

$$\mathrm{F}-\mathrm{B}(\mathrm{F})_2 \quad + \quad :\ddot{\mathrm{O}}(\mathrm{CH_3})-\mathrm{CH_3} \quad \rightleftharpoons \quad \mathrm{F}-\overset{-}{\mathrm{B}}(\mathrm{F})_2-\ddot{\mathrm{O}}^{+}(\mathrm{CH_3})-\mathrm{CH_3}$$

Boron trifluoride (a Lewis acid) | **Dimethyl ether** (a Lewis base)

$$\mathrm{Cl}-\mathrm{Al}(\mathrm{Cl})_2 \quad + \quad :\mathrm{N}(\mathrm{CH_3})_2-\mathrm{CH_3} \quad \rightleftharpoons \quad \mathrm{Cl}-\overset{-}{\mathrm{Al}}(\mathrm{Cl})_2-\mathrm{N}^{+}(\mathrm{CH_3})_2-\mathrm{CH_3}$$

Aluminum trichloride (a Lewis acid) | **Trimethylamine** (a Lewis base)

Figure 1.18 The reactions of some Lewis acids with some Lewis bases. The Lewis acids accept an electron pair; the Lewis bases donate a pair of nonbonding electrons. Note how the flow of electrons from the Lewis base to the Lewis acid is indicated by the curved arrows.

bonding electrons on oxygen, acts as a Lewis base by donating an electron pair to a proton in forming the hydronium ion, H_3O^+:

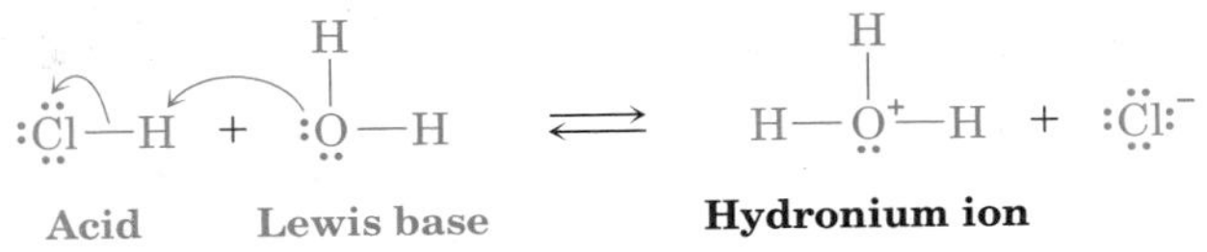

$$:\ddot{\mathrm{Cl}}-\mathrm{H} \quad + \quad :\ddot{\mathrm{O}}(\mathrm{H})-\mathrm{H} \quad \rightleftharpoons \quad \mathrm{H}-\ddot{\mathrm{O}}^{+}(\mathrm{H})-\mathrm{H} \quad + \quad :\ddot{\mathrm{Cl}}:^-$$

Acid | Lewis base | **Hydronium ion**

In a more general sense, most oxygen- and nitrogen-containing organic compounds are Lewis bases because they have lone pairs of electrons. Divalent oxygen compounds each have two lone pairs of electrons, and trivalent nitrogen compounds have one lone pair. Note in the following examples that some compounds can act both as acids and as bases depending on the circumstances, just as water can. Alcohols and carboxylic acids, for instance, act as *acids* when they donate their –OH proton but as *bases* when their oxygen atom accepts a proton.

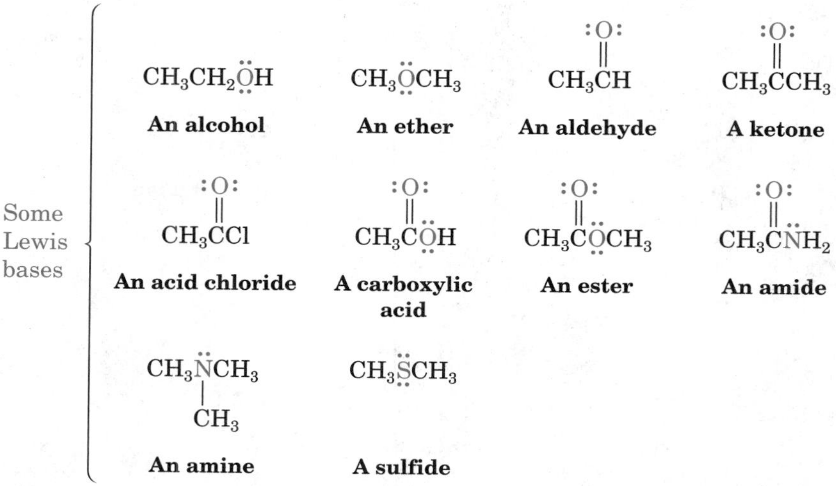

Some Lewis bases

$CH_3CH_2\ddot{O}H$ **An alcohol**	$CH_3\ddot{O}CH_3$ **An ether**	$CH_3C(=O)H$ **An aldehyde**	$CH_3C(=O)CH_3$ **A ketone**
$CH_3C(=O)Cl$ **An acid chloride**	$CH_3C(=O)\ddot{O}H$ **A carboxylic acid**	$CH_3C(=O)\ddot{O}CH_3$ **An ester**	$CH_3C(=O)\ddot{N}H_2$ **An amide**
$CH_3\ddot{N}(CH_3)CH_3$ **An amine**	$CH_3\ddot{S}CH_3$ **A sulfide**		

PRACTICE PROBLEM 1.8

Show how acetaldehyde, CH_3CHO, can act as a Lewis base.

Solution The oxygen atom of acetaldehyde has two lone pairs of electrons that it can donate to a Lewis acid such as H^+.

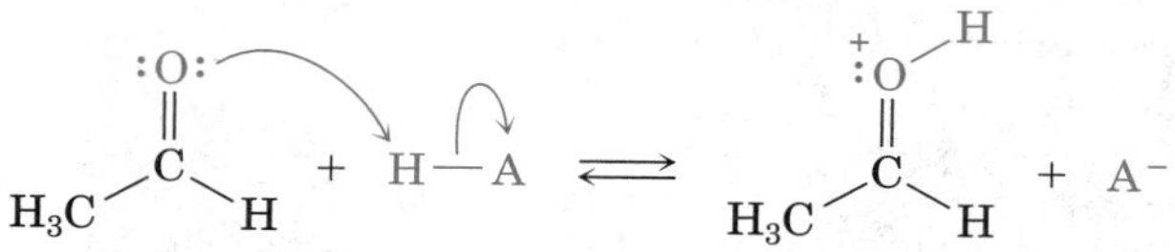

$$H_3C-C(=\ddot{O}:)-H + H-A \rightleftharpoons H_3C-C(=\overset{+}{O}-H)-H + A^-$$

Acetaldehyde

PROBLEM

1.25 Which of the following are likely to act as Lewis acids, which as Lewis bases, and which as both?

(a) $CH_3CH_2-\ddot{O}-H$ (b) $CH_3-NH-CH_3$ (c) $MgBr_2$

(d) $CH_3-B(CH_3)-CH_3$ (e) $H-\overset{+}{C}(H)-H$ (f) $CH_3-\ddot{P}(CH_3)-CH_3$

PROBLEM

1.26 Show how the species in part (a) can act as Lewis bases in their reactions with HCl, and show how the species in part (b) can act as Lewis acids in their reaction with OH^-.

(a) CH_3CH_2OH, $HN(CH_3)_2$, $P(CH_3)_3$ (b) H_3C^+, $B(CH_3)_3$, $MgBr_2$

INTERLUDE

Chemicals, Toxicity, and Risk

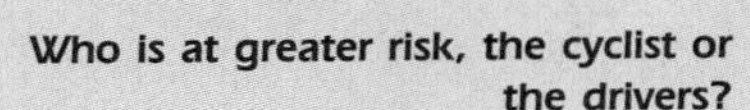

Who is at greater risk, the cyclist or the drivers?

We hear and read a lot these days about the dangers of chemicals—about pesticide residues on food, dangerous food additives, unsafe medicines, and so forth. What's a person to believe?

Life is not risk-free—we all take many risks each day. We decide to ride a bike rather than drive, even though there is a ten times greater likelihood per mile of dying in a bicycling accident than in a car. We decide to walk down stairs rather than take an elevator, even though 7000 people die from falls each year in the United States. We decide to smoke cigarettes, even though it increases our chance of getting cancer by 50%. Making judgments that affect our health is something we do every day without thinking about it.

What about risks from chemicals? Risk evaluation of chemicals is carried out by exposing test animals (usually rats) to a chemical, and then monitoring them for signs of harm. To limit the expense and time needed, the amounts administered are hundreds or thousands of times greater than those a human might normally encounter. Once the animal data are available, the interpretation of the data involves many assumptions. If a substance is harmful to animals, is it necessarily harmful to humans? How can a large dose for a small animal be translated into a small dose for a large human? The standard method for evaluating acute chemical toxicity, as opposed to long-term toxicity, is to report an *LD_{50} value,* the amount of a substance per kilogram body weight that is lethal to 50% of the test animals. The LD_{50} values of various substances are shown in Table 1.4.

Table 1.4 Some LD_{50} Values

Substance	*LD_{50} (g/kg)*	*Substance*	*LD_{50} (g/kg)*
Aflatoxin B_1	4×10^{-4}	Ethyl alcohol	10.6
Aspirin	1.1	Formaldehyde	2.4
Chloroform	3.2	Sodium cyclamate	17

(*continued*)➤

How we evaluate risk is strongly influenced by familiarity. The presence of chloroform in municipal water supplies at a barely detectable level of 0.000 000 01% has caused an outcry in many cities, yet chloroform has a lower acute toxicity than aspirin. Many foods contain natural ingredients far more toxic than synthetic food additives or pesticide residues, but the ingredients are ignored because the foods are familiar. Peanut butter, for example, contains tiny amounts of aflatoxin, a far more potent cancer threat than sodium cyclamate, an artificial sweetener that has been banned because of its "risk."

All decisions involve tradeoffs. Does the benefit of a pesticide that will increase the availability of food outweigh the health risk to 1 person in 1 million who are exposed? Do the beneficial effects of a new drug outweigh a potentially dangerous side effect to 0.001% of users? The answers aren't always obvious, but it's the responsibility of legislators and well-informed citizens to keep their responses on a factual level rather than an emotional one.

Summary and Key Words

acidity constant, K_a, 24
atomic number (Z), 3
atomic weight, 3
bond angle, 14
bond length, 11
bond strength, 11
Brønsted–Lowry acid, 23
Brønsted–Lowry base, 23
conjugate acid, 23
conjugate base, 23
covalent bond, 8
electron shell, 3
electronegativity (EN), 21
ground-state electron configuration, 4
inductive effect, 22
ionic bonding, 8
ionic solid, 8
Kekulé structure, 9
Lewis acid, 26
Lewis base, 26
Lewis structure, 8
line-bond structure, 9
lone-pair electrons, 9
mass number (A), 3

Organic chemistry is the study of carbon compounds. Although a division into inorganic and organic chemistry occurred historically, there is no scientific reason for the division.

Atoms are composed of a positively charged nucleus surrounded by negatively charged electrons that occupy specific regions of space called **orbitals.** Different orbitals have different energy levels and shapes. For example, *s* orbitals are spherical and *p* orbitals are dumbbell-shaped.

There are two fundamental kinds of chemical bonds: **ionic bonds** and **covalent bonds.** The ionic bonds commonly found in inorganic salts result from the electrical attraction of unlike charges. The covalent bonds found in organic molecules are formed by the sharing of one or more electron pairs between atoms. Electron sharing occurs when two atoms approach and their atomic orbitals overlap. Bonds that have a circular cross-section and are formed by head-on overlap of atomic orbitals are called **sigma (σ) bonds.** Bonds formed by sideways overlap of *p* orbitals are called **pi (π) bonds.**

To form bonds in organic compounds, carbon first **hybridizes** to an excited-state configuration. When forming only single bonds, carbon is sp^3-hybridized and has four equivalent **sp^3 hybrid orbitals** with tetrahedral geometry. When forming double bonds, carbon is sp^2-hybridized, has three equivalent **sp^2 orbitals** with planar geometry, and has one unhybridized *p* orbital. When form-

molecule, 8
nonbonding electron, 9
orbital, 3
organic chemistry, 2
pi (π) bond, 17
polar covalent bond, 21
sigma (σ) bond, 11
sp hybrid orbital, 19
sp^2 hybrid orbital, 16
sp^3 hybrid orbital, 13
subshell, 3
valence shell, 7

ing triple bonds, carbon is *sp*-hybridized, has two equivalent ***sp* orbitals** with linear geometry, and has two unhybridized *p* orbitals.

Organic molecules often have **polar covalent bonds** because of unsymmetrical electron sharing caused by the **electronegativity** of atoms. For example, a carbon–oxygen bond is polar because the electronegative oxygen atom attracts the bonding electrons more strongly than carbon does. Carbon–metal bonds, by contrast, are polarized in the opposite sense because carbon attracts electrons more strongly than most metals.

A **Brønsted–Lowry acid** is a substance that can donate a proton (hydrogen ion, H^+); a **Brønsted–Lowry base** is a substance that can accept a proton. The strength of an acid is expressed by its acidity constant, K_a. A **Lewis acid** is a substance that can accept an electron pair. A **Lewis base** is a substance that donates an unshared electron pair. Many organic molecules that contain oxygen and nitrogen are Lewis bases.

Working Problems

There's no better way to learn organic chemistry than by working problems. Although careful reading and rereading of the text is important, reading alone isn't enough. You must also be able to apply the information you read and be able to use your knowledge in new situations. Working problems gives you practice at doing this.

Each chapter in this book provides many problems of different sorts. The in-chapter problems are placed for immediate reinforcement of ideas just learned. The end-of-chapter problems provide additional practice and are of two types: Early problems tend to be the drill type, which provide an opportunity for you to practice your command of the fundamentals; later problems tend to be more challenging and thought-provoking.

As you study organic chemistry, take the time to work the problems. Do the ones you can and ask for help on the ones you can't. If you're stumped by a particular problem, check the accompanying *Study Guide and Solutions Manual* for an explanation that will help clarify the difficulty. Working problems takes effort, but the payoff in knowledge and understanding is immense.

ADDITIONAL PROBLEMS

1.27 How many valence (outer-shell) electrons does each of the following atoms have?
(a) Oxygen (b) Magnesium (c) Fluorine

1.28 Give the ground-state electron configuration of the following elements. For example, carbon is $1s^22s^22p^2$.
(a) Lithium (b) Sodium (c) Aluminum (d) Sulfur

1.29 What are the likely formulas of the following molecules?
(a) $AlCl_?$ (b) $CF_2Cl_?$ (c) $NI_?$

1.30 Identify the bonds in the following molecules as covalent, polar covalent, or ionic:
(a) BeF_2 (b) SiH_4 (c) CBr_4

1.31 Write Lewis (electron-dot) structures for the following molecules:
(a) H–C≡C–H (b) AlH_3 (c) CH_3OH (d) $H_2C{=}CHCl$

1.32 Write a Lewis (electron-dot) structure for acetonitrile, $CH_3C{\equiv}N$. How many electrons does the nitrogen atom have in its valence shell? How many are used for bonding, and how many are not used for bonding?

1.33 What is the hybridization of each carbon atom in acetonitrile, $CH_3C{\equiv}N$?

1.34 Fill in any unshared electrons that are missing from the following line-bond structures:

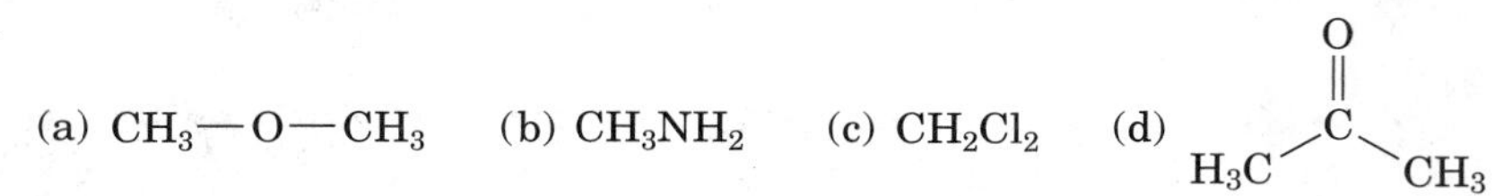

1.35 Draw both a Lewis structure and a line-bond structure for vinyl chloride, C_2H_3Cl, the starting material from which PVC [poly(vinyl chloride)] plastic is made.

1.36 There are two structures that correspond to the formula C_4H_{10}. Draw them.

1.37 Convert the following molecular formulas into line-bond structures:
(a) C_3H_8 (b) C_3H_7Br (two possibilities)
(c) C_3H_6 (two possibilities) (d) C_2H_6O (two possibilities)

1.38 Indicate the kind of hybridization (*sp*, *sp*^{2}, or *sp*^{3}) you expect for each carbon atom in the following molecules:
(a) Butane, $CH_3CH_2CH_2CH_3$ (b) 1-Butene, $CH_3CH_2CH{=}CH_2$
(c) Cyclobutene, (d) 1-Buten-3-yne, $H_2C{=}CH{-}C{\equiv}CH$

H H H H H H (cyclobutene structure: C—C, ‖ |, C—C)

1.39 What is the hybridization of each carbon atom in benzene? What shape would you expect benzene to have?

H H / C=C / H—C C—H / C—C / H H **Benzene**

1.40 Write Lewis (electron-dot) structures for the following molecules:
(a) CH_3—Be—CH_3 (b) CH_3—P—CH_3 with CH_3 on P (c) $TiCl_4$

1.41 Draw line-bond structures for the following covalent molecules:
(a) Br_2 (b) CH_3Cl (c) HF (d) CH_3CH_2OH

1.42 Indicate which of the bonds in the structures you drew for Problem 1.41 are polar covalent. Indicate bond polarity by using the symbols δ+ and δ−.

1.43 Identify the bonds in the following molecules as either ionic or covalent:
(a) NaOH (b) HOH (c) CH_3OH (d) CH_3OCH_3 (e) FF

1.44 Sodium methoxide, $NaOCH_3$, contains both ionic and covalent bonds. Indicate which is which.

1.45 Identify the most electronegative element in each of the following molecules:
(a) CH_2FCl (b) $FCH_2CH_2CH_2Br$ (c) $HOCH_2CH_2NH_2$ (d) CH_3OCH_2Li

1.46 Use the electronegativity table (Figure 1.17) to predict which of the indicated bonds in each of the following sets is more polar:
(a) $Cl–CH_3$ or $Cl–Cl$ (b) $H–CH_3$ or $H–Cl$ (c) $HO–CH_3$ or $(CH_3)_3Si–CH_3$

1.47 Indicate the direction of polarity for each bond in Problem 1.46.

1.48 Which atoms in the following structures have unshared valence electrons? Draw in these unshared electrons.

(a) CH_3SH (b) 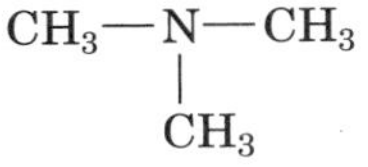 (c) CH_3CH_2Br

(d) 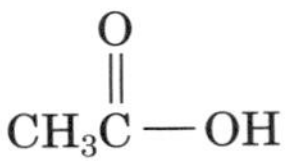 (e) 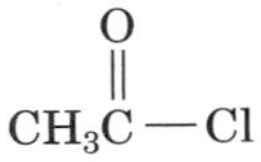

1.49 Draw a three-dimensional representation of chloroform, $CHCl_3$, using the standard convention of solid, wedged, and dashed lines. Do the same for the oxygen-bearing carbon atom in ethanol, CH_3CH_2OH.

1.50 Ammonia, $H_2N–H$, has $pK_a \approx 36$ and acetone has $pK_a \approx 19$. Will the following reaction take place? Explain.

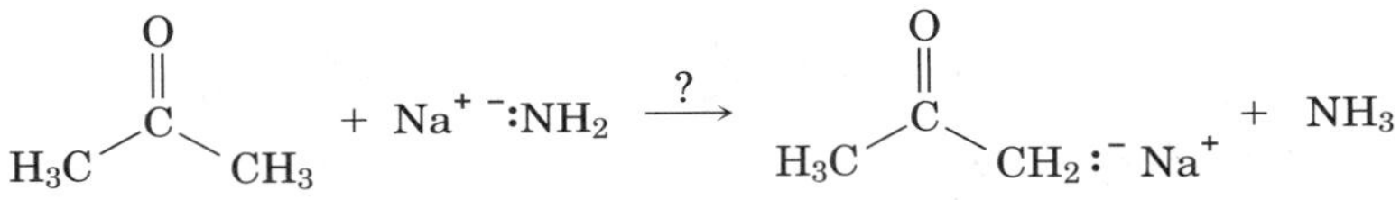

Acetone

1.51 Complete the Lewis electron-dot structure of caffeine, showing all lone-pair electrons, and identify the hybridization of the indicated atoms.

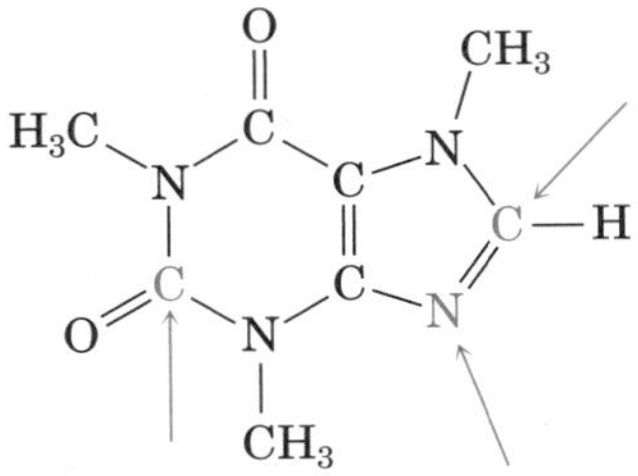

Caffeine

1.52 Which of the following substances are likely to behave as Lewis acids and which as Lewis bases?
(a) $AlBr_3$ (b) $CH_3CH_2NH_2$ (c) HF (d) CH_3SCH_3

1.53 Is the bicarbonate anion (HCO_3^-) a strong enough base to react with methanol (CH_3OH)? In other words, does the following reaction take place as written? (The pK_a of methanol is 15.5; the pK_a of H_2CO_3 is 6.4.)

$$CH_3OH + HCO_3^- \longrightarrow CH_3O^- + H_2CO_3$$

1.54 Identify the acids and bases in the following reactions:

(a) $CH_3OH + H^+ \longrightarrow CH_3\overset{+}{O}H_2$ (b) $CH_3OH + {}^-NH_2 \longrightarrow CH_3O^- + NH_3$

(c) 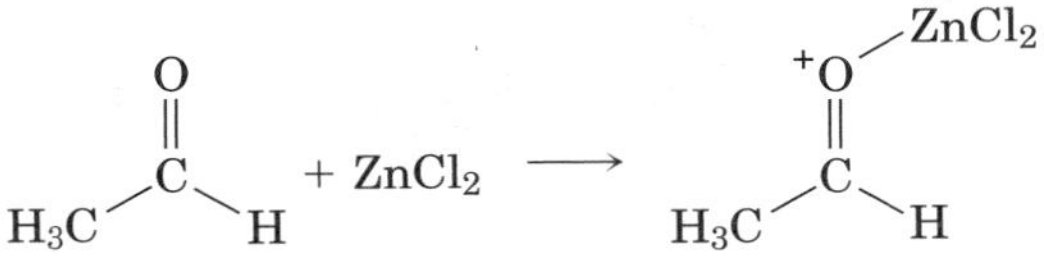

1.55 Rank the following substances in order of increasing acidity:

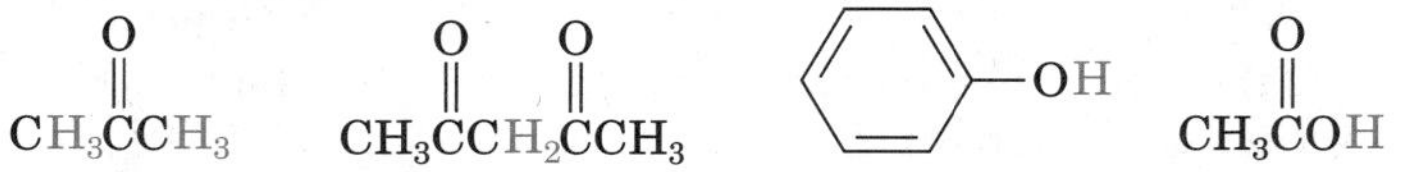

CH_3COCH_3	$CH_3COCH_2COCH_3$	C_6H_5OH	CH_3COOH
Acetone ($pK_a = 19$)	**2,4-Pentanedione** ($pK_a = 9$)	**Phenol** ($pK_a = 9.9$)	**Acetic acid** ($pK_a = 4.76$)

1.56 Which, if any, of the four substances in Problem 1.55 are strong enough acids to react almost completely with NaOH? (The pK_a of H_2O is 15.7.)

1.57 The ammonium ion (NH_4^+, $pK_a = 9.25$) has a lower pK_a than the methylammonium ion ($CH_3NH_3^+$, $pK_a = 10.66$). Which is the stronger base, ammonia (NH_3) or methylamine (CH_3NH_2)? Explain.

1.58 The ammonium ion, NH_4^+, has a geometry identical to that of methane, CH_4. What kind of hybridization do you think the nitrogen atom has? Explain.

1.59 Draw a three-dimensional representation of ethane, CH_3CH_3, using normal, dashed, and wedged lines for both carbons.

1.60 Indicate the kind of hybridization you would expect for each carbon atom in the following molecules:

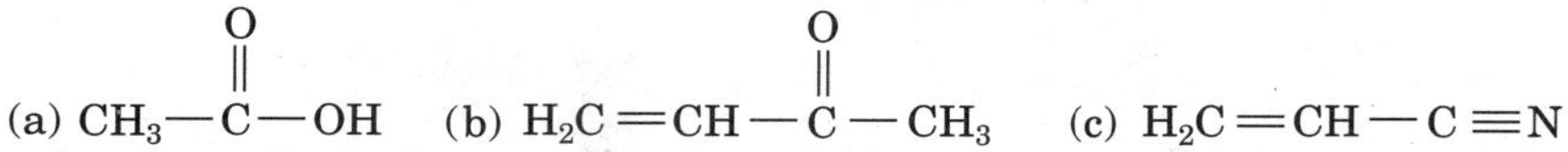

(a) $CH_3—C(=O)—OH$ **Acetic acid**

(b) $H_2C=CH—C(=O)—CH_3$ **3-Buten-2-one**

(c) $H_2C=CH—C\equiv N$ **Acrylonitrile**

1.61 Use Figure 1.17 to order the following molecules according to increasing positive character of the carbon atom:

$$CH_3F,\ CH_3OH,\ CH_3Li,\ CH_3I,\ CH_3CH_3,\ CH_3NH_2$$

1.62 Draw an orbital picture of allene, $H_2C=C=CH_2$. What hybridization must the central carbon atom have to form two double bonds? What shape does allene have?

1.63 Draw a Lewis structure and an orbital picture for carbon dioxide, CO_2. What kind of hybridization do you think the carbon atom has? What is the relationship between CO_2 and allene (Problem 1.62)?

1.64 Although most stable organic compounds have tetravalent carbon atoms, high-energy species with trivalent carbon atoms also exist. *Carbocations* are one such class of compounds. If the positively charged carbon atom has planar geometry, what hybridization do you think it has? How many valence electrons does the carbon have?

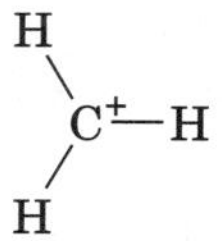

$H_2C^+—H$ **A carbocation**

1.65 Propose structures for molecules that meet the following descriptions:
(a) Contains two sp^2-hybridized carbons and two sp^3-hybridized carbons
(b) Contains only four carbons, all of which are sp^2-hybridized
(c) Contains two sp-hybridized carbons and two sp^2-hybridized carbons

1.66 There are two different substances with the formula C_4H_{10}. Draw them both, and tell how they differ.

Visualizing Chemistry

1.67 Shown below is a model of acetaminophen, a pain-reliever sold in drug stores as Tylenol. Only the connections between atoms are shown; multiple bonds are not indicated. Identify the hybridization of each carbon atom in acetaminophen, and tell which atoms have lone pairs of electrons (gray = C, red = O, blue = N, light green = H).

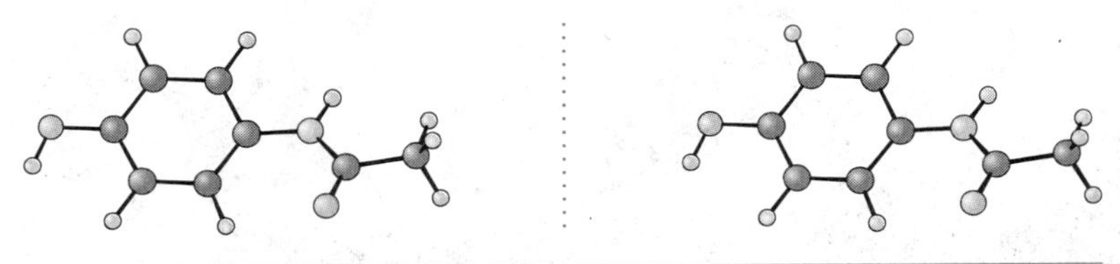

Stereo View

1.68 Shown below is a model of aspartame, $C_{14}H_{18}N_2O_5$, known commercially as Nutra-Sweet. Only the connections between atoms are shown; multiple bonds are not indicated. Complete the structure by indicating the positions of multiple bonds (gray = C, red = O, blue = N, light green = H).

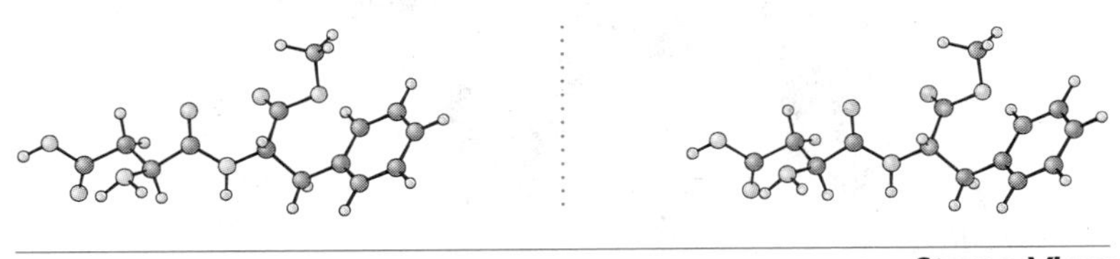

Stereo View

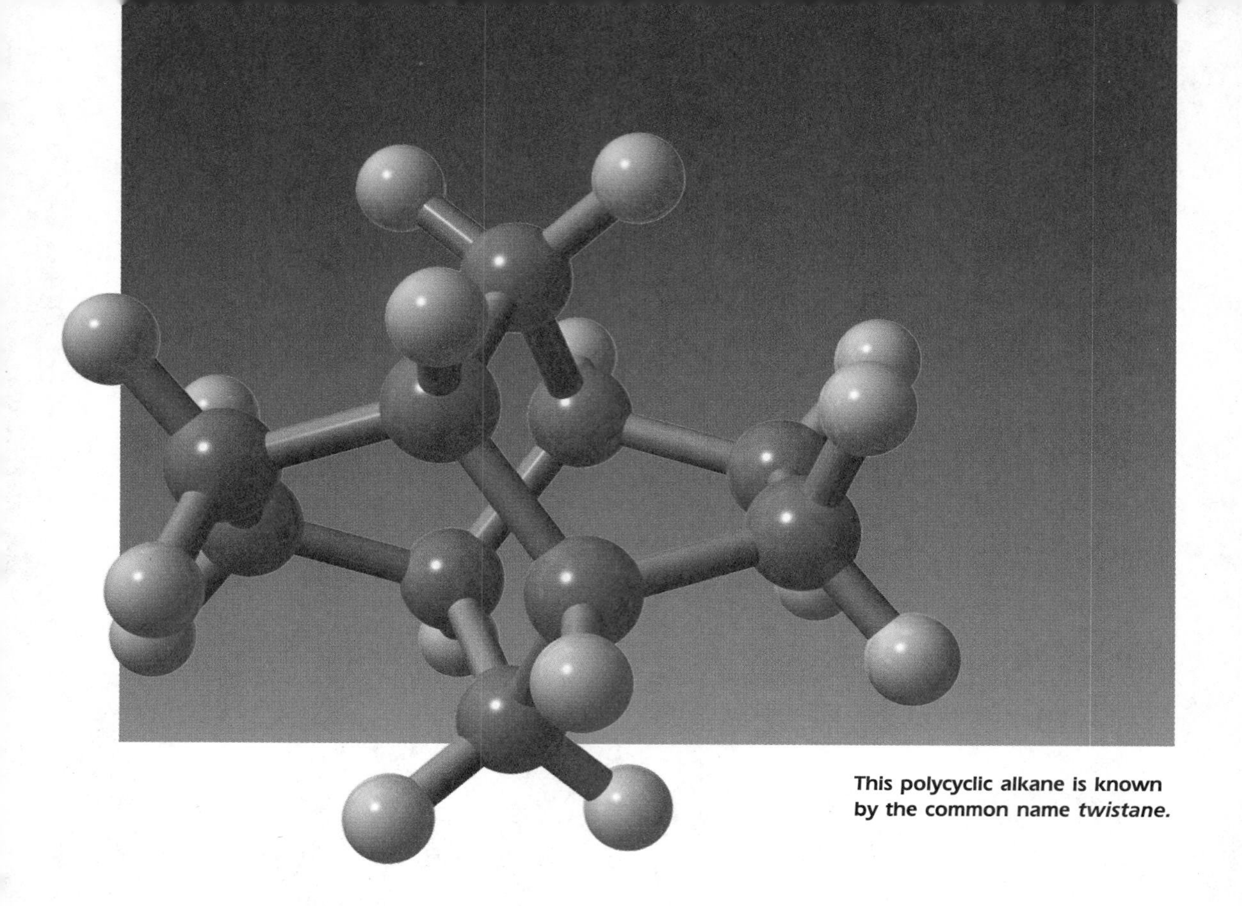

This polycyclic alkane is known by the common name *twistane.*

2 The Nature of Organic Compounds: Alkanes

There are more than *12 million* known organic compounds, each of which has its own unique physical and chemical properties. Chemists have learned through years of experience that these compounds can be classified into families according to their structural features and that the members of a given family often have similar chemical reactivity. Instead of 12 million compounds with random reactivity, there are a few dozen families of compounds whose chemistry is reasonably predictable. We'll study the chemistry of specific families of organic molecules throughout this book, beginning in this chapter with a look at the simplest family, the *alkanes.*

2.1 Functional Groups

The structural features that make it possible to classify compounds by reactivity are called *functional groups.* A **functional group** is a part of a larger molecule and is composed of an atom or group of atoms that have a characteristic chemical behavior. Chemically, a given functional group behaves almost the same way in every molecule it's in. For example, one of the simplest functional groups is the carbon–carbon double bond. Because the electronic structure of the carbon–carbon double bond remains essentially the same in all molecules where it occurs, its chemical reactivity also remains the same. Ethylene, the simplest compound with a carbon–carbon double bond, undergoes reactions that are remarkably similar to those of α-pinene, a more complicated molecule (and major component of turpentine). Both, for example, react with Br_2 to give products in which a bromine atom has added to each of the double-bond carbons (Figure 2.1).

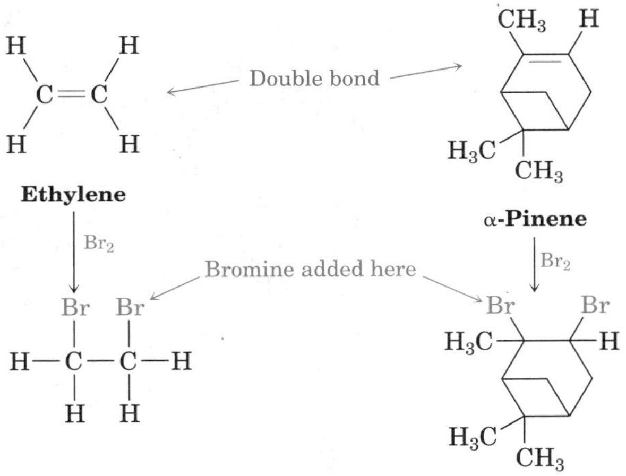

Figure 2.1 The reactions of ethylene and α-pinene with Br_2. In both cases, Br_2 reacts with the C=C double-bond functional group in exactly the same way. The size and nature of the remainder of the molecule are not important.

The example shown in Figure 2.1 is typical: *The chemistry of every organic molecule, regardless of size and complexity, is determined by the functional groups it contains.* Table 2.1 lists many of the common functional groups and gives simple examples of their occurrence. Most of the chemistry you'll be studying in the remainder of this book is the chemistry of these functional groups.

It's a good idea at this point to familiarize yourself with the structures of the functional groups shown in Table 2.1 so that you'll recognize them when you see them again. They can be grouped into several categories:

Functional Groups with Carbon–Carbon Multiple Bonds

Alkenes, alkynes, and arenes (aromatic compounds) all contain carbon–carbon multiple bonds. *Alkenes* have a double bond, *alkynes* have a triple bond, and *arenes* have three alternating double and single bonds in a six-

Table 2.1 Structure of Some Common Functional Groups

Family name	*Functional group structure*[a]	*Simple example*	*Name ending*
Alkane	(Contains only C—H and C—C single bonds)	CH_3CH_3	*-ane* Ethane
Alkene	>C=C<	$H_2C{=}CH_2$	*-ene* Ethene (Ethylene)
Alkyne	—C≡C—	H—C≡C—H	*-yne* Ethyne (Acetylene)
Arene	(six-membered ring of alternating C=C and C—C bonds, each C with one substituent bond)	(benzene ring, each C bearing H)	None Benzene
Halide	—C—X: (X = F, Cl, Br, I)	H_3C—Cl	None Chloromethane
Alcohol	—C—O—H	H_3C—O—H	*-ol* Methanol
Ether	—C—O—C—	H_3C—O—CH_3	*ether* Dimethyl ether
Amine	—C—N—H (with H on N), —C—N—H, —C—N—	H_3C—NH_2	*-amine* Methylamine
Nitrile	—C—C≡N:	H_3C—C≡N	*-nitrile* Ethanenitrile (Acetonitrile)
Sulfide	—C—S—C—	H_3C—S—CH_3	*sulfide* Dimethyl sulfide
Thiol	—C—S—H	H_3C—SH	*-thiol* Methanethiol

Family name	*Functional group structure*[a]	*Simple example*	*Name ending*
Carbonyl, $\mathrm{-C(=O)-}$	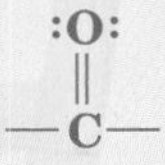		
Aldehyde	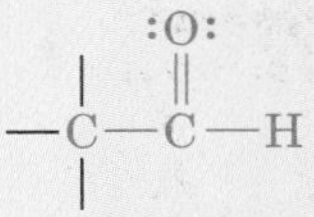 $\mathrm{-C-C(=O)-H}$	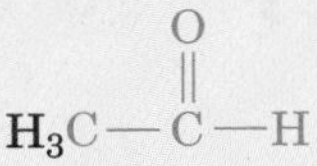 $\mathrm{H_3C-C(=O)-H}$	*-al* Ethanal (Acetaldehyde)
Ketone	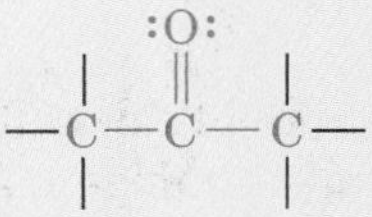 $\mathrm{-C-C(=O)-C-}$	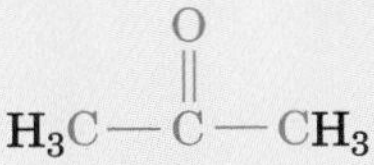 $\mathrm{H_3C-C(=O)-CH_3}$	*-one* Propanone (Acetone)
Carboxylic acid	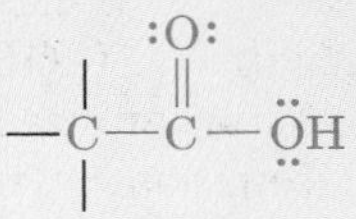 $\mathrm{-C-C(=O)-OH}$	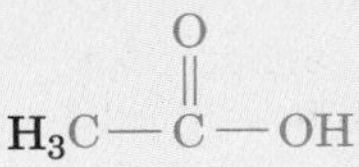 $\mathrm{H_3C-C(=O)-OH}$	*-oic acid* Ethanoic acid (Acetic acid)
Ester	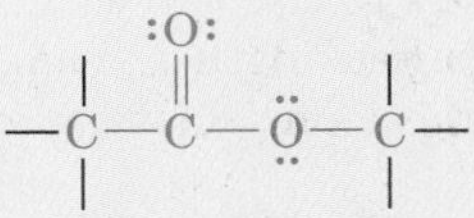 $\mathrm{-C-C(=O)-O-C-}$	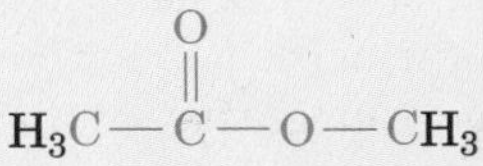 $\mathrm{H_3C-C(=O)-O-CH_3}$	*-oate* Methyl ethanoate (Methyl acetate)
Amide	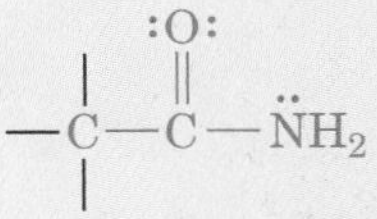 $\mathrm{-C-C(=O)-NH_2}$ $\mathrm{-C-C(=O)-N(-)-H}$ $\mathrm{-C-C(=O)-N(-)-}$	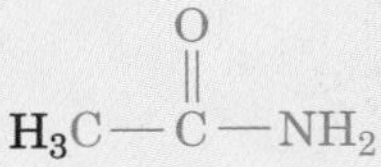 $\mathrm{H_3C-C(=O)-NH_2}$	*-amide* Ethanamide (Acetamide)
Carboxylic acid chloride	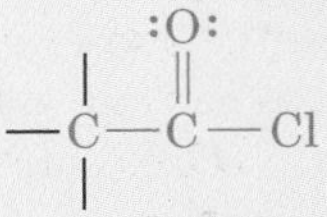 $\mathrm{-C-C(=O)-Cl}$	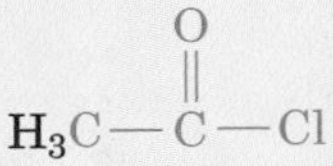 $\mathrm{H_3C-C(=O)-Cl}$	*-oyl chloride* Ethanoyl chloride (Acetyl chloride)
Carboxylic acid anhydride	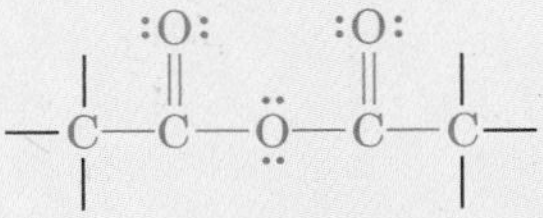 $\mathrm{-C-C(=O)-O-C(=O)-C-}$	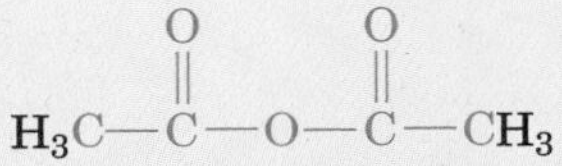 $\mathrm{H_3C-C(=O)-O-C(=O)-CH_3}$	*-oic anhydride* Ethanoic anhydride (Acetic anhydride)

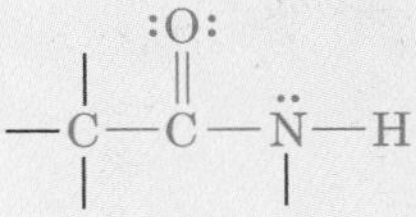

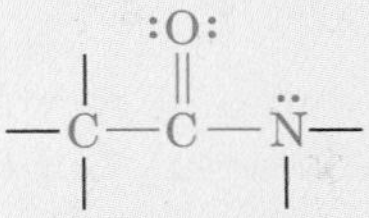

[a]The bonds whose connections aren't specified are assumed to be attached to carbon or hydrogen atoms in the rest of the molecule.

membered ring of carbon atoms. Because of their structural similarities, these compounds also have some chemical similarities.

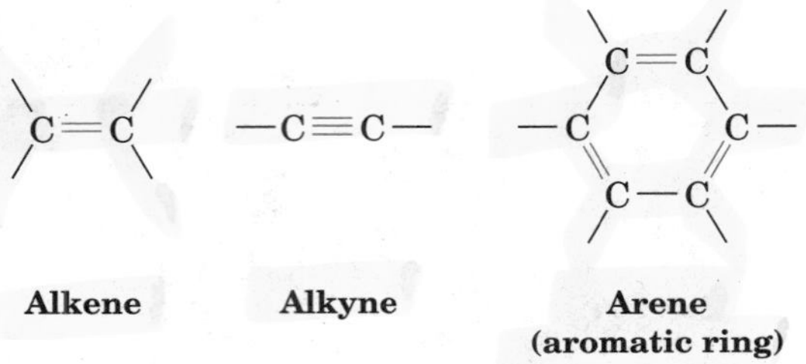

Alkene **Alkyne** **Arene (aromatic ring)**

Functional Groups with Carbon Singly Bonded to an Electronegative Atom

Alkyl halides, alcohols, ethers, amines, thiols, and sulfides all have a carbon atom singly bonded to an electronegative atom—a halogen, an oxygen, a nitrogen, or a sulfur. Alkyl halides have a carbon atom bonded to a halogen (–X), alcohols have a carbon atom bonded to a hydroxyl (–OH) group, ethers have two carbon atoms bonded to the same oxygen, amines have a carbon atom bonded to a nitrogen, thiols have a carbon atom bonded to an –SH group, and sulfides have two carbon atoms bonded to the same sulfur. In all cases, the bonds are polar, with the carbon atom bearing a partial positive charge ($\delta+$) and the electronegative atom bearing a partial negative charge ($\delta-$).

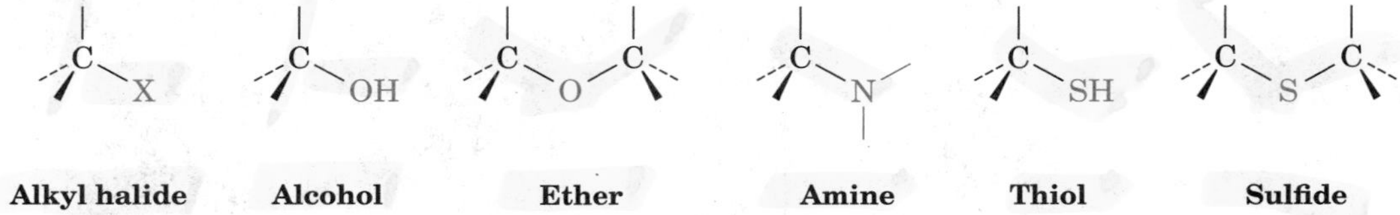

Alkyl halide **Alcohol** **Ether** **Amine** **Thiol** **Sulfide**

Functional Groups with a Carbon–Oxygen Double Bond (Carbonyl Groups)

Note particularly in Table 2.1 the different families of compounds that contain the *carbonyl group,* C=O (pronounced car-bo-**neel**). Carbon–oxygen double bonds are present in some of the most important compounds in organic chemistry. These compounds are similar in many respects but differ depending on the identity of the atoms bonded to the carbonyl-group carbon. Aldehydes have at least one hydrogen bonded to the C=O, ketones have two carbons bonded to the C=O, carboxylic acids have one carbon and one –OH group bonded to the C=O, esters have one carbon and one ether-like oxygen bonded to the C=O, amides have one carbon and one nitrogen bonded to the C=O, acid chlorides have one carbon and one chlorine bonded to the C=O, and so on.

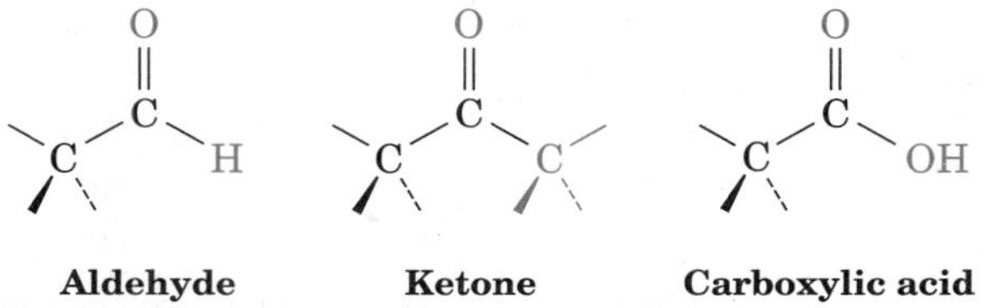

Aldehyde **Ketone** **Carboxylic acid**

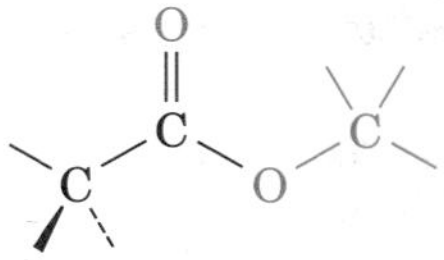

Ester 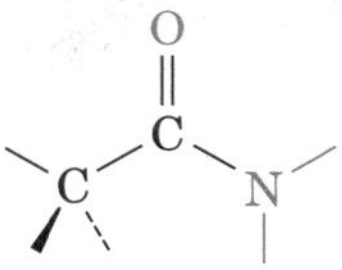 **Amide** 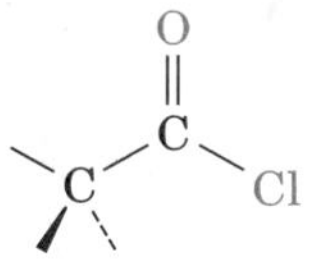 **Acid chloride**

PROBLEM

2.1 Circle and identify the functional groups in the following molecules:

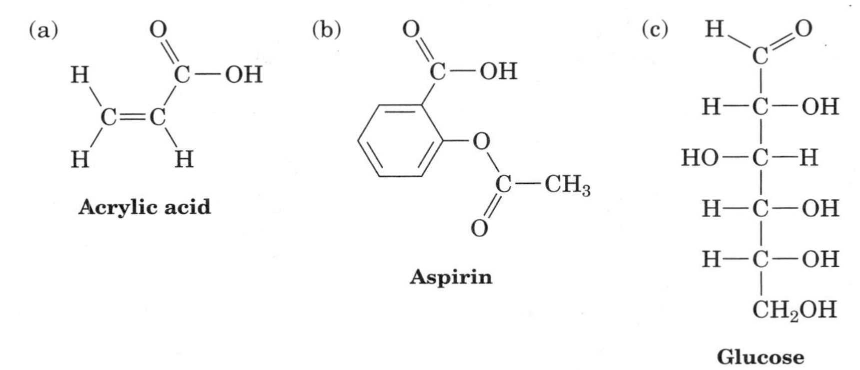

PROBLEM

2.2 Propose structures for simple molecules that contain the following functional groups:

(a) Alcohol (b) Aromatic ring (c) Carboxylic acid
(d) Amine (e) Both ketone and amine (f) Two double bonds

2.2 Alkanes and Alkyl Groups: Isomers

We saw in Section 1.8 that the C–C single bond in ethane results from σ (head-on) overlap of carbon sp^3 orbitals. If we imagine joining three, four, five, or even more carbon atoms by C–C single bonds, we can generate the large family of molecules called **alkanes.**

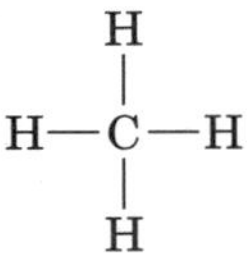

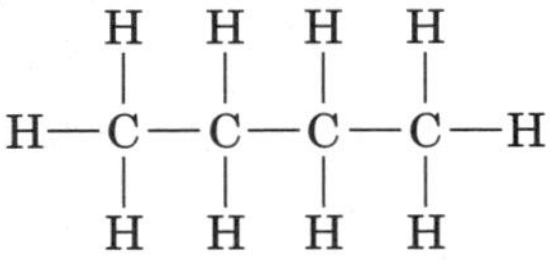

. . . and so on

Methane **Ethane** **Propane** **Butane**

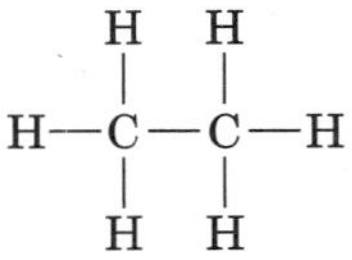

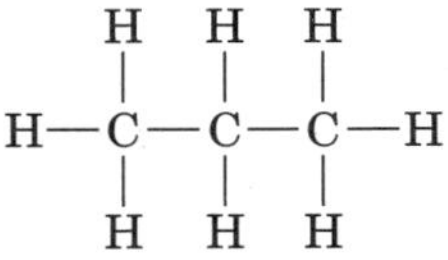

Alkanes are often described as *saturated hydrocarbons:* **hydrocarbons** because they contain only carbon and hydrogen atoms; **saturated** because they have only C–C and C–H single bonds and thus contain the maximum possible number of hydrogens per carbon. They have the general formula C_nH_{2n+2}, where n is any integer. Alkanes are also occasionally called **aliphatic** compounds, a word derived from the Greek *aleiphas,* meaning "fat." We'll see in Chapter 16 that animal fats contain long carbon chains similar to alkanes.

Think about the ways that carbon and hydrogen can combine to make alkanes. With one carbon and four hydrogens, only one structure is possible: methane, CH_4. Similarly, there is only one possible combination of two carbons with six hydrogens (ethane, CH_3CH_3) and only one possible combination of three carbons with eight hydrogens (propane, $CH_3CH_2CH_3$). If larger numbers of carbons and hydrogens combine, however, more than one kind of molecule can form. For example, there are *two* ways that molecules with the formula C_4H_{10} can form: The four carbons can be in a row (butane), or they can branch (isobutane). Similarly, there are three ways in which C_5H_{12} molecules can form, and so on for larger alkanes:

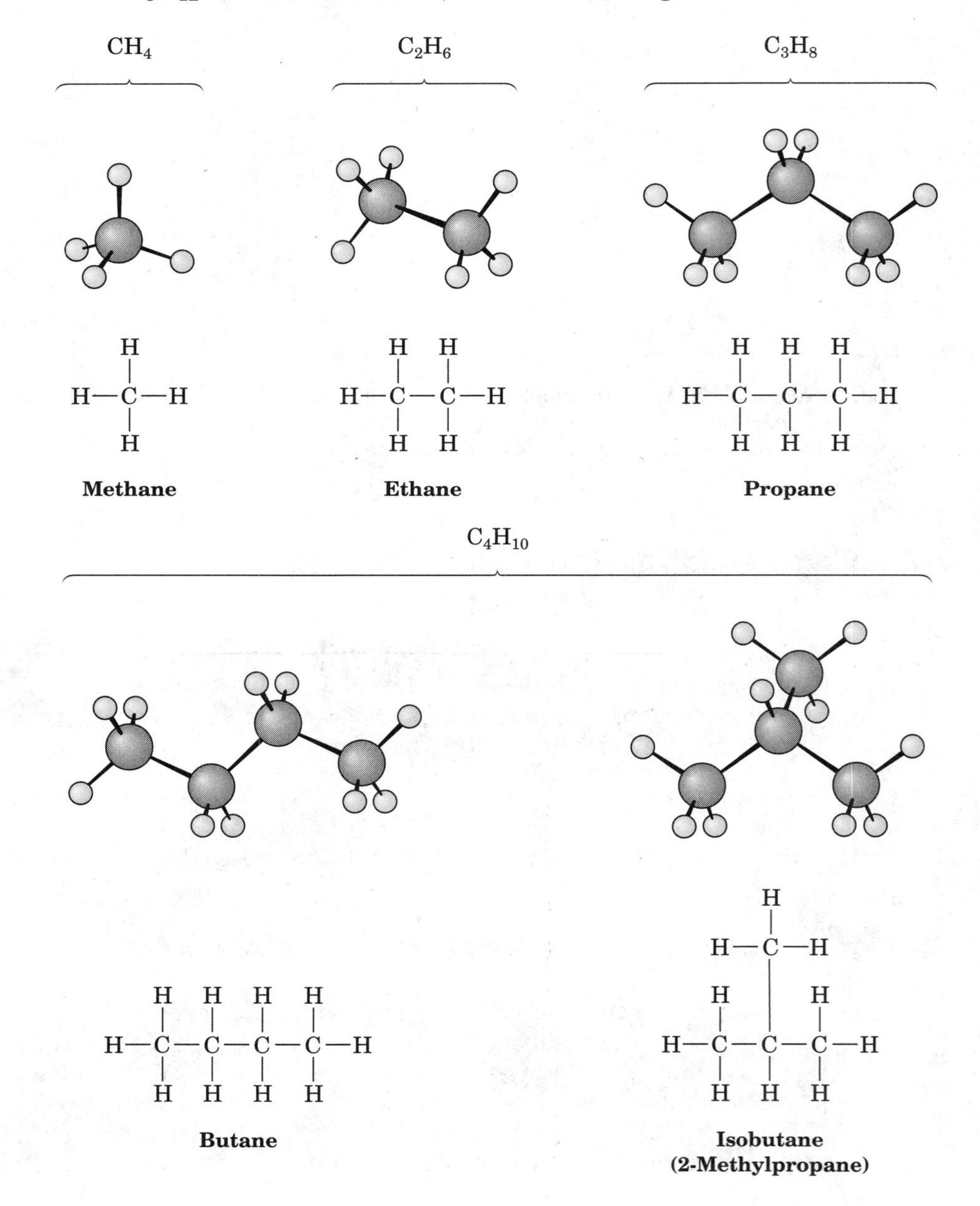

C_5H_{12}

```
H   H   H   H   H
|   |   |   |   |
H—C—C—C—C—C—H
|   |   |   |   |
H   H   H   H   H
```

Pentane

```
        H
        |
      H—C—H
H   H   |   H
|   |   |   |
H—C—C—C—C—H
|   |   |   |
H   H   H   H
```

2-Methylbutane

```
      H
      |
    H—C—H
H     |     H
|     |     |
H—C—C—C—H
|     |     |
H     |     H
    H—C—H
      |
      H
```

2,2-Dimethylpropane

Compounds like butane, whose carbons are connected in a row, are called **straight-chain alkanes,** or **normal alkanes,** whereas compounds with branched carbon chains, such as isobutane (2-methylpropane), are called **branched-chain alkanes.** The difference between the two is that you can draw a line connecting all the carbons of a straight-chain alkane without retracing your path or lifting your pencil from the paper. For a branched-chain alkane, however, you either have to retrace your path or lift your pencil from the paper to draw a line connecting all the carbons.

Compounds like the two C_4H_{10} molecules, which have the same formula but different structures, are called *isomers* from the Greek *isos* + *meros* meaning "made of the same parts." **Isomers** have the same numbers and kinds of atoms but differ in the way the atoms are arranged. Compounds like butane and isobutane, whose atoms are connected differently, are called **constitutional isomers.** We'll see shortly that other kinds of isomerism are also possible, even among compounds whose atoms are connected in the same order. As Table 2.2 shows, the number of possible alkane isomers increases dramatically as the number of carbon atoms increases.

Table 2.2 Number of Alkane Isomers

Formula	*Number of isomers*	*Formula*	*Number of isomers*
C_6H_{14}	5	$C_{10}H_{22}$	75
C_7H_{16}	9	$C_{15}H_{32}$	4,347
C_8H_{18}	18	$C_{20}H_{42}$	366,319
C_9H_{20}	35	$C_{30}H_{62}$	4,111,846,763

A given alkane can be arbitrarily shown in many ways. For example, the straight-chain, four-carbon alkane called butane can be represented by any of the structures shown in Figure 2.2. These structures aren't intended to imply any particular three-dimensional geometry for butane; they only indicate the connections among atoms. In practice, chemists rarely draw all the bonds in a molecule and usually refer to butane by the **condensed structure,** $CH_3CH_2CH_2CH_3$ or $CH_3(CH_2)_2CH_3$. In such representations, the C–C and C–H bonds are "understood" rather than shown. If a carbon has three hydrogens bonded to it, we write CH_3; if a carbon has two hydrogens bonded to it, we write CH_2, and so on. Still more simply, butane can even be represented as n-C_4H_{10}, where *n* signifies *normal,* straight-chain butane.

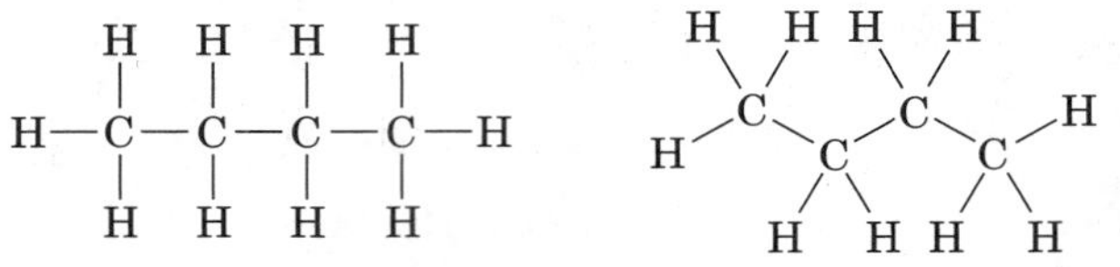

$CH_3—CH_2—CH_2—CH_3$ $CH_3CH_2CH_2CH_3$ $CH_3(CH_2)_2CH_3$

Figure 2.2 Some representations of butane (n-C_4H_{10}). The molecule is the same regardless of how it's drawn. These structures imply only that butane has a continuous chain of four carbon atoms.

Straight-chain alkanes are named according to the number of carbon atoms they contain, as shown in Table 2.3. With the exception of the first four compounds—methane, ethane, propane, and butane—whose names have historical origins, the alkanes are named based on Greek numbers, according to the number of carbons. The suffix *-ane* is added to the end of each name to identify the molecule as an alkane.

If a hydrogen atom is removed from an alkane, the partial structure that remains is called an **alkyl group.** Alkyl groups are named by replac-

Table 2.3 Names of Straight-Chain Alkanes

Number of carbons (n)	*Name*	*Formula* (C_nH_{2n+2})	*Number of carbons (n)*	*Name*	*Formula* (C_nH_{2n+2})
1	Methane	CH_4	9	Nonane	C_9H_{20}
2	Ethane	C_2H_6	10	Decane	$C_{10}H_{22}$
3	Propane	C_3H_8	11	Undecane	$C_{11}H_{24}$
4	Butane	C_4H_{10}	12	Dodecane	$C_{12}H_{26}$
5	Pentane	C_5H_{12}	13	Tridecane	$C_{13}H_{28}$
6	Hexane	C_6H_{14}	20	Icosane	$C_{20}H_{42}$
7	Heptane	C_7H_{16}	21	Henicosane	$C_{21}H_{44}$
8	Octane	C_8H_{18}	30	Triacontane	$C_{30}H_{62}$

ing the *-ane* ending of the parent alkane with an *-yl* ending. For example, removal of a hydrogen atom from methane, CH_4, generates a *methyl group,* $-CH_3$, and removal of a hydrogen atom from ethane, CH_3CH_3, generates an *ethyl group,* $-CH_2CH_3$. Similarly, removal of a hydrogen atom from the end carbon of any *n*-alkane gives the series of *n*-alkyl groups shown in Table 2.4.

Table 2.4 Some Straight-Chain Alkyl Groups

Alkane	*Name*	*Alkyl group*	*Name (abbreviation)*
CH_4	Methane	$-CH_3$	Methyl (Me)
CH_3CH_3	Ethane	$-CH_2CH_3$	Ethyl (Et)
$CH_3CH_2CH_3$	Propane	$-CH_2CH_2CH_3$	Propyl (Pr)
$CH_3CH_2CH_2CH_3$	Butane	$-CH_2CH_2CH_2CH_3$	Butyl (Bu)
$CH_3CH_2CH_2CH_2CH_3$	Pentane	$-CH_2CH_2CH_2CH_2CH_3$	Pentyl

Just as *n*-alkyl groups are generated by removing a hydrogen from an *end* carbon, branched alkyl groups are generated by removing a hydrogen atom from an *internal* carbon. Two 3-carbon alkyl groups and four 4-carbon alkyl groups are possible (Figure 2.3, p. 46).

One further word of explanation about naming alkyl groups: The prefixes used for the C_4 alkyl groups in Figure 2.3—*sec* (for secondary) and *tert* (for tertiary)—refer to the degree of alkyl substitution at the branching carbon atom. There are four possible degrees of alkyl substitution for carbon, denoted 1°, 2°, 3°, and 4°:

```
    H              H              R              R
    |              |              |              |
R — C — H      R — C — H      R — C — H      R — C — R
    |              |              |              |
    H              R              R              R
```

Primary carbon (1°) is bonded to one other carbon

Secondary carbon (2°) is bonded to two other carbons

Tertiary carbon (3°) is bonded to three other carbons

Quaternary carbon (4°) is bonded to four other carbons

The symbol **R** is used here and throughout this text to represent a *generalized* alkyl group. The R group can be methyl, ethyl, or any of a multitude of other alkyl groups. You might think of **R** as representing the **R**est of the molecule, which we aren't bothering to specify.

```
    H                                  CH3
    |                                  |
R — C — OH          CH3CH2OH        CH3CHCH2CH2OH
    |
    H
```

General class of primary alcohols, RCH_2OH

Specific examples of primary alcohols, RCH_2OH

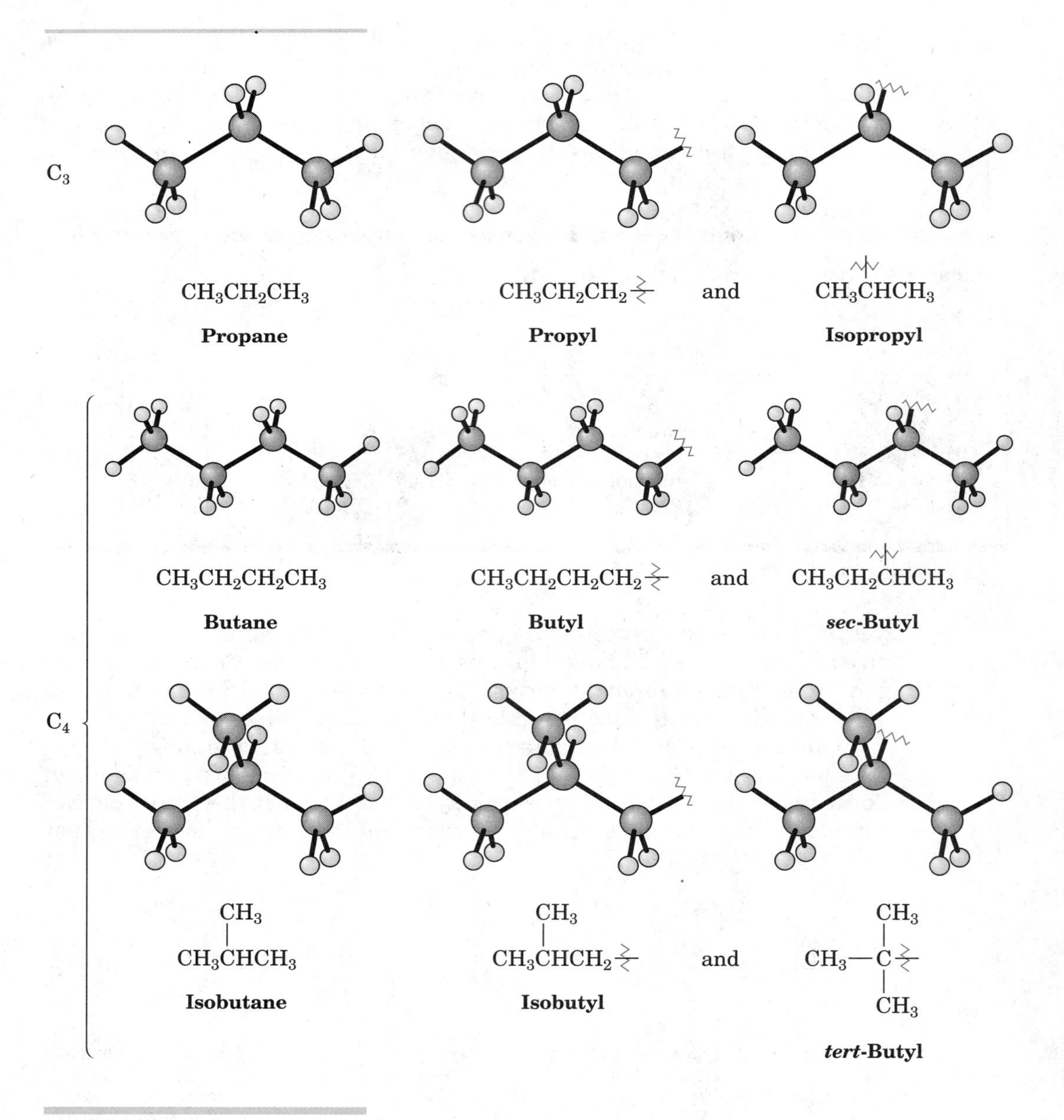

Figure 2.3 Generation of straight-chain and branched-chain alkyl groups from alkanes.

PRACTICE PROBLEM 2.1

Propose structures for two isomers with the formula C_2H_6O.

Solution We know that carbon forms four bonds, oxygen forms two, and hydrogen forms one. Putting the pieces together yields two possibilities:

$$\begin{array}{c} \text{H}\;\;\text{H} \\ |\;\;\;\;| \\ \text{H—C—C—O—H} \\ |\;\;\;\;| \\ \text{H}\;\;\text{H} \end{array} \quad \text{and} \quad \begin{array}{c} \text{H}\;\;\;\;\;\;\;\text{H} \\ |\;\;\;\;\;\;\;\;\;| \\ \text{H—C—O—C—H} \\ |\;\;\;\;\;\;\;\;\;| \\ \text{H}\;\;\;\;\;\;\;\text{H} \end{array}$$

PROBLEM

2.3 Draw structures of the five isomers of C_6H_{14}.

PROBLEM

2.4 Draw structures that meet the following descriptions:
(a) Three isomers with the formula C_8H_{18}
(b) Two isomers with the formula $C_4H_8O_2$

PROBLEM

2.5 Draw the eight possible five-carbon alkyl groups (pentyl isomers).

PROBLEM

2.6 Draw alkanes that meet the following descriptions:
(a) An alkane with two tertiary carbons
(b) An alkane that contains an isopropyl group
(c) An alkane that has one quaternary and one secondary carbon

PROBLEM

2.7 Identify the carbon atoms in the following molecules as primary, secondary, tertiary, or quaternary:

```
        CH3                       CH3CHCH3                      CH3    CH3
        |                            |                          |      |
(a) CH3CHCH2CH2CH3     (b) CH3CH2CHCH2CH3      (c) CH3CHCH2CCH3
                                                                       |
                                                                       CH3
```

2.3 Naming Branched-Chain Alkanes

In earlier times, when few pure organic chemicals were known, new compounds were named at the whim of their discoverer. Thus, urea (CH_4N_2O) is a crystalline substance isolated from urine, and barbituric acid is a tranquilizing agent named by its discoverer in honor of his friend Barbara. As the science of organic chemistry slowly grew in the nineteenth century, so too did the number of known compounds and the need for a systematic method of naming them. The system of nomenclature we'll use in this book is that devised by the International Union of Pure and Applied Chemistry (IUPAC, usually spoken as **eye**-you-pac).

A chemical name has three parts in the **IUPAC system:** prefix, parent, and suffix. The parent name selects a main part of the molecule and tells how many carbon atoms are in that part, the suffix identifies the functional-group family that the molecule belongs to, and the prefix specifies the location(s) of various substituents on the main part:

As we cover new functional groups in later chapters, the applicable IUPAC rules of nomenclature will be given. In addition, Appendix A gives an overall view of organic nomenclature and shows how compounds that contain more than one functional group are named. For the present, let's see how branched-chain alkanes are named. All but the most complex branched-chain alkanes can be named by following four steps:

Step 1 Find the parent hydrocarbon.

(a) Find the *longest continuous carbon chain* in the molecule and use the name of that chain as the parent name. The longest chain may not always be obvious; you may have to "turn corners":

$$\begin{array}{l} \qquad\qquad\quad CH_2CH_3 \\ \qquad\qquad\quad | \\ CH_3CH_2CH_2CH—CH_3 \end{array}$$

Named as a substituted hexane

(b) If two chains of equal length are present, choose the one with the larger number of branch points as the parent:

$$\begin{array}{l} \quad CH_3 \\ \quad | \\ CH_3CHCHCH_2CH_2CH_3 \\ \qquad\ | \\ \qquad\ CH_2CH_3 \end{array} \qquad NOT \qquad \begin{array}{l} \quad CH_3 \\ \quad | \\ CH_3CH—CHCH_2CH_2CH_3 \\ \qquad\qquad | \\ \qquad\qquad CH_2CH_3 \end{array}$$

Named as a hexane with *two* substituents — NOT — as a hexane with *one* substituent

Step 2 Beginning at the end *nearer the first branch point,* number each carbon atom in the parent chain:

$$\begin{array}{l} \qquad\ {}^{1}CH_3 \\ \qquad\ | \\ \qquad\ {}^{2}CH_2 \\ \qquad\ | \\ CH_3—\underset{3}{C}HCH—CH_2CH_3 \\ \qquad\qquad |_{4} \\ \qquad\qquad \underset{5}{C}H_2\underset{6}{C}H_2\underset{7}{C}H_3 \end{array} \qquad NOT \qquad \begin{array}{l} \qquad\ {}^{7}CH_3 \\ \qquad\ | \\ \qquad\ {}^{6}CH_2 \\ \qquad\ | \\ CH_3—\underset{5}{C}HCH—CH_2CH_3 \\ \qquad\qquad |_{4} \\ \qquad\qquad \underset{3}{C}H_2\underset{2}{C}H_2\underset{1}{C}H_3 \end{array}$$

The first branch occurs at C3 in the proper system of numbering, but at C4 in the improper system.

Step 3 Assign a number to each substituent according to its point of attachment on the parent chain. If there are two substituents on the same carbon, assign them both the same number. There must always be as many numbers in the name as there are substituents.

$$\begin{array}{l} \quad\ \ {}^{9}\ \ \ {}^{8} \\ \quad CH_3CH_2 \qquad\qquad CH_3 \quad CH_2CH_3 \\ \quad\ \ | \qquad\qquad\qquad\ \ | \qquad\ | \\ CH_3—\underset{7}{C}H\underset{6}{C}H_2\underset{5}{C}H_2\underset{4}{C}H—\underset{3}{C}H\underset{2}{C}H_2\underset{1}{C}H_3 \end{array}$$

Substituents:

On C3, CH_2CH_3 (3-ethyl)
On C4, CH_3 (4-methyl)
On C7, CH_3 (7-methyl)

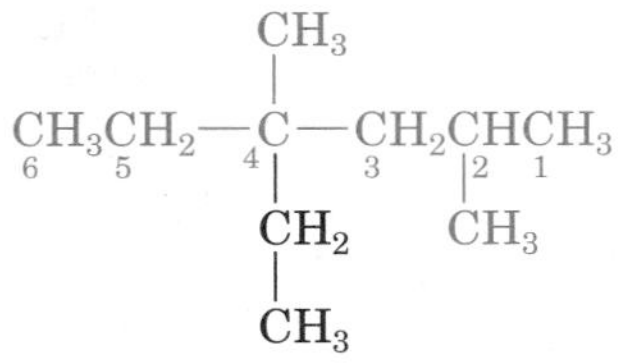

$$
\begin{array}{c}
CH_3 \\
| \\
\underset{6}{CH_3}\underset{5}{CH_2}{-}\underset{4}{C}{-}\underset{3}{CH_2}\underset{2}{CH}\underset{1}{CH_3} \\
\quad | \qquad\qquad | \\
\quad CH_2 \qquad CH_3 \\
| \\
CH_3
\end{array}
$$

Substituents:

On C2, CH_3	(2-methyl)
On C4, CH_3	(4-methyl)
On C4, CH_2CH_3	(4-ethyl)

Step 4 Write the name as a single word, using hyphens to separate the various prefixes and commas to separate numbers. If two or more different side chains are present, cite them in alphabetical order. If two or more identical side chains are present, use one of the prefixes *di-*, *tri-*, *tetra-*, and so forth. Don't use these prefixes for alphabetizing, though.

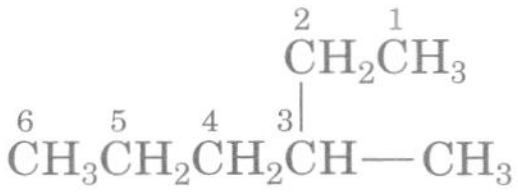

$$
\begin{array}{c}
\overset{2}{CH_2}\overset{1}{CH_3} \\
| \\
\overset{6}{CH_3}\overset{5}{CH_2}\overset{4}{CH_2}\overset{3}{CH}{-}CH_3
\end{array}
$$

3-Methylhexane

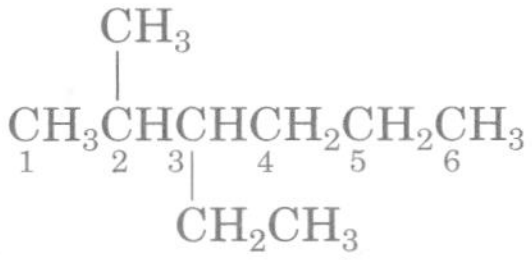

$$
\begin{array}{c}
CH_3 \\
| \\
\underset{1}{CH_3}\underset{2}{CH}\underset{3}{CH}\underset{4}{CH_2}\underset{5}{CH_2}\underset{6}{CH_3} \\
| \\
CH_2CH_3
\end{array}
$$

3-Ethyl-2-methylhexane

$$
\begin{array}{c}
{}_1CH_3 \\
| \\
{}_2CH_2 \\
| \\
CH_3{-}\underset{3}{CH}\underset{4}{CH}{-}CH_2CH_3 \\
| \\
\underset{5}{CH_2}\underset{6}{CH_2}\underset{7}{CH_3}
\end{array}
$$

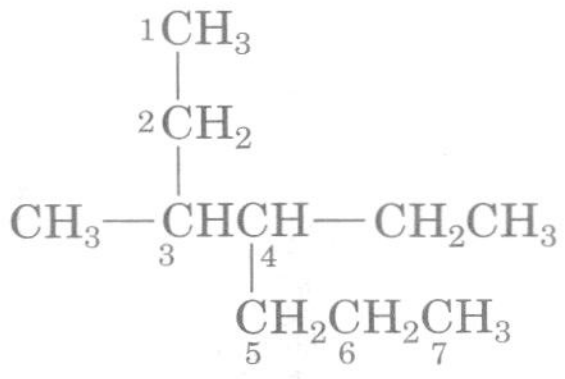

4-Ethyl-3-methylheptane

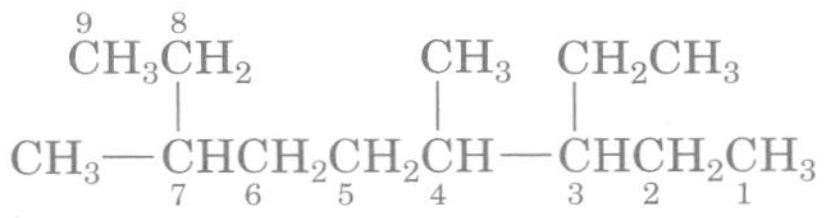

$$
\begin{array}{c}
\overset{9}{CH_3}\overset{8}{CH_2} \qquad\qquad CH_3 \quad CH_2CH_3 \\
| \qquad\qquad\qquad | \qquad\quad | \\
CH_3{-}\underset{7}{CH}\underset{6}{CH_2}\underset{5}{CH_2}\underset{4}{CH}{-}\underset{3}{CH}\underset{2}{CH_2}\underset{1}{CH_3}
\end{array}
$$

3-Ethyl-4,7-dimethylnonane

PRACTICE PROBLEM 2.2

What is the IUPAC name of the following alkane?

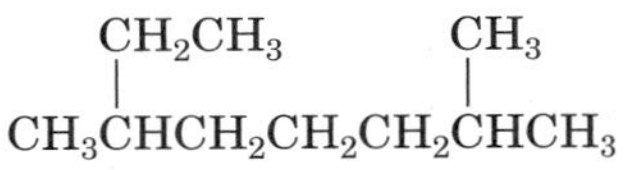

$$
\begin{array}{c}
CH_2CH_3 \qquad\qquad CH_3 \\
| \qquad\qquad\qquad | \\
CH_3CHCH_2CH_2CH_2CHCH_3
\end{array}
$$

Solution The molecule has a chain of eight carbons (octane) with two methyl substituents. Numbering from the end nearer the first methyl substituent indicates that the methyls are at C2 and C6, giving the name 2,6-dimethyloctane.

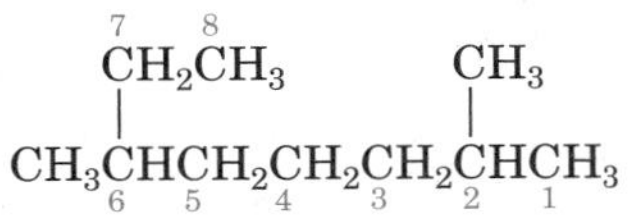

$$
\begin{array}{c}
\overset{7}{CH_2}\overset{8}{CH_3} \qquad\qquad CH_3 \\
| \qquad\qquad\qquad | \\
CH_3\underset{6}{CH}\underset{5}{CH_2}\underset{4}{CH_2}\underset{3}{CH_2}\underset{2}{CH}\underset{1}{CH_3}
\end{array}
$$

PRACTICE PROBLEM 2.3

Draw the structure of 3-isopropyl-2-methylhexane.

Solution First, look at the parent name (hexane) and draw its carbon structure:

C–C–C–C–C–C **Hexane**

Next, find the substituents (3-isopropyl and 2-methyl), and place them on the proper carbons:

```
              CH3CHCH3   <-------  An isopropyl group at C3
                 |
     C — C — C — C — C — C
     1   2|  3   4   5   6
         CH3     <-------  A methyl group at C2
```

Finally, add hydrogens to complete the structure:

```
     CH3CHCH3
        |
  CH3CHCHCH2CH2CH3      3-Isopropyl-2-methylhexane
      |
     CH3
```

PROBLEM

2.8 Give IUPAC names for the following alkanes:

(a) The three isomers of C_5H_{12}

(b)
```
             CH3
              |
  CH3CH2CHCHCH3
             |
            CH2CH3
```

(c)
```
   CH3   CH3
    |     |
CH3CHCH2CHCH3
```

(d)
```
        CH3           CH2CH3
         |              |
  CH3 — C — CH2CH2CHCH3
         |
        CH3
```

PROBLEM

2.9 Draw structures corresponding to the following IUPAC names:

(a) 3,4-Dimethylnonane (b) 3-Ethyl-4,4-dimethylheptane

(c) 2,2-Dimethyl-4-propyloctane (d) 2,2,4-Trimethylpentane

2.4 Properties of Alkanes

Alkanes are sometimes referred to as **paraffins,** a word derived from the Latin *parum affinis* meaning "slight affinity." This term aptly describes their behavior, for alkanes show little chemical affinity for other substances and are inert to most laboratory reagents. They do, however, react with oxygen, chlorine, and a few other substances under appropriate conditions.

The reaction of alkanes with O_2 occurs during combustion in an engine or furnace when the alkane is used as a fuel. Carbon dioxide and water are formed as products, and a large amount of heat is released. For example, methane (natural gas) reacts with oxygen according to the equation:

$$CH_4 + 2O_2 \longrightarrow CO_2 + 2H_2O + 890\ \text{kJ}\ (213\ \text{kcal})$$

The reaction of an alkane with Cl_2 occurs when a mixture of the two is irradiated with ultraviolet light (denoted $h\nu$, where ν is the lowercase

Greek letter nu). Depending on the relative amounts of the two reactants and on the time allowed for reaction, a sequential replacement of the alkane hydrogen atoms by chlorine occurs, leading to a mixture of chlorinated products. Methane, for instance, reacts with chlorine to yield a mixture of chloromethane (CH_3Cl), dichloromethane (CH_2Cl_2), trichloromethane ($CHCl_3$), and tetrachloromethane (CCl_4). We'll see how this chlorination reaction occurs when we take up the chemistry of alkyl halides in Chapter 7.

$$CH_4 + Cl_2 \xrightarrow{h\nu} CH_3Cl + HCl$$
$$CH_3Cl \xrightarrow{Cl_2} CH_2Cl_2 + HCl$$
$$CH_2Cl_2 \xrightarrow{Cl_2} CHCl_3 + HCl$$
$$CHCl_3 \xrightarrow{Cl_2} CCl_4 + HCl$$

Alkanes show regular increases in both boiling point and melting point as molecular weight increases (Figure 2.4). Average C–C bond parameters are nearly the same in all alkanes, with bond lengths of 1.54 ± 0.01 Å and bond strengths of 355 ± 20 kJ/mol (85 ± 5 kcal/mol). Carbon–hydrogen bond parameters are also nearly constant at 1.09 ± 0.01 Å and 400 ± 20 kJ/mol (95 ± 5 kcal/mol).

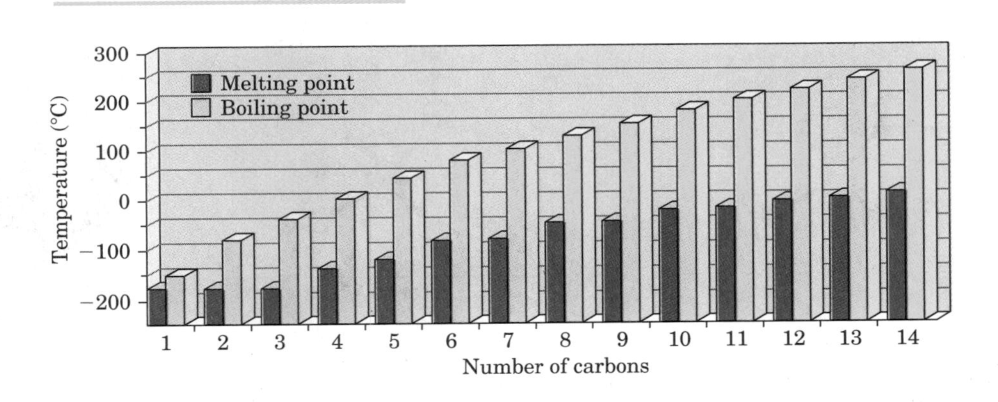

Figure 2.4 A plot of the number of carbons versus melting and boiling points for the C_1–C_{14} alkanes. There is a regular increase with molecular size.

2.5 Conformations of Ethane

We saw earlier that C–C bonds in alkanes result from σ overlap of two tetrahedrally oriented sp^3 orbitals. Let's now look into the three-dimensional consequences of such bonding. What are the spatial relationships between the hydrogens on one carbon and the hydrogens on the other?

We know that σ bonds result from the head-on overlap of two atomic orbitals and that a cross section through a σ bond is circular. Because of this circular symmetry, *rotation* is possible around carbon–carbon single bonds. Orbital overlap in the C–C bond is exactly the same for all geometric arrangements of the hydrogens (Figure 2.5). The different arrangements of atoms that result from rotation around a single bond are called **conformations,** and a specific conformation is called a **conformer** (*conform*ational iso*mer*). Unlike constitutional isomers (Section 2.2), though, different conformers usually interconvert too rapidly for them to be isolated.

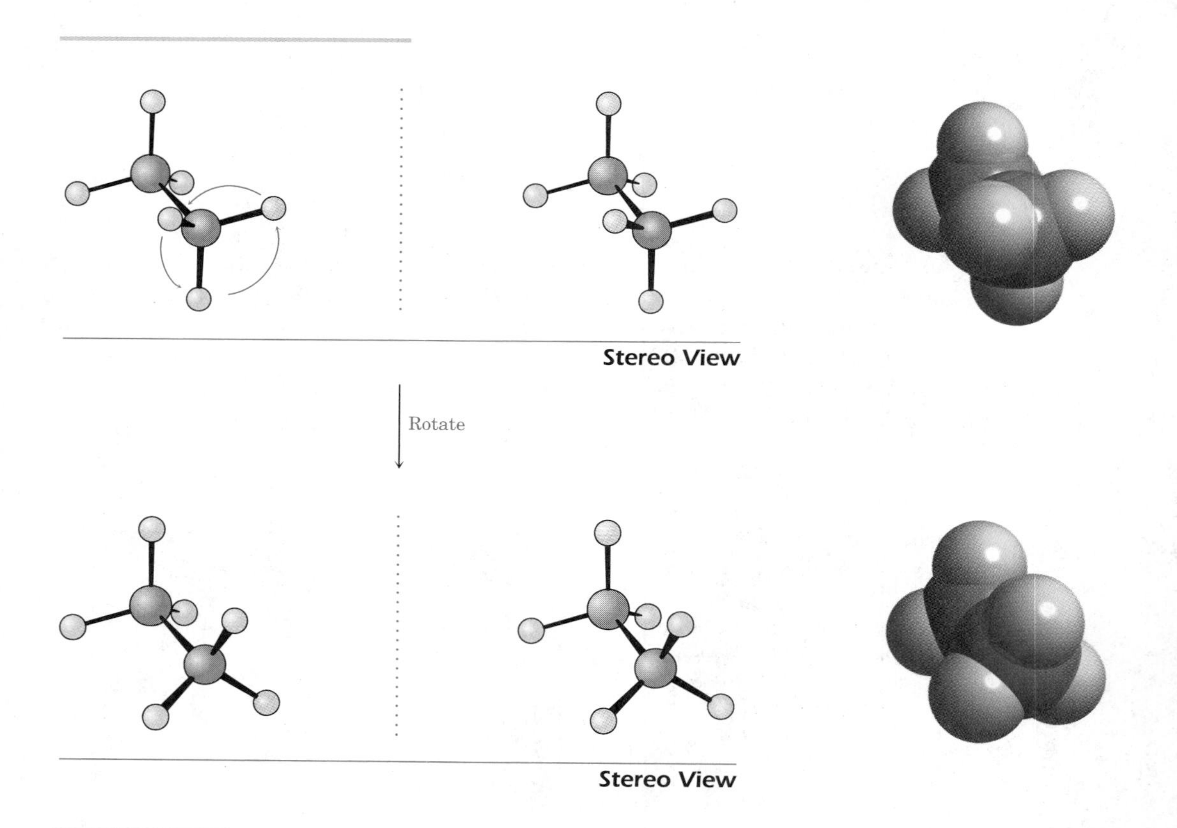

Figure 2.5 Two conformations of ethane. Rotation around the C–C single bond interconverts the different conformers.

Chemists represent conformational isomers in two ways, as shown in Figure 2.6. **Sawhorse representations** view the C–C bond from an oblique angle and indicate spatial relationships by showing all the C–H bonds. **Newman projections** view the C–C bond end-on and represent the two carbon atoms by a circle. Bonds attached to the front carbon are represented by lines to the center of the circle, and bonds attached to the rear carbon are represented by lines to the edge of the circle.

In spite of what we've just said about σ bond symmetry, we don't observe *perfectly* free rotation in ethane. Experiments show that there is a slight (12 kJ/mol; 2.9 kcal/mol) barrier to rotation and that some conformations are more stable than others. The lowest-energy, most stable conformation is the one in which all six C–H bonds are as far away from one

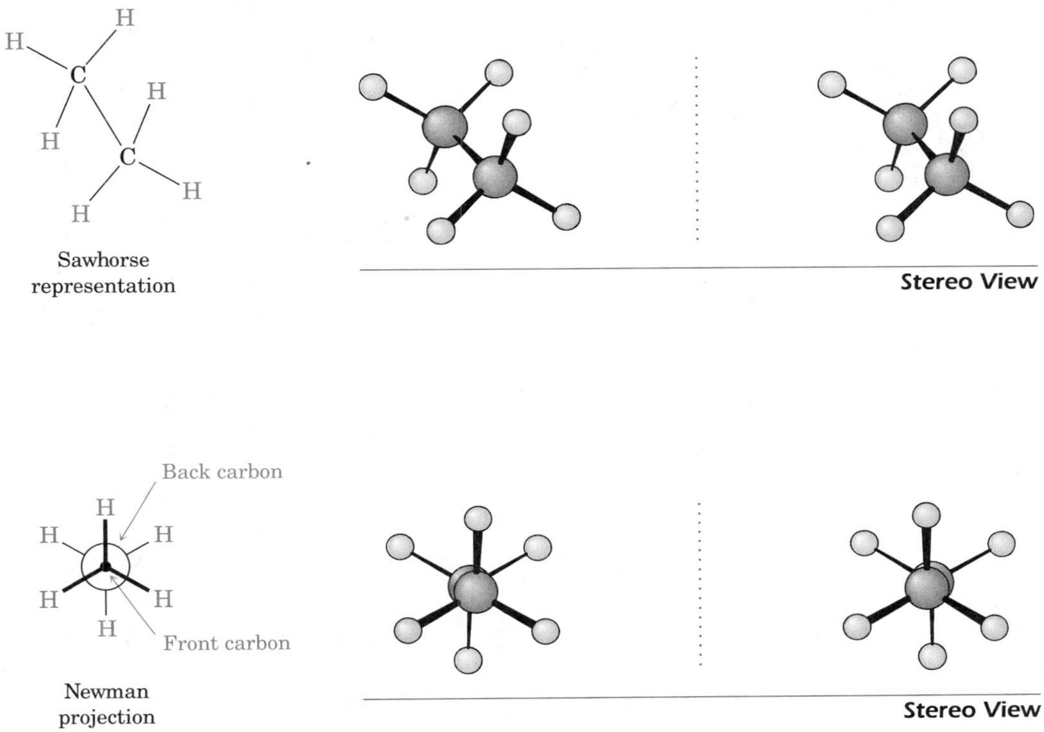

Sawhorse representation

Stereo View

Back carbon

Front carbon

Newman projection

Stereo View

Figure 2.6 A sawhorse representation and a Newman projection of ethane. The sawhorse projection views the molecule from an oblique angle, while the Newman projection views the molecule end-on.

another as possible (**staggered** when viewed end-on in a Newman projection). The highest-energy, least stable conformation is the one in which the six C–H bonds are as close as possible (**eclipsed** in a Newman projection). At any given instant, about 99% of ethane molecules have an approximately staggered conformation, and only about 1% are close to the eclipsed conformation. Both conformations are shown at the top of the next page.

The barrier to bond rotation in ethane is easier to see if you draw a graph of potential energy versus angle of bond rotation, as shown in Figure 2.7 (p. 54). The minimum energy geometries correspond to staggered conformations, and the maximum energy geometries correspond to eclipsed conformations. The barrier is caused by the slight repulsion between electron clouds in the C–H bonds as they pass by each other at close quarters in the eclipsed conformer.

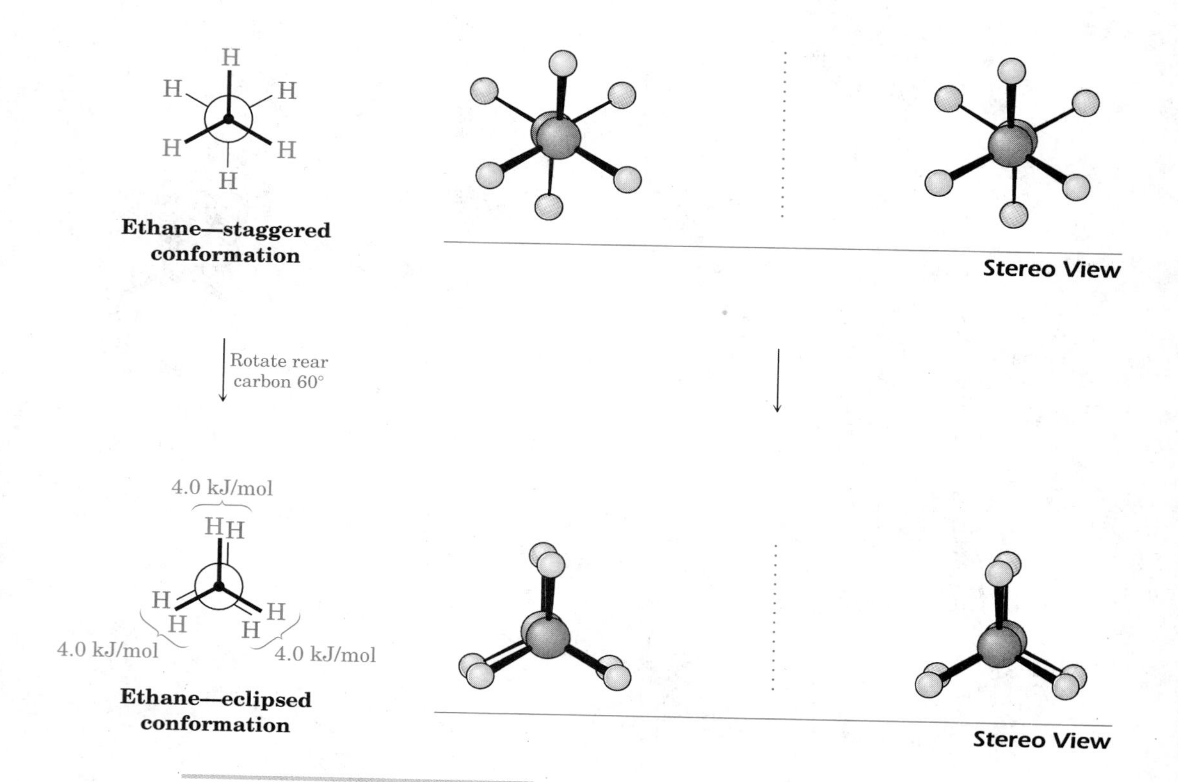

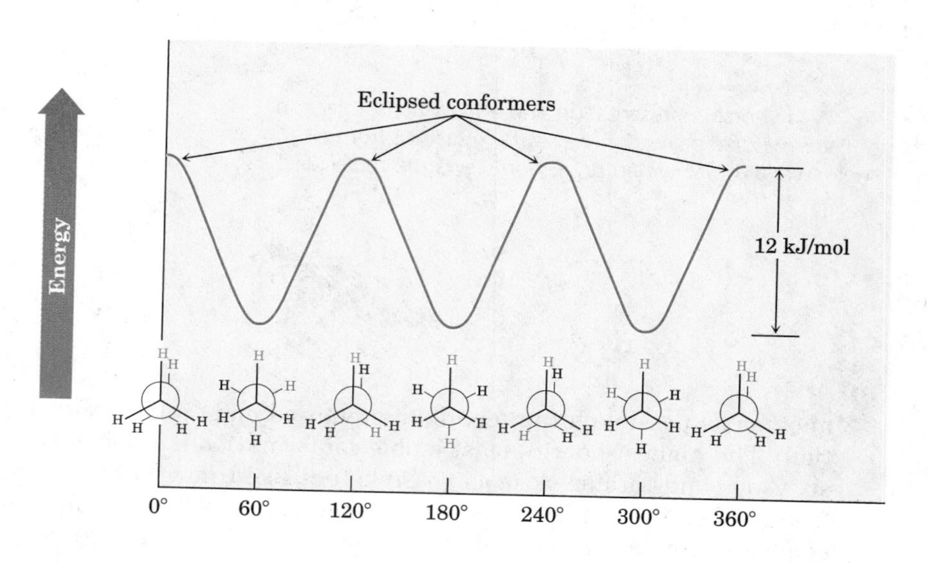

Figure 2.7 A graph of potential energy versus bond rotation in ethane. The staggered conformers are 12 kJ/mol lower in energy than the eclipsed conformers.

What is true for ethane is also true for propane, butane, and all higher alkanes. The most favored conformation for any alkane is the one in which all bonds have staggered arrangements (Figure 2.8).

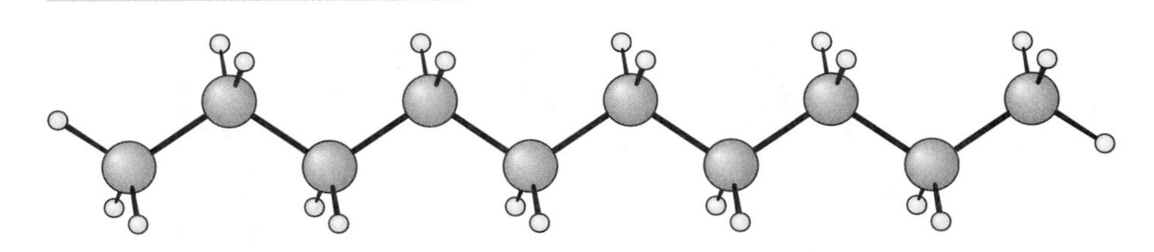

Figure 2.8 The most stable conformation of any alkane is the one in which the bonds on adjacent carbons are staggered and the carbon chain is fully extended, as in this structure of decane.

PROBLEM

2.10 Sight along a C–C bond of propane and draw a Newman projection of the most stable conformation. Draw a Newman projection of the least stable conformation.

PROBLEM

2.11 Draw a graph, similar to Figure 2.7, of energy versus angle of bond rotation for propane.

PROBLEM

2.12 Looking along the C2–C3 bond of butane, there are two different staggered conformations and two different eclipsed conformations. Draw them.

PROBLEM

2.13 Which of the butane conformations you drew in Problem 2.12 do you think is the most stable? Explain.

2.6 Drawing Chemical Structures

In the structures we've been using, a line between atoms has represented the two electrons in a covalent bond. Most chemists find themselves drawing many structures each day, and it would soon become tedious if every bond and every atom had to be indicated. Chemists have therefore devised a shorthand way of drawing **skeletal structures** that greatly simplifies matters, particularly for the cyclic compounds that we'll see shortly.

The rules for drawing skeletal structures are simple:

Rule 1 Carbon atoms usually aren't shown. Instead, a carbon atom is assumed to be at the intersection of two lines (bonds) and at the end of each line. Occasionally, a carbon atom might be indicated for emphasis or clarity.

Rule 2 Hydrogen atoms bonded to carbon aren't shown. Because carbon always has a valence of four, we mentally supply the correct number of hydrogen atoms for each carbon.

Rule 3 All atoms other than carbon and hydrogen *are* shown.

The following structures show some examples.

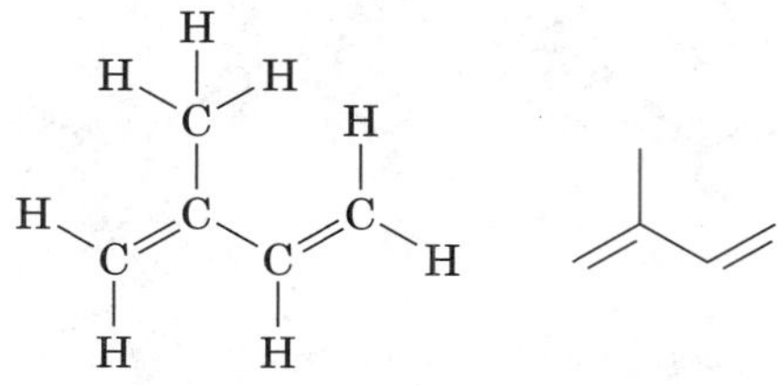

Isoprene, C_5H_8 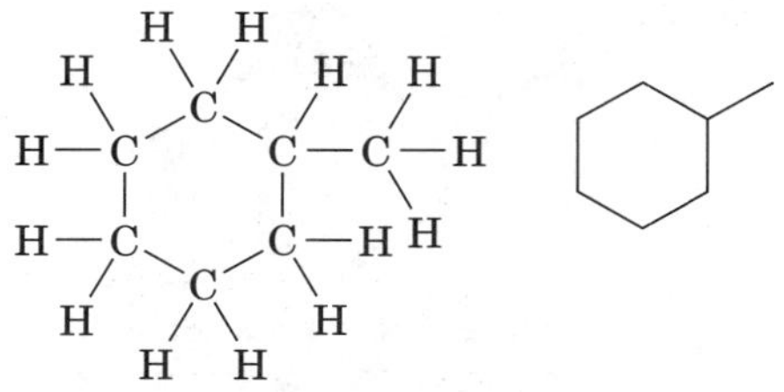 **Methylcyclohexane, C_7H_{14}**

PRACTICE PROBLEM 2.4

Convert the following skeletal structure of adrenaline into a molecular formula:

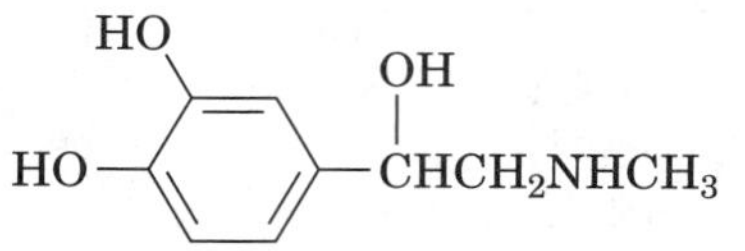

HO, HO, OH, $CHCH_2NHCH_3$ **Adrenaline**

Solution Remember that each intersection of lines is a carbon atom: $C_9H_{13}NO_3$.

PRACTICE PROBLEM 2.5

Convert the following structure into a skeletal drawing:

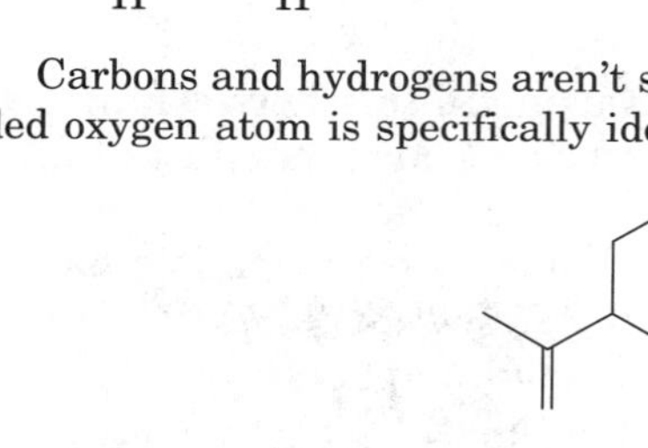

Carvone (from spearmint oil)

Solution Carbons and hydrogens aren't shown in a skeletal drawing; only the doubly bonded oxygen atom is specifically identified.

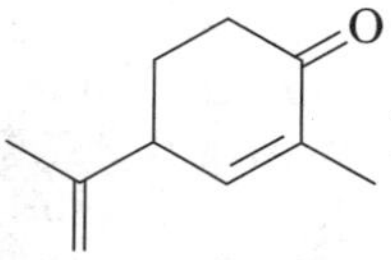

PROBLEM

2.14 Convert the following skeletal structures into molecular formulas:

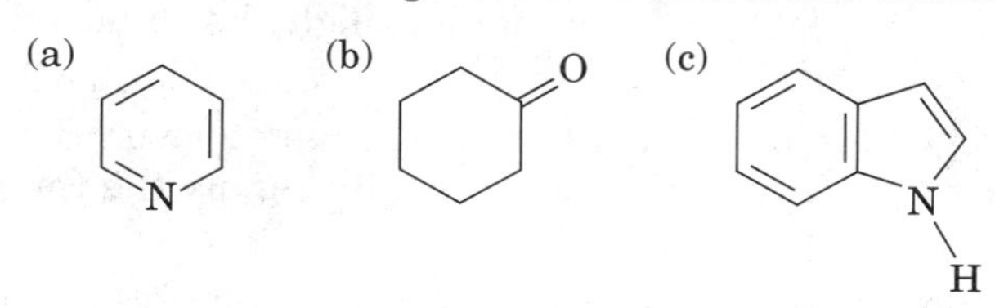

Pyridine **Cyclohexanone** **Indole**

PROBLEM

2.15 Propose skeletal structures for the following molecular formulas:
(a) C_4H_8 (b) C_3H_6O (c) C_4H_9Cl

2.7 Cycloalkanes

Though we've only discussed open-chain alkanes up to now, chemists have known for over 100 years that compounds with *rings* of carbon atoms also exist. Such compounds are called **cycloalkanes** or **alicyclic** (*ali*phatic *cyclic*) compounds. Since cycloalkanes consist of rings of $–CH_2–$ units, they have the general formula $(CH_2)_n$, or C_nH_{2n}, and are represented by polygons in skeletal drawings:

Substituted cycloalkanes are named by rules similar to those for open-chain alkanes. For most compounds, there are only two steps:

Step 1 Count the number of carbon atoms in the ring, and add the prefix *cyclo-* to the name of the corresponding alkane. If a substituent is present on the ring, the compound is named as an alkyl-substituted cycloalkane rather than as a cycloalkyl-substituted alkane.

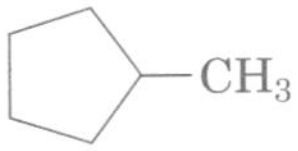

$-CH_3$ **Methylcyclopentane**

Step 2 For substituted cycloalkanes, start at a point of attachment and number the substituents on the ring so as to arrive at the lowest sum. If two or more different substituents are present, number them by alphabetical priority.

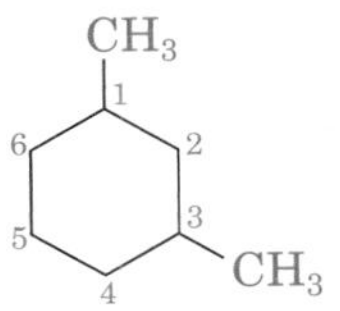

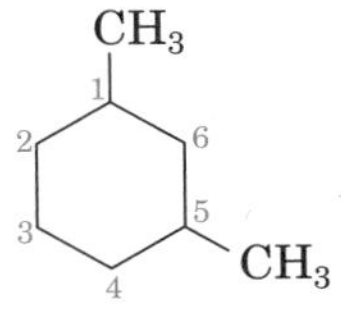

1,3-Dimethylcyclohexane (Sum: 1 + 3 = 4) *NOT* **1,5-Dimethylcyclohexane** (Sum: 1 + 5 = 6)

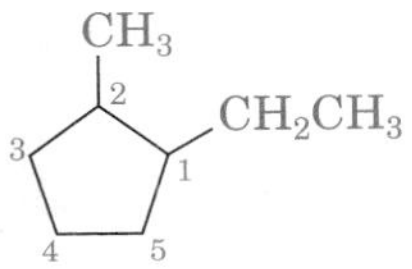

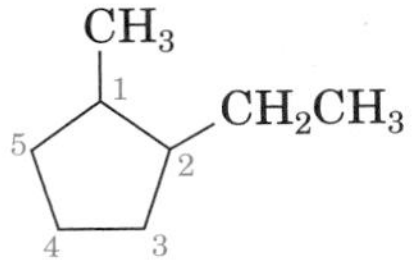

1-Ethyl-2-methylcyclopentane *NOT* **2-Ethyl-1-methylcyclopentane**

PROBLEM

2.16 Give IUPAC names for the following cycloalkanes:

(a) H_3C, CH_3 (b) CH_2CH_3, CH_3 (c) CH_3, CH_3

PROBLEM

2.17 Draw structures corresponding to the following IUPAC names:
(a) 1-*tert*-Butyl-2-methylcyclopentane (b) 1,1-Dimethylcyclobutane
(c) 1-Ethyl-4-isopropylcyclohexane

2.8 Cis–Trans Isomerism in Cycloalkanes

In many respects, the behavior of cycloalkanes is similar to that of open-chain, acyclic alkanes. Both classes of compounds are nonpolar and are chemically inert to most reagents. There are, however, some important differences.

One difference is that cycloalkanes have less conformational freedom than their open-chain counterparts. Although open-chain alkanes have nearly free rotation around their C–C single bonds, cycloalkanes are more constrained in their geometry. For example, cyclopropane is geometrically constrained to be a rigid, planar molecule. No rotation around a C–C bond is possible in cyclopropane without breaking open the ring (Figure 2.9).

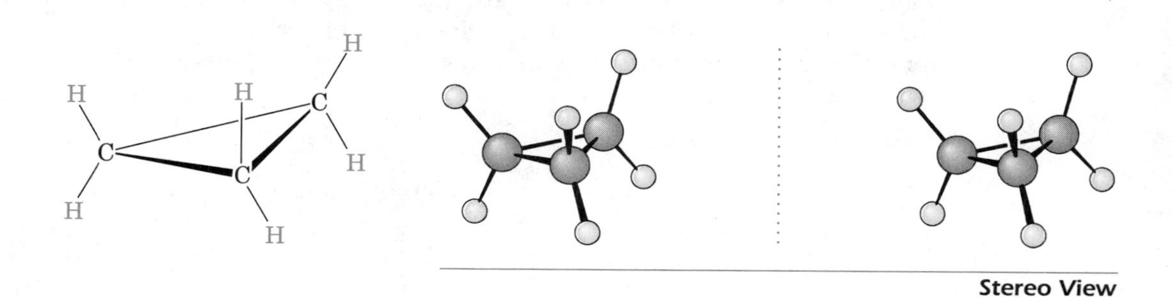

Figure 2.9 The structure of cyclopropane. No rotation is possible around the C–C bonds without breaking open the ring.

Because of their cyclic structure, cycloalkanes have two sides, a "top" side and a "bottom" side. As a result, isomerism is possible in substituted cycloalkanes. For example, there are two 1,2-dimethylcyclopropane isomers, one with the two methyl groups on the same side of the ring and one with the methyls on opposite sides. Both isomers are stable compounds and can't be interconverted without breaking bonds (Figure 2.10).

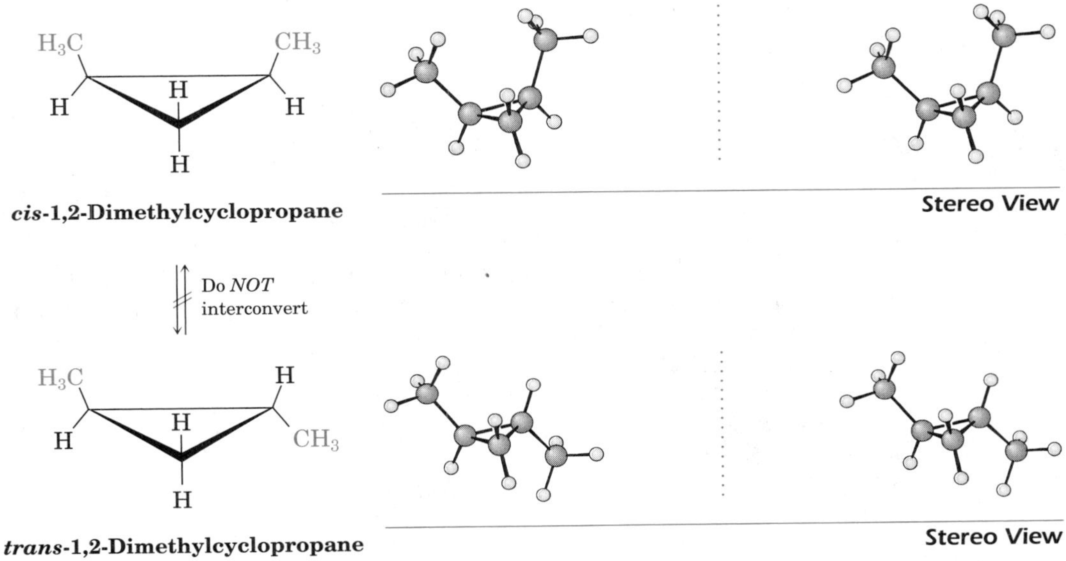

cis-1,2-Dimethylcyclopropane

Stereo View

Do *NOT* interconvert

trans-1,2-Dimethylcyclopropane

Stereo View

Figure 2.10 There are two different 1,2-dimethylcyclopropane isomers, one with the methyl groups on the same side of the ring (cis) and the other with the methyl groups on opposite sides of the ring (trans). The two isomers do not interconvert.

Unlike the constitutional isomers butane and isobutane (Section 2.2), which have different connections among atoms, the two 1,2-dimethylcyclopropanes have the *same* connections but differ in the spatial orientation of their atoms. Such compounds, which have their atoms connected in the same way but differ in three-dimensional orientation, are called **stereoisomers.** The 1,2-dimethylcyclopropanes are special kinds of stereoisomers called **cis–trans isomers.** The prefixes *cis*- (Latin, "on the same side") and *trans*- (Latin, "across") are used to distinguish between them.

PRACTICE PROBLEM 2.6

Draw *cis*-1,4-dimethylcyclohexane.

Solution *cis*-1,4-Dimethylcyclohexane contains a ring of six carbon atoms with methyl substituents on the same side of the ring at carbons 1 and 4.

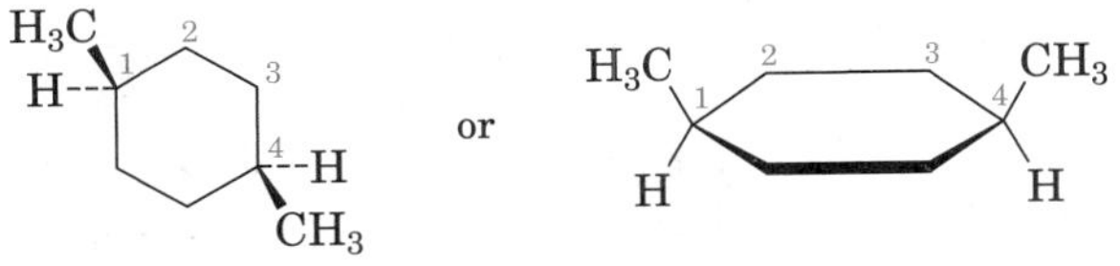

or

***cis*-1,4-Dimethylcyclohexane**

PROBLEM

2.18 Draw *cis*-1-chloro-3-methylcyclopentane.

PROBLEM

2.19 Draw both cis and trans isomers of 1,2-dibromocyclobutane.

2.9 Conformations of Some Common Cycloalkanes

In the early days of organic chemistry, cycloalkanes provoked a good deal of consternation among chemists. The problem was that if carbon prefers to have bond angles of 109.5°, how is it possible for cyclopropane and cyclobutane to exist? After all, cyclopropane must have a triangular shape with bond angles near 60°, and cyclobutane must have a square or rectangular shape with bond angles near 90°. Nonetheless, these compounds *do* exist and are stable.

Let's look at the most common cycloalkanes.

Cyclopropane

Cyclopropane is a symmetrical molecule with C–C–C bond angles of 60°, as indicated in Figure 2.11. The three carbons form an equilateral triangle, with three hydrogens protruding above and three below the plane of the carbons. All six of the C–H bonds have an eclipsed, rather than staggered, arrangement with their neighbors.

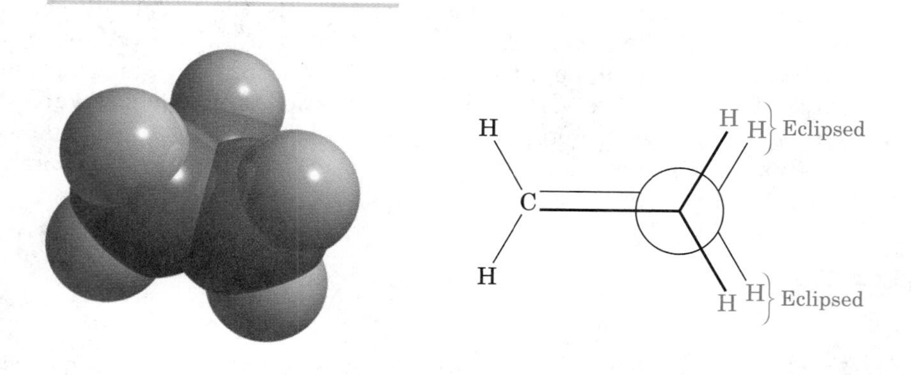

Figure 2.11 The structure of cyclopropane. The carbon atoms form an equilateral triangle, and all C–H bonds are eclipsed.

The simplest way to account for the distortion of the cyclopropane C–C–C bond angles from their ideal value of 109.5° to a value of 60° is to think of cyclopropane as having *bent bonds* (Figure 2.12). In an open-chain

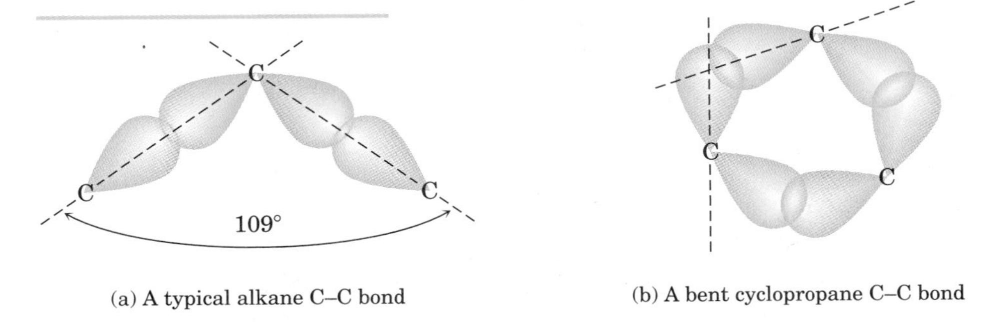

Figure 2.12 An orbital view of cyclopropane. (a) Most C–C bonds have good overlap of orbitals, but (b) cyclopropane bent bonds have poor overlap of orbitals.

alkane, maximum bonding efficiency is achieved when two atoms are located so that their overlapping orbitals point directly toward each other. In cyclopropane, however, the orbitals can't point directly toward each other but must instead overlap at a slight angle. The results of this poor overlap are that cyclopropane C–C bonds are weaker than other alkane bonds because of what is called **angle strain** and that cyclopropane is therefore more reactive than other alkanes.

Cyclobutane and Cyclopentane

Cyclobutane and cyclopentane are slightly puckered rather than flat, as indicated in Figure 2.13. This puckering makes the C–C–C bond angles a bit

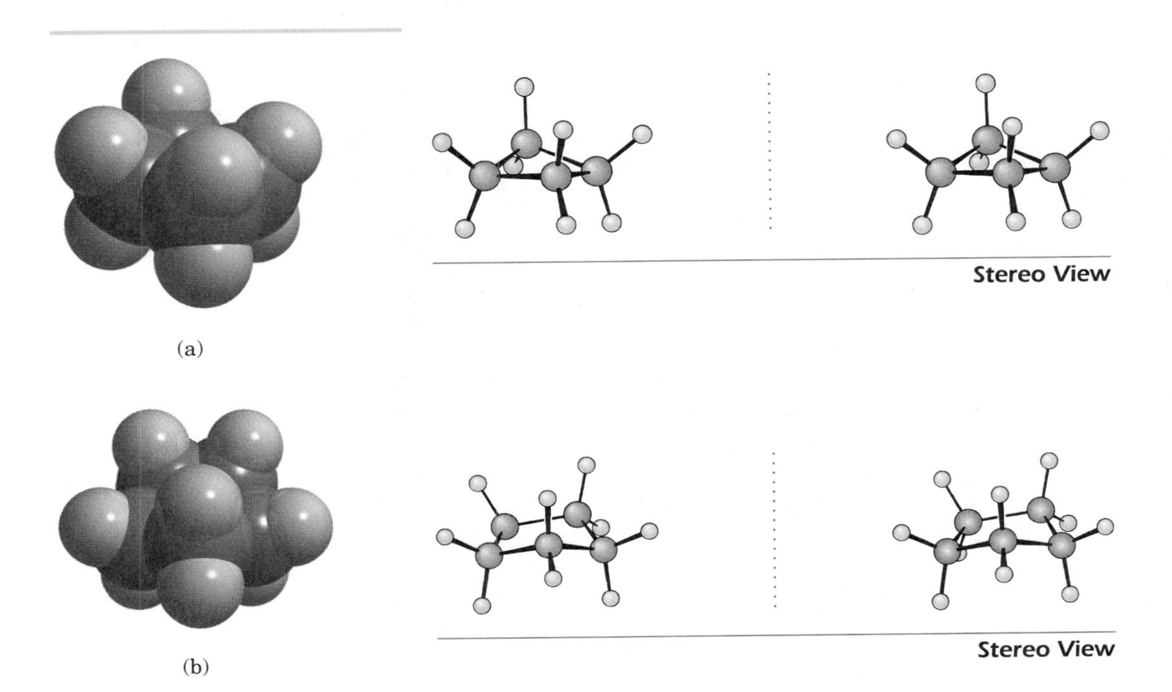

Figure 2.13 Conformations of (a) cyclobutane and (b) cyclopentane.

smaller than they would otherwise be and increases the angle strain. At the same time, though, the puckering relieves the eclipsing interactions of adjacent C–H bonds that would occur if the ring were flat.

Cyclohexane

Substituted cyclohexanes are the most common cycloalkanes because of their wide occurrence in nature. A large number of compounds, including steroids and pharmaceutical agents, have cyclohexane rings.

Cyclohexane is not flat. Rather, it is puckered into a three-dimensional shape called a **chair conformation,** in which the C–C–C bond angles are close to the ideal 109.5° tetrahedral value (Figure 2.14). In addition to being free of angle strain, chair cyclohexane is also free of all C–H eclipsing interactions because neighboring C–H bonds are staggered.

Chair conformations are drawn by following three steps:

Step 1 Draw two parallel lines, slanted downward and slightly offset from each other. These lines show that four of the cyclohexane carbon atoms lie in a plane.

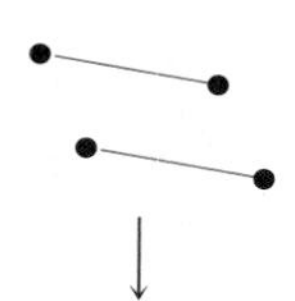

Step 2 Place the topmost carbon atom above and to the right of the plane of the other four and connect the bonds.

Step 3 Place the bottommost carbon atom below and to the left of the plane of the middle four and connect the bonds. Note that the bonds to the bottommost carbon atom are parallel to the bonds to the topmost carbon.

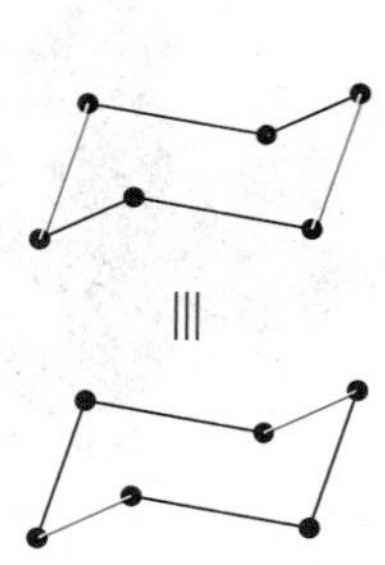

It's important to remember when viewing cyclohexane that the lower bond is in front, and the upper bond is in back. If this convention is not defined, an optical illusion can make it appear that the reverse is true.

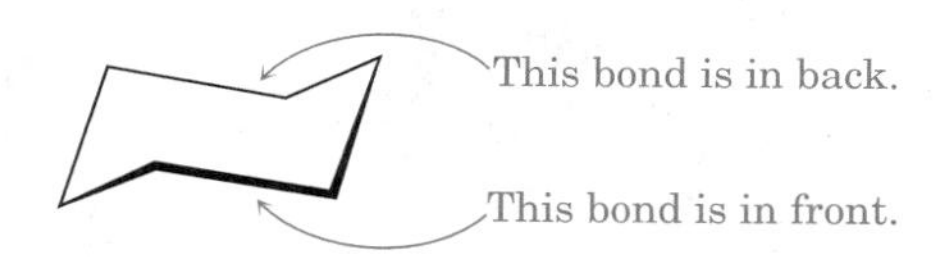

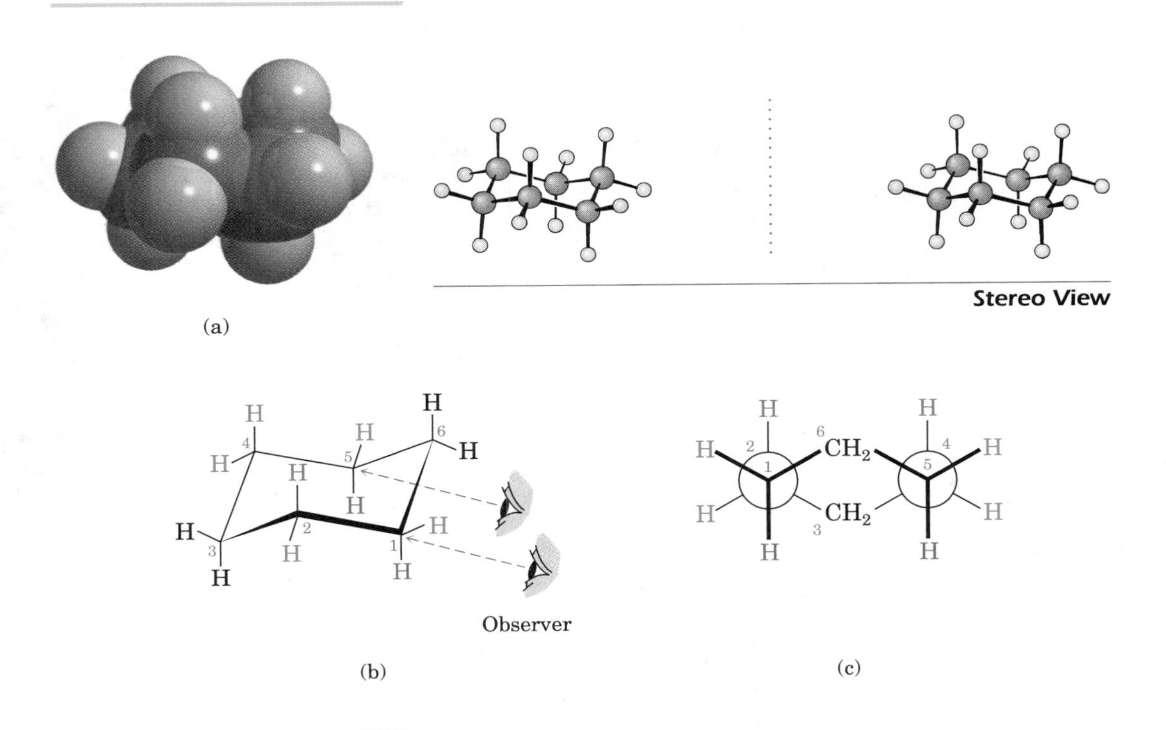

Figure 2.14 The strain-free, chair conformation of cyclohexane. All C–C–C bond angles are close to 109°, and all neighboring C–H bonds are staggered, as evident in the Newman projection (c).

2.10 Axial and Equatorial Bonds in Cyclohexane

The chair conformation of cyclohexane has many important consequences. One such consequence is that there are two kinds of positions for hydrogens on the ring—**axial positions** and **equatorial positions** (Figure 2.15, p. 64). Chair cyclohexane has six axial hydrogens that are perpendicular to the ring (parallel to the ring *axis*) and six equatorial hydrogens that are in the rough plane of the ring (around the ring *equator*).

Each side of the ring has both axial and equatorial hydrogens in an alternating arrangement. For example, if the top side of a cyclohexane ring has axial hydrogens on carbons 1, 3, and 5, then it has equatorial hydrogens on carbons 2, 4, and 6. Exactly the reverse is true for the bottom side: Carbons 1, 3, and 5 have equatorial hydrogens, but carbons 2, 4, and 6 have axial hydrogens.

Note that we haven't used the words *cis* and *trans* in this discussion of cyclohexane geometry. Two hydrogens on the same side of a ring are always cis, regardless of whether they're axial or equatorial and regardless of whether they're adjacent. Similarly, two hydrogens on opposite sides of the ring are always trans, regardless of whether they're axial or equatorial.

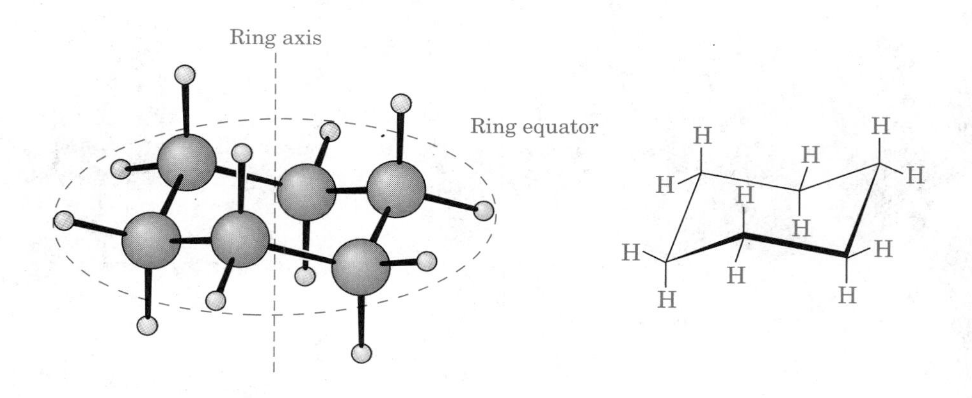

Figure 2.15 Axial (red) and equatorial (blue) hydrogen atoms in cyclohexane. The six axial C–H bonds are parallel to the ring axis, and the six equatorial C–H bonds are in a band around the ring equator.

Axial and equatorial bonds can be drawn in the following way:

Axial bonds The six axial bonds, one on each carbon, are parallel and have an alternating up–down relationship.

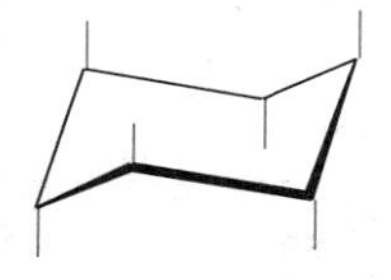

Axial bonds

Equatorial bonds The six equatorial bonds, one on each carbon, come in three sets of two parallel lines. Each set is also parallel to two ring bonds. Equatorial bonds alternate between top and bottom sides around the ring.

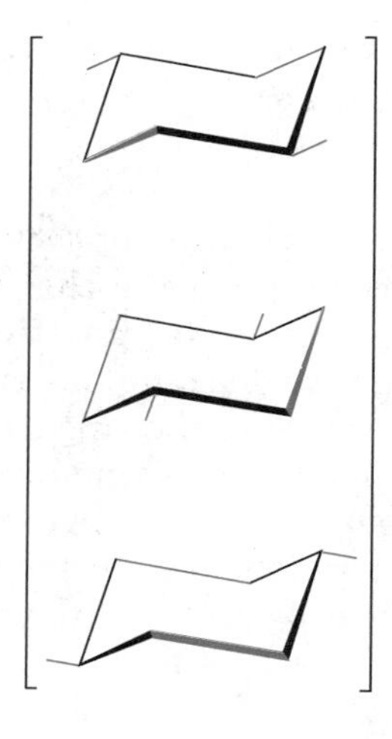

Equatorial bonds

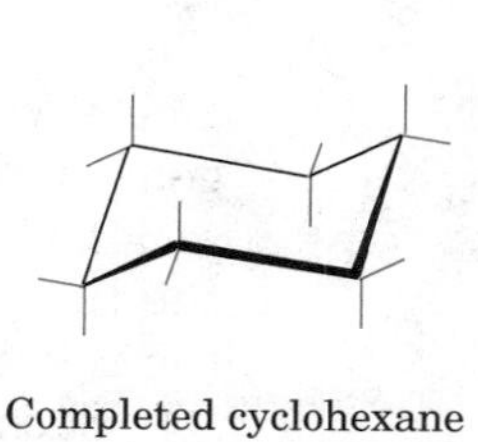

Completed cyclohexane

PROBLEM

2.20 Draw two chair structures for methylcyclohexane, one with the methyl group axial and one with the methyl group equatorial.

2.11 Conformational Mobility of Cyclohexane

Because chair cyclohexane has two kinds of positions, axial and equatorial, we might expect to find two isomeric forms of a monosubstituted cyclohexane. In fact, this expectation is wrong. There is only one methylcyclohexane, one bromocyclohexane, and so forth, because cyclohexane rings are *conformationally mobile* at room temperature. The two chair cyclohexane conformations readily interconvert, resulting in the exchange of axial and equatorial positions. This interconversion of chair conformations, usually referred to as a **ring-flip,** is shown in Figure 2.16.

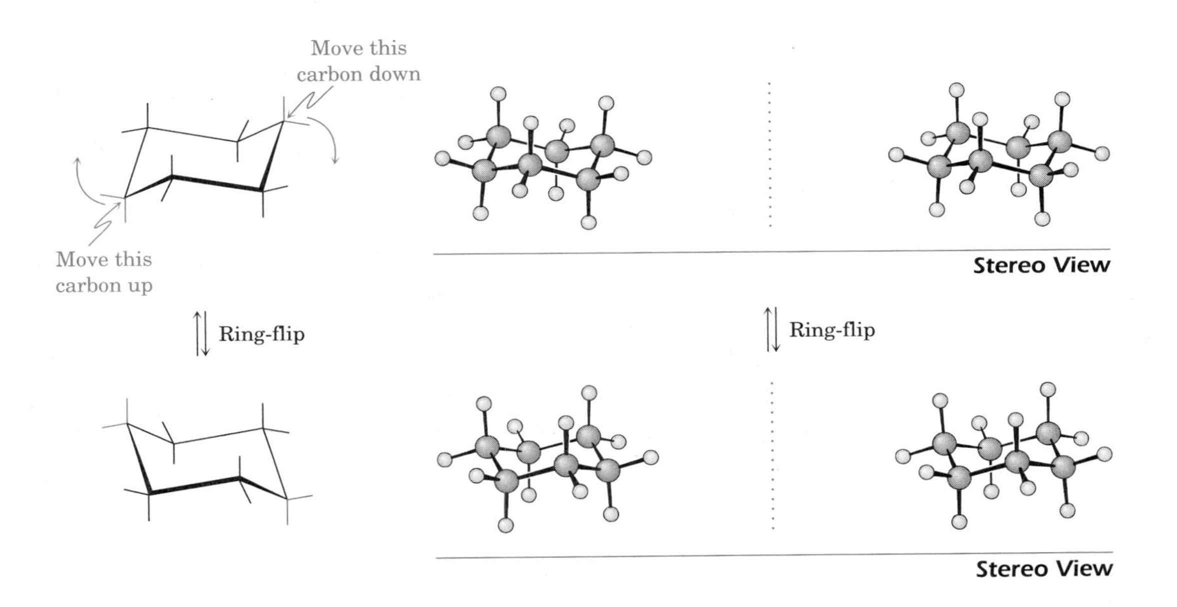

Figure 2.16 A ring-flip in chair cyclohexane interconverts axial and equatorial positions.

A chair cyclohexane can be ring-flipped by keeping the middle four carbon atoms in place while folding the two ends in opposite directions. The net result of a ring-flip is that axial and equatorial positions interconvert. An axial substituent in one chair form becomes an equatorial substituent in the ring-flipped chair form, and vice versa. For example, axial methylcyclohexane becomes equatorial methylcyclohexane after ring-flip. Since this interconversion occurs rapidly at room temperature, with an energy barrier of only 45 kJ/mol (10.8 kcal/mol), we can isolate only an interconverting mixture rather than distinct axial and equatorial isomers.

Although axial and equatorial methylcyclohexanes interconvert rapidly, they aren't equally stable. The equatorial conformer is more stable than the axial conformer by 7.6 kJ/mol (1.8 kcal/mol), meaning that about 95% of methylcyclohexane molecules have their methyl group equatorial at any given instant. The energy difference is due to an unfavorable *steric* (spatial) interaction that occurs in the axial conformer between the methyl group on carbon 1 and the axial hydrogen atoms on carbons 3 and 5 (Figure 2.17). This so-called **1,3-diaxial interaction** introduces 7.6 kJ/mol (1.8 kcal/mol) of **steric strain** into the molecule because the axial methyl group and the nearby axial hydrogen are too close together and are trying to occupy the same space.

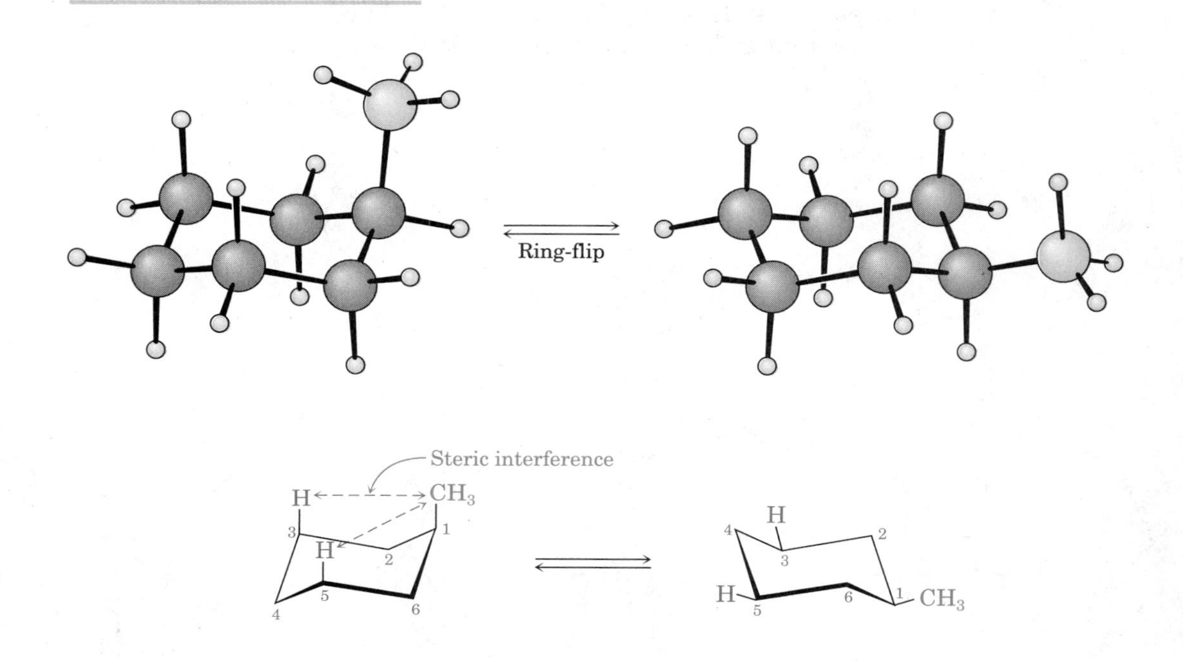

Figure 2.17 1,3-Diaxial steric interactions in axial methylcyclohexane. The equatorial conformer is more stable than the axial conformer by 7.6 kJ/mol.

What is true for methylcyclohexane is also true for all other monosubstituted cyclohexanes: A substituent is always more stable in an equatorial position than in an axial position. As you might expect, the amount of steric strain increases as the size of the axial substituent group increases.

PRACTICE PROBLEM 2.7

Draw 1,1-dimethylcyclohexane, indicating whether each methyl group is axial or equatorial.

Solution First draw a chair cyclohexane ring and then put two methyl groups on the same carbon. The methyl group in the rough plane of the ring is equatorial and the other (above or below the ring) is axial.

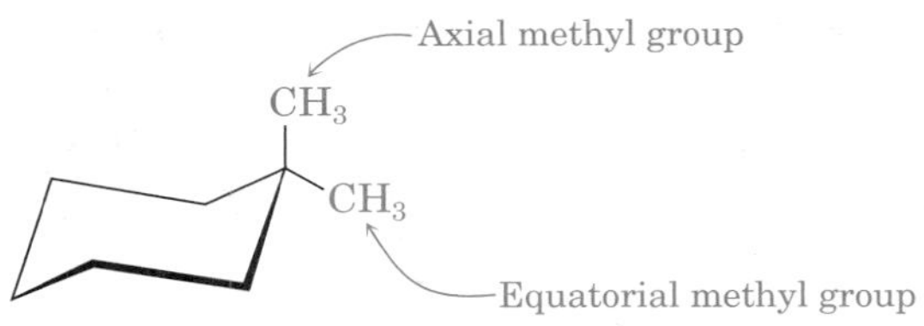

Axial methyl group

CH_3

CH_3

Equatorial methyl group

PROBLEM

2.21 Draw two different chair conformations of bromocyclohexane showing all hydrogen atoms. Label all positions as axial or equatorial. Which of the two conformations do you think is more stable?

PROBLEM

2.22 Explain why a cis-1,2-disubstituted cyclohexane such as *cis*-1,2-dichlorocyclohexane must have one group axial and one equatorial.

PROBLEM

2.23 Explain why a trans-1,2-disubstituted cyclohexane must have either both groups axial or both equatorial.

INTERLUDE

Petroleum

Many alkanes occur naturally in the plant and animal world. For example, the waxy coating on cabbage leaves contains nonacosane ($C_{29}H_{60}$), and the wood oil of the Jeffrey pine common to the Sierra Nevada mountains contains heptane (C_7H_{16}). By far the major sources of alkanes, however, are the world's natural gas and petroleum deposits. Laid down eons ago, these natural deposits are derived from the decomposition of organic matter, primarily of marine origin. *Natural gas* consists chiefly of methane but also contains ethane, propane, and butane. *Petroleum* is a complex mixture of hydrocarbons that must be separated, or *refined,* into different fractions before it can be used.

Petroleum refining begins by fractional distillation of crude petroleum into three principal cuts, according to their boiling points (bp): straight-run gasoline (bp 30–200°C), kerosene (bp 175–300°C), and gas oil (bp 275–400°C). Finally, distillation under reduced pressure yields lubricating oils and waxes, and leaves a tarry residue of asphalt.

The simple distillation of petroleum into fractions is just the beginning of the process by which automobile fuel is made. It turns out that straight-run gasoline is a poor fuel because of the phenomenon of *engine*

(*continued*)➤

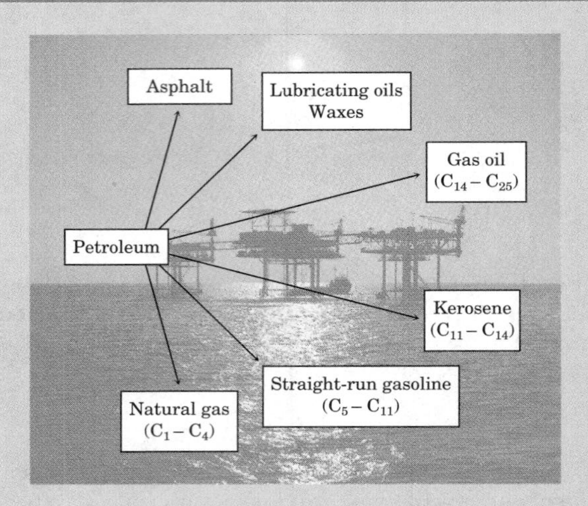

knock. In the typical four-stroke automobile engine, a piston draws a mixture of fuel and air into a cylinder on its downward stroke and compresses the mixture on its upward stroke. Just before the end of the compression, a spark plug ignites the fuel/air mix and combustion occurs, pushing the piston downward and turning the crankshaft.

Not all fuels burn equally well, though. When poor fuels are used, combustion can be initiated in an uncontrolled manner by a hot surface in the cylinder before the spark plug fires. This *preignition,* detected as an engine knock, can destroy the engine by putting irregular forces on the crankshaft and raising engine temperature.

The *octane number* of a fuel is the measure by which its antiknock properties are judged. It was recognized long ago that straight-chain alkanes are much more prone to induce engine knock than are branched-chain compounds. Heptane, a particularly bad fuel, is assigned a base value of 0 octane number, and 2,2,4-trimethylpentane (commonly known as isooctane) has a rating of 100.

$CH_3CH_2CH_2CH_2CH_2CH_2CH_3$

Heptane
(octane number = 0)

$CH_3C(CH_3)_2CH_2CH(CH_3)CH_3$

2,2,4-Trimethylpentane
(octane number = 100)

Because straight-run gasoline has a high percentage of unbranched alkanes and is therefore a poor fuel, petroleum chemists have devised several methods for producing better fuels. One of these methods, *catalytic cracking,* involves taking the kerosene cut (C_{11}–C_{14}) and "cracking" it into smaller molecules at high temperature on a silica–alumina catalyst. The major products of cracking are light hydrocarbons in the C_3–C_5 range. These small hydrocarbons are then catalytically recombined to yield C_7–C_{10} branched-chain alkanes that are suited for use as high-octane fuels.

Summary and Key Words

A **functional group** is an atom or group of atoms within a larger molecule that has a characteristic chemical reactivity. Because functional groups behave approximately the same way in all molecules where they occur, the reactions of an organic molecule are largely determined by its functional groups.

Alkanes are a class of **saturated hydrocarbons** having the general formula C_nH_{2n+2}. They contain no functional groups, are chemically rather inert, and can be either **straight-chain** or **branched.** Alkanes can be named by a series of **IUPAC** rules of nomenclature. **Isomers**—compounds that have the same chemical formula but different structures—exist for all but the simplest alkanes. Compounds such as butane and isobutane, which have the same formula but differ in the way their atoms are connected, are called **constitutional isomers.**

As a result of their symmetry, rotation is possible about C–C single bonds. Alkanes can therefore adopt any of a large number of rapidly interconverting **conformations. Staggered conformations** are more stable than **eclipsed conformations.**

Rotation

Staggered ethane **Eclipsed ethane**

Cycloalkanes contain rings of carbon atoms and have the general formula C_nH_{2n}. Because complete rotation around C–C bonds is not possible in cycloalkanes, conformational mobility is reduced and disubstituted cycloalkanes can exist as **cis–trans stereoisomers.** In a cis isomer, both substituents are on the same side of the ring, whereas in a trans isomer, the substituents are on opposite sides of the ring.

Cyclohexanes are the most common of all rings because of their wide occurrence in nature. Cyclohexane exists in a puckered, strain-free **chair conformation** in which all bond angles are near 109° and all neighboring C–H bonds are staggered. Chair cyclohexane has two kinds of bonds, axial and equatorial. **Axial bonds** are directed up and down, parallel to the ring axis; **equatorial bonds** lie in a belt around the ring equator. Chair cyclohexanes can undergo a **ring-flip** that interconverts axial and equatorial positions. Substituents on the ring are more stable in the equatorial than in the axial position.

Ring-flip

ADDITIONAL PROBLEMS

2.24 Locate and identify the functional groups in the following molecules:

(a) OH

Phenol

(b)

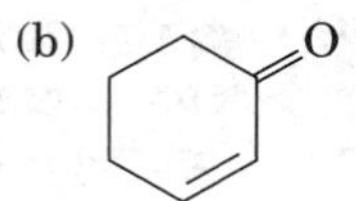

2-Cyclohexenone

(c)

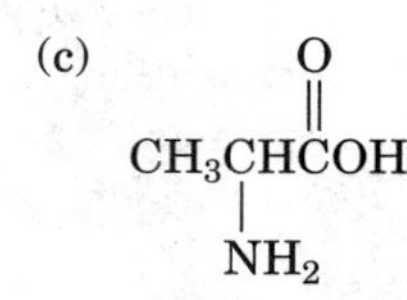

Alanine

(d)

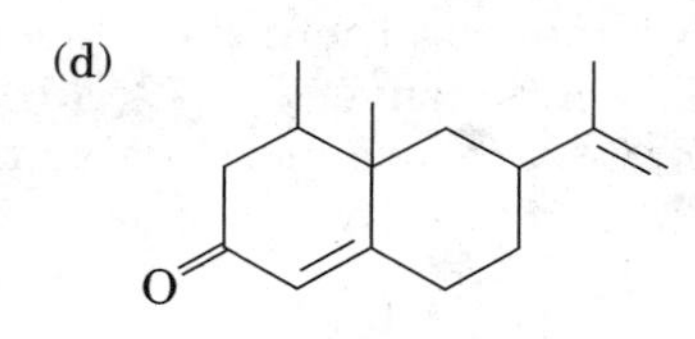

Nootkatone (from grapefruit)

(e)

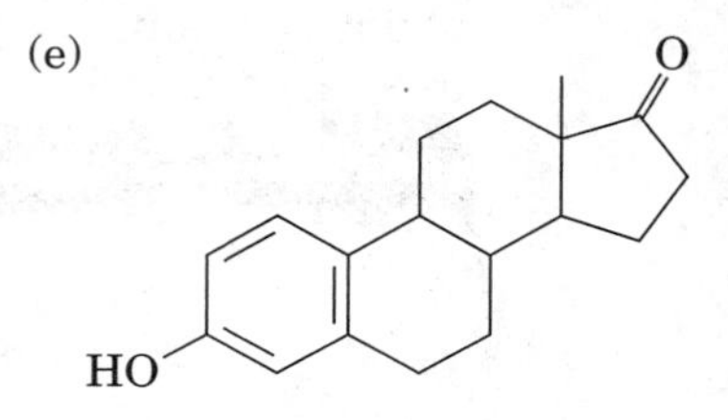

Estrone

2.25 Propose structures for molecules that fit the following descriptions:
(a) An alkene with six carbons (b) A cycloalkene with five carbons
(c) A ketone with five carbons (d) An amide with four carbons
(e) A five-carbon ester (f) An aromatic alcohol

2.26 Propose suitable structures for the following:
(a) An alkene, C_7H_{14} (b) A cycloalkene, C_3H_4 (c) A ketone, C_4H_8O
(d) A nitrile, C_5H_9N (e) A dialkene, C_5H_8 (f) A dialdehyde, $C_4H_6O_2$

2.27 Write as many structures as you can that fit the following descriptions:
(a) Alcohols with formula $C_4H_{10}O$ (b) Amines with formula $C_5H_{13}N$
(c) Ketones with formula $C_5H_{10}O$ (d) Aldehydes with formula $C_5H_{10}O$
(e) Ethers with formula $C_4H_{10}O$ (f) Esters with formula $C_4H_8O_2$

2.28 Draw all monobromo derivatives of pentane, $C_5H_{11}Br$.

2.29 Draw all monochloro derivatives of 2,5-dimethylhexane.

2.30 How many constitutional isomers are there with the formula C_3H_8O? Draw them.

2.31 Propose structures for compounds that contain the following:
(a) A quaternary carbon (b) Four methyl groups (c) An isopropyl group
(d) Two tertiary carbons (e) An amino group ($–NH_2$) bonded to a secondary carbon

2.32 What hybridization do you expect for the carbon atom in the following functional groups?
(a) Ketone (b) Nitrile (c) Ether (d) Alcohol

2.33 Which of the structures in each of the following sets represent the same compound and which represent different compounds?

(a)

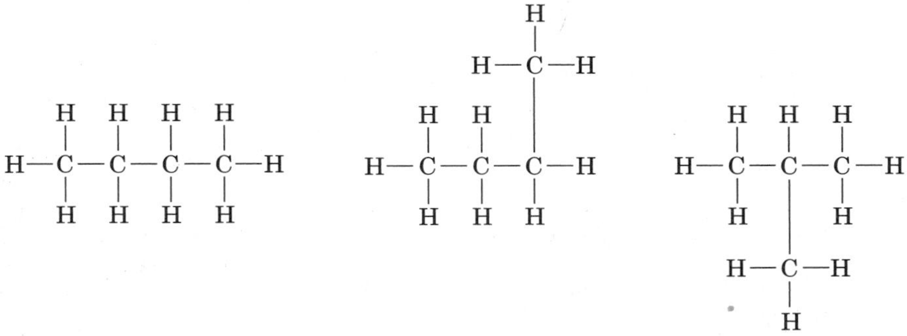

```
   H  H  H  H                 H                 H  H  H
   |  |  |  |                 |                 |  |  |
H—C—C—C—C—H              H—C—H           H—C—C—C—H
   |  |  |  |                 |                 |  |  |
   H  H  H  H           H  H  |                 H  |  H
                        |  |  |                    |
                     H—C—C—C—H               H—C—H
                        |  |  |                    |
                        H  H  H                    H
```

(b)

```
   H  H  Br H         H  Br H  H               H
   |  |  |  |         |  |  |  |               |
H—C—C—C—C—H      H—C—C—C—C—H           H—C—H
   |  |  |  |         |  |  |  |               |  H  H
   H  H  H  H         H  H  H  H               |  |  |
                                            H—C—C—C—H
                                               |  |  |
                                               H  Br H
```

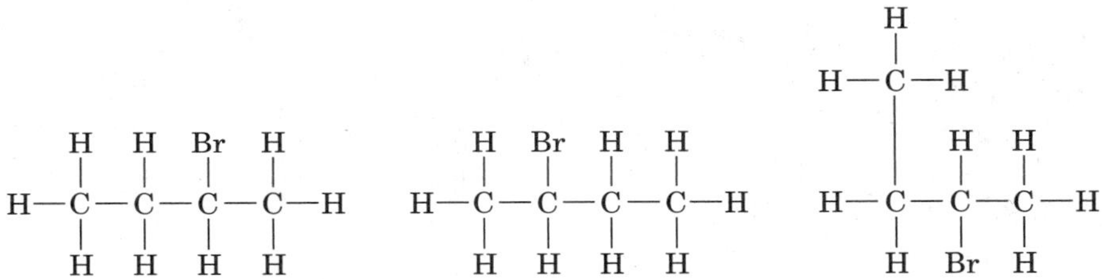

(c)

```
              CH3           CH3
              |             |
CH3CH(Br)CHCH3        CH3CHCH(Br)CH3        (CH3)2CHCH(Br)CH2CH3
```

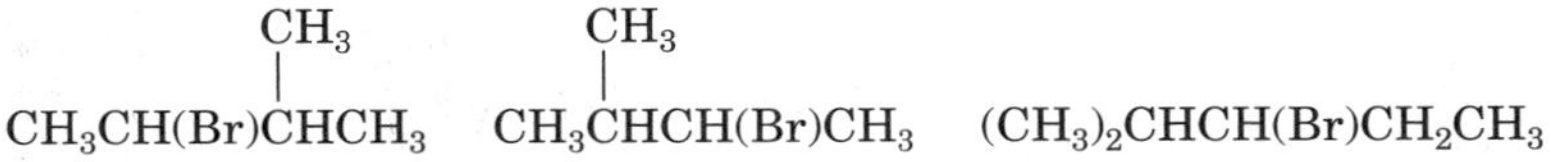

(d)

```
   OH               OH
   |                |
  [ring]—OH       [ring]            HO—[ring]
                    |               HO—
                    OH
```

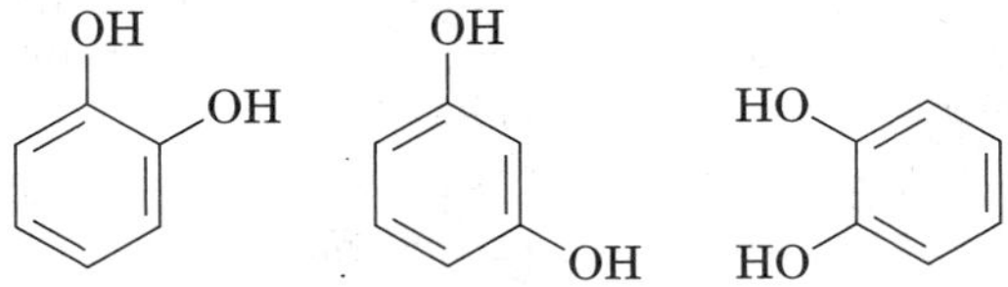

2.34 Draw structural formulas for the following substances:

(a) 2-Methylheptane
(b) 4-Ethyl-2-methylhexane
(c) 4-Ethyl-3,4-dimethyloctane
(d) 2,4,4-Trimethylheptane
(e) 1,1-Dimethylcyclopentane
(f) 4-Isopropyl-3-methylheptane

2.35 Give IUPAC names for the following alkanes:

```
                    CH3                               CH3
                    |                                 |
(a) CH3CH2CH2CHCHCH3            (b) CH3CH2CH2CHCHCH3
                 |                                 |
                 CH3                               CH2CH2CH2CH3

    CH3   CH2CH3                      CH2CH3
    |     |                           |
(c) CH3CHCH2CCH3                (d) CH3CH2CCH2CH3
          |                           |
          CH2CH3                      CH2CH3
```

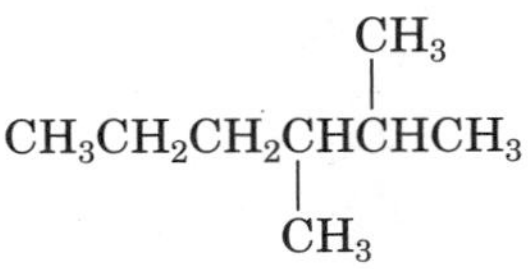

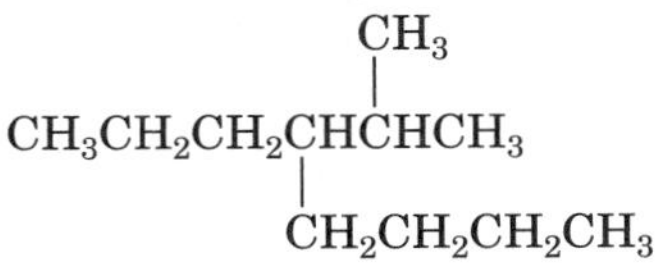

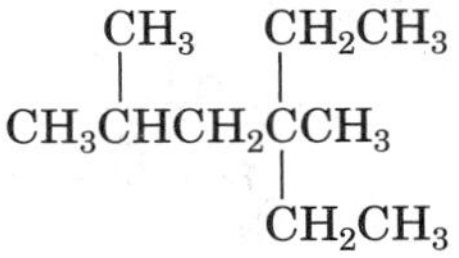

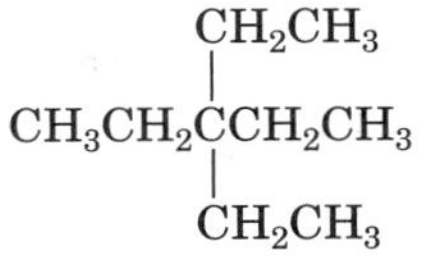

2.36 Convert the following line-bond structures into skeletal drawings:

(a) **Naphthalene**

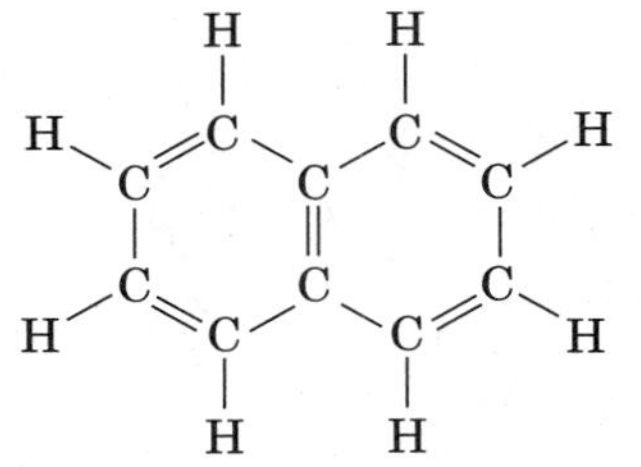

(b) **1,3-Pentadiene**

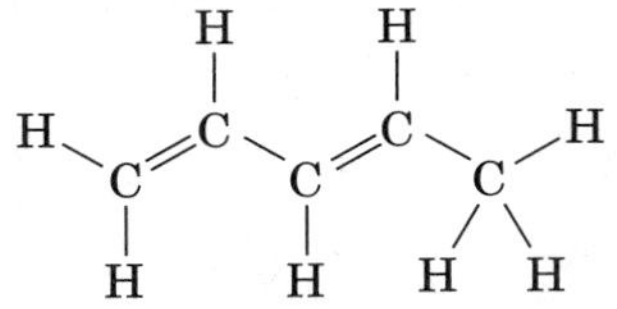

2.37 For each of the following compounds, draw a constitutional isomer having the same functional groups:

(a) $CH_3CH(CH_3)CH_2CH_2Br$ (b) cyclopentyl–OCH_3 (c) $CH_3CH_2CH_2C{\equiv}N$

(d) cyclohexyl–OH (e) $CH_3CH_2CH{=}O$ (f) phenyl–CH_2COOH

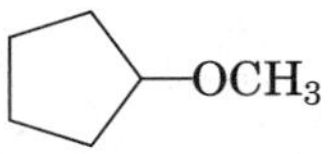

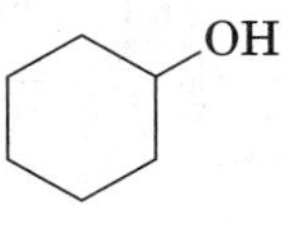

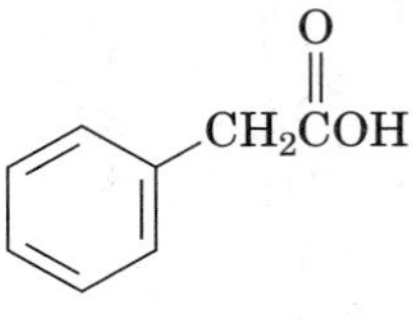

2.38 Sighting along the C2–C3 bond of 2-methylbutane, there are two different staggered conformations. Draw them both in Newman projections, tell which is more stable, and explain your choice.

2.39 Sighting along the C2–C3 bond of 2-methylbutane (see Problem 2.38), there are also two possible eclipsed conformations. Draw them both in Newman projections, tell which you think is lower in energy, and explain.

2.40 *cis*-1-*tert*-Butyl-4-methylcyclohexane exists almost exclusively in the conformation shown. What does this tell you about the relative sizes of a *tert*-butyl substituent and a methyl substituent?

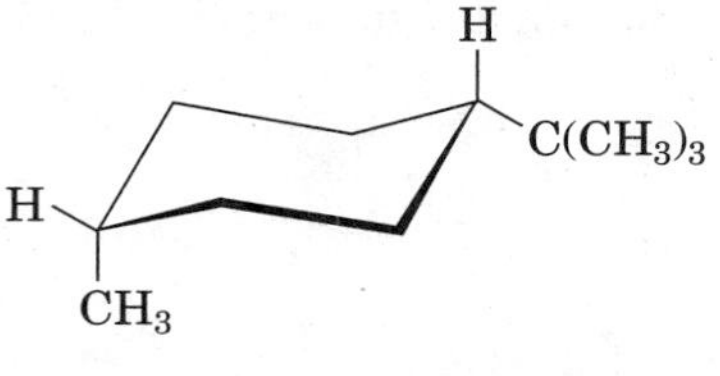

H, $C(CH_3)_3$, H, CH_3

***cis*-1-*tert*-Butyl-4-methylcyclohexane**

2.41 Give IUPAC names for the following compounds:

(a) cycloheptane with CH_3

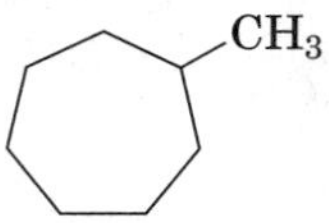

(b) H_3C, CH_3, H, H (cyclopentane)

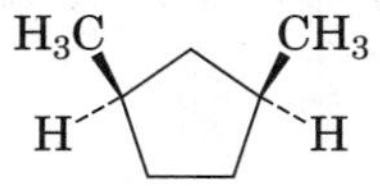

(c) CH_3, H, CH_3, H (cyclohexane)

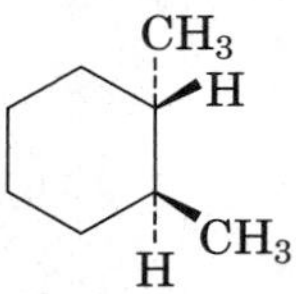

(d) cyclobutane with isopropyl and CH_3

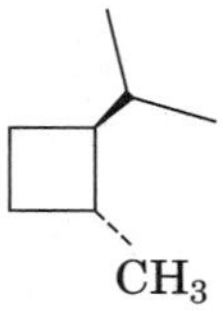

(e) H_3C–cyclohexane–CH_3, CH_3

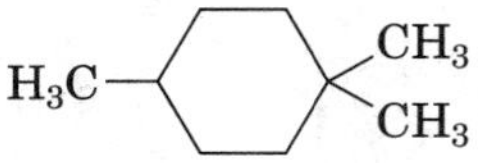

2.42 Give IUPAC names for the five isomers of C_6H_{14}.

2.43 Draw structures for the nine isomers of C_7H_{16}.

2.44 Propose structures and give correct IUPAC names for the following:
(a) A dimethyloctane
(b) A diethyldimethylhexane
(c) A cycloalkane with three methyl groups

2.45 The following names are *incorrect.* Give the proper IUPAC names.
(a) 2,2-Dimethyl-6-ethylheptane
(b) 4-Ethyl-5,5-dimethylpentane
(c) 3-Ethyl-4,4-dimethylhexane
(d) 5,5,6-Trimethyloctane

2.46 The barrier to rotation about the C–C bond in bromoethane is 15.0 kJ/mol (3.6 kcal/mol). If each hydrogen–hydrogen interaction in the eclipsed conformation is responsible for 3.8 kJ/mol (0.9 kcal/mol), how much is the hydrogen–bromine eclipsing interaction responsible for?

2.47 Make a graph of energy versus degree of bond rotation around the C–C bond in bromoethane (see Problem 2.46).

2.48 Malic acid, $C_4H_6O_5$, has been isolated from apples. Because malic acid reacts with 2 equivalents of base, it can be formulated as a dicarboxylic acid (that is, it has two –COOH groups).
(a) Draw at least five possible structures for malic acid.
(b) If malic acid is also a secondary alcohol (has an –OH group attached to a secondary carbon), what is its structure?

2.49 Cyclopropane was first prepared by reaction of 1,3-dibromopropane with sodium.
(a) Formulate the reaction.
(b) What product might the following reaction give? What geometry would you expect for the product?

$$C(CH_2Br)_4 \text{ i.e. } BrCH_2-\overset{\displaystyle CH_2Br}{\underset{\displaystyle CH_2Br}{C}}-CH_2Br \xrightarrow{4\ Na} ?$$

2.50 Tell whether the following pairs of compounds are identical, constitutional isomers, or stereoisomers.
(a) *cis*-1,3-Dibromocyclohexane and *trans*-1,4-dibromocyclohexane
(b) 2,3-Dimethylhexane and 2,5,5-trimethylpentane
(c)

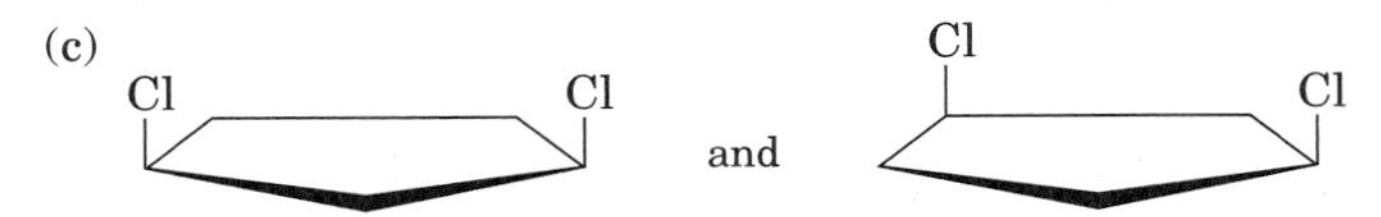

2.51 Draw two constitutional isomers of *cis*-1,2-dibromocyclopentane.

2.52 Draw a stereoisomer of *trans*-1,3-dimethylcyclobutane.

2.53 Draw *trans*-1,2-dimethylcyclohexane in its more stable chair conformation. Are the methyl groups axial or equatorial?

2.54 Draw *cis*-1,2-dimethylcyclohexane in its more stable chair conformation. Are the methyl groups axial or equatorial? Which is more stable, *cis*-1,2-dimethylcyclohexane or *trans*-1,2-dimethylcyclohexane (Problem 2.53)? Explain.

2.55 Which is more stable, *cis*-1,3-dimethylcyclohexane or *trans*-1,3-dimethylcyclohexane? Draw chair conformations of both, and explain your answer.

2.56 *N*-Methylpiperidine has the conformation shown. What does this tell you about the relative steric requirements of a methyl group versus an electron lone pair?

$\ddot{N}$ CH_3 ***N*-Methylpiperidine**

2.57 Glucose contains a six-membered ring in which all the substituents are equatorial. Draw glucose in its more stable chair conformation.

HO O CH_2OH HO OH OH **Glucose**

2.58 Draw 1,3,5-trimethylcyclohexane using a hexagon to represent the ring. How many cis–trans stereoisomers are possible?

2.59 One of the two chair structures of *cis*-1-chloro-3-methylcyclohexane is more stable than the other by 15.5 kJ/mol (3.7 kcal/mol). Which is it? What is the energy cost of a 1,3-diaxial interaction between a chlorine and a methyl group?

2.60 Amantadine is an antiviral agent that is active against influenza A infection. Draw a three-dimensional representation of amantadine showing the chair cyclohexane rings.

NH_2 **Amantadine**

2.61 Draw the three cis–trans isomers of menthol.

CH_3 OH $(CH_3)_2CH$ **Menthol**

2.62 Here's a tough one. There are two different substances named *trans*-1,2-dimethylcyclopentane. What is the relationship between them? (We'll explore this kind of isomerism in Chapter 6.)

CH_3 CH_3 and CH_3 CH_3

Visualizing Chemistry

2.63 Give IUPAC names for the following substances, and convert each drawing into a skeletal structure.

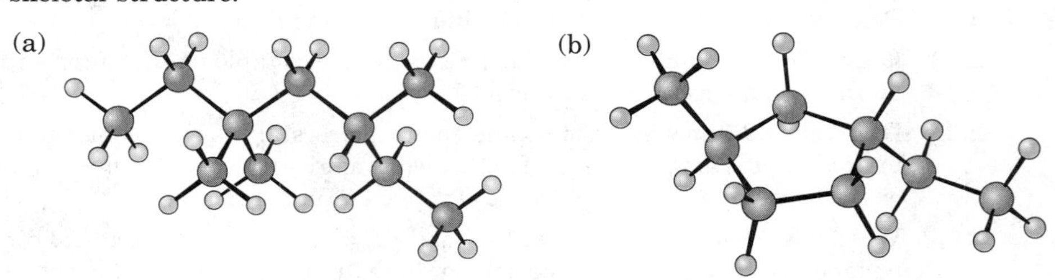

2.64 Identify the functional groups in the following substances, and convert each drawing into a molecular formula (gray = C, red = O, blue = N, light green = H).

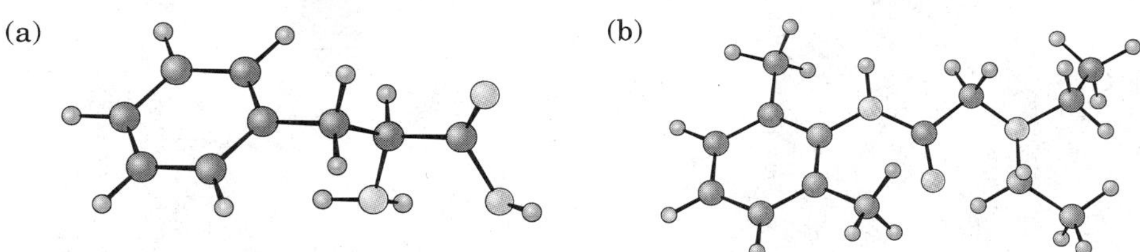

Phenylalanine

Lidocaine

2.65 The following cyclohexane derivative has three substituents—red, green, and blue. Identify each substituent as axial or equatorial, and identify each pair of relationships (red–blue, red–green, and blue–green) as cis or trans.

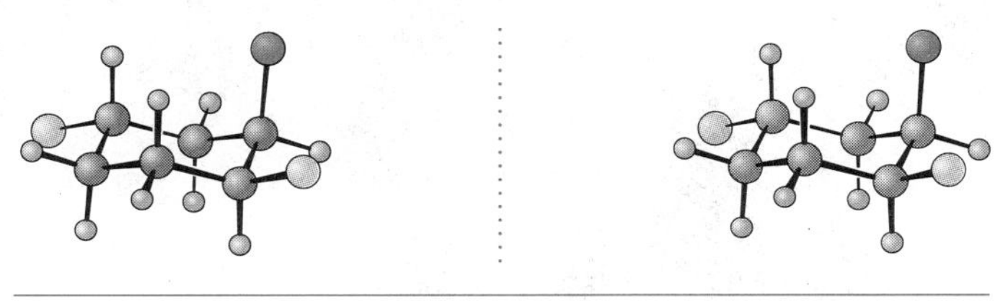

Stereo View

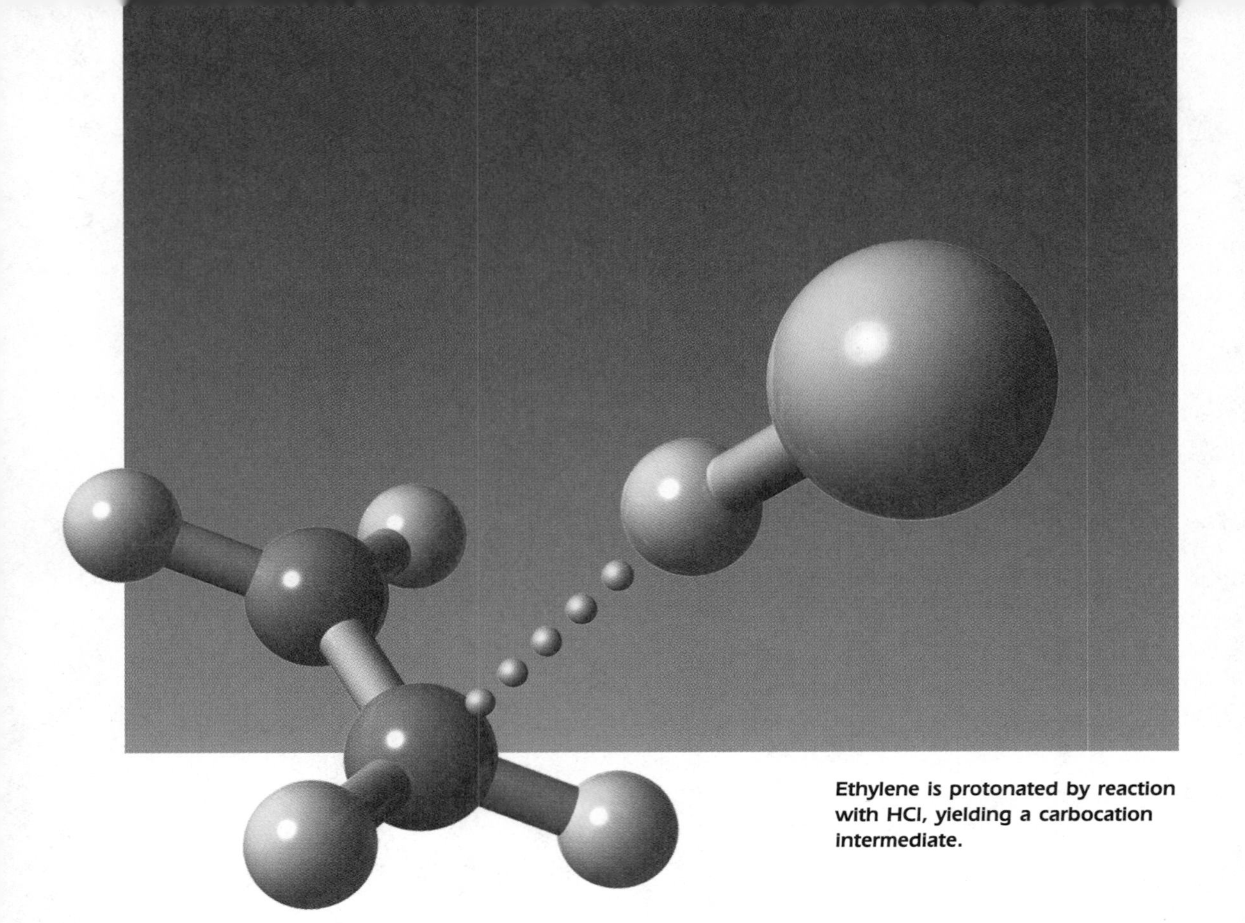

Ethylene is protonated by reaction with HCl, yielding a carbocation intermediate.

3 Alkenes: The Nature of Organic Reactions

Alkenes are hydrocarbons that contain a carbon–carbon double bond. These compounds occur abundantly in nature, and many have important biological roles. Ethylene, for example, is a plant hormone that induces ripening in fruit, and α-pinene is the major constituent of turpentine.

Ethylene ($H_2C{=}CH_2$) **α-Pinene** (H_3C, CH_3, CH_3)

We'll see in this chapter how and why alkenes behave the way they do, and we'll develop some general ideas about organic chemical reactivity that can be applied to all molecules.

3.1 Naming Alkenes

Because of their double bond, alkenes have fewer hydrogens per carbon than related alkanes and are therefore referred to as **unsaturated.** Ethylene, for example, has the formula C_2H_4 whereas ethane has the formula C_2H_6.

$$\mathrm{H_2C{=}CH_2} \qquad\qquad \mathrm{H_3C{-}CH_3}$$

Ethylene: C_2H_4
(fewer hydrogens—*unsaturated*)

Ethane: C_2H_6
(more hydrogens—*saturated*)

Alkenes are named according to a series of rules similar to those used for alkanes, with the suffix *-ene* used in place of *-ane* to identify the family. There are three steps:

Step 1 Name the parent hydrocarbon. Find the longest carbon chain that contains the double bond, and name the compound using the suffix *-ene.*

$$\mathrm{(CH_3CH_2)(CH_3CH_2CH_2)C{=}CH_2}$$

Named as a *pentene* *NOT* as a hexene, since the double bond is not contained in the six-carbon chain

Step 2 Number the carbon atoms in the chain, beginning at the end nearer the double bond. If the double bond is equidistant from the two ends, begin numbering at the end nearer the first branch point. This rule ensures that the double-bond carbons receive the lowest possible numbers:

$$\underset{6}{\mathrm{CH_3}}\underset{5}{\mathrm{CH_2}}\underset{4}{\mathrm{CH_2}}\underset{3}{\mathrm{CH}}{=}\underset{2}{\mathrm{CH}}\underset{1}{\mathrm{CH_3}} \qquad \underset{1}{\mathrm{CH_3}}\underset{2}{\mathrm{CH}}(\mathrm{CH_3})\underset{3}{\mathrm{CH}}{=}\underset{4}{\mathrm{CH}}\underset{5}{\mathrm{CH_2}}\underset{6}{\mathrm{CH_3}}$$

Step 3 Write the full name, numbering the substituents according to their position in the chain and listing them alphabetically. Indicate the position of the double bond by giving the number of the *first* alkene carbon. If more than one double bond is present, give the position of each and use one of the suffixes *-diene, -triene,* and so on.

$$\underset{6}{\mathrm{CH_3}}\underset{5}{\mathrm{CH_2}}\underset{4}{\mathrm{CH_2}}\underset{3}{\mathrm{CH}}{=}\underset{2}{\mathrm{CH}}\underset{1}{\mathrm{CH_3}} \qquad \underset{1}{\mathrm{CH_3}}\underset{2}{\mathrm{CH}}(\mathrm{CH_3})\underset{3}{\mathrm{CH}}{=}\underset{4}{\mathrm{CH}}\underset{5}{\mathrm{CH_2}}\underset{6}{\mathrm{CH_3}}$$

2-Hexene **2-Methyl-3-hexene**

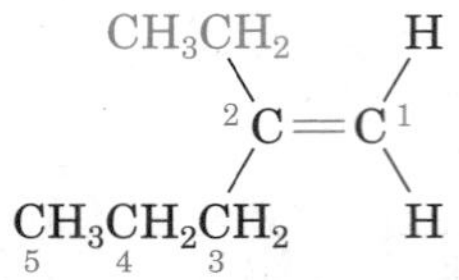

$$\underset{\substack{5\quad 4\quad 3}}{CH_3CH_2CH_2}(CH_3CH_2)\overset{2}{C}=\overset{1}{C}H_2$$

2-Ethyl-1-pentene

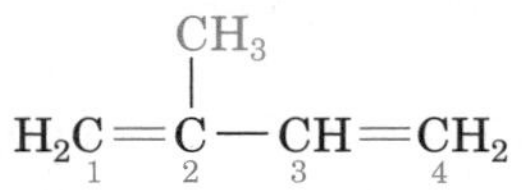

$$\underset{1}{H_2C}=\underset{2}{\overset{CH_3}{\overset{|}{C}}}-\underset{3}{CH}=\underset{4}{CH_2}$$

2-Methyl-1,3-butadiene

Cycloalkenes are named in a similar way, but because there is no chain end to begin from, we number the cycloalkene so that the double bond is between C1 and C2 and so that the first substituent has as low a number as possible. Note that it's not necessary to specify the position of the double bond in the name because it is always between C1 and C2:

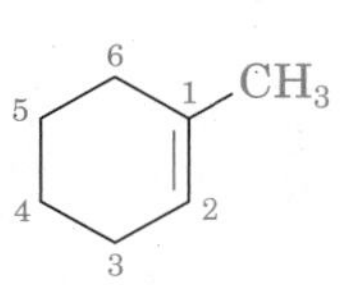

1-Methylcyclohexene 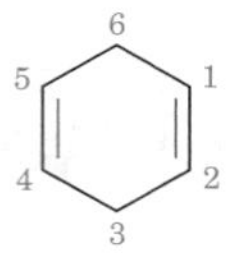 **1,4-Cyclohexadiene** 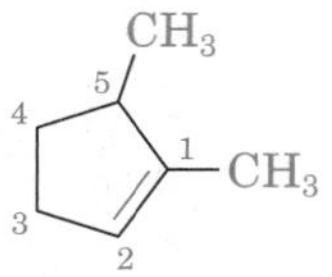 **1,5-Dimethylcyclopentene**

For historical reasons, there are a few alkenes whose names don't conform to the rules. For example, the alkene corresponding to ethane should be called *ethene,* but the name *ethylene* has been used for so long that it is accepted by IUPAC. Table 3.1 lists some other common names.

Table 3.1 Common Names of Some Alkenes[a]

Compound	*Systematic name*	*Common name*
$H_2C{=}CH_2$	Ethene	Ethylene
$CH_3CH{=}CH_2$	Propene	Propylene
$CH_3\overset{CH_3}{\overset{\vert}{C}}{=}CH_2$	2-Methylpropene	Isobutylene
$H_2C{=}\overset{CH_3}{\overset{\vert}{C}}{-}CH{=}CH_2$	2-Methyl-1,3-butadiene	Isoprene

[a]Both common and systematic names are recognized by IUPAC.

PRACTICE PROBLEM 3.1

What is the IUPAC name of the following alkene?

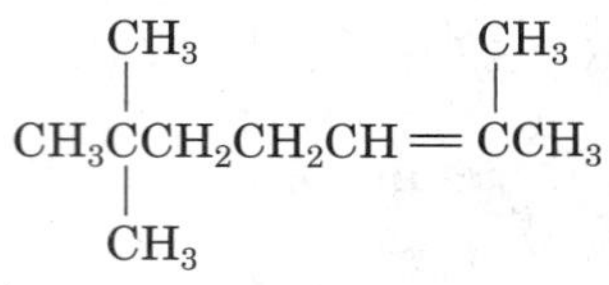

$$CH_3\underset{CH_3}{\overset{CH_3}{\overset{|}{\underset{|}{C}}}}CH_2CH_2CH=\overset{CH_3}{\overset{|}{C}}CH_3$$

Solution First, find the longest chain containing the double bond—in this case, a heptene. Next, number the chain beginning at the end nearer the double bond, and identify the substituents at each position. In this case, there are methyl groups at C2 and C6 (two):

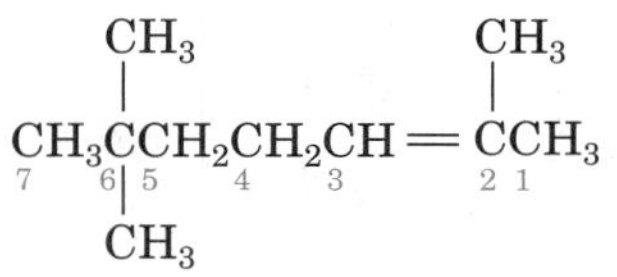

$$\overset{7}{CH_3}\overset{6}{C}(CH_3)_2\overset{5}{CH_2}\overset{4}{CH_2}\overset{3}{CH}{=}\overset{2}{C}(CH_3)\overset{1}{CH_3}$$

The full name is 2,6,6-trimethyl-2-heptene.

PROBLEM

3.1 Give IUPAC names for these compounds:

(a)  $H_2C{=}CHCH_2CH(CH_3)CH_3$ (b) $CH_3CH_2CH{=}CHCH_2CH_2CH_3$

(c) $H_2C{=}CHCH_2CH_2CH{=}CHCH_3$ (d) $CH_3CH_2CH{=}CHCH(CH_3)_2$

PROBLEM

3.2 Name the following cycloalkenes:

(a) 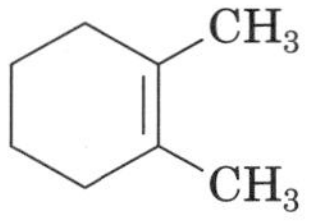 cyclohexene ring bearing CH_3 on each double-bond carbon (b) 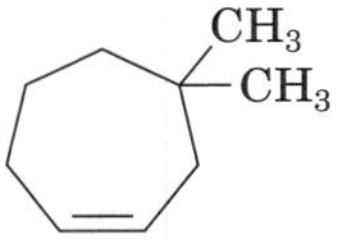 cycloheptene ring bearing two CH_3 on one carbon (c) 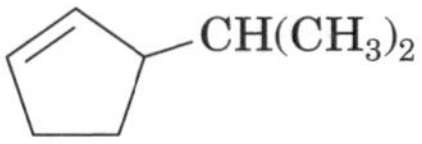 cyclopentene ring bearing $CH(CH_3)_2$

PROBLEM

3.3 Draw structures corresponding to the following IUPAC names:

(a) 2-Methyl-1-hexene (b) 4,4-Dimethyl-2-pentene

(c) 2-Methyl-1,5-hexadiene (d) 3-Ethyl-2,2-dimethyl-3-heptene

3.2 Electronic Structure of Alkenes

We saw in Section 1.9 that the carbon atoms in a double bond are sp^2-hybridized and have three equivalent orbitals, which lie in a plane at angles of 120° to one another. The fourth carbon orbital is an unhybridized p orbital perpendicular to the sp^2 plane. When two such carbon atoms approach each other, they form a σ bond by head-on overlap of sp^2 orbitals and a π bond by sideways overlap of p orbitals. The doubly bonded carbons and the four attached atoms lie in a plane, with bond angles of approximately 120° (Figure 3.1, p. 80).

We also know from Section 2.5 that rotation is possible around sir bonds, and that open-chain alkanes like ethane and propane therefor many rapidly interconverting conformations. The same is not true ble bonds, however. For rotation to take place around a double bo

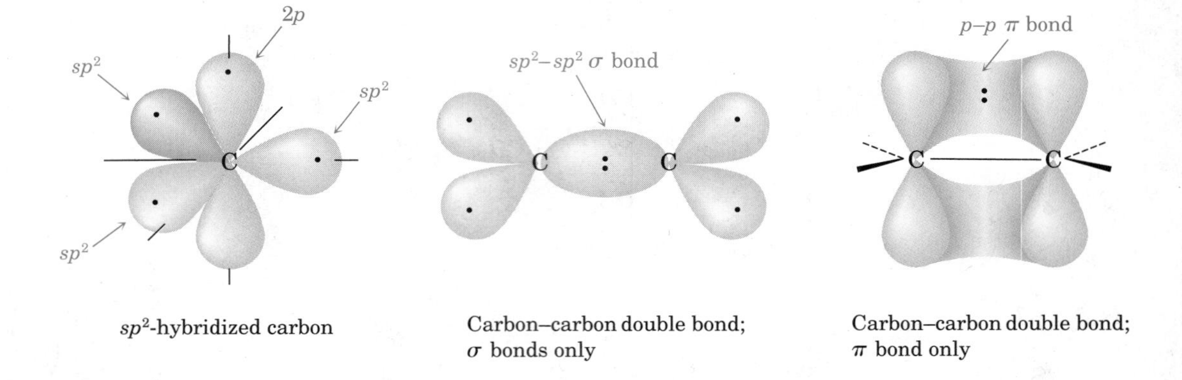

Figure 3.1 An orbital picture of the carbon–carbon double bond. The σ part is formed by head-on overlap of sp^2 orbitals, and the π part is formed by sideways overlap of p orbitals.

π part of the bond would have to break temporarily (Figure 3.2). Thus, the energy barrier to rotation around a double bond must be at least as great as the strength of the π bond itself, an estimated 235 kJ/mol (56 kcal/mol). (Recall that the rotation barrier for a single bond is only about 12 kJ/mol.)

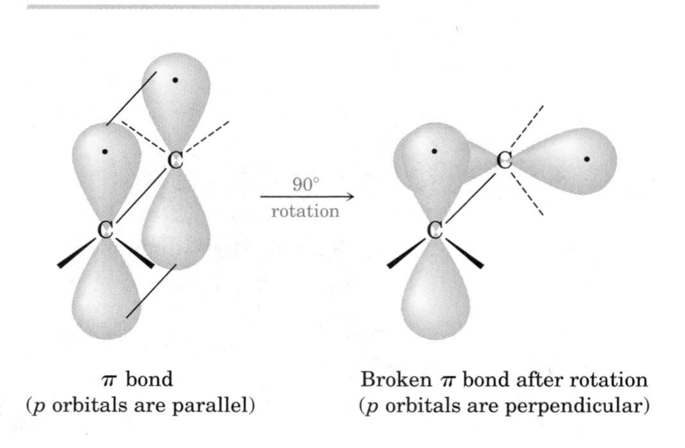

Figure 3.2 The π bond must break for rotation around a carbon–carbon double bond to take place.

3.3 Cis–Trans Isomers of Alkenes

The lack of rotation around carbon–carbon double bonds is of more than just theoretical interest; it also has chemical consequences. Imagine the situation for a disubstituted alkene such as 2-butene. (*Disubstituted* means that two substituents other than hydrogen are bonded to the double-bond carbons.) The two methyl groups in 2-butene can be either on the same side of the double bond or on opposite sides, a situation reminiscent of substituted cycloalkanes (Section 2.8). Figure 3.3 shows the two 2-butene isomers.

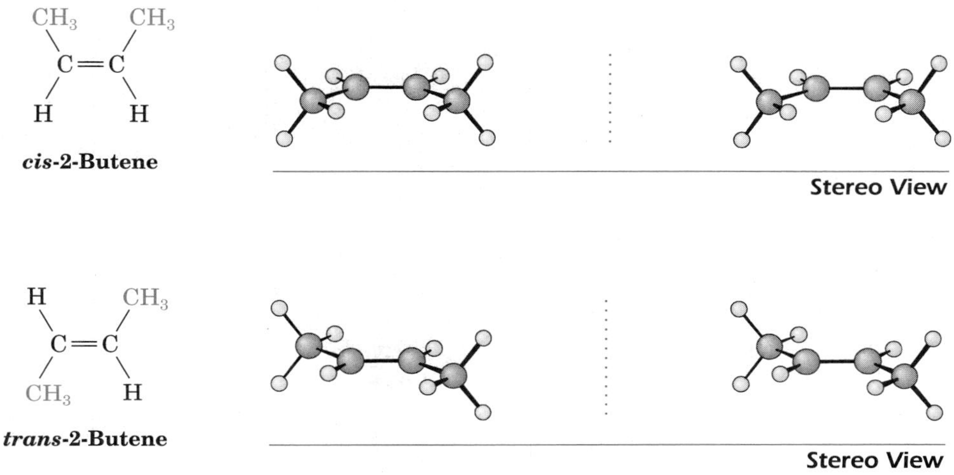

$$\begin{array}{ccc} CH_3 & & CH_3 \\ & C{=}C & \\ H & & H \end{array}$$

***cis*-2-Butene**

Stereo View

$$\begin{array}{ccc} H & & CH_3 \\ & C{=}C & \\ CH_3 & & H \end{array}$$

***trans*-2-Butene**

Stereo View

Figure 3.3 Cis and trans isomers of 2-butene. The cis isomer has the two methyl groups on the same side of the double bond, and the trans isomer has the methyl groups on opposite sides.

Since bond rotation can't occur, the two 2-butenes can't spontaneously interconvert; they are different chemical compounds. As with disubstituted cycloalkanes (Section 2.8), we call such compounds *cis–trans isomers.* The isomer with both substituents on the same side is called *cis*-2-butene, and the isomer with substituents on opposite sides is *trans*-2-butene.

Cis–trans isomerism is not limited to disubstituted alkenes. It can occur whenever both of the double-bond carbons are attached to two different groups. If one of the double-bond carbons is attached to two identical groups, however, then cis–trans isomerism is not possible (Figure 3.4).

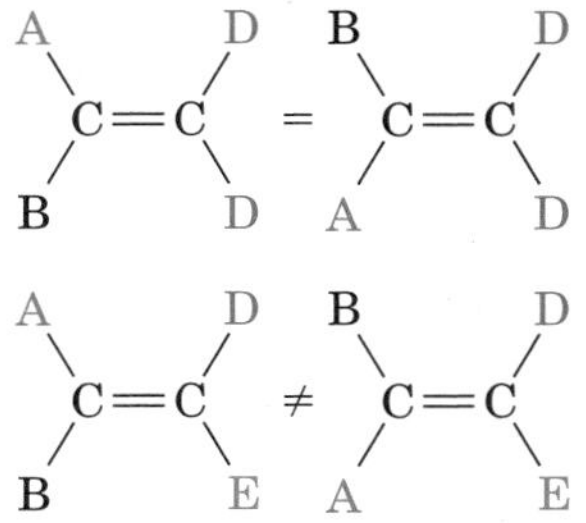

$$\begin{array}{ccc} A & & D \\ & C{=}C & \\ B & & D \end{array} = \begin{array}{ccc} B & & D \\ & C{=}C & \\ A & & D \end{array}$$

These two compounds are identical; they are not cis–trans isomers.

$$\begin{array}{ccc} A & & D \\ & C{=}C & \\ B & & E \end{array} \neq \begin{array}{ccc} B & & D \\ & C{=}C & \\ A & & E \end{array}$$

These two compounds are not identical; they are cis–trans isomers.

Figure 3.4 The requirement for cis–trans isomerism in alkenes. Both double-bond carbons must be attached to two different groups.

Although the cis–trans interconversion of alkene isomers doesn't occur spontaneously, it can be made to happen by treating the alkene with a strong acid catalyst. If we interconvert *cis*-2-butene with *trans*-2-butene and allow them to reach equilibrium, we find that they aren't of equal stability. The trans isomer is more favored than the cis isomer by a ratio of 76 (trans) to 24 (cis).

$$\underset{\textbf{Trans (76\%)}}{\text{H(H}_3\text{C)C=C(CH}_3\text{)H}} \underset{\text{catalyst}}{\overset{\text{Acid}}{\rightleftharpoons}} \underset{\textbf{Cis (24\%)}}{\text{(H}_3\text{C)HC=CH(CH}_3\text{)}}$$

Cis alkenes are less stable than their trans isomers because of steric (spatial) interference between the bulky substituents on the same side of the double bond. This is the same kind of interference, or *steric strain,* that we saw in axial methylcyclohexane (Section 2.11).

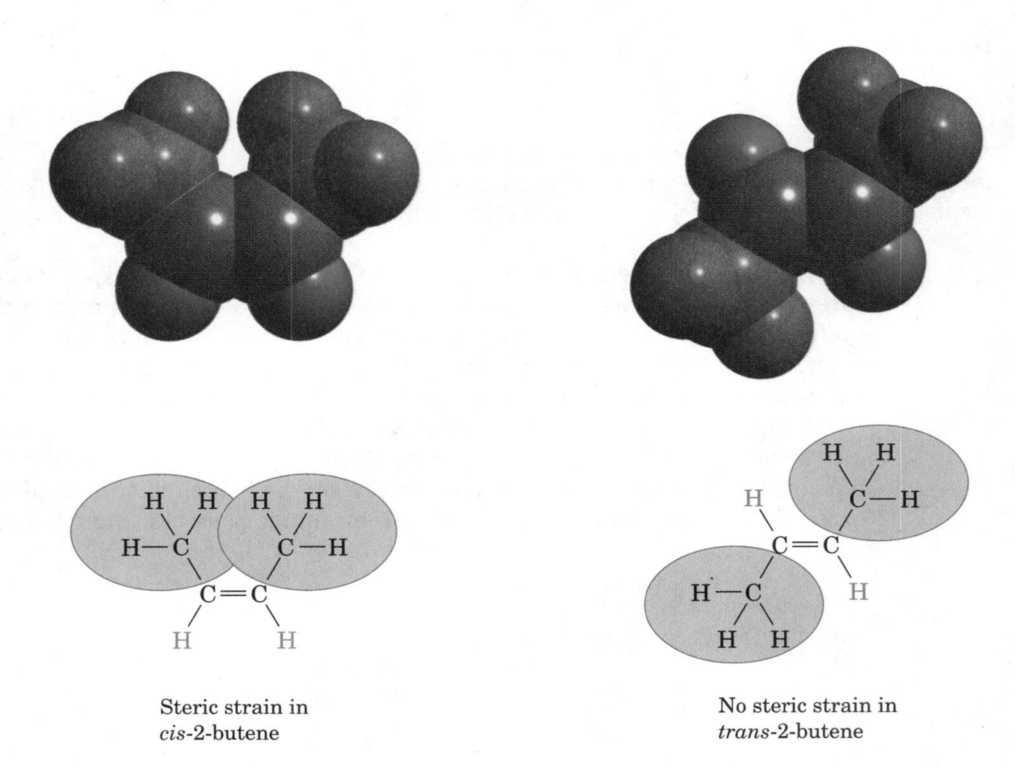

Steric strain in *cis*-2-butene

No steric strain in *trans*-2-butene

PRACTICE PROBLEM 3.2 ....

Draw the cis and trans isomers of 5-chloro-2-pentene.

Solution 5-Chloro-2-pentene is $ClCH_2CH_2CH{=}CHCH_3$. The two substituent groups are on the same side of the double bond in the cis isomer and on opposite sides in the trans isomer.

```
ClCH2CH2      CH3              H          CH3
       \     /                  \        /
        C==C                     C==C
       /     \                  /        \
      H       H          ClCH2CH2         H
```

***cis*-5-Chloro-2-pentene** ***trans*-5-Chloro-2-pentene**

PROBLEM

3.4 Which of the following compounds can exist as cis–trans isomers? Draw each cis–trans pair.
(a) $CH_3CH{=}CH_2$ (b) $(CH_3)_2C{=}CHCH_3$ (c) $ClCH{=}CHCl$
(d) $CH_3CH_2CH{=}CHCH_3$ (e) $CH_3CH_2CH{=}C(Br)CH_3$ (f) 3-Methyl-3-heptene

PROBLEM

3.5 Which is more stable, *cis*-2-methyl-3-hexene or *trans*-2-methyl-3-hexene?

PROBLEM

3.6 How can you account for the observation that cyclohexene does not show cis–trans isomerism?

3.4 Sequence Rules: The *E,Z* Designation

The cis–trans naming system used in the previous section works well for describing the geometry of disubstituted alkenes, but fails with trisubstituted and tetrasubstituted double bonds. (*Trisubstituted* means three substituents other than hydrogen on the double bond; *tetrasubstituted* means four substituents other than hydrogen.)

A more general method for describing double-bond geometry is provided by the *E,Z* system of nomenclature, which uses a set of **sequence rules** to assign priorities to the substituent groups on the double-bond carbons. Considering each of the double-bond carbons separately, we use the sequence rules to decide which of the two groups attached to each carbon is higher in priority. If the higher-priority groups are on the same side of the double bond, the alkene is designated *Z* (for the German *zusammen,* "together"). If the higher-priority groups are on opposite sides, the alkene is designated *E* (for the German *entgegen,* "opposite"). A simple way to remember which is which is to think with an accent: In the *Z* isomer, the groups are on "ze zame zide." The assignments are shown in Figure 3.5 (p. 84).

Sometimes called the *Cahn–Ingold–Prelog rules* after the chemists who proposed them, the sequence rules are as follows:

Sequence rule 1 Taking the double-bond carbons one at a time, look at the atoms directly attached to each carbon and rank them according to their atomic number. An atom with a higher atomic number is higher in priority than an atom with a lower atomic number. Thus, the

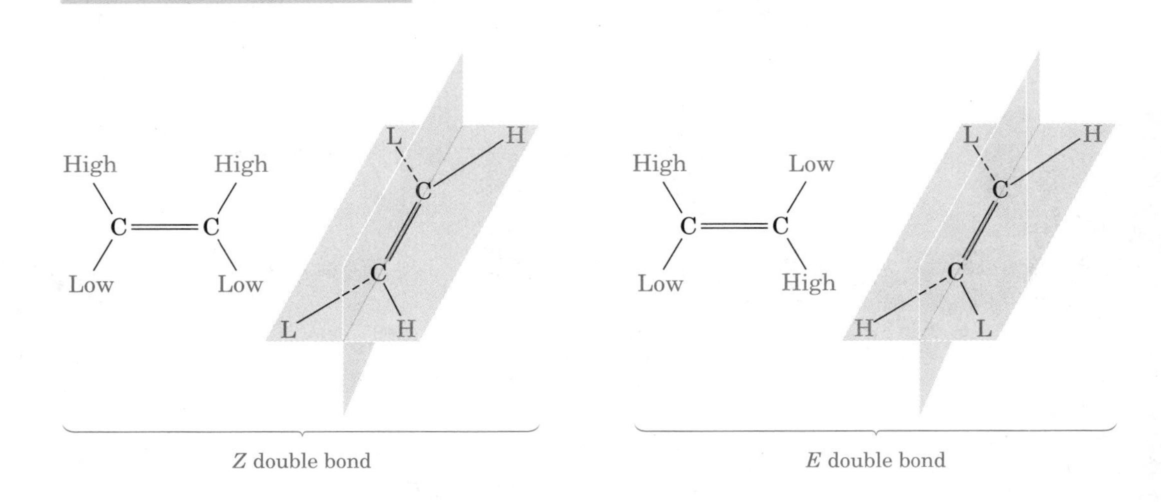

Figure 3.5 The *E,Z* system of nomenclature for substituted alkenes. The higher-priority groups are on the same side in the *Z* isomer, but are on opposite sides in the *E* isomer.

atoms that we commonly find attached to a double-bond carbon are assigned the following priorities:

$$\overset{35}{\text{Br}} > \overset{17}{\text{Cl}} > \overset{8}{\text{O}} > \overset{7}{\text{N}} > \overset{6}{\text{C}} > \overset{1}{\text{H}}$$

For example:

Low priority H, High priority Cl; C=C; High priority CH_3, Low priority CH_3

(a) (*E*)-2-Chloro-2-butene

Low priority H, Low priority CH_3; C=C; High priority CH_3, High priority Cl

(b) (*Z*)-2-Chloro-2-butene

Because chlorine has a higher atomic number than carbon, it receives higher priority than a methyl (CH_3) group. Methyl receives higher priority than hydrogen, however, and isomer (a) is therefore assigned *E* geometry (high-priority groups on opposite sides of the double bond). Isomer (b) has *Z* geometry (high-priority groups on "ze zame zide" of the double bond).

Sequence rule 2 If a decision can't be reached by ranking the first atoms in the substituents, look at the second, third, or fourth atoms away from the double-bond carbons until the first difference is found. Thus, an ethyl substituent, $-CH_2CH_3$, and a methyl substituent, $-CH_3$, are equivalent by rule 1 because both have carbon as the first atom. By rule 2, however, ethyl receives higher priority than methyl because its *second* atoms are C, H, H rather than H, H, H. Look at the following examples to see how the rule works:

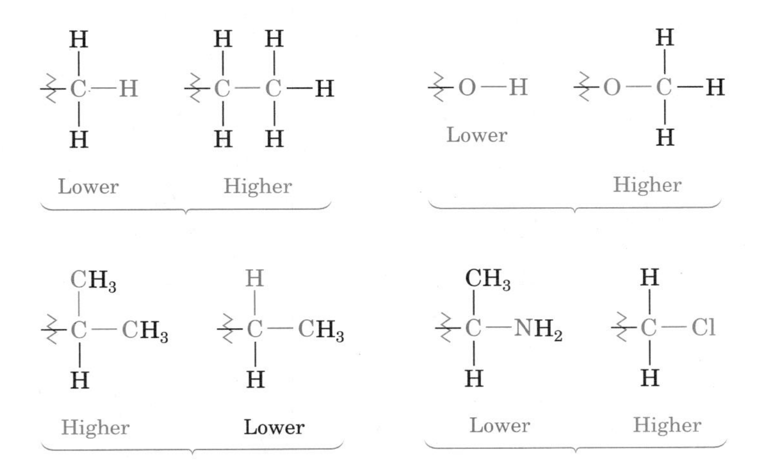

Sequence rule 3 Multiple-bonded atoms are equivalent to the same number of single-bonded atoms. For example, an aldehyde substituent (–CH=O), which has a carbon atom *doubly* bonded to *one* oxygen, is equivalent to a substituent with a carbon atom *singly* bonded to *two* oxygens:

is equivalent to

This carbon is bonded to H, O, O

This oxygen is bonded to C, C

This carbon is bonded to H, O, O

This oxygen is bonded to C, C

As further examples, the following pairs are equivalent:

is equivalent to

This carbon is bonded to H, C, C

This carbon is bonded to H, H, C, C

This carbon is bonded to H, C, C

This carbon is bonded to H, H, C, C

is equivalent to

This carbon is bonded to C, C, C

This carbon is bonded to H, C, C, C

This carbon is bonded to C, C, C

This carbon is bonded to H, C, C, C

PRACTICE PROBLEM 3.3

Assign *E* or *Z* configuration to the double bond in the following compound:

```
H           CH(CH3)2
  \        /
   C == C
  /        \
H3C         CH2OH
```

Solution Look at the two double-bond carbons individually. The left-hand carbon has two substituents, $-H$ and $-CH_3$, of which $-CH_3$ receives higher priority by rule 1. The right-hand carbon also has two substituents, $-CH(CH_3)_2$ and $-CH_2OH$, which are equivalent by rule 1. By rule 2, however, $-CH_2OH$ receives higher priority than $-CH(CH_3)_2$ because $-CH_2OH$ has an *oxygen* and two hydrogens as the next atoms, whereas $-CH(CH_3)_2$ has two *carbons* and a hydrogen. Thus, the two high-priority groups are on the same side of the double bond, and we assign *Z* configuration.

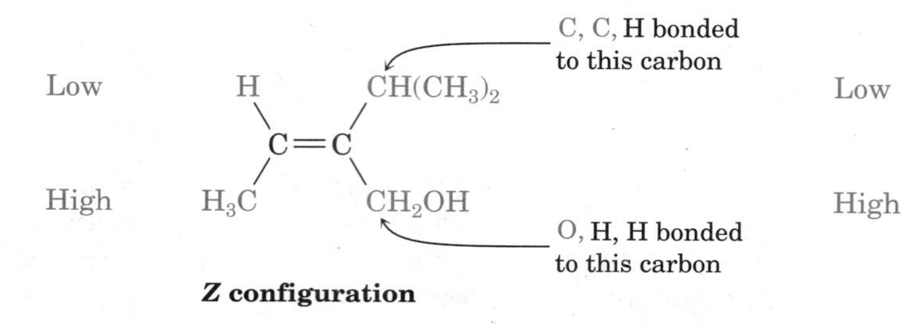

***Z* configuration**

PROBLEM

3.7 Which member in each of the following sets is higher in priority?
(a) $-H$ or $-Br$ (b) $-Cl$ or $-Br$ (c) $-CH_3$ or $-CH_2CH_3$
(d) $-NH_2$ or $-OH$ (e) $-CH_2OH$ or $-CH_3$ (f) $-CH_2OH$ or $-CH{=}O$

PROBLEM

3.8 Which is higher in priority, $-\overset{\overset{O}{\|}}{C}-OH$ or $-\overset{\overset{O}{\|}}{C}-OCH_3$? Explain.

PROBLEM

3.9 Which is higher in priority, isopropyl or *n*-octyl? Explain.

PROBLEM

3.10 Assign *E* or *Z* configuration to the following compounds:

```
                                           O
                                           ‖
(a) CH3O        Cl        (b) H3C          C—OCH3
        \      /                  \       /
         C == C                    C == C
        /      \                  /       \
       H        CH3              H         OCH3
```

3.5 Kinds of Organic Reactions

"RASE"

Now that we know something about alkenes and the double-bond functional group, it's time to learn about their chemical reactivity. As an introduction, we'll first look at some of the basic principles that underlie all organic reactions. In particular, we'll develop some general notions about why compounds react the way they do, and we'll see some methods that have been developed to help understand how reactions take place.

Organic chemical reactions can be organized either by *what kinds* of reactions occur or by *how* reactions occur. We'll begin by looking at the kinds of reactions that take place. There are four particularly important kinds of organic reactions: *additions, eliminations, substitutions,* and *rearrangements.*

Addition reactions occur when two reactants add together to form a single new product with no atoms "left over." We can generalize the process as:

These reactants add together . . . $A + B \longrightarrow C$ . . . to give this single product.

As an example of an important addition reaction that we'll be studying soon, alkenes react with HCl to yield alkyl chlorides:

These two reactants . . . H—Cl + $H_2C{=}CH_2$ ⟶ CH_3CH_2Cl (H—C(H)(H)—C(H)(Cl)—H) . . . add to give this product.

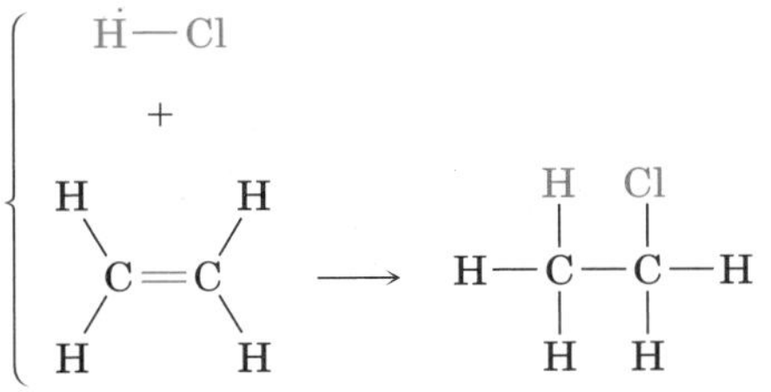

Ethylene (an alkene) **Chloroethane (an alkyl halide)**

Elimination reactions are, in a sense, the opposite of addition reactions. Eliminations occur when a single reactant splits into two products, a process we can generalize as:

This one reactant . . . $A \longrightarrow B + C$ . . . splits apart to give these two products.

As an example of an important elimination reaction, alkyl halides split apart into an acid and an alkene when treated with base:

This one reactant . . . CH_3CH_2Cl (H—C(H)(H)—C(H)(Cl)—H) $\xrightarrow{NaOH}$ $H_2C{=}CH_2$ + H—Cl . . . gives these two products.

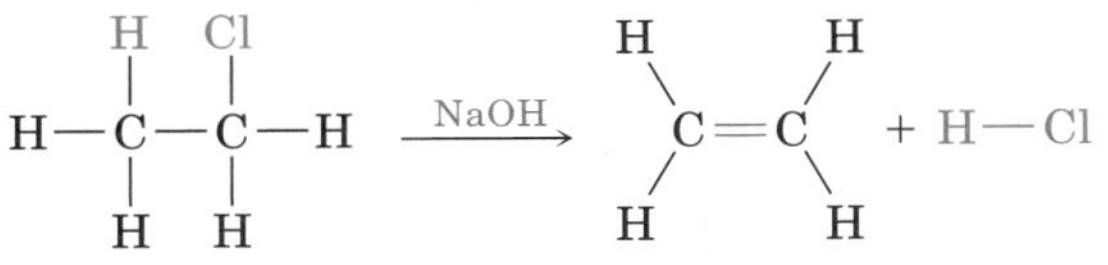

Chloroethane (an alkyl halide) **Ethylene (an alkene)**

Substitution reactions occur when two reactants exchange parts to give two new products, a process we can generalize as:

These two reactants exchange parts . . . $A{-}B + C{-}D \longrightarrow A{-}C + B{-}D$ . . . to give these two new products.

As an example of a substitution reaction, we saw in Section 2.4 that alkanes react with Cl_2 in the presence of ultraviolet light to yield alkyl chlorides. A –Cl group substitutes for the –H group of the alkane, and two new products result:

These two reactants . . . $CH_4 + Cl{-}Cl \xrightarrow{\text{Light}} CH_3{-}Cl + H{-}Cl$ . . . give these two products.

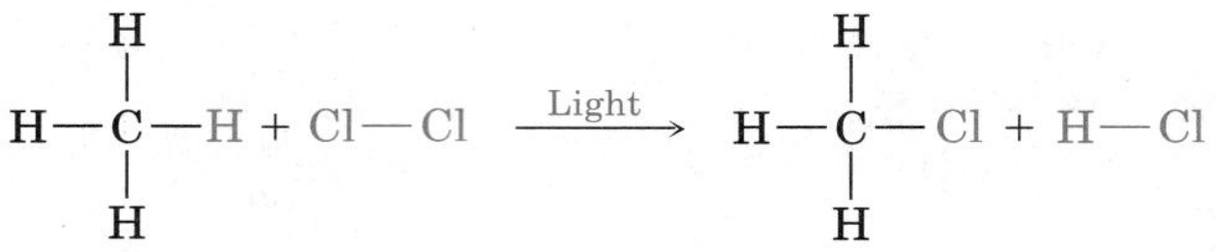

Methane (an alkane) → **Chloromethane (an alkyl halide)**

Rearrangement reactions occur when a single reactant undergoes a reorganization of bonds and atoms to yield a single isomeric product, a process we can generalize as:

This single reactant . . . $A \longrightarrow B$ . . . gives this isomeric product.

As an example of a rearrangement reaction, we saw in Section 3.3 that *cis*-2-butene can be converted into its isomer *trans*-2-butene by treatment with an acid catalyst:

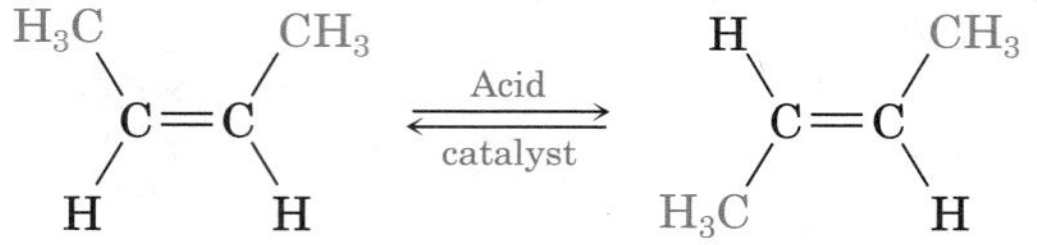

$$\text{cis-}H_3C{-}CH{=}CH{-}CH_3 \underset{\text{catalyst}}{\overset{\text{Acid}}{\rightleftharpoons}} \text{trans-}H_3C{-}CH{=}CH{-}CH_3$$

***cis*-2-Butene (24%)** ⇌ ***trans*-2-Butene (76%)**

PROBLEM

3.11 Classify the following reactions as additions, eliminations, substitutions, or rearrangements:

(a) $CH_3Br + KOH \longrightarrow CH_3OH + KBr$ (b) $CH_3CH_2OH \longrightarrow H_2C{=}CH_2 + H_2O$

(c) $H_2C{=}CH_2 + H_2 \longrightarrow CH_3CH_3$

3.6 How Reactions Occur: Mechanisms

Having looked at the kinds of reactions that take place, let's now see *how* reactions occur. An overall description of how a reaction occurs is called a **reaction mechanism.** A mechanism describes in detail exactly what takes place at each stage of a chemical transformation—which bonds are broken and in what order, which bonds are formed and in what order, and what the relative rate of each step is.

All chemical reactions involve bond breaking and bond making. When two reactants come together, react, and yield products, specific chemical bonds in the reactants are broken, and specific bonds in the products are formed. Fundamentally, there are two ways in which a covalent two-electron bond can break. A bond can break in an electronically *symmetrical* way so that one electron remains with each product fragment, or a bond can break in an electronically *unsymmetrical* way so that both electrons remain with one product fragment, leaving the other fragment with a vacant orbital. The symmetrical cleavage is said to be **homolytic,** and the unsymmetrical cleavage is said to be **heterolytic.**

$$\mathrm{A{:}B} \longrightarrow \mathrm{A\cdot} + \mathrm{\cdot B} \qquad \text{Homolytic bond breaking (radical)}$$

$$\mathrm{A{:}B} \longrightarrow \mathrm{A^+} + \mathrm{{:}B^-} \qquad \text{Heterolytic bond breaking (polar)}$$

Conversely, there are two ways in which a covalent two-electron bond can form: in an electronically symmetrical (**homogenic**) way when one electron is donated to the new bond by each reactant, or in an electronically unsymmetrical (**heterogenic**) way when both bonding electrons are donated to the new bond by one reactant.

$$\mathrm{A\cdot} + \mathrm{\cdot B} \longrightarrow \mathrm{A{:}B} \qquad \text{Homogenic bond making (radical)}$$

$$\mathrm{A^+} + \mathrm{{:}B^-} \longrightarrow \mathrm{A{:}B} \qquad \text{Heterogenic bond making (polar)}$$

Processes that involve symmetrical bond breaking and making are called **radical reactions.** A **radical** (sometimes called a "free radical") is a chemical species that contains an *odd* number of valence electrons and thus has an orbital with only one electron. Processes that involve unsymmetrical bond breaking and making are called **polar reactions.** Polar reactions involve species that contain an *even* number of valence electrons and have only electron pairs in their orbitals. Polar processes are the more common reaction type in organic chemistry, and much of this book is devoted to their description.

To see how polar reactions occur, we need to recall the discussion of polar covalent bonds in Section 1.11 and look more deeply into the effects of bond polarity on organic molecules. We've seen that certain bonds in a molecule, particularly the bonds in functional groups, are often polar. When a carbon atom bonds to an electronegative atom such as chlorine or oxygen, the bond is polarized so that the carbon bears a partial positive charge ($\delta+$) and the electronegative atom bears a partial negative charge ($\delta-$). Conversely, when carbon bonds to an atom that is less electronegative than itself, the opposite polarity results. Such is the case with most carbon–metal (*organometallic*) bonds.

$\mathrm{Y}^{\delta-}$–$\mathrm{C}^{\delta+}$ where Y = O, N, Cl, Br, I

$\mathrm{M}^{\delta+}$–$\mathrm{C}^{\delta-}$ where M = a metal such as Mg or Li

What effect does bond polarity have on chemical reactions? *Because unlike charges attract each other, the fundamental characteristic of all polar reactions is that electron-rich sites in one molecule react with electron-poor sites in another molecule.* Bonds are made when an electron-rich reactant donates a pair of electrons to an electron-poor reactant, and bonds are broken when one of the two product fragments leaves with the electron pair.

Chemists usually indicate the electron movement that occurs during a polar reaction by using curved arrows. A *curved arrow shows where electrons move during the reaction.* It means that an electron pair moves *from* the atom or bond at the tail of the arrow *to* the atom or bond at the head of the arrow.

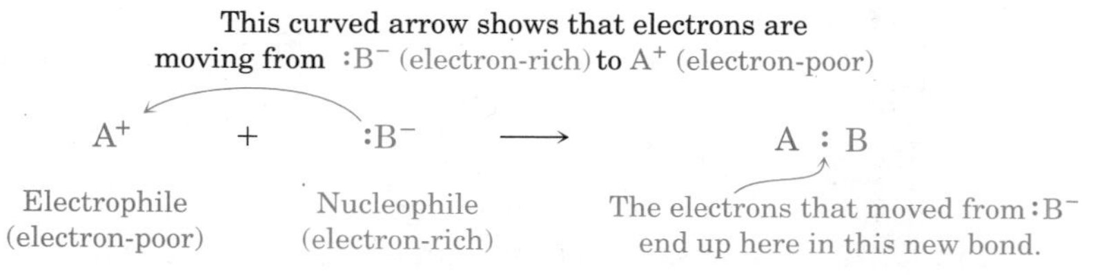

In referring to polar reactions, chemists have coined the words *nucleophile* and *electrophile.* A **nucleophile** is a reagent that is "nucleus loving"; a nucleophile has an electron-rich atom and can form a bond by donating an electron pair to an electron-poor atom. Nucleophiles often have lone pairs of electrons and, frequently, are negatively charged. An **electrophile,** by contrast, is "electron-loving"; an electrophile has an electron-poor atom and can form a bond by accepting an electron pair from a nucleophile. Electrophiles are often, though not always, positively charged.

PRACTICE PROBLEM 3.4

What is the direction of bond polarity in the amine functional group, $C{-}NH_2$?

Solution Nitrogen is more electronegative than carbon according to Figure 1.17, so an amine is polarized with carbon $\delta+$ and nitrogen $\delta-$.

PROBLEM

3.12 What is the direction of bond polarity in the following functional groups? (See Figure 1.17 for electronegativity values.)
(a) Ketone (b) Alkyl chloride (c) Alcohol (d) Alkyllithium, R–Li

PROBLEM

3.13 Identify the functional groups, and show the direction of bond polarity in each of the following molecules:

(a) Acetone, $CH_3\overset{\overset{\displaystyle O}{\|}}{C}CH_3$
(b) Chloroethane, CH_3CH_2Cl
(c) Methanethiol, CH_3SH
(d) Tetraethyllead, $(CH_3CH_2)_4Pb$ (the "lead" in gasoline)

PROBLEM

3.14 Which of the following are most likely to behave as electrophiles, and which as nucleophiles? Explain.
(a) H^+ (b) HO^- (c) Br^+ (d) NH_3 (e) H–C≡C–H

3.7 An Example of a Polar Reaction: Addition of HCl to Ethylene

Let's look in detail at a typical polar reaction, the reaction of ethylene with HCl. When ethylene is treated with hydrogen chloride at room temperature, chloroethane is produced. Overall, the reaction can be formulated as

$$H_2C{=}CH_2 + H{-}Cl \longrightarrow H{-}CH_2{-}CH_2Cl$$

Ethylene (nucleophile)	Hydrogen chloride (electrophile)	Chloroethane

This reaction, an example of a general polar reaction type known as an *electrophilic addition,* can be understood using the general concepts discussed in the previous section. We'll begin by looking at the nature of the two reactants.

What do we know about ethylene? We know from Sections 1.9 and 3.2 that a carbon–carbon double bond results from orbital overlap of two sp^2-hybridized carbon atoms. The σ part of the double bond results from sp^2–sp^2 overlap, and the π part results from p–p overlap.

What kind of chemical reactivity might we expect of a C=C bond? We know that *alkanes* are relatively inert because their valence electrons are tied up in strong, nonpolar C–C and C–H bonds. Furthermore, the bonding electrons in alkanes are inaccessible to external reagents because they are sheltered in σ bonds between nuclei. The situation for *alkenes* is quite different, however. For one thing, double bonds have greater electron density than single bonds—four electrons in a double bond versus only two electrons in a single bond. Also, the electrons in the π bond are accessible to external reagents because they are located above and below the plane of the double bond rather than between the nuclei (Figure 3.6, p. 92).

Both electron richness and electron accessibility lead to the prediction that C=C bonds should behave as *nucleophiles.* That is, the chemistry of alkenes should involve reaction of the electron-rich double bond with electron-poor reagents. This is exactly what we find: The most important reaction of alkenes is their reaction with electrophiles.

What about HCl? As a strong acid, HCl is a powerful proton (H^+) donor. Because a proton is positively charged and electron-poor, it is a good electrophile. Thus, the reaction of H^+ with ethylene is a typical electrophile–nucleophile combination, characteristic of all polar reactions. (Although chemists often talk about "H^+" when referring to acids, there is really no such free species. Protons are always associated with another molecule for stability—for example, with water in H_3O^+.)

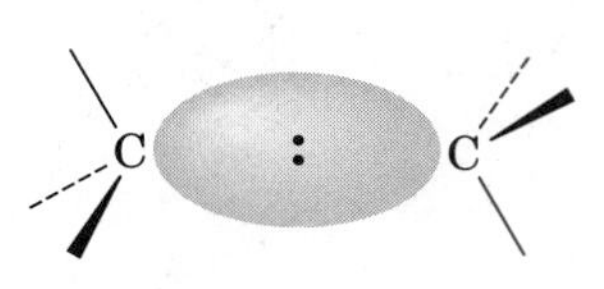

Carbon–carbon σ bond:
stronger; less accessible bonding electrons

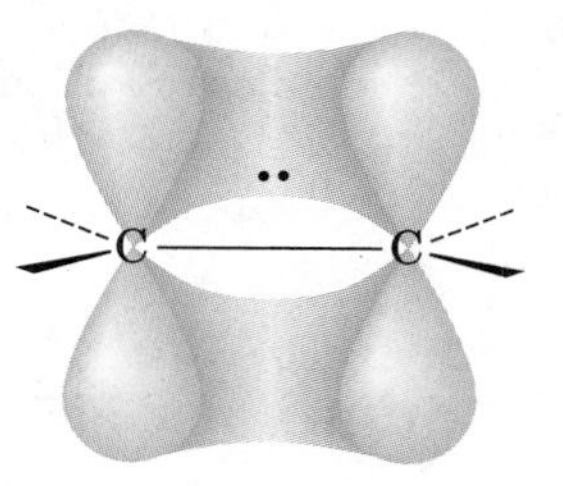

Carbon–carbon π bond:
weaker; more accessible electrons

Figure 3.6 A comparison of carbon–carbon single and double bonds. A double bond is both more electron-rich (more nucleophilic) and more accessible to reaction with external reagents than a single bond.

3.8 The Mechanism of an Organic Reaction: Addition of HCl to Ethylene

We can view the electrophilic addition reaction between ethylene and HCl as proceeding by the mechanism shown in Figure 3.7. The reaction takes place in two steps, beginning when the alkene reacts with H^+. Two electrons from the ethylene π bond move to form a new σ bond between the H^+ and one of the ethylene carbon atoms, as shown by following the path of the curved arrow in Figure 3.7. (*Remember:* A curved arrow is used to indicate how an electron pair moves in a polar reaction.) The other ethylene carbon atom, having lost its share of the π electrons, is now left with a vacant *p* orbital, has only six valence electrons, and carries a positive charge. In the second step, this positively charged species—a carbon cation, or **carbocation**—is itself an electrophile that accepts an electron pair from the nucleophilic Cl^- anion to form a C–Cl bond and give the neutral addition product. Once again, a curved arrow in Figure 3.7 shows the path of the electron-pair movement from Cl^- ion to carbon.

PRACTICE PROBLEM 3.5

What product would you expect from reaction of HCl with cyclohexene?

Solution HCl adds to the double-bond functional group in cyclohexene in exactly the same way it adds to ethylene, yielding an addition product.

H H + HCl ⟶ H Cl H H

Cyclohexene **Chlorocyclohexane**

The electrophile H^+ is attacked by the π electrons of the double bond, and a new C–H σ bond is formed. This leaves the other carbon atom with a + charge and a vacant *p* orbital.

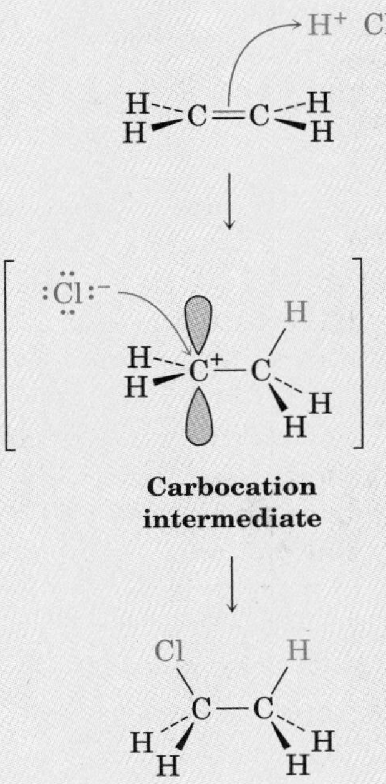

Carbocation intermediate

Cl^- donates an electron pair to the positively charged carbon atom, forming a C–Cl σ bond and yielding the neutral addition product.

Figure 3.7 The mechanism of the electrophilic addition of HCl to ethylene. The reaction takes place in two steps and involves an intermediate carbocation.

PROBLEM

3.15 Reaction of HCl with 2-methylpropene yields 2-chloro-2-methylpropane. What is the structure of the carbocation formed during the reaction? Show the mechanism of the reaction.

$$(CH_3)_2C{=}CH_2 + HCl \longrightarrow (CH_3)_3C{-}Cl$$

PROBLEM

3.16 Reaction of HCl with 2-pentene yields a mixture of two addition products. Write the reaction, and show the two products.

3.9 Describing a Reaction: Rates and Equilibria

Every chemical reaction can go both forward and backward. Reactants can give products, and products can revert to reactants. The position of the resultant chemical equilibrium is expressed by an equation in which K_{eq}, the

equilibrium constant, is equal to the concentration of products, divided by the concentration of reactants.

For the generalized reaction: $aA + bB \rightleftharpoons cC + dD$

we have $$K_{eq} = \frac{[\text{Products}]}{[\text{Reactants}]} = \frac{[C]^c[D]^d}{[A]^a[B]^b}$$

The value of the equilibrium constant tells which side of the reaction arrow is favored. If K_{eq} is larger than 1, then the product concentration term $[C]^c[D]^d$ is larger than the reactant concentration term $[A]^a[B]^b$, and the reaction proceeds as written from left to right. If K_{eq} is smaller than 1, the reaction does not take place as written but instead goes from right to left.

What the equilibrium equation does *not* tell is the *rate* of the reaction, or how fast the equilibrium is established. Some reactions are extremely slow even though they have favorable equilibrium constants. Gasoline is stable at room temperature, for example, because the rate of its reaction with oxygen is slow at 298 K. At higher temperatures, however, such as occur in contact with a lighted match, gasoline reacts rapidly with O_2 and undergoes complete conversion to water and carbon dioxide. Rates (*how fast* a reaction occurs) and equilibria (*how much* a reaction occurs) are entirely different.

Rate ⟶ Is reaction fast or slow?

Equilibrium ⟶ In what direction does reaction proceed?

What determines whether a reaction occurs? For a reaction to have a favorable equilibrium constant, the energy level of the products must be lower than the energy level of the reactants. In other words, energy must be given off.[1] Such reactions are said to be **exothermic** (from the Greek *exo,* "outside," and *therme,* "heat"). If the energy level of the products is higher than the energy level of the reactants, then the equilibrium constant for the reaction is unfavorable, and energy must be added to make the reaction take place. Such reactions are said to be **endothermic** (Greek *endon,* "within").

A good analogy for the relationship between energy and chemical reactivity (stability) is that of a rock poised near the top of a hill. The rock, in its unstable position, has stored the energy that was required to raise it. When it rolls downhill, the rock releases its energy until it reaches a stable, low-energy position at the bottom of the hill. In the same way, the energy level in a chemical reaction goes downhill as the energy stored in the chemical bonds of a reactant is released and a more stable product is formed (Figure 3.8).

---

[1]The "energy" released in a favorable reaction is actually the Gibbs free energy (ΔG), which is the sum of two contributions, an *enthalpy* contribution (ΔH) that measures the heat change in the reaction and a temperature-dependent *entropy* contribution ($T\Delta S$) that measures the change in molecular disorder during the reaction ($\Delta G = \Delta H - T\Delta S$). The entropy contribution is often small, however, so that $\Delta G \approx \Delta H$.

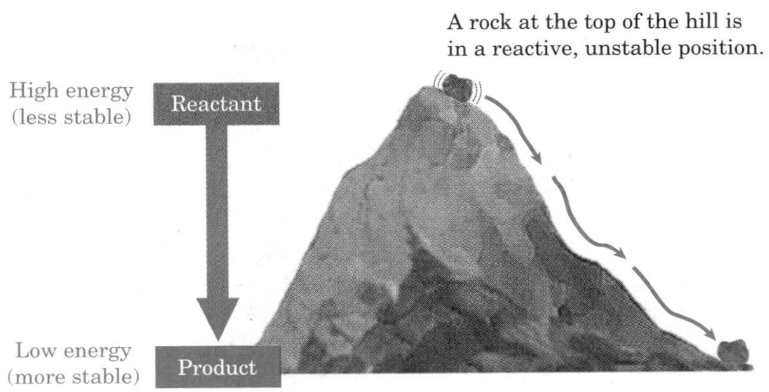

Figure 3.8 The relationship between energy and stability. Like a rock near the top of a hill, high-energy substances are unstable. They release their energy when they drop "downhill" to form low-energy, stable products.

The amount of heat released in an exothermic reaction or absorbed in an endothermic reaction is called the *enthalpy of reaction,* or **heat of reaction, ΔH** (Δ is the capital Greek letter delta and is used to indicate "change"). By convention, ΔH has a negative value in an exothermic reaction because heat is released, and a positive value in an endothermic reaction because heat is absorbed. Since ΔH is a measure of the difference in energy between reactants and products, the size of ΔH determines the size of the equilibrium constant, K_{eq}. Favorable reactions with large K_{eq}'s are highly exothermic and have negative heats of reaction, whereas unfavorable reactions with small K_{eq}'s are endothermic and have positive heats of reaction.

$$A + B \rightleftharpoons C + D$$

$$K_{eq} = \frac{[C][D]}{[A][B]} \qquad \begin{array}{l} \text{Exothermic if } K_{eq} > 1\text{; negative value of } \Delta H \\ \text{Endothermic if } K_{eq} < 1\text{; positive value of } \Delta H \end{array}$$

PRACTICE PROBLEM 3.6

Which reaction is more favorable, one with $\Delta H = -60$ kJ/mol or one with $\Delta H = +60$ kJ/mol?

Solution Reactions with negative ΔH are exothermic and thus are favorable, but reactions with positive ΔH are endothermic and unfavorable.

PROBLEM

3.17 Which reaction is more exothermic, one with $\Delta H = -10$ kJ/mol or one with $\Delta H = +10$ kJ/mol?

PROBLEM

3.18 Which reaction is more exothermic, one with $K_{eq} = 1000$ or one with $K_{eq} = 0.001$?

3.10 Describing a Reaction: Reaction Energy Diagrams and Transition States

For a reaction to take place, reactant molecules must collide, and reorganization of atoms and bonds must occur. Let's look again at the addition reaction between ethylene and HCl. As the reaction proceeds, ethylene and HCl approach each other, the C=C π bond breaks, a new C–H bond forms in the first step, and a new C–Cl bond forms in the second step.

$$H_2C{=}CH_2 + H^+ \longrightarrow [H{-}CH_2{-}\overset{+}{C}H_2] \xrightarrow{:\ddot{Cl}:^-} H{-}CH_2{-}CH_2{-}Cl$$

Ethylene (nucleophile) **A carbocation** **Chloroethane**

Over the years, chemists have developed a method for depicting the energy changes that occur during a reaction by using **reaction energy diagrams** of the sort shown in Figure 3.9. The vertical axis of the diagram represents the total energy of all reactants, and the horizontal axis represents the progress of the reaction from beginning (left) to end (right).

At the beginning of the reaction, ethylene and HCl have the total amount of energy indicated by the reactant level on the left side of the diagram. As the two molecules crowd together, their electron clouds repel each other, causing the energy level to rise. If the collision has occurred with sufficient force and proper orientation, the reactants continue to approach each other despite the repulsion until the new C–H bond starts to form and the H–Cl bond starts to break. At some point, a structure of maximum energy is reached, a point we call the **transition state.**

The transition state represents the highest-energy structure involved in this step of the reaction and can't be isolated. Nevertheless, we can imagine it to be a kind of activated complex of the two reactants in which the C=C π bond is partially broken and the new C–H bond is partially formed (Figure 3.10).

The energy difference between reactants and transition state, called the **activation energy, E_{act},** measures how rapidly the reaction occurs. A large activation energy results in a slow reaction because few of the reacting molecules collide with enough energy to reach the transition state. A small activation energy results in a rapid reaction because almost all reacting molecules are energetic enough to climb to the transition state. As

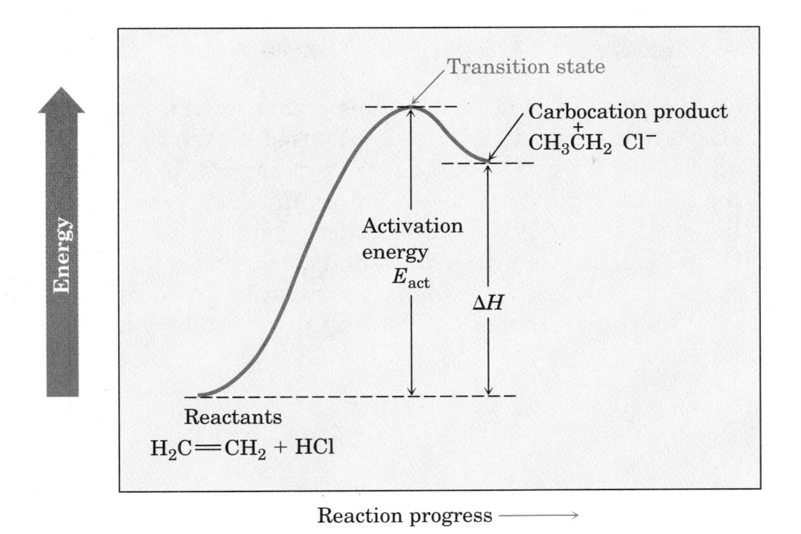

Figure 3.9 A reaction energy diagram for the first step in the reaction of ethylene with HCl. The energy difference between reactants and transition state, E_{act}, determines the reaction rate. The energy difference between reactants and carbocation product, ΔH, determines the position of the equilibrium.

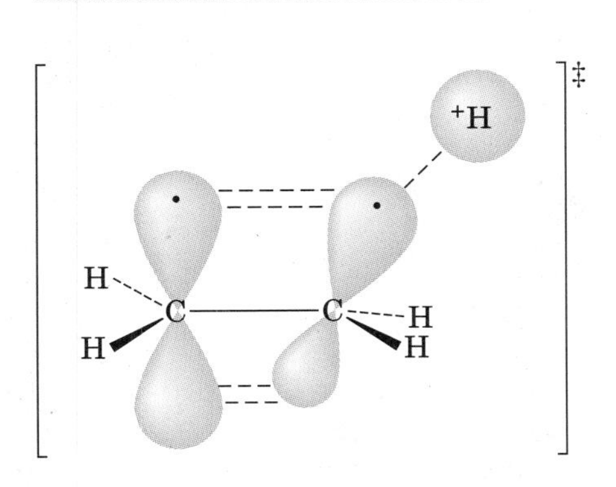

Figure 3.10 A hypothetical transition-state structure for the first step of the reaction of ethylene with HCl. The C=C π bond is just beginning to break, and the C–H bond is just beginning to form.

an analogy, think about hikers climbing over a mountain pass. If the pass is a high one, the hikers need a lot of energy and surmount the barrier slowly. If the pass is low, however, the hikers need less energy and reach the top quickly.

Most organic reactions have activation energies in the range 40–125 kJ/mol (10–30 kcal/mol). Reactions with activation energies less than 80 kJ/mol take place spontaneously at room temperature or below, whereas reactions with higher activation energies normally require heating. Heat gives the colliding molecules enough energy to climb the activation barrier.

Once the transition state has been reached, the reaction proceeds to yield carbocation product. Energy is released as the new C–H bond forms fully, and the curve in the reaction energy diagram in Figure 3.9 therefore turns downward until it reaches a minimum. This minimum point represents the energy level of the carbocation product of the first step. The energy change, ΔH, between reactants and carbocation product is simply the difference between the two levels in the diagram.[2] Since the carbocation is less stable than the alkene reactant, the first step is endothermic, and energy is absorbed.

PROBLEM

3.19 Which reaction is faster, one with $E_{act} = 60$ kJ/mol or one with $E_{act} = 80$ kJ/mol? Is it possible to predict which of the two has the larger K_{eq}?

3.11 Describing a Reaction: Intermediates

How can we describe the carbocation structure formed in the first step of the reaction of ethylene with HCl? The carbocation is clearly different from the reactants, yet it isn't a transition state and it isn't a final product.

$$H_2C{=}CH_2 + H^+ \longrightarrow \left[H{-}\overset{H}{\underset{H}{C}}{-}\overset{+}{\underset{H}{C}}{-}H \right] \xrightarrow{Cl^-} H{-}\overset{H}{\underset{H}{C}}{-}\overset{Cl}{\underset{H}{C}}{-}H$$

Ethylene — Reaction intermediate — **Chloroethane**

We call the carbocation, which is formed transiently during the course of the multistep reaction, a reaction **intermediate.** As soon as the intermediate is formed in the first step, it immediately reacts with Cl^- in a second step to give the final product, chloroethane. This second step has its own activation energy, E_{act}, its own transition state, and its own energy change, ΔH. We can view the second transition state as an activated complex between the electrophilic carbocation intermediate and nucleophilic Cl^- anion, a complex in which the new C–Cl bond is just starting to form.

A complete energy diagram for the overall reaction of ethylene with HCl is shown in Figure 3.11. In essence, diagrams for the two individual

[2] As noted previously, this argument assumes that $\Delta G \approx \Delta H$.

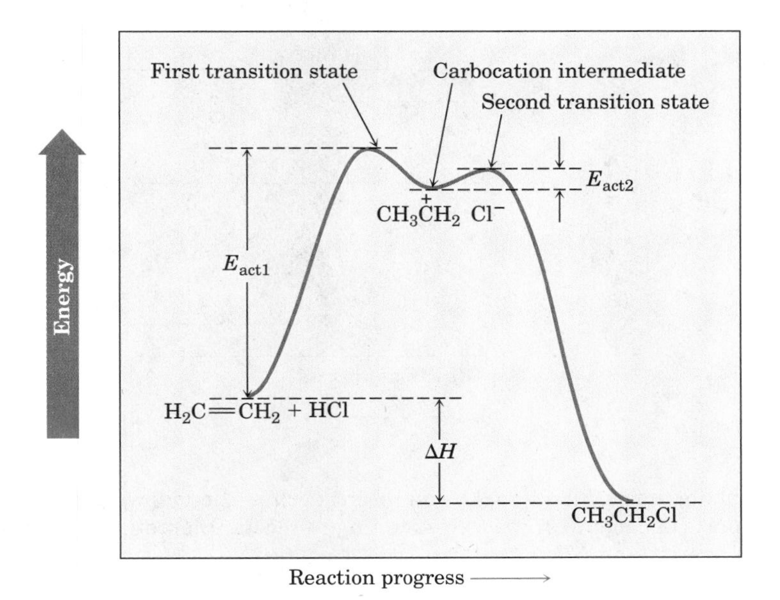

Figure 3.11 An overall reaction energy diagram for the reaction of ethylene with HCl. Two separate steps are involved, each with its own activation energy, transition state, and energy change. The energy minimum between the two steps represents the carbocation reaction intermediate.

steps are joined in the middle so that the *product* of step 1 (the carbocation) is the *reactant* for step 2. As indicated in Figure 3.11, the carbocation intermediate lies at an energy minimum between steps 1 and 2. Since the energy level of this intermediate is higher than the level of either the initial reactants (ethylene + HCl) or the final product (chloroethane), the intermediate is reactive and can't be isolated. It is, however, more stable than either of the two transition states.

Each step in a multistep process can be considered separately. Each step has its own E_{act} (rate) and its own ΔH (energy change). The overall ΔH of the reaction, however, is the energy difference between initial reactants (far left) and final products (far right). This is always true, regardless of the shape of the reaction energy curve. Note, for example, that the energy diagram for the reaction of HCl with ethylene in Figure 3.11 shows that the energy level of the final product is lower than the energy level of the reactants. Thus, the overall reaction is exothermic.

PRACTICE PROBLEM 3.7

Sketch a reaction energy diagram for a one-step reaction that is fast and highly exothermic.

Solution A fast reaction has a low E_{act}, and a highly exothermic reaction has a large negative ΔH. Thus, the diagram will look like this:

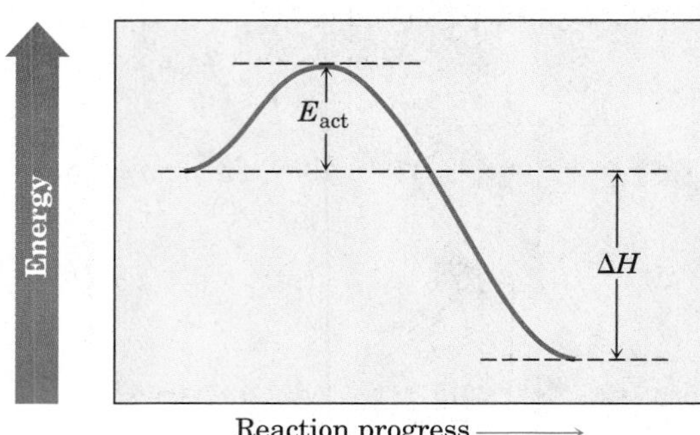

PROBLEM

3.20 Sketch reaction energy diagrams to represent each of the following situations, and label the parts of each diagram corresponding to reactant, product, transition state, activation energy, and ΔH.
(a) An exothermic reaction that takes place in one step
(b) An endothermic reaction that takes place in one step

PROBLEM

3.21 Draw a reaction energy diagram for a two-step reaction with an endothermic first step and an exothermic second step. Label the intermediate.

INTERLUDE

Carrots, Alkenes, and the Chemistry of Vision

The pigments responsible for the striking colors of these bird feathers are derived from β-carotene.

Your mother may have told you when you were growing up that eating carrots is good for your eyes. Although that's probably not true for healthy adults on a proper diet (your mother lied), the chemistry of carrots and the chemistry of vision are nevertheless related. Alkenes are important in both.

(continued)
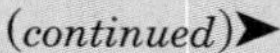

Carrots are rich in β-carotene, a purple-orange alkene that is an excellent dietary source of vitamin A. β-Carotene is converted to vitamin A by enzymes in the liver, oxidized to an aldehyde called all-*trans*-retinal, and then isomerized by a change in geometry of the C11–C12 double bond to produce 11-*cis*-retinal, the light-sensitive pigment on which the visual systems of all living things are based.

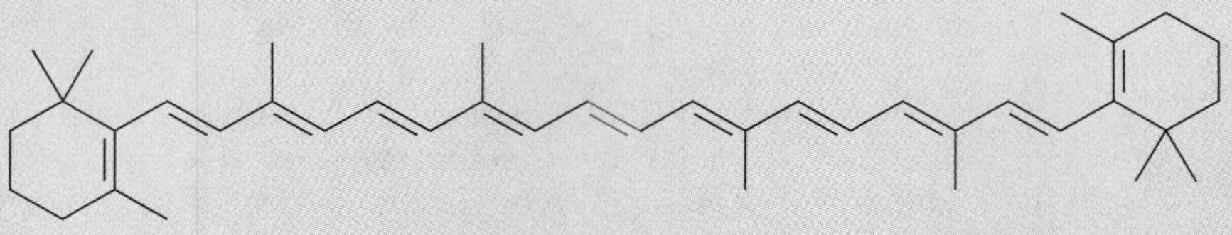

β-Carotene

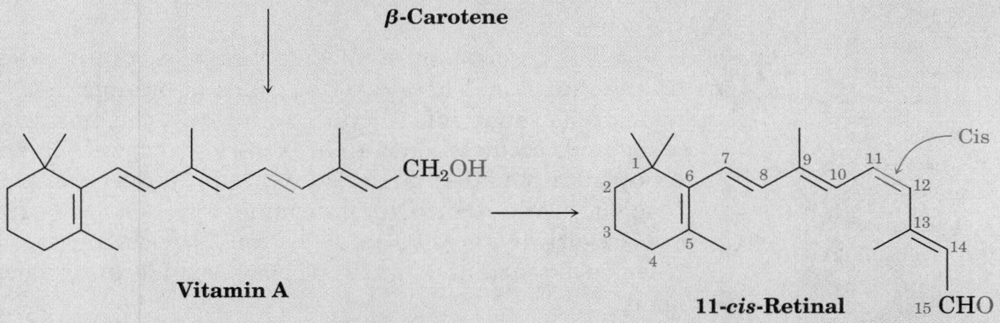

CH_2OH

Vitamin A

Cis

1 2 3 4 5 6 7 8 9 10 11 12 13 14

15 CHO

11-*cis*-Retinal

The retina of the eye contains two types of light-sensitive receptor cells, *rod cells* and *cone cells.* The three million or so rod cells are primarily responsible for seeing in dim light, whereas the hundred million cone cells are responsible for seeing in bright light and for the perception of colors. In the rod cells of the eye, 11-*cis*-retinal is converted into *rhodopsin,* a light-sensitive substance formed from the protein *opsin* and 11-*cis*-retinal. When light strikes the rod cell, isomerization of the C11–C12 double bond occurs, and 11-*trans*-rhodopsin, also called *metarhodopsin II,* is produced. This cis–trans isomerization is accompanied by a change in molecular geometry that causes a nerve impulse to be sent to the brain where it is perceived as vision.

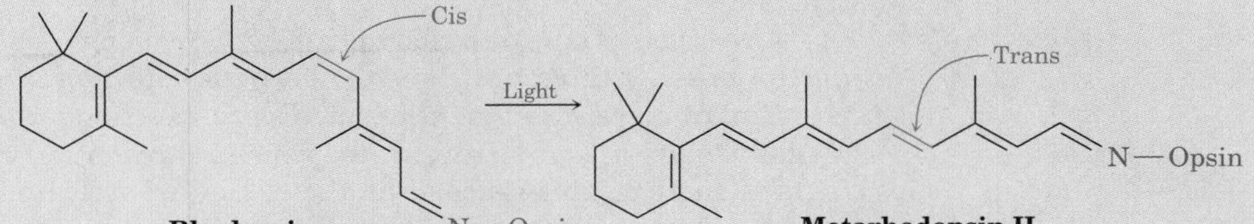

Cis

Light

Trans

N—Opsin

N—Opsin

Rhodopsin

Metarhodopsin II

Metarhodopsin II is then recycled into rhodopsin by a multistep sequence involving cleavage into all-*trans*-retinal and cis–trans isomerization back to 11-*cis*-retinal.

Summary and Key Words

activation energy (E_{act}), 96
addition reaction, 87
alkene, 76
carbocation, 92
electrophile, 90
elimination reaction, 87
endothermic, 94
equilibrium constant (K_{eq}), 93–94
exothermic, 94
heat of reaction (ΔH), 95
heterogenic, 89
heterolytic, 89
homogenic, 89
homolytic, 89
intermediate, 98
nucleophile, 90
polar reaction, 89
radical, 89
radical reaction, 89
reaction energy diagram, 96
reaction mechanism, 88
rearrangement reaction, 88
sequence rules, 83
substitution reaction, 88
transition state, 96
unsaturated, 77

Alkenes are hydrocarbons that contain one or more carbon–carbon double bonds. Because they contain fewer hydrogens than related alkanes, alkenes are often referred to as **unsaturated.**

A double bond consists of two parts: a σ bond formed by head-on overlap of two sp^2 orbitals and a π bond formed by sideways overlap of two *p* orbitals. Because rotation around the double bond is not possible, substituted alkenes can exist as **cis–trans stereoisomers.** The geometry of a double bond can be described as either ***Z*** (*zusammen*) or ***E*** (*entgegen*) by application of a series of sequence rules.

A full description of how a reaction occurs is called its **mechanism.** There are two kinds of organic mechanisms: polar and radical. **Polar reactions,** the most common kind, occur when an electron-rich reagent, or **nucleophile,** donates an electron pair to an electron-poor reagent, or **electrophile,** in forming a new bond. **Radical reactions** involve odd-electron species and occur when each reactant donates one electron in forming a new bond.

$$A^+ + :B^- \longrightarrow A:B \quad \textbf{Polar reaction}$$

$$A\cdot + \cdot B \longrightarrow A:B \quad \textbf{Radical reaction}$$

The energy change that takes place during a reaction can be described by considering both **rates** (how fast a reaction occurs) and **equilibria** (how much the reaction occurs). The equilibrium position of a reaction is determined by ΔH, the energy change that takes place during the reaction. If the reaction is **exothermic,** energy is given off, and the reaction has a favorable equilibrium constant. If the reaction is **endothermic,** energy is absorbed and the reaction has an unfavorable equilibrium constant.

A reaction can be described pictorially by using a **reaction energy diagram,** which follows the course of the reaction from reactant to product. Every reaction proceeds through a **transition state,** which represents the highest-energy point reached and is a kind of activated complex between reactants. The amount of energy needed by reactants to reach the transition state is the **activation energy, E_{act}.** The larger the activation energy, the slower the reaction.

Many reactions take place in more than one step and involve the formation of an **intermediate.** An intermediate is a species that is formed during the course of a multistep reaction and that lies in an energy minimum between two transition states. Intermediates are more stable than transition states but are often too reactive to be isolated.

ADDITIONAL PROBLEMS

3.22 Identify the functional groups in each of the following molecules:

(a) $CH_3CH_2C{\equiv}N$ (b) 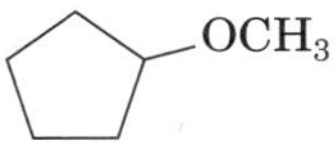 OCH_3

(c) 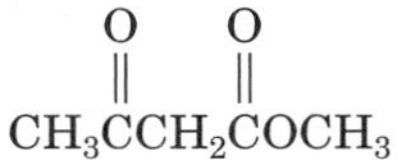 $CH_3\overset{O}{\overset{\|}{C}}CH_2\overset{O}{\overset{\|}{C}}OCH_3$ (d) 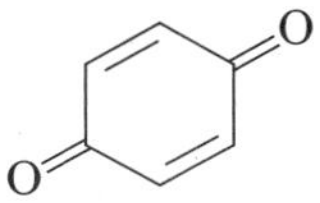 O O

(e) 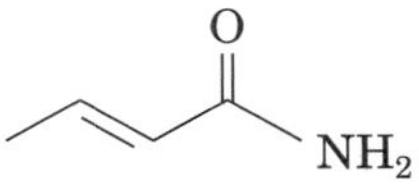 O NH_2 (f) 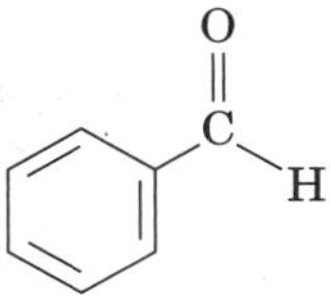 O C H

3.23 Predict the direction of polarization of the functional groups you identified in Problem 3.22.

3.24 Identify the functional groups in each of the following molecules:

(a) 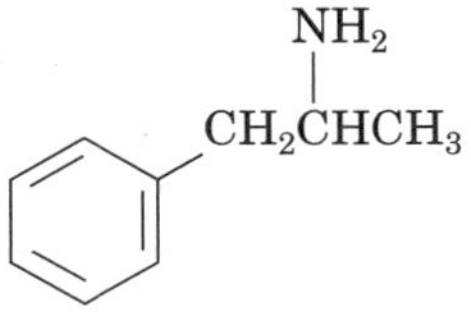 NH_2 CH_2CHCH_3

Amphetamine

(b) 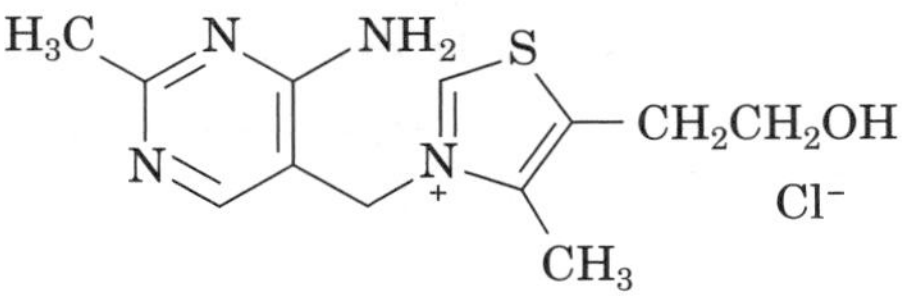 H_3C N NH_2 S N N + CH_2CH_2OH Cl^- CH_3

Thiamine

3.25 Explain the differences between addition, elimination, substitution, and rearrangement reactions.

3.26 Which of the following are most likely to behave as electrophiles, and which as nucleophiles?
(a) Cl^- (b) NH_3 (c) Mg^{2+} (d) CN^- (e) $CH_3{}^+$

3.27 Give IUPAC names for the following alkenes:

(a) $CH_3CH{=}CHCH(CH_3)CH_2CH_3$ (b) $CH_3CH{=}CHCH(CH_2CH_2CH_3)CH_2CH_2CH_3$

(c) $H_2C{=}C(CH_2CH_3)CH_2CH_3$ (d) $H_2C{=}C{=}CHCH_3$

3.28 Name the following cycloalkenes by IUPAC rules:

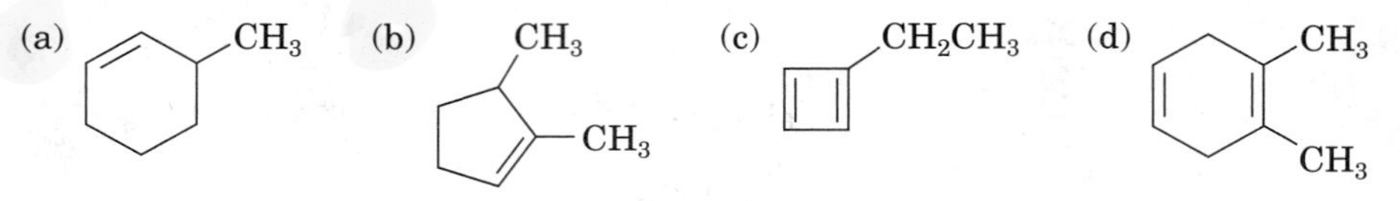

3.29 Draw structures corresponding to the following IUPAC names:
(a) 3-Propyl-2-heptene (b) 2,4-Dimethyl-2-hexene
(c) 1,5-Octadiene (d) 4-Methyl-1,3-pentadiene
(e) *cis*-4,4-Dimethyl-2-hexene (f) (*E*)-3-Methyl-3-heptene

3.30 Draw the structures of the following cycloalkenes:
(a) *cis*-4,5-Dimethylcyclohexene (b) 3,3,4,4-Tetramethylcyclobutene

3.31 The following names are incorrect. Draw each molecule, and give its correct name.
(a) 1-Methyl-2-cyclopentene (b) 1-Methyl-1-pentene
(c) 6-Ethylcycloheptene (d) 3-Methyl-2-ethylcyclohexene

3.32 Which of the following molecules show cis–trans isomerism?

(a) $CH_3C(CH_3){=}CHCH_2CH_3$ (b) $ClCH_2CH_2C(CH_3){=}C(CH_3)CH_2CH_2Cl$

(c) HO

3.33 Draw and name molecules that meet the following descriptions:
(a) An alkene, C_6H_{12}, that does not show cis–trans isomerism
(b) The *E* isomer of a trisubstituted alkene, C_6H_{12}
(c) A cycloalkene, C_7H_{12}, with a tetrasubstituted double bond

3.34 Neglecting cis–trans isomers, there are five substances with the formula C_4H_8. Draw and name them.

3.35 Which of the molecules you drew in Problem 3.34 show cis–trans isomerism? Draw and name their cis–trans isomers.

3.36 Draw four possible structures for each of the following formulas:
(a) C_6H_{10} (b) C_8H_8O (c) $C_7H_{10}Cl_2$

3.37 How can you explain the fact that cyclohexene does not show cis–trans isomerism but cyclodecene does?

3.38 Rank the following sets of substituents in order of priority according to the sequence rules:
(a) $-CH_3$, $-Br$, $-H$, $-I$
(b) $-OH$, $-OCH_3$, $-H$, $-COOH$
(c) $-CH_3$, $-COOH$, $-CH_2OH$, $-CHO$
(d) $-CH_3$, $-CH{=}CH_2$, $-CH_2CH_3$, $-CH(CH_3)_2$

3.39 Assign *E* or *Z* configuration to the following alkenes:

(a) 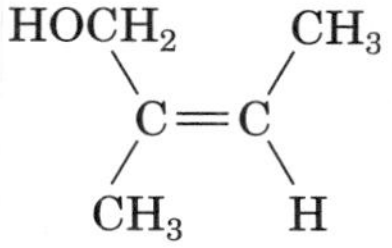 $HOCH_2$ / CH_3 — $C{=}C$ — CH_3 / H (b) 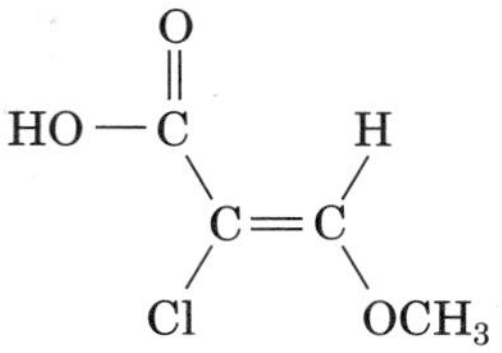 HO—C(=O) / H — $C{=}C$ — Cl / OCH_3

3.40 Draw and name the five C_5H_{10} alkene isomers. Ignore cis–trans isomers.

3.41 Menthene, a hydrocarbon found in mint plants, has the IUPAC name 1-isopropyl-4-methylcyclohexene. What is the structure of menthene?

3.42 Name the following cycloalkenes:

(a) 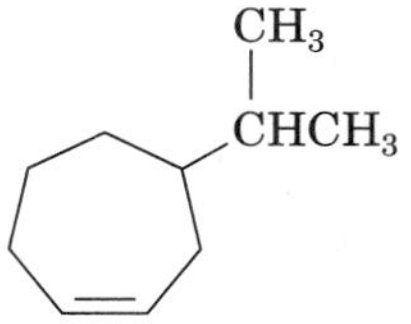 CH_3, $CHCH_3$ (b) 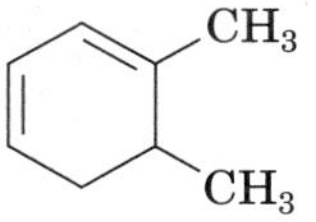 CH_3, CH_3 (c) 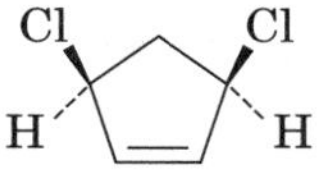 Cl, Cl, H, H

3.43 Which of the following are most likely to behave as electrophiles, and which as nucleophiles?

(a) Zn^{2+} (b) $CH_3\ddot{N}H_2$ (c) 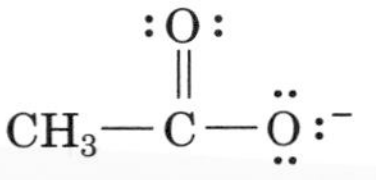 $CH_3—C(=\ddot{O}:)—\ddot{O}:^-$ (d) $H\ddot{S}:^-$

3.44 α-Farnesene is a constituent of the natural waxy coating found on apples. What is its IUPAC name?

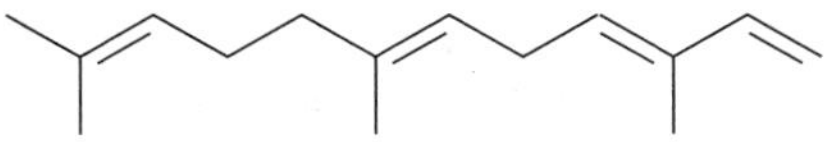

α-Farnesene

3.45 Indicate *E* or *Z* configuration for each of the double bonds in α-farnesene (see Problem 3.44).

3.46 Reaction of 2-methylpropene with HCl might, in principle, lead to a mixture of two products. Draw them.

3.47 Give an example of each of the following:
(a) An electrophile (b) A nucleophile
(c) An oxygen-containing functional group

3.48 If a reaction has $K_{eq} = 0.001$, is it likely to be exothermic or endothermic? Explain.

3.49 If a reaction has $E_{act} = 15$ kJ/mol, is it likely to be fast or slow at room temperature? Explain.

3.50 If a reaction has $\Delta H = 25$ kJ/mol, is it exothermic or endothermic? Is it likely to be fast or slow at room temperature? Explain.

3.51 Draw a reaction energy diagram for a two-step exothermic reaction whose first step is faster than its second step. Label the parts of the diagram corresponding to reactants, products, transition states, intermediate, activation energies, and overall ΔH.

3.52 Draw a reaction energy diagram for a two-step reaction whose second step is faster than its first step.

3.53 Draw a reaction energy diagram for a reaction with $K_{eq} = 1$.

3.54 Describe the difference between a transition state and a reaction intermediate.

3.55 Consider the reaction energy diagram shown, and answer the following questions:

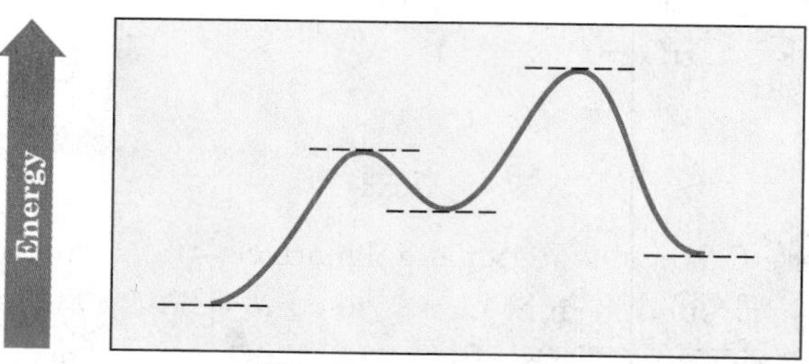

(a) Indicate the overall ΔH for the reaction. Is it positive or negative?
(b) How many steps are involved in the reaction?
(c) Which step is faster?
(d) How many transition states are there? Label them.

3.56 When isopropylidenecyclohexane is treated with strong acid at room temperature, isomerization occurs by the mechanism shown below to yield 1-isopropylcyclohexene:

Acid catalyst

Isopropylidenecyclohexane **1-Isopropylcyclohexene**

At equilibrium, the product mixture contains about 30% isopropylidenecyclohexane and about 70% 1-isopropylcyclohexene.
(a) What kind of reaction is occurring? Is the mechanism polar or radical?
(b) Draw curved arrows to indicate electron flow in each step.
(c) Calculate K_{eq} for the reaction.

3.57 We'll see in the next chapter that the stability of carbocations depends on the number of alkyl groups attached to the positively charged carbon—the more alkyl groups, the more stable the cation. Draw the two possible carbocation intermediates that might be formed in the reaction of HCl with 2-methylpropene (Problem 3.46), tell which is more stable, and predict which product will form.

Visualizing Chemistry

3.58 Give IUPAC names for the following alkenes, and convert each drawing into a skeletal structure.

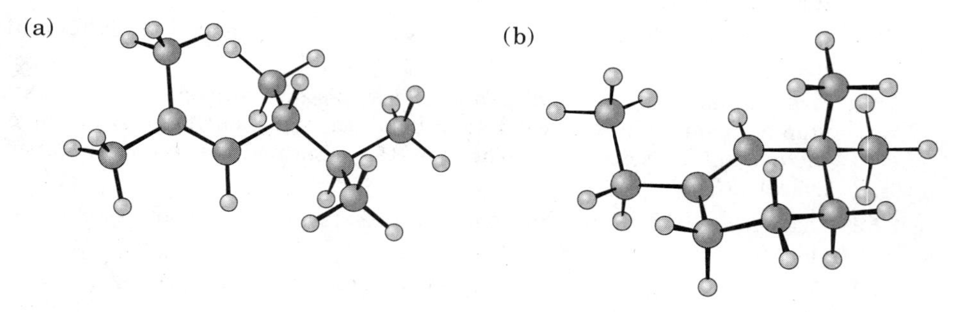

3.59 Assign stereochemistry (E or Z) to each of the following alkenes, and convert each drawing into a skeletal structure (gray = C, red = O, yellow-green = Cl, light green = H).

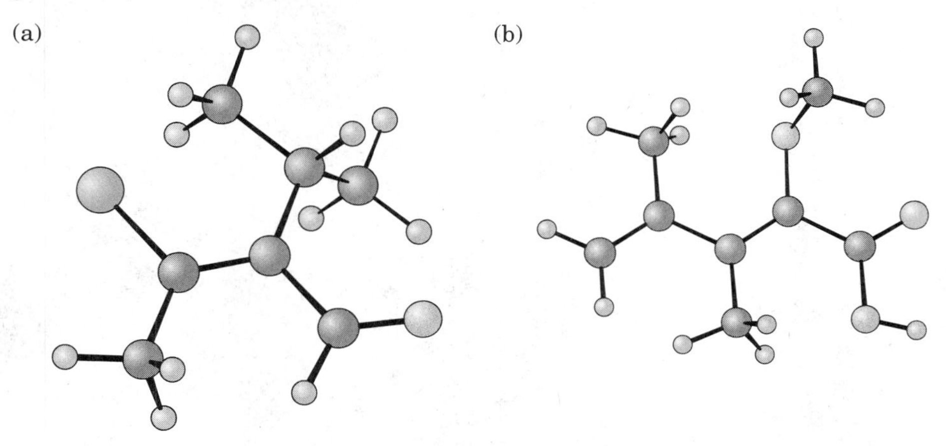

3.60 The following drawing does *not* represent a stable molecule. Why not?

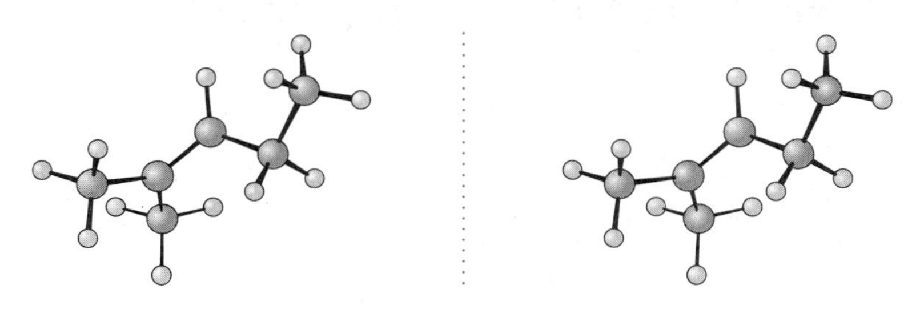

The isopropyl cation has a planar, trivalent carbon atom with a vacant *p* orbital.

4 Alkenes and Alkynes

We saw in the previous chapter how organic reactions can be classified, and we developed some general ideas about how reactions can be described. In this chapter, we'll apply those ideas to a systematic study of the alkene and alkyne families of compounds. In particular, we'll see that the most important reaction of these two functional groups is the addition to the C=C and C≡C multiple bonds of various reagents X–Y to yield saturated products:

$$\gt C{=}C\lt \; + \; X{-}Y \longrightarrow -\overset{X}{\underset{|}{\overset{|}{C}}}-\overset{Y}{\underset{|}{\overset{|}{C}}}-$$

An alkene — An addition product

4.1 Addition of HX to Alkenes: Hydrohalogenation

We saw in Section 3.8 that alkenes react with HCl to yield alkyl chloride addition products. For example, ethylene reacts with HCl to give chloroethane. The reaction takes place in two steps and involves a carbocation intermediate.

$$H_2C{=}CH_2 + H^+ \longrightarrow \left[H{-}\overset{H}{\underset{H}{C}}{-}\overset{+}{\underset{H}{C}}{-}H \right] \xrightarrow{Cl^-} H{-}\overset{H}{\underset{H}{C}}{-}\overset{Cl}{\underset{H}{C}}{-}H$$

Ethylene — Carbocation intermediate — **Chloroethane**

The addition of halogen acids, HX, to alkenes is a general reaction that allows chemists to prepare a variety of halo-substituted alkane products. Thus, HCl, HBr, and HI all add to alkenes:[1]

$$(CH_3)_2C{=}CH_2 + HCl \xrightarrow{Ether} CH_3{-}\overset{Cl}{\underset{CH_3}{C}}{-}CH_3$$

2-Methylpropene — **2-Chloro-2-methylpropane**

[1]Organic reaction equations can be written in different ways to emphasize different points. For example, the reaction of ethylene with HCl might be written in the format A + B ⟶ C to emphasize that both reactants are equally important for the purposes of the discussion. The solvent and notes about other reaction conditions such as temperature are usually written either above or below the reaction arrow:

$$H_2C{=}CH_2 + HBr \xrightarrow[25°C]{Ether} CH_3CH_2Br$$

(Ether = Solvent)

Alternatively, we might choose to write the same reaction in the format

$$A \xrightarrow{B} C$$

to emphasize that reactant A is the organic starting material whose chemistry is of greater interest. Reactant B is then placed above the reaction arrow, together with notes about solvent and reaction conditions. For example:

$$H_2C{=}CH_2 \xrightarrow[Ether,\ 25°C]{HBr} CH_3CH_2Br$$

(HBr = Reagent; Ether = Solvent)

Both reaction formats are frequently used in chemistry, and you sometimes have to look carefully at the overall transformation to identify the different roles of the substances shown next to the reaction arrows.

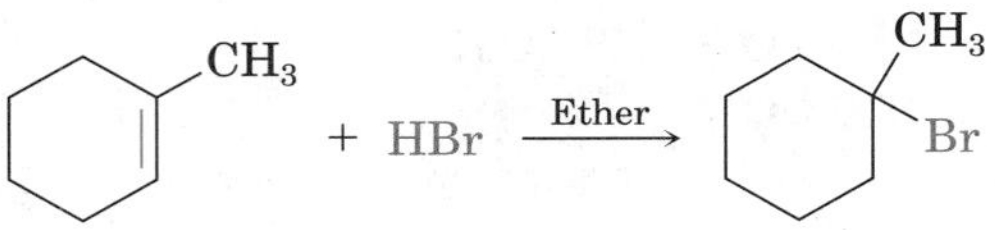

$$\text{1-Methylcyclohexene} + HBr \xrightarrow{\text{Ether}} \text{1-Bromo-1-methylcyclohexane}$$

1-Methylcyclohexene 1-Bromo-1-methylcyclohexane

$$CH_3CH_2CH_2CH{=}CH_2 + HI \xrightarrow{\text{Ether}} CH_3CH_2CH_2CH(I)CH_3$$

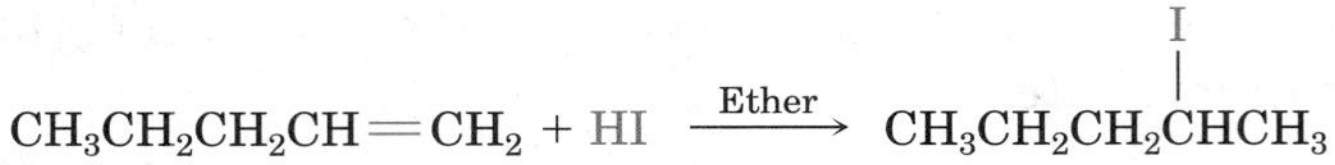

1-Pentene 2-Iodopentane

4.2 Orientation of Alkene Addition Reactions: Markovnikov's Rule

Look carefully at the reactions shown in the previous section. In each case, an unsymmetrically substituted alkene has given a single addition product rather than the mixture that might have been expected. For example, 2-methylpropene *might* have reacted with HCl to give 1-chloro-2-methylpropane in addition to 2-chloro-2-methylpropane, but it didn't. We say that reactions are **regiospecific** (**ree**-jee-oh-specific) when only one of the two possible directions of addition occurs.[2]

A regiospecific reaction:

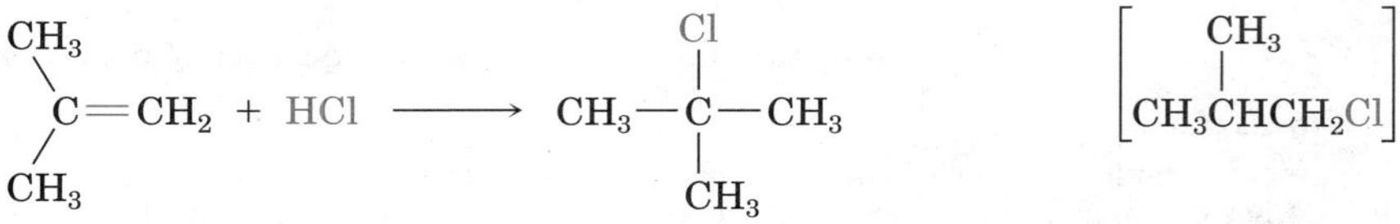

$$(CH_3)_2C{=}CH_2 + HCl \longrightarrow CH_3{-}C(Cl)(CH_3){-}CH_3 \qquad \left[CH_3CH(CH_3)CH_2Cl\right]$$

2-Methylpropene 2-Chloro-2-methylpropane (*sole product*) 1-Chloro-2-methylpropane (*NOT* formed)

After looking at many such reactions, the Russian chemist Vladimir Markovnikov proposed in 1869 what has become known as **Markovnikov's rule:**

> ***Markovnikov's rule*** In the addition of HX to an alkene, the H attaches to the carbon with fewer alkyl substituents, and the X attaches to the carbon with more alkyl substituents.

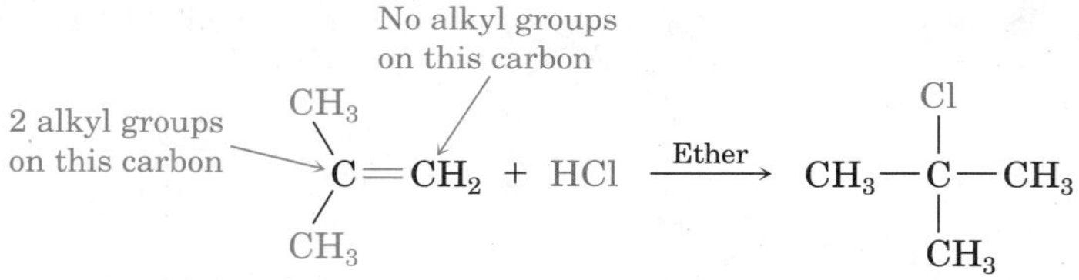

$$(CH_3)_2C{=}CH_2 + HCl \xrightarrow{\text{Ether}} CH_3{-}C(Cl)(CH_3){-}CH_3$$

2 alkyl groups on this carbon; No alkyl groups on this carbon

2-Methylpropene 2-Chloro-2-methylpropane

---

[2]The term *regioselective* rather than regiospecific is used when one product predominates but is not formed exclusively.

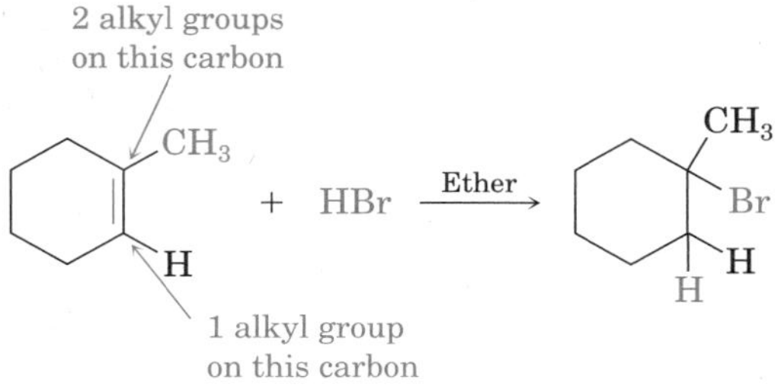

2 alkyl groups on this carbon

1 alkyl group on this carbon

1-Methylcyclohexene + HBr $\xrightarrow{\text{Ether}}$ 1-Bromo-1-methylcyclohexane

When both double-bond carbon atoms have the same degree of substitution, a mixture of addition products results:

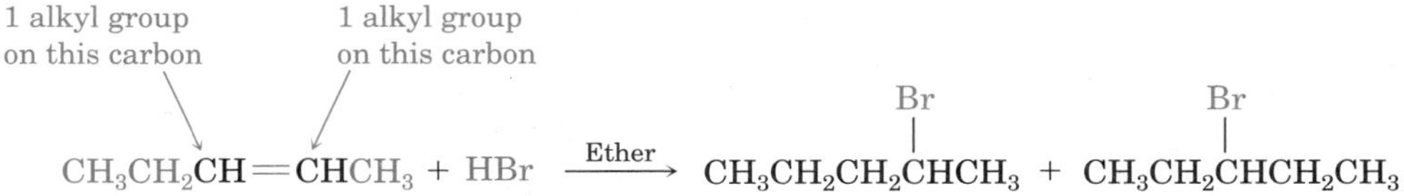

1 alkyl group on this carbon — 1 alkyl group on this carbon

$$CH_3CH_2CH{=}CHCH_3 + HBr \xrightarrow{\text{Ether}} CH_3CH_2CH_2CHBrCH_3 + CH_3CH_2CHBrCH_2CH_3$$

2-Pentene → **2-Bromopentane** + **3-Bromopentane**

Since carbocations are involved as intermediates in these reactions (Section 3.11), another way to express Markovnikov's rule is to say that, in the addition of HX to alkenes, the more highly substituted carbocation intermediate is formed rather than the less highly substituted one. For example, addition of H^+ to 2-methylpropene yields the intermediate *tertiary* carbocation rather than the alternative primary carbocation. Why should this be?

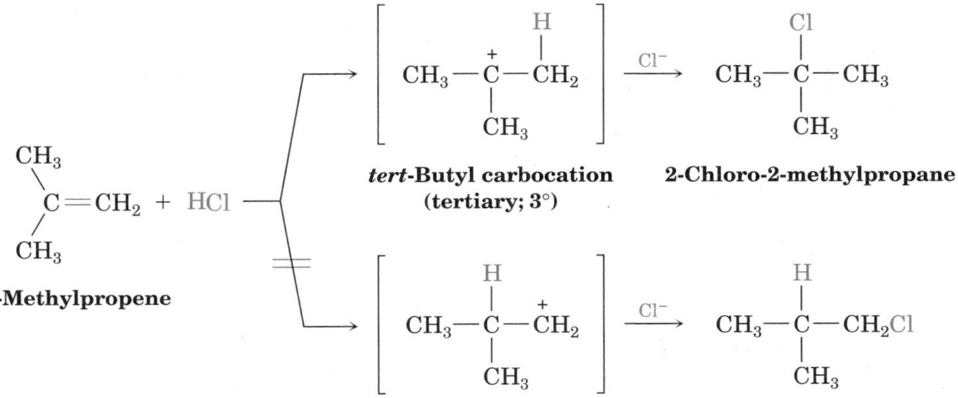

$$(CH_3)_2C{=}CH_2 + HCl \rightarrow \left[CH_3-\overset{+}{C}(CH_3)-CH_2H\right] \xrightarrow{Cl^-} CH_3-CCl(CH_3)-CH_3$$

2-Methylpropene → ***tert*-Butyl carbocation (tertiary; 3°)** → **2-Chloro-2-methylpropane**

$$(CH_3)_2C{=}CH_2 + HCl \nrightarrow \left[CH_3-CH(CH_3)-\overset{+}{C}H_2\right] \xrightarrow{Cl^-} CH_3-CH(CH_3)-CH_2Cl$$

Isobutyl carbocation (primary; 1°) → **1-Chloro-2-methylpropane** (*NOT formed*)

PRACTICE PROBLEM 4.1

What product would you expect from the reaction of HCl with 1-ethylcyclopentene?

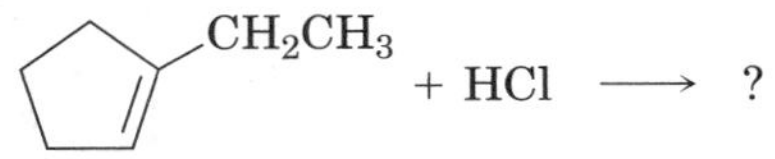

$$\text{1-ethylcyclopentene } (CH_2CH_3) + HCl \longrightarrow ?$$

Solution Markovnikov's rule predicts that H^+ will add to the double-bond carbon that has one alkyl group (C2 on the ring), and the Cl will add to the double-bond carbon that has two alkyl groups (C1 on the ring). The expected product is 1-chloro-1-ethylcyclopentane.

2 alkyl groups on this carbon

CH_2CH_3 + HCl ⟶ CH_2CH_3, Cl

1 alkyl group on this carbon

1-Chloro-1-ethylcyclopentane

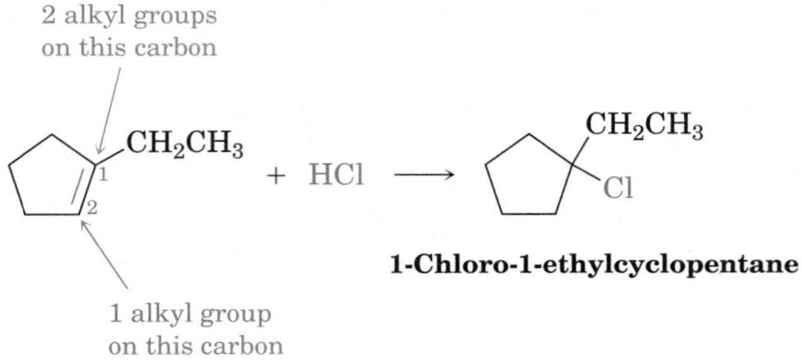

PROBLEM

4.1 Predict the products of the following reactions:

(a) $CH_3CH_2CH{=}CH_2 + HCl \longrightarrow ?$

(b) $CH_3\overset{CH_3}{\overset{|}{C}}{=}CHCH_2CH_3 + HI \longrightarrow ?$

(c) 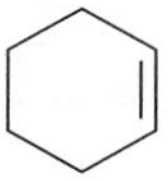 Cyclohexene + HCl ⟶ ?

PROBLEM

4.2 What alkenes would you start with to prepare the following alkyl halides?

(a) Bromocyclopentane

(b) $CH_3CH_2\overset{Br}{\overset{|}{C}}HCH_2CH_2CH_3$

(c) 1-Iodo-1-isopropylcyclohexane

(d) 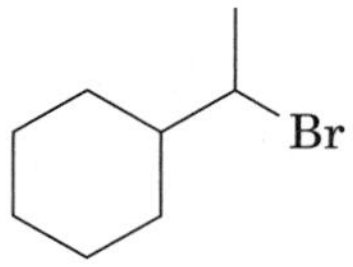 Br

4.3 Carbocation Structure and Stability

To understand why Markovnikov's rule works, we need to learn more about the structure and stability of substituted carbocations. Regarding structure, evidence shows that carbocations are *planar*. The positively charged carbon atom is sp^2-hybridized, and the three substituents bonded to it are oriented to the corners of an equilateral triangle (Figure 4.1). Because there are only six electrons in the carbon valence shell, and because all six are used in the three σ bonds, the p orbital extending above and below the plane is vacant.

Regarding stability, measurements show that carbocation stability increases with increasing alkyl substitution: More highly substituted carbo-

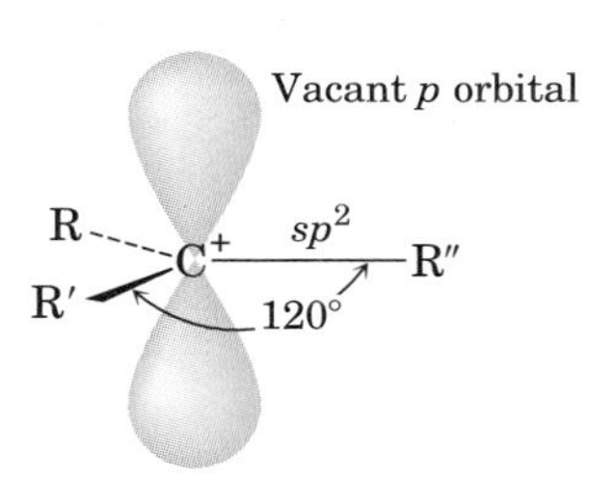

Figure 4.1 The electronic structure of a carbocation. The carbon is sp^2-hybridized and has a vacant *p* orbital.

cations are more stable than less highly substituted ones because alkyl groups tend to donate electrons to the positively charged carbon atom. The more alkyl groups there are, the more electron donation there is, and the more stable the carbocation.

R_3C^+ $\quad$ R_2CH^+ $\quad$ RCH_2^+ $\quad$ CH_3^+

Tertiary (3°) > Secondary (2°) > Primary (1°) > Methyl

More stable ← Less stable

With the above information, we can now explain Markovnikov's rule. In the reaction of 1-methylcyclohexene with HBr, for example, the intermediate carbocation might have either *three* alkyl substituents (a tertiary cation, 3°) or *two* alkyl substituents (a secondary cation, 2°). Because the tertiary cation is more stable than the secondary one, it's the tertiary cation that forms as the reaction intermediate, thus leading to the observed tertiary alkyl bromide product.

1-Methylcyclohexene + HBr → **(A tertiary carbocation)** —Br^-→ **1-Bromo-1-methylcyclohexane**

1-Methylcyclohexene + HBr ⇸ **(A secondary carbocation)** —Br^-→ **1-Bromo-2-methylcyclohexane** (*NOT formed*)

PROBLEM

4.3 Show the structures of the carbocation intermediates you would expect in the following reactions:

(a) 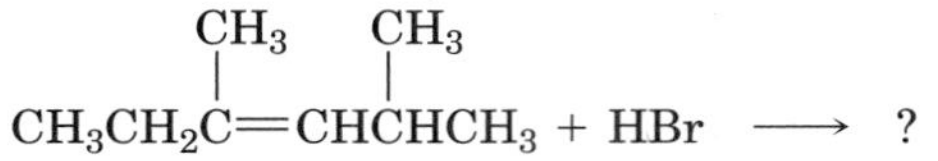 $CH_3CH_2C(CH_3){=}CHCH(CH_3)CH_3 + HBr \longrightarrow ?$

(b) 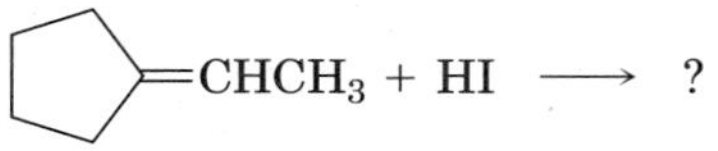 (cyclopentylidene)$=CHCH_3 + HI \longrightarrow ?$

4.4 Addition of H_2O to Alkenes: Hydration

Water adds to simple alkenes to yield alcohols, ROH, a process called **hydration.** Industrially, more than 300,000 tons of ethanol are produced each year in the United States by this method:

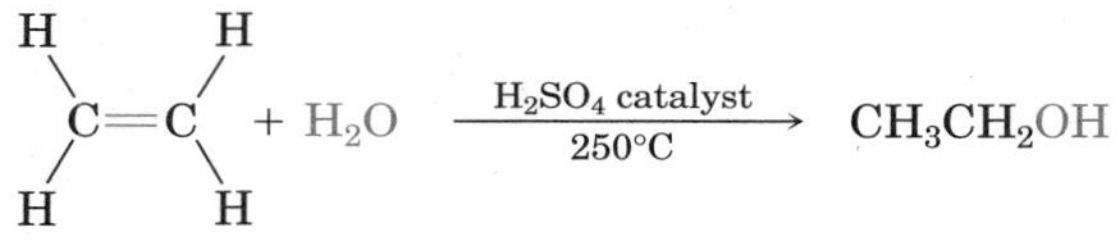

$$H_2C{=}CH_2 + H_2O \xrightarrow[250°C]{H_2SO_4\ \text{catalyst}} CH_3CH_2OH$$

Ethylene **Ethanol**

Hydration takes place on reaction of an alkene with aqueous acid by a mechanism similar to that of HX addition. Reaction of the alkene double bond with H^+ yields a carbocation intermediate, which then reacts with water as nucleophile to yield a protonated alcohol (ROH_2^+) product. Loss of H^+ from the protonated alcohol gives the neutral alcohol and regenerates the acid catalyst (Figure 4.2). The addition of water to an unsymmetrical alkene follows Markovnikov's rule just as addition of HX does, giving the more highly substituted alcohol as product.

Unfortunately, the reaction conditions required for hydration are so severe that molecules are sometimes destroyed by the high temperatures and strongly acidic conditions. For example, the hydration of ethylene to produce ethanol requires a sulfuric acid catalyst and reaction temperatures of up to 250°C.

PRACTICE PROBLEM 4.2

What product would you expect from addition of water to methylenecyclopentane?

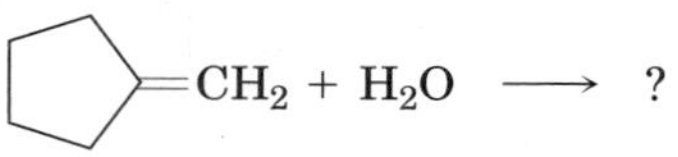

$$\text{(cyclopentylidene)}{=}CH_2 + H_2O \longrightarrow ?$$

Methylenecyclopentane

The alkene double bond reacts with H^+ to yield a carbocation intermediate.

Water acts as a nucleophile to donate a pair of electrons to form a carbon–oxygen bond and produce a protonated alcohol intermediate.

Loss of H^+ from the protonated alcohol intermediate then gives the neutral alcohol product and regenerates the acid catalyst.

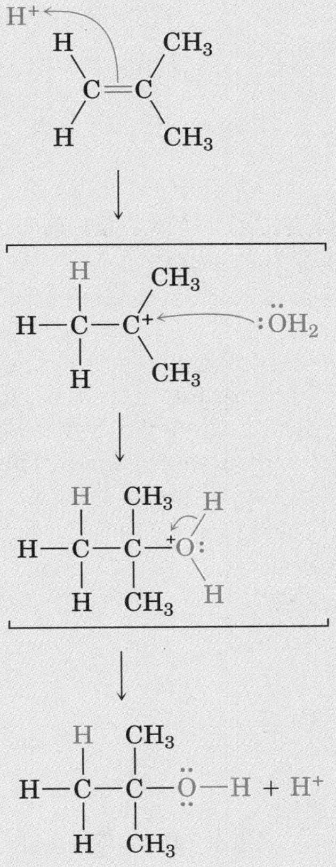

Figure 4.2 Mechanism of the acid-catalyzed hydration of an alkene. Protonation of the alkene gives a carbocation intermediate, which reacts with water.

Solution According to Markovnikov's rule, H^+ adds to the carbon that already has more hydrogens (the CH_2 carbon), and OH adds to the carbon that has fewer hydrogens (the ring carbon). Thus, the product will be a tertiary alcohol.

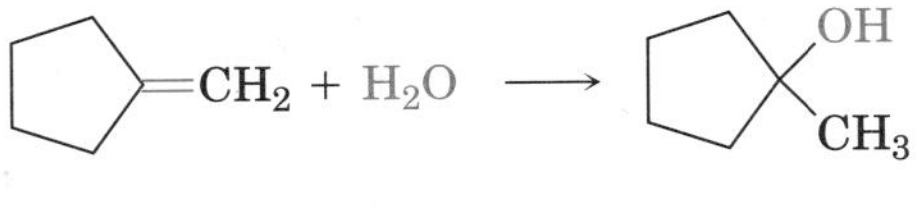

PROBLEM

4.4 What product would you expect to obtain from addition of water to the following alkenes?

(a) $CH_3CH_2C(CH_3){=}CHCH_2CH_3$ (b) 1-Methylcyclopentene
(c) 2,5-Dimethyl-2-heptene

PROBLEM

4.5 What alkenes might the following alcohols be made from?

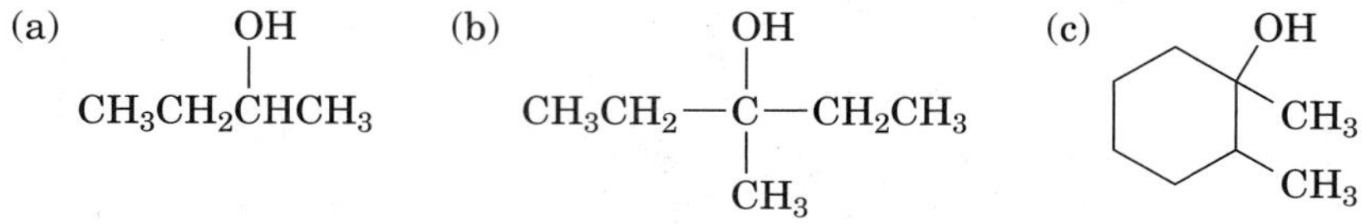

(a) $CH_3CH_2CH(OH)CH_3$ (b) $CH_3CH_2C(OH)(CH_3)CH_2CH_3$ (c) 1-methyl-2-methylcyclohexanol (OH and CH_3 on one ring carbon, CH_3 on the adjacent carbon)

4.5 Addition of X_2 to Alkenes: Halogenation

Many other reagents besides HX and H_2O add to alkenes. Bromine and chlorine, for instance, add readily to yield 1,2-dihaloalkanes. More than 7 million tons of 1,2-dichloroethane (also called ethylene dichloride) are synthesized each year in the United States by addition of Cl_2 to ethylene. The product is used both as a solvent and as a starting material for the synthesis of poly(vinyl chloride), PVC.

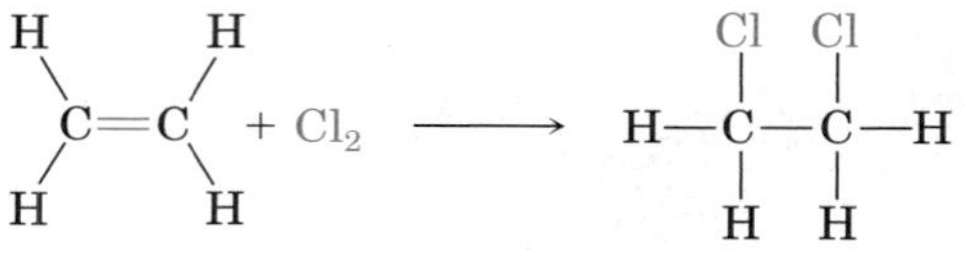

$$H_2C{=}CH_2 + Cl_2 \longrightarrow ClH_2C{-}CH_2Cl$$

Ethylene → **1,2-Dichloroethane (Ethylene dichloride)**

Addition of Br_2 also serves as a simple and rapid laboratory test for unsaturation. A sample of unknown structure is dissolved in dichloromethane, CH_2Cl_2, and several drops of Br_2 are added. Immediate disappearance of the reddish Br_2 color signals a positive test and indicates that the sample is an alkene.

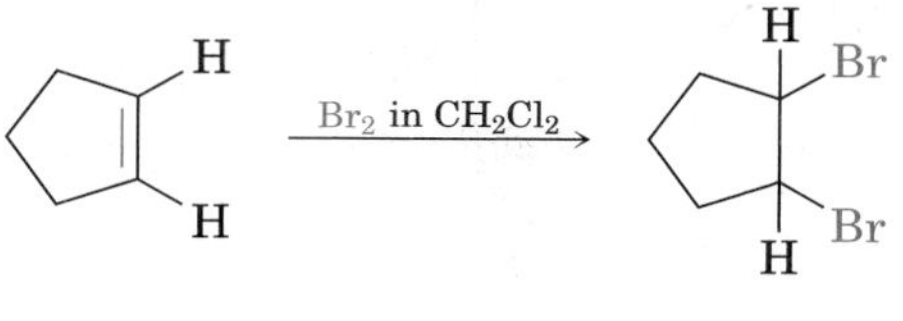

$$\text{Cyclopentene} \xrightarrow{Br_2 \text{ in } CH_2Cl_2} \text{1,2-Dibromocyclopentane (95\%)}$$

Cyclopentene **1,2-Dibromocyclopentane (95%)**

Based on what we've seen thus far, a possible mechanism for the reaction of Br_2 or Cl_2 with an alkene is shown in Figure 4.3. As a Br_2 molecule approaches an alkene, the π electrons of the alkene form a bond to one Br atom, displacing the other atom as Br^- anion. The net result is that electrophilic Br^+ adds to the alkene in much the same way that H^+ does, giving a carbocation intermediate that can react further with Br^- to yield the dibromo addition product.

Although the mechanism shown in Figure 4.3 looks reasonable, it's not completely consistent with known facts. In particular, the mechanism

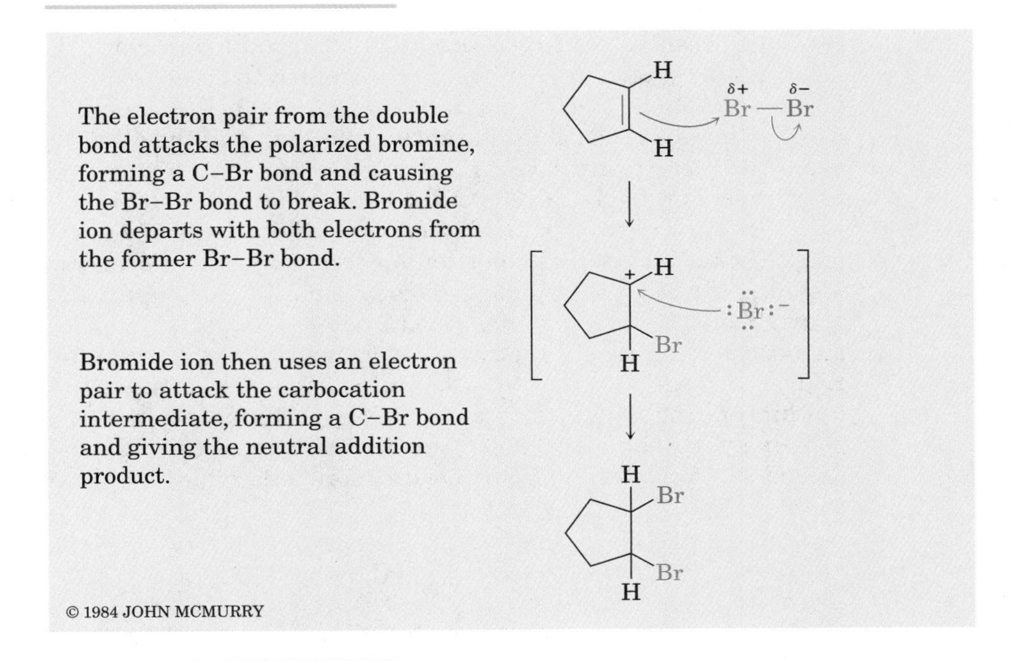

The electron pair from the double bond attacks the polarized bromine, forming a C–Br bond and causing the Br–Br bond to break. Bromide ion departs with both electrons from the former Br–Br bond.
Bromide ion then uses an electron pair to attack the carbocation intermediate, forming a C–Br bond and giving the neutral addition product.
H
δ+
δ−
Br
Br
H
H
:Br:⁻
Br
H
H
Br
Br
H

Figure 4.3 A possible mechanism for the addition of Br_2 to cyclopentene.

doesn't explain the *stereochemistry* of halogen addition. That is, the mechanism doesn't explain what product stereoisomers (Section 2.8) are formed in the reaction.

Let's look again at the reaction of Br_2 with cyclopentene and assume that Br^+ adds from the bottom side of the molecule to form the carbocation intermediate shown in Figure 4.4. (The addition could just as well occur

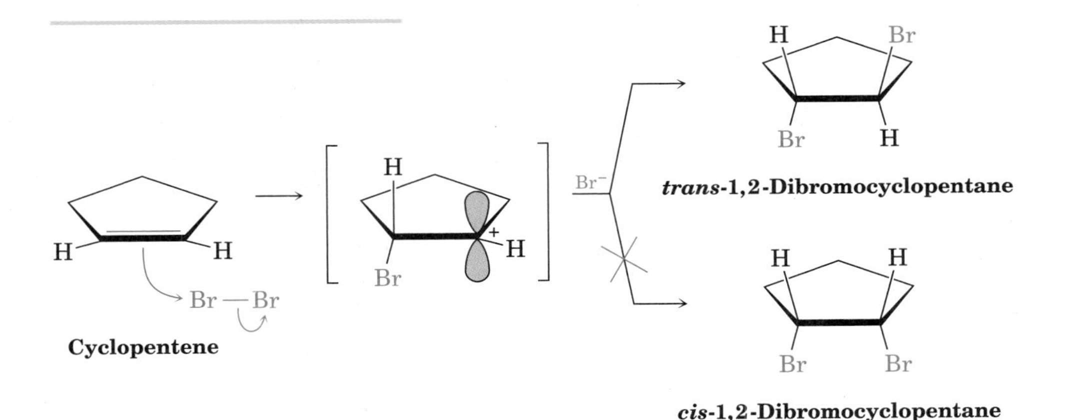

Figure 4.4 Stereochemistry of the addition of bromine to cyclopentene. Only the trans product is formed.

from the top side, but we'll consider only one possibility for simplicity.) Because the positively charged carbon is planar and sp^2-hybridized, it could be attacked by Br^- anion in the second step of the reaction from either the top or the bottom side to give a *mixture* of products. One product has the two Br atoms on the same side of the ring (cis), and the other has them on opposite sides (trans). We find, however, that only *trans*-1,2-dibromocyclopentane is produced. None of the cis product is formed.

Because the two Br atoms add to opposite faces of the cyclopentene double bond, we say that the reaction occurs with **anti stereochemistry,** meaning that the two bromines come from directions approximately 180° apart. This result is best explained by imagining that the reaction intermediate is not a true carbocation but is instead a cyclic **bromonium ion,** R_2Br^+, formed by overlap of bromine lone-pair electrons with the vacant *p* orbital of the neighboring carbon (Figure 4.5). Since the bromine atom effectively "shields" one side of the molecule, reaction with Br^- ion in the second step occurs from the opposite, more accessible side to give the anti product.

PROBLEM

4.6 What product would you expect to obtain from addition of Br_2 to 1,2-dimethylcyclohexene? Show the stereochemistry of the product.

PROBLEM

4.7 Show the structure of the intermediate bromonium ion formed in Problem 4.6.

4.6 Addition of H_2 to Alkenes: Hydrogenation

Addition of H_2 to the C=C bond occurs when alkenes are exposed to an atmosphere of hydrogen gas in the presence of a catalyst. We describe the result by saying that the double bond is **hydrogenated,** or *reduced.* (The word **reduction** in organic chemistry usually refers to the addition of hydrogen or removal of oxygen from a molecule.) For most alkene hydrogenations, either palladium metal or platinum (as PtO_2) is used as the catalyst.

$$\text{C=C} + \text{H—H} \xrightarrow{\text{Catalyst}} \text{H—C—C—H}$$

An alkene **An alkane**

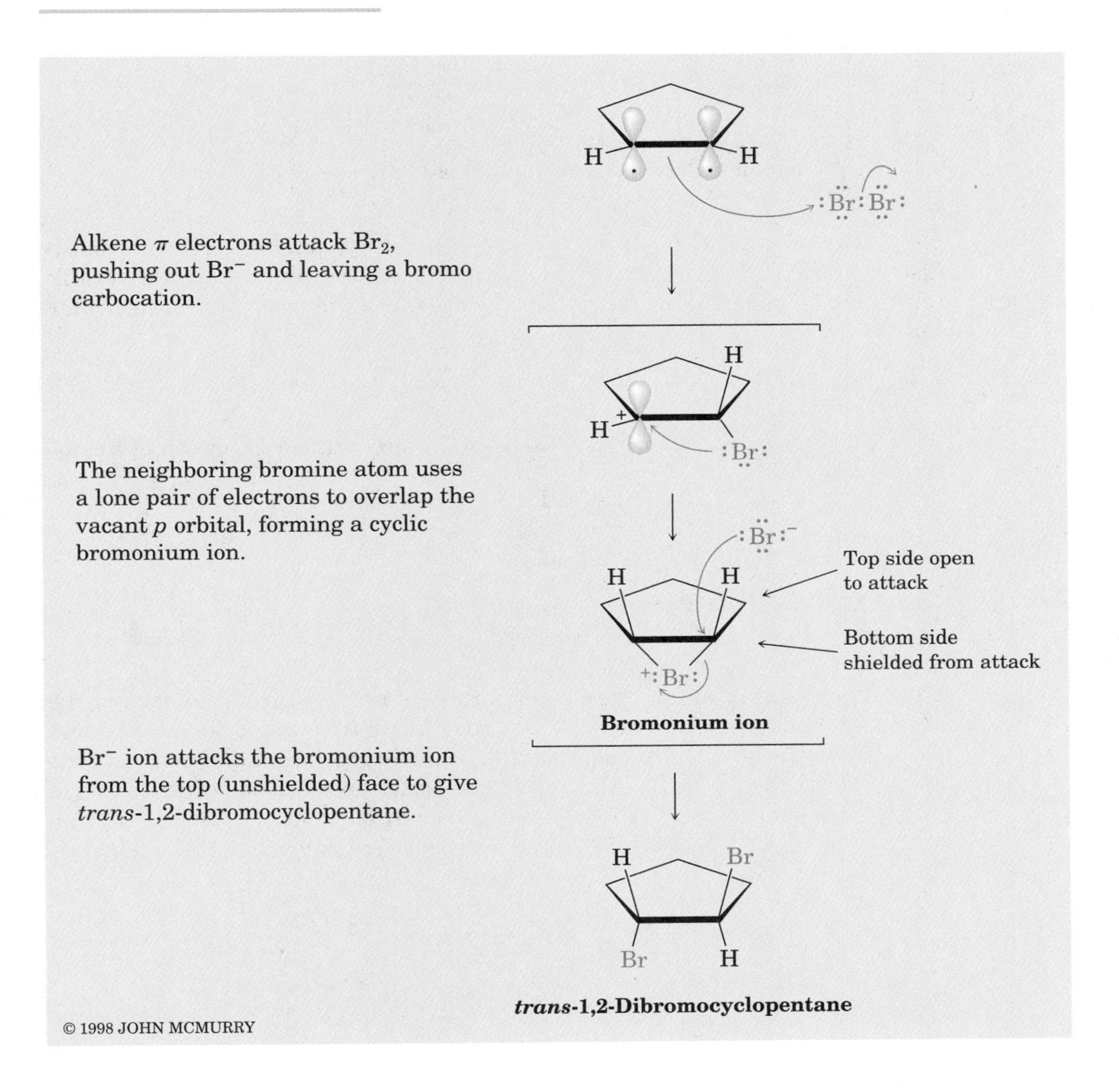

Figure 4.5 Mechanism of the addition of Br_2 to an alkene. A cyclic bromonium ion intermediate is formed, shielding one face of the double bond and resulting in formation of anti stereochemistry for the addition product.

Catalytic hydrogenation of alkenes, unlike most other organic reactions, is a *heterogeneous* process, rather than a homogeneous one. That is, the hydrogenation reaction occurs on the surface of solid catalyst particles rather than in solution. The reaction occurs with **syn stereochemistry** (the opposite of *anti*), meaning that both hydrogens add to the double bond from the same side.

$$\text{1,2-Dimethylcyclohexene} \xrightarrow[CH_3CO_2H]{H_2,\ PtO_2} \textit{cis}\text{-1,2-Dimethylcyclohexane (82\%)}$$

1,2-Dimethylcyclohexene **_cis_-1,2-Dimethylcyclohexane** **(82%)**

In addition to its usefulness in the laboratory, alkene hydrogenation is also of great commercial value. In the food industry, unsaturated vegetable oils are catalytically hydrogenated on a vast scale to produce the saturated fats used in margarine.

PROBLEM

4.8 What product would you expect to obtain from catalytic hydrogenation of the following alkenes?
(a) $(CH_3)_2C{=}CHCH_2CH_3$ (b) 3,3-Dimethylcyclopentene

4.7 Oxidation of Alkenes

Hydroxylation of an alkene—the addition of an –OH group to each of the alkene carbons—can be carried out by reaction of the alkene with potassium permanganate, $KMnO_4$, in basic solution. Since oxygen is added to the alkene during the reaction, we call this an **oxidation.** The reaction occurs with syn stereochemistry and yields a 1,2-dialcohol, or **diol,** product (also called a *glycol*). For example, cyclohexene gives *cis*-1,2-cyclohexanediol in 37% yield.

$$\text{Cyclohexene} + KMnO_4 \xrightarrow[NaOH]{H_2O} \textit{cis}\text{-1,2-Cyclohexanediol (37\%)}$$

Cyclohexene **_cis_-1,2-Cyclohexanediol** **(37%)**

When oxidation of the alkene is carried out with $KMnO_4$ in either neutral or acidic solution, *cleavage* of the double bond occurs and carbonyl-containing products are obtained. If the double bond is tetrasubstituted, the two carbonyl-containing products are ketones; if a hydrogen is present on the double bond, one of the carbonyl-containing products is a carboxylic acid; and if two hydrogens are present on one carbon, CO_2 is formed:

$$\text{Isopropylidenecyclohexane} + KMnO_4 \xrightarrow{H_2O} \text{Cyclohexanone} + O{=}C(CH_3)_2$$

Isopropylidenecyclohexane **Cyclohexanone** **Acetone**
(two ketones)

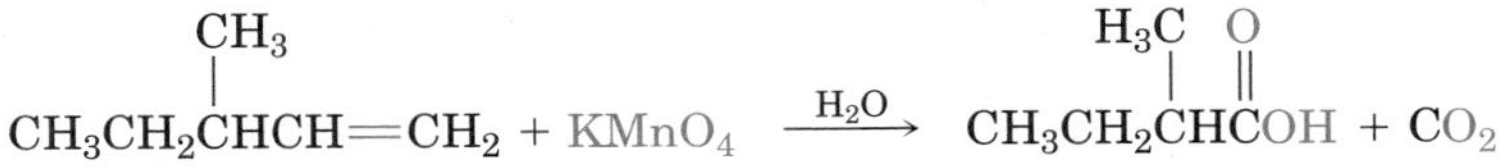

$$\mathrm{CH_3CH_2\underset{\substack{|\\ }}{\overset{\substack{CH_3\\|}}{C}}HCH{=}CH_2 + KMnO_4 \xrightarrow{H_2O} CH_3CH_2\overset{\substack{H_3C\\|}}{C}H\overset{\substack{O\\\|}}{C}OH + CO_2}$$

3-Methyl-1-pentene → **2-Methylbutanoic acid (45%)**

PRACTICE PROBLEM 4.3

Predict the product of reaction of 2-pentene with aqueous acidic $KMnO_4$.

Solution Reaction of an alkene with acidic $KMnO_4$ yields carbonyl-containing products in which the double bond is broken and the two fragments have a C=O group in place of the original alkene C=C. If a hydrogen is present on the double bond, a carboxylic acid is produced. Thus, 2-pentene gives the following reaction:

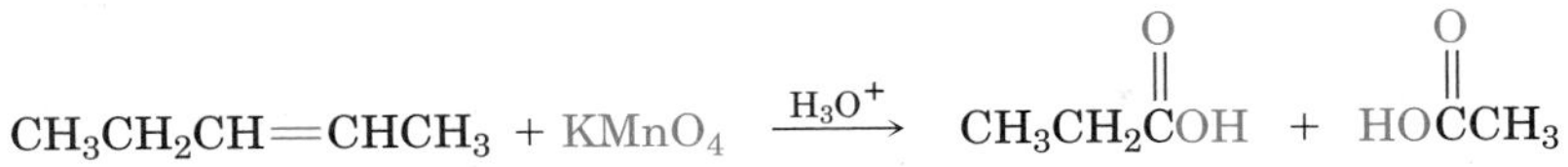

$$\mathrm{CH_3CH_2CH{=}CHCH_3 + KMnO_4 \xrightarrow{H_3O^+} CH_3CH_2\overset{\substack{O\\\|}}{C}OH + HO\overset{\substack{O\\\|}}{C}CH_3}$$

2-Pentene → **Propanoic acid** + **Acetic acid**

PRACTICE PROBLEM 4.4

What alkene gives a mixture of acetone and propanoic acid on reaction with acidic $KMnO_4$?

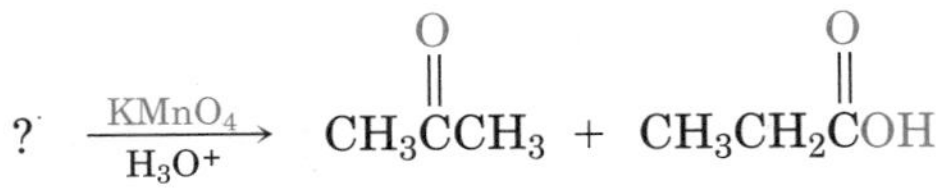

$$? \xrightarrow[H_3O^+]{KMnO_4} \mathrm{CH_3\overset{\substack{O\\\|}}{C}CH_3 + CH_3CH_2\overset{\substack{O\\\|}}{C}OH}$$

Acetone + **Propanoic acid**

Solution To find the starting alkene that gives the cleavage products shown, remove the oxygen atoms from the two products, join the fragments with a double bond, and replace the –OH by –H:

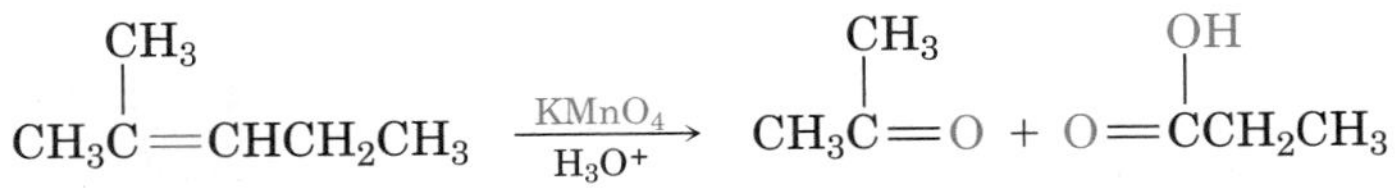

$$\mathrm{CH_3\overset{\substack{CH_3\\|}}{C}{=}CHCH_2CH_3 \xrightarrow[H_3O^+]{KMnO_4} CH_3\overset{\substack{CH_3\\|}}{C}{=}O + O{=}\overset{\substack{OH\\|}}{C}CH_2CH_3}$$

2-Methyl-2-pentene → **Acetone** + **Propanoic acid**

PROBLEM

4.9 Predict the product of the reaction of 1,2-dimethylcyclohexene with the following:
(a) $KMnO_4$, H_3O^+ (b) $KMnO_4$, OH^-, H_2O

PROBLEM

4.10 Propose structures for alkenes that yield the following products on treatment with acidic $KMnO_4$:
(a) $(CH_3)_2C{=}O + CO_2$ (b) 2 equiv CH_3CH_2COOH

4.8 Alkene Polymers

No other group of synthetic chemicals has had as great an impact on our day-to-day lives as the synthetic *polymers.* From carpeting to clothing to foam coffee cups, we are surrounded by polymers.

A **polymer** is a large (sometimes *very* large) molecule built up by repetitive bonding together of many smaller molecules, called **monomers.** As we'll see in later chapters, nature makes wide use of biological polymers. Cellulose, for example, is a polymer built of repeating sugar units; proteins are polymers built of repeating amino acid units; and nucleic acids are polymers built of repeating nucleotide units. Although synthetic polymers are chemically much simpler than biopolymers, there is an immense diversity to the structures and properties of synthetic polymers, depending on the nature of the monomers and on the reaction conditions used for polymerization.

Radical Polymerization of Alkenes

Many simple alkenes undergo rapid polymerization when treated with a small amount of a radical catalyst. Ethylene, for example, yields polyethylene, an enormous alkane that may have up to several thousand monomer units incorporated into its long hydrocarbon chain. Ethylene polymerization is usually carried out at high pressure (1000–3000 atm) and high temperature (100–250°C) with a radical catalyst such as benzoyl peroxide.

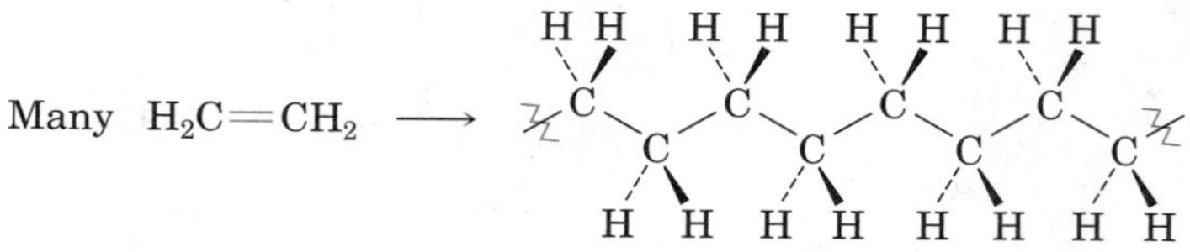

$$\text{Many } H_2C{=}CH_2 \longrightarrow \text{—}(CH_2CH_2CH_2CH_2CH_2CH_2CH_2CH_2)\text{—}$$

Ethylene **A section of polyethylene**

Radical polymerizations of alkenes involve three kinds of steps: *initiation, propagation,* and *termination.*

Step 1 *Initiation* occurs when small amounts of radicals are generated by the catalyst. For example,when benzoyl peroxide is used as initiator, the O–O bond is broken on heating to yield benzoyloxy radicals.

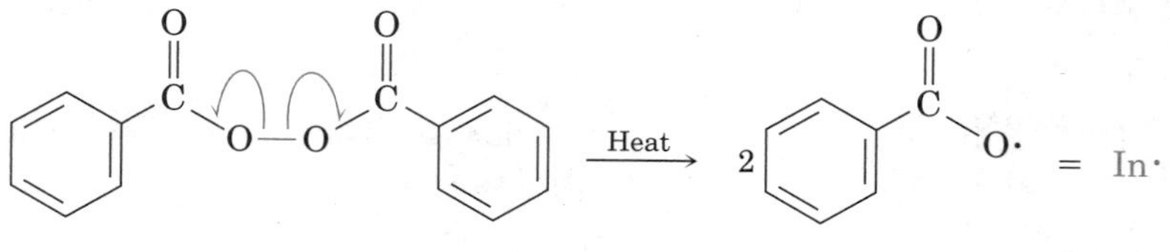

$$C_6H_5C(=O)O{-}OC(=O)C_6H_5 \xrightarrow{\text{Heat}} 2\ C_6H_5C(=O)O\cdot \quad = \text{In}\cdot$$

Benzoyl peroxide **Benzoyloxy radical**

Step 2 The benzoyloxy radical produced in step 1 adds to the C=C bond of ethylene to generate a carbon radical in a process similar to what occurs when an electrophile such as H^+ adds to a C=C bond to generate a carbocation (Section 3.8). Note that this radical addition step results in formation of a bond between the initiator and the ethylene molecule in which one electron has been contributed by each partner. The remaining electron from the ethylene π bond remains on carbon as the radical site.

$$In\cdot + H_2C{=}CH_2 \longrightarrow In{-}CH_2CH_2\cdot$$

Step 3 *Propagation* of the polymerization reaction occurs when the carbon radical formed in step 2 adds to another ethylene molecule. Repetition of this step for hundreds or thousands of times builds the polymer chain.

$$In{-}CH_2CH_2\cdot + H_2C{=}CH_2 \longrightarrow In{-}CH_2CH_2CH_2CH_2\cdot$$

$$\xrightarrow[\text{many times}]{\text{Repeat}} In{-}(CH_2CH_2)_nCH_2CH_2\cdot$$

Step 4 *Termination* of the polymer chain eventually occurs by a reaction that consumes the radical. For example, combination of two chains by chance meeting is a possible chain-terminating reaction:

$$2\,R{-}CH_2CH_2\cdot \longrightarrow R{-}CH_2CH_2CH_2CH_2{-}R$$

Polymerization of Substituted Ethylenes

Many substituted ethylenes, called *vinyl monomers,* undergo radical-initiated polymerization, yielding polymers with substituent groups regularly spaced along the polymer backbone. Propylene, for example, yields polypropylene when polymerized.

$$\underset{\textbf{Propylene}}{H_2C{=}CHCH_3} \longrightarrow \underset{\textbf{Polypropylene}}{\left(CH_2\overset{CH_3}{\overset{|}{C}H}CH_2\overset{CH_3}{\overset{|}{C}H}CH_2\overset{CH_3}{\overset{|}{C}H}CH_2\overset{CH_3}{\overset{|}{C}H}\right)}$$

Table 4.1 (p. 124) shows some commercially important vinyl monomers and lists some industrial uses of the different polymers that result.

PRACTICE PROBLEM 4.5

Show the structure of poly(vinyl chloride), a polymer made from $H_2C{=}CHCl$, by drawing several repeating units.

Solution The general structure of poly(vinyl chloride) is

$$\underset{\textbf{Vinyl chloride}}{H_2C{=}\overset{Cl}{\overset{|}{C}H}} \longrightarrow \underset{\textbf{Poly(vinyl chloride)}}{\left(CH_2\overset{Cl}{\overset{|}{C}H}CH_2\overset{Cl}{\overset{|}{C}H}CH_2\overset{Cl}{\overset{|}{C}H}\right)}$$

PROBLEM

4.11 Show the structure of Teflon by drawing several repeating units. The monomer unit is tetrafluoroethylene, $F_2C{=}CF_2$.

Table 4.1 Some Alkene Polymers and Their Uses

Monomer name	*Formula*	*Trade or common name of polymer*	*Uses*
Ethylene	$H_2C{=}CH_2$	Polyethylene	Packaging, bottles, cable insulation, films and sheets
Propene (propylene)	$H_2C{=}CHCH_3$	Polypropylene	Automotive moldings, rope, carpet fibers
Chloroethylene (vinyl chloride)	$H_2C{=}CHCl$	Poly(vinyl chloride), Tedlar	Insulation, films, pipes
Styrene	$H_2C{=}CHC_6H_5$	Polystyrene, Styron	Foam and molded articles
Tetrafluoroethylene	$F_2C{=}CF_2$	Teflon	Valves and gaskets, coatings
Acrylonitrile	$H_2C{=}CHCN$	Orlon, Acrilan	Fibers
Methyl methacrylate	$H_2C{=}C(CH_3)CO_2CH_3$	Plexiglas, Lucite	Molded articles, paints
Vinyl acetate	$H_2C{=}CHOCOCH_3$	Poly(vinyl acetate)	Paints, adhesives

4.9 Preparation of Alkenes: Elimination Reactions

Just as addition reactions account for most of the chemistry that alkenes undergo, *elimination reactions* account for most of the ways used to prepare alkenes. Additions and eliminations are, in many respects, two sides of the same coin:

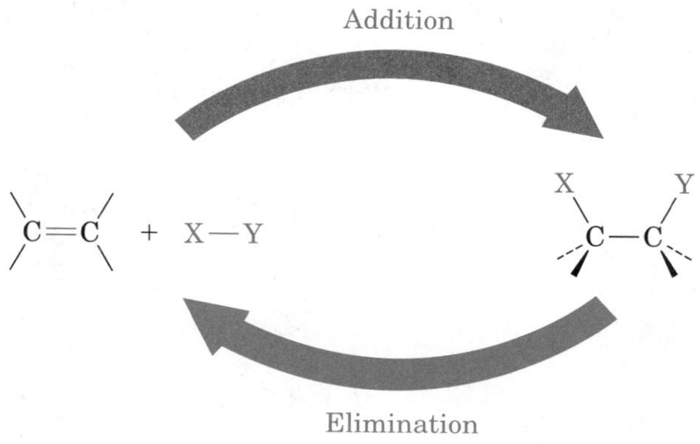

Addition

$$\mathrm{C{=}C} + \mathrm{X{-}Y} \rightleftharpoons \mathrm{X{-}C{-}C{-}Y}$$

Elimination

The two most common kinds of elimination reactions are the **dehydrohalogenation** of an alkyl halide (elimination of HX) and the **dehydration** of an alcohol (elimination of water, H_2O). We'll return for a closer look at how these reactions take place in Chapter 7.

Elimination of HX from Alkyl Halides: Dehydrohalogenation

Alkyl halides can be synthesized by addition of HX to alkenes. Conversely, alkenes can be synthesized by elimination of HX from alkyl halides. Dehydrohalogenation is usually effected by treating the alkyl halide with a strong base. Thus, bromocyclohexane yields cyclohexene when treated with potassium hydroxide in alcohol solution:

$$\text{Bromocyclohexane} \xrightarrow[CH_3CH_2OH]{KOH} \text{Cyclohexene (81\%)} + KBr + H_2O$$

Bromocyclohexane **Cyclohexene (81%)**

Elimination reactions are a bit more complex than addition reactions because of the regioselectivity problem: What product results from the dehydrohalogenation of an unsymmetrical halide? In fact, elimination reactions usually give mixtures of alkene products, and the best we can do is to predict which product will be major.

According to **Zaitsev's rule,** a predictive rule formulated by the Russian chemist Alexander Zaitsev in 1875, base-induced elimination reactions generally give the more highly substituted alkene product. For example, on treatment of 2-bromobutane with KOH in ethanol, 2-butene (disubstituted; two alkyl group substituents on the double-bond carbons) predominates over 1-butene (monosubstituted; one alkyl group substituent on the double-bond carbons).

$$CH_3CH_2CH(Br)CH_3 \xrightarrow[CH_3CH_2OH]{KOH} CH_3CH{=}CHCH_3 + CH_3CH_2CH{=}CH_2$$

2-Bromobutane **2-Butene (81%)** **1-Butene (19%)**

Zaitsev's rule: In the elimination of HX from an alkyl halide, the more highly substituted alkene product predominates.

PRACTICE PROBLEM 4.6

What product would you expect from reaction of 1-chloro-1-methylcyclohexane with KOH in ethanol?

Solution Treatment of an alkyl halide with a strong base such as KOH causes dehydrohalogenation and yields an alkene. To find the products in a specific case, draw the structure of the starting material and locate the hydrogen atoms on each neighboring carbon. Then generate the potential alkene products by removing HX in as many ways as possible. The major product will be the one that has the most highly substituted double bond—in this case, 1-methylcyclohexene.

$$\text{1-Chloro-1-methylcyclohexane} + KOH \xrightarrow{CH_3CH_2OH} \text{1-Methylcyclohexene (major)} + \text{Methylenecyclohexane (minor)}$$

1-Chloro-1-methylcyclohexane **1-Methylcyclohexene** (major) **Methylenecyclohexane** (minor)

PROBLEM

4.12 What products would you expect from the reaction of 2-bromo-2-methylbutane with KOH in ethanol? Which will be major?

PROBLEM

4.13 What alkyl halide starting materials might the following alkenes have been made from?

(a) $CH_3CH(CH_3)CH_2CH_2CH(CH_3)CH{=}CH_2$ (b) 3,4-dimethylcyclopentene (ring with CH_3 and CH_3 substituents)

Elimination of H_2O from Alcohols: Dehydration

The dehydration of an alcohol is one of the most useful methods of alkene synthesis, and many ways of carrying out the reaction have been devised. A method that works particularly well for secondary and tertiary alcohols is by treatment with a strong acid. For example, when 1-methylcyclohexanol is treated with aqueous sulfuric acid, dehydration occurs to yield 1-methylcyclohexene:

$$\text{1-Methylcyclohexanol} \xrightarrow[50°C]{H_2SO_4,\ H_2O} \text{1-Methylcyclohexene} + H_2O$$

1-Methylcyclohexanol **1-Methylcyclohexene (91%)**

Acid-catalyzed dehydrations usually follow Zaitsev's rule and yield the more highly substituted alkene as major product. Thus, 2-methyl-2-butanol gives primarily 2-methyl-2-butene (trisubstituted) rather than 2-methyl-1-butene (disubstituted):

$$CH_3CH_2{-}C(OH)(CH_3){-}CH_3 \xrightarrow[25°C]{H_2SO_4,\ H_2O} CH_3CH{=}C(CH_3)CH_3 + CH_3CH_2C(CH_3){=}CH_2$$

2-Methyl-2-butanol **2-Methyl-2-butene** (major) **2-Methyl-1-butene** (minor)

PRACTICE PROBLEM 4.7

Predict the major product of the following reaction:

$$\underset{\text{CH}_3\text{CH}_2\text{CHCHCH}_3}{\text{H}_3\text{C}\ \ \text{OH}} \xrightarrow{\text{H}_2\text{SO}_4,\ \text{H}_2\text{O}} ?$$

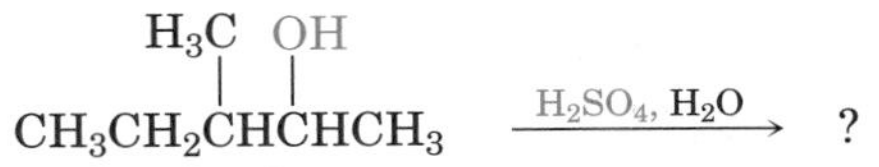

Solution Treatment of an alcohol with H_2SO_4 leads to dehydration and formation of the more highly substituted alkene product (Zaitsev's rule). Thus, dehydration of 3-methyl-2-pentanol yields 3-methyl-2-pentene as the major product rather than 3-methyl-1-pentene.

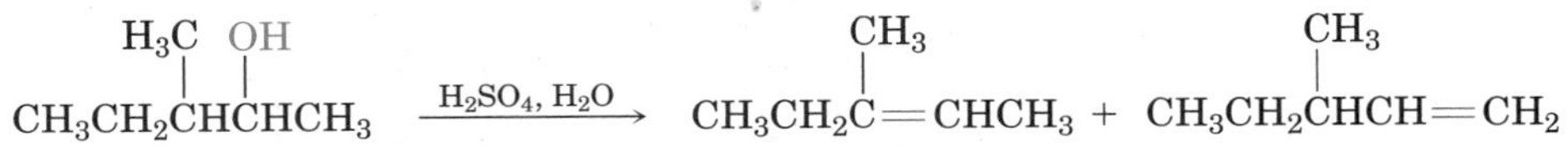

$$\underset{\text{CH}_3\text{CH}_2\text{CHCHCH}_3}{\text{H}_3\text{C}\ \ \text{OH}} \xrightarrow{\text{H}_2\text{SO}_4,\ \text{H}_2\text{O}} \underset{\text{CH}_3\text{CH}_2\text{C}{=}\text{CHCH}_3}{\text{CH}_3} + \underset{\text{CH}_3\text{CH}_2\text{CHCH}{=}\text{CH}_2}{\text{CH}_3}$$

3-Methyl-2-pentanol — **3-Methyl-2-pentene** (major) — **3-Methyl-1-pentene** (minor)

PROBLEM

4.14 Predict the products you would expect from the following reactions. Indicate the major product in each case.

(a) 2-Bromo-2-methylpentane + KOH $\longrightarrow$?

(b) 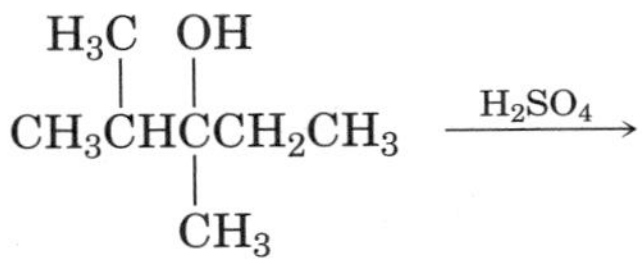 $\text{CH}_3\text{CHCCH}_2\text{CH}_3$ (with H_3C on C2, OH and CH_3 on C3) $\xrightarrow{\text{H}_2\text{SO}_4}$

PROBLEM

4.15 What alcohols might the following alkenes have been made from?

(a) 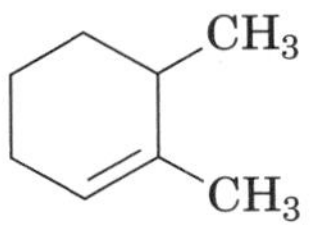 1,6-dimethylcyclohexene ring with substituents CH$_3$, CH$_3$

(b) $\text{CH}_3\text{CH}_2\text{CH}{=}\text{CHCH}_2\text{CH}_2\text{CH}_3$

4.10 Conjugated Dienes

Multiple bonds that alternate with single bonds are said to be **conjugated.** Thus, 1,3-butadiene is a **conjugated diene,** whereas 1,4-pentadiene is a nonconjugated diene with isolated double bonds.

$$\text{H}_2\text{C}{=}\text{CH}{-}\text{CH}{=}\text{CH}_2 \qquad \text{H}_2\text{C}{=}\text{CH}{-}\text{CH}_2{-}\text{CH}{=}\text{CH}_2$$

1,3-Butadiene (conjugated; alternating double and single bonds)

1,4-Pentadiene (nonconjugated; nonalternating double and single bonds)

What's so special about conjugated dienes that we need to look at them separately? The orbital view of 1,3-butadiene shown in Figure 4.6 provides a clue to the answer: *There is an electronic interaction between the two*

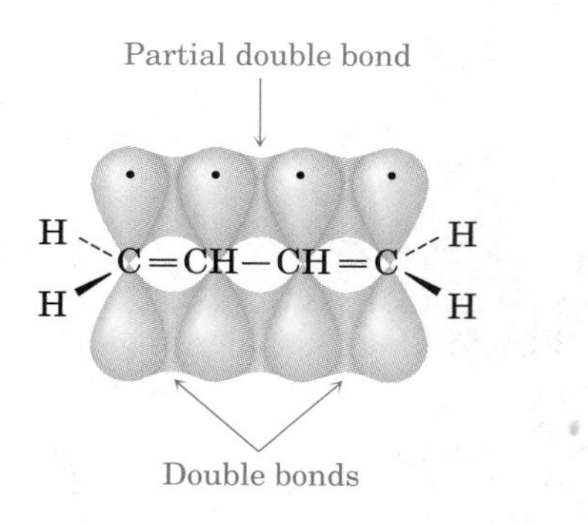

Figure 4.6 An orbital view of 1,3-butadiene. Each of the four carbon atoms has a *p* orbital, allowing for an electronic interaction across the C2–C3 single bond.

double bonds of a conjugated diene because of *p* orbital overlap across the central single bond. This interaction of *p* orbitals across a single bond gives conjugated dienes some unusual properties.

Although much of the chemistry of conjugated dienes and other alkenes is similar, there is a striking difference in their addition reactions with electrophiles like HX and X_2. When HX adds to an isolated alkene, Markovnikov's rule usually predicts the formation of a single product. When HX adds to a conjugated diene, though, mixtures of products are usually obtained. For example, reaction of HBr with 1,3-butadiene yields two products:

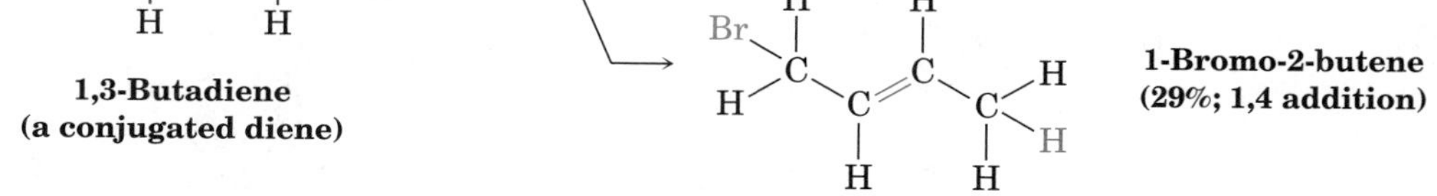

1,3-Butadiene (a conjugated diene) + HBr →

3-Bromo-1-butene (71%; 1,2 addition)

1-Bromo-2-butene (29%; 1,4 addition)

3-Bromo-1-butene is the typical product of Markovnikov addition, but 1-bromo-2-butene appears unusual. The double bond in this product has moved to a position between C2 and C3, while HBr has added to C1 and C4, a result described as **1,4 addition.**

How can we account for the formation of the 1,4-addition product? The answer is that an *allylic carbocation* is involved as an intermediate in the reaction. (The word **allylic** means "next to a double bond.") When H^+ adds to an electron-rich π bond of 1,3-butadiene, two carbocation intermediates are possible—a primary nonallylic carbocation and a secondary allylic carbocation. Allylic carbocations are very stable and therefore form faster than less stable, nonallylic carbocations.

1,3-Butadiene + HBr

→ Secondary, allylic carbocation + Br^-

⇸ Primary carbocation (*NOT formed*) + Br^-

4.11 Stability of Allylic Carbocations: Resonance

Why are allylic carbocations particularly stable? To find the answer, look at the orbital picture of an allylic carbocation in Figure 4.7. The positively charged carbon atom has a vacant *p* orbital that can overlap the *p* orbitals of the neighboring double bond.

From an orbital viewpoint, an allylic carbocation is symmetrical. All three carbon atoms are sp^2-hybridized, and each has a *p* orbital. Thus, the *p* orbital on the central carbon can overlap equally well with *p* orbitals on *either* of the two neighboring carbons, and the two electrons are free to move about over the entire three-orbital array.

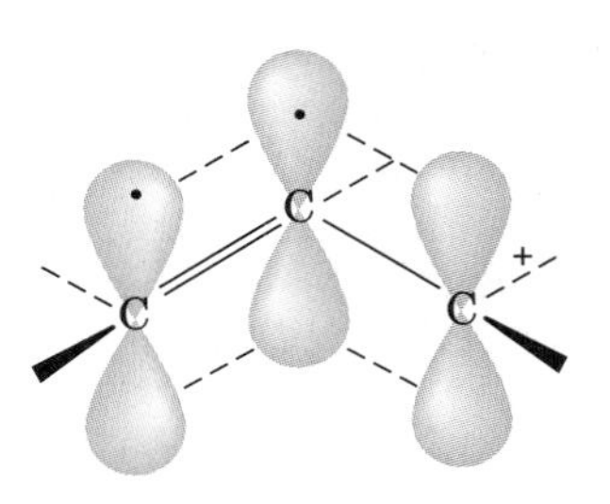

Figure 4.7 An orbital picture of an allylic carbocation. The vacant *p* orbital on the positively charged carbon can overlap the double-bond *p* orbitals.

One consequence of this orbital picture is that there are two ways to draw an allylic carbocation. We can draw it with the vacant p orbital on the left and the double bond on the right, or we can draw it with the vacant p orbital on the right and the double bond on the left. *Neither structure is correct by itself; the true structure of the allylic carbocation is somewhere in between the two.*

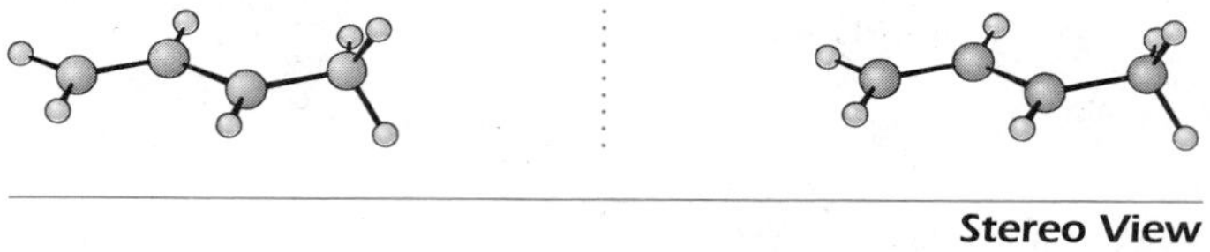

Stereo View

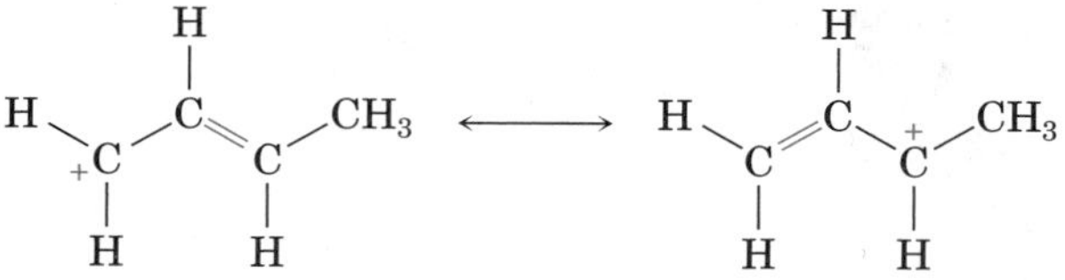

Two resonance forms of an allylic carbocation

The two individual structures of the allylic carbocation are called **resonance forms,** and their special relationship is indicated by the double-headed arrow between them. The only difference between the resonance forms is the position of the bonding electrons. The atoms themselves remain in exactly the same place in both resonance forms.

The best way to think about resonance is to realize that a species like an allylic carbocation is no different from any other organic substance. An allylic carbocation doesn't jump back and forth between two resonance forms, spending part of its time looking like one and the rest of its time looking like the other. Rather, an allylic carbocation has a single, unchanging structure that we call a **resonance hybrid.** A useful analogy is to think of a resonance hybrid as being like a mixed-breed dog. Just as a dog that's a mixture of dachshund and German shepherd doesn't change back and forth from one to the other, a resonance hybrid doesn't change back and forth between forms.

The difficulty in understanding resonance hybrids is visual, because we can't draw an accurate single picture of a resonance hybrid by using familiar kinds of structures. The line-bond structures that work so well to represent most organic molecules just don't work well for resonance hybrids like allylic carbocations because the two C–C bonds are equivalent and each is midway between single and double.

One of the most important consequences of resonance theory is that *the greater the number of possible resonance forms, the greater the stability.* Because an allylic carbocation is a resonance hybrid of two forms, it is more stable than a typical nonallylic carbocation, which has only one form. This stability is due to the fact that the π electrons can be spread out, or *delocalized,* over an extended p orbital network rather than remain centered in only one bond.

In addition to its effect on stability, the resonance picture of an allylic carbocation also has chemical consequences. When the allylic carbocation produced by protonation of 1,3-butadiene reacts with Br^- ion to complete the addition, reaction can occur at either C1 or C3, because both share the positive charge. The result is a mixture of 1,2- and 1,4-addition products.

Br^-

1,4 addition (29%)

1,2 addition (71%)

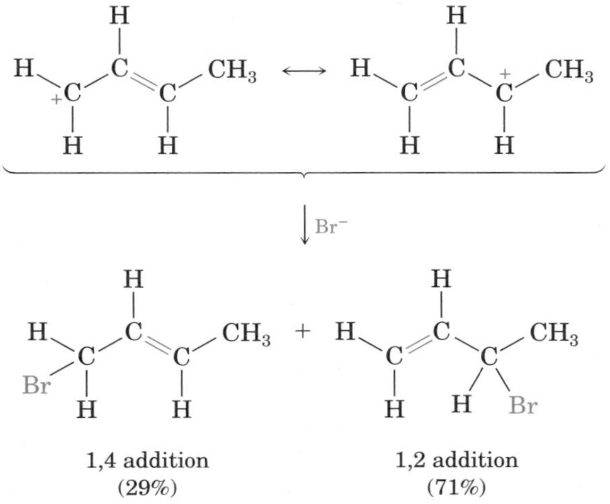

PROBLEM

4.16 1,3-Butadiene reacts with Br_2 to yield a mixture of 1,2- and 1,4-addition products. Show the structure of each.

4.12 Drawing and Interpreting Resonance Forms

Resonance is an extremely useful concept for explaining a variety of chemical phenomena. In inorganic chemistry, for example, the carbonate ion (CO_3^{2-}) has identical bond lengths for its three C–O bonds. Although there is no single line-bond structure that can account for this equivalence of C–O bonds, resonance theory accounts for it nicely. The carbonate ion is simply a resonance hybrid of three resonance forms. The three oxygens share the π electrons and the negative charges equally:

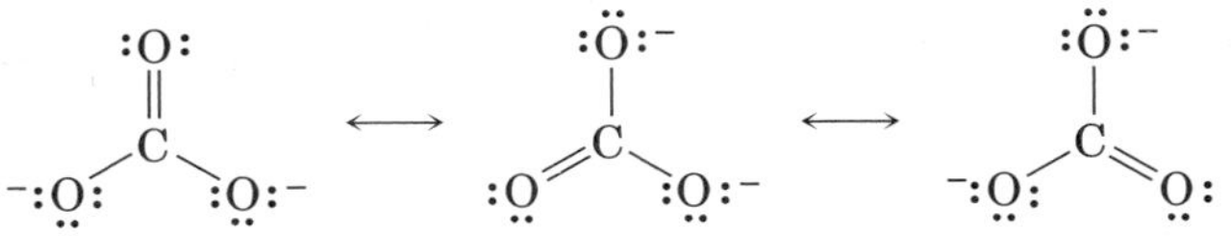

Carbonate ion (three resonance forms)

As an example from organic chemistry, we'll see in the next chapter that the six carbon–carbon bonds in aromatic compounds like benzene are equivalent because benzene is a resonance hybrid of two forms. Each form has alternating single and double bonds, and neither form is correct by itself. The true benzene structure is a hybrid of the two forms.

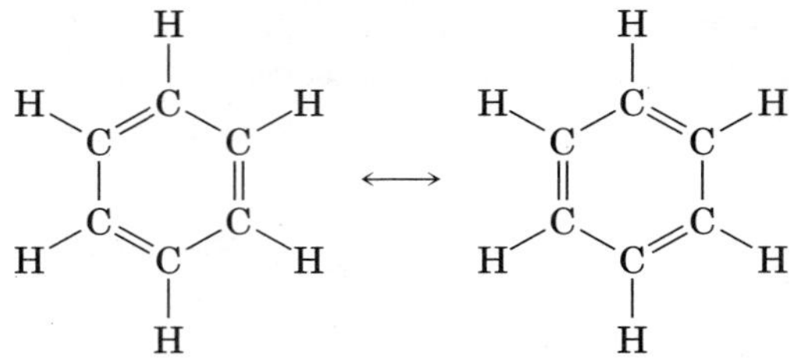

Benzene (two resonance forms)

When first dealing with resonance theory, it's often useful to have a set of guidelines that describe how to draw and interpret resonance forms. The following rules should prove helpful:

Rule 1 *Resonance forms are imaginary.* The real structure is a composite, or hybrid, of the different forms. Substances like the allylic carbocation, the carbonate ion, and benzene are no different from any other substance. They have single, unchanging structures, and they do not switch back and forth between resonance forms. The only difference between these and other substances is in the way they must be represented on paper.

Rule 2 *Resonance forms differ from each other only in the placement of their π or nonbonding electrons.* Neither the position nor the hybridization of atoms changes from one resonance form to another. In benzene, for example, the π electrons in the double bonds move, but the six carbon and six hydrogen atoms remain in place:

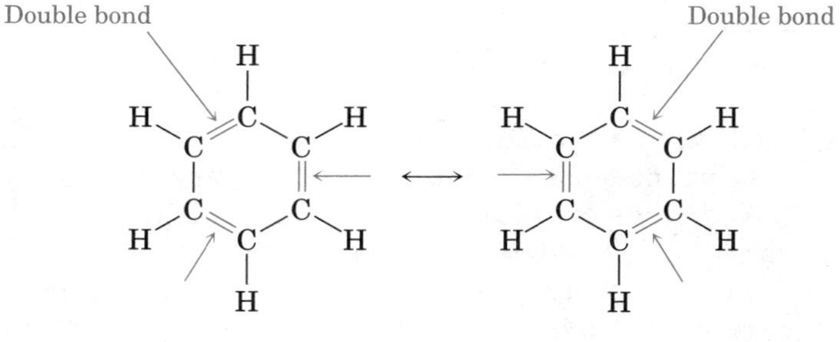

By contrast, two structures such as 1,3-cyclohexadiene and 1,4-cyclohexadiene are *not* resonance structures, because their hydrogen atoms don't occupy the same positions. Instead, the two dienes are constitutional isomers.

Constitutional isomers, *NOT* resonance forms

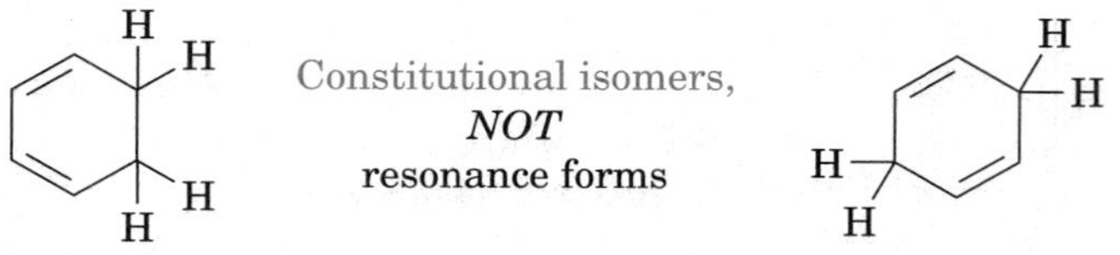

1,3-Cyclohexadiene **1,4-Cyclohexadiene**

Rule 3 Different resonance forms of a substance don't have to be equivalent. For example, the allylic carbocation obtained by reaction of 1,3-butadiene with H^+ is unsymmetrical. One end of the delocalized π electron system has a methyl substituent, and the other end is unsubstituted. Even though the two resonance forms aren't equivalent, they both contribute to the overall resonance hybrid.

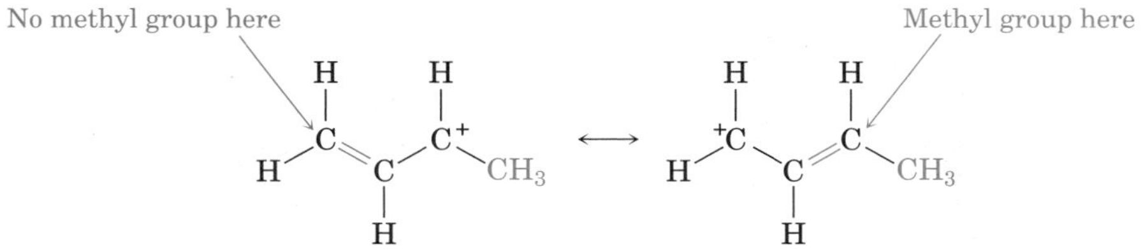

No methyl group here

Methyl group here

In general, when two resonance forms are nonequivalent, the actual structure of the resonance hybrid is closer to the more stable form than to the less stable form. Thus, we might expect the butenyl carbocation to look a bit more like a secondary carbocation than like a primary one.

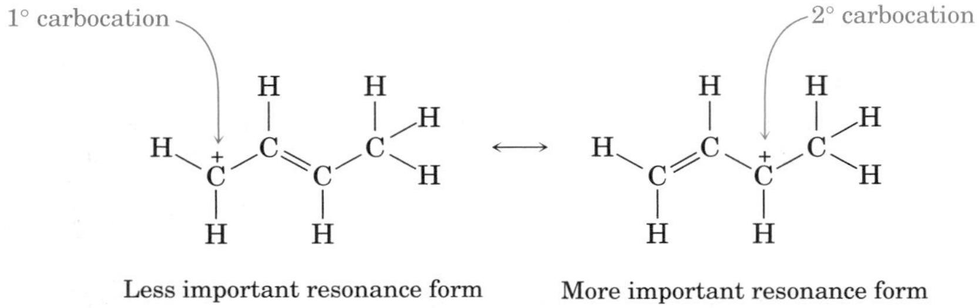

1° carbocation

2° carbocation

Less important resonance form

More important resonance form

Rule 4 Resonance forms must be valid Lewis structures and obey normal rules of valency. A resonance form is like any other structure: The octet rule still applies. For example, one of the following structures for the acetate ion is not a valid resonance form because the carbon atom has five bonds and ten electrons:

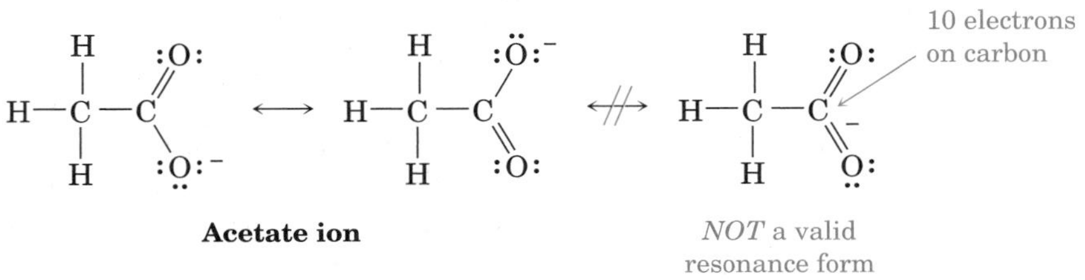

10 electrons on carbon

Acetate ion

NOT a valid resonance form

Rule 5 The resonance hybrid is more stable than any single resonance form. In other words, resonance leads to stability. The greater the number of resonance forms, the more stable the substance. We've already seen, for example, that an allylic carbocation is more stable than a nonallylic one. In a similar manner, we'll see in the next chapter that a benzene ring is more stable than a cyclic alkene.

PRACTICE PROBLEM 4.8

Use resonance structures to explain why the two C–O bonds of sodium formate are equivalent.

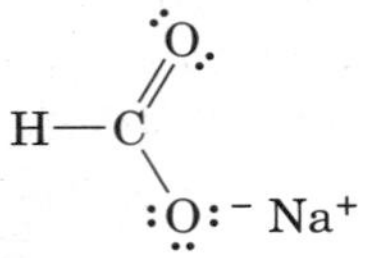

$$\mathrm{H{-}C({=}\ddot{O}{:})({-}\ddot{O}{:}^{-}\ Na^{+})}$$

Sodium formate

Solution The formate anion is a resonance hybrid of two equivalent resonance forms, which can be drawn by showing the double bond either to the top oxygen or to the bottom oxygen. Only the positions of the electrons are different in the two forms.

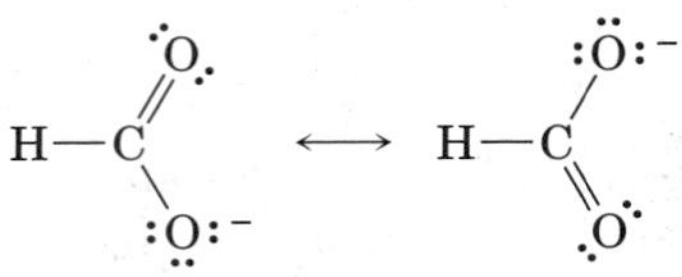

$$\mathrm{H{-}C({=}\ddot{O}{:})({-}\ddot{O}{:}^{-})} \longleftrightarrow \mathrm{H{-}C({-}\ddot{O}{:}^{-})({=}\ddot{O}{:})}$$

PROBLEM

4.17 Give the structure of all possible monoadducts of HCl and 1,3-pentadiene.

PROBLEM

4.18 Look at the possible carbocation intermediates produced during addition of HCl to 1,3-pentadiene (Problem 4.17) and predict which is the most stable.

PROBLEM

4.19 Draw as many resonance structures as you can for the following species:

(a) cyclohexenyl–$\overset{+}{C}H_2$ (b) $CH_3-\overset{O}{\overset{\|}{C}}-\ddot{C}H_2^{-}$ (c) cyclohexadienyl cation (+)

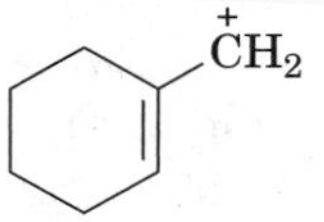

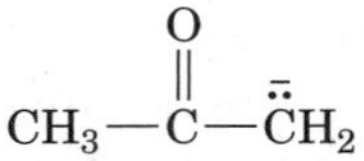

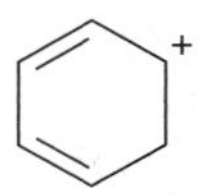

4.13 Alkynes

Alkynes are hydrocarbons that contain a carbon–carbon triple bond. Since four hydrogens must be removed from an alkane, C_nH_{2n+2}, to produce a triple bond, the general formula for an alkyne is C_nH_{2n-2}.

A C≡C triple bond results from the overlap of two *sp*-hybridized carbon atoms (Section 1.10). The two *sp* hybrid orbitals of carbon lie at an angle of 180° to each other along an axis that is perpendicular to the axes of the unhybridized $2p_y$ and $2p_z$ orbitals. When two *sp*-hybridized carbons approach each other, one *sp*–*sp* σ bond and two *p*–*p* π bonds form—a net triple bond (Figure 4.8). The two remaining *sp* orbitals form bonds to other atoms at an angle of 180° from the carbon–carbon σ bond, making acetylene, H–C≡C–H, a linear molecule.

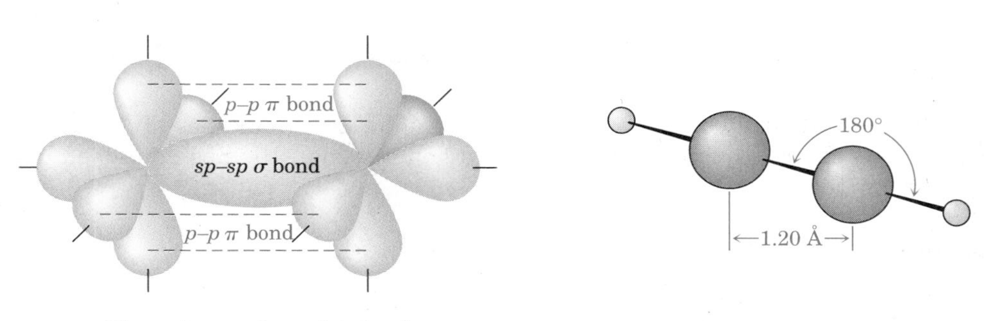

Figure 4.8 The electronic structure of a carbon–carbon triple bond.

Alkynes are named by general rules similar to those used for alkanes (Section 2.3) and alkenes (Section 3.1). The suffix *-yne* is used in the parent hydrocarbon name to denote an alkyne, and the position of the triple bond is indicated by its number in the chain. Numbering begins at the chain end nearer the triple bond so that the triple bond receives as low a number as possible.

$$\overset{8}{C}H_3\overset{7}{C}H_2\overset{6}{C}H(CH_3)\overset{5}{C}H_2\overset{4}{C}\equiv\overset{3}{C}\overset{2}{C}H_2\overset{1}{C}H_3$$

Begin numbering at the end nearer the triple bond.

6-Methyl-3-octyne

Compounds containing both double and triple bonds are called *enynes* (not ynenes). Numbering of the hydrocarbon chain starts from the end nearer the first multiple bond, whether double or triple. If there is a choice in numbering, double bonds receive lower numbers than triple bonds. For example,

$$\underset{7}{HC}\equiv\underset{6}{C}\underset{5}{C}H_2\underset{4}{C}H_2\underset{3}{C}H_2\underset{2}{C}H=\underset{1}{C}H_2$$

1-Hepten-6-yne

$$\underset{1}{HC}\equiv\underset{2}{C}\underset{3}{C}H_2\underset{4}{C}H(CH_3)\underset{5}{C}H_2\underset{6}{C}H_2\underset{7}{C}H=\underset{8}{C}H\underset{9}{C}H_3$$

4-Methyl-7-nonen-1-yne

PROBLEM

4.20 Give IUPAC names for the following compounds:

(a) $CH_3CH_2C\equiv CCH_2CH(CH_3)CH_3$

(b) $HC\equiv CC(CH_3)_2CH_3$

(c) $CH_3CH(CH_3)CH_2C\equiv CCH_3$

(d) $CH_3CH=CHCH_2C\equiv CCH_3$

4.14 Reactions of Alkynes: Addition of H_2, HX, and X_2

Based on their structural similarity, we might expect alkynes and alkenes to have chemical similarities also. As a general rule, alkynes do indeed behave much the same as alkenes.

Addition of H_2 to Alkynes

Alkynes are easily converted into alkanes by reduction with 2 equivalents of H_2 over a palladium catalyst. The reaction proceeds through an alkene intermediate, and the reaction can be stopped at the alkene stage if the right catalyst is used. The catalyst most often used for this purpose is the Lindlar catalyst, a specially prepared form of palladium metal. Because hydrogenation occurs with syn stereochemistry, alkynes give cis alkenes when reduced. For example,

$$CH_3(CH_2)_3C{\equiv}C(CH_2)_3CH_3 \xrightarrow[Pd/C]{2\ H_2} CH_3(CH_2)_8CH_3$$

5-Decyne → **Decane (96%)**

$$CH_3(CH_2)_3C{\equiv}C(CH_2)_3CH_3 \xrightarrow[\text{Lindlar catalyst}]{H_2} (H)(CH_3(CH_2)_3)C{=}C(H)((CH_2)_3CH_3)$$

5-Decyne → ***cis*-5-Decene (96%)**

Addition of HX to Alkynes

Alkynes give addition products on reaction with HCl, HBr, and HI. Although the reactions can usually be stopped after addition of 1 equivalent of HX to yield a *vinylic* halide (**vinylic** means "on the C=C double bond"), an excess of HX leads to formation of a dihalide product. As the following example indicates, the regioselectivity of addition to monosubstituted alkynes follows Markovnikov's rule. The H atom adds to the terminal carbon of the triple bond, and the X atom adds to the internal, more highly substituted carbon:

$$CH_3CH_2CH_2CH_2C{\equiv}CH + HBr \longrightarrow CH_3CH_2CH_2CH_2C(Br){=}CH_2$$

1-Hexyne → **2-Bromo-1-hexene**

Addition of X_2 to Alkynes

Bromine and chlorine add to alkynes to give dihalide addition products with anti stereochemistry:

$$CH_3CH_2CH_2CH_2C{\equiv}CH + Br_2 \xrightarrow{CCl_4} (CH_3CH_2CH_2CH_2)(Br)C{=}C(Br)(H)$$

1-Hexyne → **(*E*)-1,2-Dibromo-1-hexene**

PROBLEM

4.21 What products would you expect from the following reactions?

(a) $CH_3CH_2CH_2C{\equiv}CH + 1\ \text{equiv}\ Cl_2 \longrightarrow ?$

(b) $CH_3CH_2CH_2C{\equiv}CCH_2CH_3 + 1\ \text{equiv HBr} \longrightarrow ?$

(c) $CH_3CH(CH_3)CH_2C{\equiv}CCH_2CH_3 + H_2 \xrightarrow[\text{catalyst}]{\text{Lindlar}} ?$

4.15 Addition of H_2O to Alkynes

Addition of water takes place when an alkyne is treated with aqueous sulfuric acid in the presence of mercuric sulfate catalyst:

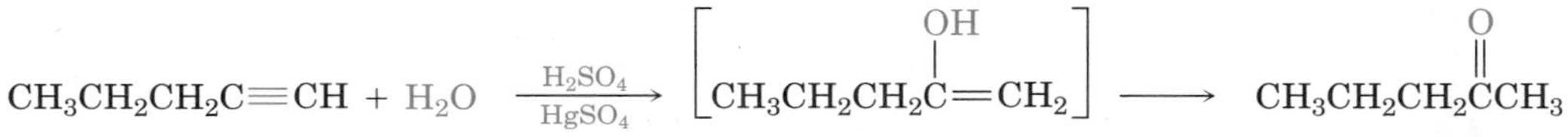

$$\underset{\textbf{1-Pentyne}}{CH_3CH_2CH_2C{\equiv}CH} + H_2O \xrightarrow[HgSO_4]{H_2SO_4} \left[\underset{\text{An enol}}{CH_3CH_2CH_2C(OH){=}CH_2}\right] \longrightarrow \underset{\textbf{2-Pentanone (78\%)}}{CH_3CH_2CH_2C(=O)CH_3}$$

Markovnikov regioselectivity is found for the hydration reaction, with the H attaching to the less substituted carbon and the OH attaching to the more substituted carbon. Interestingly, though, the expected vinylic alcohol, or *enol* (*ene* = alkene; *ol* = alcohol), is not isolated. Instead, the enol rearranges to a more stable isomer, a ketone ($R_2C{=}O$). It turns out that enols and ketones rapidly interconvert—a process called *tautomerism,* which we'll discuss in more detail in Section 11.1. With few exceptions, the keto–enol equilibrium heavily favors the ketone. Enols are almost never isolated.

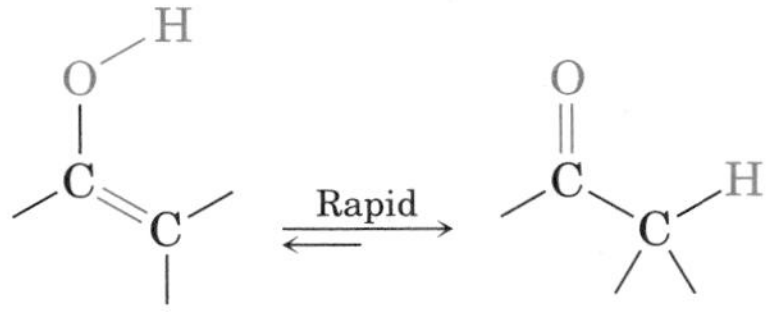

$$\underset{\substack{\textbf{Enol tautomer}\\\textbf{(less favored)}}}{C(OH){=}C} \underset{}{\overset{\text{Rapid}}{\rightleftharpoons}} \underset{\substack{\textbf{Keto tautomer}\\\textbf{(more favored)}}}{C(=O){-}C{-}H}$$

A mixture of both possible ketones results when an internal alkyne (R–C≡C–R′) is hydrated, but only a single product is formed from reaction of a terminal alkyne (R–C≡C–H).

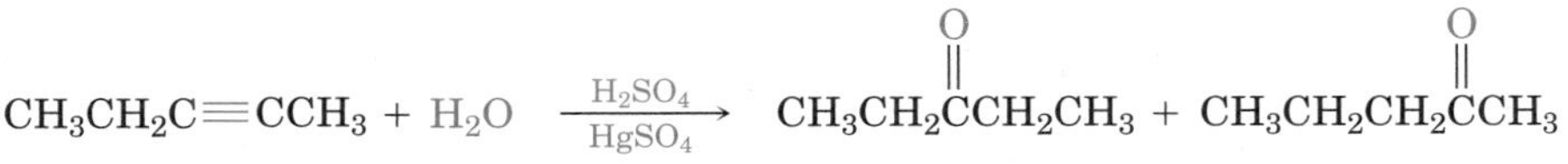

$$\underset{\substack{\textbf{2-Pentyne}\\\textbf{(an internal alkyne)}}}{CH_3CH_2C{\equiv}CCH_3} + H_2O \xrightarrow[HgSO_4]{H_2SO_4} \underset{\textbf{3-Pentanone}}{CH_3CH_2C(=O)CH_2CH_3} + \underset{\textbf{2-Pentanone}}{CH_3CH_2CH_2C(=O)CH_3}$$

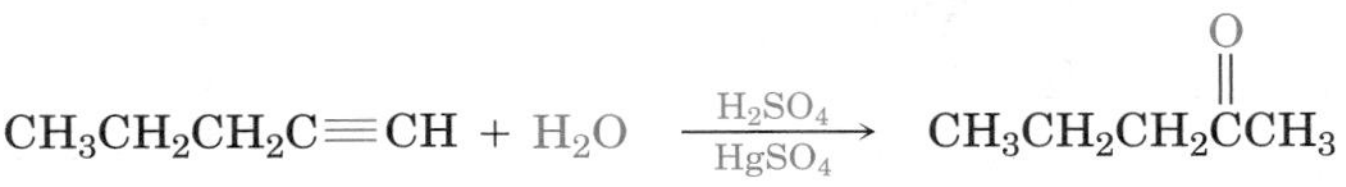

$$\underset{\substack{\textbf{1-Pentyne}\\\textbf{(a terminal alkyne)}}}{CH_3CH_2CH_2C{\equiv}CH} + H_2O \xrightarrow[HgSO_4]{H_2SO_4} \underset{\textbf{2-Pentanone}}{CH_3CH_2CH_2C(=O)CH_3}$$

PRACTICE PROBLEM 4.9

What product would you obtain by hydration of 4-methyl-1-hexyne?

Solution Addition of water to 4-methyl-1-hexyne according to Markovnikov's rule should yield a product with the OH group attached to C2 rather than C1. This enol then isomerizes to yield a ketone:

$$\underset{\textbf{4-Methyl-1-hexyne}}{CH_3CH_2\overset{\overset{CH_3}{|}}{C}HCH_2C{\equiv}CH} + H_2O \xrightarrow[HgSO_4]{H_2SO_4} \left[CH_3CH_2\overset{\overset{CH_3}{|}}{C}HCH_2\overset{\overset{OH}{|}}{C}{=}CH_2\right]$$

$$\longrightarrow \underset{\textbf{4-Methyl-2-hexanone}}{CH_3CH_2\overset{\overset{CH_3}{|}}{C}HCH_2\overset{\overset{O}{\|}}{C}CH_3}$$

PROBLEM

4.22 What product would you obtain by hydration of 4-octyne?

PROBLEM

4.23 What alkynes would you start with to prepare the following ketones by a hydration reaction?

(a) $CH_3CH_2CH_2\overset{\overset{O}{\|}}{C}CH_3$ (b) $CH_3CH_2CH_2\overset{\overset{O}{\|}}{C}CH_2CH_3$

4.16 Alkyne Acidity: Formation of Acetylide Anions

The most striking difference between the chemistry of alkenes and alkynes is that terminal alkynes (R–C≡C–H) are weakly acidic, with $pK_a \approx 25$ (Section 1.12). Alkenes, by contrast, have $pK_a \approx 44$. When a terminal alkyne is treated with a strong base such as sodium amide, $NaNH_2$, the terminal hydrogen is removed, and an acetylide anion is formed:

$$R{-}C{\equiv}C{-}H + :\ddot{N}H_2\ Na^+ \longrightarrow \underset{\textbf{Acetylide anion}}{R{-}C{\equiv}C:^-\ Na^+} + :NH_3$$

The presence of an unshared electron pair on the negatively charged alkyne carbon makes acetylide anions both basic and nucleophilic. As a result, acetylide anions react with alkyl halides such as bromomethane to substitute for the halogen and yield a new alkyne product. We won't study the mechanism of this substitution reaction until Chapter 7, but will note for the present that it is a very useful method for preparing larger alkynes from simpler precursors. Terminal alkynes can be prepared by reaction of acetylene itself, and internal alkynes can be prepared by further reaction of a terminal alkyne:

$$\underset{\textbf{Acetylene}}{HC{\equiv}CH} \xrightarrow{NaNH_2} HC{\equiv}C^- \ Na^+ \xrightarrow{RCH_2Br} \underset{\textbf{A terminal alkyne}}{HC{\equiv}CCH_2R'}$$

$$\underset{\textbf{A terminal alkyne}}{RC{\equiv}CH} \xrightarrow{NaNH_2} RC{\equiv}C^- \ Na^+ \xrightarrow{R'CH_2Br} \underset{\textbf{An internal alkyne}}{RC{\equiv}CCH_2R'}$$

As an example, conversion of 1-hexyne into its anion, followed by reaction with 1-bromobutane, yields 5-decyne:

$$\underset{\textbf{1-Hexyne}}{CH_3CH_2CH_2CH_2C{\equiv}CH} \xrightarrow[2.\ CH_3CH_2CH_2CH_2Br]{1.\ NaNH_2,\ NH_3} \underset{\textbf{5-Decyne (76\%)}}{CH_3CH_2CH_2CH_2C{\equiv}CCH_2CH_2CH_2CH_3}$$

The one limitation to the reaction of an acetylide anion with an alkyl halide is that only primary alkyl halides, RCH_2X, can be used, for reasons that will be discussed in Chapter 7. If a secondary or tertiary alkyl halide is used instead, the basic acetylide anion causes dehydrohalogenation of the alkyl halide, giving an alkene (Section 4.9).

PRACTICE PROBLEM 4.10

What alkyne and what alkyl halide would you use to prepare 1-pentyne?

Solution Draw the structure of the target molecule, and identify the alkyl group(s) attached to the triple-bonded carbons. In the present case, one of the alkyne carbons has a propyl group attached to it, and the other has a hydrogen attached to it. Thus, 1-pentyne could be prepared by treatment of acetylene with $NaNH_2$ to yield sodium acetylide, followed by reaction with 1-bromopropane:

$$\underset{\textbf{Acetylene}}{H{-}C{\equiv}C{-}H} + :\ddot{N}H_2^- \ Na^+ \longrightarrow \underset{\textbf{Sodium acetylide}}{H{-}C{\equiv}C:^- \ Na^+} + :NH_3$$

$$H{-}C{\equiv}C:^- \ Na^+ + \underset{\textbf{1-Bromopropane}}{CH_3CH_2CH_2Br} \longrightarrow \underset{\textbf{1-Pentyne}}{H{-}C{\equiv}C{-}CH_2CH_2CH_3}$$

This propyl group comes from 1-bromopropane

PROBLEM

4.24 Show the alkyne and alkyl halide from which the following products can be obtained. Where two routes look feasible, list both.
(a) 5-Methyl-1-hexyne (b) 2-Hexyne (c) 4-Methyl-2-pentyne

INTERLUDE

Natural Rubber

Crude rubber is harvested from the rubber tree, *Hevea brasiliensis*.

Rubber—an unusual name for an unusual substance—is a naturally occurring alkene polymer produced by more than 400 different plants. The major source, however, is the so-called rubber tree, *Hevea brasiliensis,* from which the crude material is harvested as it drips from a slice made through the bark. The name *rubber* was coined by Joseph Priestley, the discoverer of oxygen and early researcher of rubber chemistry, for the simple reason that one of rubber's early uses was to rub out pencil marks on paper.

Unlike polyethylene and other simple alkene polymers, natural rubber is a polymer of a conjugated diene, *isoprene,* or 2-methyl-1,3-butadiene. The polymerization takes place by 1,4 addition (Section 4.10) of each isoprene monomer unit to the growing chain, leading to formation of a polymer that still contains double bonds spaced regularly at four-carbon intervals. As the following structure shows, these double bonds have Z stereochemistry:

Z geometry

Many isoprene units → **A segment of natural rubber**

Crude rubber, called latex, is collected from the tree as an aqueous dispersion that is washed, dried, and coagulated by warming in air to give a polymer with chains that average about 5000 monomer units in length and have molecular weights of 200,000–500,000. This crude coagulate is too soft and tacky to be useful until it is hardened by heating with elemental sulfur, a process called *vulcanization.* By mechanisms that are still not fully understood, vulcanization cross-links the rubber chains together by forming carbon–sulfur bonds between them, thereby hardening and stiffening the polymer. The exact degree of hardening can

(*continued*)➤

be varied, yielding material soft enough for automobile tires or hard enough for bowling balls (*ebonite*).

The remarkable ability of rubber to stretch and then contract to its original shape is due to the irregular shapes of the polymer chains caused by the double bonds. These double bonds introduce bends and kinks into the polymer chains, thereby preventing neighboring chains from nestling together into tightly packed, semicrystalline regions. When stretched, the randomly coiled chains straighten out and orient along the direction of the pull, but are kept from sliding over each other by the cross-links. When the stretch is released, the polymer reverts to its original random state.

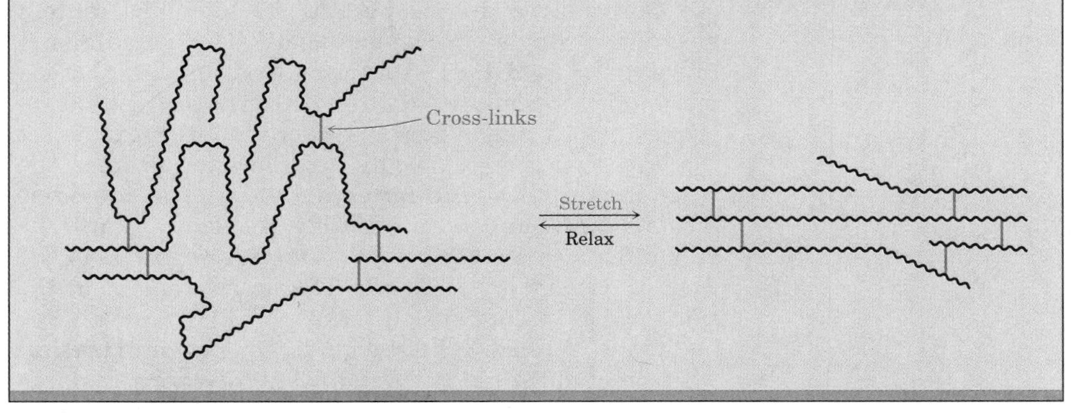

Summary and Key Words

1,4 addition, 128
alkyne, 134
allylic, 128
anti stereochemistry, 117
bromonium ion, 117
conjugated diene, 127
conjugation, 127
dehydration, 125
dehydrohalogenation, 125
diol, 120
hydration, 114
hydrogenation, 118
hydroxylation, 120
Markovnikov's rule, 110
monomer, 122
oxidation, 120
polymer, 122
reduction, 118

The chemistry of alkenes is dominated by addition reactions of electrophiles. When HX reacts with an alkene, **Markovnikov's rule** predicts that the hydrogen will add to the carbon that has fewer alkyl substituents and the X group will add to the carbon that has more alkyl substituents. For example,

$$(H_3C)_2C{=}CH_2 + HCl \longrightarrow H_3C{-}C(Cl)(CH_3){-}CH_3$$

Many other electrophiles besides HX add to alkenes. Thus, Br_2 and Cl_2 add to give 1,2-dihalide addition products having **anti stereochemistry.** Addition of H_2O (**hydration**) takes place on reaction of the alkene with aqueous acid, and H_2 can be added to alkenes (**hydrogenation**) by reaction in the presence of a metal catalyst such as platinum or palladium.

Oxidation of alkenes is carried out using potassium permanganate, $KMnO_4$. Under basic conditions, $KMnO_4$ reacts with alkenes to yield cis **1,2-diols.** Under neutral or acidic conditions, $KMnO_4$ cleaves double bonds to yield carbonyl-containing products.

Alkenes are prepared from alkyl halides and alcohols by elimination reactions. Treatment of an alkyl halide with a strong base effects **dehydrohalogenation,** and treatment of an alcohol with acid effects **dehydration.** These elimination reactions usually give a mixture of alkene products in which the more highly substituted alkene predominates (**Zaitsev's rule**).

Conjugated dienes, such as 1,3-butadiene, contain alternating single and double bonds. Conjugated dienes undergo **1,4 addition** of electrophiles through the formation of a resonance-stabilized **allylic** carbocation intermediate. No single line-bond representation can depict the true structure of an allylic carbocation. Rather, the true structure is a **resonance hybrid** somewhere intermediate between two contributing resonance forms. The only difference between two **resonance forms** is in the location of bonding electrons. The atoms remain in the same places in both structures.

Many simple alkenes undergo **polymerization** when treated with a radical catalyst. **Polymers** are large molecules built up by the repetitive bonding together of many small **monomer** units.

Alkynes are hydrocarbons that contain carbon–carbon triple bonds. Much of the chemistry of alkynes is similar to that of alkenes. For example, alkynes react with 1 equiv of HBr and HCl to yield **vinylic** halides, and with 1 equiv of Br_2 and Cl_2 to yield 1,2-dihalides. Alkynes can also be hydrated by reaction with aqueous sulfuric acid in the presence of mercuric sulfate catalyst. The reaction leads to an intermediate enol that immediately isomerizes to a ketone. Alkynes can also be hydrogenated with the Lindlar catalyst to yield a cis alkene. Terminal alkynes are weakly acidic and can be converted into **acetylide anions** by treatment with a strong base. Reaction of the acetylide anion with a primary alkyl halide then gives an internal alkyne.

Summary of Reactions

Note: No stereochemistry is implied unless specifically stated or indicated with wedged, solid, and dashed lines.

1. Reactions of alkenes

(a) Addition of HX, where X = Cl, Br, or I (Sections 4.1–4.2)

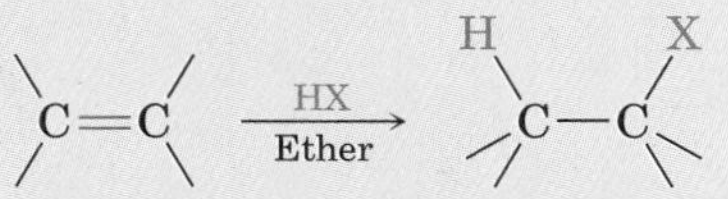

$$\text{C=C} \xrightarrow[\text{Ether}]{\text{HX}} \text{H–C–C–X}$$

Markovnikov's rule: H adds to the less highly substituted carbon, and X adds to the more highly substituted one.

(b) Addition of H_2O (Section 4.4)

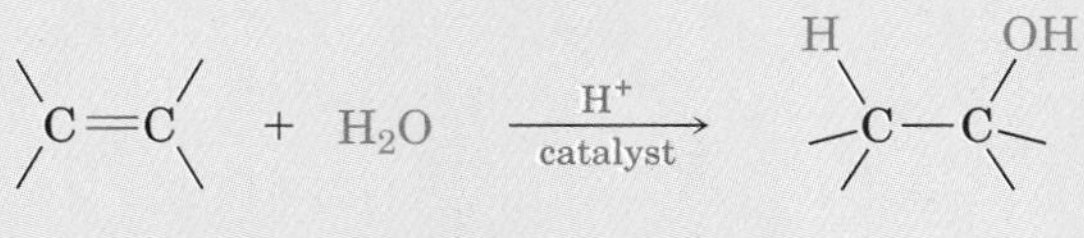

$$\text{C=C} + H_2O \xrightarrow[\text{catalyst}]{H^+} \text{H–C–C–OH}$$

Markovnikov's rule: H adds to the less highly substituted carbon, and OH adds to the more highly substituted one.

(c) Addition of X_2, where X = Cl or Br (Section 4.5)

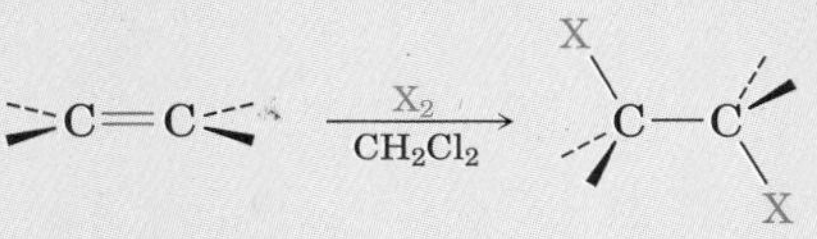

$$\text{C=C} \xrightarrow[CH_2Cl_2]{X_2} \text{X–C–C–X}$$

Anti addition

(d) Addition of H_2 (hydrogenation; Section 4.6)

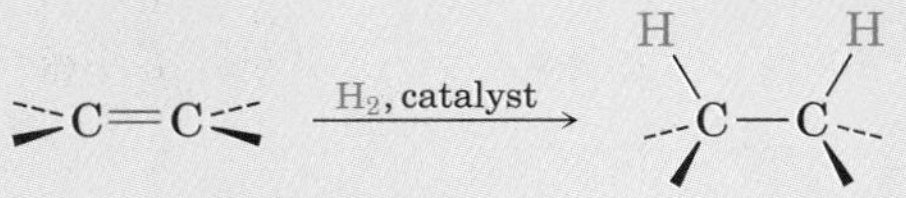

$$\text{C=C} \xrightarrow{H_2\text{, catalyst}} \text{H–C–C–H}$$

Syn addition

(e) Hydroxylation with $KMnO_4$ (Section 4.7)

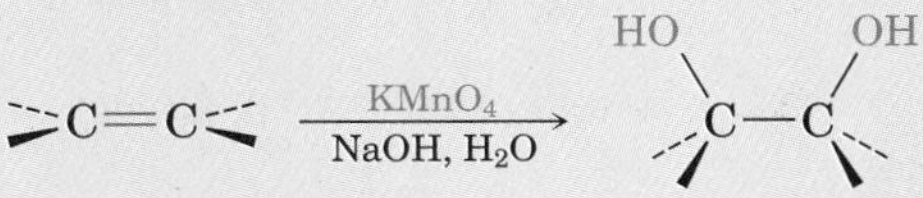

$$\text{C=C} \xrightarrow[\text{NaOH, } H_2O]{KMnO_4} \text{HO–C–C–OH}$$

Syn addition

(f) Oxidative cleavage of alkenes with acidic $KMnO_4$ (Section 4.7)

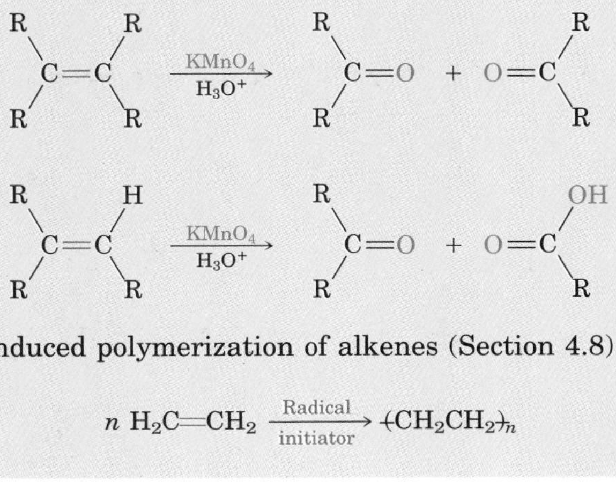

$$R_2C{=}CR_2 \xrightarrow[H_3O^+]{KMnO_4} R_2C{=}O + O{=}CR_2$$

$$R_2C{=}CHR \xrightarrow[H_3O^+]{KMnO_4} R_2C{=}O + O{=}C(OH)R$$

(g) Radical-induced polymerization of alkenes (Section 4.8)

$$n\ H_2C{=}CH_2 \xrightarrow[\text{initiator}]{\text{Radical}} \text{+}CH_2CH_2\text{+}_n$$

2. Synthesis of alkenes by elimination reactions
   (a) Dehydrohalogenation of alkyl halides (Section 4.9)

$$\text{H–C–C–X} \xrightarrow{\text{Base}} \text{C=C}$$

Zaitsev's rule: Major product formed is the alkene with the more highly substituted double bond.

   (b) Dehydration of alcohols (Section 4.9)

$$\text{H–C–C–OH} \xrightarrow[\text{H}_2\text{O}]{\text{H}_2\text{SO}_4} \text{C=C} + \text{H}_2\text{O}$$

Zaitsev's rule: Major product formed is the alkene with the more highly substituted double bond.

3. Reactions of alkynes
   (a) Addition of H_2 (hydrogenation; Section 4.14)

$$\text{R–C}\equiv\text{C–R}' \xrightarrow[\text{Lindlar catalyst}]{\text{H}_2} \text{(H)(R)C=C(H)(R}')$$

A cis alkene

Syn addition

   (b) Addition of HX, where X = Cl, Br, or I (Section 4.14)

$$\text{–C}\equiv\text{C–} + \text{HX} \longrightarrow \text{(H)C=C(X)}$$

Markovnikov's rule: H adds to the less highly substituted carbon, and X adds to the more highly substituted one.

   (c) Addition of X_2, where X = Cl or Br (Section 4.14)

$$\text{–C}\equiv\text{C–} + \text{X}_2 \longrightarrow \text{(X)C=C(X)}$$

Anti addition

   (d) Addition of H_2O to yield ketones (Section 4.15)

$$\text{–C}\equiv\text{C–} + \text{H}_2\text{O} \xrightarrow[\text{HgSO}_4]{\text{H}_2\text{SO}_4} \left[\text{(OH)C=C(H)}\right] \longrightarrow \text{(O=)C–C(H)–}$$

   (e) Acidity: conversion into acetylide anions (Section 4.16)

$$\text{R–C}\equiv\text{C–H} \xrightarrow{\text{NaNH}_2} \text{R–C}\equiv\text{C:}^-\ \text{Na}^+ + \text{NH}_3$$

   (f) Reaction of acetylide ions with alkyl halides (Section 4.16)

$$\text{R–C}\equiv\text{C:}^-\ \text{Na}^+ + \text{R}'\text{CH}_2\text{X} \longrightarrow \text{R–C}\equiv\text{C–CH}_2\text{R}' + \text{NaX}$$

ADDITIONAL PROBLEMS

4.25 Give IUPAC names for the following compounds:

(a) $CH_3CH{=}CH\overset{CH_3}{\overset{|}{C}}{=}CHCH_3$ (b) $CH_3CH{=}CH\overset{CH_2CH_2CH_3}{\overset{|}{C}H}CH_2C{\equiv}CH$

(c) $H_2C{=}C{=}\overset{CH_3}{\overset{|}{C}}CH_3$ (d) $HC{\equiv}CCH_2C{\equiv}C\overset{CH_3}{\overset{|}{C}H}CH_3$

4.26 Draw structures corresponding to the following IUPAC names:
(a) 3-Ethyl-1-heptyne (b) 3,5-Dimethyl-4-hexen-1-yne
(c) 1,5-Heptadiyne (d) 1-Methyl-1,3-cyclopentadiene

4.27 Draw three possible structures for each of the following formulas:
(a) C_6H_8 (b) C_6H_8O

4.28 Name the following alkynes according to IUPAC rules:
(a) $CH_3CH_2C{\equiv}CCH_2CH_2CH_3$ (b) $CH_3CH_2C{\equiv}CC(CH_3)_3$
(c) $CH_3C{\equiv}CCH_2C{\equiv}CCH_2CH_3$ (d) $H_2C{=}CHCH{=}CHC{\equiv}CH$

4.29 Draw structures corresponding to the following IUPAC names:
(a) 3-Heptyne (b) 3,3-Dimethyl-4-octyne
(c) 3,4-Dimethylcyclodecyne (d) 2,2,5,5-Tetramethyl-3-hexyne

4.30 Draw and name all the possible pentyne isomers, C_5H_8.

4.31 Draw and name the six possible diene isomers of formula C_5H_8. Which of the six are conjugated dienes?

4.32 The following two hydrocarbons have been isolated from plants in the sunflower family. Name them according to IUPAC rules.
(a) $CH_3CH{=}CHC{\equiv}CC{\equiv}CCH{=}CHCH{=}CHCH{=}CH_2$ (all trans)
(b) $CH_3C{\equiv}CC{\equiv}CC{\equiv}CC{\equiv}CC{\equiv}CCH{=}CH_2$

4.33 Predict the products of the following reactions. Indicate regioselectivity where relevant. (The aromatic ring is inert to all the indicated reagents.)

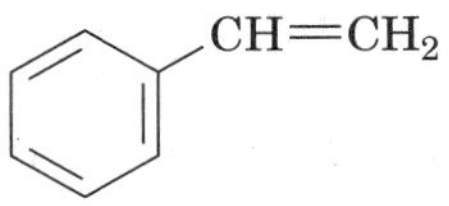

$C_6H_5CH{=}CH_2$ **Styrene**

(a) Styrene + H_2 $\xrightarrow{Pd}$? (b) Styrene + Br_2 $\longrightarrow$?

(c) Styrene + HBr $\longrightarrow$? (d) Styrene + $KMnO_4$ $\xrightarrow{NaOH,\ H_2O}$?

4.34 Suggest structures for alkenes that give the following reaction products. There may be more than one answer for some cases.

(a) ? $\xrightarrow{H_2/Pd\ catalyst}$ 2-Methylhexane

(b) ? $\xrightarrow{Br_2\ in\ CH_2Cl_2}$ 2,3-Dibromo-5-methylhexane

(c) ? $\xrightarrow{HBr}$ 2-Bromo-3-methylheptane

(d) ? $\xrightarrow[H_2O]{KMnO_4,OH^-}$ $CH_3\overset{CH_3}{\overset{|}{C}H}CH_2\overset{HO}{\overset{|}{C}H}\overset{OH}{\overset{|}{C}H}CH_2CH_3$

4.35 Using an oxidative cleavage reaction, explain how you would distinguish between these two isomeric cyclohexadienes:

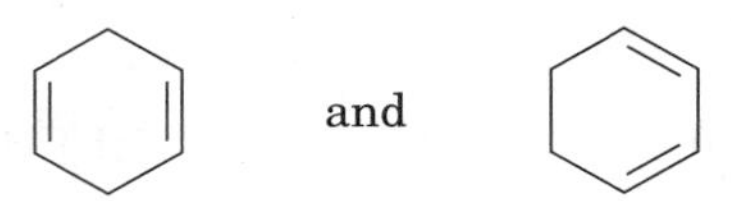

4.36 Formulate the reaction of cyclohexene with Br_2, showing the reaction intermediate and the final product with correct stereochemistry.

4.37 What products would you expect to obtain from reaction of 1,3-cyclohexadiene with each of the following?

(a) 1 mol Br_2 in CH_2Cl_2 (b) 1 mol HCl

(c) 1 mol DCl (D = deuterium, ^2H) (d) H_2 over a Pd catalyst

4.38 Draw the structure of a hydrocarbon that reacts with only 1 equiv of H_2 on catalytic hydrogenation and gives only pentanoic acid, $CH_3CH_2CH_2CH_2COOH$, on treatment with acidic $KMnO_4$. Write the reactions involved.

4.39 Give the structure of an alkene that yields the following keto acid on reaction with $KMnO_4$ in aqueous acid:

$$? \xrightarrow[H_3O^+]{KMnO_4} \text{HO}\overset{\text{O}}{\overset{\|}{\text{C}}}\text{CH}_2\text{CH}_2\text{CH}_2\text{CH}_2\overset{\text{O}}{\overset{\|}{\text{C}}}\text{CH}_3$$

4.40 What alkenes would you hydrate to obtain the following alcohols?

(a) $CH_3CH_2\overset{\text{OH}}{\overset{|}{\text{C}}}\text{HCH}_3$ (b) OH (c) OH CH3

4.41 What alkynes would you hydrate to obtain the following ketones?

(a) $CH_3\overset{\text{CH}_3}{\overset{|}{\text{C}}}\text{HCH}_2\overset{\text{O}}{\overset{\|}{\text{C}}}\text{CH}_3$ (b) O CH3

4.42 Draw the structure of a hydrocarbon that reacts with 2 equiv of H_2 on catalytic hydrogenation and gives only succinic acid on reaction with acidic $KMnO_4$.

$$\text{HO}\overset{\text{O}}{\overset{\|}{\text{C}}}\text{CH}_2\text{CH}_2\overset{\text{O}}{\overset{\|}{\text{C}}}\text{OH}$$ **Succinic acid**

4.43 Predict the products of the following reactions on 1-hexyne:

(a) $\xrightarrow{\text{1 equiv HBr}}$? (b) $\xrightarrow{\text{1 equiv Cl}_2}$? (c) $\xrightarrow{\text{H}_2\text{, Lindlar catalyst}}$?

4.44 Predict the products of the following reactions on 5-decyne:

(a) $\xrightarrow{\text{H}_2\text{, Lindlar catalyst}}$? (b) $\xrightarrow{\text{2 equiv Br}_2}$? (c) $\xrightarrow{\text{H}_2\text{O, H}_2\text{SO}_4\text{, HgSO}_4}$?

4.45 In planning the synthesis of a compound, it's as important to know what *not* to do as to know what to do. What is wrong with each of the following reactions?

(a)
$$\underset{\displaystyle CH_3C{=}CHCH_3}{\overset{\displaystyle CH_3}{|}} \xrightarrow{HBr} \overset{\displaystyle CH_3}{|}\,CH_3CHCHCH_3\,\underset{\displaystyle Br}{|}$$

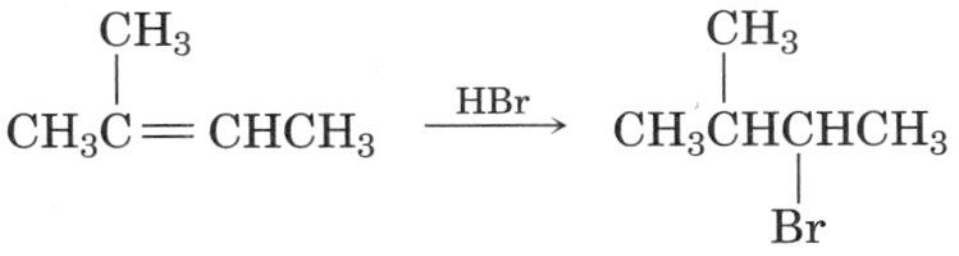

(b)
$$\text{(cyclohexene)} \xrightarrow[H_2O,\ OH^-]{KMnO_4} \text{(cyclohexane with H, OH on one carbon and OH, H on the adjacent carbon)}$$

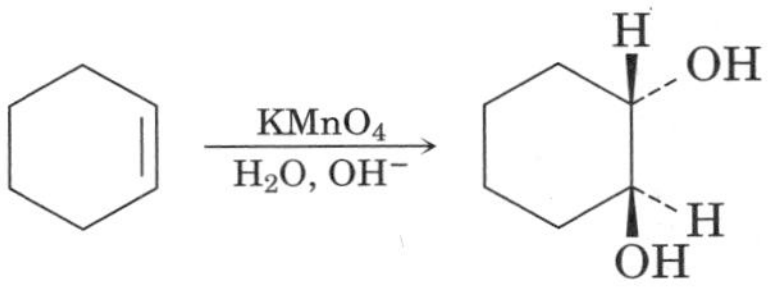

(c)
$$CH_3CH_2\overset{\displaystyle CH_3}{\overset{|}{C}}HCH_2C{\equiv}CH \xrightarrow[HgSO_4]{H_2O,\ H_2SO_4} CH_3CH_2\overset{\displaystyle CH_3}{\overset{|}{C}}HCH_2CH_2\overset{\displaystyle O}{\overset{\|}{C}}H$$

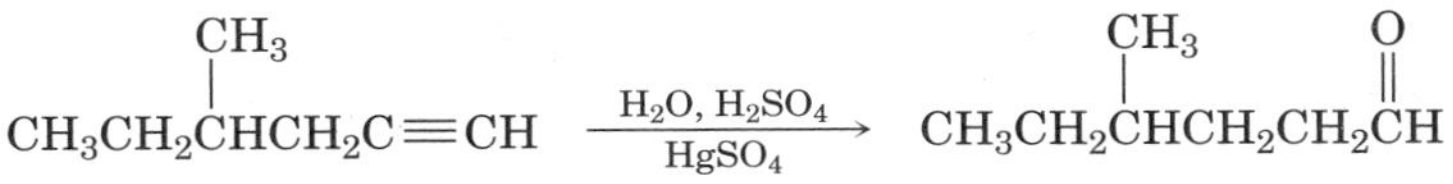

4.46 Acrylonitrile, $H_2C{=}CHC{\equiv}N$, contains a carbon–carbon double bond and a carbon–nitrogen triple bond. Sketch the orbitals involved in the multiple bonding in acrylonitrile, and indicate the hybridization of the carbons. Is acrylonitrile conjugated?

4.47 How would you prepare *cis*-2-butene starting from 1-propyne, an alkyl halide, and any other reagents needed? (This problem can't be worked in a single step. You'll have to carry out more than one reaction.)

4.48 Using 1-butyne as the only organic starting material, along with any inorganic reagents needed, how would you synthesize the following compounds? (More than one step may be needed.)
(a) Butane
(b) 1,1,2,2-Tetrachlorobutane
(c) 2-Bromobutane
(d) 2-Butanone ($CH_3CH_2COCH_3$)

4.49 Give the structure of an alkene that provides only acetone, $(CH_3)_2C{=}O$, on reaction with acidic $KMnO_4$.

4.50 Compound A has the formula C_8H_8. It reacts rapidly with acidic $KMnO_4$ but reacts with only 1 equiv of H_2 over a palladium catalyst. On hydrogenation under conditions that reduce aromatic rings, A reacts with 4 equiv of H_2, and hydrocarbon B, C_8H_{16}, is produced. The reaction of A with $KMnO_4$ gives CO_2 and a carboxylic acid C, $C_7H_6O_2$. What are the structures of A, B, and C? Write all the reactions.

4.51 The sex attractant of the common housefly is a hydrocarbon named *muscalure*, $C_{23}H_{46}$. On treatment of muscalure with aqueous acidic $KMnO_4$, two products are obtained, $CH_3(CH_2)_{12}COOH$ and $CH_3(CH_2)_7COOH$. Propose a structure for muscalure.

4.52 How would you synthesize muscalure (Problem 4.51) starting from acetylene and any alkyl halides needed? (The double bond in muscalure is cis.)

4.53 Draw a reaction energy diagram for the addition of HBr to 1-pentene. Let one curve on your diagram show the formation of 1-bromopentane product and another curve on the same diagram show the formation of 2-bromopentane product. Label the positions for all reactants, intermediates, and products.

4.54 Make sketches of what you imagine the transition-state structures to look like in the reaction of HBr with 1-pentene (Problem 4.53).

4.55 Methylenecyclohexane, on treatment with strong acid, isomerizes to yield 1-methylcyclohexene. Propose a mechanism by which the reaction might occur.

$$\text{(cyclohexane)}{=}CH_2 \xrightarrow{H^+} \text{(cyclohexene)}{-}CH_3$$

Methylenecyclohexane **1-Methylcyclohexene**

4.56 α-Terpinene, $C_{10}H_{16}$, is a pleasant-smelling hydrocarbon that has been isolated from oil of marjoram. On hydrogenation over a palladium catalyst, α-terpinene reacts with 2 mol equiv of hydrogen to yield a new hydrocarbon, $C_{10}H_{20}$. On reaction with acidic $KMnO_4$, α-terpinene yields oxalic acid and 6-methyl-2,5-heptanedione. Propose a structure for α-terpinene.

$$HO(O{=})C{-}C({=}O)OH \qquad CH_3\overset{O}{\overset{\|}{C}}CH_2CH_2\overset{O}{\overset{\|}{C}}\underset{CH_3}{\underset{|}{C}H}CH_3$$

Oxalic acid **6-Methyl-2,5-heptanedione**

4.57 Explain the observation that hydroxylation of *cis*-2-butene with basic $KMnO_4$ yields a different product than hydroxylation of *trans*-2-butene. First draw the structure and show the stereochemistry of each product, and then make molecular models. We'll explore the stereochemistry of the products in more detail in Chapter 6.

Visualizing Chemistry

4.58 Name the following alkenes, and predict the products of their reaction with (i) $KMnO_4$ in aqueous acid and (ii) $KMnO_4$ in aqueous NaOH.

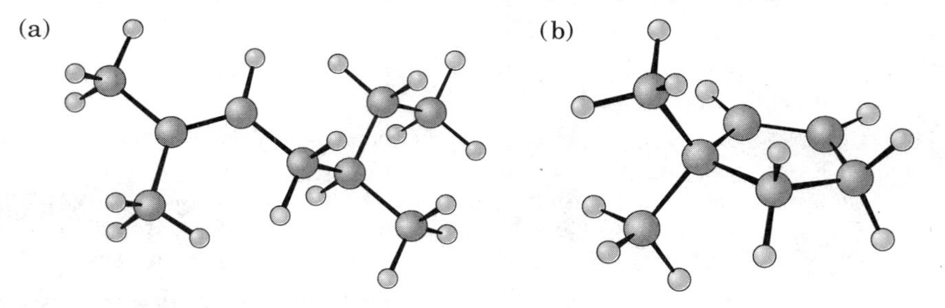

4.59 Name the following alkynes, and predict the products of their reaction with (i) H_2 in the presence of a Lindlar catalyst and (ii) H_3O^+ in the presence of $HgSO_4$.

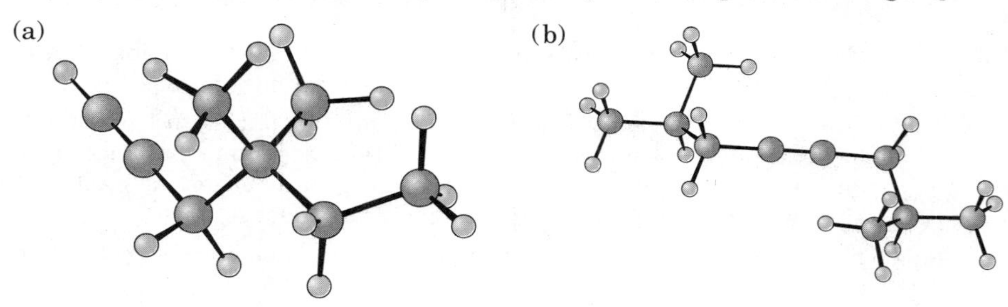

4.60 The following drawing of 4-methyl-1,3-pentadiene represents a high-energy conformation rather than a low-energy conformation. Explain.

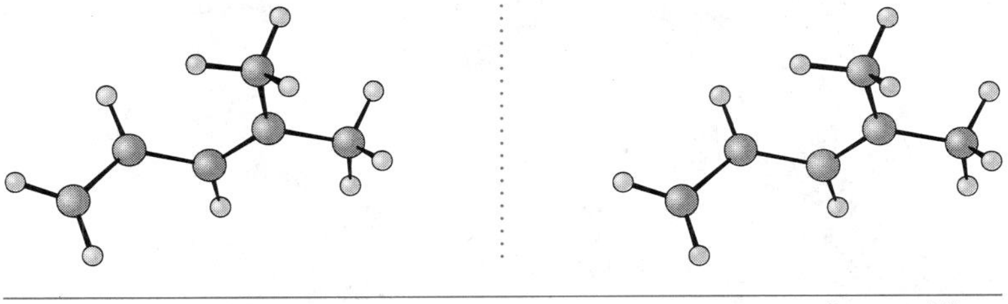

Stereo View

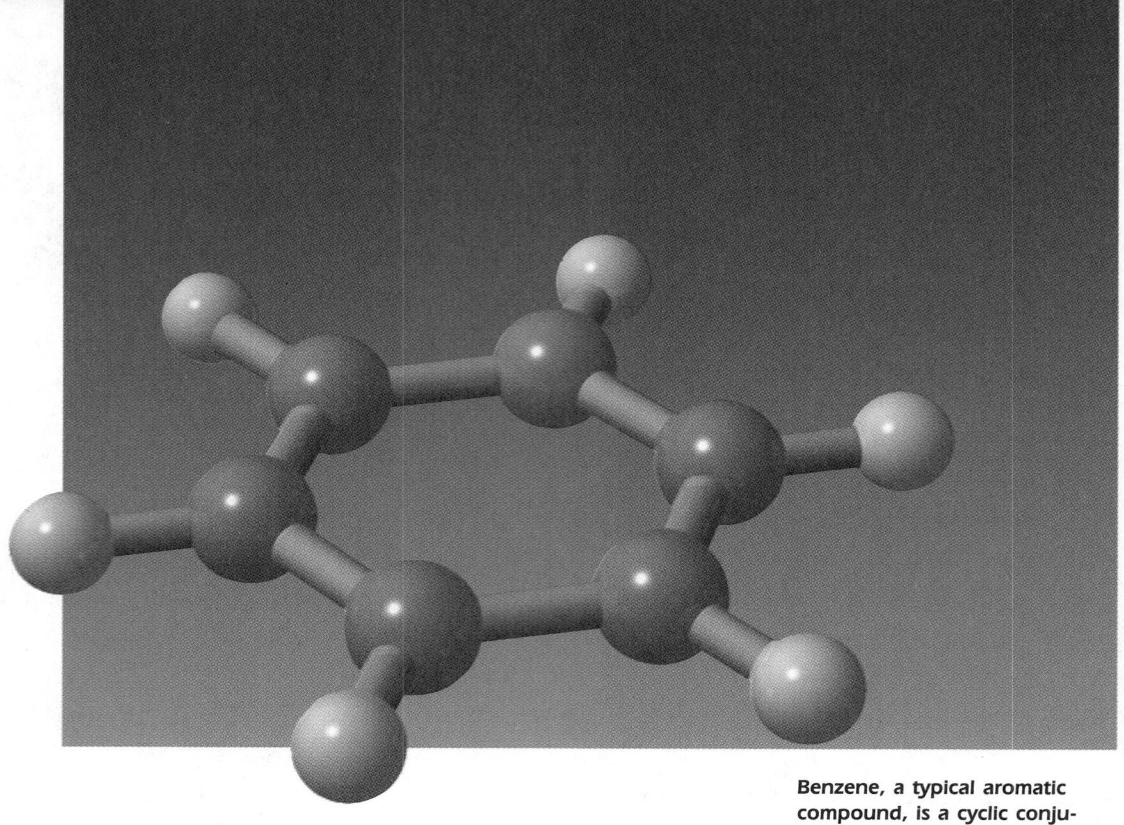

Benzene, a typical aromatic compound, is a cyclic conjugated molecule.

5 Aromatic Compounds

In the early days of organic chemistry, the word *aromatic* was used to describe fragrant substances such as benzaldehyde (from cherries, peaches, and almonds), toluene (from tolu balsam), and benzene (from coal distillate). It was soon realized, however, that the substances grouped as aromatic differ from other compounds in their chemical behavior.

Today, we use the word **aromatic** to refer to the class of compounds that contain benzene-like, six-membered rings with three double bonds. Many important compounds are aromatic in part, including the steroidal hormone estrone and the analgesic ibuprofen. Benzene itself causes a depressed white blood cell count (leukopenia) on prolonged exposure and should not be used as a laboratory solvent. We'll see in this chapter how aromatic substances behave and why they're different from the alkanes, alkenes, and alkynes we've studied up to this point.

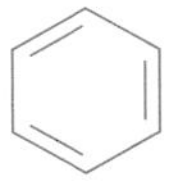

Benzene

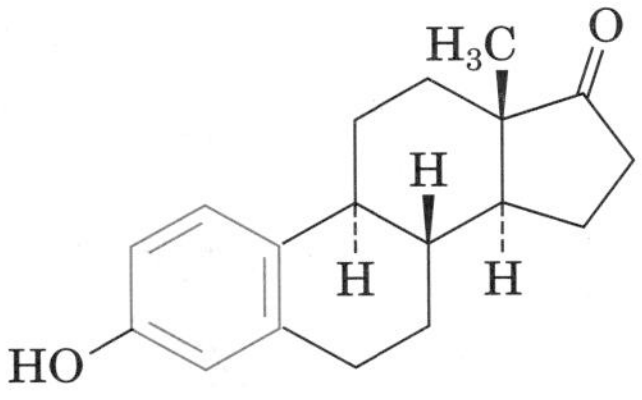

Estrone

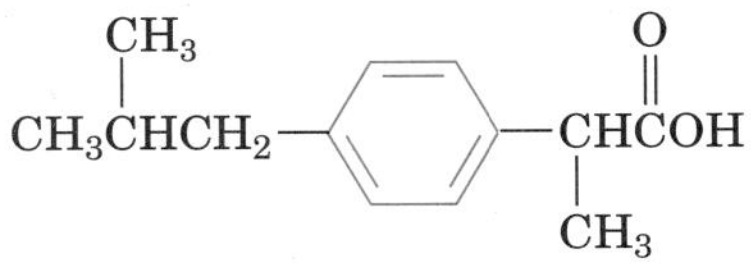

Ibuprofen

5.1 Structure of Benzene: The Kekulé Proposal

By the mid-1800s, benzene was known to have the molecular formula C_6H_6, and its chemistry was being actively explored. It was known that, although benzene is relatively unreactive toward most reagents that attack alkenes, it reacts with Br_2 in the presence of iron to give the *substitution* product C_6H_5Br rather than the *addition* product $C_6H_6Br_2$. Furthermore, only one monobromo substitution product was known; no isomers had been prepared.

$$C_6H_6 + Br_2 \xrightarrow{Fe} C_6H_5Br + HBr \qquad [C_6H_6Br_2]$$

Benzene — **Bromobenzene** (substitution product) — (Addition product —*NOT formed*)

On the basis of these and other results, August Kekulé proposed in 1865 that benzene contains a ring of carbon atoms and can be formulated as 1,3,5-cyclohexatriene. Kekulé reasoned that this structure would readily account for the isolation of only a single monobromo substitution product, because all six carbon atoms and all six hydrogens in 1,3,5-cyclohexatriene are equivalent.

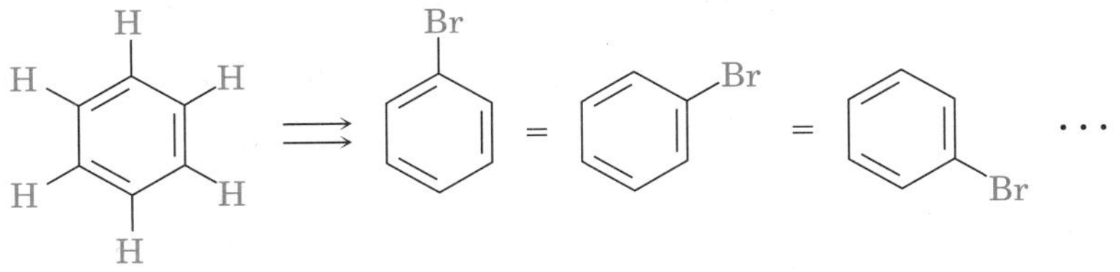

All six hydrogens are equivalent

Only one possible monobromo substitution product

Kekulé's proposal was widely criticized at the time. Although it satisfactorily accounts for the correct number of monosubstituted benzene isomers, it fails to answer two critical questions: Why is benzene unreactive compared with other alkenes, and why does benzene give a substitution product rather than an addition product on reaction with Br_2?

PROBLEM

5.1 How many dibromobenzene derivatives, $C_6H_4Br_2$, are possible according to Kekulé's theory? Draw them.

5.2 Stability of Benzene

The unusual stability of benzene was a great puzzle to early chemists. Although its formula, C_6H_6, indicates that unsaturation must be present, benzene shows none of the behavior characteristic of alkenes. For example, alkenes readily react with $KMnO_4$ to give 1,2-diols; they react with aqueous acid to give alcohols; and they react with HCl to give chloroalkanes. Benzene does none of these things. *Benzene does not undergo electrophilic addition reactions.*

Cyclohexene $\xrightarrow{KMnO_4,\ H_2O}$ (cyclohexane-1,2-diol: OH, OH)

Cyclohexene $\xrightarrow{H_3O^+}$ (cyclohexanol: OH)

Cyclohexene $\xrightarrow{HCl,\ Ether}$ (chlorocyclohexane: Cl)

Benzene $\xrightarrow{KMnO_4}$ No reaction

Benzene $\xrightarrow{H_3O^+}$ No reaction

Benzene $\xrightarrow{HCl,\ Ether}$ No reaction

Further evidence for the unusual nature of benzene is that all carbon–carbon bonds in benzene have the same length, intermediate between a typical single bond and a typical double bond. Most C–C single bonds have lengths near 1.54 Å, and most C=C double bonds are about 1.34 Å long, but the carbon–carbon bonds in benzene are 1.39 Å long (Figure 5.1).

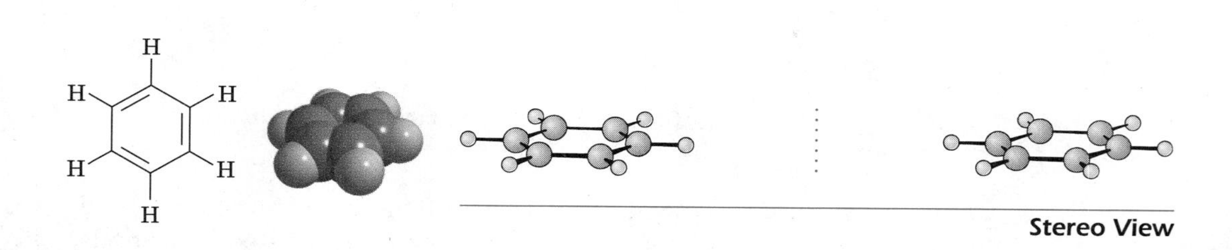

Figure 5.1 The structure of benzene. All six carbon–carbon bonds are 1.39 Å in length.

5.3 Structure of Benzene: The Resonance Proposal

How can we account for benzene's properties, and how can we best represent its structure? To answer these questions, we need to look again at resonance theory. We saw in Sections 4.11 and 4.12 that an allylic carbocation is best described as a resonance hybrid of two contributing forms. Neither resonance form is correct by itself; the true structure of an allylic carbocation is intermediate between the two forms:

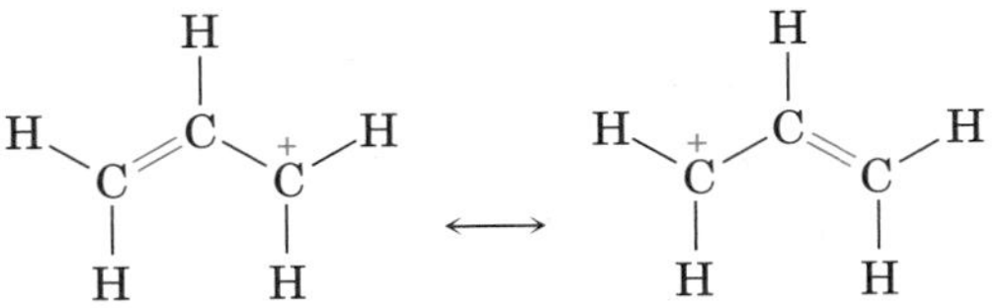

In the same way, resonance theory says that benzene can't be described satisfactorily by a single line-bond structure but is instead a resonance hybrid of two forms. Benzene doesn't oscillate back and forth between two forms; its true structure is somewhere between the two. Each carbon–carbon connection is an average of 1.5 bonds, midway between a single bond and a double bond.

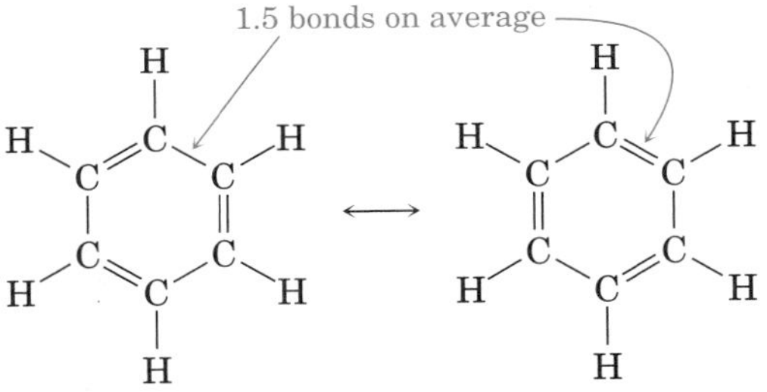

1.5 bonds on average

An orbital view of benzene shows the situation more clearly, emphasizing the cyclic conjugation of the benzene molecule and the equivalence of the six carbon–carbon bonds. Benzene is a flat, symmetrical molecule with the shape of a regular hexagon. All C–C–C bond angles are 120°, each carbon atom is sp^2-hybridized, and each carbon has a p orbital perpendicular to the plane of the six-membered ring. Since all six p orbitals are equivalent, it's impossible to define three localized alkene π bonds in which a given p orbital overlaps only one neighboring p orbital. Rather, each p orbital overlaps equally well with *both* neighboring p orbitals, leading to a structure for benzene in which the π electrons are delocalized around the ring in two doughnut-shaped clouds (Figure 5.2, p. 154).

We can now see why benzene is unusually stable. According to resonance theory, the more resonance forms a substance has, the more stable it is. Benzene, with two resonance forms of equal energy, is more stable than a normal alkene.

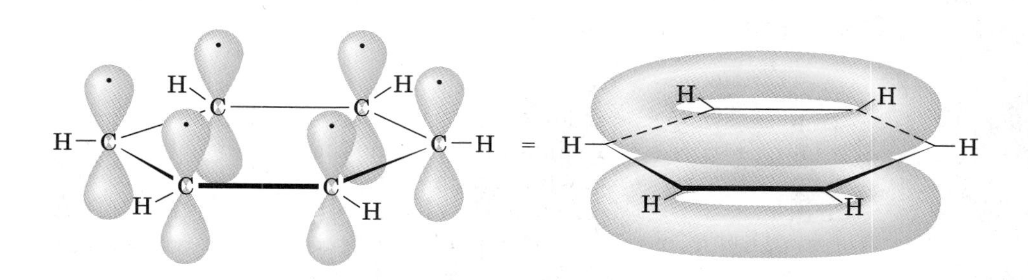

Figure 5.2 An orbital picture of benzene. Each of the six carbon atoms has a *p* orbital that can overlap equally well with neighboring *p* orbitals on both sides. The π electrons are thus delocalized around the ring in two doughnut-shaped clouds.

PROBLEM

5.2 How does resonance theory account for the fact that there is only one 1,2-dibromobenzene rather than the two isomers that Kekulé's theory would suggest?

5.4 Naming Aromatic Compounds

Aromatic substances, more than any other class of organic compounds, have acquired a large number of common names. Although the use of such names is discouraged, IUPAC rules allow for those shown in Table 5.1 to be retained. Thus, methylbenzene is commonly known as toluene, hydroxybenzene as phenol, aminobenzene as aniline, and so on.

Monosubstituted benzenes are systematically named in the same manner as other hydrocarbons, with *-benzene* as the parent name. Thus, C_6H_5Br is bromobenzene, and $C_6H_5CH_2CH_3$ is ethylbenzene. The name **phenyl** (**fen**-nil) is used for the $-C_6H_5$ unit when the benzene ring is considered as a substituent, and the name **benzyl** is used for the $C_6H_5CH_2-$ group.

Br | CH_2CH_3 | CH_2

Bromo**benzene** Ethyl**benzene** **A phenyl group** **A benzyl group**

Disubstituted benzenes are named using one of the prefixes *ortho-* (*o*), *meta-* (*m*), or *para-* (*p*). An ortho-disubstituted benzene has its two substituents in a 1,2 relationship on the ring; a meta-disubstituted benzene has its two substituents in a 1,3 relationship; and a para-disubstituted benzene has its substituents in a 1,4 relationship:

Table 5.1 Common Names of Some Aromatic Compounds

Formula	*Name*	*Formula*	*Name*
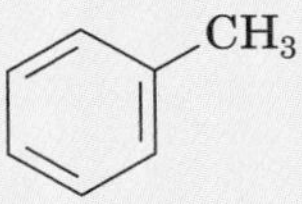 CH_3	Toluene (bp 111°C)	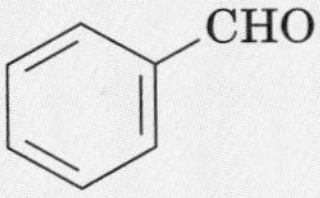 CHO	Benzaldehyde (bp 178°C)
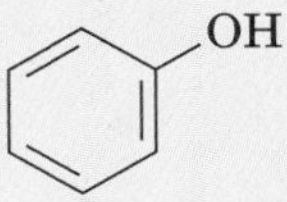 OH	Phenol (mp 43°C)	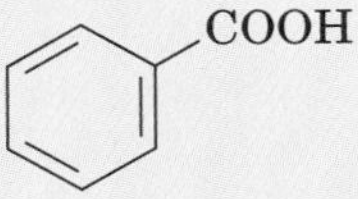 COOH	Benzoic acid (mp 122°C)
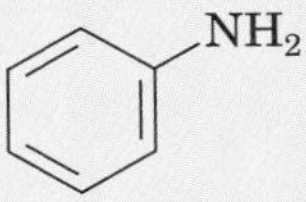 NH_2	Aniline (bp 184°C)	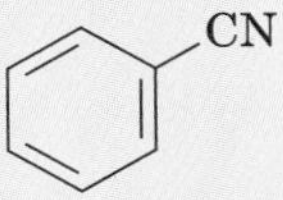 CN	Benzonitrile (bp 191°C)
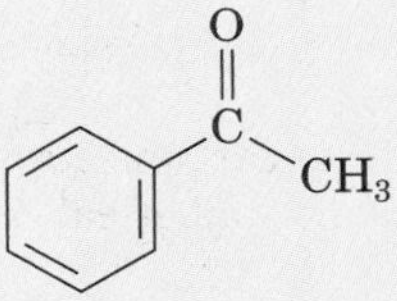 O, C, CH_3	Acetophenone (mp 21°C)	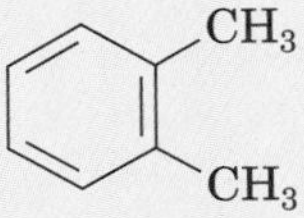 CH_3, CH_3	*ortho*-Xylene (bp 144°C)

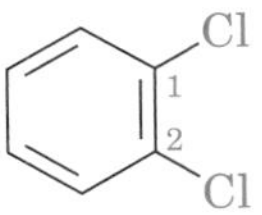

ortho-**Dichlorobenzene**
1,2 disubstituted

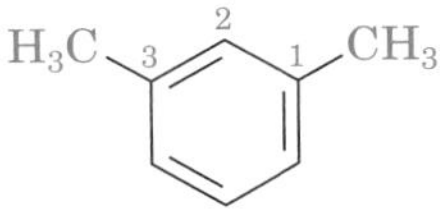

meta-**Xylene**
1,3 disubstituted

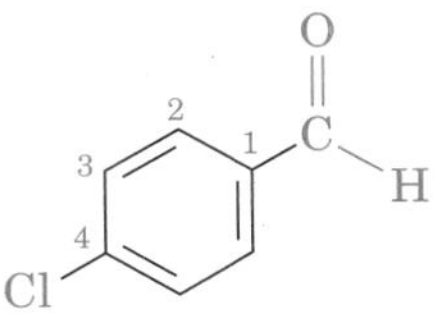

para-**Chlorobenzaldehyde**
1,4 disubstituted

Benzenes with more than two substituents are named by numbering the position of each substituent on the ring so that the lowest possible numbers are used. The substituents are listed alphabetically when writing the name.

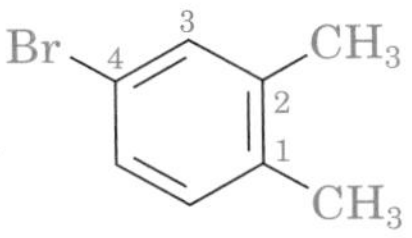

4-Bromo-1,2-dimethyl**benzene**

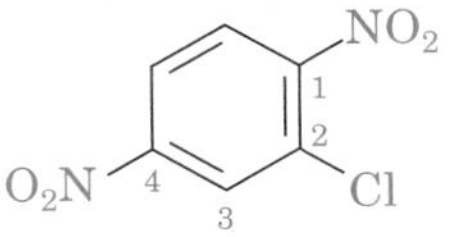

2-Chloro-1,4-dinitro**benzene**

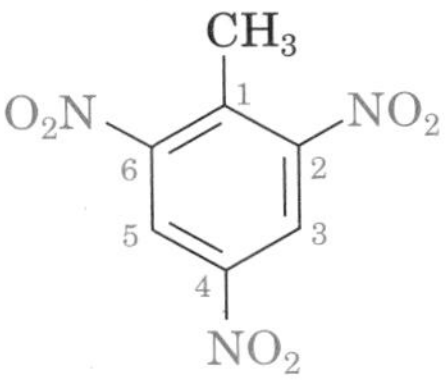

2,4,6-Trinitro**toluene (TNT)**

In the third example shown, note that *-toluene* is used as the parent name rather than *-benzene*. Any of the monosubstituted aromatic compounds shown in Table 5.1 can serve as a parent name, with the principal substituent (–CH_3 in toluene, for example) assumed to be on carbon 1. The following two examples further illustrate this practice:

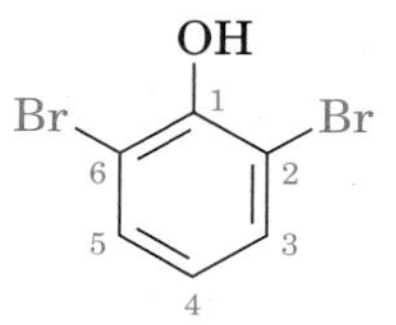

OH
Br 1 Br Cl O
6 2 C OH
5 3
4

2,6-Dibromo***phenol*** 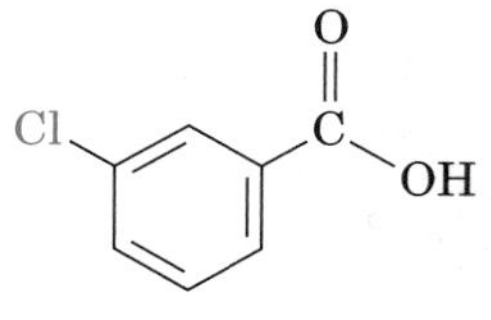 *m*-Chloro***benzoic acid***

PRACTICE PROBLEM 5.1

What is the IUPAC name of the following compound?

Cl NO_2

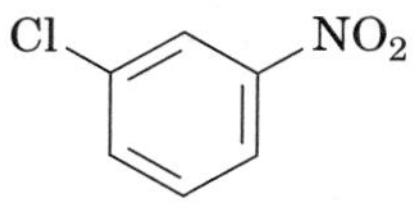

Solution Because the nitro group (–NO_2) and chloro group are on carbons 1 and 3, they have a meta relationship. Citing the two substituents in alphabetical order gives the IUPAC name *m*-chloronitrobenzene.

PROBLEM

5.3 Tell whether the following compounds are ortho, meta, or para disubstituted:

(a) 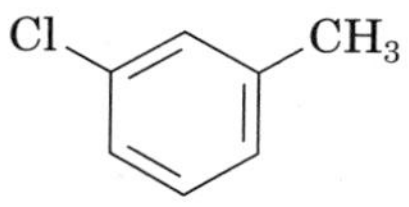 Cl CH_3 (b) 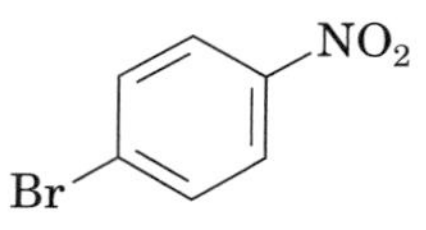 NO_2 Br (c) 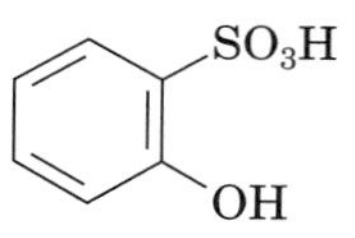 SO_3H OH

PROBLEM

5.4 Give IUPAC names for the following compounds:

(a) 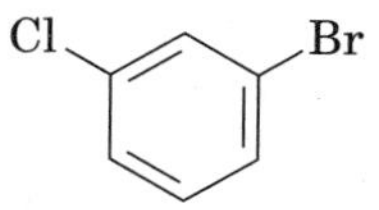 Cl Br (b) 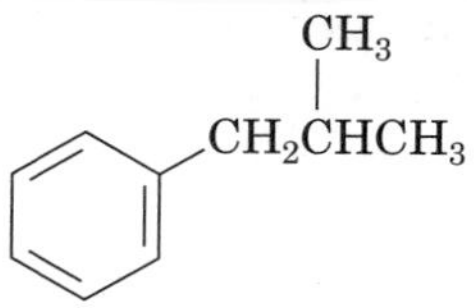 CH_3 CH_2CHCH_3 (c) 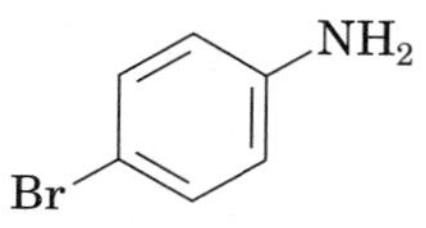 NH_2 Br

PROBLEM

5.5 Draw structures corresponding to the following IUPAC names:

(a) *p*-Bromochlorobenzene (b) *p*-Bromotoluene
(c) *m*-Chloroaniline (d) 1-Chloro-3,5-dimethylbenzene

5.5 Chemistry of Benzene: Electrophilic Aromatic Substitution Reactions

The most important reaction of aromatic compounds is **electrophilic aromatic substitution.** That is, an electron-poor reagent (an electrophile, E^+) reacts with an aromatic ring and substitutes for one of the ring hydrogens:

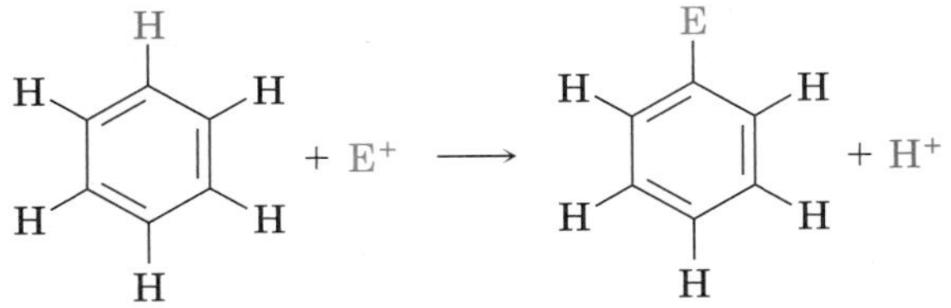

Many different substituents can be introduced onto the aromatic ring by electrophilic substitution reactions. By choosing the proper reagents, it's possible to *halogenate* the aromatic ring (substitute a halogen: –F, –Cl, –Br, or –I), *nitrate* it (substitute a nitro group: $-NO_2$), *sulfonate* it (substitute a sulfonic acid group: $-SO_3H$), *alkylate* it (substitute an alkyl group: –R), or *acylate* it (substitute an acyl group: –COR). Starting with only a few simple materials, we can prepare many thousands of substituted aromatic compounds (Figure 5.3).

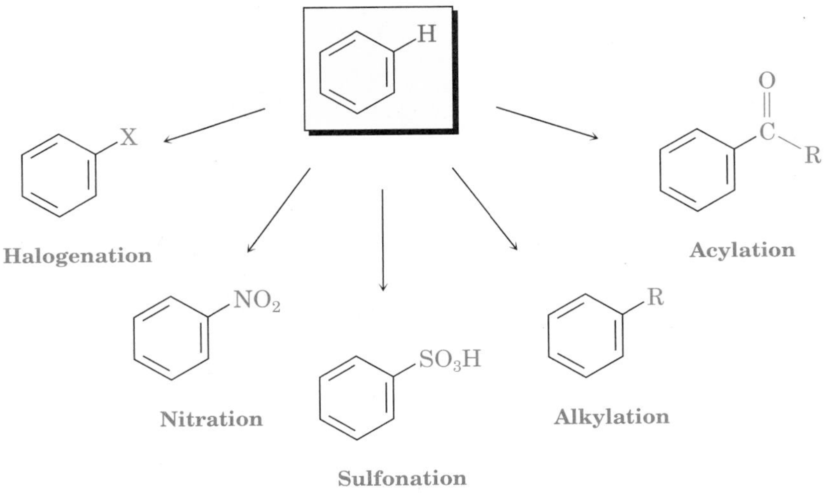

Figure 5.3 Some electrophilic aromatic substitution reactions.

All these reactions—and many more as well—take place by a similar mechanism. Let's begin a study of this fundamental reaction type by looking at one reaction in detail, the bromination of benzene.

PROBLEM

5.6 There are three products that might form on bromination of toluene. Draw and name them.

5.6 Bromination of Benzene

Benzene reacts with Br_2 in the presence of $FeBr_3$ as catalyst to yield the substitution product bromobenzene:

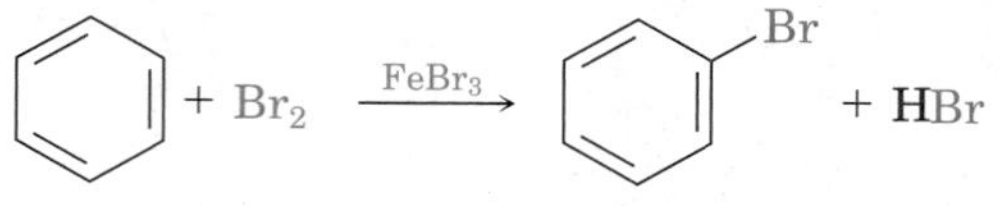

$$\text{Benzene} + Br_2 \xrightarrow{FeBr_3} \text{Bromobenzene (80\%)} + HBr$$

Benzene **Bromobenzene (80%)**

Before seeing how this electrophilic *substitution* reaction occurs, let's briefly recall what was said in Sections 3.7–3.11 about electrophilic *addition* reactions of alkenes. When a reagent such as HCl adds to an alkene, the electrophilic H^+ approaches the *p* orbitals of the double bond and forms a bond to one carbon, leaving a positive charge on the other carbon. The carbocation intermediate is then attacked by the nucleophile Cl^- ion to yield the addition product (Figure 5.4).

H—Cl

Alkene Carbocation intermediate Addition product

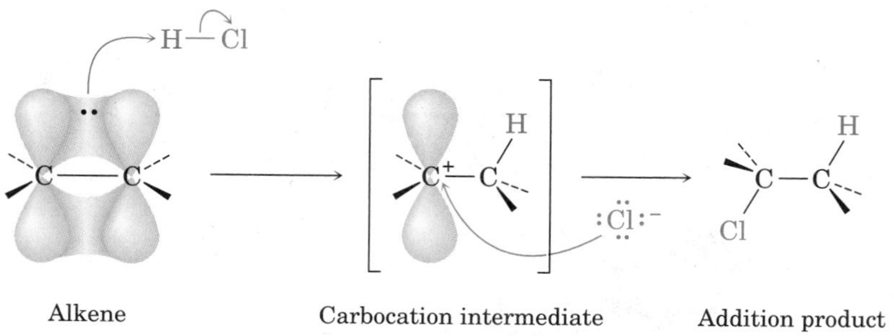

Figure 5.4 The mechanism of an electrophilic addition reaction of an alkene.

An electrophilic aromatic *substitution* reaction begins in a similar way, but there are a number of differences. One difference is that aromatic rings are less reactive toward electrophiles than alkenes. For example, Br_2 in CH_2Cl_2 solution reacts instantly with most alkenes but does not react with benzene. For bromination of benzene to take place, a catalyst such as $FeBr_3$ is needed. The catalyst makes the Br_2 molecule more electrophilic by reacting with it to give $FeBr_4^-$ and Br^+.

$$FeBr_3 + Br_2 \longrightarrow FeBr_4^- + Br^+$$

The electrophilic Br^+ then reacts with the electron-rich (nucleophilic) benzene ring to yield a nonaromatic carbocation intermediate. This carbocation is allylic (Section 4.11) and is a hybrid of three resonance forms:

Although more stable than a typical nonallylic carbocation, the intermediate in electrophilic aromatic substitution is nevertheless much less stable than the starting aromatic reactant. Thus, reaction of an electrophile with a benzene ring is endothermic, has a relatively high activation energy, and is rather slow. Figure 5.5 gives reaction energy diagrams that compare the reaction of an electrophile E^+ with an alkene and with benzene. The benzene reaction is slower (has a higher E_{act}) because the starting material is so stable.

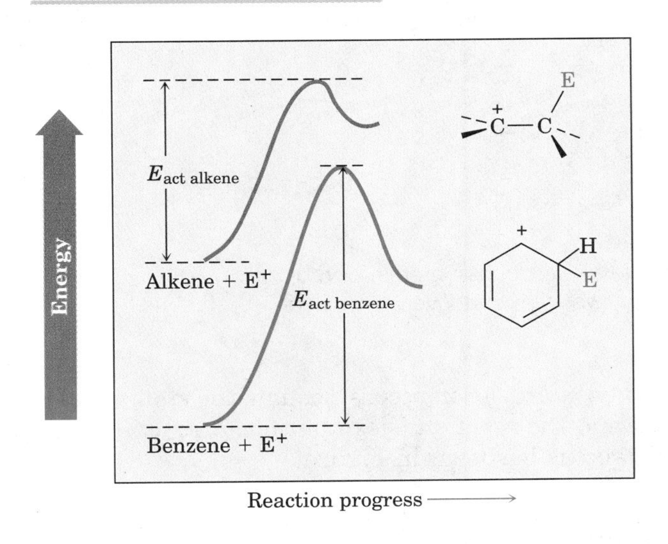

Figure 5.5 A comparison of the reactions of an electrophile (E^+) with an alkene and with benzene: E_{act} (alkene) $< E_{act}$ (benzene).

A second difference between alkene addition reactions and aromatic substitution reactions occurs after the electrophile has added to the benzene ring to give the carbocation intermediate. Instead of Br^- adding to the carbocation intermediate to yield an addition product, a base removes H^+ from the bromine-bearing carbon to yield the neutral aromatic substitution product. The net effect is the substitution of Br^+ for H^+ by the overall mechanism shown in Figure 5.6 (p. 160).

Why does the reaction of Br_2 with benzene take a different course than its reaction with an alkene? The answer is simple: If *addition* occurred, the stability of the aromatic ring would be lost and the overall reaction would be

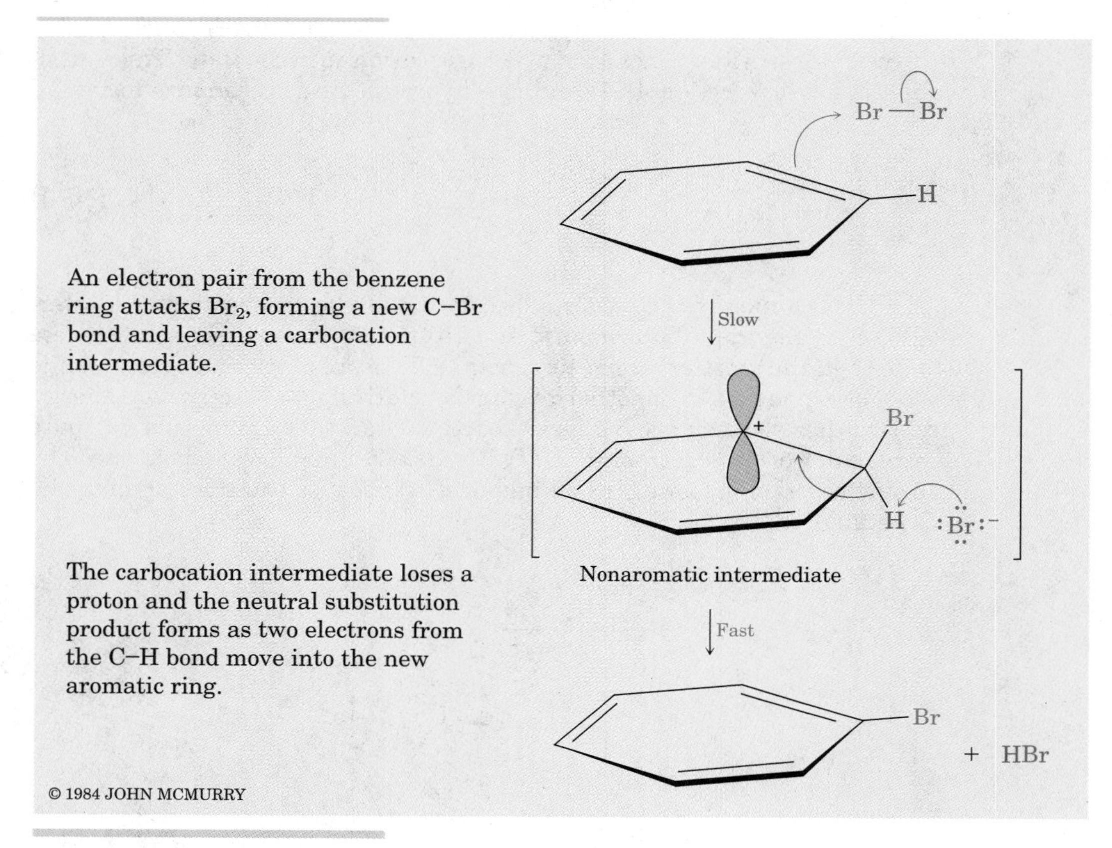

Br — Br
H
An electron pair from the benzene ring attacks Br_2, forming a new C–Br bond and leaving a carbocation intermediate.
Slow
+
Br
H
:Br:⁻
The carbocation intermediate loses a proton and the neutral substitution product forms as two electrons from the C–H bond move into the new aromatic ring.
Nonaromatic intermediate
Fast
Br
+ HBr
© 1984 JOHN MCMURRY

Figure 5.6 The mechanism of the electrophilic bromination of benzene. The reaction occurs in two steps and involves a carbocation intermediate.

endothermic. When *substitution* occurs, though, the stability of the aromatic ring is retained and the reaction is exothermic. A reaction energy diagram for the overall process is shown in Figure 5.7.

5.7 Other Electrophilic Aromatic Substitution Reactions

Many electrophilic aromatic substitutions besides bromination occur by the same general mechanism. Let's look briefly at some of these other reactions.

Chlorination

Aromatic rings react with Cl_2 in the presence of $FeCl_3$ catalyst to yield chlorobenzenes. This kind of reaction is used in the synthesis of numerous pharmaceutical agents, including the tranquilizer diazepam (Valium).

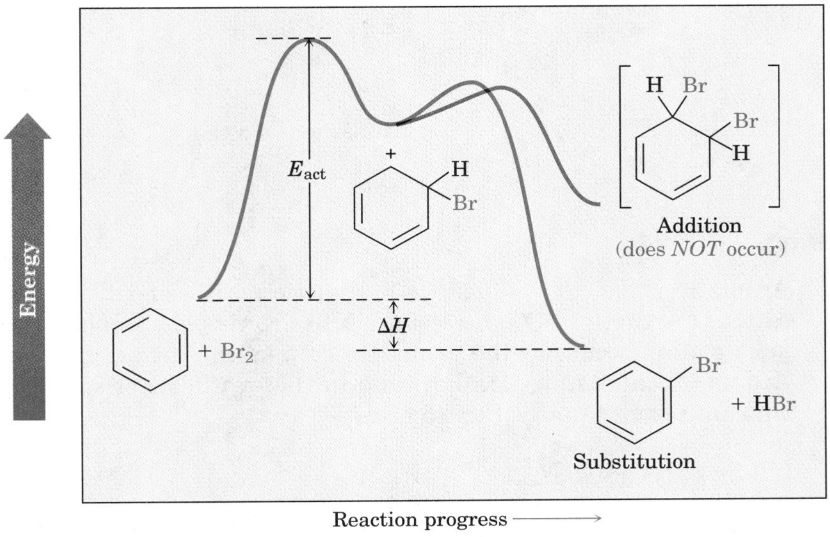

Figure 5.7 A reaction energy diagram for the electrophilic bromination of benzene.

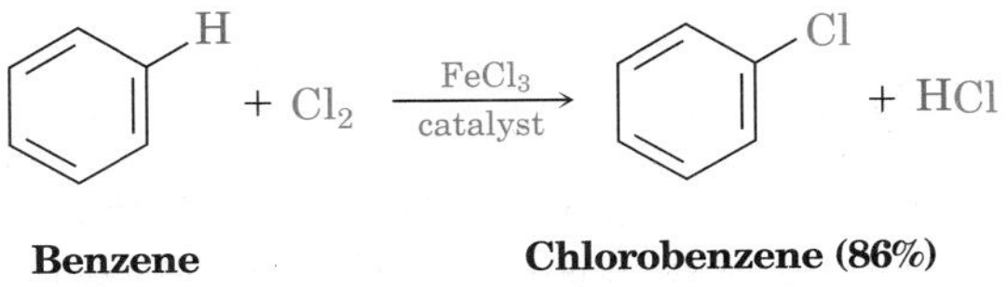

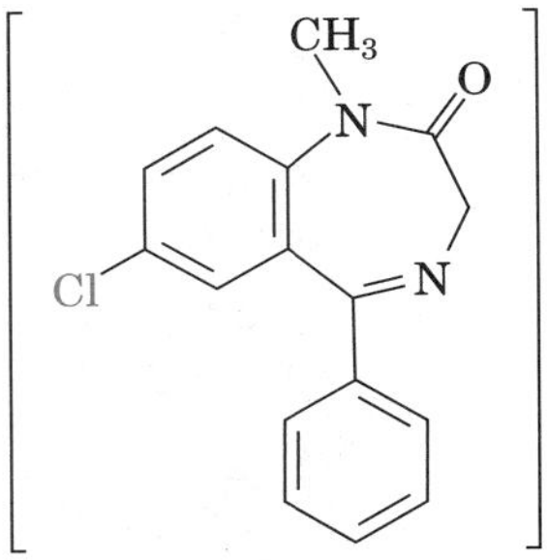

Diazepam

Nitration

Aromatic rings are nitrated by reaction with a mixture of concentrated nitric and sulfuric acids. The electrophile is the nitronium ion, NO_2^+, which is formed by reaction of HNO_3 with H^+ followed by loss of water and which reacts with benzene in much the same way Br^+ does. Nitration of aromatic rings is a key step in the synthesis of explosives such as TNT (2,4,6-trinitrotoluene), dyes, and many pharmaceutical agents.

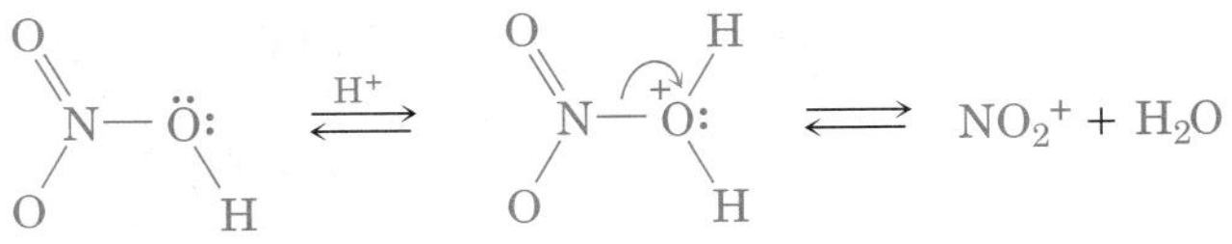

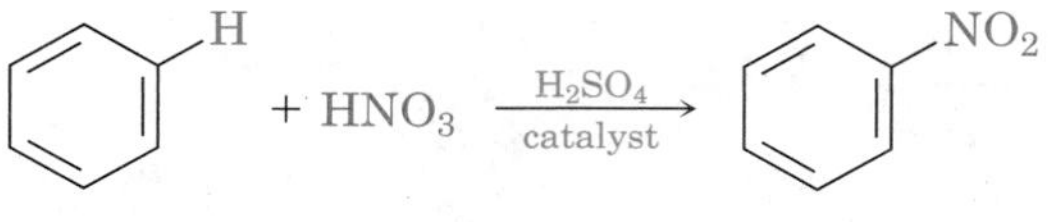

$$C_6H_5H + HNO_3 \xrightarrow[\text{catalyst}]{H_2SO_4} C_6H_5NO_2$$

Benzene **Nitrobenzene (85%)**

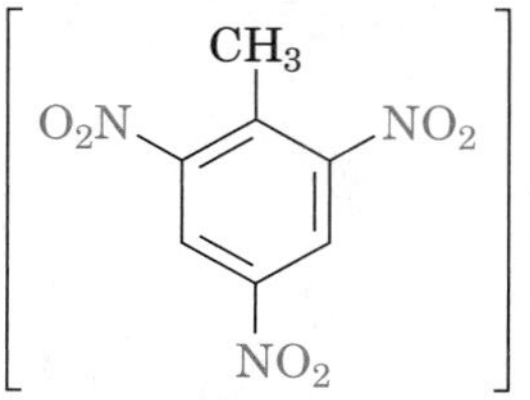

Trinitrotoluene (TNT)

Sulfonation

Aromatic rings are sulfonated by reaction with so-called *fuming sulfuric acid,* a mixture of SO_3 and H_2SO_4. The reactive electrophile is HSO_3^+, and substitution occurs by the usual two-step mechanism seen for bromination. Aromatic sulfonation is a key step in the synthesis of such compounds as the sulfa drug family of antibiotics.

$$SO_3 + H_2SO_4 \longrightarrow SO_3H^+ + HSO_4^-$$

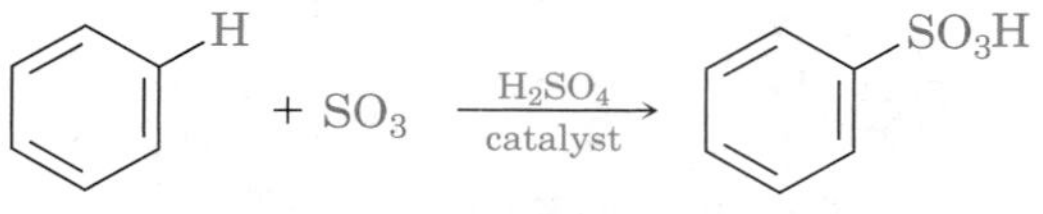

$$C_6H_5H + SO_3 \xrightarrow[\text{catalyst}]{H_2SO_4} C_6H_5SO_3H$$

Benzene **Benzenesulfonic acid (95%)**

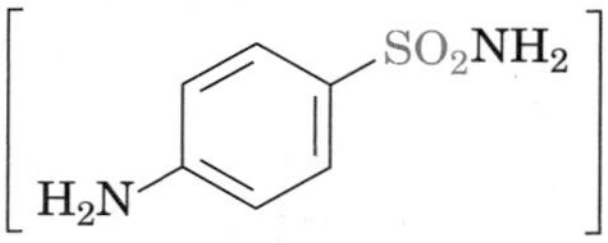

Sulfanilimide (a sulfa drug)

PRACTICE PROBLEM 5.2

Show the mechanism of the reaction of benzene with fuming sulfuric acid to yield benzenesulfonic acid.

Solution The electrophile in sulfonation reactions is HSO_3^+, and the reaction occurs by the same two-step process common to all electrophilic aromatic substitutions.

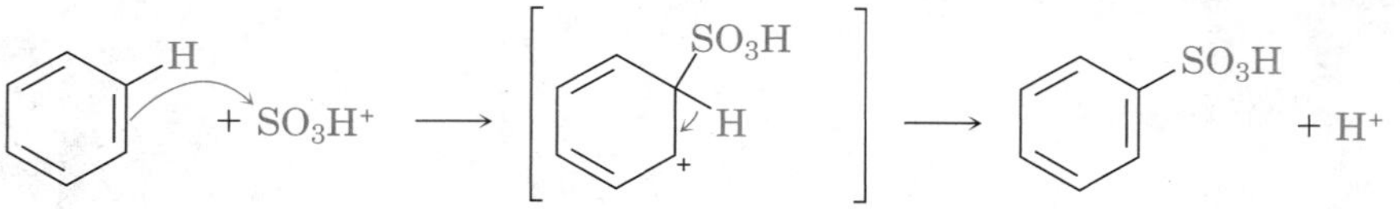

$$C_6H_5H + SO_3H^+ \longrightarrow [\text{Carbocation intermediate}] \longrightarrow C_6H_5SO_3H + H^+$$

Carbocation intermediate

PROBLEM

5.7 Show the mechanism of the reaction of benzene with nitric acid and sulfuric acid to yield nitrobenzene.

PROBLEM

5.8 Chlorination of *o*-xylene (*o*-dimethylbenzene) yields a mixture of two products, but chlorination of *p*-xylene yields a single product. Explain.

PROBLEM

5.9 How many products might be formed on chlorination of *m*-xylene?

PROBLEM

5.10 How can you account for the fact that deuterium (D, ^2H) slowly replaces hydrogen (^1H) in the aromatic ring when benzene is treated with D_2SO_4?

5.8 The Friedel–Crafts Alkylation and Acylation Reactions

An alkyl group is attached to an aromatic ring on reaction with an alkyl chloride, RCl, in the presence of $AlCl_3$ catalyst, a process called the **Friedel–Crafts alkylation reaction.** For example, benzene reacts with 2-chloropropane in the presence of $AlCl_3$ to yield isopropylbenzene (also called cumene):

$$C_6H_6 + CH_3CH(Cl)CH_3 \xrightarrow{AlCl_3} C_6H_5CH(CH_3)CH_3 + HCl$$

Benzene **2-Chloropropane** **Cumene (85%) (Isopropylbenzene)**

The Friedel–Crafts alkylation reaction is an aromatic substitution in which the electrophile is a carbocation, R^+. Aluminum chloride catalyzes the reaction by helping the alkyl chloride to ionize, in much the same way that $FeBr_3$ helps Br_2 to ionize (Section 5.6). The overall Friedel–Crafts mechanism for the synthesis of isopropylbenzene is shown in Figure 5.8 (p. 164).

Though extremely useful, the Friedel–Crafts alkylation reaction has several important limitations. For example, only *alkyl* halides can be used; aryl halides such as chlorobenzene don't react. In addition, Friedel–Crafts reactions don't succeed on aromatic rings that are already substituted by the groups $-NO_2$, –C≡N, $-SO_3H$, or –COR. Such aromatic rings are much less reactive than benzene for reasons we'll discuss in Section 5.9.

Closely related to the Friedel–Crafts alkylation reaction is the **Friedel–Crafts acylation reaction.** When an aromatic compound is treated with a carboxylic acid chloride, RCOCl, in the presence of $AlCl_3$, an **acyl** (**a**-sil) **group,** –COR, is introduced onto the ring. For example, reaction of benzene with acetyl chloride yields acetophenone, a ketone:

$$C_6H_6 + CH_3C(=O)Cl \xrightarrow[80°C]{AlCl_3} C_6H_5C(=O)CH_3 + HCl$$

Benzene **Acetyl chloride** **Acetophenone (95%)**

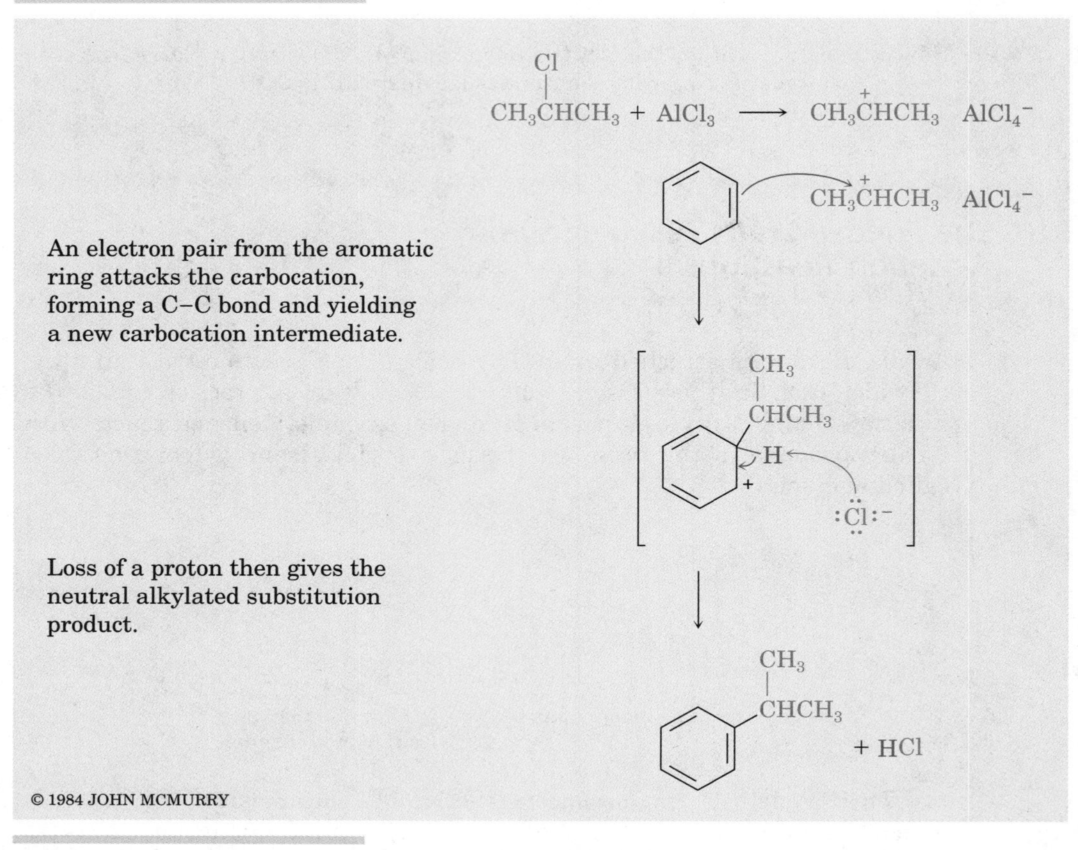
Cl
CH3CHCH3 + AlCl3 ⟶ CH3CHCH3 AlCl4−
CH3CHCH3 AlCl4−
An electron pair from the aromatic ring attacks the carbocation, forming a C–C bond and yielding a new carbocation intermediate.
CH3
CHCH3
H
:Cl:−
Loss of a proton then gives the neutral alkylated substitution product.
CH3
CHCH3
+ HCl
© 1984 JOHN MCMURRY

Figure 5.8 Mechanism of the Friedel–Crafts alkylation reaction in the synthesis of isopropylbenzene. The electrophile is a carbocation generated by $AlCl_3$-assisted ionization of an alkyl chloride.

PROBLEM

5.11 What products would you expect to obtain from the reaction of the following compounds with chloroethane and $AlCl_3$?
(a) Benzene (b) *p*-Xylene

PROBLEM

5.12 What products would you expect to obtain from the reaction of benzene with the following reagents?
(a) $(CH_3)_3CCl$, $AlCl_3$ (b) CH_3CH_2COCl, $AlCl_3$

5.9 Substituent Effects in Electrophilic Aromatic Substitution

Only one product can form when an electrophilic substitution occurs on benzene, but what would happen if we were to carry out an electrophilic substitution reaction on a ring that already has a substituent? Substituents already present on an aromatic ring have two effects:

1. Substituents affect the *reactivity* of the aromatic ring. Some groups activate the ring for further electrophilic substitution, and some deactivate it. An –OH group activates the ring, for instance, making it 1000 times more reactive than benzene toward nitration. An $-NO_2$ group deactivates the ring, however, making it 20 million times less reactive than benzene.

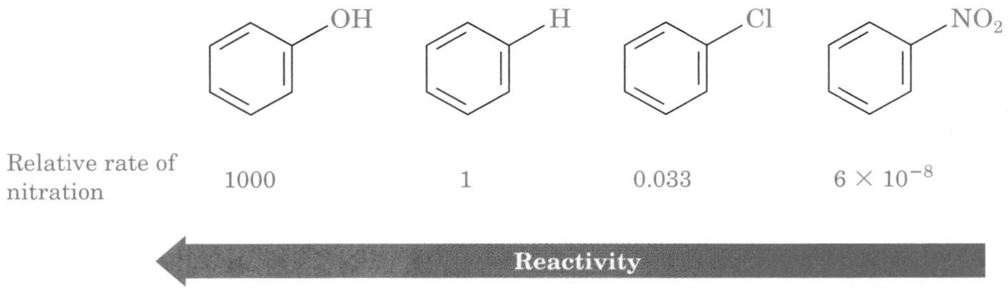

	OH	H	Cl	NO_2
Relative rate of nitration	1000	1	0.033	6×10^{-8}

← Reactivity

2. Substituents affect the *orientation* of the reaction. The three possible disubstituted products—ortho, meta, and para—are usually not formed in equal amounts. Instead, the nature of the substituent already present on the ring determines the position of the second substitution. An –OH group directs further substitution toward the ortho and para positions, for instance, while a –CN group directs further substitution primarily toward the meta position.

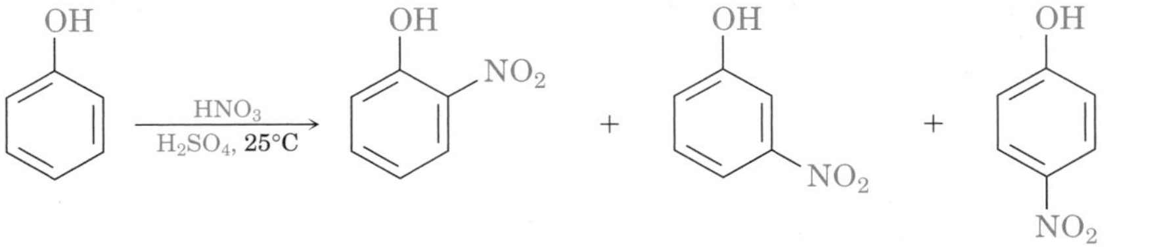

Phenol $\xrightarrow[H_2SO_4,\ 25°C]{HNO_3}$ *o*-Nitrophenol (50%) + *m*-Nitrophenol (0%) + *p*-Nitrophenol (50%)

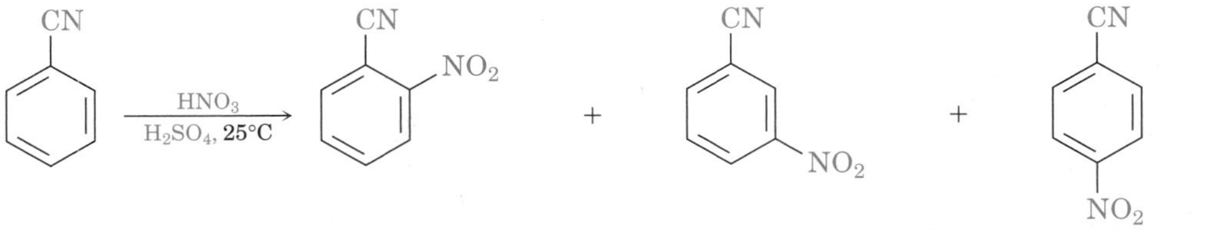

Benzonitrile $\xrightarrow[H_2SO_4,\ 25°C]{HNO_3}$ *o*-Nitrobenzonitrile (17%) + *m*-Nitrobenzonitrile (81%) + *p*-Nitrobenzonitrile (2%)

Substituents can be classified into three groups, as shown in Figure 5.9 (p. 166): ortho- and para-directing activators, ortho- and para-directing deactivators, and meta-directing deactivators. No meta-directing activators are known. Note how the directing effects of the groups correlate with their reactivities. All meta-directing groups are deactivating, and most ortho- and para-directing groups are activating. The halogens are unique in being ortho- and para-directing deactivators.

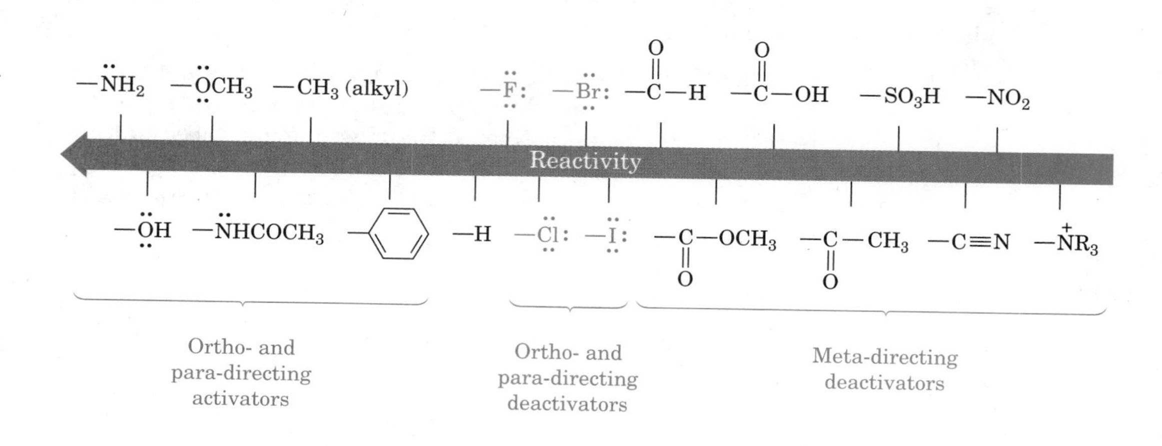

Figure 5.9 Classification of substituent effects in electrophilic aromatic substitutions.

PRACTICE PROBLEM 5.3

Which would you expect to react faster in an electrophilic aromatic substitution reaction, chlorobenzene or ethylbenzene? Explain.

Solution According to Figure 5.9, a chloro substituent is deactivating, whereas an alkyl group is activating. Thus, ethylbenzene is more reactive than chlorobenzene.

PROBLEM

5.13 Use Figure 5.9 to rank the compounds in each of the following groups in order of their reactivity to electrophilic aromatic substitution:
(a) Nitrobenzene, phenol (hydroxybenzene), toluene
(b) Phenol, benzene, chlorobenzene, benzoic acid
(c) Benzene, bromobenzene, benzaldehyde, aniline (aminobenzene)

5.10 An Explanation of Substituent Effects

Activating and Deactivating Effects in Aromatic Rings

What makes a group either activating or deactivating? *The common feature of all activating groups is that they donate electrons to the ring,* thereby stabilizing the carbocation intermediate and lowering the activation energy for its formation. *The common feature of all deactivating groups, by contrast, is that they withdraw electrons from the ring,* thereby destabilizing the carbocation intermediate and raising the activation energy for its formation.

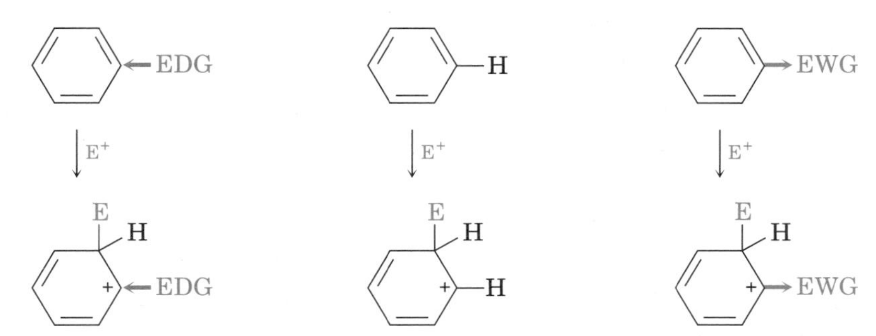

Electron-donating group (EDG); carbocation intermediate is more stabilized, so ring is more reactive.

Electron-withdrawing group (EWG); carbocation intermediate is less stabilized, so ring is less reactive.

The electron donation or withdrawal may occur either inductively (Section 1.11) or by resonance. Inductive effects are due to electronegativity differences between the ring and the attached substituent, while resonance effects are due to orbital overlap between a *p* orbital on the ring and a *p* orbital on the substituent.

Orienting Effects in Aromatic Rings: Ortho and Para Directors

Let's look at the nitration of phenol as an example of how ortho/para-directing substituents work. In the first step, attack on the electrophilic nitronium ion (NO_2^+) can occur either ortho, meta, or para to the –OH group, giving the carbocation intermediates shown in Figure 5.10 (p. 168). The ortho and para intermediates are more stable than the meta intermediate because each has a resonance form that allows the positive charge to be stabilized by the substituent oxygen atom. Since the ortho and para intermediates are more stable than the meta intermediate, they are formed faster.

In general, any substituent that has a lone pair of electrons on the atom directly bonded to the aromatic ring allows an electron-donating resonance interaction to occur, and thus acts as an ortho/para director:

Ortho/para directors: $C_6H_5\ddot{O}H$, $C_6H_5\ddot{N}H_2$, $C_6H_5\ddot{Cl}:$, $C_6H_5\ddot{Br}:$

Orienting Effects in Aromatic Rings: Meta Directors

The influence of meta-directing substituents can be explained using the same kinds of arguments used for ortho/para directors. Look at the chlorination of benzaldehyde, for example (Figure 5.11, p. 169). Of the three possible carbocation intermediates, those produced by reaction at ortho and para positions are least stable. In both cases, the unfavorable resonance forms

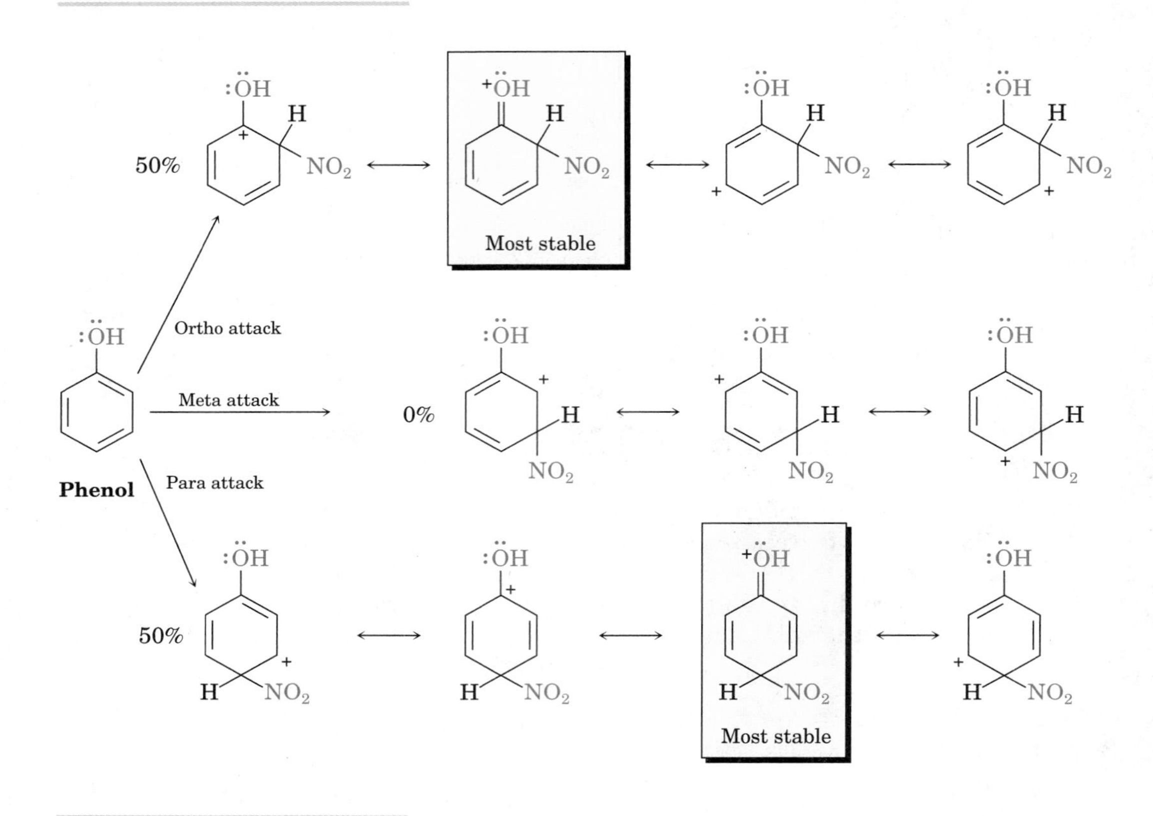

Figure 5.10 Intermediates in the nitration of phenol. The ortho and para intermediates are more stable than the meta intermediate because of a resonance form involving the oxygen atom.

indicated in Figure 5.11 place the positive charge directly on the carbon that bears the aldehyde group, where it is disfavored by a repulsive interaction with the positively polarized carbon atom of the C=O group. Hence, the meta intermediate is most favored and is formed faster than the ortho and para intermediates.

In general, any substituent that has a positively polarized atom ($\delta+$) directly attached to the ring allows a destabilizing, electron-withdrawing resonance interaction to occur at ortho and para positions, and thus acts as a meta director:

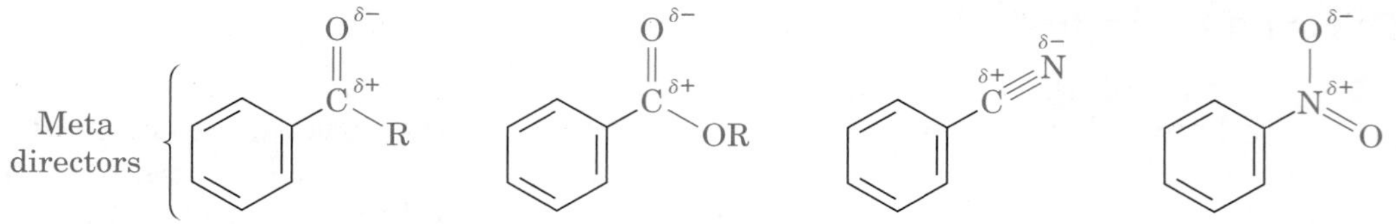

Meta directors

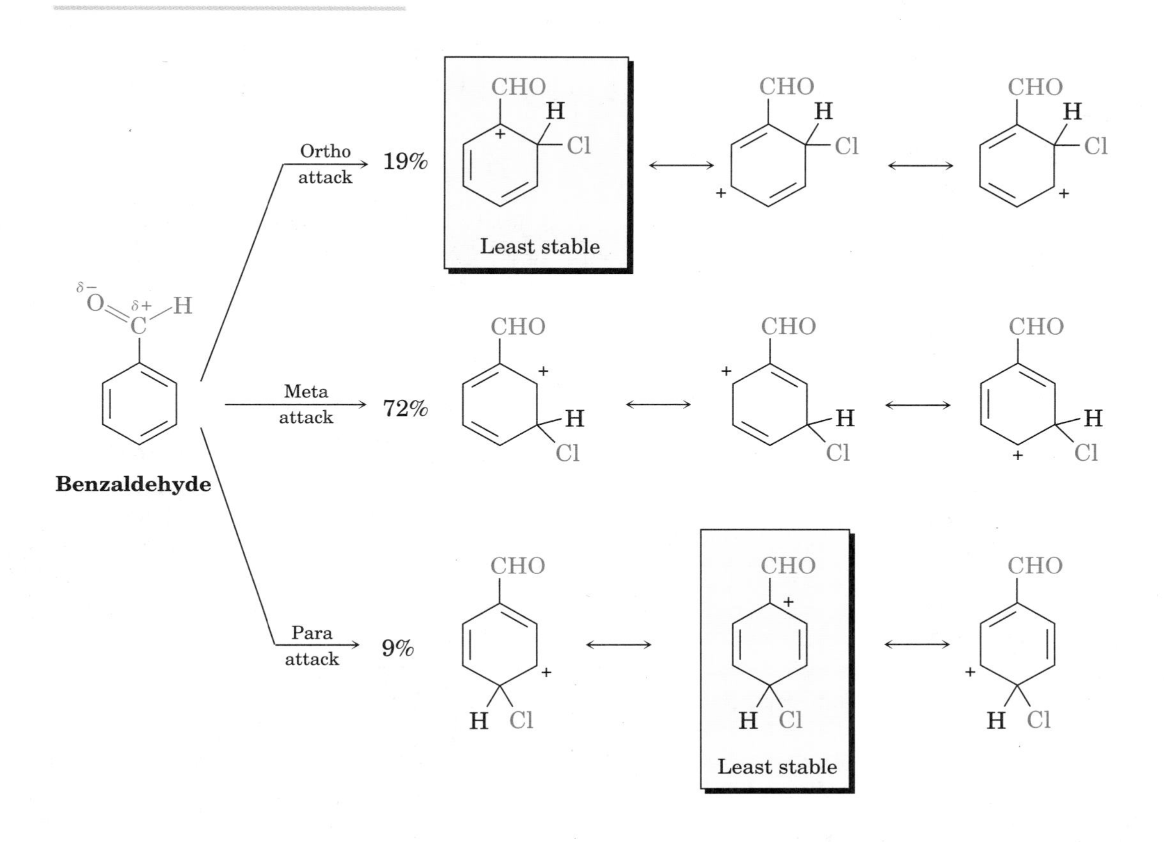

Figure 5.11 Intermediates in the chlorination of benzaldehyde. The ortho and para intermediates are less stable than the meta intermediate because they have unfavorable resonance forms.

PRACTICE PROBLEM 5.4

What product(s) would you expect from bromination of aniline, $C_6H_5NH_2$?

Solution Figure 5.9 indicates that an amino group, $-NH_2$, is ortho/para directing. We therefore expect to obtain a mixture of *o*-bromoaniline and *p*-bromoaniline.

NH_2 $\xrightarrow{Br_2}$ NH_2, Br + Br, NH_2

Aniline **o-Bromoaniline** **p-Bromoaniline**

PROBLEM

5.14 What product(s) would you expect from nitration of the following compounds?
(a) Nitrobenzene (b) Bromobenzene (c) Toluene
(d) Benzoic acid (e) *p*-Xylene

PROBLEM

5.15 Draw resonance structures of the three possible carbocation intermediates to show how a methoxyl group ($-OCH_3$) directs bromination toward ortho and para positions.

PROBLEM

5.16 Draw resonance structures of the three possible carbocation intermediates to show how an acetyl group, $CH_3C{=}O$, directs bromination toward the meta position.

5.11 Oxidation and Reduction of Aromatic Compounds

Despite its unsaturation, a benzene ring is normally inert to strong oxidizing agents such as $KMnO_4$. (Recall from Section 4.7 that $KMnO_4$ cleaves alkene C=C bonds.) Alkyl groups attached to the aromatic ring are readily attacked by oxidizing agents, however, and are converted into carboxyl groups (–COOH). For example, butylbenzene is oxidized by $KMnO_4$ to give benzoic acid. The mechanism of this reaction is not fully understood but probably involves attack on the side-chain C–H bonds at the position next to the aromatic ring (the **benzylic** position) to give radical intermediates.

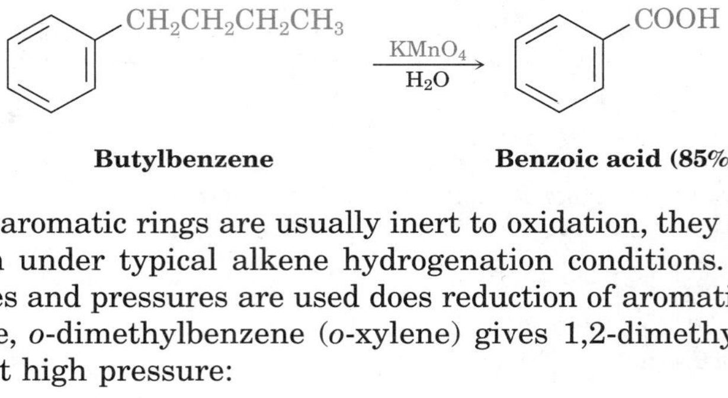

$CH_2CH_2CH_2CH_3$ $\xrightarrow[H_2O]{KMnO_4}$ COOH

Butylbenzene **Benzoic acid (85%)**

Just as aromatic rings are usually inert to oxidation, they are also inert to reduction under typical alkene hydrogenation conditions. Only if high temperatures and pressures are used does reduction of aromatic rings occur. For example, *o*-dimethylbenzene (*o*-xylene) gives 1,2-dimethylcyclohexane if reduced at high pressure:

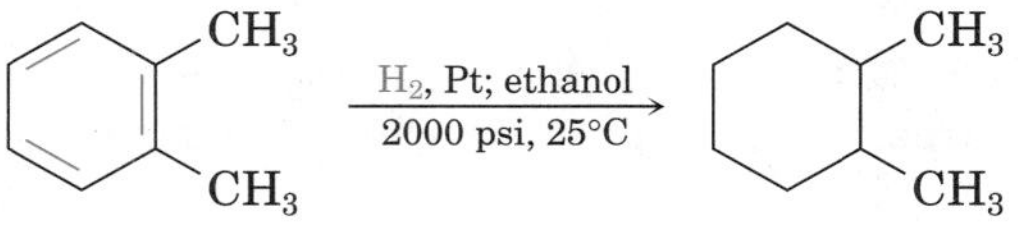

CH_3, CH_3 $\xrightarrow[2000\ psi,\ 25°C]{H_2,\ Pt;\ ethanol}$ CH_3, CH_3

***o*-Xylene** **1,2-Dimethylcyclohexane (100%)**

PROBLEM

5.17 What aromatic products do you expect to obtain from oxidation of the following substances with $KMnO_4$?

(a) *m*-Chloroethylbenzene (b) 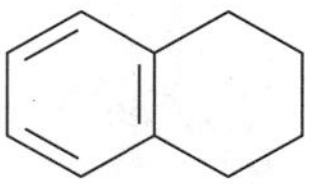 **Tetralin**

5.12 Polycyclic Aromatic Hydrocarbons

The concept of aromaticity—the unusual chemical stability that arises in cyclic conjugated molecules like benzene—can be extended beyond simple monocyclic compounds to include **polycyclic aromatic compounds.** Naphthalene, with two benzene-like rings fused together, and anthracene, with three fused rings, are two of the simplest polycyclic aromatic molecules.

Naphthalene **Anthracene**

Naphthalene and other polycyclic aromatic hydrocarbons show many of the chemical properties we associate with aromaticity. Both, for example, react with electrophilic reagents such as Br_2 to give substitution products rather than double-bond addition products.

$\xrightarrow[\Delta]{Br_2,\ Fe}$ Br + HBr

Naphthalene **1-Bromonaphthalene (75%)**

We'll see in Chapter 12 that nitrogen-containing compounds like pyridine and pyrrole are also aromatic, even though they don't contain benzene rings.

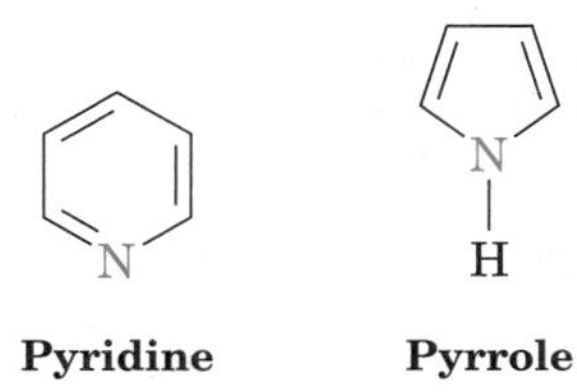

Pyridine **Pyrrole**

PROBLEM

5.18 There are three resonance structures of naphthalene, of which only one is shown. Draw the other two.

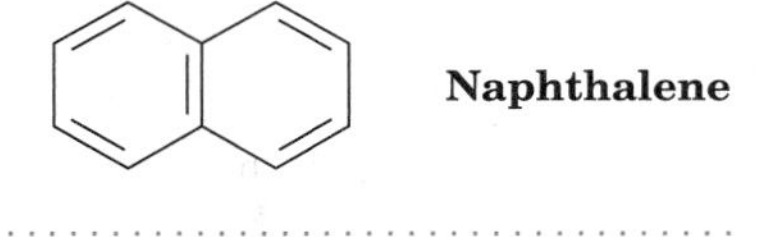

Naphthalene

5.13 Organic Synthesis

The laboratory synthesis of organic molecules from simple precursors might be carried out for many reasons. In the pharmaceutical industry, new organic molecules are often designed and synthesized for evaluation as medicines. In the chemical industry, syntheses are often undertaken to devise more economical routes to known compounds. In this book, too, we'll sometimes devise syntheses of complex molecules from simpler precursors, but the purpose here is simply to help you learn organic chemistry. Devising a route for the synthesis of an organic molecule requires that you approach chemical problems in a logical way, draw on all your knowledge of organic reactivity, and organize that knowledge into a workable plan.

The only trick to devising an organic synthesis is to *work backward.* Look at the product and ask yourself, "What is the immediate precursor of that product?" Having found an immediate precursor, work backward again, one step at a time, until a suitable starting material is found. Let's try some examples.

PRACTICE PROBLEM 5.5 . . . .

Synthesize *m*-chloronitrobenzene starting from benzene.

Solution Ask, "What is an immediate precursor of *m*-chloronitrobenzene?"

? ⟶ [structure: NO_2, Cl]

***m*-Chloronitrobenzene**

There are two substituents on the ring, a –Cl group, which is ortho/para-directing, and an $-NO_2$ group, which is meta-directing. We can't nitrate chlorobenzene, because the wrong isomers (*o*- and *p*-chloronitrobenzenes) would result, but chlorination of nitrobenzene should give the desired product.

Chlorobenzene —HNO_3, H_2SO_4 (crossed out)→ ***m*-Chloronitrobenzene**

Nitrobenzene —Cl_2, $FeCl_3$→ ***m*-Chloronitrobenzene**

"What is an immediate precursor of nitrobenzene?" Benzene, which can be nitrated. Thus, in two steps, we've solved the problem.

$$\text{Benzene} \xrightarrow{HNO_3,\ H_2SO_4} \text{Nitrobenzene} \xrightarrow{Cl_2,\ FeCl_3} m\text{-Chloronitrobenzene}$$

Benzene **Nitrobenzene** ***m*-Chloronitrobenzene**

PRACTICE PROBLEM 5.6 ....

Synthesize *p*-bromobenzoic acid starting from benzene.

Solution Ask, "What is an immediate precursor of *p*-bromobenzoic acid?"

$$? \longrightarrow Br\text{–}C_6H_4\text{–}COOH$$

***p*-Bromobenzoic acid**

There are two substituents on the ring, a –COOH group, which is meta-directing, and a –Br atom, which is ortho/para-directing. We can't brominate benzoic acid, because the wrong isomer (*m*-bromobenzoic acid) would be formed. We've seen, however, that oxidation of alkylbenzene side chains yields benzoic acids. An immediate precursor of our target molecule might therefore be *p*-bromotoluene.

$$Br\text{–}C_6H_4\text{–}CH_3 \xrightarrow[H_2O]{KMnO_4} Br\text{–}C_6H_4\text{–}COOH$$

***p*-Bromotoluene** ***p*-Bromobenzoic acid**

"What is an immediate precursor of *p*-bromotoluene?" Perhaps toluene, because the methyl group would direct bromination to the ortho and para positions, and we could then separate isomers. Alternatively, bromobenzene might be an immediate precursor because we could carry out a Friedel–Crafts alkylation and obtain the para product. Both methods are satisfactory.

$$\text{Toluene} \xrightarrow[FeBr_2]{Br_2} Br\text{–}C_6H_4\text{–}CH_3 + \text{Ortho isomer}$$

$$\text{Bromobenzene} \xrightarrow[AlCl_3]{CH_3Cl} Br\text{–}C_6H_4\text{–}CH_3$$

Toluene

Bromobenzene

***p*-Bromotoluene**
(separate and purify)

"What is an immediate precursor of toluene?" Benzene, which can be methylated in a Friedel–Crafts reaction:

$$\text{Benzene} \xrightarrow[AlCl_3]{CH_3Cl} \text{Toluene}$$

Benzene **Toluene**

"Alternatively, what is an immediate precursor of bromobenzene?" Benzene, which can be brominated:

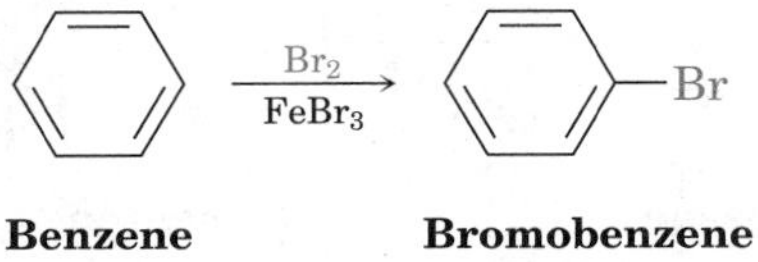

$$\text{Benzene} \xrightarrow[FeBr_3]{Br_2} \text{Bromobenzene}$$

Benzene **Bromobenzene**

Our backward synthetic (*retrosynthetic*) analysis has provided two workable routes from benzene to *p*-bromobenzoic acid.

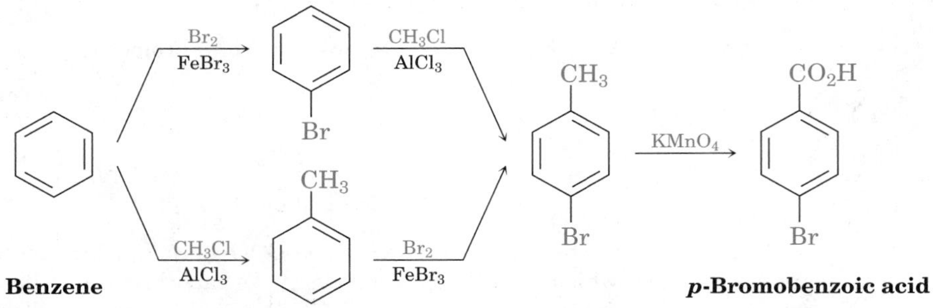

$$\text{Benzene} \xrightarrow[FeBr_3]{Br_2} \text{Bromobenzene} \xrightarrow[AlCl_3]{CH_3Cl} p\text{-Bromotoluene}$$

$$\text{Benzene} \xrightarrow[AlCl_3]{CH_3Cl} \text{Toluene} \xrightarrow[FeBr_3]{Br_2} p\text{-Bromotoluene}$$

$$p\text{-Bromotoluene} \xrightarrow{KMnO_4} p\text{-Bromobenzoic acid}$$

Benzene **p-Bromobenzoic acid**

PROBLEM

5.19 Propose a synthesis of each of the following substances from benzene:

(a) H_3C–C_6H_4–$C(=O)CH_3$ (para) (b) Cl–C_6H_4–NO_2 (para)

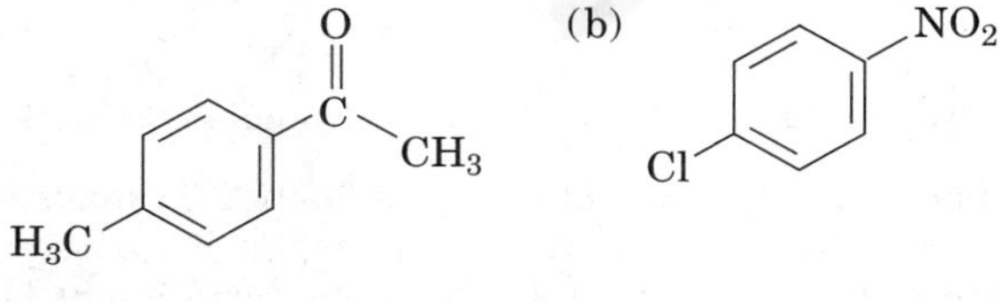

PROBLEM

5.20 Synthesize the following substances from benzene:
(a) *o*-Bromotoluene (b) 2-Bromo-1,4-dimethylbenzene

PROBLEM

5.21 How would you prepare *m*-chlorobenzoic acid from benzene?

INTERLUDE

Aspirin and Other Aromatic NSAID's

Long-distance runners sometimes call ibuprofen "the fifth basic food group" because of its usefulness in controlling aches and pains.

Whatever the cause—tennis elbow, a sprained ankle, or a wrenched knee—pain and inflammation seem to go together. They are, however, different in their origin, and powerful drugs are available for treating each separately. Codeine, for example, is a powerful *analgesic,* or pain reliever, while cortisone and related steroids are potent *anti-inflammatory* agents often used for treating arthritis. For minor pains and inflammation, though, both problems are often treated using a common, over-the-counter medication called an *NSAID,* for *nonsteroidal anti-inflammatory drug.*

The most common NSAID is aspirin, or acetylsalicylic acid, whose use goes back to the late 1800s. It has been known since before 400 BC that fevers can be lowered by chewing the bark of willow trees. The active agent in willow bark was found in 1827 to be an aromatic compound called *salicin,* which could be converted by reaction with water (*hydrolysis*) into salicyl alcohol and then oxidized to give salicylic acid. Salicylic acid turned out to be even more effective than salicin for reducing fevers and also to have both analgesic and anti-inflammatory properties. Unfortunately, salicylic acid is too corrosive to the walls of the stomach for everyday use. Conversion of the phenol –OH group into an ester, however, yielded acetylsalicylic acid, which proved just as potent as salicylic acid but less corrosive to the stomach.

CH_2OH, OH → $COOH$, OH → $COOH$, $OCCH_3$ (C=O)

Salicyl alcohol **Salicylic acid** **Acetylsalicylic acid (Aspirin)**

Though extraordinary in its powers, aspirin is also more dangerous than commonly believed. Only about 15 g can be fatal to a small child, and aspirin can cause stomach bleeding and allergic reactions in long-term users. Even more serious is a condition called *Reye's syndrome,* a potentially fatal reaction to aspirin sometimes seen in children recover-

(*continued*)➤

ing from the flu. As a result of these problems, numerous other NSAID's have been developed in the last two decades, most notably ibuprofen and naproxen.

Like aspirin, both ibuprofen and naproxen are relatively simple aromatic compounds containing a side-chain carboxylic acid group. Ibuprofen, sold under the names Advil, Motrin, Nuprin, and others, has roughly the same potency as aspirin but is less prone to cause stomach upset. Naproxen, sold under the names Naprosyn and Aleve, also has about the same potency as aspirin, but remains active in the body six times longer. NSAID's are a godsend to arthritis sufferers and weekend athletes, but they should always be treated with respect and not be overused.

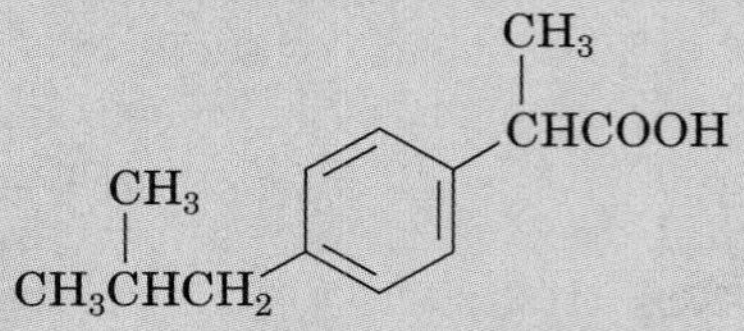

CH_3 CH_3CHCH_2 — $CHCOOH$ (CH_3)

Ibuprofen
(Advil, Motrin, Nuprin)

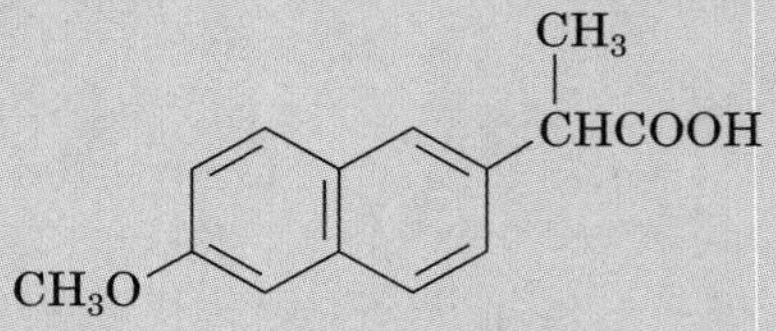

CH_3O — $CHCOOH$ (CH_3)

Naproxen
(Naprosyn, Aleve)

Summary and Key Words

acyl group, 163
aromatic, 150
benzyl, 154
benzylic position, 170
electrophilic aromatic substitution, 157
Friedel–Crafts acylation reaction, 163
Friedel–Crafts alkylation reaction, 163
phenyl, 154
polycyclic aromatic compound, 171

The word **aromatic** refers to the class of compounds structurally related to benzene. Aromatic compounds are named according to IUPAC rules, with disubstituted benzenes referred to as either **ortho** (1,2 disubstituted), **meta** (1,3 disubstituted), or **para** (1,4 disubstituted). Benzene is a resonance hybrid of two equivalent forms, neither of which is correct by itself. The true structure of benzene is intermediate between the two.

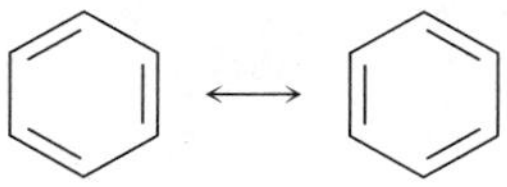

$\longleftrightarrow$

The most common reaction of aromatic compounds is **electrophilic aromatic substitution.** In this two-step polar reaction, the π electrons of the aromatic ring first attack the electrophile to yield a resonance-stabilized carbocation intermediate, which then loses H^+ to give a substituted aromatic product. Bromination, chlorination, iodination, nitration, sulfonation, **Friedel–Crafts alkylation,** and **Friedel–Crafts acylation** can all be carried out with the proper choice of reagent. Friedel–Crafts alkylation is particularly useful for preparing a variety of alkylbenzenes but is limited because only alkyl halides can be used and strongly deactivated rings do not react.

Substituents on the benzene ring affect both the reactivity of the ring toward further substitution and the orientation of that further substitution. Substituents can be classified either as **activators** or **deactivators,** and either as **ortho/para directors** or as **meta directors.**

The side chains of alkylbenzenes have unique reactivity because of the neighboring aromatic ring. Thus, an alkyl group attached to the aromatic ring can be degraded to a carboxyl group (–COOH) by oxidation with aqueous $KMnO_4$. In addition, aromatic rings can be reduced to yield cyclohexanes on catalytic hydrogenation at high pressure.

Summary of Reactions

1. Electrophilic aromatic substitution
   (a) Bromination (Section 5.6)

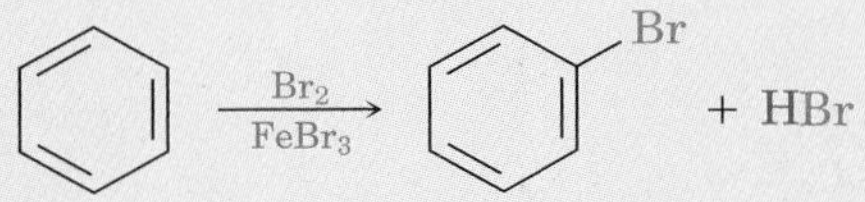

$$\text{Benzene} \xrightarrow[FeBr_3]{Br_2} \text{C}_6\text{H}_5\text{Br} + HBr$$

   (b) Chlorination (Section 5.7)

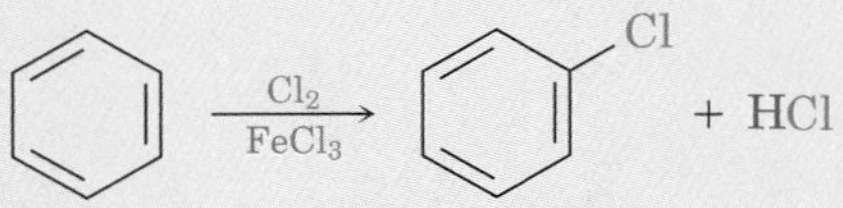

$$\text{Benzene} \xrightarrow[FeCl_3]{Cl_2} \text{C}_6\text{H}_5\text{Cl} + HCl$$

   (c) Nitration (Section 5.7)

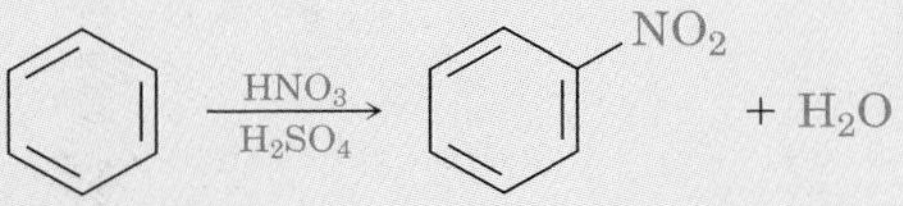

$$\text{Benzene} \xrightarrow[H_2SO_4]{HNO_3} \text{C}_6\text{H}_5\text{NO}_2 + H_2O$$

   (d) Sulfonation (Section 5.7)

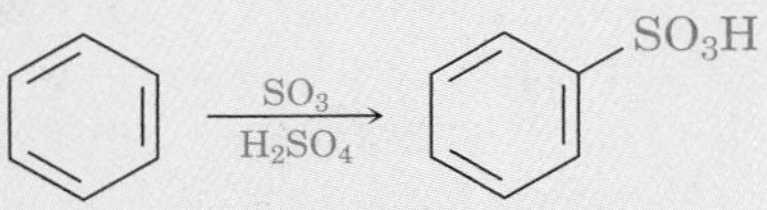

$$\text{Benzene} \xrightarrow[H_2SO_4]{SO_3} \text{C}_6\text{H}_5\text{SO}_3\text{H}$$

   (e) Friedel–Crafts alkylation (Section 5.8)

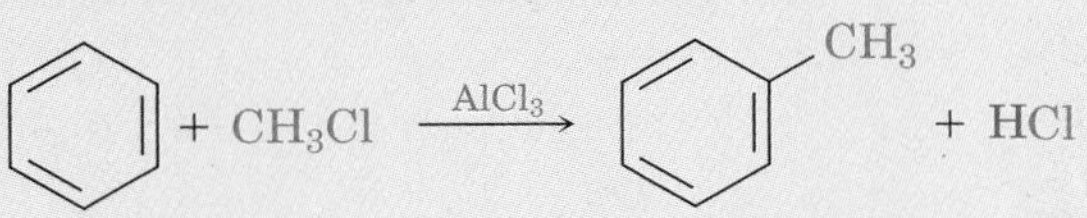

$$\text{Benzene} + CH_3Cl \xrightarrow{AlCl_3} \text{C}_6\text{H}_5\text{CH}_3 + HCl$$

(f) Friedel–Crafts acylation (Section 5.8)

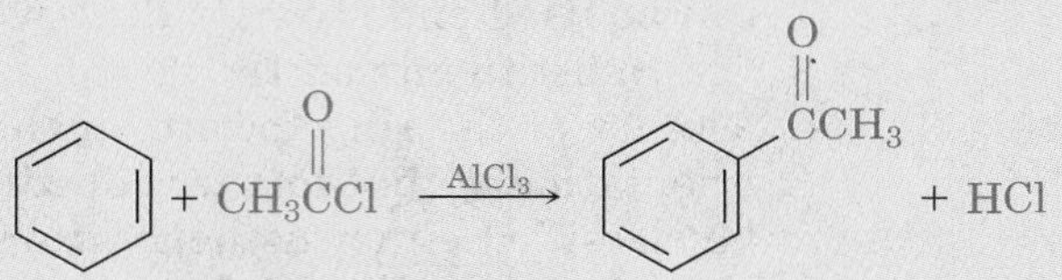

$$C_6H_6 + CH_3\overset{O}{\overset{\|}{C}}Cl \xrightarrow{AlCl_3} C_6H_5\overset{O}{\overset{\|}{C}}CH_3 + HCl$$

2. Oxidation of aromatic side chains (Section 5.11)

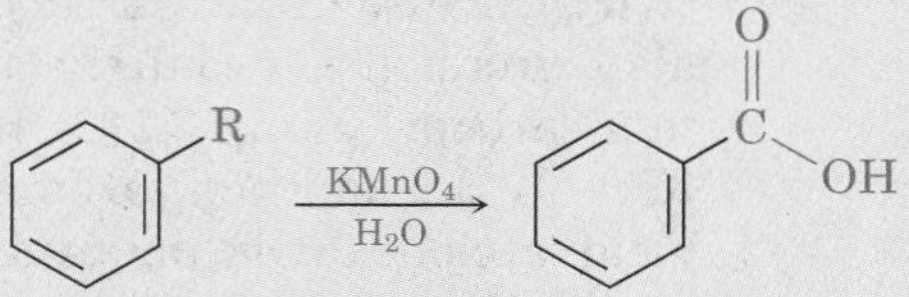

$$C_6H_5R \xrightarrow[H_2O]{KMnO_4} C_6H_5\overset{O}{\overset{\|}{C}}OH$$

3. Hydrogenation of aromatic rings (Section 5.11)

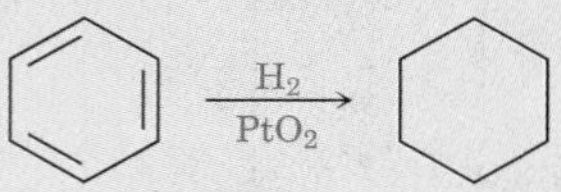

$$C_6H_6 \xrightarrow[PtO_2]{H_2} C_6H_{12}$$

ADDITIONAL PROBLEMS

5.22 Give IUPAC names for the following compounds:

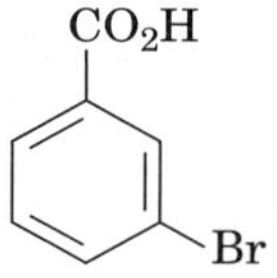

(a) $C_6H_5CH_2CH_2CH_2CH(CH_3)CH_3$

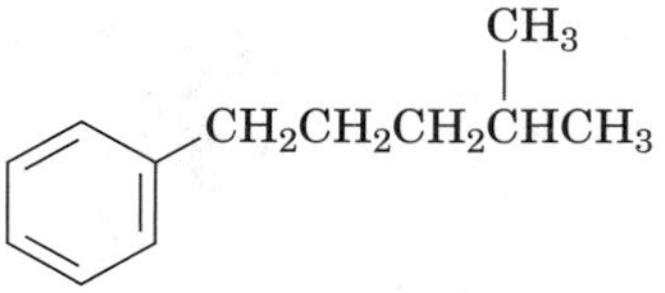

(b) benzene ring with CO_2H and Br (meta)

(c) benzene ring with Br, H_3C, and CH_3 (1,3,5-positions)

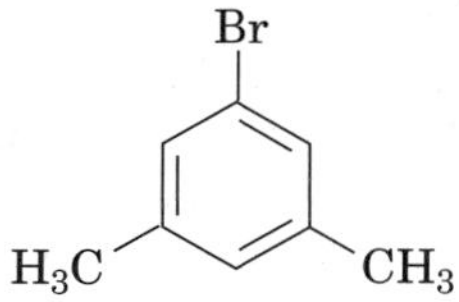

(d) benzene ring with Br and $CH_2CH_2CH_3$ (ortho)

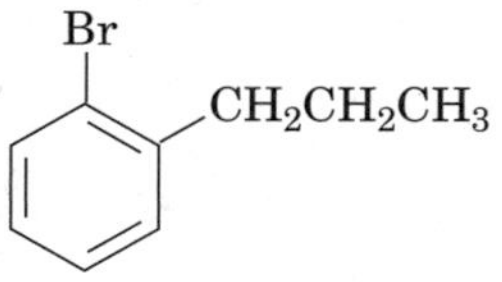

5.23 Draw structures corresponding to the following names:
(a) *m*-Bromophenol (b) 1,3,5-Benzenetriol
(c) *p*-Iodonitrobenzene (d) 2,4,6-Trinitrotoluene (TNT)
(e) *o*-Aminobenzoic acid (f) 3-Methyl-2-phenylhexane

5.24 Draw and name all aromatic compounds with the formula C_7H_7Cl.

5.25 Draw and name all isomeric:
(a) Dinitrobenzenes (b) Bromodimethylbenzenes

5.26 Propose structures for aromatic hydrocarbons meeting the following descriptions:
(a) C_9H_{12}; can give only one product on aromatic bromination
(b) C_8H_{10}; can give three products on aromatic chlorination
(c) $C_{10}H_{14}$; can give two products on aromatic nitration

5.27 Formulate the reaction of benzene with 2-chloro-2-methylpropane in the presence of $AlCl_3$ catalyst to give *tert*-butylbenzene.

5.28 Predict the major product(s) of mononitration of the following substances:
(a) Bromobenzene (b) Benzonitrile (cyanobenzene) (c) Benzoic acid
(d) Nitrobenzene (e) Phenol (f) Benzaldehyde

5.29 Which of the substances listed in Problem 5.28 react faster than benzene and which react slower?

5.30 Rank the compounds in each group according to their reactivity toward electrophilic substitution:
(a) Chlorobenzene, *o*-dichlorobenzene, benzene
(b) *p*-Bromonitrobenzene, nitrobenzene, phenol
(c) Fluorobenzene, benzaldehyde, *o*-dimethylbenzene

5.31 Show the steps involved in the Friedel–Crafts reaction of benzene with CH_3Cl.

5.32 The orientation of electrophilic aromatic substitution on a disubstituted benzene ring is usually controlled by whichever of the two groups already on the ring is the more powerful activator. Name and draw the structure(s) of the major product(s) of electrophilic chlorination of these substances:
(a) *m*-Nitrophenol (b) *o*-Methylphenol (c) *p*-Chloronitrobenzene

5.33 Predict the major product(s) you would expect to obtain from sulfonation of the following substances (see Problem 5.32):
(a) Bromobenzene (b) *m*-Bromophenol (c) *p*-Nitrotoluene

5.34 Rank the following aromatic compounds in the expected order of their reactivity toward Friedel–Crafts acylation. Which compounds are unreactive?
(a) Bromobenzene (b) Toluene (c) Anisole ($C_6H_5OCH_3$)
(d) Nitrobenzene (e) *p*-Bromotoluene

5.35 What is the structure of the compound with formula C_8H_9Br that gives *p*-bromobenzoic acid on oxidation with $KMnO_4$?

5.36 Draw the four resonance structures of anthracene.

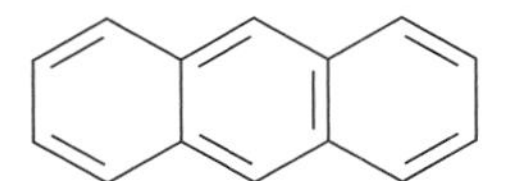

Anthracene

5.37 Draw the five resonance structures of phenanthrene.

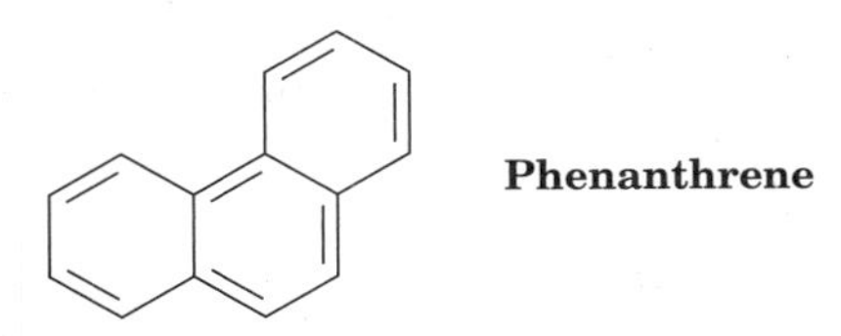

Phenanthrene

5.38 Explain why bromination of biphenyl occurs at the ortho and para positions rather than at the meta positions. Use resonance structures of the carbocation intermediates in your explanation.

Biphenyl

5.39 In light of your answer to Problem 5.38, at what position and on which ring would you expect nitration of 4-bromobiphenyl to occur?

—Br **4-Bromobiphenyl**

5.40 Starting with benzene, how would you synthesize the following substances? Assume that you can separate ortho and para isomers if necessary.
(a) *m*-Bromobenzenesulfonic acid
(b) *o*-Chlorobenzenesulfonic acid
(c) *p*-Chlorotoluene

5.41 Starting from any aromatic hydrocarbon of your choice, how would you synthesize the following substances? Ortho and para isomers can be separated if necessary.
(a) *o*-Nitrobenzoic acid
(b) *p*-*tert*-Butylbenzoic acid

5.42 Explain by drawing resonance structures of the intermediate carbocations why naphthalene undergoes electrophilic aromatic substitution at C1 rather than at C2.

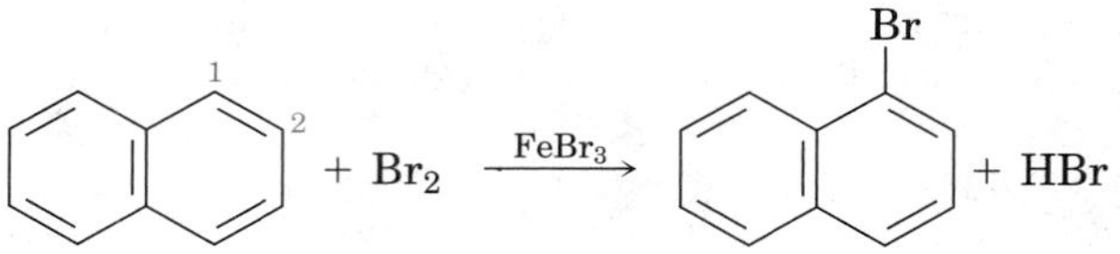

$$\text{Naphthalene} + Br_2 \xrightarrow{FeBr_3} \text{1-Bromonaphthalene} + HBr$$

5.43 We said in Section 4.11 that allylic carbocations are stabilized by resonance. Draw resonance structures to account for a similar stabilization of benzylic carbocations.

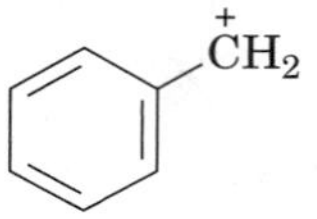

$C_6H_5\overset{+}{C}H_2$ A benzylic carbocation

5.44 Addition of HBr to 1-phenylpropene yields (1-bromopropyl)benzene as the exclusive product. Propose a mechanism for the reaction, and explain why none of the other regioisomer is produced (see Problem 5.43).

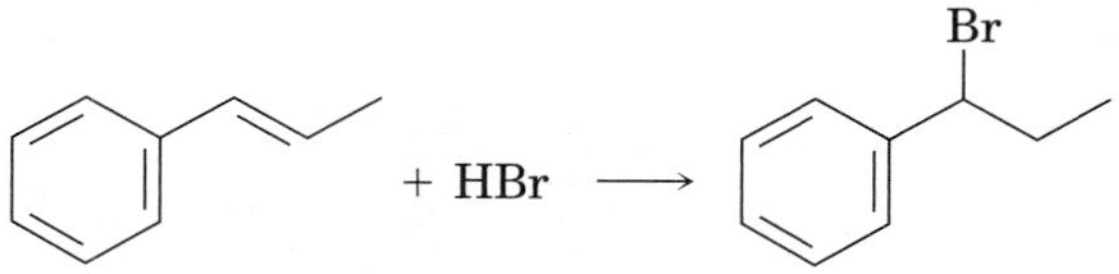

$$C_6H_5CH{=}CHCH_3 + HBr \longrightarrow C_6H_5CHBrCH_2CH_3$$

5.45 The following syntheses have flaws in them. What is wrong with each?

(a) Toluene (CH_3) $\xrightarrow{\text{1. } Cl_2,\ FeCl_3;\ \text{2. } KMnO_4}$ *m*-chlorobenzoic acid (COOH, Cl)

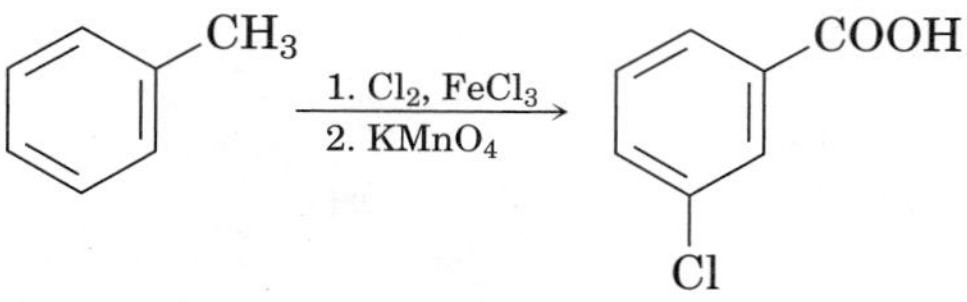

(b) Chlorobenzene (Cl) $\xrightarrow{\text{1. } (CH_3)_3CCl,\ AlCl_3;\ \text{2. } KMnO_4,\ H_2O}$ *m*-chlorobenzoic acid (Cl, COOH)

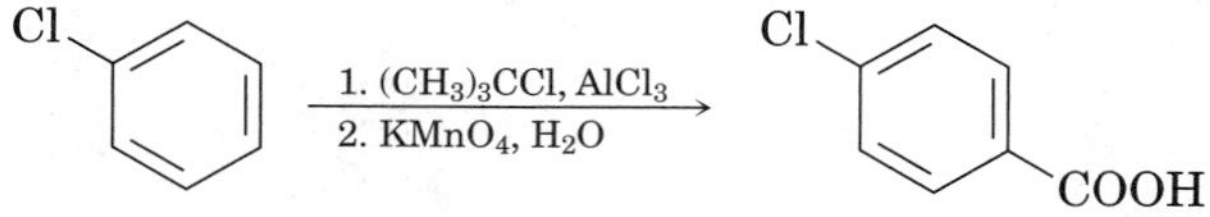

5.46 Pyridine is a cyclic nitrogen-containing compound that shows many of the properties associated with aromaticity. For example, pyridine undergoes electrophilic substitution reactions. Draw an orbital picture of pyridine and account for its aromatic properties.

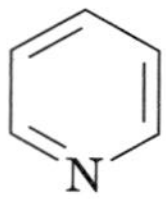

Pyridine

5.47 Would you expect the trimethylammonium group to be an activating or deactivating substituent? Explain.

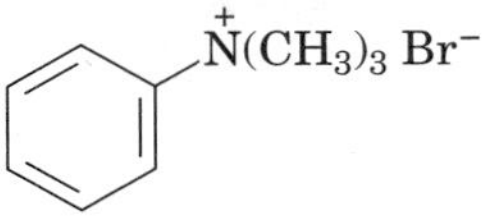

$\overset{+}{N}(CH_3)_3\ Br^-$

Phenyltrimethylammonium bromide

5.48 Starting with toluene, how would you synthesize the three nitrobenzoic acids?

5.49 Carbocations generated by reaction of an alkene with a strong acid catalyst can react with aromatic rings in a Friedel–Crafts reaction. Propose a mechanism to account for the industrial synthesis of the food preservative BHT from *p*-cresol and 2-methylpropene:

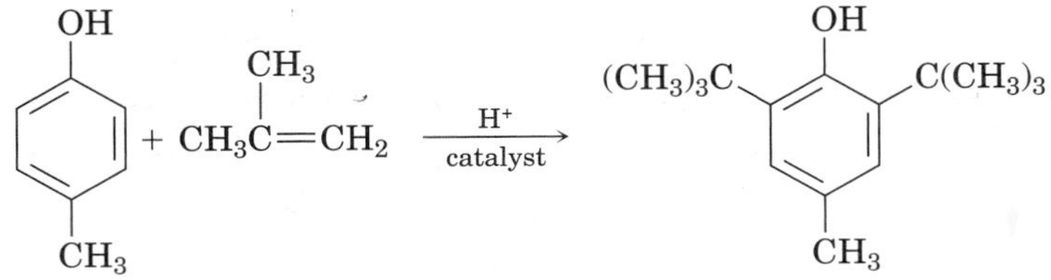

OH, CH_3 + $CH_3\overset{CH_3}{C}{=}CH_2$ $\xrightarrow[\text{catalyst}]{H^+}$ OH, $(CH_3)_3C$, $C(CH_3)_3$, CH_3

***p*-Cresol** **BHT**

5.50 You know the mechanism of HBr addition to alkenes, and you know the effects of various substituent groups on aromatic substitution. Use this knowledge to predict which of the following two alkenes reacts faster with HBr. Explain your answer by drawing resonance structures of the carbocation intermediates.

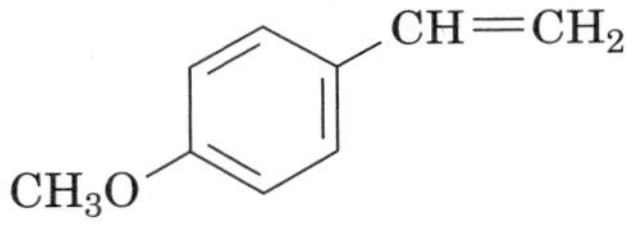

$CH{=}CH_2$, CH_3O and $CH{=}CH_2$, O_2N

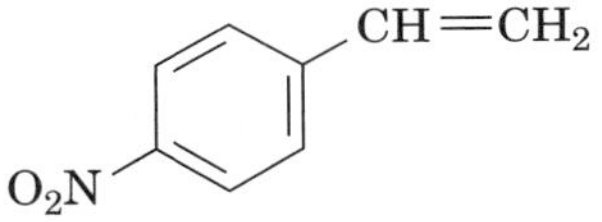

5.51 Identify the reagents represented by the letters a–d in the following scheme:

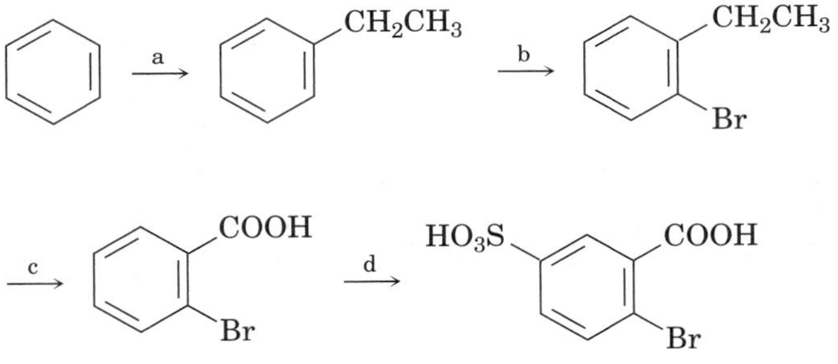

$\xrightarrow{a}$ CH_2CH_3 $\xrightarrow{b}$ CH_2CH_3, Br

$\xrightarrow{c}$ COOH, Br $\xrightarrow{d}$ HO_3S, COOH, Br

Visualizing Chemistry

5.52 Give IUPAC names for the following substances (gray = C, red = O, blue = N, light green = H).

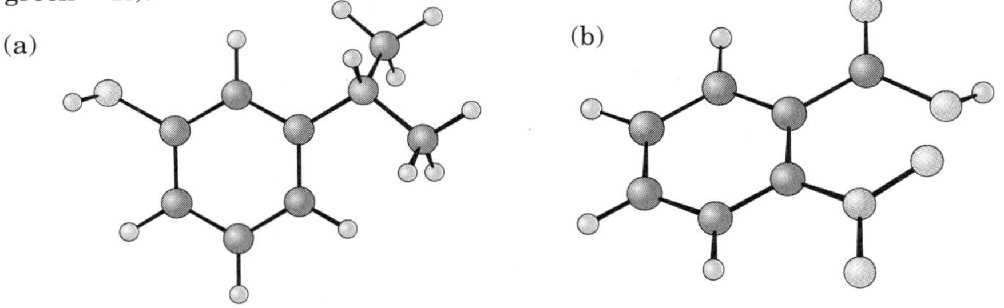

5.53 Draw and name the product from reaction of each of the following substances with (i) Br_2, $FeBr_3$ and (ii) CH_3COCl, $AlCl_3$ (gray = C, red = O, light green = H). (See Problem 5.32.)

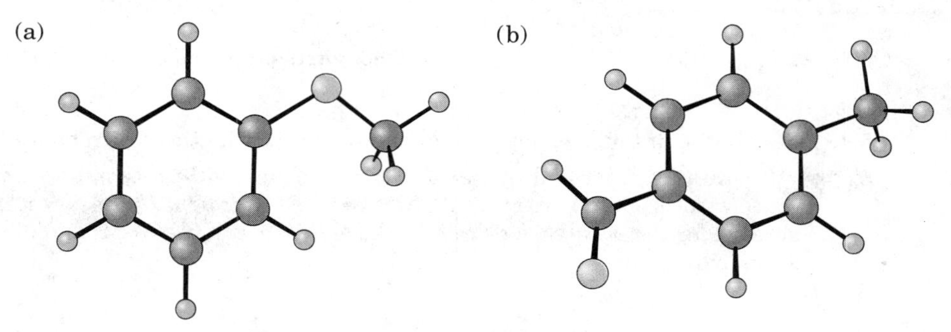

5.54 Draw two resonance structures for the following carbocation, indicating the positions of the double bonds.

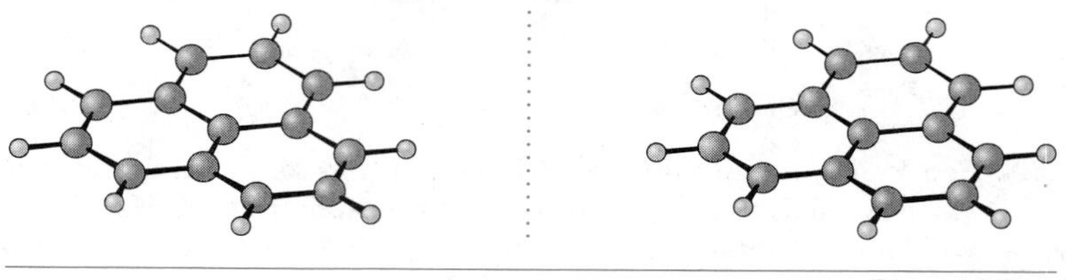

Stereo View

These two models of alanine [$CH_3CH(NH_2)COOH$] are *enantiomers,* or mirror images.

6 Stereochemistry

Up to this point, we've been concerned only with the general nature of chemical reactions and with the specific chemistry of hydrocarbon functional groups. Although we took a brief look at constitutional isomers of alkanes in Section 2.2 and cis–trans stereoisomers of cycloalkanes in Section 2.8, we've given little thought to any chemical consequences that might arise from the spatial arrangements of atoms in molecules. It's now time to look more deeply into these consequences. **Stereochemistry** is the branch of chemistry concerned with the three-dimensional nature of molecules.

6.1 Stereochemistry and the Tetrahedral Carbon

Are you right-handed or left-handed? Though most of us don't often think about it, handedness plays a surprisingly large part in our daily activities. Musical instruments such as oboes and clarinets have a handedness to them, the last available softball glove always fits the wrong hand, and left-handed people write in a "funny" way. The fundamental reason for these difficulties is that our hands aren't identical—they're *mirror images.* When you hold your *left* hand up to a mirror, the reflection looks like a *right* hand. Try it.

Handedness is also important in organic chemistry and is crucial in biochemistry, where carbohydrates, amino acids, nucleic acids, and many other naturally occurring molecules are handed. To see how molecular handedness arises, look at the molecules shown in Figure 6.1. On the left of Figure 6.1 are three molecules, and on the right are their images reflected in

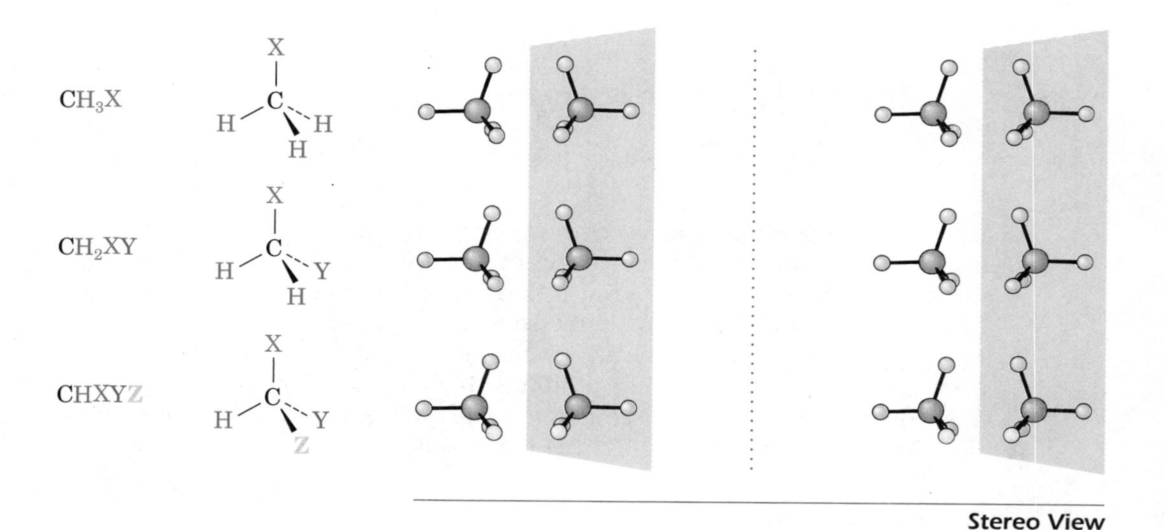

Figure 6.1 Three tetrahedral carbon atoms and their mirror images. Molecules of the type CH_3X and CH_2XY are identical to their mirror images, but a molecule of the type CHXYZ is not. A CHXYZ molecule is related to its mirror image in the same way that a right hand is related to a left hand.

a mirror. The CH_3X and CH_2XY molecules are identical to their mirror images and thus are not handed. If you make molecular models of each molecule and of its mirror image, you find that they are identical and that you can superimpose one on the other. By contrast, the CHXYZ molecule is *not* identical to its mirror image. You can't superimpose a model of the molecule on a model of its mirror image for the same reason that you can't superimpose a left hand on a right hand: They simply aren't the same. You might superimpose *two* of the substituents, X and Y for example, but H and Z would be reversed. If the H and Z substituents were superimposed, X and Y would be reversed.

A molecule that is not identical to its mirror image is a special kind of stereoisomer called an **enantiomer** (e-**nan**-tee-o-mer; Greek *enantio,* "opposite"). Enantiomers, which are related to each other as a right hand is related to a left hand, result whenever a tetrahedral carbon atom is bonded to four different substituents (one need not be H). For example, lactic acid (2-hydroxypropanoic acid) exists as a pair of enantiomers because there are four different groups (–H, –OH, $–CH_3$, –COOH) bonded to the central carbon atom:

H
CH3—C—COOH
OH

H
X—C—Z
Y

Lactic acid: a molecule of general formula CHXYZ

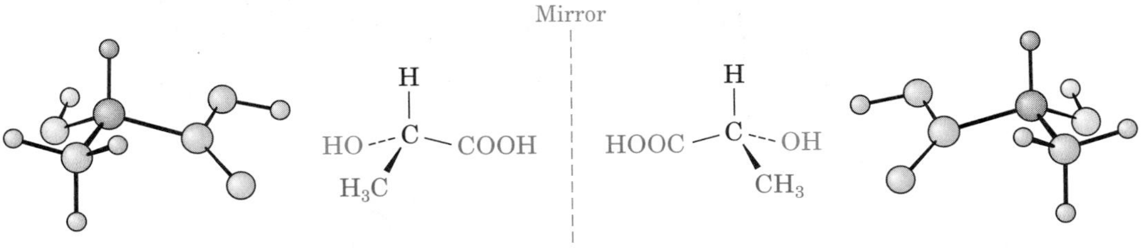

Mirror

H
HO---C—COOH
H3C

H
HOOC—C---OH
CH3

(+)-Lactic acid
$[\alpha]_D$ = +3.82°

(−)-Lactic acid
$[\alpha]_D$ = −3.82°

No matter how hard you try, you can't superimpose a molecule of "right-handed" lactic acid on top of a molecule of "left-handed" lactic acid; the two molecules aren't identical, as shown in Figure 6.2 (p. 186).

6.2 The Reason for Handedness in Molecules: Chirality

Compounds that are not identical to their mirror images and thus exist in two enantiomeric forms are said to be **chiral** (**ky**-ral; Greek *cheir,* "hand"). You can't take a chiral molecule and its mirror image (enantiomer) and place one on top of the other so that all atoms coincide.

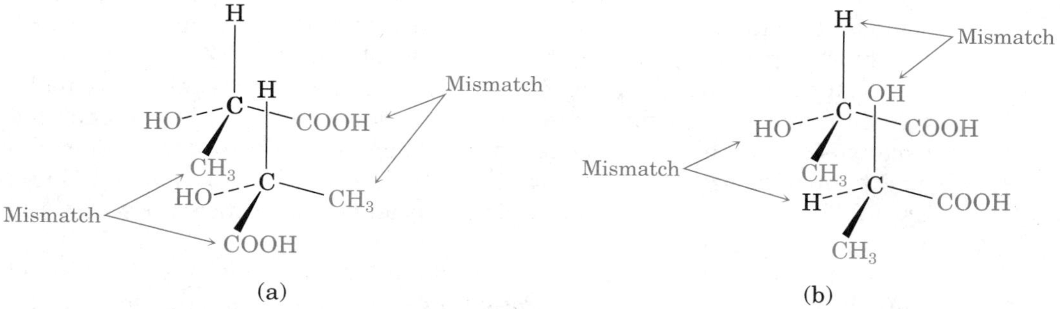

H
H
HO C COOH
Mismatch
CH_3
HO C CH_3
Mismatch
COOH
(a)

H
Mismatch
OH
HO C COOH
Mismatch
CH_3
H C COOH
CH_3
(b)

Figure 6.2 Attempts at superimposing the mirror-image forms of lactic acid: (a) When the –H and –OH substituents match up, the –COOH and $–CH_3$ substituents don't. (b) When –COOH and $–CH_3$ match up, –H and –OH don't. Regardless of how the molecules are oriented, they aren't identical.

How can you predict whether a given molecule is or is not chiral? *A compound is not chiral if it contains a plane of symmetry.* A **plane of symmetry** is a plane that cuts through the middle of an object (or molecule) so that one half of the object is an exact mirror image of the other half. A laboratory flask, for example, has a plane of symmetry. If you were to cut the flask in half, one half would be an exact mirror image of the other half. A hand, however, does not have a plane of symmetry. One "half" of a hand is not a mirror image of the other "half" (Figure 6.3).

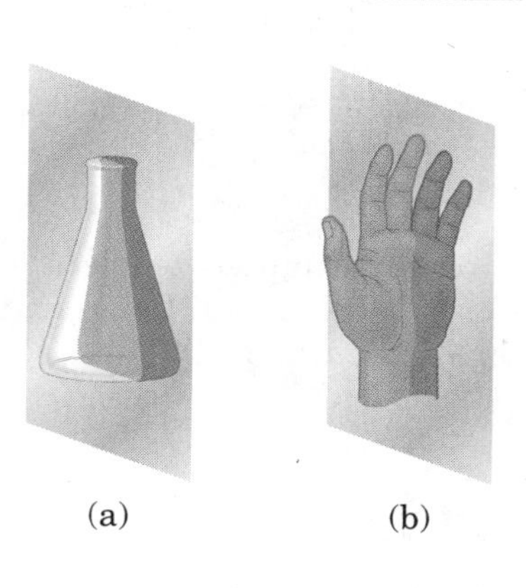

Figure 6.3 The meaning of *symmetry plane.* An object like the flask (a) has a plane of symmetry passing through it, making the right and left halves mirror images. An object like a hand (b) has no symmetry plane; the right "half" of a hand is not a mirror image of the left "half."

A molecule that has a plane of symmetry in any of its possible conformations must be identical to its mirror image and hence must be nonchiral, or **achiral** (a-**ky**-ral). Thus, propanoic acid, CH_3CH_2COOH, contains a plane of symmetry when lined up as shown in Figure 6.4 and is therefore achiral. Lactic acid, $CH_3CH(OH)COOH$, however, has no plane of symmetry and is chiral.

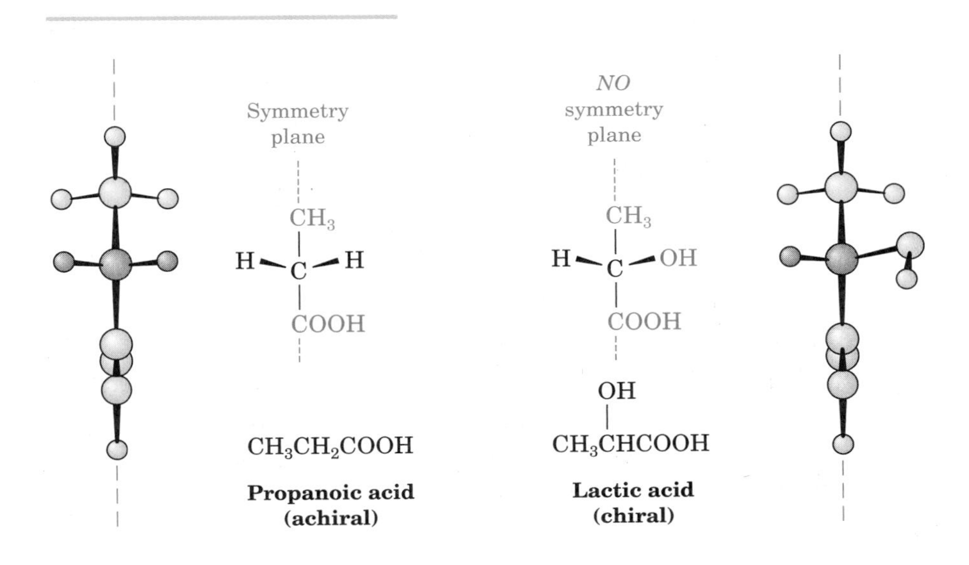

Figure 6.4 The achiral propanoic acid molecule versus the chiral lactic acid molecule. Propanoic acid has a plane of symmetry that makes one side of the molecule a mirror image of the other side. Lactic acid has no such symmetry plane.

The most common (although not the only) cause of chirality in an organic molecule is the presence of a carbon atom bonded to four different groups—for example, the central carbon atom in lactic acid. Such carbons are referred to as *asymmetric centers,* or **stereocenters.** (Note that *chirality* is a property of the entire molecule, whereas a stereocenter is the *cause* of chirality.)

Detecting stereocenters in a complex molecule takes practice, because it's not always immediately apparent that four different groups are bonded to a given carbon. The differences don't necessarily appear right next to the stereocenter. For example, 5-bromodecane is a chiral molecule because four different groups are bonded to C5 (marked by an asterisk):

$$CH_3CH_2CH_2CH_2CH_2\overset{\displaystyle Br}{\underset{\displaystyle H}{\overset{|}{\underset{|}{C^*}}}}CH_2CH_2CH_2CH_3$$

5-Bromodecane (chiral)

Substituents on carbon 5

—H

—Br

—$CH_2CH_2CH_2CH_3$ (butyl)

—$CH_2CH_2CH_2CH_2CH_3$ (pentyl)

A butyl substituent is very *similar* to a pentyl substituent, but it isn't identical. The difference isn't apparent until four carbons away from the stereocenter, but there's still a difference.

In the examples of chiral molecules shown below, check for yourself that the labeled atoms are indeed stereocenters. (When checking for stereocenters, it's helpful to note that CH_2, CH_3, C=C, C≡C, and C=O carbons *can't* be stereocenters, because they have at least two identical bonds.)

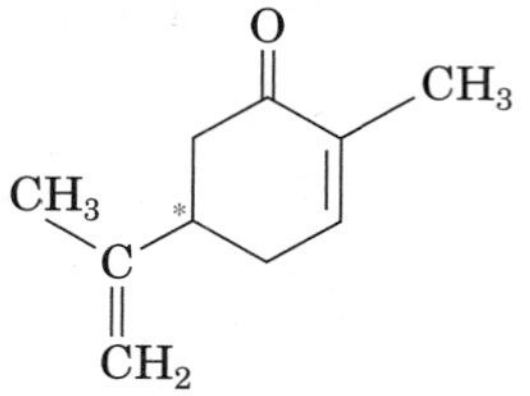

Carvone (spearmint oil) 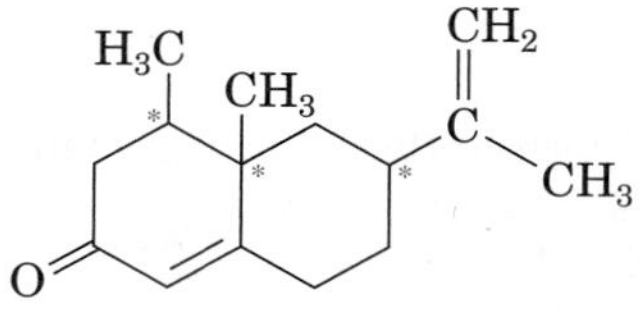 **Nootkatone (grapefruit oil)**

PRACTICE PROBLEM 6.1

Draw the structure of a chiral alcohol.

Solution An alcohol is a compound that contains the –OH functional group. To make an alcohol chiral, we need to have four different groups bonded to a single carbon atom, say –H, –OH, $–CH_3$, and $–CH_2CH_3$:

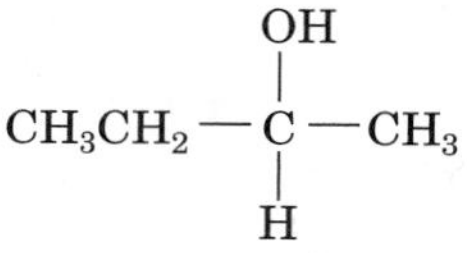

$$CH_3CH_2-\overset{\displaystyle OH}{\underset{\displaystyle H}{C}}-CH_3$$

2-Butanol

PRACTICE PROBLEM 6.2

Is 3-methylhexane chiral?

Solution Draw the structure of 3-methylhexane and cross out all the CH_2 and CH_3 carbons because they can't be stereocenters. Then look closely at any carbon that remains to see if it's bonded to four different groups. Since C3 is bonded to –H, $–CH_3$, $–CH_2CH_3$, and $–CH_2CH_2CH_3$, the molecule is chiral.

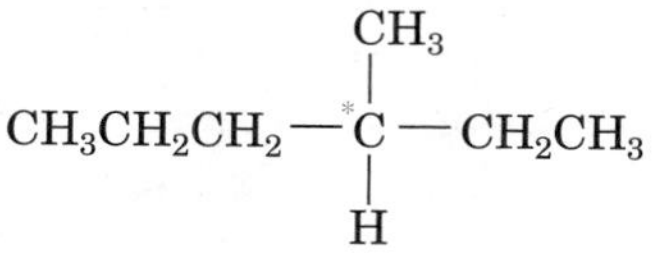

$$CH_3CH_2CH_2-\overset{\displaystyle CH_3}{\underset{\displaystyle H}{C^{*}}}-CH_2CH_3$$

3-Methylhexane (chiral)

PRACTICE PROBLEM 6.3

Is 2-methylcyclohexanone chiral?

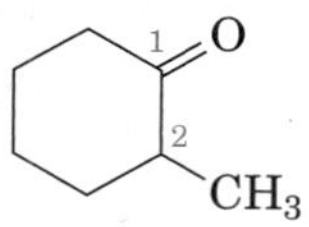

2-Methylcyclohexanone

Solution Ignoring the CH_3 carbon, the four CH_2 carbons in the ring, and the C=O carbon, look carefully at C2. Carbon 2 is bonded to four different groups: a $-CH_3$ group, an –H atom, a –C=O carbon in the ring, and a $-CH_2-$ ring carbon. Thus, 2-methylcyclohexanone is chiral.

PROBLEM

6.1 Which of the following objects are chiral (handed)?
(a) Bean stalk (b) Screwdriver (c) Screw (d) Shoe

PROBLEM

6.2 Which of the following compounds are chiral?
(a) 3-Bromopentane (b) 1,3-Dibromopentane
(c) 3-Methyl-1-hexene (d) *cis*-1,4-Dimethylcyclohexane

PROBLEM

6.3 Which of the following molecules are chiral? Identify the stereocenter(s) in each.

(a) CH_3

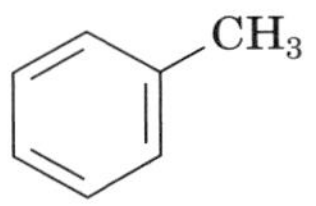

Toluene

(b) $CH_2CH_2CH_3$, N, H

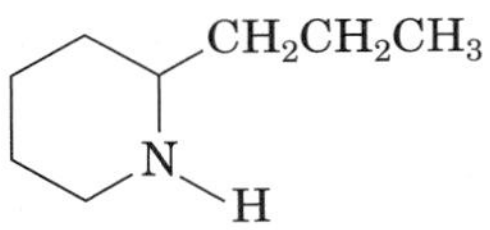

Coniine
(from poison hemlock)

(c) CH_3CH_2, O, O, N, N, H, H, O

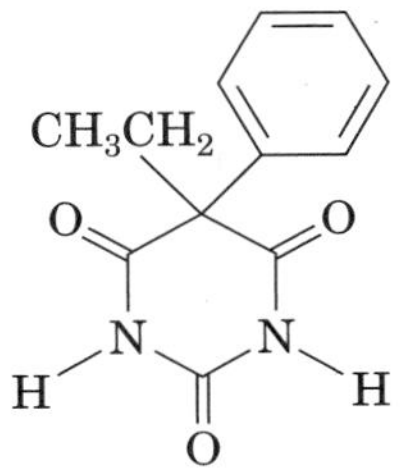

Phenobarbital
(tranquilizer)

PROBLEM

6.4 Place asterisks at the stereocenters in the following molecules:

(a)

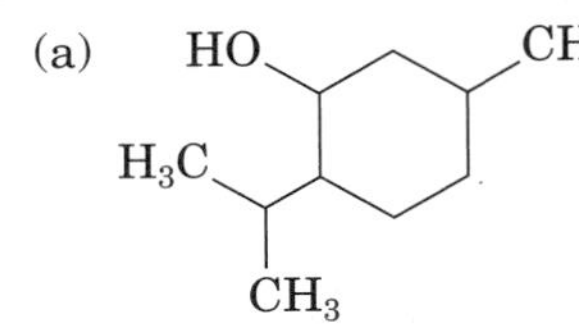

Menthol

(b) H_3C, CH_3, CH_3, O

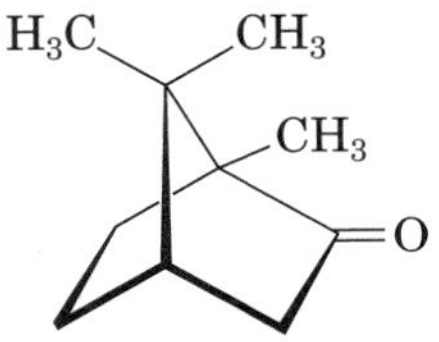

Camphor

(c) CH_3O, NCH_3

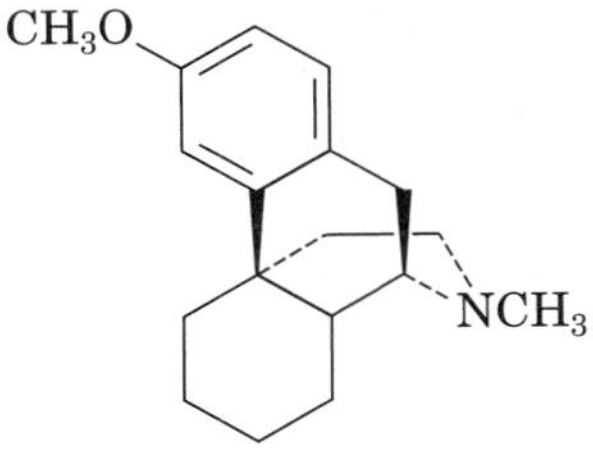

Dextromethorphan
(a cough suppressant)

PROBLEM

6.5 Alanine, an amino acid found in proteins, is a chiral molecule. Use the standard convention of wedged, solid, and dashed lines to draw the two enantiomers of alanine.

$$\begin{array}{c} NH_2 \\ | \\ CH_3CHCOOH \end{array} \quad \textbf{Alanine}$$

6.3 Optical Activity

The study of stereochemistry originated in the early nineteenth century during investigations of the French physicist Jean Baptiste Biot into the nature of *plane-polarized light.* A beam of ordinary light consists of electromagnetic waves that oscillate in an infinite number of planes at right angles to the direction of light travel. When a beam of ordinary light passes through a device called a *polarizer,* though, only the light waves oscillating in a *single* plane pass through, and the light is said to be plane-polarized. Light waves in all other planes are blocked out (Figure 6.5).

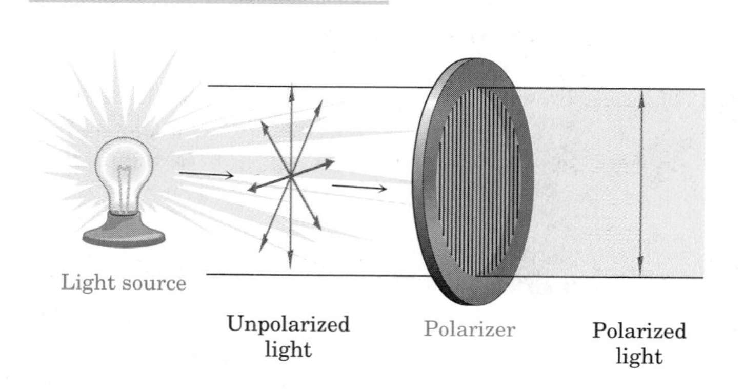

Figure 6.5 The nature of plane-polarized light. Only electromagnetic waves that oscillate in a single plane pass through the polarizer.

Biot made the remarkable observation that, when a beam of plane-polarized light passes through a solution of certain organic molecules, such as sugar or camphor, the plane of polarization is *rotated.* Not all organic molecules exhibit this property, but those that do are said to be **optically active.**

The amount of rotation can be measured with an instrument called a *polarimeter,* represented in Figure 6.6. Optically active organic molecules are placed in a sample tube, plane-polarized light is passed through the tube, and rotation of the plane occurs. The light then goes through a second polarizer called the *analyzer.* By rotating the analyzer until light passes through *it,* we can find the new plane of polarization and can tell to what

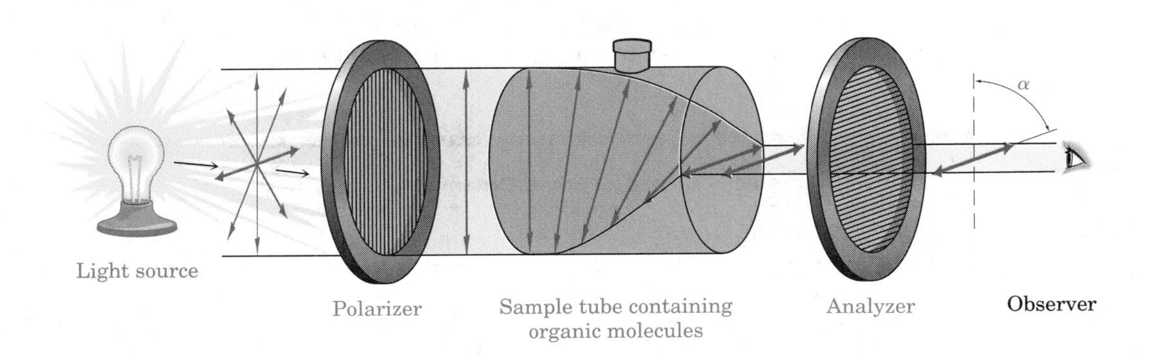

Figure 6.6 Schematic representation of a polarimeter. Plane-polarized light passes through a solution of optically active molecules, which rotate the plane of polarization.

extent rotation has occurred. The amount of rotation observed is denoted by α (Greek alpha) and is expressed in degrees.

In addition to determining the extent of rotation, we can also find the direction. From the vantage point of the observer looking at the analyzer, some optically active molecules rotate plane-polarized light to the left (counterclockwise) and are said to be **levorotatory,** whereas other molecules rotate light to the right (clockwise) and are said to be **dextrorotatory.** By convention, rotation to the left is given a minus sign (−), and rotation to the right is given a plus sign (+). For example, (−)-morphine is levorotatory and (+)-sucrose is dextrorotatory.

6.4 Specific Rotation

The amount of rotation observed in a polarimetry experiment depends on the structure of the sample molecules and on the number of molecules encountered by the light beam. The number of molecules encountered depends, in turn, on sample concentration and sample path length. If the concentration of the sample in a tube is doubled, the observed rotation is doubled. If the concentration is kept constant but the length of the sample tube is doubled, the observed rotation is doubled.

To express optical rotation data so that comparisons can be made, we have to choose standard conditions. The **specific rotation, $[\alpha]_D$,** of a compound is defined as the observed rotation when light of 589.6 nanometer (nm; 1 nm = 10^{-9} m) wavelength is used with a sample path length l of 1 decimeter (1 dm = 10 cm) and a sample concentration C of 1 g/mL. (Light of 589.6 nm, the so-called sodium D line, is the yellow light emitted from common sodium lamps.)

$$[\alpha]_D = \frac{\text{Observed rotation (degrees)}}{\text{Path length, } l \text{ (dm)} \times \text{Concentration, } C \text{ (g/mL)}} = \frac{\alpha}{l \times C}$$

When optical rotation data are expressed in this standard way, the specific rotation $[\alpha]_D$ is a physical constant characteristic of a given optically active compound. Some examples are listed in Table 6.1.

Table 6.1 Specific Rotations of Some Organic Molecules

Compound	$[\alpha]_D$ *(degrees)*	*Compound*	$[\alpha]_D$ *(degrees)*
Camphor	+44.26	Penicillin V	+233
Morphine	−132	Monosodium glutamate	+25.5
Sucrose	+66.47	Benzene	0
Cholesterol	−31.5	Acetic acid	0

PRACTICE PROBLEM 6.4

A 1.20 g sample of cocaine, $[\alpha]_D = -16°$, was dissolved in 7.50 mL of chloroform and placed in a sample tube having a path length of 5.00 cm. What was the observed rotation?

Solution Observed rotation, α, is equal to specific rotation $[\alpha]_D$ times sample concentration C times path length l:

$$\alpha = [\alpha]_D \times C \times l$$

where $[\alpha]_D = -16°$, $l = 5.00$ cm $= 0.500$ dm, and $C = 1.20$ g/7.50 mL $= 0.160$ g/mL.
Thus, $\alpha = -16° \times 0.500 \times 0.160 = -1.3°$.

PROBLEM

6.6 Is cocaine (Practice Problem 6.4) dextrorotatory or levorotatory?

PROBLEM

6.7 A 1.50 g sample of coniine, the toxic extract of poison hemlock, was dissolved in 10.0 mL of ethanol and placed in a sample tube with a path length of 5.00 cm. The observed rotation at the sodium D line was +1.21°. Calculate the specific rotation $[\alpha]_D$ for coniine.

6.5 Pasteur's Discovery of Enantiomers

Little was done after Biot's discovery of optical activity until 1848, when Louis Pasteur began work on a study of crystalline tartaric acid salts derived from wine. On recrystallizing a concentrated solution of sodium ammonium tartrate below 28°C, Pasteur made the surprising observation that two distinct kinds of crystals precipitated. Furthermore, the two kinds of crystals were *mirror images* and were related in the same way that a right hand is related to a left hand.

Working carefully with a pair of tweezers, Pasteur was able to separate the crystals into two piles, one of "right-handed" crystals and one of "left-handed" crystals, like those shown in Figure 6.7. Although the original sample (a 50:50 mixture of right and left) was optically inactive, *solutions of crystals from each of the sorted piles were optically active,* and their specific rotations were equal in amount but opposite in sign.

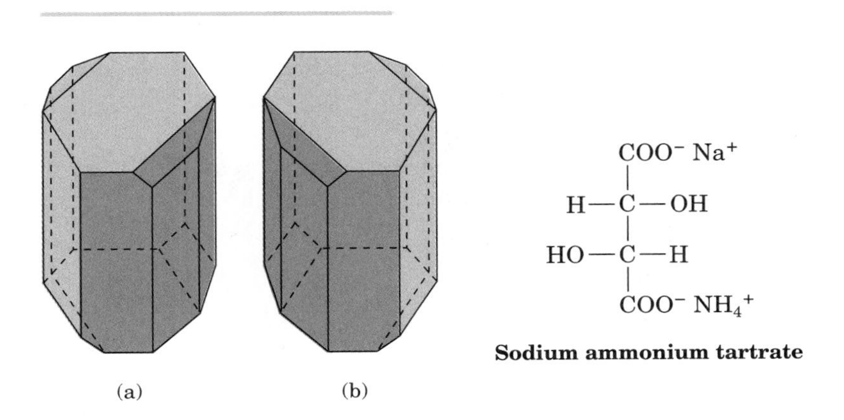

Figure 6.7 Crystals of sodium ammonium tartrate. One of the crystals is dextrorotatory in solution, and the other is levorotatory. The drawings are taken from Pasteur's original sketches.

Pasteur was far ahead of his time. Although the structural theory of Kekulé had not yet been proposed, Pasteur explained his results by speaking of the molecules themselves, saying, "There is no doubt that [in the *dextro* tartaric acid] there exists an asymmetric arrangement having a nonsuperimposable image. It is no less certain that the atoms of the *levo* acid possess precisely the inverse asymmetric arrangement." Pasteur's vision was extraordinary, for it was not until 25 years later that his theories regarding the asymmetry of chiral molecules were confirmed.

Today, we would describe Pasteur's work by saying that he had discovered the phenomenon of enantiomerism. Enantiomers (also called *optical isomers*) have identical physical properties, such as melting points and boiling points, but differ in the direction in which their solutions rotate plane-polarized light.

6.6 Sequence Rules for Specifying Configuration

Although drawings provide visual representations of stereochemistry, they are difficult to translate into words. Thus, a verbal method for specifying the three-dimensional arrangement (the **configuration**) of substituents around a stereocenter is also necessary. The method used employs the same sequence rules given in Section 3.4 for specifying *E* and *Z* alkene stereochemistry. Let's briefly review these sequence rules and see how they're used to specify the configuration of a stereocenter. For a more thorough review, you should reread Section 3.4.

Rule 1 Look at the four atoms directly attached to the stereocenter, and assign priorities in order of decreasing atomic number. The atom with highest atomic number is ranked first; the atom with lowest atomic number is ranked fourth.

Rule 2 If a decision about priority can't be reached by applying rule 1, compare atomic numbers of the second atoms in each substituent, continuing on as necessary through the third or fourth atoms until a point of difference is reached.

Rule 3 Multiple-bonded atoms are considered equivalent to the same number of single-bonded atoms. For example:

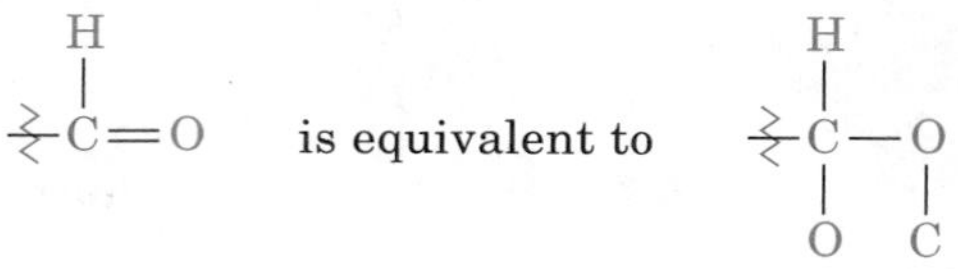

H
|
—C=O is equivalent to —C(H)(O)—O—C

Having assigned priorities to the four groups attached to a stereocenter, we describe the stereochemical configuration around the carbon by comparing the four groups to the thumb and fingers of a hand. Hold up one of your hands, with fingers curled and thumb outstretched, and orient it so that the thumb points from the carbon to the group of lowest priority (4). If you are using your *right* hand and the fingers curl in the direction of decreasing priority (1 → 2 → 3) for the remaining three groups, we say that the stereocenter has the ***R* configuration** (Latin *rectus,* "right"). If it's the fingers of your *left* hand that curl in the 1 → 2 → 3 direction, the stereocenter has the ***S* configuration** (Latin *sinister,* "left").

Look at (−)-lactic acid to see an example of how a configuration is assigned. Sequence rule 1 says that –OH has priority 1 and –H has priority 4, but it doesn't let us distinguish between $-CH_3$ and –COOH because both groups have carbon as their first atom. Sequence rule 2, however, says that –COOH is higher priority than $-CH_3$ because oxygen outranks hydrogen (the second atom in each group).

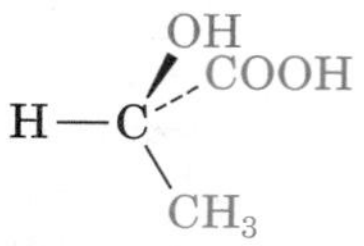

H—C(OH)(COOH)(CH₃)

(−)-Lactic acid

Priorities

4	—H	(Low)
3	$-CH_3$	
2	—C(=O)—OH	
1	—OH	(High)

When a right hand is held so that the thumb points from the carbon to the fourth-priority group (–H), the fingers of the hand curl in the 1 → 2 → 3 direction, so we assign *R* configuration to (−)-lactic acid. Applying the same procedure to (+)-lactic acid leads to the opposite assignment (Figure 6.8).

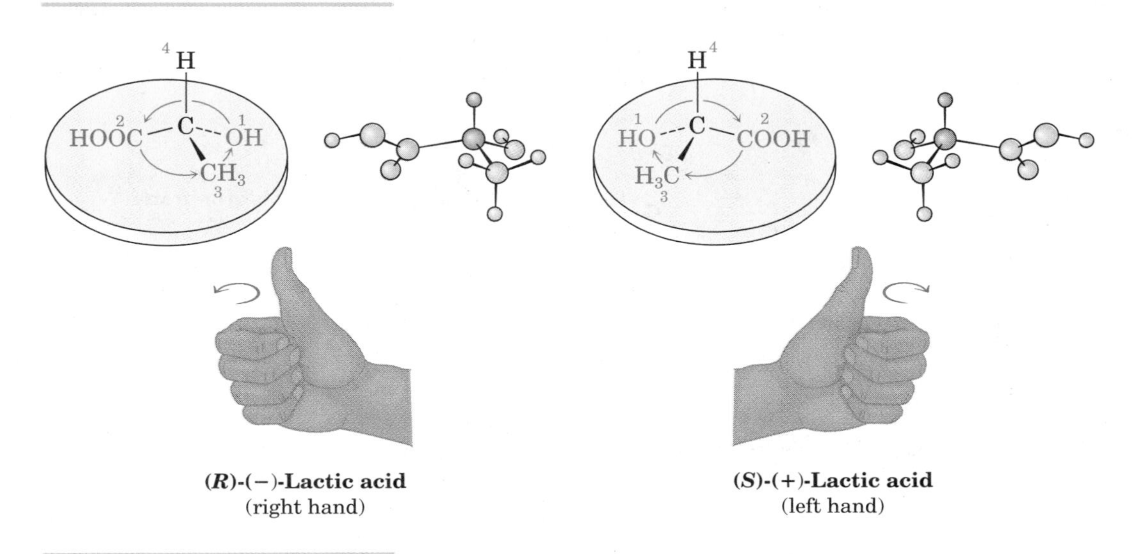

Figure 6.8 Assignment of configuration to (*R*)-(−)-lactic acid and (*S*)-(+)-lactic acid.

Further examples are provided by naturally occurring (−)-glyceraldehyde and (+)-alanine, which have the *S* configurations shown in Figure 6.9. *Note that the sign of optical rotation, (+) or (−), is not related to the R,S designation.* (*S*)-Alanine happens to be dextrorotatory (+), and (*S*)-glyceraldehyde happens to be levorotatory (−), but there is no simple correlation between *R,S* configuration and direction of optical rotation.

(*S*)-Glyceraldehyde
[(*S*)-(−)-2,3-Dihydroxypropanal]
$[\alpha]_D = -8.7°$

(*S*)-Alanine
[(*S*)-(+)-2-Aminopropanoic acid]
$[\alpha]_D = +8.5°$

Figure 6.9 Assignment of configuration to (−)-glyceraldehyde and (+)-alanine. Both happen to have the *S* configuration, although one is levorotatory and the other is dextrorotatory.

PRACTICE PROBLEM 6.5

Draw a tetrahedral representation of (*R*)-2-chlorobutane.

Solution The four substituents bonded to the stereocenter of (*R*)-2-chlorobutane can be assigned the following priorities: (1) –Cl, (2) $-CH_2CH_3$, (3) $-CH_3$, (4) –H. To draw a tetrahedral representation of the molecule, it's easiest to orient the low-priority –H group toward the top and then arrange the other three groups so that the direction of travel from 1 → 2 → 3 corresponds to the curl of your right hand.

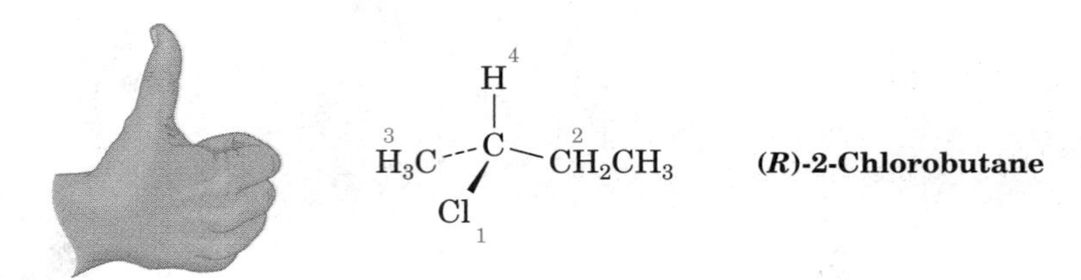

Using molecular models is a great help in working problems of this sort.

PROBLEM

6.8 Assign priorities to the substituents in each of the following sets:
(a) –H, –Br, $-CH_2CH_3$, $-CH_2CH_2OH$ (b) –COOH, $-COOCH_3$, $-CH_2OH$, –OH
(c) –Br, $-CH_2Br$, –Cl, $-CH_2Cl$

PROBLEM

6.9 Assign *R,S* configurations to the following molecules:

(a) Br on top of C; H_3C (dashed, left), H (wedge, bottom), COOH (right)

(b) H on top of C; HO (dashed, left), CH_3 (wedge, bottom), COOH (right)

(c) NH_2 on top of C; NC (dashed, left), H (wedge, bottom), CH_3 (right)

PROBLEM

6.10 Draw a tetrahedral representation of (*S*)-2-hydroxypentane (2-pentanol).

6.7 Diastereomers

Molecules like lactic acid and glyceraldehyde are relatively simple to deal with because each has only one stereocenter and only two enantiomeric forms. The situation becomes more complex, however, for molecules that have more than one stereocenter. Take the amino acid threonine (2-amino-3-hydroxybutanoic acid) as an example. Since threonine has two stereocenters (C2 and C3), there are four possible stereoisomers, as shown in Figure 6.10. (Check for yourself that the *R,S* configurations are correct.)

The four stereoisomers of 2-amino-3-hydroxybutanoic acid can be classified into two pairs of enantiomers (mirror images). The 2*S*,3*S* stereoiso-

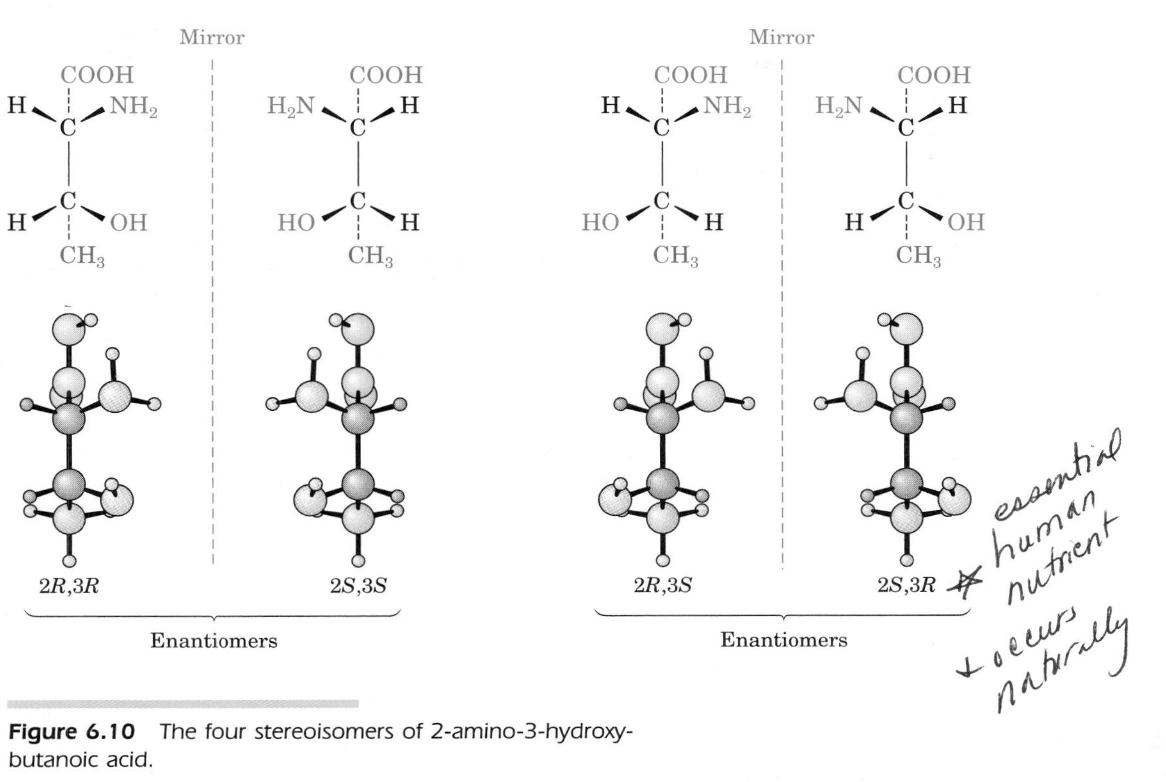

Figure 6.10 The four stereoisomers of 2-amino-3-hydroxy-butanoic acid.

mer is the mirror image of 2*R*,3*R*, and the 2*S*,3*R* stereoisomer is the mirror image of 2*R*,3*S*. But what is the relationship between any two configurations that are not mirror images? What, for example, is the relationship between the 2*R*,3*R* isomer and the 2*R*,3*S* isomer? These two compounds are stereoisomers, yet they aren't enantiomers. To describe such a relationship, we need a new term—*diastereomer.*

Diastereomers are stereoisomers that are not mirror images. Since we used the right-hand/left-hand analogy to describe the relationship between two enantiomers, we might extend the analogy by saying that the relationship between diastereomers is that of hands from two different people. Your hand and your friend's hand look *similar,* but they aren't identical and they aren't mirror images. The same is true of diastereomers; they're similar, but not identical and not mirror images.

Note carefully the difference between enantiomers and diastereomers: Enantiomers have opposite configurations at *all* stereocenters; diastereomers have opposite configurations at *some* (one or more) stereocenters, but the same configuration at others. A full description of the four threonine stereoisomers is given in Table 6.2 (p. 198).

Of the four threonine stereoisomers, only the 2*S*,3*R* isomer, $[\alpha]_D = -28.3°$, occurs naturally and is an essential human nutrient. Most biologically important organic molecules are chiral, and usually only one stereoisomer is found in nature.

Table 6.2 Relationships Among the Four Stereoisomers of Threonine

Stereoisomer	*Enantiomeric with*	*Diastereomeric with*
2*R*,3*R*	2*S*,3*S*	2*R*,3*S* and 2*S*,3*R*
2*S*,3*S*	2*R*,3*R*	2*R*,3*S* and 2*S*,3*R*
2*R*,3*S*	2*S*,3*R*	2*R*,3*R* and 2*S*,3*S*
2*S*,3*R*	2*R*,3*S*	2*R*,3*R* and 2*S*,3*S*

PROBLEM

6.11 Assign *R* or *S* configuration to each stereocenter in the following molecules:

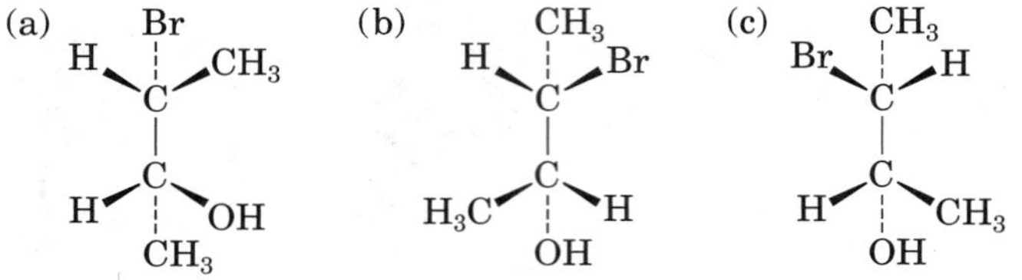

(a) Br, H, CH_3, C, C, H, OH, CH_3

(b) CH_3, H, Br, C, C, H_3C, H, OH

(c) CH_3, Br, H, C, C, H, CH_3, OH

PROBLEM

6.12 Which of the compounds in Problem 6.11 are enantiomers, and which are diastereomers?

PROBLEM

6.13 Chloramphenicol is a powerful antibiotic isolated from the *Streptomyces venezuelae* bacterium. It is active against a broad spectrum of bacterial infections and is particularly valuable against typhoid fever. Assign *R* or *S* configuration to the stereocenters in chloramphenicol.

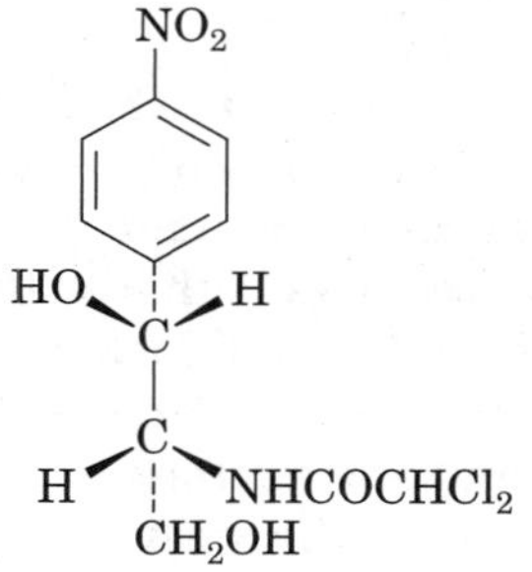

NO_2, HO, H, C, C, H, $NHCOCHCl_2$, CH_2OH

Chloramphenicol
$[\alpha]_D = +18.6°$

6.8 Meso Compounds

Let's look at one more example of a compound with two stereocenters: the tartaric acid used by Pasteur. The four stereoisomers can be drawn as follows:

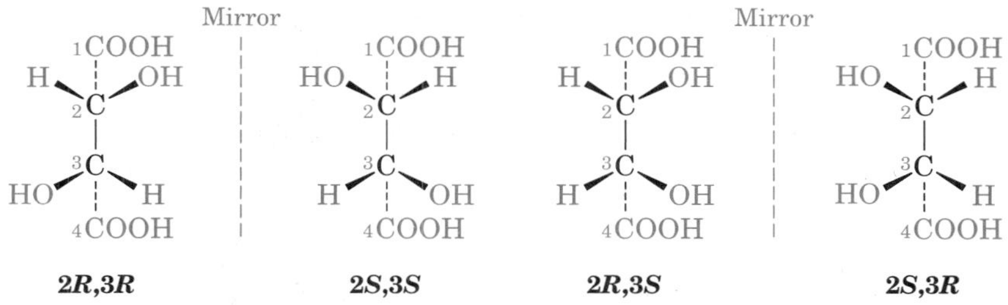

Mirror Mirror

1COOH, H, OH, 2C, 3C, HO, H, 4COOH — **2R,3R**

1COOH, HO, H, 2C, 3C, H, OH, 4COOH — **2S,3S**

1COOH, H, OH, 2C, 3C, H, OH, 4COOH — **2R,3S**

1COOH, HO, H, 2C, 3C, HO, H, 4COOH — **2S,3R**

The mirror-image 2*R*,3*R* and 2*S*,3*S* structures are not identical and therefore represent an enantiomeric pair. A careful look, however, shows that the 2*R*,3*S* and 2*S*,3*R* structures *are* identical, as can be seen by rotating one structure 180°:

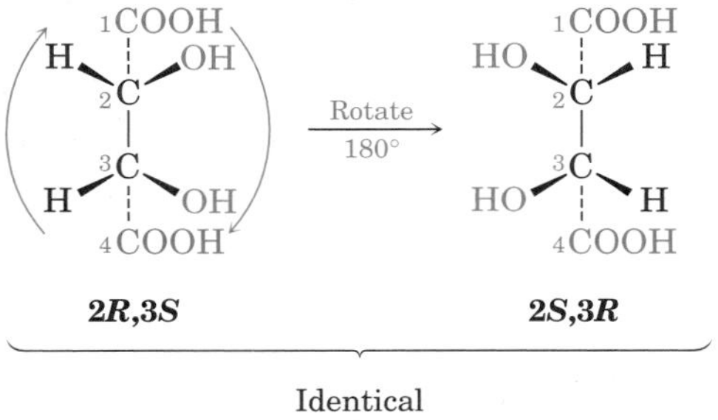

1COOH, H, OH, 2C, 3C, H, OH, 4COOH — **2R,3S**

Rotate 180°

1COOH, HO, H, 2C, 3C, HO, H, 4COOH — **2S,3R**

Identical

The 2*R*,3*S* and 2*S*,3*R* structures are identical because the molecule has a plane of symmetry and is therefore achiral. The symmetry plane cuts through the C2–C3 bond, making one half of the molecule a mirror image of the other half (Figure 6.11).

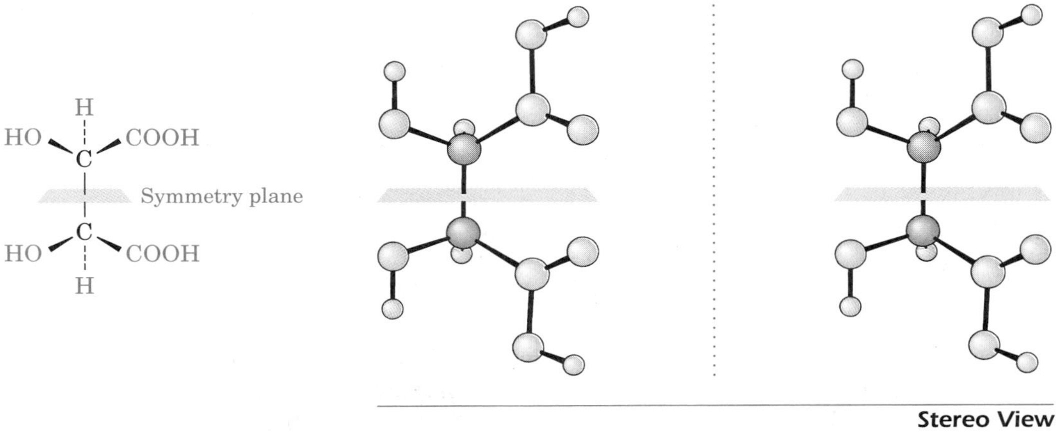

H, HO, C, COOH, Symmetry plane, HO, C, COOH, H

Stereo View

Figure 6.11 A symmetry plane cutting through the C2–C3 bond of *meso*-tartaric acid makes the molecule achiral.

Because of the plane of symmetry, the tartaric acid stereoisomer shown in Figure 6.11 is achiral, despite the fact that it has two stereocenters. Such compounds that are achiral, yet contain stereocenters, are called **meso compounds** (**me**-zo). Thus, tartaric acid exists in three stereoisomeric forms: two enantiomers and one meso form.

PRACTICE PROBLEM 6.6

Does *cis*-1,2-dimethylcyclobutane have any stereocenters? Is it a chiral molecule?

Solution Looking at the structure of *cis*-1,2-dimethylcyclobutane, we see that both of the methyl-bearing ring carbons (C1 and C2) are stereocenters. Overall, though, the compound is achiral because there is a symmetry plane bisecting the ring between C1 and C2. Thus, *cis*-1,2-dimethylcyclobutane is a meso compound.

Symmetry plane

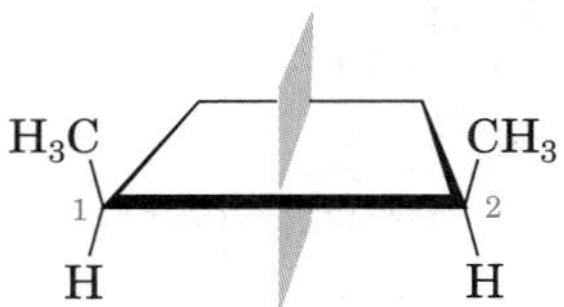

PROBLEM

6.14 Which of the following substances have meso forms?
(a) 2,3-Dibromobutane (b) 2,3-Dibromopentane (c) 2,4-Dibromopentane

PROBLEM

6.15 Which of the following structures represent meso compounds?

(a) (b) 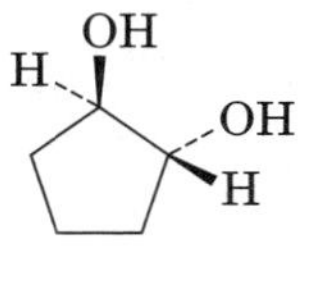 (c)

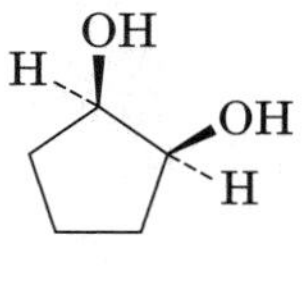

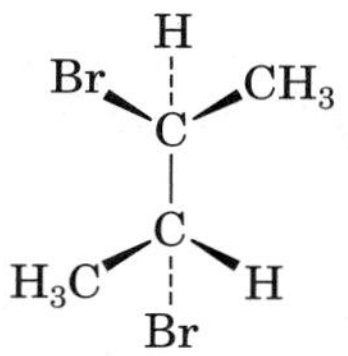

6.9 Molecules with More Than Two Stereocenters

One stereocenter gives rise to two stereoisomers (one pair of enantiomers), and two stereocenters give rise to a maximum of four stereoisomers (two pairs of enantiomers). In general, a molecule with n stereocenters has a maximum of 2^n stereoisomers (2^{n-1} pairs of enantiomers). For example, cholesterol has eight stereocenters. Thus, $2^8 = 256$ stereoisomers of cholesterol, or 128 pairs of enantiomers, are possible in principle, though many would be too strained to exist. Only one, however, is produced in nature.

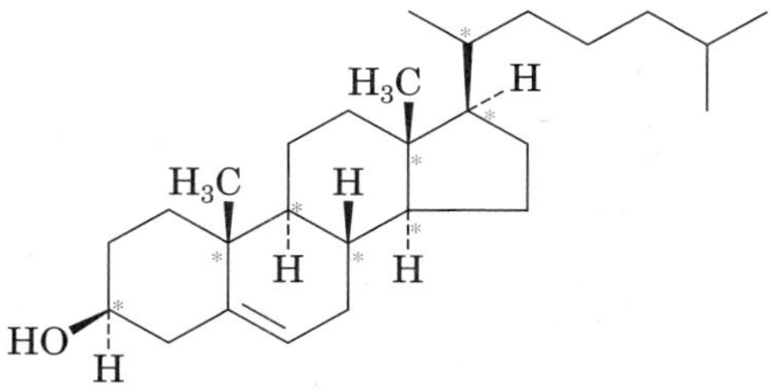

Cholesterol
(eight stereocenters)

PROBLEM

6.16 Nandrolone is an anabolic steroid used by some athletes to build muscle mass. How many stereocenters does nandrolone have? How many stereoisomers of nandrolone are possible in principle?

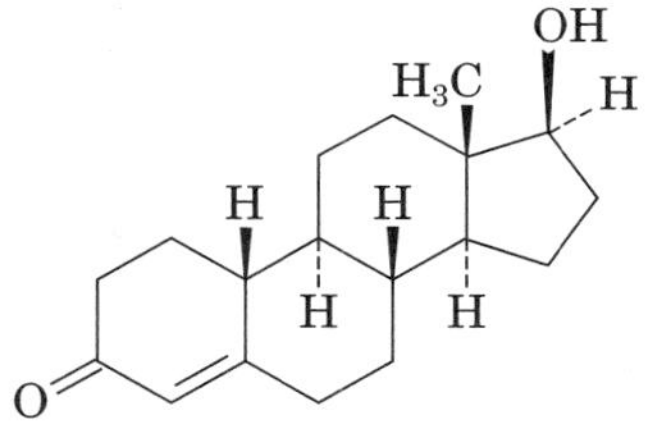

Nandrolone

6.10 Racemic Mixtures and the Resolution of Enantiomers

To conclude this discussion of stereoisomerism, let's return for a final look at Pasteur's pioneering work. Pasteur took an optically inactive tartaric acid salt and found that he could crystallize from it two optically active forms having the 2*R*,3*R* and 2*S*,3*S* configurations. But what was the optically inactive form he started with? It couldn't have been *meso*-tartaric acid, because *meso*-tartaric acid is a different compound and can't interconvert with the two chiral enantiomers without breaking and re-forming bonds.

The answer is that Pasteur started with a 50:50 *mixture* of the two chiral tartaric acid enantiomers. Such a mixture is called a **racemic** (ray-**see**-mic) **mixture,** or **racemate.** Racemic mixtures, often denoted by the symbol (±), show no optical activity because they contain equal amounts of (+) and (−) forms. The (+) rotation from one enantiomer exactly cancels the (−) rotation from the other. Through good fortune, Pasteur was able to separate, or **resolve,** (±)-tartaric acid into its (+) and (−) enantiomers. Unfortunately, the fractional crystallization technique he used doesn't work for most racemic mixtures, and other methods are required.

The most common method of resolution uses an acid–base reaction between a racemic mixture of chiral carboxylic acids and an amine. Just as inorganic acids such as HCl react with ammonia (NH_3) to yield ammonium salts, so organic carboxylic acids (RCOOH) react with amines (RNH_2):

$$HCl + NH_3 \longrightarrow NH_4^+ \ Cl^-$$

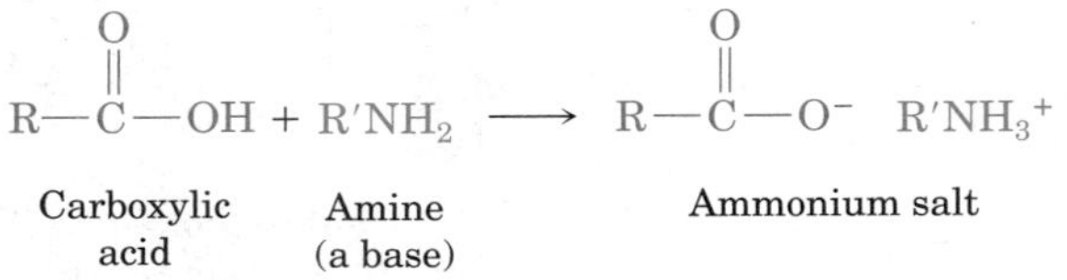

$$R{-}\overset{O}{\overset{\|}{C}}{-}OH + R'NH_2 \longrightarrow R{-}\overset{O}{\overset{\|}{C}}{-}O^- \ R'NH_3^+$$

Carboxylic acid — Amine (a base) — Ammonium salt

To understand how this method of resolution works, let's see what happens when a racemic mixture of chiral acids, such as (+)- and (−)-lactic acids, reacts with an achiral amine base such as methylamine (Figure 6.12). Stereochemically, the situation is analogous to what happens when left and right hands (chiral) pick up a ball (achiral). Both left and right hands pick up the ball equally well, and the products—ball in right hand versus ball in left hand—are mirror images. In the same way, both (+)- and (−)-lactic acid react with methylamine equally well, and the product is a racemic mixture of methylammonium (+)-lactate and methylammonium (−)-lactate.

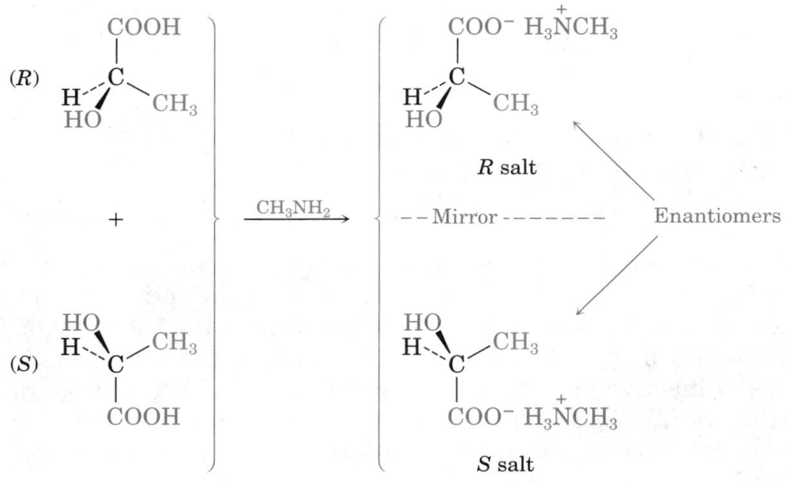

Figure 6.12 Reaction of racemic lactic acid with achiral methylamine leads to a racemic mixture of ammonium salts.

Now let's see what happens when the racemic mixture of (+)- and (−)-lactic acids reacts with a *single* enantiomer of a chiral amine base, such as (*R*)-1-phenylethylamine (Figure 6.13). Stereochemically, this situation is analogous to what happens when a hand (a chiral reagent) puts on a glove (*also a chiral reagent*). *Left and right hands do not put on the same glove in the same way.* The products—right hand in right glove versus left hand in right glove—are not mirror images; they're altogether different.

In the same way, (+)- and (−)-lactic acid react with (*R*)-1-phenylethylamine to give different products. (*R*)-Lactic acid reacts with (*R*)-1-phenylethylamine to give the *R,R* salt, whereas (*S*)-lactic acid reacts with the same *R* amine to give the *S,R* salt. *The two salts are diastereomers* (Section 6.7). They are different compounds and have different chemical and physical properties. It may therefore be possible to separate them by crystallization or some other means. Once separated, acidification of the two diastereomeric salts with HCl then allows us to isolate the two pure enantiomers of lactic acid and to recover the chiral amine for further use.

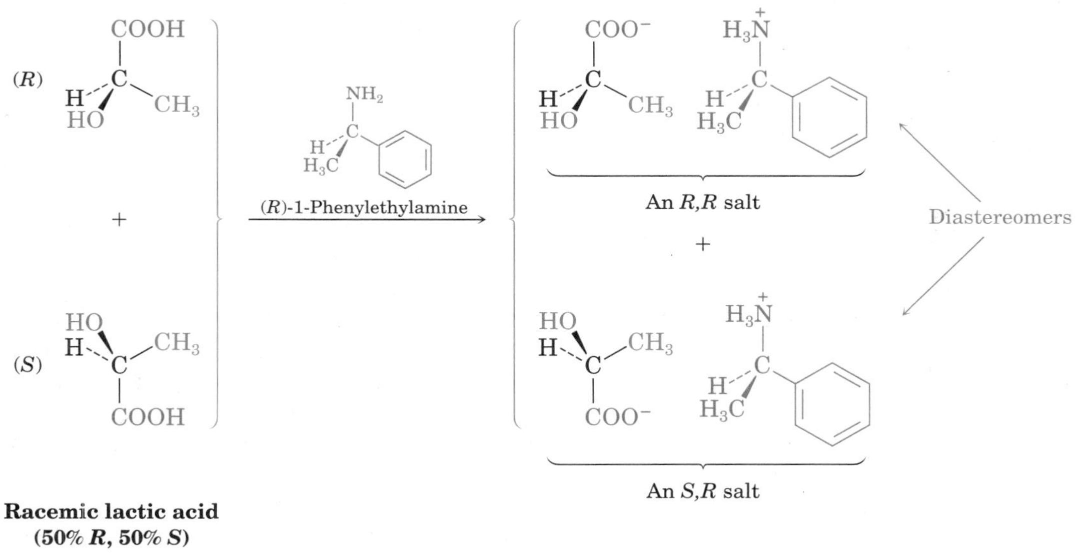

(*R*)-1-Phenylethylamine

Racemic lactic acid (50% *R*, 50% *S*)

An *R,R* salt

An *S,R* salt

Diastereomers

Figure 6.13 Reaction of racemic lactic acid with optically pure (*R*)-1-phenylethylamine leads to a mixture of diastereomeric salts, which have different properties and can, in principle, be separated.

PRACTICE PROBLEM 6.7

We'll see in Section 10.6 that carboxylic acids (RCOOH) react with alcohols (R′OH) to form esters (RCOOR′). Suppose that racemic lactic acid reacts with methanol to form the ester, methyl lactate. What stereochemistry would you expect the product(s) to have? What is the relationship of the products?

$$CH_3CH(OH)COOH + CH_3OH \longrightarrow CH_3CH(OH)COOCH_3 + H_2O$$

Lactic acid **Methanol** **Methyl lactate**

Solution Reaction of a racemic acid with an achiral alcohol such as methanol yields a racemic mixture of mirror-image (enantiomeric) products:

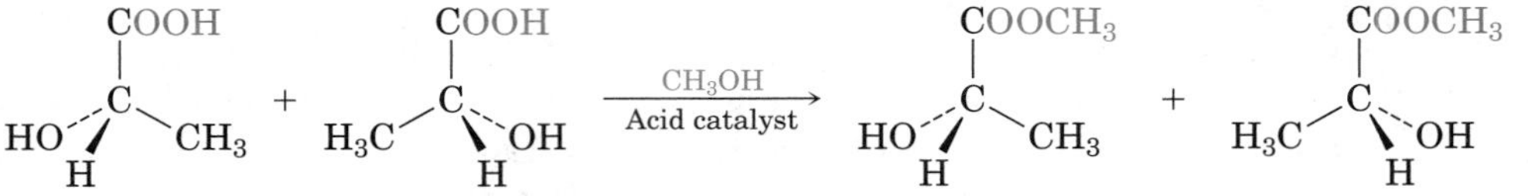

$$\underset{\textbf{(S)-Lactic acid}}{HO\text{-}CH(CH_3)\text{-}COOH} + \underset{\textbf{(R)-Lactic acid}}{H_3C\text{-}CH(OH)\text{-}COOH} \xrightarrow[\text{Acid catalyst}]{CH_3OH} \underset{\textbf{Methyl (S)-lactate}}{HO\text{-}CH(CH_3)\text{-}COOCH_3} + \underset{\textbf{Methyl (R)-lactate}}{H_3C\text{-}CH(OH)\text{-}COOCH_3}$$

PROBLEM

6.17 Suppose that acetic acid (CH_3COOH) reacts with (*S*)-2-butanol to form an ester (see Practice Problem 6.7). What stereochemistry would you expect the product(s) to have? What is the relationship of the products?

$$\underset{\textbf{Acetic acid}}{CH_3\overset{O}{\overset{\|}{C}}OH} + \underset{\textbf{2-Butanol}}{HO\overset{CH_3}{\overset{|}{C}}HCH_2CH_3} \longrightarrow \underset{\textbf{\textit{sec}-Butyl acetate}}{CH_3\overset{O}{\overset{\|}{C}}O\overset{CH_3}{\overset{|}{C}}HCH_2CH_3} + H_2O$$

6.11 Physical Properties of Stereoisomers

If such seemingly simple compounds as tartaric acid can exist in different stereoisomeric configurations, the question arises whether the different stereoisomers have different physical properties. The answer is yes, they do.

Some physical properties of the three stereoisomers of tartaric acid and of the racemic mixture are shown in Table 6.3. The (+) and (−) enantiomers have identical melting points, solubilities, and densities. They differ only in the sign of their rotation of plane-polarized light. The meso isomer, by contrast, is diastereomeric with the (+) and (−) forms. It is therefore a different compound altogether, and has different physical properties. The racemic mixture is different still. Though a mixture of enantiomers, racemates act as though they were pure compounds, different from either enantiomer or from the meso form.

Table 6.3 Some Properties of the Stereoisomers of Tartaric Acid

Stereoisomer	*Melting point (°C)*	$[\alpha]_D$ *(degrees)*	*Density* (g/cm^3)	*Solubility at 20°C* ($g/100\ mL\ H_2O$)
(+)	168–170	+12	1.7598	139.0
(−)	168–170	−12	1.7598	139.0
Meso	146–148	0	1.6660	125.0
(±)	206	0	1.7880	20.6

6.12 A Brief Review of Isomerism

As noted on several previous occasions, isomers are compounds that have the same chemical formula but different structures. We've seen several kinds of isomers in the past few chapters, and it's a good idea at this point to see how they relate to one another (Figure 6.14).

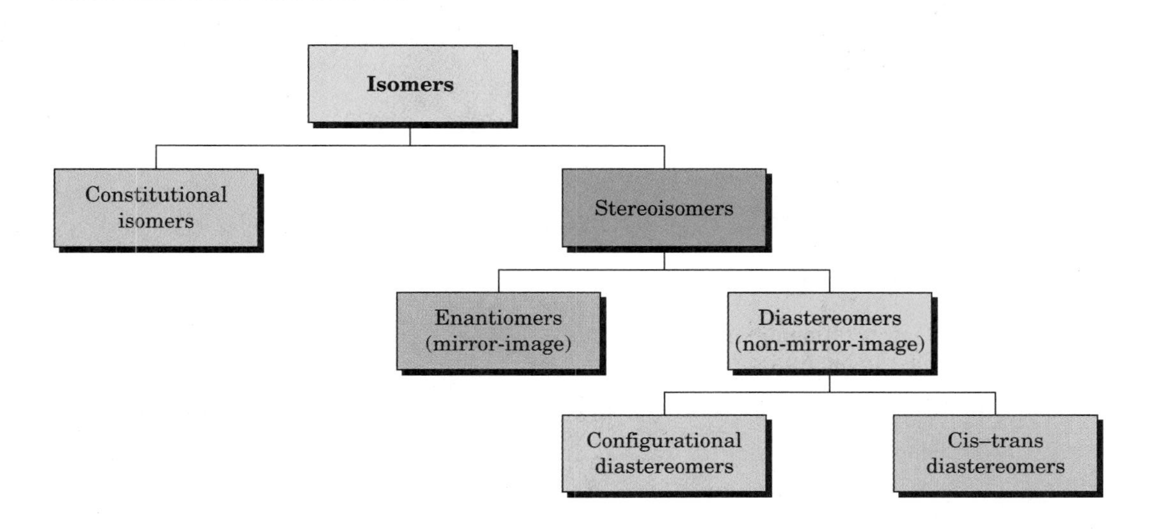

Figure 6.14 A summary of the different kinds of isomers.

There are two fundamental types of isomerism, both of which we've now encountered: constitutional isomerism and stereoisomerism.

Constitutional isomers (Section 2.2) are compounds whose atoms are connected differently. Among the kinds of constitutional isomers we've seen are skeletal, functional, and positional isomers.

Constitutional isomers—different connections among atoms:

Different carbon skeletons:	$CH_3CH(CH_3)CH_3$ (CH₃ on the central carbon) **Isobutane**	and	$CH_3CH_2CH_2CH_3$ **Butane**
Different functional groups:	CH_3CH_2OH **Ethyl alcohol**	and	CH_3OCH_3 **Dimethyl ether**
Different position of functional groups:	$CH_3CH(NH_2)CH_3$ **Isopropylamine**	and	$CH_3CH_2CH_2NH_2$ **Propylamine**

Stereoisomers (Section 2.8) are compounds whose atoms are connected in the same way but with a different geometry. Among the kinds of stereoisomers we've seen are enantiomers, diastereomers, and cis–trans isomers (both in alkenes and in cycloalkanes). In fact, though, cis–trans isomers are really just another kind of diastereomers because they are non-mirror-image stereoisomers.

Stereoisomers—same connections among atoms but different geometry:

Enantiomers (nonsuperimposable mirror-image stereoisomers)

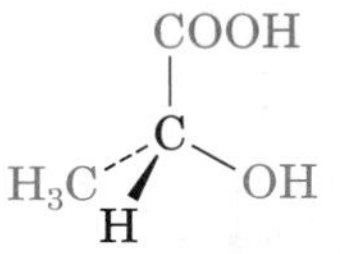

COOH, H_3C, H, OH — **(*R*)-Lactic acid**

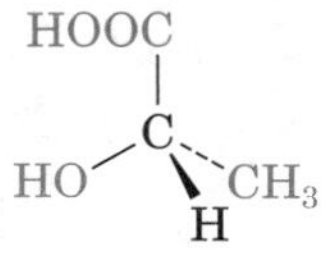

HOOC, HO, H, CH_3 — **(*S*)-Lactic acid**

Diastereomers (nonsuperimposable, non-mirror-image stereoisomers)

Configurational diastereomers

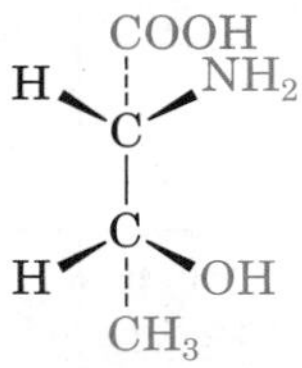

COOH, H, NH_2, H, OH, CH_3 — **2*R*,3*R*-2-Amino-3-hydroxybutanoic acid**

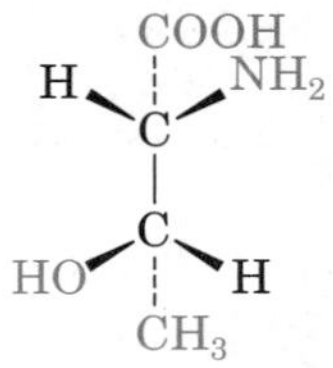

COOH, H, NH_2, HO, H, CH_3 — **2*R*,3*S*-2-Amino-3-hydroxybutanoic acid**

Cis–trans diastereomers (substituents on same side or opposite side of double bond or ring)

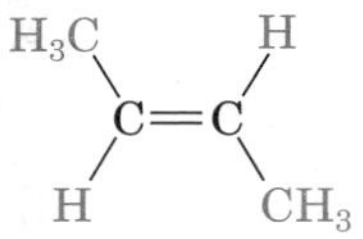

H_3C, H, C=C, H, CH_3 — ***trans*-2-Butene** and H_3C, CH_3, C=C, H, H — ***cis*-2-Butene**

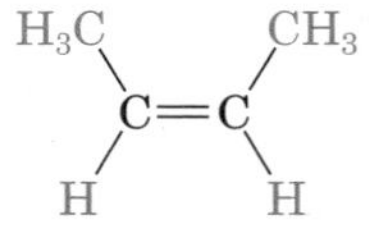

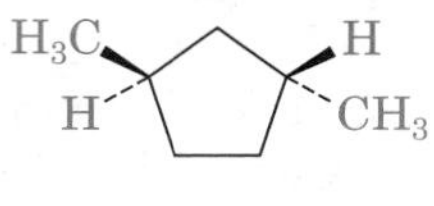

H_3C, H, H, CH_3 — ***trans*-1,3-Dimethylcyclopentane** and H_3C, H, CH_3, H — ***cis*-1,3-Dimethylcyclopentane**

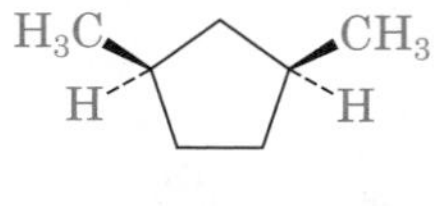

PROBLEM

6.18 What kinds of isomers are the following pairs?
(a) (*S*)-5-Chloro-2-hexene and chlorocyclohexane
(b) (2*R*,3*R*)-Dibromopentane and (2*S*,3*R*)-dibromopentane

6.13 Stereochemistry of Reactions: Addition of HBr to Alkenes

Many organic reactions, including some that we've studied, yield products with stereocenters. For example, addition of HBr to 1-butene yields 2-bromobutane, a chiral molecule. What predictions can we make about the stereochemistry of this chiral product? If a single enantiomer is formed, is it *R* or *S*? If a mixture of enantiomers is formed, how much of each? In fact, the 2-bromobutane produced is a racemic mixture of *R* and *S* enantiomers. Let's see why.

$$CH_3CH_2CH{=}CH_2 \xrightarrow[\text{Ether}]{\text{HBr}} CH_3CH_2\overset{\text{Br}}{\overset{|}{C}}HCH_3$$

1-Butene (achiral) → **(±)-2-Bromobutane (chiral)**

To understand why a racemic product results from the reaction of HBr with 1-butene, think about what happens during the reaction. 1-Butene is first protonated to yield an intermediate secondary carbocation. Since the trivalent carbon is sp^2-hybridized, the cation has no stereocenters, has a plane of symmetry, and is achiral. As a result, it can be attacked by Br^- equally well from either the top or the bottom. Attack from the top leads to (*S*)-2-bromobutane, and attack from the bottom leads to (*R*)-2-bromobutane. Since both pathways occur with equal probability, a racemic product mixture results (Figure 6.15).

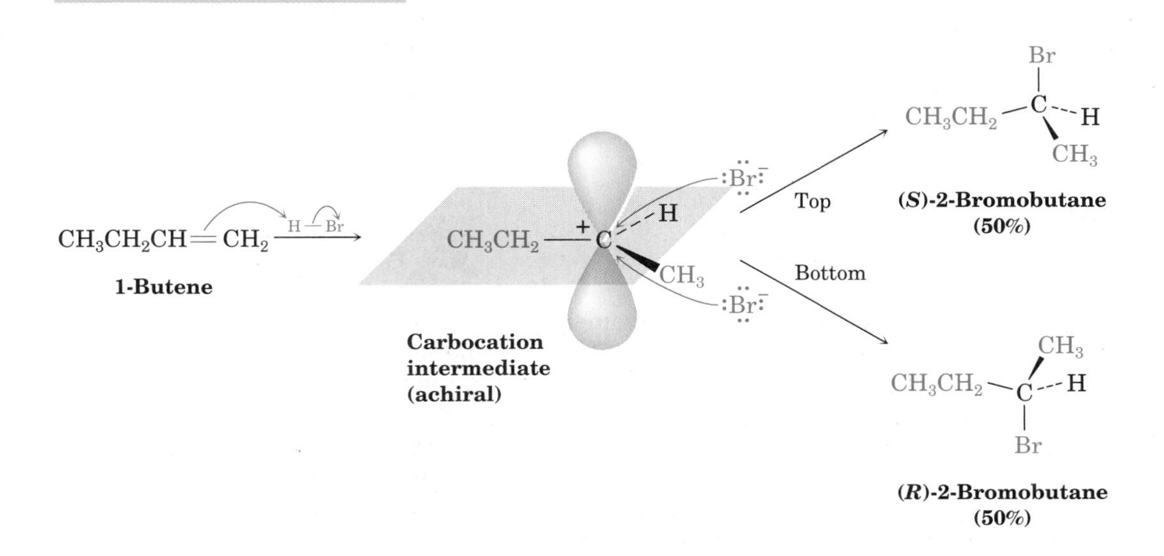

Figure 6.15 Stereochemistry of the addition of HBr to 1-butene. The intermediate carbocation is attacked equally well from both top and bottom, giving a racemic mixture of products that is 50% *R* and 50% *S*.

What is true for the reaction of 1-butene with HBr is also true for all other reactions: *Reactions between achiral reactants always lead to optically inactive products.* Optically active products can't be produced from optically inactive reactants.

6.14 Chirality in Nature

Just as different stereoisomeric forms of a chiral molecule have different physical properties, they usually have different biological properties as well. For example, the (+) enantiomer of carvone has the odor of caraway seeds, and the (−) enantiomer has the odor of spearmint.

(+)-Carvone (in caraway seeds)

(−)-Carvone (in spearmint oil)

Another example of how a change in chirality can affect the biological properties of a molecule is found in the amino acid, dopa. More properly named 2-amino-3-(3,4-dihydroxyphenyl)propanoic acid, dopa has a single stereocenter and can thus exist in two stereoisomeric forms. The dextrorotatory enantiomer, D-dopa, has no physiological effect on humans, but the levorotatory enantiomer, L-dopa, is widely used for its potent activity against Parkinson's disease, a chronic malady of the central nervous system.

D-Dopa (no biological effect)

L-Dopa (anti-Parkinsonian agent)

Why do different stereoisomers have different biological properties? To exert its biological action, a chiral molecule must fit into a chiral receptor at a target site, much as a hand fits into a glove. But just as a right hand can fit only into a right-hand glove, so a particular stereoisomer can fit only into a receptor having the proper complementary shape. Any other stereoisomer will be a misfit, like a right hand in a left-handed glove. A schematic representation of the interaction between a chiral molecule and a chiral biological receptor is shown in Figure 6.16. One enantiomer fits the receptor perfectly, but the other does not.

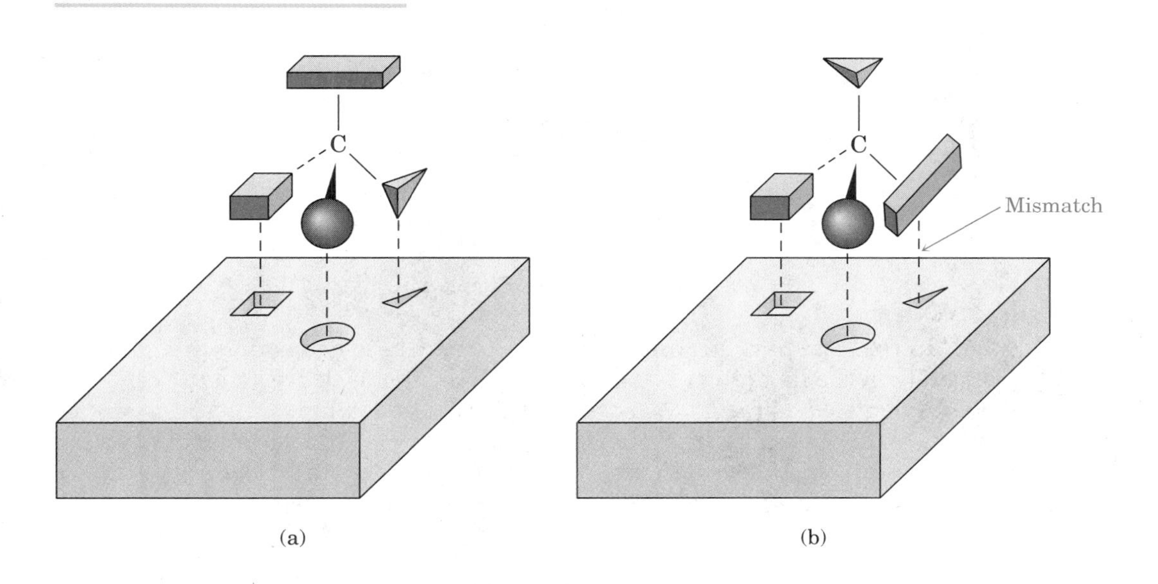

Figure 6.16 (a) One enantiomer fits easily into a chiral receptor site to exert its biological effect, but (b) the other enantiomer can't fit into the same receptor.

INTERLUDE

Chiral Drugs

The *S* enantiomer of ibuprofen soothes the aches and pains of athletic injuries. The *R* enantiomer has no effect.

The hundreds of different pharmaceutical agents approved for use by the U.S. Food and Drug Administration come from many sources. Many are isolated directly from plants or bacteria, others are made by chemical modification of naturally occurring compounds, and still others are made entirely in the laboratory and have no relatives in nature.

Those drugs that come from natural sources, either directly or after chemical modification, are usually chiral and are generally found only as a single enantiomer rather than as a racemic mixture. Penicillin V, for example, an antibiotic isolated from the *Penicillium* mold, has a 2*S*,5*R*,6*R* configuration. Its enantiomer, which does not occur naturally but can be made in the laboratory, has essentially no biological activity.

(*continued*)➤

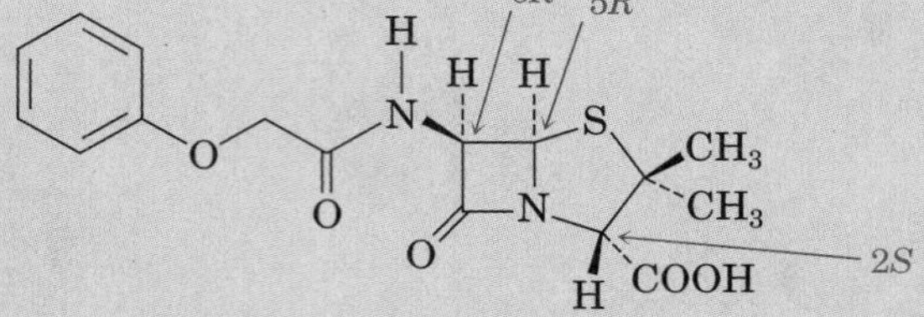

Penicillin V (2*S*,5*R*,6*R* configuration)

In contrast to drugs from natural sources, those drugs that are made entirely in the laboratory either are achiral or, if chiral, are generally produced and sold as racemic mixtures. Ibuprofen, for example, contains one stereocenter, but only the *S* enantiomer is an analgesic/anti-inflammatory agent useful in treating rheumatism. Even though the *R* enantiomer of ibuprofen is inactive, the substance marketed under such trade names as Advil, Nuprin, and Motrin is a racemic mixture of *R* and *S*.

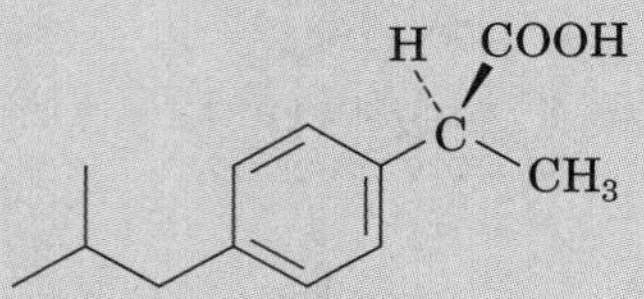

(*S*)-Ibuprofen
(an active analgesic agent)

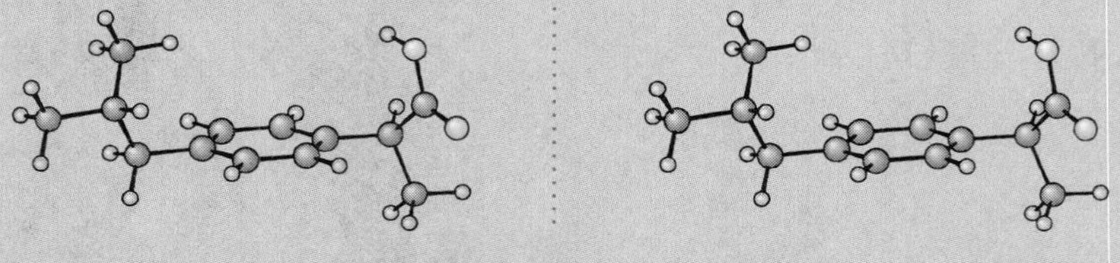

Stereo View

Not only is it wasteful to synthesize and administer a physiologically inactive enantiomer, many examples are now known where the presence of the "wrong" enantiomer in a racemic mixture either affects the body's ability to utilize the "right" enantiomer or has unintended effects of its own. The presence of (*R*)-ibuprofen in the racemic mixture, for instance, seems to slow down substantially the rate at which the *S* enantiomer takes effect.

To get around this problem, pharmaceutical companies are now devising methods of so-called *asymmetric synthesis,* which allows them to prepare only a single enantiomer rather than a racemic mixture. Viable methods have already been developed for the preparation of (*S*)-ibuprofen, and the time may not be far off when television ads show famous tennis players talking about the advantages of chiral drugs.

Summary and Key Words

achiral, 187
chiral, 185
configuration, 193
dextrorotatory, 191
diastereomer, 197
enantiomer, 185
levorotatory, 191
meso compound, 200
optical activity, 190
plane of symmetry, 186
R configuration, 194
racemic mixture, 201
resolution, 201
S configuration, 194
specific rotation, $[\alpha]_D$, 191
stereocenter, 187
stereochemistry, 183

A molecule that is not identical to its mirror image is said to be **chiral,** meaning "handed." A chiral molecule is one that does not contain a **plane of symmetry.** The usual cause of chirality is the presence of a tetrahedral carbon atom bonded to four different groups—a so-called **stereocenter.** Chiral compounds can exist as a pair of mirror-image stereoisomers called **enantiomers,** which are related to each other as a right hand is related to a left hand. When a beam of plane-polarized light is passed through a solution of a pure enantiomer, the plane of polarization is rotated, and the compound is said to be **optically active.**

The three-dimensional **configuration** of a stereocenter is specified as either ***R*** or ***S*.** Sequence rules are used to assign priorities to the four substituents on the chiral carbon, and the atom is then compared with a hand, oriented so that the thumb points from the carbon to the group of lowest priority (4). If the fingers of a *right* hand curl in the direction of decreasing priority for the remaining three groups, the stereocenter has the *R* configuration. If the fingers of a *left* hand curl in the $1 \rightarrow 2 \rightarrow 3$ direction, the stereocenter has the *S* configuration.

Some molecules have more than one stereocenter. Enantiomers have opposite configurations at all stereocenters, whereas **diastereomers** have the same configuration in at least one center but opposite configurations at the others. **Meso compounds** contain stereocenters but are achiral overall because they contain a plane of symmetry. **Racemates** are 50:50 mixtures of (+) and (−) enantiomers. Racemic mixtures and individual diastereomers differ in both their physical properties and their biological properties.

ADDITIONAL PROBLEMS

6.19 Cholic acid, the major steroid found in bile, was found to have a rotation of +2.22° when a 3.00 g sample was dissolved in 5.00 mL of alcohol in a sample tube with a 1.00 cm path length. Calculate $[\alpha]_D$ for cholic acid.

6.20 Polarimeters are so sensitive that they can measure rotations to the thousandth of a degree, an important advantage when only small amounts of a sample are available. For example, when 7.00 mg of ecdysone, an insect hormone that controls molting in the silkworm moth, was dissolved in 1.00 mL of chloroform in a cell with a 2.00 cm path length, an observed rotation of +0.087° was found. Calculate $[\alpha]_D$ for ecdysone.

6.21 Define the following terms:

(a) Chirality (b) Stereocenter (c) Diastereomer
(d) Racemate (e) Meso compound (f) Enantiomer

6.22 Which of the following objects are chiral?
(a) A basketball (b) A wine glass (c) An ear
(d) A snowflake (e) A coin (f) Scissors

6.23 Which of the following compounds are chiral?
(a) 2,4-Dimethylheptane (b) 5-Ethyl-3,3-dimethylheptane
(c) *cis*-1,3-Dimethylcyclohexane

6.24 Penicillin V is a broad-spectrum antibiotic that contains three stereocenters. Identify them with asterisks.

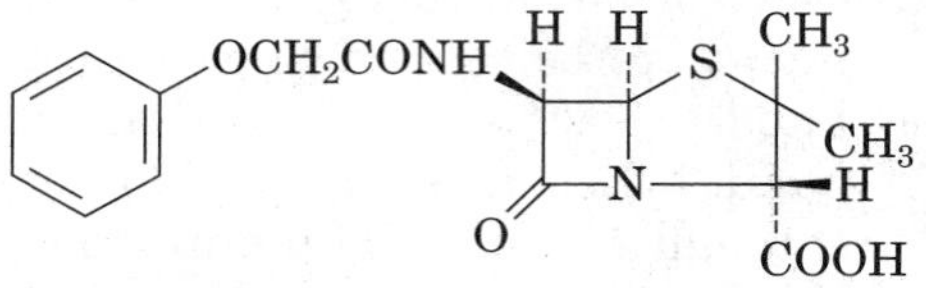

OCH$_2$CONH H H S CH$_3$ CH$_3$ N H O COOH

Penicillin V (antibiotic)

6.25 Draw chiral molecules that meet the following descriptions:
(a) A chloroalkane, $C_5H_{11}Cl$ (b) An alcohol, $C_6H_{14}O$
(c) An alkene, C_6H_{12} (d) An alkane, C_8H_{18}

6.26 Which of the following compounds are chiral? Label all stereocenters.

(a) H$_3$C CH$_3$ CH$_3$CH$_2$CH—C—CH$_2$CH$_3$ CH$_3$

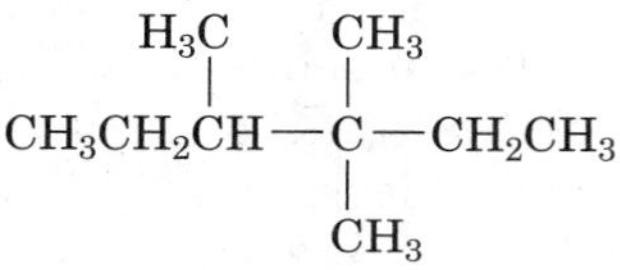

(b) O CH$_3$

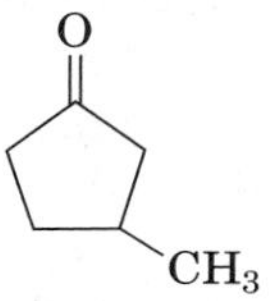

(c) H$_3$C O

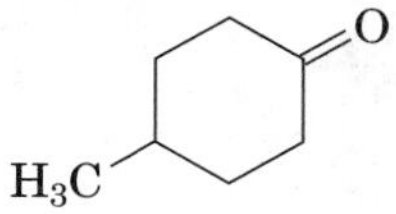

(d) BrCH$_2$CHCHCH$_2$Br

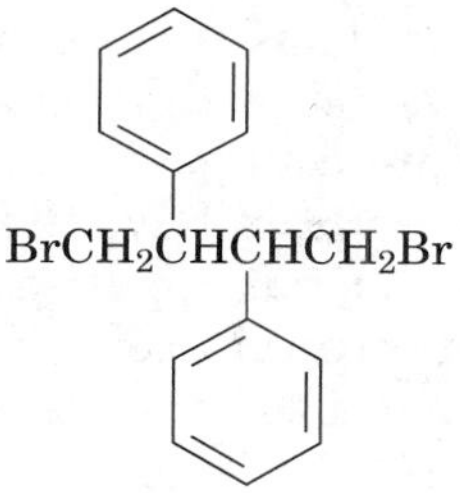

(e) CH$_3$ H H$_3$C H

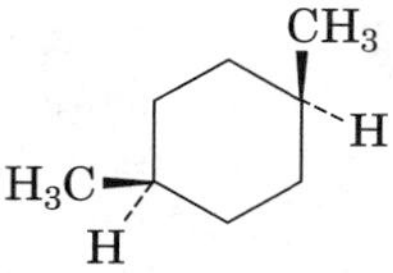

6.27 There are eight alcohols with the formula $C_5H_{12}O$. Draw them, and tell which are chiral.

6.28 Propose structures for compounds that meet the following descriptions:
(a) A chiral alcohol with four carbons (b) A chiral carboxylic acid
(c) A compound with two stereocenters

6.29 Assign priorities to the substituents in each of the following sets:
(a) –H, –OH, –OCH$_3$, –CH$_3$ (b) –Br, –CH$_3$, –CH$_2$Br, –Cl
(c) –CH=CH$_2$, –CH(CH$_3$)$_2$, –C(CH$_3$)$_3$, –CH$_2$CH$_3$
(d) –COOCH$_3$, –COCH$_3$, –CH$_2$OCH$_3$, –OCH$_3$

6.30 One enantiomer of lactic acid is shown below. Is it *R* or *S*? Draw its mirror image in the standard tetrahedral representation.

COOH C H OH H$_3$C

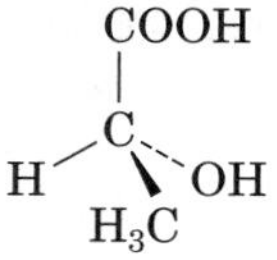

6.31 Draw tetrahedral representations of both enantiomers of the amino acid serine. Tell which of your structures is *S* and which is *R*.

$$\mathrm{HOCH_2CH(NH_2)COOH}$$ **Serine**

6.32 If naturally occurring (*S*)-serine has $[\alpha]_D = -6.83°$, what specific rotation do you expect for (*R*)-serine?

6.33 Assign *R* or *S* configuration to the stereocenters in the following molecules:

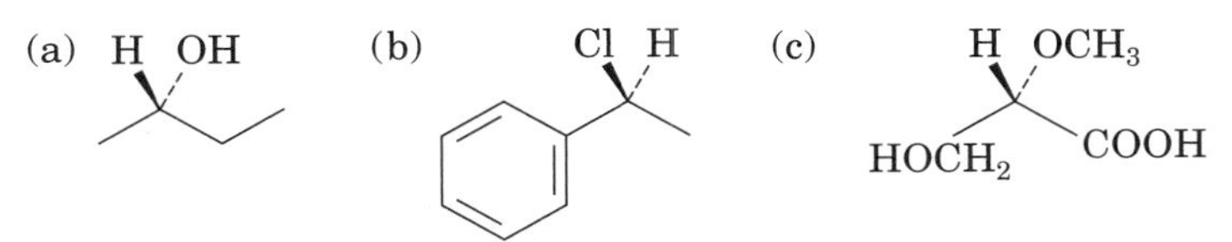

6.34 What is the relationship between the specific rotations of (2*R*,3*R*)-dihydroxypentane and (2*S*,3*S*)-dihydroxypentane? Between (2*R*,3*S*)-dihydroxypentane and (2*R*,3*R*)-dihydroxypentane?

6.35 What is the stereochemical configuration of the enantiomer of (2*S*,4*R*)-dibromooctane?

6.36 What are the stereochemical configurations of the two diastereomers of (2*S*,4*R*)-dibromooctane?

6.37 Draw examples of the following:
(a) A meso compound with the formula C_8H_{18}
(b) A compound with two stereocenters, one *R* and the other *S*

6.38 Draw a tetrahedral representation of (*S*)-2-butanol, $CH_3CH_2CH(OH)CH_3$.

6.39 Tell whether the following Newman projection of 2-chlorobutane is *R* or *S*. (You might want to review Section 2.5.)

Cl
H CH3
H3C H
H

6.40 Draw a Newman projection that is enantiomeric with the one shown in Problem 6.39.

6.41 Draw a Newman projection of *meso*-tartaric acid.

6.42 Draw Newman projections of (2*R*,3*R*)- and (2*S*,3*S*)-tartaric acid, and compare them to the projection you drew in Problem 6.41 for the meso form.

6.43 Glucose has four stereocenters. How many stereoisomers of glucose are possible?

6.44 Draw a tetrahedral representation of (*R*)-3-chloro-1-pentene.

6.45 Draw all the stereoisomers of 1,2-dimethylcyclopentane. Assign *R*,*S* configurations to the stereocenters in all isomers, and indicate which stereoisomers are chiral and which, if any, are meso.

6.46 Assign *R* or *S* configuration to each stereocenter in the following molecules:

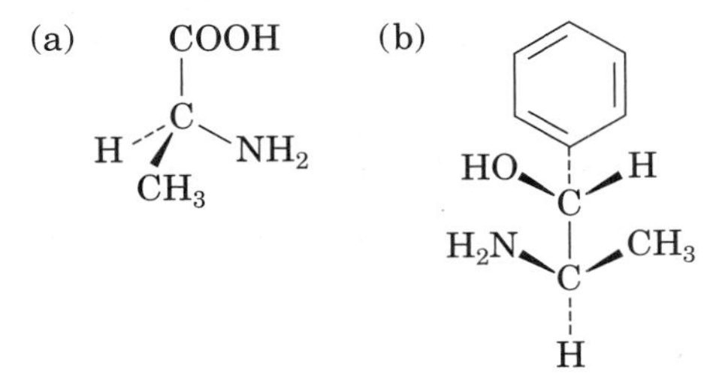

6.47 Hydroxylation of *cis*-2-butene with $KMnO_4$ yields 2,3-butanediol. What is the stereochemistry of the product? (Review Section 4.7.)

6.48 Answer Problem 6.47 for *trans*-2-butene.

6.49 How many stereoisomers of 2,4-dibromo-3-chloropentane are there? Draw them, and indicate which are optically active.

6.50 Alkenes undergo reaction with peroxycarboxylic acids (RCO_3H) to give compounds called *epoxides.* For example, *cis*-2-butene gives 2,3-epoxybutane:

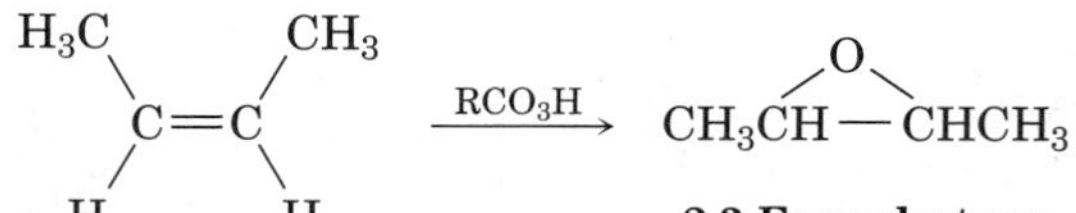

$$\underset{H\quad\quad H}{\overset{H_3C\quad\quad CH_3}{C{=}C}} \xrightarrow{RCO_3H} CH_3\overset{\overset{O}{\frown}}{CH{-}CH}CH_3$$

2,3-Epoxybutane

Assuming that both C–O bonds form from the same side of the molecule (syn stereochemistry), show the stereochemistry of the product. Is the epoxide chiral? How many stereocenters does it have? How would you describe the product stereochemically?

6.51 Answer Problem 6.50 assuming that the epoxidation was carried out on *trans*-2-butene.

6.52 Ribose, an essential part of ribonucleic acid (RNA), has the following structure:

H H H OH
HO CHO
HO H HO H

Ribose

How many stereocenters does ribose have? Identify them with asterisks. How many stereoisomers of ribose are there?

6.53 Draw the structure of the enantiomer of ribose (see Problem 6.52).

6.54 Draw the structure of a diastereomer of ribose (see Problem 6.52).

6.55 On catalytic hydrogenation over a platinum catalyst, ribose (see Problem 6.52) is converted into ribitol. Is ribitol optically active or inactive? Explain.

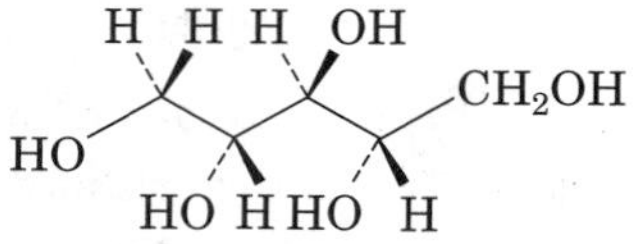

H H H OH
HO CH_2OH
HO H HO H

Ribitol

6.56 Draw the two enantiomers of the amino acid cysteine, $HSCH_2CH(NH_2)COOH$, and identify each as *R* or *S*.

6.57 Draw the structure of (*R*)-2-methylcyclohexanone.

6.58 Compound A, C_7H_{14}, is optically active. On catalytic reduction over a palladium catalyst, 1 equiv of H_2 is absorbed, yielding compound B, C_7H_{16}. On cleavage of A with acidic $KMnO_4$, two fragments are obtained. One fragment can be identified as acetic acid, CH_3COOH, and the other fragment, C, is an optically active carboxylic acid. Formulate the reactions, and propose structures for A, B, and C.

6.59 *Allenes* are compounds with adjacent C=C bonds. Even though they don't contain stereocenters, many allenes are chiral. For example, mycomycin, an antibiotic isolated from the bacterium *Nocardia acidophilus,* is chiral and has $[\alpha]_D = -130°$. Can you explain why mycomycin is chiral? Making a molecular model should be helpful.

$$HC{\equiv}C{-}C{\equiv}C{-}CH{=}C{=}CH{-}CH{=}CH{-}CH{=}CH{-}CH_2COOH$$

Mycomycin (an allene)

Visualizing Chemistry

6.60 Which of the following structures are identical? (Gray = C, red = O, yellow-green = Cl, light green = H.)

(a)

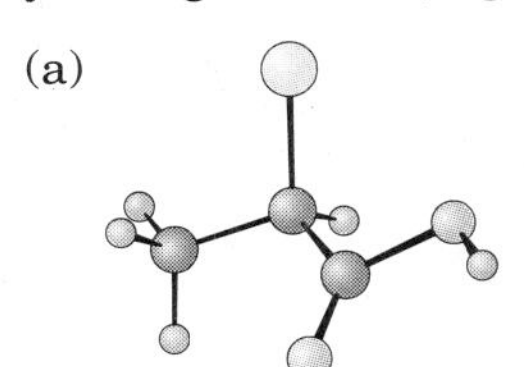

(b)

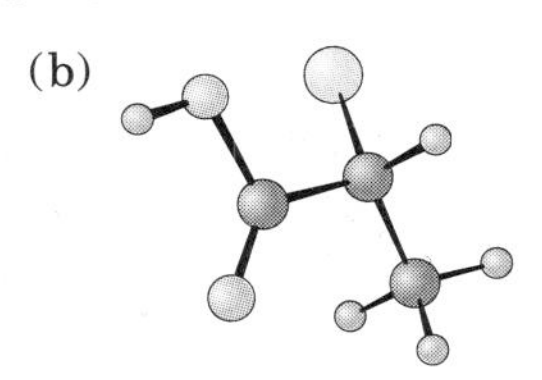

(c)

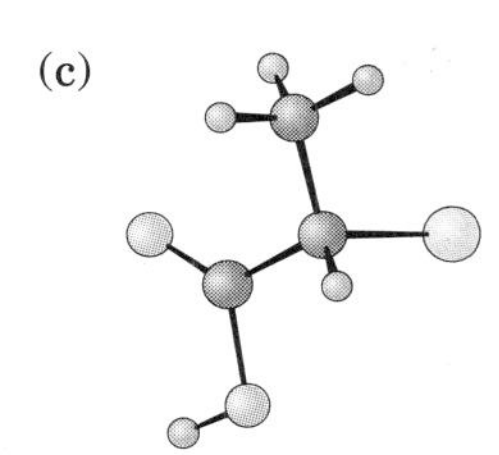

(d)

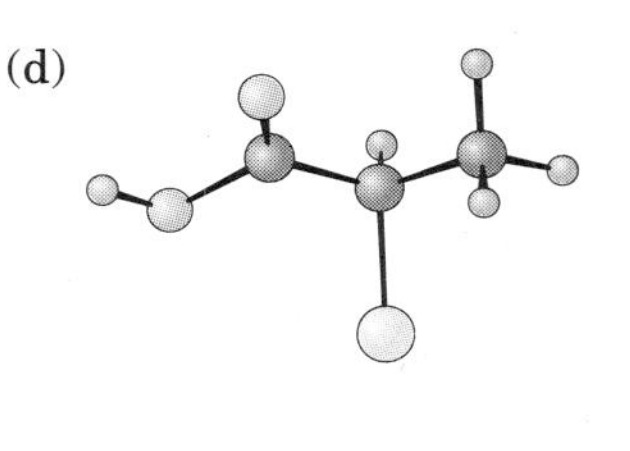

6.61 Assign *R* or *S* configuration to the following molecules (gray = C, red = O, blue = N, light green = H).

(a)

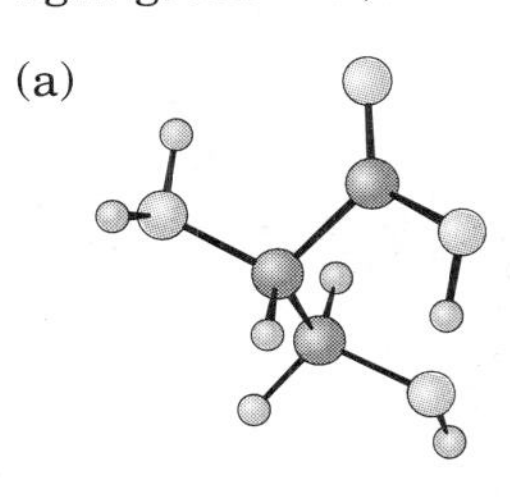

Serine

(b)

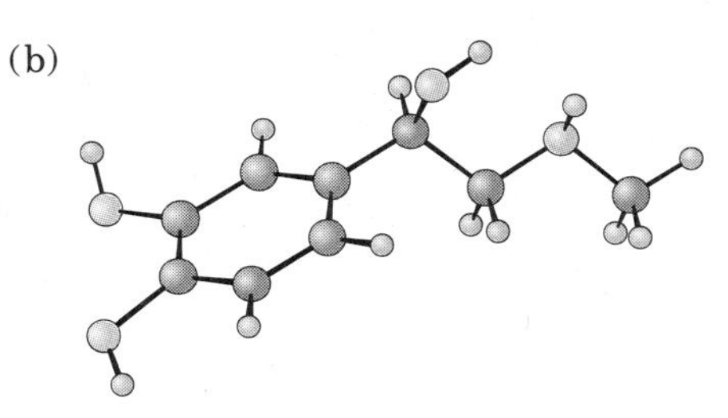

Adrenaline

6.62 Which, if any, of the following structures represent meso compounds? (Gray = C, red = O, blue = N, yellow-green = Cl, light green = H.)

(a)

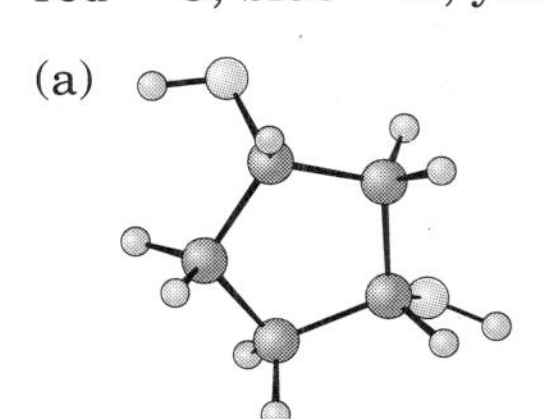

(b)

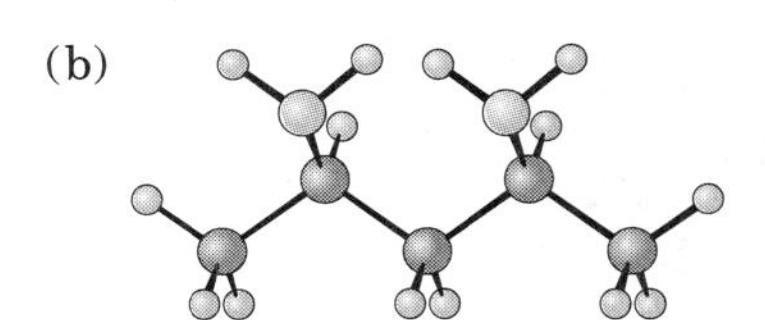

(c)

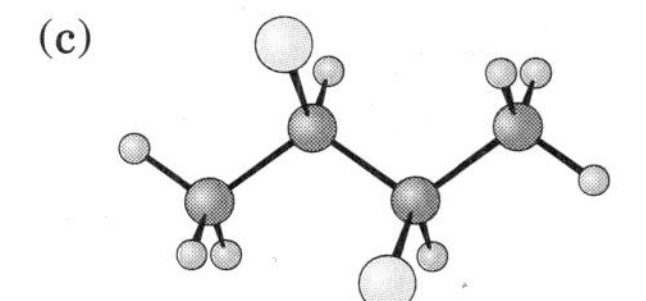

6.63 Assign *R* or *S* configuration to each stereocenter in pseudoephedrine, an over-the-counter decongestant found in cold remedies (gray = C, red = O, blue = N, light green = H).

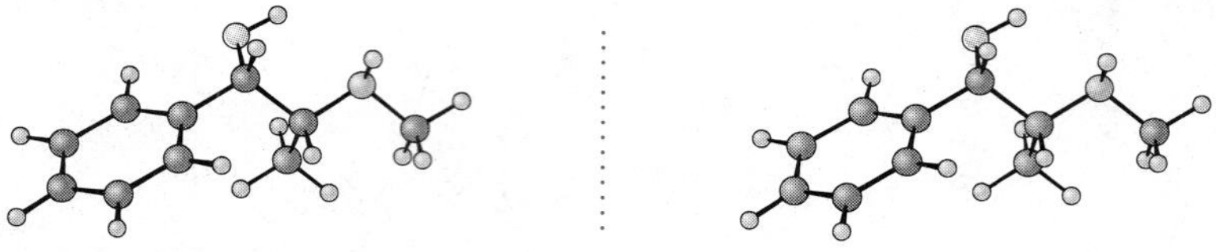

Stereo View

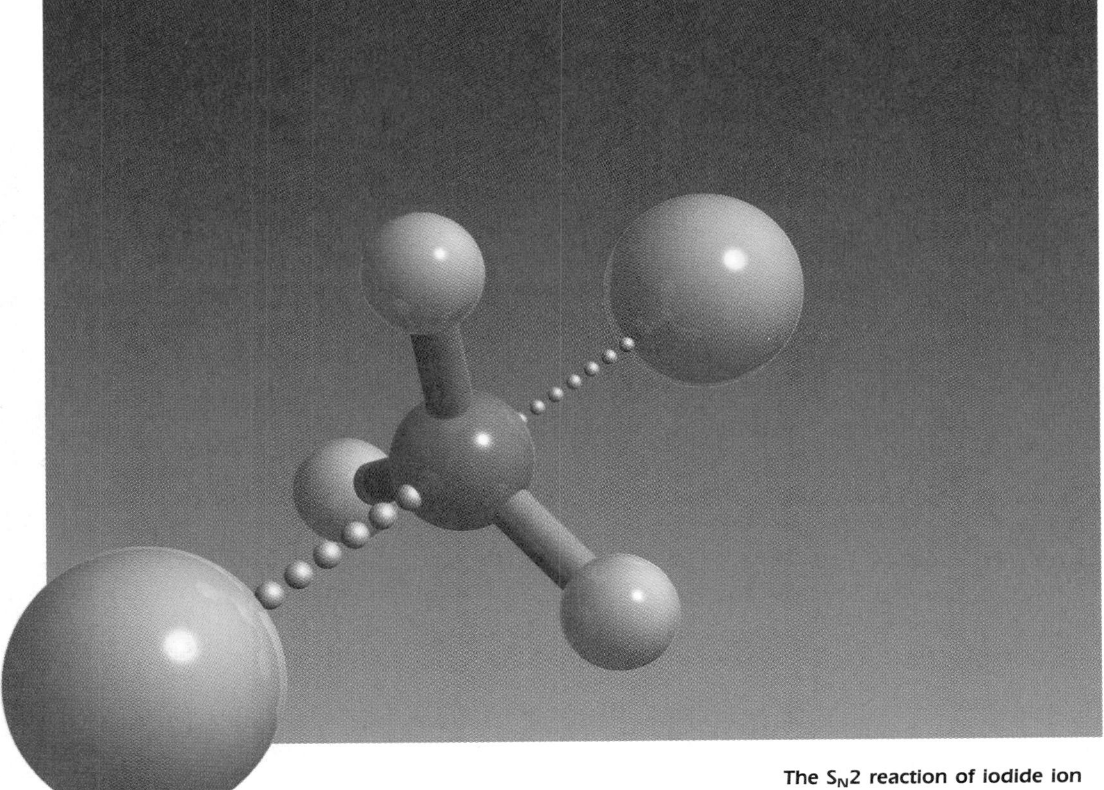

The S_N2 reaction of iodide ion on bromomethane occurs in a single step through a planar transition state.

7 Alkyl Halides

It would be difficult to study organic chemistry for long without becoming aware of halo-substituted alkanes, or **alkyl halides.** Among their many applications, alkyl halides are used as industrial solvents, inhaled anesthetics in medicine, pesticides, and refrigerants.

```
   Cl  Cl              F   Br
   |   |               |   |
H—C—C—H           F—C—C—H
   |   |               |   |
   H   H               F   Cl
```

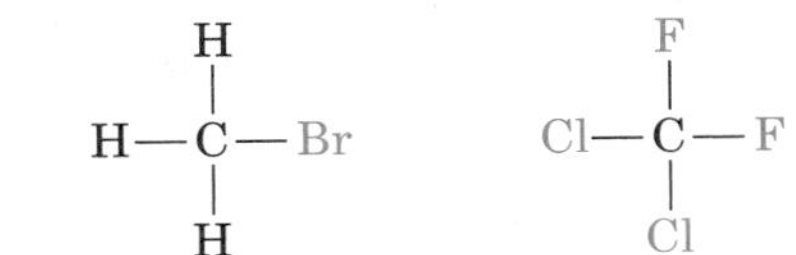

1,2-Dichloroethane (a solvent)

Halothane (an inhaled anesthetic)

Bromomethane (a fumigant)

Dichlorodifluoromethane (a refrigerant)

7.1 Naming Alkyl Halides

Alkyl halides are named in the same way as alkanes (Section 2.3), by considering the halogen as a substituent on the parent alkane chain. There are three steps:

Step 1 Find the longest chain, and name it as the parent. If a multiple bond is present, the parent chain must contain it.

Step 2 Number the carbons of the parent chain beginning at the end nearer the first substituent, regardless of whether it is alkyl or halo. Assign each substituent a number according to its position on the chain. If there are substituents the same distance from both ends, begin numbering at the end nearer the substituent with alphabetical priority.

CH_3 (on C2), Br (on C5), CH_3 (on C4)

$CH_3CHCH_2CHCHCH_2CH_3$ (carbons numbered 1 2 3 4 5 6 7)

5-Bromo-2,4-dimethylheptane

Br (on C2), CH_3 (on C5), CH_3 (on C4)

$CH_3CHCH_2CHCHCH_2CH_3$ (carbons numbered 1 2 3 4 5 6 7)

2-Bromo-4,5-dimethylheptane

Step 3 Write the name, listing all substituents in alphabetical order and using one of the prefixes *di-*, *tri-*, and so forth if more than one of the same substituent is present.

Cl Cl (on C2, C3), CH_3 (on C4)

$CH_3CHCHCHCH_2CH_3$ (carbons numbered 1 2 3 4 5 6)

2,3-Dichloro-4-methylhexane

In addition to their systematic names, many simple alkyl halides are also named by identifying first the alkyl group and then the halogen. For example, CH_3I can be called either iodomethane or methyl iodide.

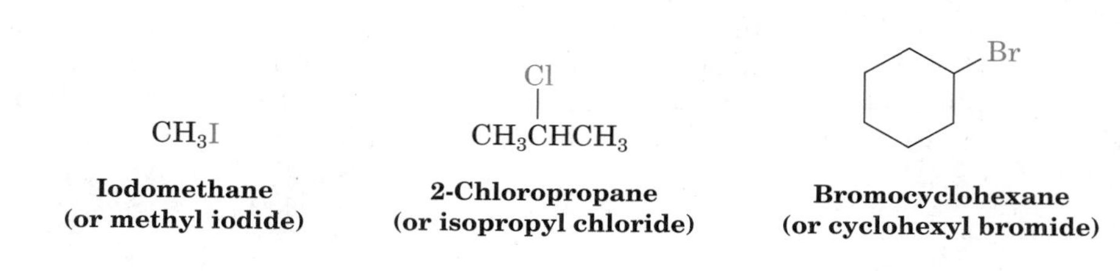

Iodomethane (or methyl iodide) **2-Chloropropane (or isopropyl chloride)** **Bromocyclohexane (or cyclohexyl bromide)**

PROBLEM

7.1 Give the IUPAC names of the following alkyl halides:

(a) $CH_3CH_2CHCH_3$ with Br on C2

(b) $CH_3CH_2CHCHCH_3$ with Cl and CH_3 substituents

(c) $CH_3CHCH_2CH_2Cl$ with CH_3 substituent

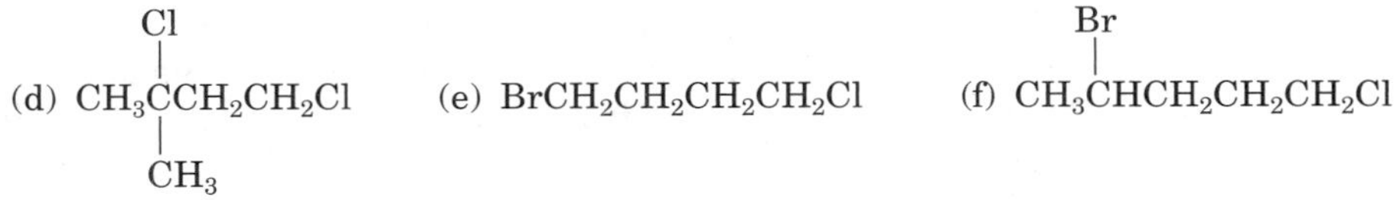

(d) $CH_3C(Cl)(CH_3)CH_2CH_2Cl$ (e) $BrCH_2CH_2CH_2CH_2Cl$ (f) $CH_3CH(Br)CH_2CH_2CH_2Cl$

PROBLEM

7.2 Draw structures corresponding to the following names:
(a) 2-Chloro-3,3-dimethylhexane (b) 3,3-Dichloro-2-methylhexane
(c) 3-Bromo-3-ethylpentane (d) 2-Bromo-5-chloro-3-methylhexane

7.2 Preparation of Alkyl Halides: Radical Chlorination of Alkanes

We've already seen several methods for preparing alkyl halides, including the addition reactions of HX and X_2 with alkenes (Sections 4.1 and 4.5):

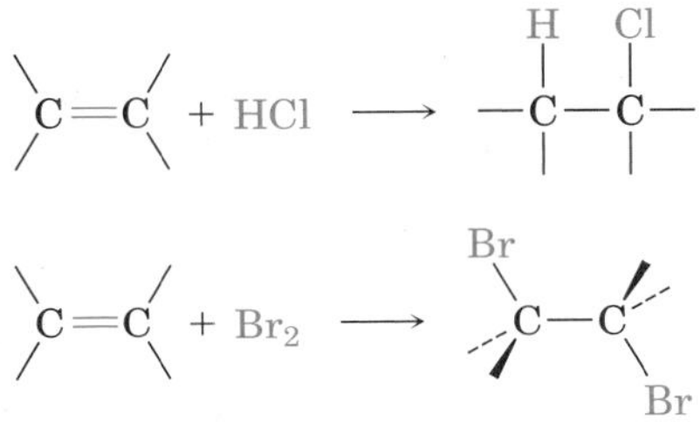

$$\text{C=C} + HCl \longrightarrow -\overset{H}{C}-\overset{Cl}{C}-$$

$$\text{C=C} + Br_2 \longrightarrow \overset{Br}{C}-\underset{Br}{C}$$

Another method of alkyl halide synthesis is the reaction of an alkane with Cl_2 or Br_2. Although inert to most reagents, alkanes react readily with Cl_2 in the presence of ultraviolet light ($h\nu$) to give chlorinated alkane products. For example, methane reacts with Cl_2 to give chloromethane:

$$CH_4 + Cl_2 \xrightarrow{h\nu} CH_3Cl + HCl$$

Methane **Chloromethane**

The chlorination of methane is a typical **radical substitution reaction** rather than a polar reaction of the sort we've been studying until now. Recall from Section 3.6 that radical reactions involve substances that have an odd number of electrons. Bonds are formed in radical reactions when each partner donates one electron to the new bond, and bonds are broken when each fragment leaves with one electron.

$$A\cdot + \cdot B \longrightarrow A{:}B \quad \text{Radical (homogenic) bond making}$$

$$A{:}B \longrightarrow A\cdot + \cdot B \quad \text{Radical (homolytic) bond breaking}$$

Radical substitution reactions normally require three kinds of steps: an *initiation* step, *propagation* steps, and *termination* steps. As its name implies, the initiation step starts the reaction by producing reactive radicals. In the chlorination reaction, for instance, the relatively weak Cl–Cl bond is broken by irradiation with ultraviolet light to give two chlorine radicals, Cl·.

Once an initiation step has occurred and a small number of radicals have been produced, the reaction continues in a self-sustaining cycle of two propagation steps. In the first propagation step, Cl· abstracts an H atom from methane to produce HCl and a methyl radical (·CH_3). In the second propagation step, the methyl radical abstracts a Cl atom from Cl_2 to yield chloromethane and a new chlorine radical, which then cycles back into the first propagation step, making the overall process a *chain reaction.*

Occasionally, two radicals might collide and combine to form a stable product. When this happens, the propagation cycle is interrupted and the chain reaction is terminated. The overall mechanism of methane chlorination is shown in Figure 7.1.

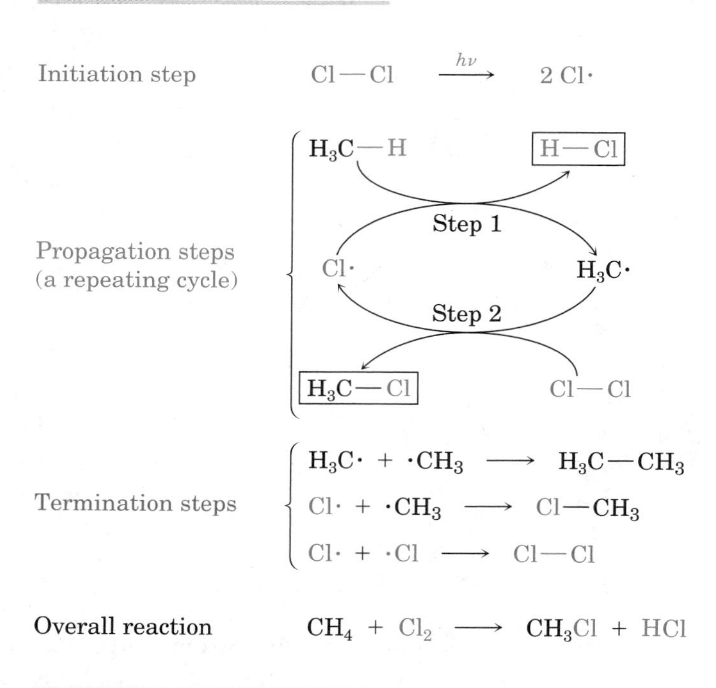

Figure 7.1 Mechanism of the radical chlorination of methane. Three kinds of steps are required: *initiation, propagation,* and *termination.*

Though interesting from a mechanistic point of view, alkane chlorination is a poor method for preparing most alkyl chlorides because mixtures of products usually result. Chlorination of methane does not stop cleanly at the monochlorinated stage, but continues on, giving a mixture of dichloro, trichloro, and even tetrachloro products that must be separated:

$$CH_4 + Cl_2 \xrightarrow{h\nu} CH_3Cl + HCl$$
$$\xrightarrow{Cl_2} CH_2Cl_2 + HCl$$
$$\xrightarrow{Cl_2} CHCl_3 + HCl$$
$$\xrightarrow{Cl_2} CCl_4 + HCl$$

The situation is even worse for chlorination of alkanes that have more than one kind of hydrogen. Chlorination of butane, for instance, gives two monochlorinated products as well as several dichlorobutanes, trichlorobutanes, and so on. Of the monochloro product, 30% is 1-chlorobutane and 70% is 2-chlorobutane:

$$\underset{\textbf{Butane}}{CH_3CH_2CH_2CH_3} + Cl_2 \xrightarrow{h\nu} \underbrace{\underset{\textbf{1-Chlorobutane}}{CH_3CH_2CH_2CH_2Cl} + \underset{\textbf{2-Chlorobutane}}{CH_3CH_2\overset{\displaystyle Cl}{\overset{|}{C}}HCH_3}}_{30:70} + \text{Dichloro-, trichloro-, tetrachloro-, and so on}$$

PRACTICE PROBLEM 7.1

Draw all the monochloro products you might get from radical chlorination of 2-methylbutane.

Solution Draw the structure of the starting material and begin systematically replacing each kind of hydrogen by chlorine. In this example, there are four possibilities.

$$\underset{\textbf{2-Methylbutane}}{CH_3{-}CH_2{-}\overset{\displaystyle CH_3}{\overset{|}{\underset{\displaystyle H}{\underset{|}{C}}}}{-}CH_3} \xrightarrow[h\nu]{Cl_2} CH_3CH_2\overset{\displaystyle CH_3}{\overset{|}{C}}HCH_2Cl + CH_3CH_2{-}\overset{\displaystyle CH_3}{\overset{|}{\underset{\displaystyle Cl}{\underset{|}{C}}}}{-}CH_3 + CH_3CH(Cl)\overset{\displaystyle CH_3}{\overset{|}{C}}HCH_3 + CH_2(Cl)CH_2\overset{\displaystyle CH_3}{\overset{|}{C}}HCH_3$$

PROBLEM

7.3 Draw and name all monochloro products you might obtain from radical chlorination of 3-methylpentane. Which, if any, are chiral?

PROBLEM

7.4 Radical chlorination of pentane is a poor way to prepare 1-chloropentane, but radical chlorination of 2,2-dimethylpropane is a good way to prepare 1-chloro-2,2-dimethylpropane. Explain.

7.3 Alkyl Halides from Alcohols

The most general method for preparing alkyl halides is to make them from alcohols, a reaction carried out simply by treating the alcohol with HX. 1-Methylcyclohexanol, for example, is converted into 1-chloro-1-methylcyclohexane by treating with HCl:

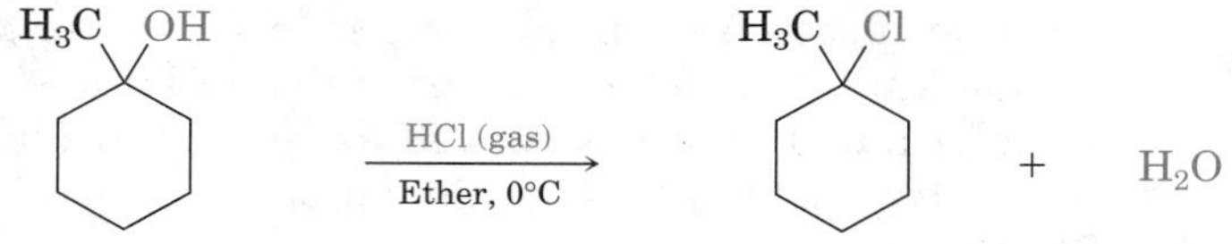

$$\text{1-Methylcyclohexanol} \xrightarrow[\text{Ether, 0°C}]{\text{HCl (gas)}} \text{1-Chloro-1-methylcyclohexane (90\%)} + H_2O$$

1-Methylcyclohexanol **1-Chloro-1-methylcyclohexane (90%)**

For reasons that will be discussed in Section 7.8, the reaction works best with tertiary alcohols. Primary and secondary alcohols react much more slowly.

$$R{-}OH + HX \longrightarrow R{-}X + H_2O$$

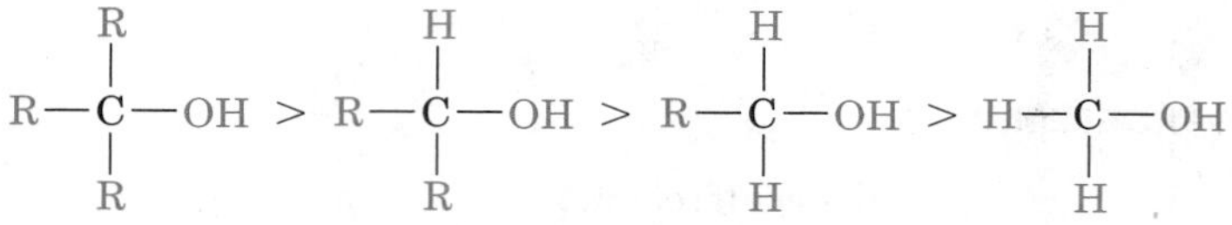

$$R_3C{-}OH > R_2CH{-}OH > RCH_2{-}OH > H_3C{-}OH$$

3° 2° 1° Methyl

More reactive ← Reactivity — Less reactive

Primary and secondary alcohols are best converted into alkyl halides by treatment with either thionyl chloride ($SOCl_2$) or phosphorus tribromide (PBr_3). These reactions normally take place in high yield.

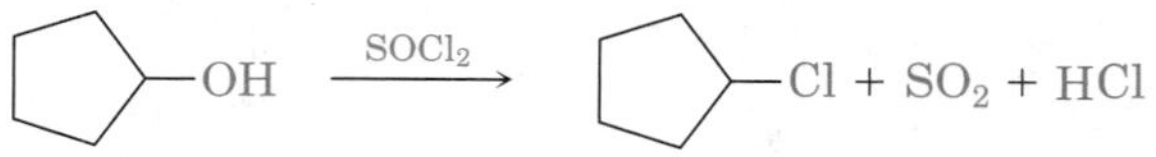

$$\text{Cyclopentanol} \xrightarrow{SOCl_2} \text{Chlorocyclopentane} + SO_2 + HCl$$

Cyclopentanol **Chlorocyclopentane**

$$3\ CH_3CH_2CH(OH)CH_3 \xrightarrow[\text{Ether, 35°C}]{PBr_3} 3\ CH_3CH_2CH(Br)CH_3 + P(OH)_3$$

2-Butanol **2-Bromobutane (86%)**

PRACTICE PROBLEM 7.2

Predict the product of the following reaction:

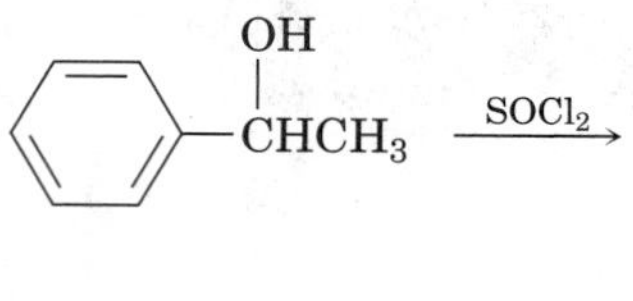

$$C_6H_5{-}CH(OH)CH_3 \xrightarrow{SOCl_2}$$

Solution Alcohols yield alkyl chlorides on treatment with $SOCl_2$:

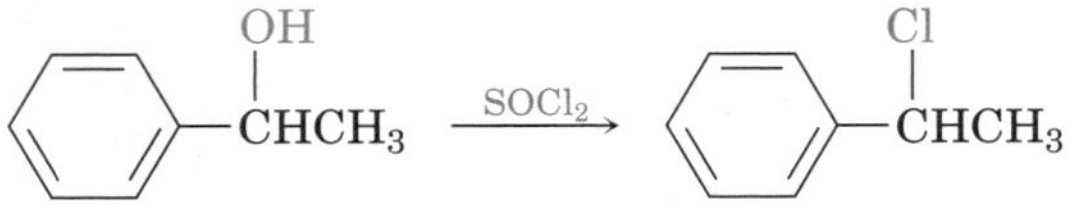

$$C_6H_5\text{-}CH(OH)CH_3 \xrightarrow{SOCl_2} C_6H_5\text{-}CH(Cl)CH_3$$

PROBLEM

7.5 How would you prepare the following alkyl halides from the appropriate alcohols?

(a) 2-Chloro-2-methylpropane

(b) 2-Bromo-4-methylpentane

(c) $BrCH_2CH_2CH_2CH_2CH(CH_3)CH_3$

(d) $CH_3CH_2CH(CH_3)CH_2C(Cl)(CH_3)CH_3$

PROBLEM

7.6 Predict the products of the following reactions:

(a) $CH_3CH_2CH(OH)CH_2CH(CH_3)CH_3 + PBr_3 \longrightarrow ?$

(b) 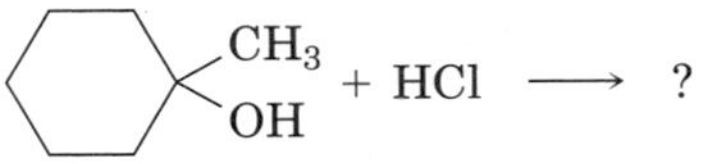 1-Methylcyclohexanol (CH_3, OH on the same ring carbon) $+ HCl \longrightarrow ?$

(c) 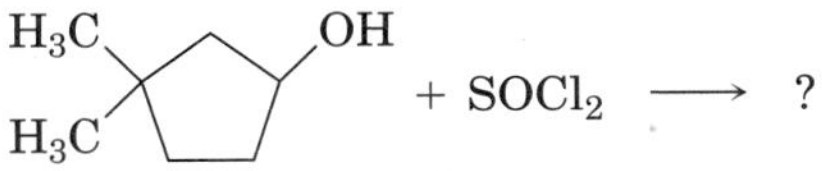 3,3-Dimethylcyclopentanol (H_3C, H_3C on one ring carbon; OH) $+ SOCl_2 \longrightarrow ?$

7.4 Reactions of Alkyl Halides: Grignard Reagents

Alkyl halides react with magnesium metal in ether solvent to yield organomagnesium halides, called **Grignard reagents** after their discoverer, Victor Grignard. Grignard reagents contain a carbon–metal bond and are thus *organometallic compounds.*

$$R\text{—}X + Mg \xrightarrow{\text{Ether}} R\text{—}Mg\text{—}X$$

where R = 1°, 2°, or 3° alkyl, aryl, or alkenyl

X = Cl, Br, or I

For example,

$$\text{C}_6\text{H}_5\text{–Br} \xrightarrow[\text{Ether}]{\text{Mg}} \text{C}_6\text{H}_5\text{–MgBr}$$

Bromobenzene → **Phenylmagnesium bromide**

$$\text{CH}_3\text{CH}_2\text{CH(Cl)CH}_3 \xrightarrow[\text{Ether}]{\text{Mg}} \text{CH}_3\text{CH}_2\text{CH(MgCl)CH}_3$$

2-Chlorobutane → ***sec*-Butylmagnesium chloride**

Grignard reagents are extraordinarily useful and versatile compounds. As you might expect from the discussion of electronegativity and bond polarity in Section 1.11, a carbon–magnesium bond is strongly polarized, making the organic part both nucleophilic and basic.

$\delta+$ MgX
$\delta-$ C ← Basic and nucleophilic site

Because of their nucleophilicity, Grignard reagents react with a wide variety of electrophiles; because of their basicity, Grignard reagents react with acids. For example, they react with acids such as HCl or H_2O to yield hydrocarbons. The overall sequence, R–X → R–MgX → R–H, is a useful method for converting an organic halide into a hydrocarbon:

$$\text{CH}_3(\text{CH}_2)_8\text{CH}_2\text{Br} \xrightarrow[\text{2. H}_2\text{O}]{\text{1. Mg}} \text{CH}_3(\text{CH}_2)_8\text{CH}_3$$

1-Bromodecane → **Decane (85%)**

PRACTICE PROBLEM 7.3 ....

By using several reactions in sequence, you can accomplish transformations that can't be done in a single step. How would you prepare the alkane methylcyclohexane from the alcohol 1-methylcyclohexanol?

$$\text{1-Methylcyclohexanol (C(OH)(CH}_3\text{))} \xrightarrow{?} \text{Methylcyclohexane (–CH}_3\text{)}$$

1-Methylcyclohexanol → **Methylcyclohexane**

Solution We know that alcohols can be converted into alkyl halides and that alkyl halides can be converted into alkanes. Carrying out the two reactions sequentially thus converts 1-methylcyclohexanol into methylcyclohexane.

$$\text{C(OH)(CH}_3\text{)} \xrightarrow{\text{HBr}} \text{C(Br)(CH}_3\text{)} \xrightarrow[\text{2. H}_2\text{O}]{\text{1. Mg, ether}} \text{–CH}_3$$

1-Methylcyclohexanol → **1-Bromo-1-methylcyclohexane** → **Methylcyclohexane**

PROBLEM

7.7 An advantage to preparing an alkane from a Grignard reagent is that deuterium (D; the ^{2}H isotope of hydrogen) can be placed at a specific site in a molecule. How might you convert 2-bromobutane into 2-deuteriobutane?

$$CH_3\underset{\displaystyle |}{\overset{\displaystyle Br}{C}}HCH_2CH_3 \xrightarrow{?} CH_3\overset{\displaystyle D}{C}HCH_2CH_3$$

PROBLEM

7.8 How could you convert 4-methyl-1-pentanol into 2-methylpentane?

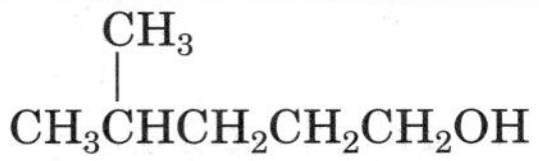

$$CH_3\overset{\displaystyle CH_3}{C}HCH_2CH_2CH_2OH \quad \textbf{4-Methyl-1-pentanol}$$

7.5 Nucleophilic Substitution Reactions: The Discovery

In 1896, the German chemist Paul Walden made a remarkable discovery. He found that (+)- and (−)-malic acids could be interconverted by a series of simple reactions. When Walden treated (−)-malic acid with PCl_5, he isolated (+)-chlorosuccinic acid. This, on reaction with wet Ag_2O, gave (+)-malic acid. Similarly, reaction of (+)-malic acid with PCl_5 gave (−)-chlorosuccinic acid, which was converted into (−)-malic acid when treated with wet Ag_2O. The full cycle of reactions reported by Walden is shown in Figure 7.2.

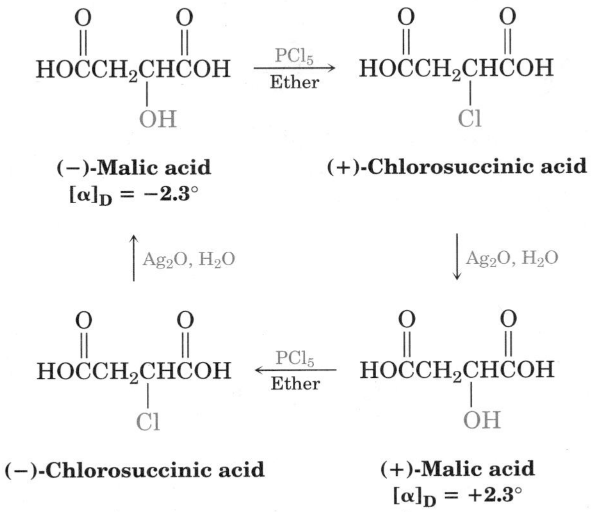

$$HO\overset{O}{\overset{\|}{C}}CH_2\underset{OH}{CH}\overset{O}{\overset{\|}{C}}OH \xrightarrow[\text{Ether}]{PCl_5} HO\overset{O}{\overset{\|}{C}}CH_2\underset{Cl}{CH}\overset{O}{\overset{\|}{C}}OH$$

(−)-Malic acid $[\alpha]_D = -2.3°$ **(+)-Chlorosuccinic acid**

↑ Ag_2O, H_2O ↓ Ag_2O, H_2O

$$HO\overset{O}{\overset{\|}{C}}CH_2\underset{Cl}{CH}\overset{O}{\overset{\|}{C}}OH \xleftarrow[\text{Ether}]{PCl_5} HO\overset{O}{\overset{\|}{C}}CH_2\underset{OH}{CH}\overset{O}{\overset{\|}{C}}OH$$

(−)-Chlorosuccinic acid **(+)-Malic acid** $[\alpha]_D = +2.3°$

Figure 7.2 Walden's cycle of reactions interconverting (+)- and (−)-malic acids.

At the time, the results were astonishing. Since (−)-malic acid was converted into (+)-malic acid, *some reactions in the cycle must have occurred with an inversion, or change, in the configuration of the stereocenter.* But which ones, and how?

Today we refer to the transformations taking place in Walden's cycle as **nucleophilic substitution reactions,** because each step involves the substitution of one nucleophile (chloride ion, Cl^-, or hydroxide ion, OH^-) for another. Nucleophilic substitution reactions are one of the most common and versatile reaction types in organic chemistry.

7.6 Kinds of Nucleophilic Substitution Reactions

Following the work of Walden, a series of investigations was undertaken during the 1920s and 1930s to clarify the mechanism of nucleophilic substitution reactions and to find out how inversions of configuration occur. We now know that nucleophilic substitutions occur by two major pathways, named the *S_N1 reaction* and the *S_N2 reaction.* In both cases, the "S_N" part of the name stands for "substitution, nucleophilic." The meaning of the 1 and the 2 will be discussed in Sections 7.7 and 7.8.

Regardless of mechanism, the overall change during all nucleophilic substitution reactions is the same: A *nucleophile* (Nu: or $Nu:^-$) reacts with a *substrate* R–X and substitutes for $X:^-$ (the **leaving group**) to yield the product R–Nu. If the nucleophile is neutral (Nu:), then the product is positively charged to maintain charge conservation; if the nucleophile is negatively charged ($Nu:^-$), the product is neutral.

Negatively charged $Nu:^-$ $\quad Nu:^- + R{-}X \longrightarrow R{-}Nu + X:^-$

Neutral Nu: $\quad Nu: + R{-}X \longrightarrow R{-}Nu^+ + X:^-$

Because of the wide scope of nucleophilic substitution reactions, many products can be prepared. In fact, we've already seen a number of nucleophilic substitution reactions in previous chapters. The reaction of acetylide anions with alkyl halides (Section 4.16), for example, is an S_N2 reaction. Table 7.1 lists others.

PRACTICE PROBLEM 7.4

What is the substitution product from reaction of 1-chloropropane with sodium hydroxide?

Solution Write the two starting materials, and identify the nucleophile (in this instance, OH^-) and the leaving group (in this instance, Cl^-). Then replace the –Cl group by –OH and write the complete equation.

$$CH_3CH_2CH_2Cl + Na^+ \ {}^-OH \longrightarrow CH_3CH_2CH_2OH + Na^+ \ {}^-Cl$$

1-Chloropropane **1-Propanol**

Table 7.1 Some Nucleophilic Substitution Reactions with Bromomethane:
$Nu:^- + CH_3Br \longrightarrow Nu{-}CH_3 + :\ddot{Br}:^-$

Attacking nucleophile		*Product*	
Formula	*Name*	*Formula*	*Name*
$H:^-$	Hydride	CH_4	Methane
$CH_3\ddot{S}:^-$	Methanethiolate	CH_3SCH_3	Dimethyl sulfide
$H\ddot{S}:^-$	Hydrosulfide	$HSCH_3$	Methanethiol
$N{\equiv}C:^-$	Cyanide	$N{\equiv}CCH_3$	Acetonitrile
$:\ddot{I}:^-$	Iodide	ICH_3	Iodomethane
$H\ddot{O}:^-$	Hydroxide	$HOCH_3$	Methanol
$CH_3\ddot{O}:^-$	Methoxide	CH_3OCH_3	Dimethyl ether
$N{=}N{=}\ddot{N}:^-$	Azide	N_3CH_3	Azidomethane
$:\ddot{Cl}:^-$	Chloride	$ClCH_3$	Chloromethane
$CH_3CO_2:^-$	Acetate	$CH_3CO_2CH_3$	Methyl acetate
$H_3N:$	Ammonia	$H_3\overset{+}{N}CH_3\ Br^-$	Methylammonium bromide
$(CH_3)_3N:$	Trimethylamine	$(CH_3)_3\overset{+}{N}CH_3\ Br^-$	Tetramethylammonium bromide

PRACTICE PROBLEM 7.5

How would you prepare 1-propanethiol, $CH_3CH_2CH_2SH$, using a nucleophilic substitution reaction?

Solution Since the product contains an –SH group, it might be prepared by reaction of SH^- (hydrosulfide ion) with 1-bromopropane:

$$\underset{\textbf{1-Bromopropane}}{CH_3CH_2CH_2Br} + Na^+\ ^-SH \longrightarrow \underset{\textbf{1-Propanethiol}}{CH_3CH_2CH_2SH} + Na^+\ ^-Br$$

PROBLEM

7.9 What substitution products would you expect to obtain from the following reactions?

(a) $CH_3CH_2\overset{\overset{\large Br}{|}}{C}HCH_3 + LiI \longrightarrow\ ?$

(b) $CH_3\overset{\overset{\large CH_3}{|}}{C}HCH_2Cl + HS^- \longrightarrow\ ?$

(c) 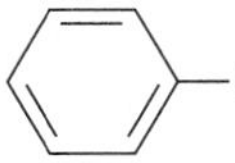 $C_6H_5{-}CH_2Br + NaCN \longrightarrow\ ?$ (benzene ring)

PROBLEM

7.10 How might you prepare the following substances by using nucleophilic substitution reactions?
(a) $CH_3CH_2CH_2CH_2OH$ (b) $(CH_3)_2CHCH_2CH_2N_3$

7.7 The S_N2 Reaction

An **S_N2 reaction** takes place in a single step without intermediates when the entering nucleophile attacks the substrate from a direction 180° away from the leaving group. As the nucleophile comes in on one side of the molecule, an electron pair on the nucleophile Nu:$^-$ forces out the leaving group X:$^-$, which departs from the other side of the molecule and takes with it the electron pair from the C–X bond. In the transition state for the reaction, the new Nu–C bond is partially forming at the same time the old C–X bond is partially breaking, and the negative charge is shared by both the incoming nucleophile and the outgoing leaving group. The mechanism is shown in Figure 7.3 for the reaction of OH^- with (*S*)-2-bromobutane.

The nucleophile ^-OH uses its lone-pair electrons to attack the alkyl halide carbon 180° away from the departing halogen. This leads to a transition state with a partially formed C–OH bond and a partially broken C–Br bond.

The stereochemistry at carbon is inverted as the C–OH bond forms fully and the bromide ion departs with the electron pair from the former C–Br bond.

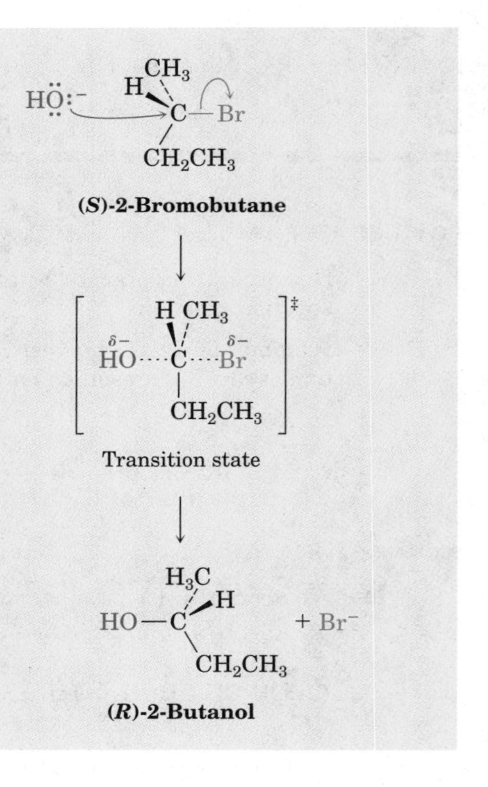

Figure 7.3 The mechanism of the S_N2 reaction. The reaction takes place in a single step when the incoming nucleophile (OH^-) approaches from a direction 180° away from the leaving group (Br^-), thereby inverting the stereochemistry at carbon.

Let's see what evidence there is for this mechanism and what the chemical consequences are.

Rates of S_N2 Reactions

The exact speed with which a reaction occurs is called the **reaction rate** and is a quantity that can often be measured. The determination of reaction rates and of how those rates depend on reactant concentrations is a powerful tool for probing mechanisms. As an example, let's look at the effect of reactant concentrations on the rate of the S_N2 reaction of OH^- with CH_3Br to yield CH_3OH:

$$HO:^- + CH_3{-}Br: \longrightarrow HO{-}CH_3 + :Br:^-$$

The S_N2 reaction of CH_3Br with OH^- takes place when substrate and nucleophile collide and react in a single step. At a given concentration of reactants, the reaction takes place at a certain rate. If we double the concentration of OH^-, the frequency of encounter between the two reactants doubles, and we therefore find that the reaction rate also doubles. Similarly, if we double the concentration of CH_3Br, the reaction rate doubles. Thus, the origin of the "2" in S_N2: S_N2 reactions are said to be **bimolecular** because the rate of the reaction depends on the concentration of *two* substances, alkyl halide and nucleophile.

PROBLEM

7.11 What effects would the following changes have on the rate of the S_N2 reaction between CH_3I and sodium acetate?
(a) The CH_3I concentration is tripled.
(b) Both CH_3I and CH_3CO_2Na concentrations are doubled.

Stereochemistry of S_N2 Reactions

Look carefully at the mechanism of the S_N2 reaction shown in Figure 7.3. As the incoming nucleophile attacks the substrate and begins pushing out the leaving group on the opposite side, the stereochemistry of the molecule *inverts.* (*S*)-2-Bromobutane gives (*R*)-2-butanol, for example, by an inversion of configuration that occurs through a planar transition state (Figure 7.4, p. 230).

PRACTICE PROBLEM 7.6

What product would you expect to obtain from the S_N2 reaction of (*S*)-2-iodooctane with sodium cyanide, NaCN?

Solution Table 7.1 shows that cyanide ion is a nucleophile in the S_N2 reaction. We therefore expect it to displace iodide ion from (*S*)-2-iodooctane with inversion of configuration to yield (*R*)-2-methyloctanenitrile.

I H — $Na^+\ CN^-$ → H CN + NaI

(*S*)-2-Iodooctane **(*R*)-2-Methyloctanenitrile**

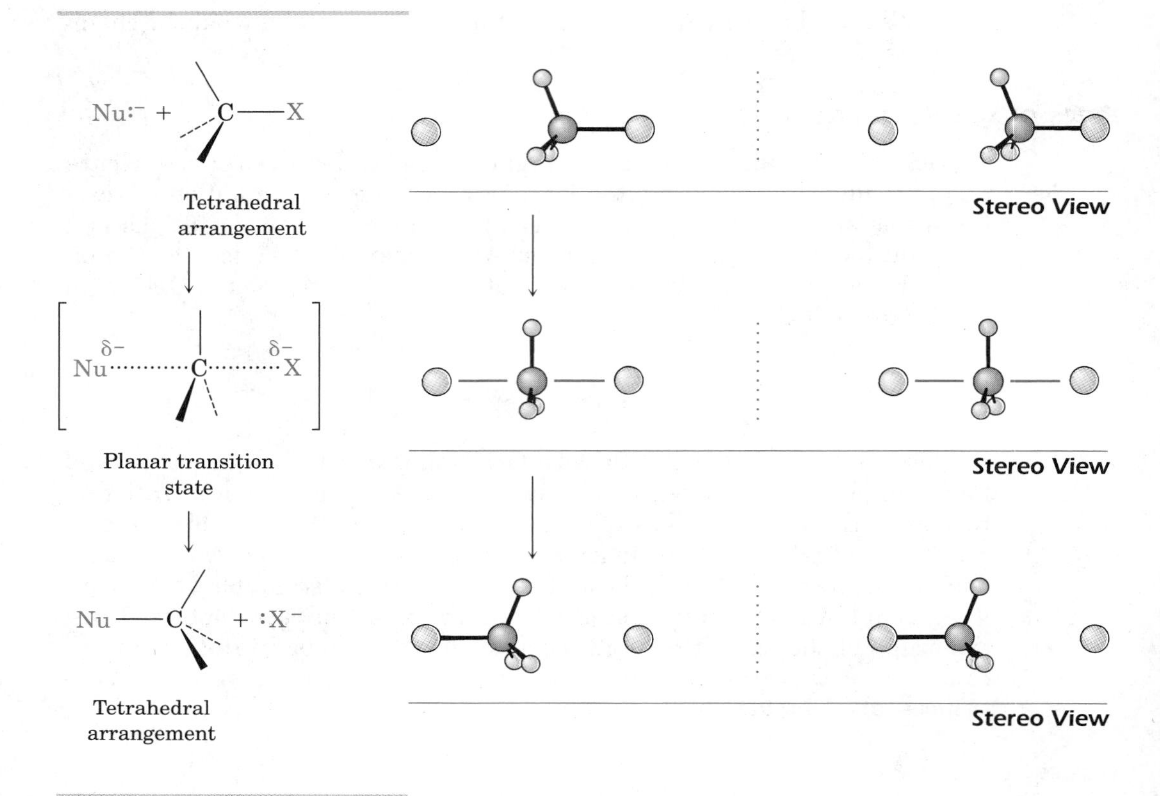

Figure 7.4 The transition state of the S_N2 reaction has a planar arrangement of the carbon atom and the three attached groups.

PROBLEM

7.12 What product would you expect to obtain from the S_N2 reaction of (*S*)-2-bromohexane with sodium acetate, CH_3CO_2Na? Show the stereochemistry of both product and reactant.

PROBLEM

7.13 How can you explain the fact that treatment of (*R*)-2-bromohexane with NaBr yields *racemic* 2-bromohexane?

Steric Effects in S_N2 Reactions

The ease with which a nucleophile can approach a substrate to carry out an S_N2 reaction depends on steric accessibility. Bulky substrates, in which the halide-bearing carbon atom is shielded from attack, react more slowly than those in which the carbon is more accessible (Figure 7.5).

Methyl halides (CH_3X) are the most reactive substrates, followed by primary alkyls (RCH_2–X) such as ethyl and propyl. Alkyl branching next to the leaving group slows the reaction greatly for secondary halides (R_2CH–X), and further branching effectively halts the reaction for tertiary halides (R_3C–X).

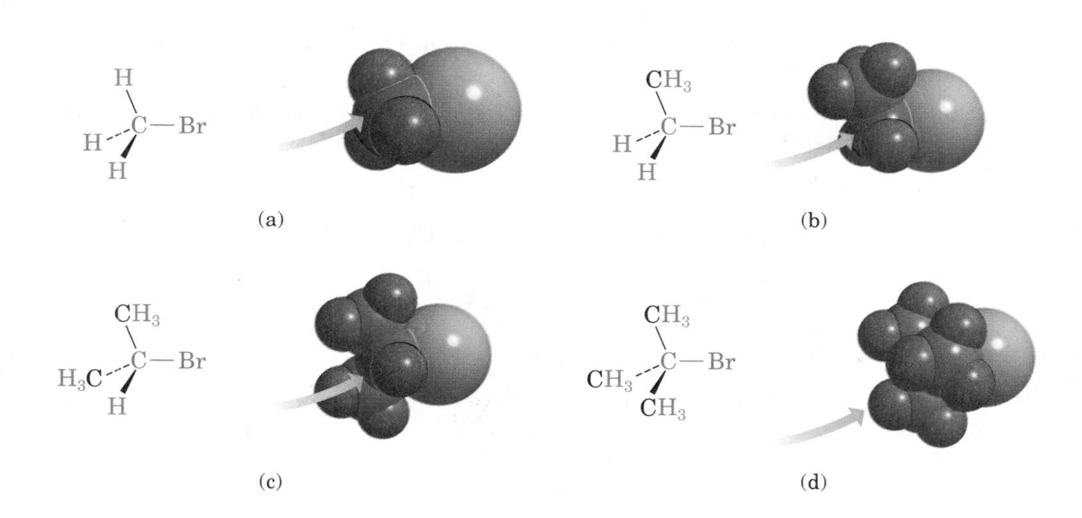

Figure 7.5 Steric hindrance to the S_N2 reaction. As these computer-generated models show, the carbon atom in (a) bromomethane is readily accessible, resulting in a fast S_N2 reaction, but the carbon atoms in (b) bromoethane, (c) 2-bromopropane, and (d) 2-bromo-2-methylpropane are successively less accessible, resulting in successively slower S_N2 reactions.

$$R{-}Br + Cl^- \longrightarrow R{-}Cl + Br^-$$

	$H{-}CH_2{-}Br$	$H_3C{-}CH_2{-}Br$	$H_3C{-}CH(CH_3){-}Br$	$H_3C{-}C(CH_3)_2{-}CH_2{-}Br$	$H_3C{-}C(CH_3)_2{-}Br$
	(Methyl)	(Primary)	(Secondary)	(Neopentyl)	(Tertiary)
Relative reactivity	2,000,000	40,000	500	1	<1

More reactive ← **Reactivity as substrate** — Less reactive

Vinylic ($R_2C{=}CRX$) and aryl (Ar–X) halides are not shown on this reactivity list because they are completely unreactive toward S_N2 displacements. This lack of reactivity is due to steric hindrance, because the incoming nucleophile would have to burrow through part of the molecule to carry out a back-side displacement.

$R_2C{=}C(Cl)R$ + $Nu{:}^-$ ⟶̸ No reaction

Vinylic halide

$Ar{-}Cl$ + $:Nu^-$ ⟶̸ No reaction

Aryl halide

PRACTICE PROBLEM 7.7

Which would you expect to be faster, the S_N2 reaction of OH^- ion with 1-bromopentane or with 2-bromopentane?

Solution Since 1-bromopentane is a 1° halide and 2-bromopentane is a 2° halide, reaction with the less hindered 1-bromopentane should be faster.

$CH_3CH_2CH_2CH_2CH_2Br$ (Primary) **1-Bromopentane**

$CH_3CH_2CH_2CH(Br)CH_3$ (Secondary) **2-Bromopentane**

PROBLEM

7.14 Which of the following S_N2 reactions would you expect to be faster?
(a) Reaction of CN^- (cyanide ion) with $CH_3CH(Br)CH_3$ or with $CH_3CH_2CH_2Br$?
(b) Reaction of I^- with $(CH_3)_2CHCH_2Cl$ or with $H_2C{=}CHCl$?

The Leaving Group in S_N2 Reactions

Another variable that can affect the S_N2 reaction is the identity of the leaving group displaced by the attacking nucleophile. Because the leaving group is expelled with a negative charge in most S_N2 reactions, the best leaving groups are those that give the most stable anions. A halide ion (I^-, Br^-, or Cl^-) is the most common leaving group, though others are also possible. Such anions as F^-, OH^-, OR^-, and $NH_2{}^-$ are rarely found as leaving groups.

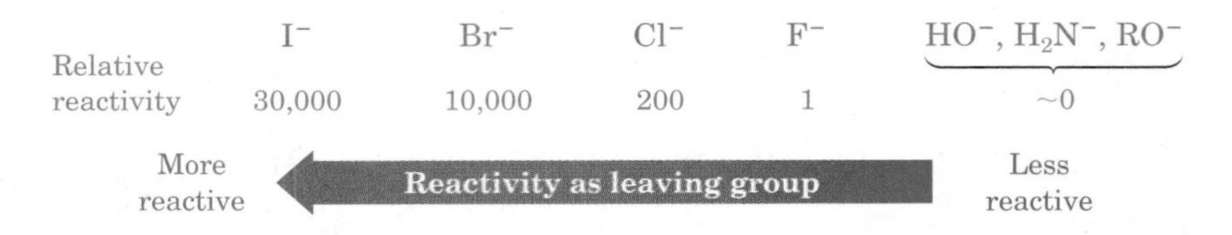

PROBLEM

7.15 Rank the following compounds in order of their expected reactivity toward S_N2 reaction: CH_3I, CH_3F, CH_3Br.

7.8 The S_N1 Reaction

Most nucleophilic substitutions take place by the S_N2 pathway, but an alternative called the **S_N1 reaction** can also occur. In general, S_N1 reactions take place only on *tertiary* substrates and only under neutral or acidic conditions in a hydroxylic solvent such as water or alcohol. We saw in Section 7.3, for example, that alkyl halides can be prepared from alcohols by treatment with HCl or HBr. Tertiary alcohols react rapidly, but primary and secondary alcohols are much slower.

$$R_3COH >> R_2CHOH > RCH_2OH > CH_3OH$$

What's going on here? Clearly, a nucleophilic substitution reaction is taking place—a halogen is replacing a hydroxyl group—yet the reactivity order 3° > 2° > 1° is backward from the normal S_N2 order. Furthermore, a hydroxyl group is being replaced, although we said in the previous section that OH^- is a poor leaving group. What's going on is that this is *not* an S_N2 reaction; it is an S_N1 reaction, whose mechanism is shown in Figure 7.6.

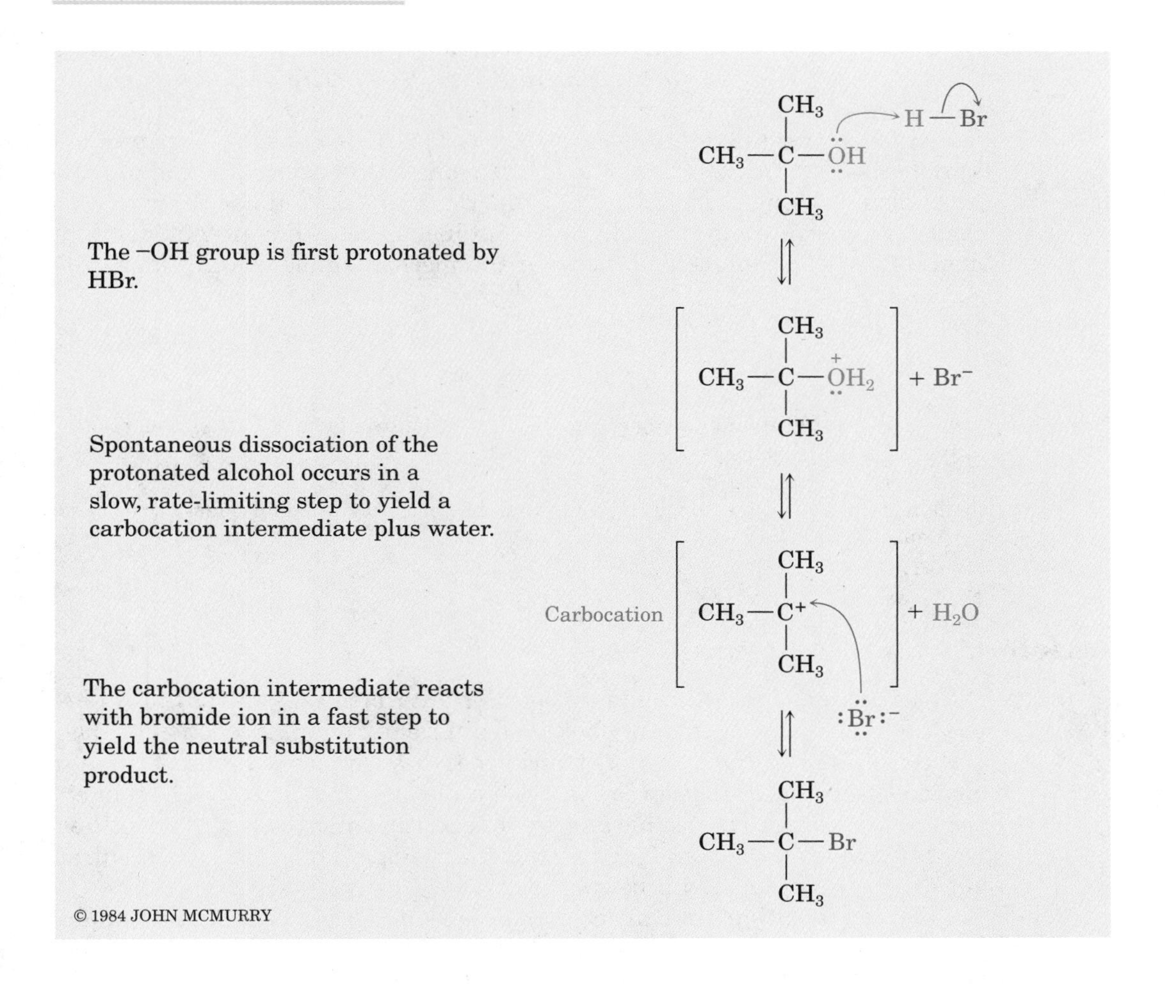

Figure 7.6 The mechanism of the S_N1 reaction of *tert*-butyl alcohol with HBr to yield an alkyl halide. Neutral H_2O is the leaving group.

Unlike what occurs in an S_N2 reaction, where the leaving group is displaced *at the same time* that the incoming nucleophile approaches, an S_N1 reaction occurs by spontaneous loss of the leaving group *before* the incoming nucleophile approaches. Loss of the leaving group gives a carbocation intermediate, which then reacts with nucleophile in a second step to yield the substitution product.

This two-step mechanism explains why tertiary alcohols react with HBr so much more rapidly than primary or secondary ones do: S_N1 reactions can occur only when stable carbocation intermediates are formed. The more stable the carbocation intermediate, the faster the S_N1 reaction. Thus, the reactivity order of alcohols with HBr is the same as the stability order of carbocations (Section 4.3).

Rates of S_N1 Reactions

Unlike the rate of an S_N2 reaction, which depends on the concentrations of both substrate and nucleophile, *the rate of an S_N1 reaction depends only on the concentration of the substrate and is independent of the nucleophile concentration.* Thus, the origin of the "1" in S_N1: S_N1 reactions are **unimolecular** because the rate of the reaction depends on the concentration of only one substance. The observation that S_N1 reactions are unimolecular means that the substrate must undergo a spontaneous reaction without involvement of the nucleophile, exactly what the mechanism shown in Figure 7.6 accounts for.

PROBLEM

7.16 What effect would the following changes have on the rate of the S_N1 reaction of *tert*-butyl alcohol with HBr?
(a) The HBr concentration is tripled.
(b) The HBr concentration is halved, and the *tert*-butyl alcohol concentration is doubled.

Stereochemistry of S_N1 Reactions

If S_N1 reactions occur through carbocation intermediates as shown in Figure 7.6, their stereochemistry should be different from S_N2 reactions. Since carbocations are planar and sp^2-hybridized, they are achiral. The positively charged carbon can therefore react with a nucleophile equally well from either face, leading to a racemic mixture of enantiomers (Figure 7.7). In other words, if we carry out an S_N1 reaction on a single enantiomer of a chiral substrate, the product will be racemic.

The expectation that an S_N1 reaction on a chiral substrate should lead to a racemic product is exactly what is found. For example, reaction of (*R*)-1-phenyl-1-butanol with HCl gives a racemic alkyl chloride product:

$CH_3CH_2CH_2$–C(H)(Ph)–OH + HCl ⟶ $CH_3CH_2CH_2$–C(H)(Ph)–Cl + Cl–C(H)(Ph)–$CH_2CH_2CH_3$

(*R*)-1-Phenyl-1-butanol | **(*R*)-1-Phenyl-1-chlorobutane** (50%, retention) | **(*S*)-1-Phenyl-1-chlorobutane** (50%, inversion)

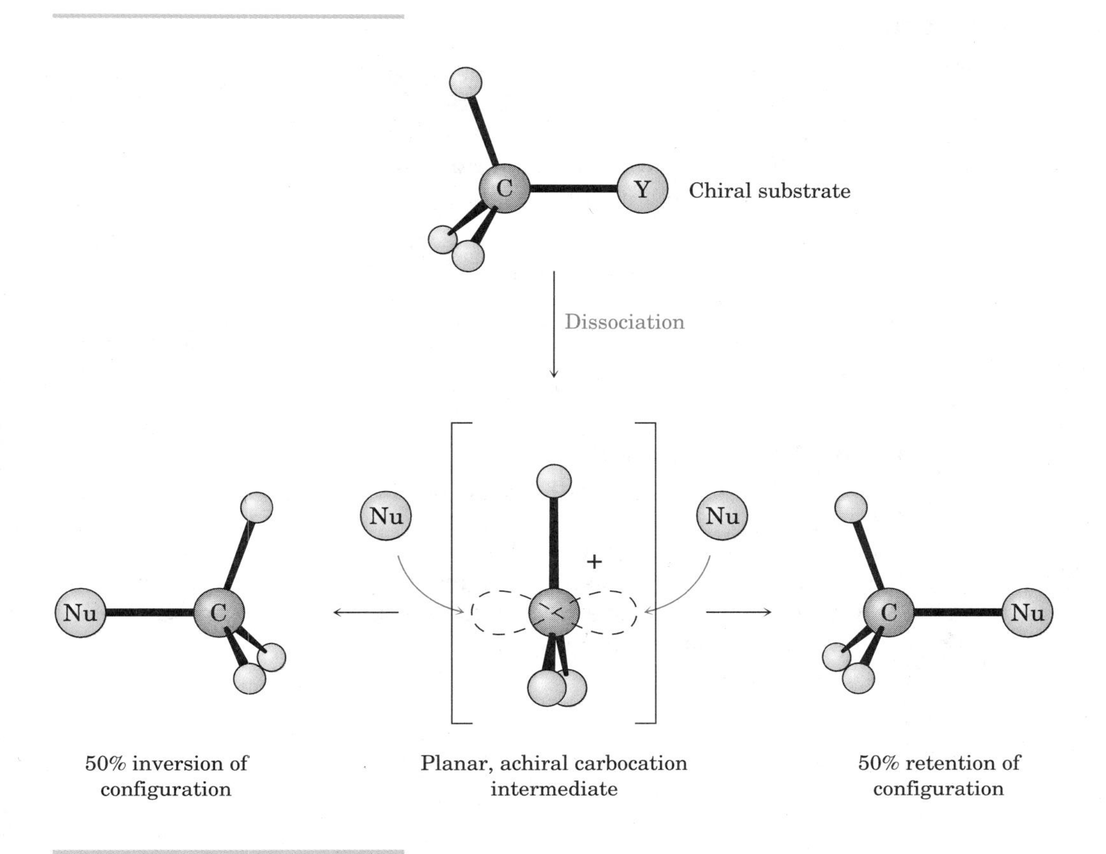

Figure 7.7 An S_N1 reaction on a chiral substrate. An optically active starting material gives a racemic product.

PRACTICE PROBLEM 7.8

What stereochemistry would you expect for the S_N1 reaction of (*R*)-3-bromo-3-methylhexane with methanol to yield 3-methoxy-3-methylhexane?

Solution First draw the starting alkyl halide, showing its correct stereochemistry. Then replace the –Br with a methoxy group (–OCH_3) to give the racemic product.

H_3C Br

+ CH_3OH ⟶

(*R*)-3-Bromo-3-methylhexane

CH_3O CH_3

(*S*)-3-Methoxy-3-methylhexane (50%)

H_3C OCH_3

(*R*)-3-Methoxy-3-methylhexane (50%)

PROBLEM

7.17 What product would you expect to obtain from the S_N1 reaction of (*S*)-3-methyl-3-octanol [(*S*)-3-hydroxy-3-methyloctane] with HBr? Show the stereochemistry of both starting material and product.

The Leaving Group in S_N1 Reactions

The best leaving groups in S_N1 reactions are those that are most stable, just as in S_N2 reactions. Note that if an S_N1 reaction is carried out under acidic conditions, as occurs when a tertiary alcohol reacts with HX to yield an alkyl halide (Figure 7.6), neutral water can be the leaving group. The S_N1 reactivity order of leaving groups is:

$$I^- > Br^- > H_2O = Cl^- >> F^-$$

Better leaving group ← Leaving group ability — Worse leaving group

7.9 Eliminations: The E2 Reaction

Two kinds of reactions are possible when a nucleophile/base reacts with an alkyl halide. The nucleophile/base can substitute for the leaving group in an S_N1 or S_N2 reaction, or the nucleophile/base can cause elimination of HX, leading to formation of an alkene:

Substitution H–C–C–Br + OH^- ⟶ H–C–C–OH + Br^-

Elimination H–C–C–Br + OH^- ⟶ C=C + H_2O + Br^-

We saw in Section 4.9 that the elimination of HX from alkyl halides is an extremely useful method for preparing alkenes. The topic is complex, though, because eliminations can take place by several different mechanistic pathways, just as substitutions can.

The **E2 reaction** (elimination, bimolecular) takes place when an alkyl halide is treated with a strong base, such as hydroxide ion or alkoxide ion (RO^-). The mechanism is shown in Figure 7.8.

Like the S_N2 reaction, the E2 reaction takes place in one step without intermediates. As the attacking base begins to abstract H^+ from a carbon atom next to the leaving group, the C–H and C–X bonds begin to break and the C=C double bond begins to form. When the leaving group departs, it takes with it the two electrons from the former C–X bond.

One good piece of evidence supporting this mechanism comes from measurements of reaction rates. Because both base and alkyl halide are involved in the one-step mechanism, E2 reactions show the same bimolecular behavior that S_N2 reactions do. A second piece of evidence involves the stereochemistry of E2 reactions. Eliminations almost always occur from an **anti periplanar geometry,** meaning that all reacting atoms lie in the same

Base (B:) attacks a neighboring hydrogen and begins to remove the H at the same time as the alkene double bond starts to form and the X group starts to leave.

Transition state

Neutral alkene is produced when the C–H bond is fully broken and the X group has departed with the C–X bond electron pair.

Figure 7.8 The mechanism of the E2 reaction. The reaction takes place in a single step through a transition state in which the double bond begins to form at the same time the H and X groups are leaving. (Dotted lines indicate partial bonding in the transition state.)

plane (*periplanar*) and that the H and X depart from opposite sides (*anti*) of the molecule:

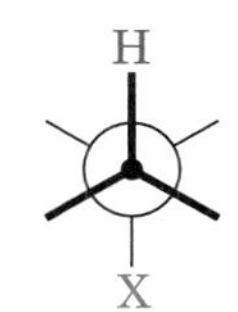

Anti periplanar geometry (staggered, lower energy)

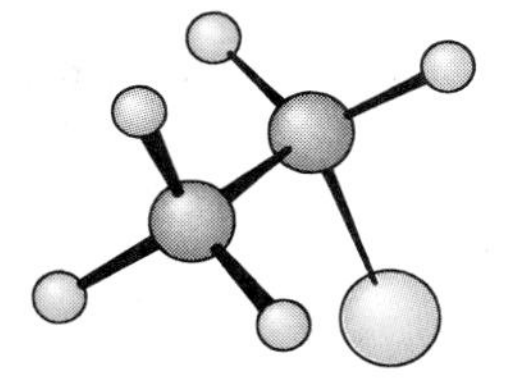

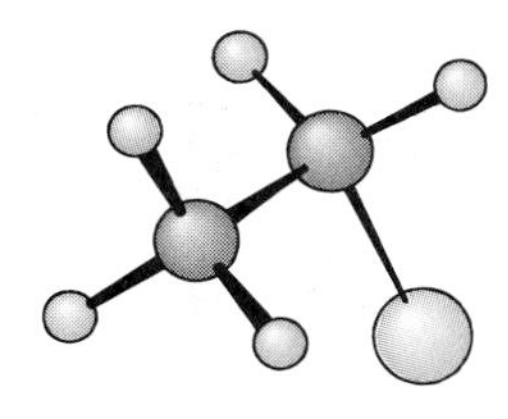

Stereo View

What's so special about anti periplanar geometry? Because the sp^3 orbitals in the C–H and C–X bonds of the reactant must overlap and become *p* orbitals in the alkene product, *they must also overlap in the transition state.* This can only occur if the orbitals are in the same plane (periplanar) to begin with (Figure 7.9, p. 238).

Anti periplanar geometry for E2 reactions has stereochemical consequences that provide strong evidence for the proposed mechanism. To cite just one example, *meso*-1,2-dibromo-1,2-diphenylethane undergoes E2 elimination on treatment with base to give the pure *E* alkene, rather than a mixture of *E* and *Z* alkenes:

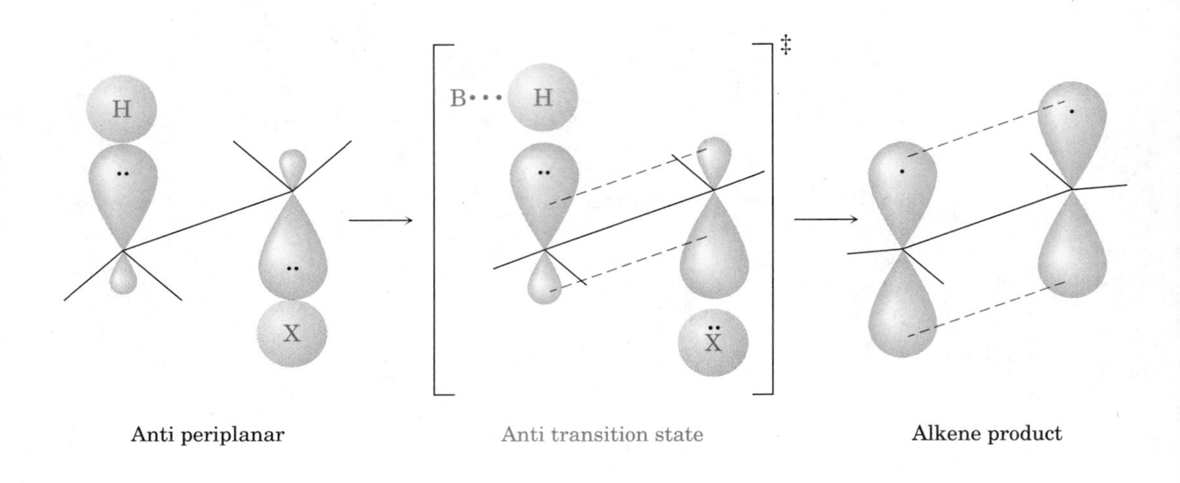

Figure 7.9 The transition state for an E2 reaction of an alkyl halide. Overlap of developing *p* orbitals in the transition state requires periplanar geometry in the reactant.

Base: → KOH / Ethanol →

[Ph = phenyl]

***meso*-1,2-Dibromo-1,2-diphenylethane** → **(*E*)-1-Bromo-1,2-diphenylethylene**

PRACTICE PROBLEM 7.9

What stereochemistry (*E* or *Z*) do you expect for the alkene obtained by E2 elimination of (1*S*,2*S*)-1,2-dibromo-1,2-diphenylethane?

Solution First, draw (1*S*,2*S*)-1,2-dibromo-1,2-diphenylethane so that you can see its stereochemistry and so that the –H and –Br groups to be eliminated are anti periplanar (molecular models are extremely helpful here). Keeping all substituents in approximately their same positions, eliminate HBr and see what alkene results. The product is (*Z*)-1-bromo-1,2-diphenylethylene.

B: →

PROBLEM

7.18 Ignoring double-bond stereochemistry, what elimination products would you expect from the following reactions? (Remember Zaitsev's rule; Section 4.9.)

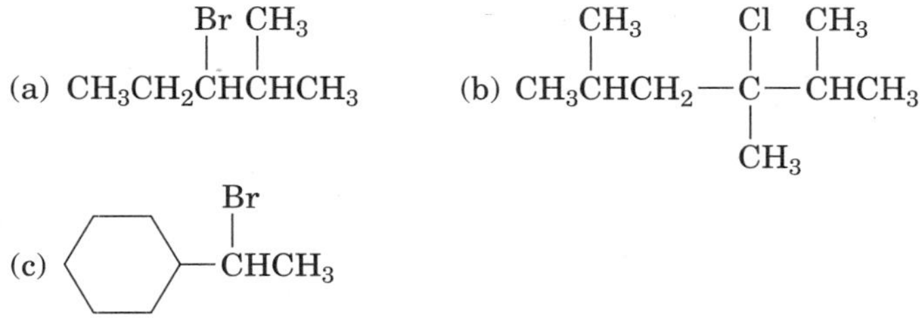

(a) $CH_3CH_2\overset{\overset{\displaystyle Br}{|}}{C}H\overset{\overset{\displaystyle CH_3}{|}}{C}HCH_3$

(b) $CH_3\overset{\overset{\displaystyle CH_3}{|}}{C}HCH_2-\underset{\underset{\displaystyle CH_3}{|}}{\overset{\overset{\displaystyle Cl}{|}}{C}}-\overset{\overset{\displaystyle CH_3}{|}}{C}HCH_3$

(c) Cyclohexyl$-\overset{\overset{\displaystyle Br}{|}}{C}HCH_3$

PROBLEM

7.19 What stereochemistry (*E* or *Z*) do you expect for the alkene obtained by E2 elimination of (1*R*,2*R*)-1,2-dibromo-1,2-diphenylethane? Draw a Newman projection of the reacting conformation.

7.10 Eliminations: The E1 Reaction

Just as the S_N2 reaction has an analog in the E2 reaction, the S_N1 reaction has an analog in the **E1 reaction** (elimination, unimolecular). Both S_N1 and E1 are unimolecular reactions that occur by spontaneous dissociation of the substrate to produce an intermediate carbocation, which then undergoes further reaction.

We saw in Section 4.9 that alcohols can be dehydrated by treatment with H_2SO_4 to give alkenes, a reaction that occurs by the E1 mechanism shown in Figure 7.10 (p. 240). Acid first protonates the –OH group of the alcohol, the protonated alcohol spontaneously loses water to yield a carbocation intermediate, and the carbocation then loses H^+ from a neighboring carbon to give alkene product. Tertiary alcohols react much faster than primary or secondary ones because they lead to more stable carbocation intermediates.

PROBLEM

7.20 What effect on the rate of an E1 dehydration of 2-methyl-2-propanol would you expect if the concentration of the alcohol were tripled?

7.11 A Summary of Reactivity: S_N1, S_N2, E1, E2

Now that we've seen four different kinds of nucleophilic substitution/elimination reactions, you may well wonder how to predict what will take place in any given case. Will substitution or elimination occur? Will the reaction

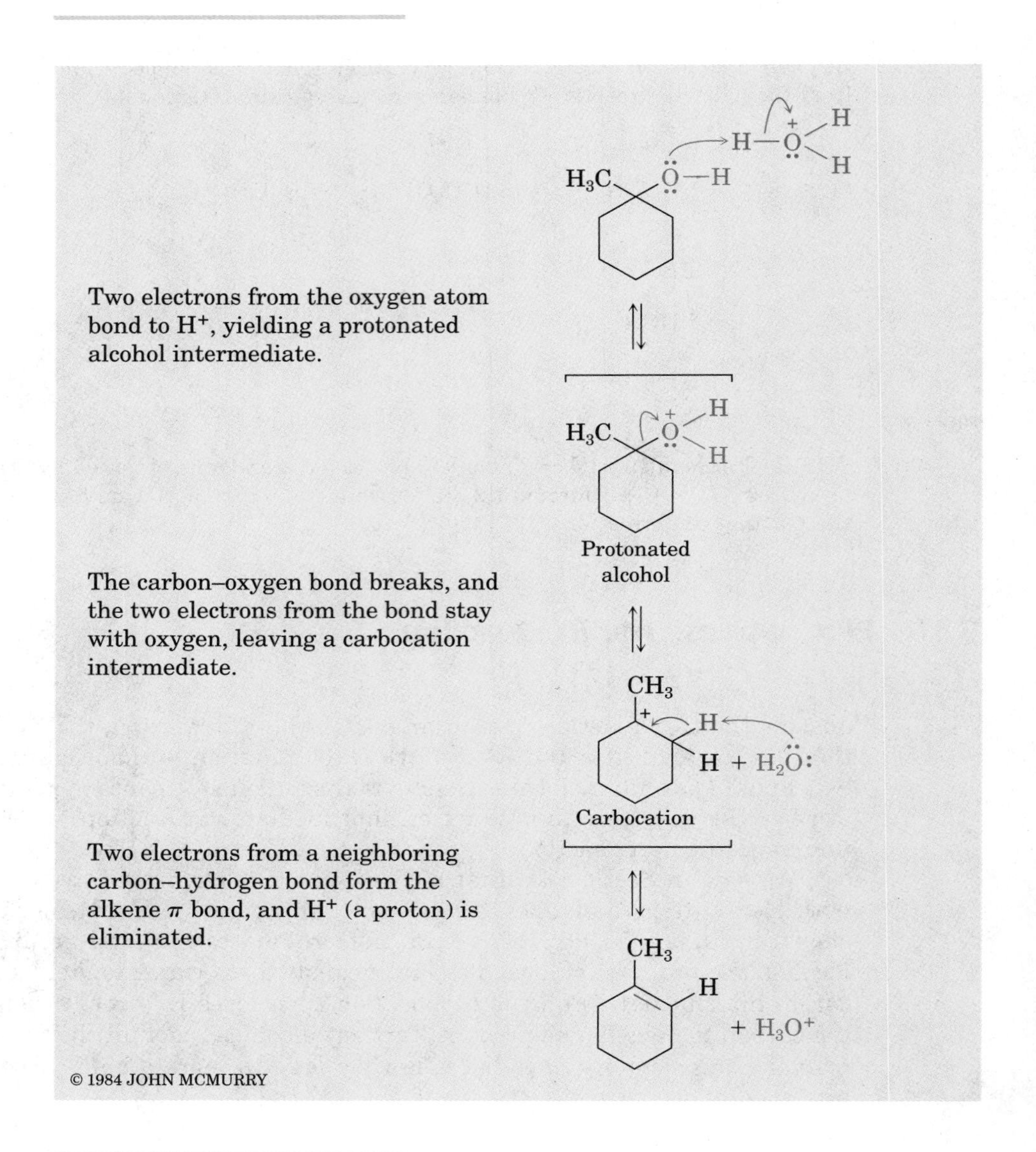
Two electrons from the oxygen atom bond to H+, yielding a protonated alcohol intermediate.
The carbon–oxygen bond breaks, and the two electrons from the bond stay with oxygen, leaving a carbocation intermediate.
Two electrons from a neighboring carbon–hydrogen bond form the alkene π bond, and H+ (a proton) is eliminated.
H3C
O—H
H—O
H
H
Protonated alcohol
CH3
H
H + H2O:
Carbocation
+ H3O+
© 1984 JOHN MCMURRY

Figure 7.10 Mechanism of the E1 reaction in the acid-catalyzed dehydration of a tertiary alcohol. Two steps are involved, and a carbocation intermediate is formed.

be unimolecular or bimolecular? There are no rigid answers to these questions, but it's possible to make some broad generalizations.

1. *A primary substrate* normally reacts by an S_N2 pathway because it is unhindered and because its dissociation would give an unstable, primary carbocation. If a primary alkyl halide is treated with a good nucleophile such as I^-, Br^-, RS^-, NH_3, or CN^-, only S_N2 substitution occurs. If a strong base such as hydroxide ion or an alkoxide ion (RO^-) is used, some E2 elimination might also occur.

2. *A secondary substrate* reacts by either an S_N2 or an E2 pathway. If a secondary alkyl halide is treated with a good nucleophile, S_N2 substitution predominates. If a strong base such as OH^- is used, E2 elimination predominates.
3. *A tertiary substrate* can react by any of three pathways, S_N1, E1, or E2. If a tertiary alcohol is treated with HCl, HBr, or HI, then S_N1 substitution occurs; if the alcohol is treated with H_2SO_4, then E1 elimination occurs; and if a tertiary alkyl halide is treated with a strong base OH^-, then E2 elimination occurs.

These generalizations are summarized in Table 7.2.

Table 7.2 A Summary of Substitution and Elimination Reactions

Halide type	S_N1	S_N2	E1	E2
RCH_2X (primary)	Does not occur	Highly favored	Does not occur	Occurs when strong bases are used
R_2CHX (secondary)	Can occur	Occurs in competition with E2 reaction	Can occur	Favored when strong bases are used
R_3CX (tertiary)	Favored in hydroxylic solvents	Does not occur	Occurs in competition with S_N1 reaction	Favored when bases are used

PRACTICE PROBLEM 7.10

Tell what kind of reaction this is:

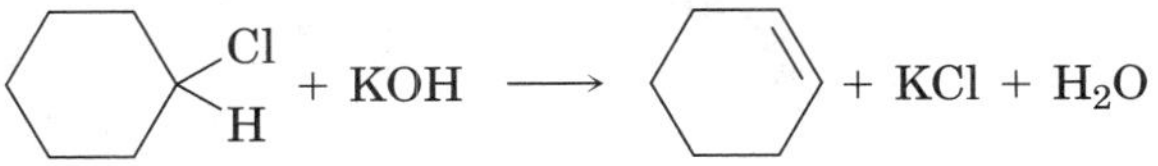

$$\text{(cyclohexyl-CHCl: } C_6H_{10}(Cl)(H)) + KOH \longrightarrow \text{cyclohexene} + KCl + H_2O$$

Solution Because KOH is a strong base and a secondary alkyl halide is undergoing elimination of HCl, this is an E2 reaction.

PROBLEM

7.21 Tell whether the following reactions are S_N1, S_N2, E1, or E2:

(a) 1-Bromobutane + NaN_3 $\longrightarrow$ 1-Azidobutane

(b) $CH_3CH_2CH(Cl)CH_2CH_3 + KOH \longrightarrow CH_3CH_2CH{=}CHCH_3$

(c) 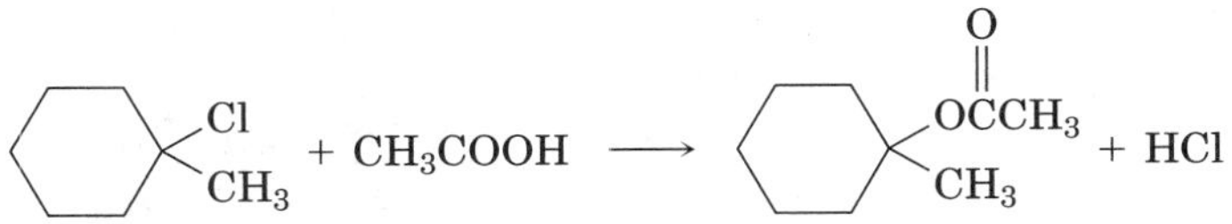 1-chloro-1-methylcyclohexane ($C_6H_{10}(Cl)(CH_3)$) $+ CH_3COOH \longrightarrow$ $C_6H_{10}(OCOCH_3)(CH_3)$ (with $\overset{O}{\overset{\|}{OCCH_3}}$) $+ HCl$

7.12 Substitution Reactions in Living Organisms

Chemical reactions in living organisms take place by reaction pathways analogous to those carried out in the laboratory. Thus, a number of processes that occur in living organisms involve nucleophilic substitution reactions. Perhaps the most common of all biological substitution reactions is *methylation*—the transfer of a methyl group from a donor molecule to a nucleophile.

A laboratory chemist would probably use CH_3I for a methylation reaction, but living organisms operate more subtly. The large and complex molecule *S*-adenosylmethionine is the biological methyl-group donor. Since the sulfur atom of *S*-adenosylmethionine has a positive charge (a *sulfonium ion*), it is an excellent leaving group for S_N2 displacements on the methyl carbon. An example of the action of *S*-adenosylmethionine in biological methylations takes place in the adrenal medulla during the formation of adrenaline from norepinephrine (Figure 7.11).

Norepinephrine: HO, HO (ring) $-CH(OH)CH_2NH_2$ + *S*-Adenosylmethionine: $HOOCCH(NH_2)CH_2CH_2\overset{+}{S}(CH_3)CH_2$–(ribose–adenine)

$\downarrow$ S_N2 reaction

Adrenaline: HO, HO (ring) $-CH(OH)CH_2NHCH_3$ + $HOOCCH(NH_2)CH_2CH_2SCH_2$–(ribose–adenine)

Figure 7.11 The biological methylation of norepinephrine by reaction with *S*-adenosylmethionine to yield adrenaline.

After having dealt only with simple alkyl halides like CH_3I up to this point, it's something of a surprise to encounter a molecule as complex as *S*-adenosylmethionine. From the chemical standpoint, though, iodomethane and *S*-adenosylmethionine do exactly the same thing: Both transfer a methyl group in an S_N2 reaction. The same chemical principles apply to both.

INTERLUDE

Naturally Occurring Organohalogen Compounds

Marine corals secrete organohalogen compounds that act as a feeding deterrent to starfish.

As recently as 1968, only about 30 naturally occurring organohalogen compounds were known. It was simply assumed that chloroform, halogenated phenols, chlorinated aromatic compounds called PCB's, and other such substances found in the environment were industrial "pollutants." Now, less than a quarter century later, the situation is quite different. More than 2000 organohalogen compounds have been found to occur naturally, and many thousands more surely exist. From compounds as simple as chloromethane to those as complex as nostocyclophane D and chartelline A, an extraordinarily diverse range of organohalogen compounds exists in plants, bacteria, and animals.

Nostocyclophane D
(from the blue-green alga
Nostoc linckia)

Chartelline A
(from the bryozoan
Chartella papyracea)

Some naturally occurring organohalogen compounds are produced in massive quantities. Forest fires, volcanoes, and marine kelp release up to 5 *million* tons of CH_3Cl per year, for example, while annual industrial emissions total only about 26,000 tons. A detailed examination of one species of Okinawan acorn worm in a 1 km^2 study area showed that they released nearly 100 pounds per day of halogenated phenols, compounds previously thought to be nonnatural pollutants.

Why do organisms produce organohalogen compounds, many of which are undoubtedly toxic? The answer seems to be that many organisms use organohalogen compounds for self-defense, either as feeding deterrents, as irritants to predators, or as natural pesticides. Marine

(*continued*)➤

sponges, coral, and sea hares, for example, release foul-tasting organohalogen compounds that deter fish, starfish, and other predators from eating them. More remarkably, even humans appear to produce halogenated compounds as part of their defense against infection. The human immune system contains a peroxidase enzyme capable of carrying out halogenation reactions on fungi and bacteria, thereby killing the pathogen.

Much remains to be learned—only a few hundred of the more than 500,000 known species of marine organisms have been examined—but it is already clear that organohalogen compounds are an integral part of the world around us.

Summary and Key Words

alkyl halide, 217
anti periplanar geometry, 236
bimolecular, 228
E1 reaction, 239
E2 reaction, 236
Grignard reagent, 223
leaving group, 226
nucleophilic substitution reaction, 226
radical substitution reaction, 219
reaction rate, 228
S_N1 reaction, 232
S_N2 reaction, 228
unimolecular, 234

Alkyl halides are usually prepared from alcohols by treatment either with HX (for tertiary alcohols) or with $SOCl_2$ or PBr_3 (for primary and secondary alcohols). Alkyl halides react with magnesium metal to form organomagnesium halides, or **Grignard reagents.** These organometallic compounds react with acids to yield the corresponding alkanes.

Treatment of an alkyl halide with a nucleophile/base results either in substitution or in elimination. **Nucleophilic substitution reactions** occur by two mechanisms: S_N2 and S_N1. In the **S_N2 reaction,** the entering nucleophile attacks the substrate from a direction 180° away from the **leaving group,** resulting in an umbrella-like inversion of configuration at the carbon atom. S_N2 reactions are strongly inhibited by increasing steric bulk of the reagents and are favored only for primary substrates. In the **S_N1 reaction,** the substrate spontaneously dissociates to a carbocation, which then reacts with a nucleophile in a second step. In consequence, S_N1 reactions take place with racemization of configuration at the carbon atom and are favored only for tertiary substrates.

Elimination reactions also occur by two mechanisms: E2 and E1. In the **E2 reaction,** a base abstracts a proton at the same time that the leaving group departs. The E2 reaction takes place with **anti periplanar** geometry and occurs when a substrate is treated with a strong base. In the **E1 reaction,** the substrate spontaneously dissociates to form a carbocation, which can subsequently lose H^+ from a neighboring carbon. The reaction occurs on tertiary substrates in neutral or acidic hydroxylic solvents.

Summary of Reactions

1. Synthesis of alkyl halides
   (a) Radical chlorination of alkanes (Section 7.2)

$$-\overset{|}{\underset{|}{C}}-H + Cl_2 \xrightarrow{h\nu} -\overset{|}{\underset{|}{C}}-Cl + HCl$$

Reaction is very unselective

   (b) Alkyl halides from alcohols (Section 7.3)
      (1) Reaction of tertiary alcohols with HX, where X = Cl, Br

$$R_3C{-}OH \xrightarrow[\text{Ether}]{HX} R_3C{-}X + H_2O$$

      (2) Reaction of primary and secondary alcohols with PBr_3 and $SOCl_2$

$$ROH + PBr_3 \longrightarrow RBr$$

$$ROH + SOCl_2 \longrightarrow RCl$$

2. Reactions of alkyl halides
   (a) Formation and protonation of Grignard reagents (Section 7.4)

$$RX + Mg \longrightarrow RMgX$$

$$RMgX \longrightarrow RH$$

   (b) S_N2 reaction: back-side attack of nucleophile on alkyl halide (Sections 7.6 and 7.7)

$$Nu{:}^- + {-}C{-}X \longrightarrow Nu{-}C{-} + X{:}^-$$

Substrate must be primary or secondary

   (c) S_N1 reaction: carbocation intermediate is involved (Section 7.8)

$$R_3C{-}X \longrightarrow [R_3C^+] \xrightarrow{:Nu^-} R_3C{-}Nu + :X^-$$

Substrate must be tertiary or (occasionally) secondary

   (d) E2 reaction (Section 7.9)

$$\text{Base:} + H{-}C{-}C{-}Br \xrightarrow[\text{Ethanol}]{KOH} C{=}C$$

Anti periplanar geometry is required

(e) E1 reaction (Section 7.10)

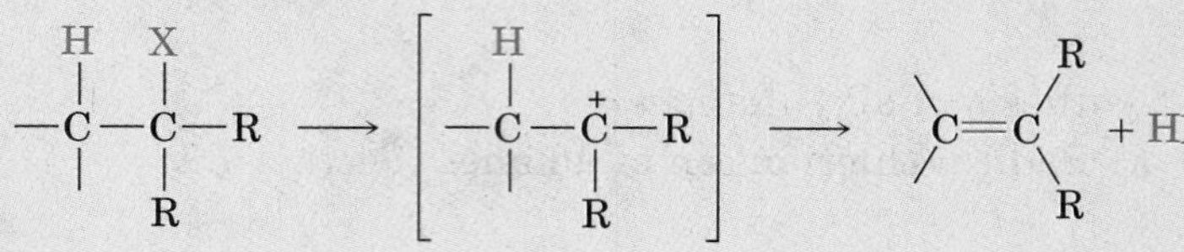

$$-\overset{\displaystyle H}{\underset{|}{\overset{|}{C}}}-\overset{\displaystyle X}{\underset{\displaystyle R}{\overset{|}{\underset{|}{C}}}}-R \longrightarrow \left[-\overset{\displaystyle H}{\underset{|}{\overset{|}{C}}}-\overset{+}{\underset{\displaystyle R}{\underset{|}{C}}}-R\right] \longrightarrow \;\; >C=C<^{R}_{R} \;+ HX$$

Best for tertiary substrates in neutral or acidic solvents.
Carbocation intermediate is involved.

ADDITIONAL PROBLEMS

7.22 Name the following alkyl halides according to IUPAC rules:

(a) 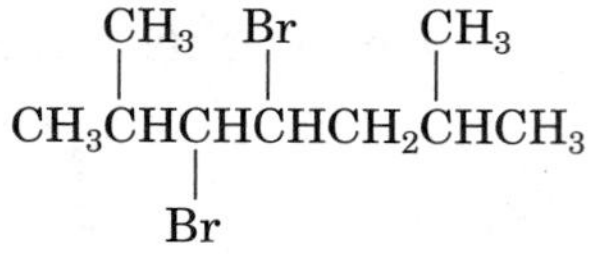 $CH_3CH(CH_3)CH(Br)CH(Br)CH_2CH(CH_3)CH_3$ (drawn as $CH_3\overset{CH_3}{C}H\overset{Br}{C}H\underset{Br}{C}HCH_2\overset{CH_3}{C}HCH_3$)

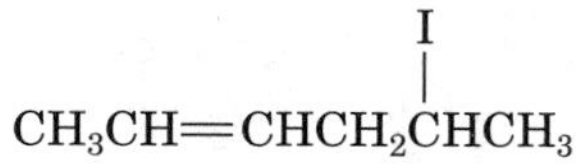

(b) $CH_3CH{=}CHCH_2\overset{I}{C}HCH_3$

(c) $CH_3\overset{Br}{\underset{CH_3}{C}}CH_2CH_2\overset{Cl}{C}H\overset{CH_3}{C}HCH_3$

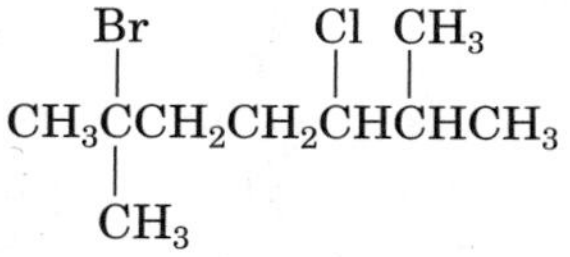

(d) 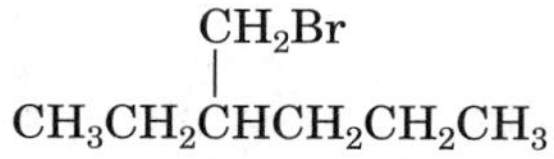 $CH_3CH_2\overset{CH_2Br}{C}HCH_2CH_2CH_3$

7.23 Draw structures corresponding to the following IUPAC names:
(a) 2,3-Dichloro-4-methylhexane (b) 4-Bromo-4-ethyl-2-methylhexane
(c) 3-Iodo-2,2,4,4-tetramethylpentane

7.24 Draw and name the monochlorination products you might obtain by radical chlorination of 2-methylpentane. Which of the products are chiral? Are any of the products optically active?

7.25 Describe the effects of the following variables on both S_N2 and S_N1 reactions:
(a) Substrate structure (b) Leaving group

7.26 How would you prepare the following compounds, starting with cyclopentene and any other reagents needed?
(a) Chlorocyclopentane (b) Cyclopentanol
(c) Cyclopentylmagnesium chloride (d) Cyclopentane

7.27 Which reagent in each of the following pairs is a better leaving group?
(a) F^- or Br^- (b) Cl^- or NH_2^- (c) OH^- or I^-

7.28 Predict the product(s) of the following reactions:

(a) 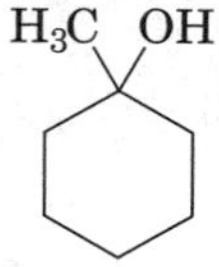 1-methylcyclohexanol (H_3C, OH on same ring carbon) $\xrightarrow[\text{Ether}]{\text{HBr}}$?

(b) $CH_3CH_2CH_2CH_2OH \xrightarrow{SOCl_2}$?

(c) 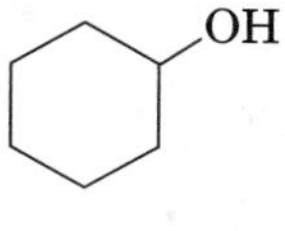 cyclohexanol (ring with OH) $\xrightarrow[\text{Ether}]{PBr_3}$?

(d) $CH_3CH_2CH(Br)CH_3 \xrightarrow[\text{Ether}]{\text{Mg}}$ A $\xrightarrow{H_2O}$ B

7.29 Which alkyl halide in each of the following pairs will react faster in an S_N2 reaction with OH^-?
(a) Bromobenzene or benzyl bromide, $C_6H_5CH_2Br$
(b) CH_3Cl or $(CH_3)_3CCl$
(c) $CH_3CH{=}CHBr$ or $H_2C{=}CHCH_2Br$

7.30 How might you prepare the following molecules using a nucleophilic substitution reaction at some step?

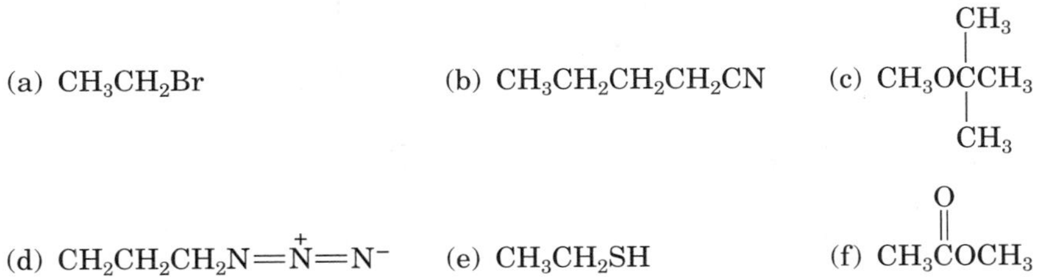

(a) CH_3CH_2Br (b) $CH_3CH_2CH_2CH_2CN$ (c) $CH_3O\underset{\underset{CH_3}{|}}{\overset{\overset{CH_3}{|}}{C}}CH_3$

(d) $CH_2CH_2CH_2N{=}\overset{+}{N}{=}N^-$ (e) CH_3CH_2SH (f) $CH_3\overset{\overset{O}{\|}}{C}OCH_3$

7.31 What products do you expect from reaction of 1-bromopropane with the following reagents?
(a) NaI (b) NaCN (c) NaOH (d) Mg, then H_2O (e) $NaOCH_3$

7.32 Order the following compounds with respect to both S_N1 and S_N2 reactivity:

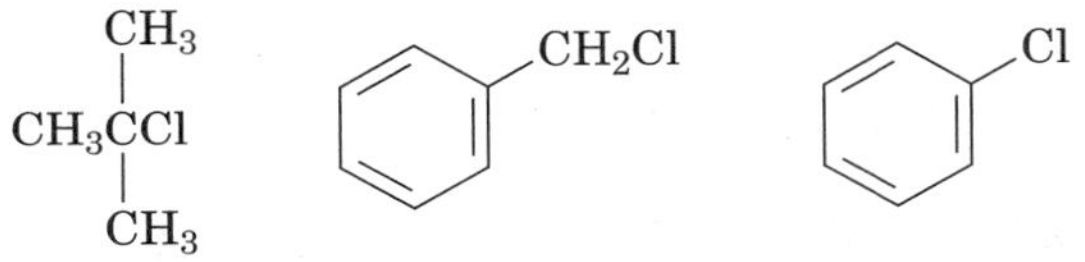

$CH_3\underset{\underset{CH_3}{|}}{\overset{\overset{CH_3}{|}}{C}}Cl$ (benzene ring)–CH_2Cl (benzene ring)–Cl

7.33 Order each set of compounds with respect to S_N2 reactivity:
(a) $(CH_3)_3CCl$, $CH_3CH_2CH_2Cl$, $CH_3CH_2CH(Cl)CH_3$
(b) $(CH_3)_2CHCH(Br)CH_3$, $(CH_3)_2CHCH_2Br$, CH_3Br

7.34 What is wrong with each of the following reactions?

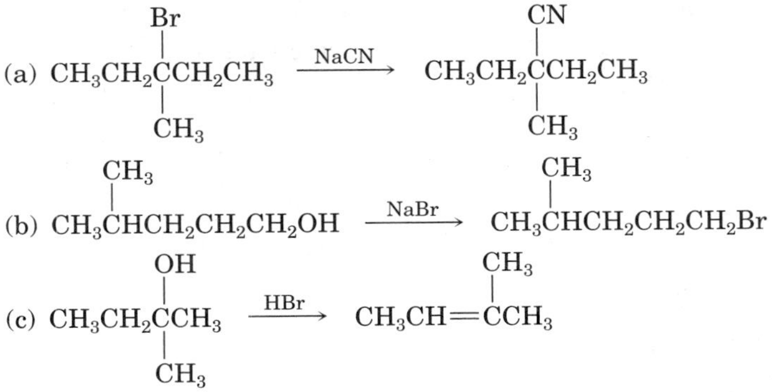

(a) $CH_3CH_2\underset{\underset{CH_3}{|}}{\overset{\overset{Br}{|}}{C}}CH_2CH_3 \xrightarrow{NaCN} CH_3CH_2\underset{\underset{CH_3}{|}}{\overset{\overset{CN}{|}}{C}}CH_2CH_3$

(b) $CH_3\overset{\overset{CH_3}{|}}{C}HCH_2CH_2CH_2OH \xrightarrow{NaBr} CH_3\overset{\overset{CH_3}{|}}{C}HCH_2CH_2CH_2Br$

(c) $CH_3CH_2\underset{\underset{CH_3}{|}}{\overset{\overset{OH}{|}}{C}}CH_3 \xrightarrow{HBr} CH_3CH{=}\overset{\overset{CH_3}{|}}{C}CH_3$

7.35 Predict the product and give the stereochemistry of reactions of the following nucleophiles with (*R*)-2-bromooctane:
(a) CN^- (b) $CH_3CO_2^-$ (c) Br^-

7.36 Draw all isomers of C_4H_9Br, name them, and arrange them in order of decreasing reactivity in the S_N2 reaction.

7.37 Although radical chlorination of alkanes is usually unselective, chlorination of propene, $CH_3CH{=}CH_2$, occurs almost exclusively on the methyl group rather than on the double bond. Draw resonance structures of the allyl radical $CH_2{=}CHCH_2\cdot$ to account for this result.

7.38 Draw resonance structures of the benzyl radical $C_6H_5CH_2\cdot$ to account for the fact that radical chlorination of toluene occurs exclusively on the methyl group rather than on the aromatic ring.

7.39 Ethers can be prepared by S_N2 reaction of an alkoxide ion with an alkyl halide: $R–O^- + R'–Br \rightarrow R–O–R' + Br^-$. Suppose you wanted to prepare cyclohexyl methyl ether. Which route would be better, reaction of methoxide ion, CH_3O^-, with bromocyclohexane or reaction of cyclohexoxide ion with bromomethane? Explain.

7.40 How could you prepare diethyl ether, $CH_3CH_2OCH_2CH_3$, starting from ethyl alcohol and any inorganic reagents needed? (See Problem 7.39.)

7.41 How could you prepare cyclohexane starting from 3-bromocyclohexene?

7.42 The S_N2 reaction can occur *intramolecularly* (within the same molecule). What product would you expect from treatment of 4-bromo-1-butanol with base?

$$BrCH_2CH_2CH_2CH_2OH \xrightarrow{\text{Base}} [BrCH_2CH_2CH_2CH_2O^-\ Na^+] \longrightarrow ?$$

7.43 In light of your answer to Problem 7.42, propose a synthesis of 1,4-dioxane from 1,2-dibromoethane.

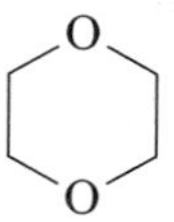

1,4-Dioxane

7.44 Propose a structure for an alkyl halide that can give a mixture of three alkenes on E2 reaction.

7.45 Heating either *tert*-butyl chloride or *tert*-butyl bromide with ethanol yields the same reaction mixture: about 80% *tert*-butyl ethyl ether [$(CH_3)_3COCH_2CH_3$] and 20% 2-methylpropene. Explain.

7.46 What effect would you expect the following changes to have on the rate of the reaction of 1-iodo-2-methylbutane with CN^-?

(a) CN^- concentration is halved and 1-iodo-2-methylbutane concentration is doubled.

(b) Both CN^- and 1-iodo-2-methylbutane concentrations are tripled.

7.47 What effect would you expect on the rate of reaction of ethyl alcohol with 2-iodo-2-methylbutane if the concentration of the alkyl halide is tripled?

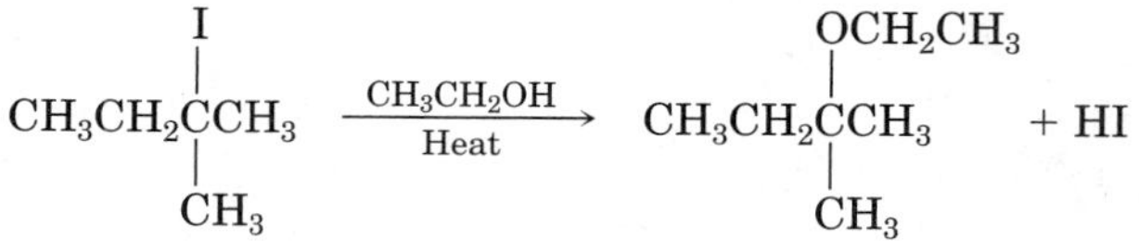

$$CH_3CH_2C(I)(CH_3)CH_3 \xrightarrow[\text{Heat}]{CH_3CH_2OH} CH_3CH_2C(OCH_2CH_3)(CH_3)CH_3 + HI$$

2-Iodo-2-methylbutane

7.48 Identify the following reactions as either S_N1, S_N2, E1 or E2:

(a) 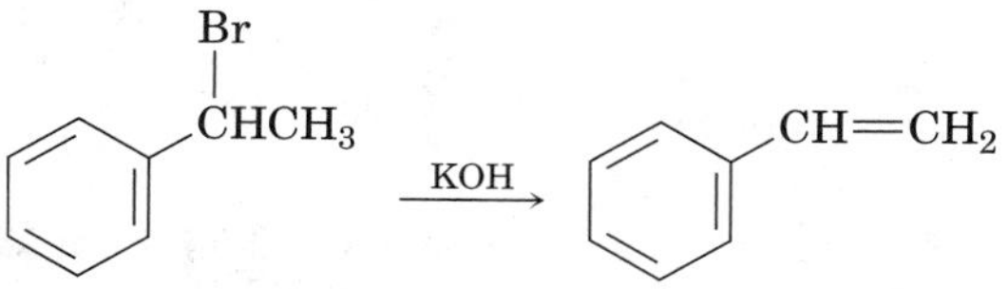 $C_6H_5CH(Br)CH_3 \xrightarrow{KOH} C_6H_5CH{=}CH_2$

(b) 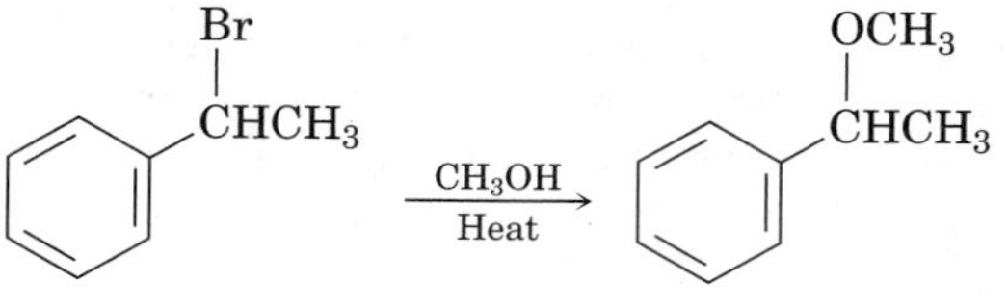 $C_6H_5CH(Br)CH_3 \xrightarrow[\text{Heat}]{CH_3OH} C_6H_5CH(OCH_3)CH_3$

7.49 How can you explain the fact that *trans*-1-bromo-2-methylcyclohexane yields the non-Zaitsev elimination product 3-methylcyclohexene on treatment with base?

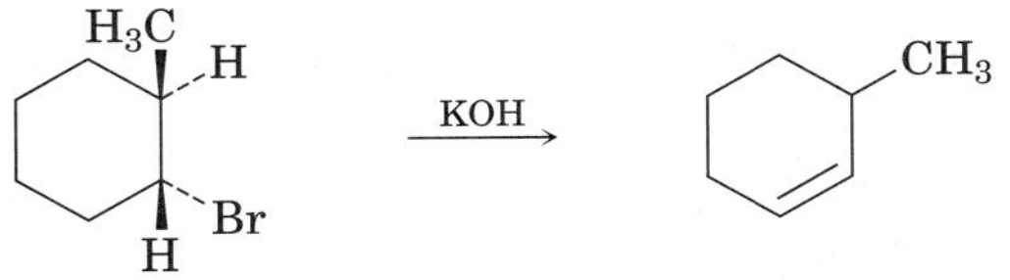

$$\xrightarrow{\text{KOH}}$$

trans-1-Bromo-2-methylcyclohexane **3-Methylcyclohexene**

7.50 Propose a structure for an alkyl halide that gives (*Z*)-2,3-diphenyl-2-butene on E2 reaction.

7.51 Predict the major alkene product from the following eliminations:

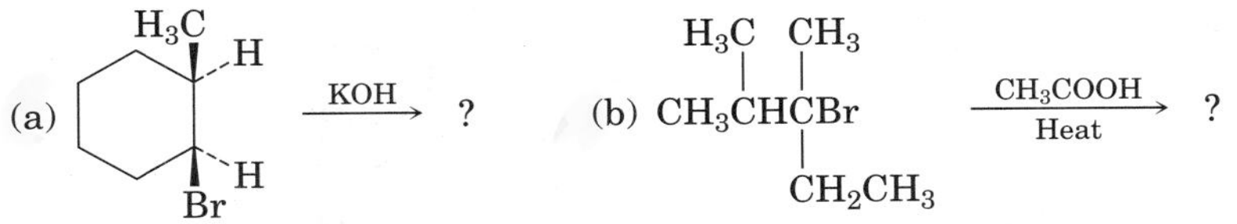

(a) $\xrightarrow{\text{KOH}}$?

(b) $CH_3CHC(CH_3)(CH_2CH_3)Br$ with H_3C on CH $\xrightarrow[\text{Heat}]{CH_3COOH}$?

7.52 (2*R*,3*S*)-2-Bromo-3-phenylbutane undergoes E2 reaction on treatment with sodium ethoxide to yield (*Z*)-2-phenyl-2-butene:

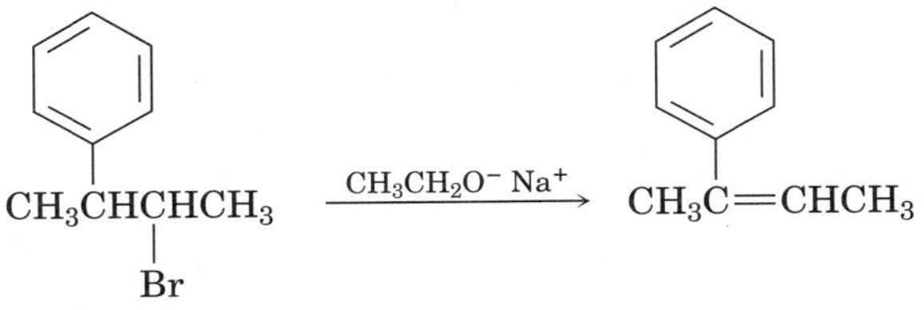

$$CH_3CH(C_6H_5)CH(Br)CH_3 \xrightarrow{CH_3CH_2O^- \ Na^+} CH_3C(C_6H_5){=}CHCH_3$$

Formulate the reaction showing the proper stereochemistry, and explain the observed result using Newman projections.

7.53 In light of your answer to Problem 7.52, which alkene, *E* or *Z*, would you expect from the E2 reaction of (2*R*,3*R*)-2-bromo-3-phenylbutane?

7.54 Reaction of HBr with (*R*)-3-methyl-3-hexanol yields (±)-3-bromo-3-methylhexane. Explain.

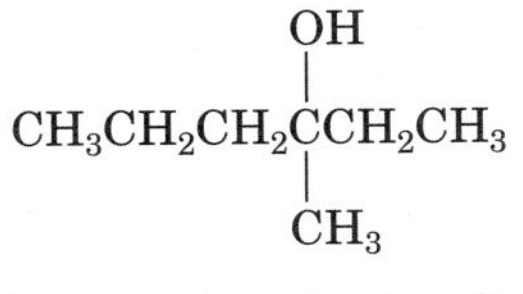

$$CH_3CH_2CH_2C(OH)(CH_3)CH_2CH_3 \quad \textbf{3-Methyl-3-hexanol}$$

7.55 (*S*)-2-Butanol slowly racemizes on standing in dilute sulfuric acid. Propose a mechanism to account for this observation.

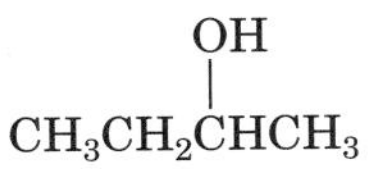

$$CH_3CH_2CH(OH)CH_3 \quad \textbf{2-Butanol}$$

7.56 (*S*)-3-Methylhexane undergoes radical chlorination to yield 3-chloro-3-methylhexane as the major product. Is the product chiral? Is it optically active? What stereoisomers are produced and in what ratio?

7.57 Draw the eight diastereomers of 1,2,3,4,5,6-hexachlorocyclohexane. One isomer loses HCl in an E2 reaction nearly 1000 times more slowly than the others. Which isomer reacts so slowly, and why?

7.58 Compound A is optically inactive and has the formula $C_{16}H_{16}Br_2$. On treatment with a strong base, A gives hydrocarbon B, $C_{16}H_{14}$, which absorbs 2 equiv of H_2 when reduced over a palladium catalyst. Hydrocarbon B also reacts with acidic $KMnO_4$ to give two carbonyl-containing products. One product, C, is a carboxylic acid with the formula $C_7H_6O_2$. The other product is oxalic acid, HO_2CCO_2H. Formulate the reactions involved, and suggest structures for A, B, and C. What is the stereochemistry of A?

Visualizing Chemistry

7.59 Write the product you would expect from reaction of each of the following molecules with (i) $Na^+ \ ^-SCH_3$ and (ii) NaOH (gray = C, yellow-green = Cl, light green = H).

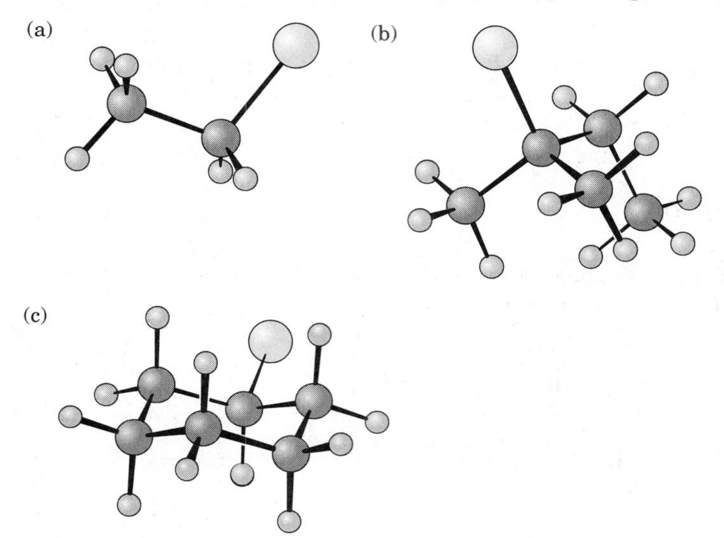

7.60 Assign *R* or *S* configuration to the following molecule, write the product you would expect from S_N2 reaction with NaCN, and assign *R* or *S* configuration to the product (gray = C, red = O, yellow-green = Cl, light green = H).

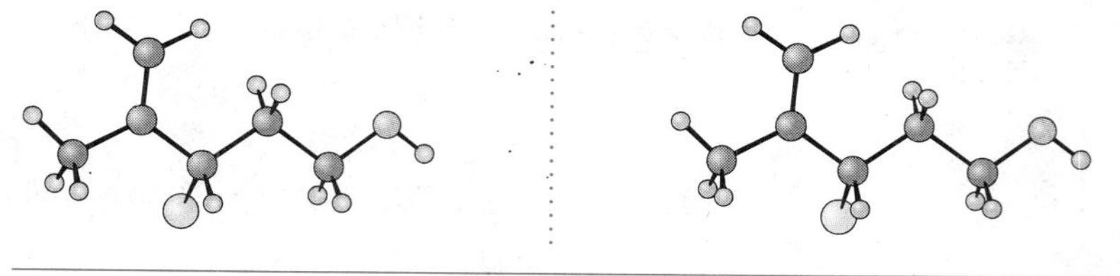

Stereo View

7.61 Draw the structure and assign *Z* or *E* stereochemistry to the product you expect from E2 reaction of the following molecule with NaOH (gray = C, yellow-green = Cl, light green = H).

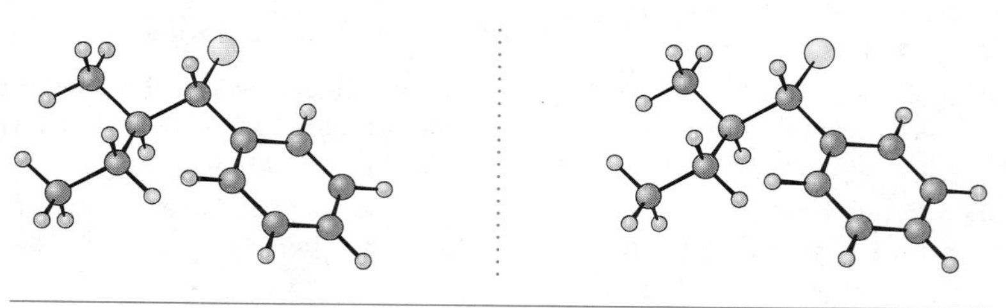

Stereo View

Ethanol and methanol are the two most widely used alcohols.

8 Alcohols, Ethers, and Phenols

An **alcohol** is a compound that has a hydroxyl group (–OH) bonded to a saturated, sp^3-hybridized carbon atom; a **phenol** has a hydroxyl group bonded to an aromatic ring; and an **ether** has an oxygen atom bonded to two organic groups. All three classes of compounds can be thought of as organic derivatives of water in which one or both of the water hydrogens is replaced by an organic substituent (H–O–H becomes R–O–H or R–O–R′).

CH_3CH_2OH — **Ethanol**

(benzene ring–OH) — **Phenol**

CH_3OCH_3 — **Dimethyl ether**

Alcohols, phenols, and ethers occur widely in nature and have many industrial, pharmaceutical, and biological applications. Ethanol, for instance, is a fuel additive, an industrial solvent, and a beverage; menthol is a flavoring agent; BHT (butylated hydroxytoluene) is an antioxidant food additive; and diethyl ether (the familiar "ether" of medical use) was once popular as an anesthetic agent.

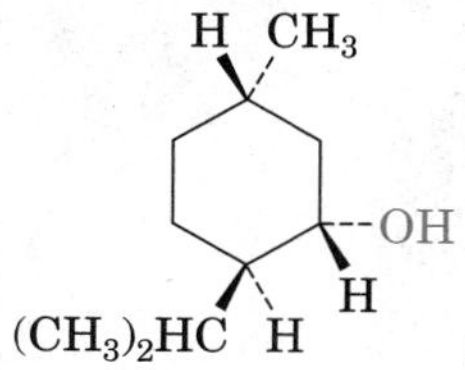

Menthol

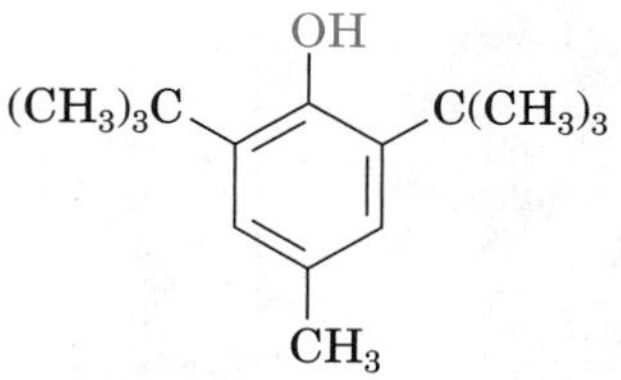

BHT

$CH_3CH_2OCH_2CH_3$

Diethyl ether

8.1 Naming Alcohols, Phenols, and Ethers

Alcohols

Alcohols are classified as primary (1°), secondary (2°), or tertiary (3°), depending on the number of carbon substituents bonded to the hydroxyl-bearing carbon:

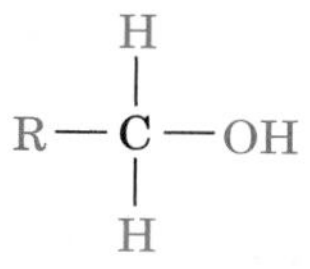

A primary alcohol (1°)

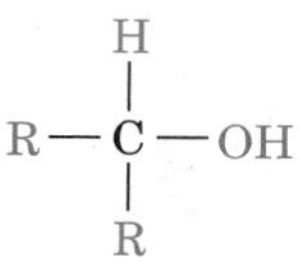

A secondary alcohol (2°)

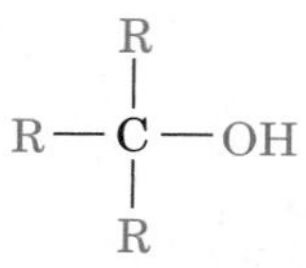

A tertiary alcohol (3°)

Simple alcohols are named in the IUPAC system as derivatives of the parent alkane, using the suffix *-ol.*

Step 1 Select the longest carbon chain containing the hydroxyl group, and replace the *-e* ending of the corresponding alkane with *-ol.*

Step 2 Number the carbons of the parent chain beginning at the end nearer the hydroxyl group.

Step 3 Number all substituents according to their position on the chain, and write the name listing the substituents in alphabetical order.

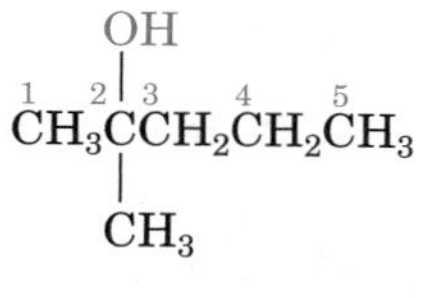

2-Methyl-2-pentanol

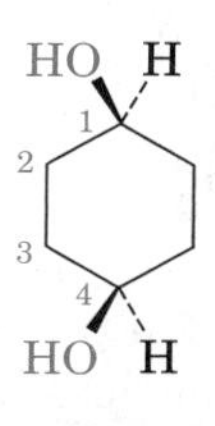

***cis*-1,4-Cyclohexanediol**

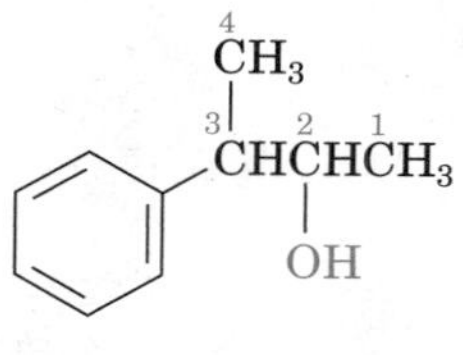

3-Phenyl-2-butanol

Some well-known alcohols also have common names. For example,

$C_6H_5CH_2OH$ — **Benzyl alcohol (Phenylmethanol)**

$CH_3C(CH_3)_2OH$ — ***tert*-Butyl alcohol (2-Methyl-2-propanol)**

$HOCH_2CH_2OH$ — **Ethylene glycol (1,2-Ethanediol)**

$HOCH_2CH(OH)CH_2OH$ — **Glycerol (1,2,3-Propanetriol)**

Phenols

The word *phenol* is used both as the name of a specific substance (hydroxybenzene) and as the family name for hydroxy-substituted aromatic compounds. Phenols are named as substituted aromatic compounds according to the rules discussed in Section 5.4. Note that *-phenol* is used as the parent name rather than *-benzene.*

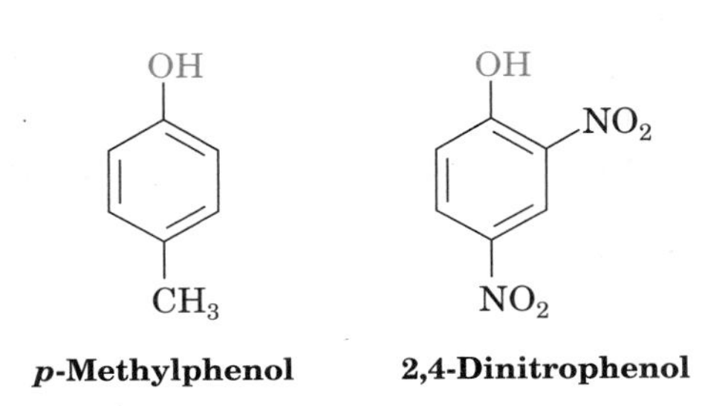

***p*-Methylphenol** **2,4-Dinitrophenol**

Ethers

Simple ethers that contain no other functional groups are named by identifying the two organic groups and adding the word *ether.*

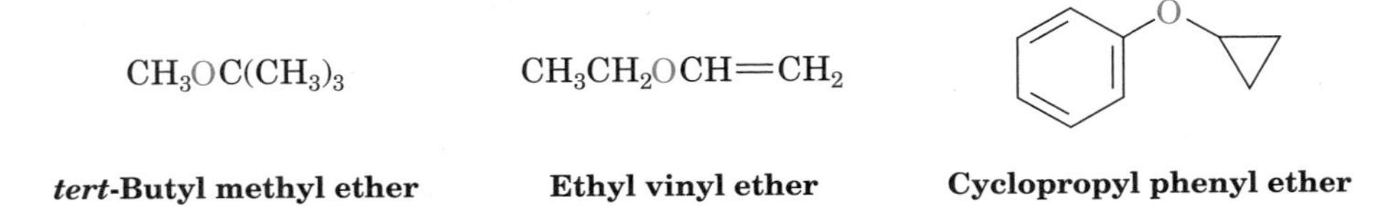

***tert*-Butyl methyl ether** **Ethyl vinyl ether** **Cyclopropyl phenyl ether**

If more than one ether linkage is present or if other functional groups are present, the ether part is named as an *alkoxy* substituent on the parent compound.

$CH_3O-C_6H_4-OCH_3$ — *p*-Dimethoxy**benzene**

$C_6H_9-O-C(CH_3)_3$ (ring positions 1, 2, 3, 4) — 4-*tert*-Butoxy-**1-cyclohexene**

PROBLEM

8.1 Give IUPAC names for the following alcohols:

(a) 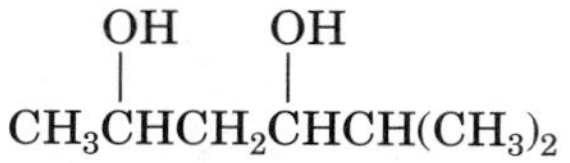 $CH_3CH(OH)CH_2CH(OH)CH(CH_3)_2$

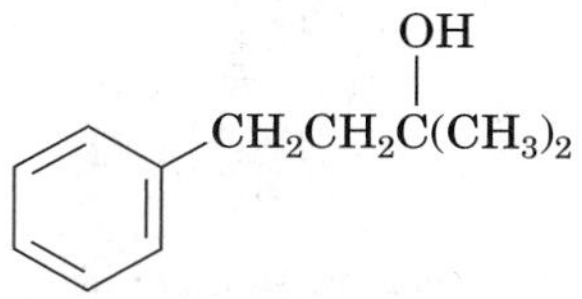

(b) $C_6H_5CH_2CH_2C(OH)(CH_3)_2$

(c) 4,4-dimethyl-substituted cyclohexane bearing OH (structure: OH on ring carbon; H_3C and CH_3 on the opposite ring carbon)

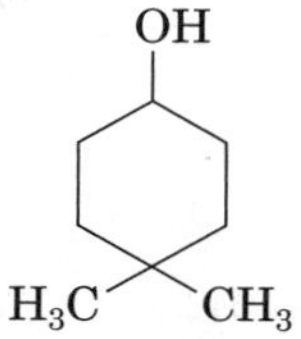

(d) 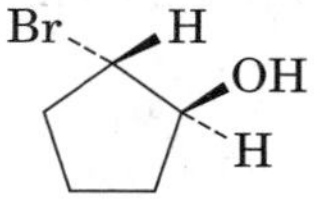 Br, H, OH, H on cyclopentane ring (stereochemical structure)

PROBLEM

8.2 Identify the alcohols in Problem 8.1 as primary, secondary, or tertiary.

PROBLEM

8.3 Draw structures corresponding to the following IUPAC names:
(a) 2-Methyl-2-hexanol
(b) 1,5-Hexanediol
(c) 2-Ethyl-2-buten-1-ol
(d) 3-Cyclohexen-1-ol
(e) *o*-Bromophenol
(f) 2,4,6-Trinitrophenol

PROBLEM

8.4 Name the following ethers by IUPAC rules:

(a) 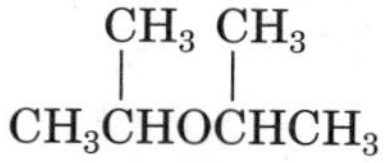 $CH_3CH(CH_3)OCH(CH_3)CH_3$

(b) 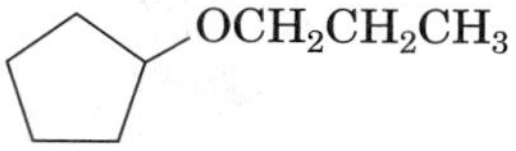 cyclopentyl–$OCH_2CH_2CH_3$

(c) 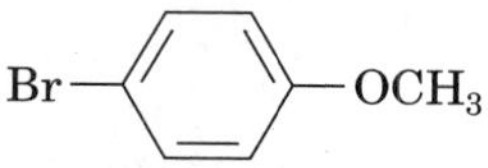 Br–C_6H_4–OCH_3

(d) $(CH_3)_2CHCH_2OCH_2CH_3$

8.2 Properties of Alcohols, Phenols, and Ethers: Hydrogen Bonding

Alcohols, phenols, and ethers can be thought of as organic derivatives of water in which one or both of the hydrogens have been replaced by organic parts. Thus, all three classes of compounds have nearly the same geometry as water. The R–O–H or R–O–R′ bonds have an approximately tetrahedral bond angle—for example, 112° in dimethyl ether—and the oxygen atom is sp^3-hybridized.

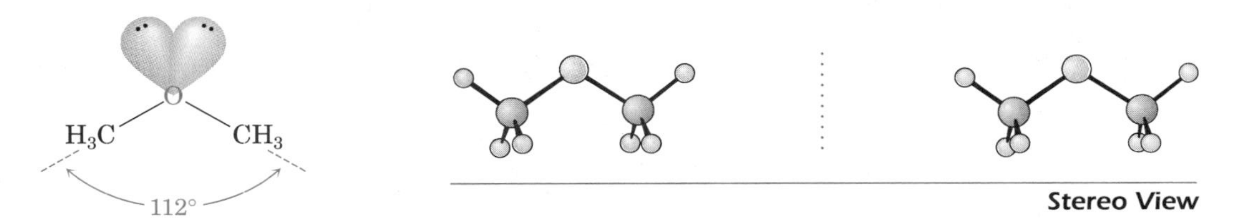

Alcohols and phenols differ significantly from the hydrocarbons and alkyl halides we've studied thus far. As shown in Figure 8.1, alcohols have higher boiling points than alkanes or haloalkanes of similar molecular weight. For example, the molecular weights of 1-propanol (mol wt = 60), butane (mol wt = 58), and chloroethane (mol wt = 65) are similar, but 1-propanol boils at 97.4°C, compared with −0.5°C for the alkane and 12.3°C for the chloroalkane. Similarly, phenols have higher boiling points than aromatic hydrocarbons. Phenol itself, for example, boils at 182°C, whereas toluene boils at 110.6°C.

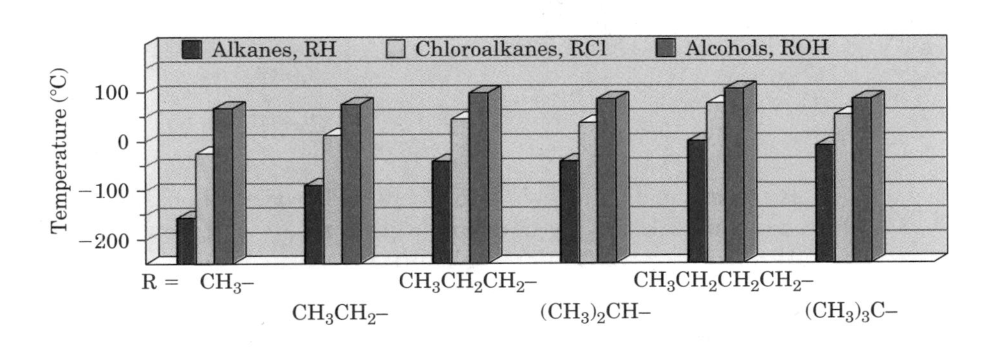

Figure 8.1 A comparison of boiling points for some alkanes, chloroalkanes, and alcohols. Alcohols generally have higher boiling points.

Alcohols and phenols have unusually high boiling points because, like water, they form hydrogen bonds. The positively polarized –OH hydrogen of one molecule is attracted to the negatively polarized oxygen of another molecule (Figure 8.2, p. 256). Although hydrogen bonds have a strength of only about 20–40 kJ/mol (5–10 kcal/mol) versus 400–420 kJ/mol for a typical O–H covalent bond, the presence of a great many hydrogen bonds in solution means that extra energy is required to break them during the boiling process. Ethers, because they lack hydroxyl groups, can't form hydrogen bonds and therefore have lower boiling points.

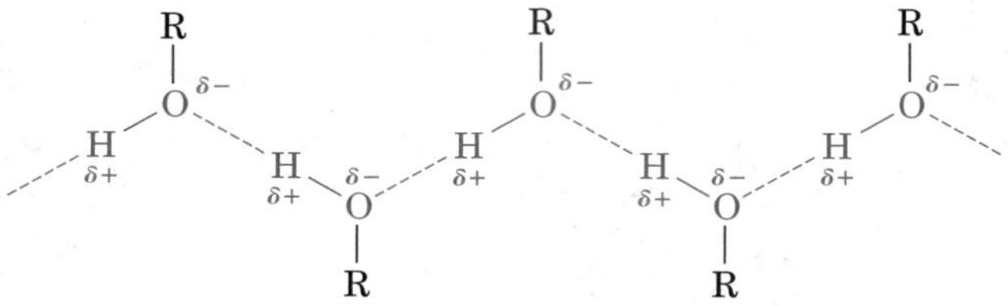

Figure 8.2 Hydrogen bonding in alcohols and phenols. The weak attraction between a positively polarized –OH hydrogen and a pair of nonbonding electrons on oxygen holds the molecules together. Note that the O–H····O bond angle is approximately 180°.

8.3 Properties of Alcohols and Phenols: Acidity

Alcohols and phenols, like water, are both weakly basic and weakly acidic. As weak Lewis bases, alcohols and phenols are reversibly protonated by strong acids to yield oxonium ions, ROH_2^+:

$$R—\ddot{O}—H + HX \rightleftharpoons R—\overset{H}{\overset{|}{O^+}}—H \quad X^-$$

As weak acids, alcohols and phenols dissociate to a slight extent in dilute aqueous solution by donating a proton to water, generating H_3O^+ and an **alkoxide ion (RO^-)** or a **phenoxide ion (ArO^-).**

$$R\ddot{O}—H + H_2\ddot{O}: \rightleftharpoons R\ddot{O}:^- + H_3O:^+$$

Table 8.1 gives the pK_a values of some common alcohols and phenols in comparison with water and HCl. (You might want to review Section 1.12 to brush up on the behavior of acids.)

The data in Table 8.1 show that alcohols are about as acidic as water. They are generally much less acidic than carboxylic acids or mineral acids, they don't react with weak bases like bicarbonate ion, and they react only to a limited extent with metal hydroxides such as NaOH. Alcohols do, however, react with alkali metals such as sodium and potassium to yield alkoxide salts that are themselves strong bases.

$$2\ CH_3OH + 2\ Na \longrightarrow 2\ CH_3O^-\ Na^+ + H_2$$

Methanol **Sodium methoxide**

Table 8.1 Acidity Constants of Some Alcohols and Phenols

Alcohol or phenol	pK_a	
$(CH_3)_3COH$	18.00	Weaker acid
CH_3CH_2OH	16.00	
[HOH, water][a]	[15.74]	
CH_3OH	15.54	
p-Methylphenol	10.17	
Phenol	9.89	
p-Bromophenol	9.35	
p-Nitrophenol	7.15	
[HCl, hydrochloric acid][a]	[−7.00]	Stronger acid

[a]Values for water and hydrochloric acid are shown for reference.

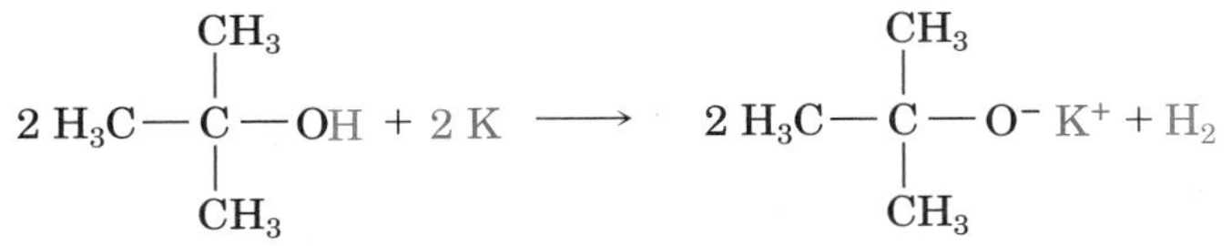

$$2\ H_3C-C(CH_3)_2-OH + 2\ K \longrightarrow 2\ H_3C-C(CH_3)_2-O^-\ K^+ + H_2$$

***tert*-Butyl alcohol** → **Potassium *tert*-butoxide**

Phenols are much more acidic than alcohols. In fact, some nitro-substituted phenols approach or surpass the acidity of carboxylic acids. One practical consequence of this acidity is that phenols are soluble in dilute aqueous NaOH.

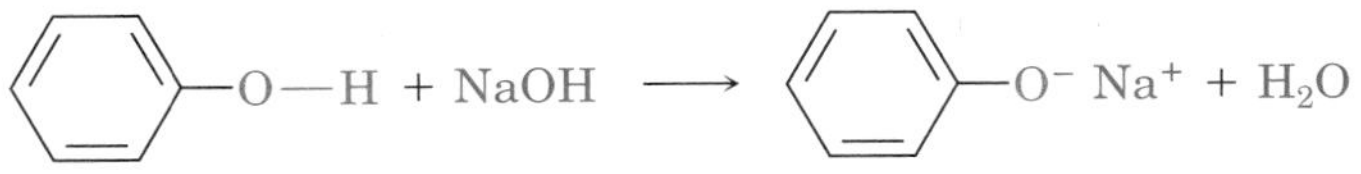

$$C_6H_5-O-H + NaOH \longrightarrow C_6H_5-O^-\ Na^+ + H_2O$$

Phenol → **Sodium phenoxide**

Phenols are more acidic than alcohols because the phenoxide anion is resonance-stabilized by the aromatic ring. Sharing the negative charge over the ring increases the stability of the phenoxide anion and thus increases the tendency of the corresponding phenol to dissociate.

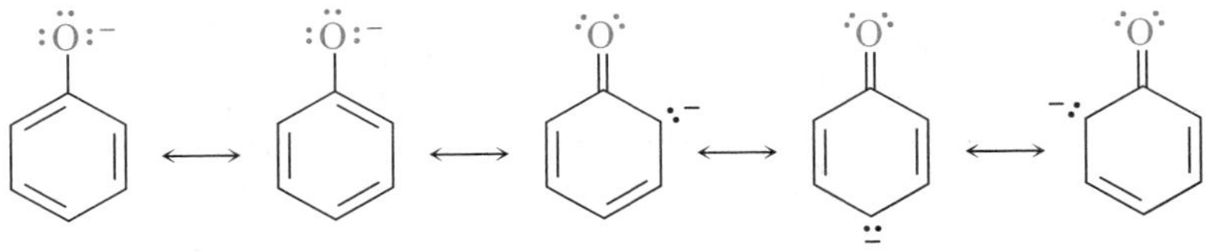

Substituted phenols can be either more or less acidic than phenol itself. Phenols with a substituent that is electron-withdrawing are more acidic because the substituent stabilizes the corresponding phenoxide anion and makes dissociation more favorable. Phenols with a substituent that is electron-donating are less acidic because the substituent destabilizes the phenoxide anion and makes dissociation less favorable. Note that the acidifying effect of a substituent on a phenol is related to its activating or deactivating effect in electrophilic aromatic substitution (Section 5.10). Substituents that make a phenol more acidic deactivate a ring toward electrophilic substitution, while substituents that make a phenol less acidic activate a ring.

Electron-withdrawing groups (EWG) stabilize phenoxide anion, resulting in increased phenol acidity

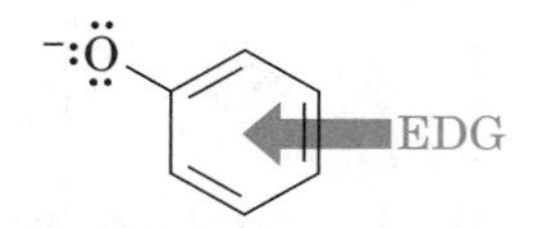

Electron-donating groups (EDG) destabilize phenoxide anion, resulting in decreased phenol acidity

PRACTICE PROBLEM 8.1

Which would you expect to be more acidic, *p*-methylphenol or *p*-cyanophenol?

Solution We know from their effects on aromatic substitution (Section 5.9) that methyl is an activating group (electron donor) whereas cyano is a deactivating group (electron acceptor). Thus, *p*-cyanophenol is more acidic.

PROBLEM

8.5 Rank the compounds in each group in order of increasing acidity.
(a) Methanol, phenol, *p*-nitrophenol, *p*-methylphenol
(b) Benzyl alcohol, *p*-bromophenol, 2,4-dibromophenol, *p*-methoxyphenol

PROBLEM

8.6 Draw as many resonance structures as you can for the anion of *p*-cyanophenol.

8.4 Synthesis of Alcohols

Alcohols occupy a central position in organic chemistry. They can be prepared from many other kinds of compounds (alkenes, alkyl halides, ketones, aldehydes, and esters, among others), and they can be transformed into an equally wide assortment of compounds. Let's review briefly some of the methods of alcohol synthesis we've already seen.

Alcohols can be prepared by hydration of alkenes (Section 4.4). Treatment of the alkene with water and an acid catalyst leads to the Markovnikov product:

$$\text{1-Methylcyclohexene} \xrightarrow[H_2SO_4\ \text{catalyst}]{H_2O} \text{1-Methylcyclohexanol}$$

1-Methylcyclohexene **1-Methylcyclohexanol**

1,2-Diols can be prepared by direct hydroxylation of an alkene with basic $KMnO_4$ (Section 4.7). The reaction takes place with syn stereochemistry:

$$\text{1-Methylcyclohexene} \xrightarrow[H_2O,\ NaOH]{KMnO_4} \text{1-Methyl-}cis\text{-1,2-cyclohexanediol}$$

1-Methylcyclohexene **1-Methyl-*cis*-1,2-cyclohexanediol**

8.5 Alcohols from Carbonyl Compounds

The most general method for preparing alcohols is by reduction of carbonyl compounds—the formal addition of H_2 to a C=O double bond. For the moment, we'll only list the *kinds* of carbonyl reductions that can be carried out, but we'll return for a look at the reduction mechanism in the next chapter (Section 9.6).

$$C{=}O \xrightarrow{[H]} H{-}C{-}OH$$

A carbonyl compound An alcohol

where [H] is a generalized reducing agent

Reduction of Aldehydes and Ketones

Aldehydes and ketones are easily reduced to yield alcohols. Aldehydes are converted into primary alcohols, and ketones are converted into secondary alcohols.

$$RCHO \xrightarrow{[H]} RCH_2OH \qquad RCOR' \xrightarrow{[H]} RR'CHOH$$

An aldehyde A primary alcohol A ketone A secondary alcohol

Many reducing reagents are available, but sodium borohydride, $NaBH_4$, is usually chosen because of its safety and ease of handling.

Aldehyde reduction

$$CH_3CH_2CH_2\overset{O}{\overset{\|}{C}}H \xrightarrow[2.\ H_3O^+]{1.\ NaBH_4,\ ethanol} CH_3CH_2CH_2\underset{H}{\underset{|}{\overset{OH}{\overset{|}{C}}}}H$$

Butanal **1-Butanol (85%)**
(a 1° alcohol)

Ketone reduction

$$\text{Dicyclohexyl ketone} \xrightarrow[2.\ H_3O^+]{1.\ NaBH_4,\ ethanol} \text{Dicyclohexylmethanol}$$

Dicyclohexyl ketone **Dicyclohexylmethanol (88%)**
(a 2° alcohol)

Lithium aluminum hydride, $LiAlH_4$, is another reducing agent that is sometimes used to reduce ketones and aldehydes. Although it is far more powerful and reactive than $NaBH_4$, $LiAlH_4$ is also far more dangerous. It reacts violently with water, and it decomposes explosively when heated above 120°C.

Reduction of Esters and Carboxylic Acids

Esters and carboxylic acids are reduced to give primary alcohols:

$$R-\overset{O}{\overset{\|}{C}}-OH \quad \text{or} \quad R-\overset{O}{\overset{\|}{C}}-OR' \xrightarrow{[H]} R-\underset{H}{\underset{|}{\overset{OH}{\overset{|}{C}}}}-H$$

A carboxylic acid An ester A primary alcohol

These reactions are more difficult than the corresponding reductions of ketones and aldehydes, so $LiAlH_4$ is used rather than $NaBH_4$. Note that only one hydrogen is added to the carbonyl carbon atom during the reduction of a ketone or aldehyde, but two hydrogens are added to the carbonyl carbon during reduction of an ester or carboxylic acid.

Carboxylic acid reduction

$$CH_3(CH_2)_7CH=CH(CH_2)_7\overset{O}{\overset{\|}{C}}OH \xrightarrow[2.\ H_3O^+]{1.\ LiAlH_4,\ ether} CH_3(CH_2)_7CH=CH(CH_2)_7CH_2OH$$

9-Octadecenoic acid
(Oleic acid)

9-Octadecen-1-ol (87%)

Ester reduction

$$\underset{\textbf{Methyl 2-pentenoate}}{CH_3CH_2CH{=}CH\overset{\overset{O}{\|}}{C}OCH_3} \xrightarrow[2.\ H_3O^+]{1.\ LiAlH_4,\ ether} \underset{\textbf{2-Penten-1-ol (91\%)}}{CH_3CH_2CH{=}CHCH_2OH} + CH_3OH$$

PRACTICE PROBLEM 8.2

Predict the product of the following reaction:

$$CH_3CH_2CH_2\overset{\overset{O}{\|}}{C}CH_2CH_3 \xrightarrow{NaBH_4} ?$$

Solution Ketones are reduced by treatment with $NaBH_4$ to yield secondary alcohols. Thus, reduction of 3-hexanone yields 3-hexanol.

$$\underset{\textbf{3-Hexanone}}{CH_3CH_2CH_2\overset{\overset{O}{\|}}{C}CH_2CH_3} \xrightarrow{NaBH_4} \underset{\textbf{3-Hexanol}}{CH_3CH_2CH_2\overset{\overset{OH}{|}}{C}HCH_2CH_3}$$

PROBLEM

8.7 How would you carry out the following reactions?

(a) $CH_3\overset{\overset{O}{\|}}{C}CH_2CH_2\overset{\overset{O}{\|}}{C}OCH_3 \xrightarrow{?} CH_3\overset{\overset{OH}{|}}{C}HCH_2CH_2\overset{\overset{O}{\|}}{C}OCH_3$

(b) $CH_3\overset{\overset{O}{\|}}{C}CH_2CH_2\overset{\overset{O}{\|}}{C}OCH_3 \xrightarrow{?} CH_3\overset{\overset{OH}{|}}{C}HCH_2CH_2CH_2OH$

PROBLEM

8.8 What carbonyl compounds give the following alcohols on reduction with $LiAlH_4$? Show all possibilities.

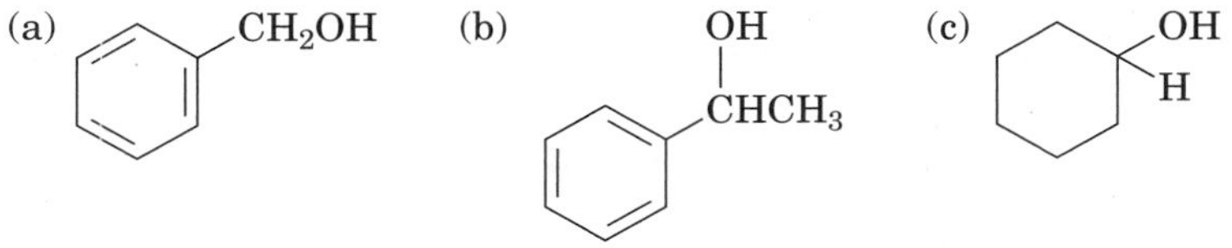

(a) $C_6H_5CH_2OH$ (benzene ring–CH_2OH) (b) $C_6H_5\overset{\overset{OH}{|}}{C}HCH_3$ (c) cyclohexane ring bearing OH and H

8.6 Ethers from Alcohols: The Williamson Ether Synthesis

Alkoxides react with alkyl halides to yield ethers, a reaction known as the **Williamson ether synthesis.** Although discovered more than 100 years ago, the Williamson synthesis is still the best method for preparing both symmetrical and unsymmetrical ethers.

$$\text{C}_5\text{H}_9\text{–}\ddot{\text{O}}\text{:}^- + \text{CH}_3\text{–I} \longrightarrow \text{C}_5\text{H}_9\text{–O–CH}_3 + \text{I}^-$$

Cyclopentoxide ion **Cyclopentyl methyl ether (74%)**

The alkoxide ion needed in the reaction is usually prepared by reaction of an alcohol with sodium or potassium metal (Section 8.3):

$$2\ \text{ROH} + 2\ \text{Na} \longrightarrow 2\ \text{RO}^-\ \text{Na}^+ + \text{H}_2$$

Mechanistically, the Williamson synthesis is an S_N2 reaction (Section 7.7) that occurs by nucleophilic substitution of halide ion by the alkoxide ion. Thus, the reaction is subject to all the normal S_N2 limitations. Primary alkyl halides work best because competitive E2 elimination of HX can occur with more hindered substrates. Unsymmetrical ethers are therefore best prepared by reaction of the more hindered alkoxide partner with the less hindered alkyl halide partner, rather than vice versa. For example, *tert*-butyl methyl ether is best synthesized by reaction of *tert*-butoxide ion with iodomethane, rather than by reaction of methoxide ion with 2-chloro-2-methylpropane.

S_N2 reaction

$$(\text{CH}_3)_3\text{C–}\ddot{\text{O}}\text{:}^- + \text{CH}_3\text{–I} \longrightarrow (\text{CH}_3)_3\text{C–O–CH}_3 + \text{I}^-$$

***tert*-Butoxide ion** **Iodomethane** ***tert*-Butyl methyl ether**

E2 reaction

$$\text{CH}_3\ddot{\text{O}}\text{:}^- + \text{H–CH}_2\text{–C(CH}_3)_2\text{–Cl} \longrightarrow \text{H}_2\text{C=C(CH}_3)_2 + \text{CH}_3\text{OH} + \text{Cl}^-$$

Methoxide ion **2-Chloro-2-methylpropane** **2-Methylpropene**

PROBLEM

8.9 Treatment of cyclohexanol with sodium metal gives an alkoxide ion that undergoes reaction with iodoethane to yield cyclohexyl ethyl ether. Write the reaction, showing all the steps.

PROBLEM

8.10 How would you prepare the following ethers?
(a) Methyl propyl ether (b) Anisole (methyl phenyl ether)

(c) $C_6H_5CH_2OCH(CH_3)_2$

PROBLEM

8.11 Rank the following compounds in order of their expected reactivity toward an alkoxide ion in the Williamson ether synthesis: bromoethane, 2-bromopropane, chloroethane, 2-chloro-2-methylpropane.

8.7 Reactions of Alcohols

Dehydration

Alcohols can be dehydrated to give alkenes (Section 4.9). Tertiary alcohols lose water easily when treated with H_2SO_4, but primary and secondary alcohols require higher temperatures. As we saw in Section 7.10, the mechanism of this dehydration is simply an E1 reaction. Strong acid protonates the alcohol oxygen, the protonated intermediate spontaneously loses water to generate a carbocation, and loss of H^+ from a neighboring carbon atom then yields the alkene product:

$$(CH_3)_3C\text{-}OH \xrightarrow{H_2SO_4} \left[H_3C\text{-}\overset{+}{C}(CH_3)\text{-}CH_3 \right] + H_2O \xrightarrow{-H^+} H_2C{=}C(CH_3)_2$$

tert-Butyl alcohol + H_2O **2-Methylpropene (97%)**

Conversion into Alkyl Halides and Ethers

Alcohols can be converted into both alkyl halides (Section 7.3) and ethers (Section 8.6). Tertiary alcohols are readily transformed into alkyl halides by an S_N1 mechanism on treatment with either HCl or HBr. Primary and secondary alcohols are much more resistant to reaction with acids, however, and are best converted into halides by treatment with either $SOCl_2$ or PBr_3.

Oxidation of Alcohols

One of the most valuable reactions of alcohols is their oxidation to yield carbonyl compounds by a formal loss of H_2: CH–OH → C=O. Primary alcohols yield aldehydes or carboxylic acids, and secondary alcohols yield ketones, but tertiary alcohols don't normally react with oxidizing agents.

Primary alcohol

$$RCH_2OH \xrightarrow{[O]} RCHO \xrightarrow{[O]} RCOOH$$

An aldehyde A carboxylic acid

Secondary alcohol

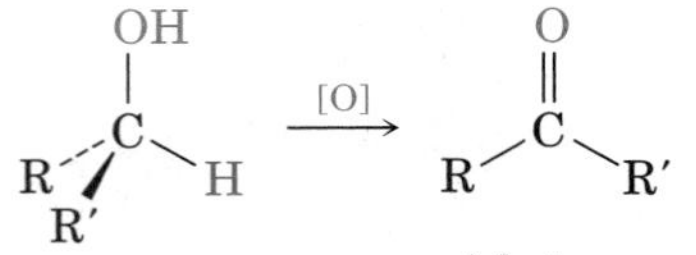

$$R(R')C(OH)H \xrightarrow{[O]} RC(=O)R'$$

A ketone

Primary alcohols are oxidized either to aldehydes or to carboxylic acids, depending on the reagents chosen. Probably the best method for preparing an aldehyde from a primary alcohol on a laboratory scale (as opposed to an industrial scale) is by use of pyridinium chlorochromate (PCC), $C_5H_6NCrO_3Cl$, in dichloromethane solvent. This reagent is too expensive for large-scale use in industry, however.

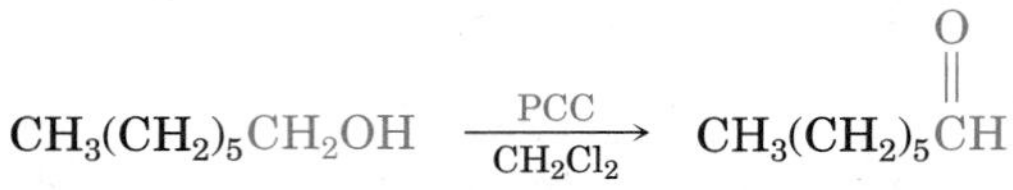

$$CH_3(CH_2)_5CH_2OH \xrightarrow[CH_2Cl_2]{PCC} CH_3(CH_2)_5CH{=}O$$

1-Heptanol → **Heptanal (78%)**

Many oxidizing agents, such as chromium trioxide (CrO_3) and sodium dichromate ($Na_2Cr_2O_7$) in aqueous acid solution, oxidize primary alcohols to carboxylic acids. Although aldehydes are intermediates in these oxidations, they usually can't be isolated because further oxidation takes place too rapidly.

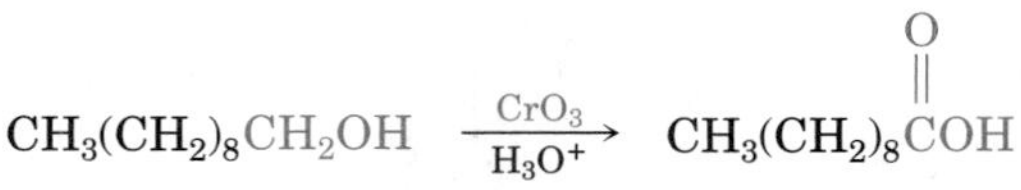

$$CH_3(CH_2)_8CH_2OH \xrightarrow[H_3O^+]{CrO_3} CH_3(CH_2)_8C(=O)OH$$

1-Decanol → **Decanoic acid (93%)**

Secondary alcohols are oxidized easily to produce ketones. Sodium dichromate in aqueous acetic acid is often used as the oxidant.

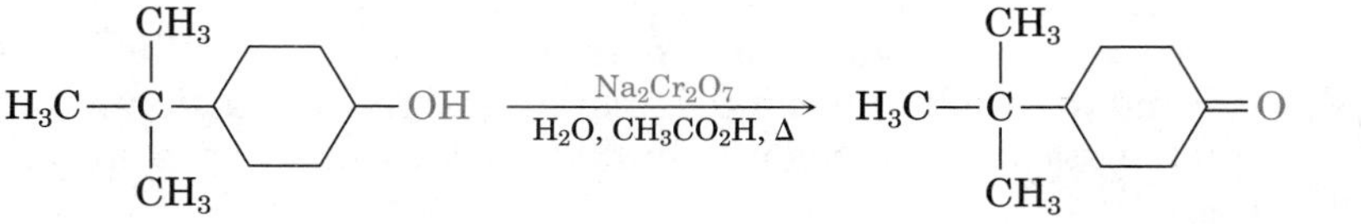

$$(CH_3)_3C\text{-}C_6H_{10}\text{-}OH \xrightarrow[H_2O,\ CH_3CO_2H,\ \Delta]{Na_2Cr_2O_7} (CH_3)_3C\text{-}C_6H_{10}{=}O$$

4-*tert*-Butylcyclohexanol → **4-*tert*-Butylcyclohexanone (91%)**

PRACTICE PROBLEM 8.3

What product would you expect from reaction of benzyl alcohol with CrO_3?

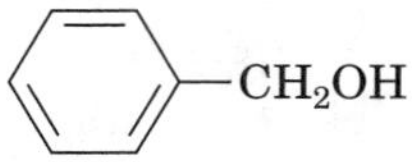

$C_6H_5CH_2OH$ **Benzyl alcohol**

Solution Treatment of a primary alcohol with CrO_3 yields a carboxylic acid. Thus, oxidation of benzyl alcohol should yield benzoic acid.

$$C_6H_5CH_2OH \xrightarrow[H_3O^+]{CrO_3} C_6H_5COOH$$

Benzyl alcohol **Benzoic acid**

PROBLEM

8.12 What alcohols would give the following products on oxidation?

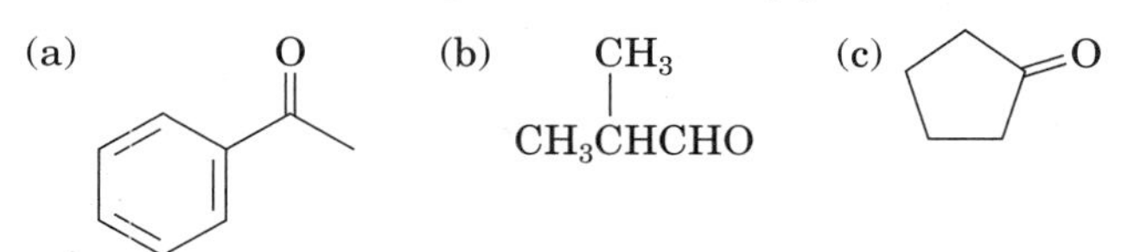

PROBLEM

8.13 What products would you expect to obtain from oxidation of the following alcohols with CrO_3?
(a) Cyclohexanol (b) 1-Hexanol (c) 2-Hexanol

PROBLEM

8.14 What products would you expect to obtain from oxidation of the alcohols in Problem 8.13 with pyridinium chlorochromate (PCC)?

8.8 Synthesis and Reactions of Phenols

Phenols can be synthesized from aromatic starting materials by a two-step sequence. The starting compound is first sulfonated by treatment with SO_3/H_2SO_4, and the arenesulfonic acid product is then converted into a phenol by high-temperature reaction with NaOH.

$$CH_3C_6H_5 \xrightarrow[H_2SO_4]{SO_3} p\text{-}CH_3C_6H_4SO_3H \xrightarrow[300^\circ C\text{; 2. } H_3O^+]{1.\ NaOH} p\text{-}CH_3C_6H_4OH$$

Toluene ***p*-Toluenesulfonic acid** ***p*-Methylphenol (72%)**

PROBLEM

8.15 *p*-Cresol (*p*-methylphenol) is used industrially both as an antiseptic and as a starting material to prepare the food additive BHT. Show how you could synthesize *p*-cresol from benzene.

Alcohol-Like Reactions

Phenols and alcohols are very different in spite of the fact that both have –OH groups. Phenols can't be dehydrated by treatment with acid and can't be converted into halides by treatment with HX. Phenols can, however, be converted into ethers by S_N2 reaction with alkyl halides in the presence of base. Williamson ether synthesis with phenols occurs easily because phenols are more acidic than alcohols and are therefore more easily converted into their anions.

$$o\text{-Nitrophenol} + CH_3CH_2CH_2CH_2Br \xrightarrow[\text{Acetone}]{K_2CO_3} \text{Butyl } o\text{-nitrophenyl ether } (80\%)$$

o-Nitrophenol **1-Bromobutane** **Butyl o-nitrophenyl ether (80%)**

Electrophilic Aromatic Substitution Reactions

The –OH group is an activating, ortho/para-directing substituent in electrophilic aromatic substitution reactions (Sections 5.9 and 5.10). As a result, phenols are reactive substrates for electrophilic halogenation, nitration, and sulfonation.

Phenol (OH) → Br_2, $FeBr_3$ → o-bromophenol (OH, Br); → HNO_3, H_2SO_4 → o-nitrophenol (OH, NO_2); → SO_3, H_2SO_4 → o-hydroxybenzenesulfonic acid (OH, SO_3H)

Oxidation of Phenols: Quinones

Treatment of a phenol with a strong oxidizing agent such as sodium dichromate yields a cyclohexadienedione, or **quinone:**

$$\text{Phenol (–OH)} \xrightarrow[H_3O^+]{Na_2Cr_2O_7} O{=}C_6H_4{=}O$$

Phenol **Benzoquinone**

Quinones are an interesting and valuable class of compounds because of their oxidation–reduction properties. They can be easily reduced to **hydroquinones** (*p*-dihydroxybenzenes) by $NaBH_4$ or $SnCl_2$, and hydroquinones can be easily oxidized back to quinones by $Na_2Cr_2O_7$. Hydroquinone is used, among other things, as a photographic developer, because it reduces Ag^+ on film to metallic silver.

$$O{=}C_6H_4{=}O \underset{Na_2Cr_2O_7}{\overset{NaBH_4}{\rightleftharpoons}} HO{-}C_6H_4{-}OH$$

Benzoquinone **Hydroquinone**

8.9 Reactions of Ethers: Acidic Cleavage

Ethers are unreactive to many common reagents, a property that accounts for their frequent use as reaction solvents. Halogens, mild acids, bases, and nucleophiles have no effect on most ethers. In fact, ethers undergo only one general reaction—they are cleaved by strong acids. Aqueous HI is the usual reagent for cleaving ethers, although aqueous HBr also works.

Acidic ether cleavages are typical nucleophilic substitution reactions. They take place by either an S_N1 or S_N2 pathway, depending on the structure of the ether. Primary and secondary alkyl ethers react by an S_N2 pathway, in which nucleophilic iodide ion attacks the protonated ether at the less highly substituted site. The ether oxygen atom stays with the more hindered alkyl group, and the iodide bonds to the less hindered group. For example, ethyl isopropyl ether yields isopropyl alcohol and iodoethane on cleavage by HI:

$$CH_3CH(CH_3){-}\ddot{O}{-}CH_2CH_3 \xrightarrow{H{-}I} \left[CH_3CH(CH_3){-}\overset{+}{O}(H){-}CH_2CH_3\right] \xrightarrow[S_N2]{:\ddot{I}:^-} CH_3CH(CH_3)OH + CH_3CH_2I$$

Ethyl isopropyl ether **Isopropyl alcohol** **Iodoethane**

Tertiary ethers cleave by an S_N1 mechanism because they can produce stable intermediate carbocations. In such reactions, the ether oxygen atom stays with the *less* hindered alkyl group and the halide bonds to the tertiary group. Like most S_N1 reactions, the cleavage is fast and often takes place at room temperature or below.

Secondary Tertiary

$$C_6H_{11}{-}O{-}C(CH_3)_3 \xrightarrow[S_N1]{HBr,\ H_2O} C_6H_{11}OH + CH_3CBr(CH_3)_2$$

***tert*-Butyl cyclohexyl ether** **Cyclohexanol** **2-Bromo-2-methylpropane**

PRACTICE PROBLEM 8.4 ....

What products would you expect from the reaction of methyl cyclopentyl ether with HI?

Solution Iodide ion attacks the less hindered methyl group rather than the more hindered cyclopentyl group in an S_N2 reaction, giving iodomethane and cyclopentanol:

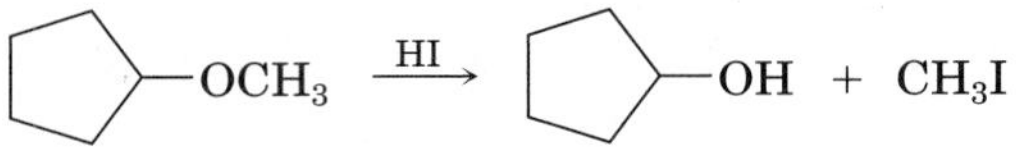

$$\text{C}_5\text{H}_9\text{—OCH}_3 \xrightarrow{\text{HI}} \text{C}_5\text{H}_9\text{—OH} + CH_3I$$

PROBLEM

8.16 What products do you expect from the reaction of the following ethers with HI?
(a) $CH_3CH_2OCH_2CH_3$ (b) Cyclohexyl ethyl ether (c) $(CH_3)_3COCH_2CH_3$

8.10 Cyclic Ethers: Epoxides

For the most part, cyclic ethers behave like acyclic ethers. The chemistry of the ether functional group is the same whether it's in an open chain or in a ring. Thus, common cyclic ethers such as tetrahydrofuran (THF) and dioxane are often used as solvents because of their inertness.

1,4-Dioxane **Tetrahydrofuran**

The only cyclic ethers that behave differently from open-chain ethers are the three-membered-ring compounds called **epoxides,** or **oxiranes.** The strain of the three-membered ring gives epoxides unique chemical reactivity.

Epoxides are prepared by reaction of an alkene with a *peroxyacid,* RCO_3H. *m*-Chloroperoxybenzoic acid is often used because it is more stable and more easily handled than most other peroxyacids.

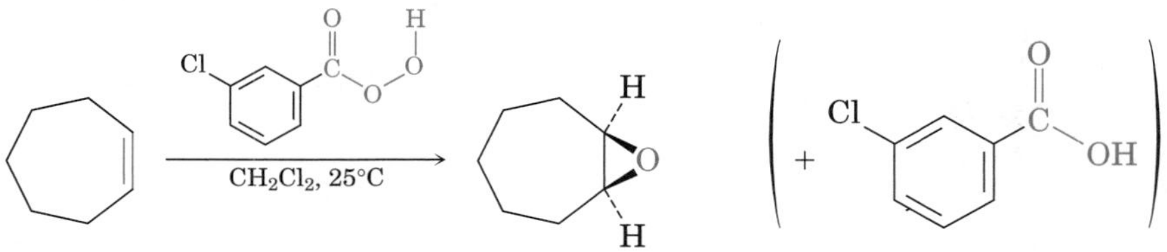

CH_2Cl_2, 25°C

Cycloheptene **1,2-Epoxycycloheptane (78%)**

PROBLEM

8.17 What product do you expect from reaction of *cis*-2-butene with *m*-chloroperoxybenzoic acid, assuming syn stereochemistry?

PROBLEM

8.18 Reaction of *trans*-2-butene with *m*-chloroperoxybenzoic acid yields a different epoxide from that obtained by reaction of the cis isomer (see Problem 8.17). Explain.

8.11 Ring-Opening Reactions of Epoxides

Epoxide rings are cleaved by treatment with acid just as other ethers are cleaved. The major difference is that epoxides react under much milder conditions because of the strain of the three-membered ring. Dilute aqueous acid at room temperature converts an epoxide to a 1,2-diol, also called a *vicinal glycol*. (The word *vicinal* means "adjacent," and a *glycol* is a diol.) Two million tons of ethylene glycol, most of it used as automobile antifreeze, are produced every year in the United States by acid-catalyzed hydration of ethylene oxide. (Note that the name *ethylene glycol* refers to the glycol derived *from* ethylene. Similarly, *ethylene oxide* is the epoxide derived from ethylene.)

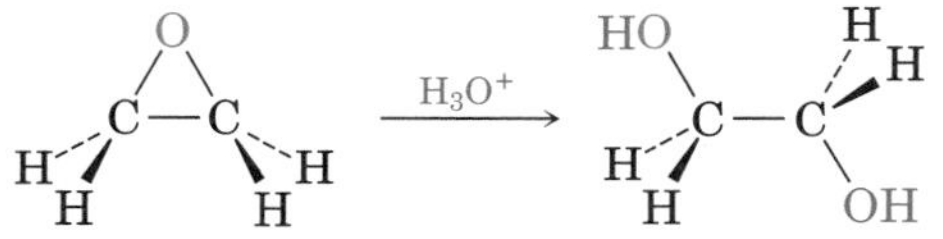

Ethylene oxide $\xrightarrow{H_3O^+}$ **Ethylene glycol (1,2-Ethanediol)**

Acid-catalyzed epoxide cleavage takes place by S_N2 attack of H_2O on the protonated epoxide, in a manner analogous to the final step of alkene bromination, where a three-membered-ring bromonium ion is opened by nucleophilic attack (Section 4.5).

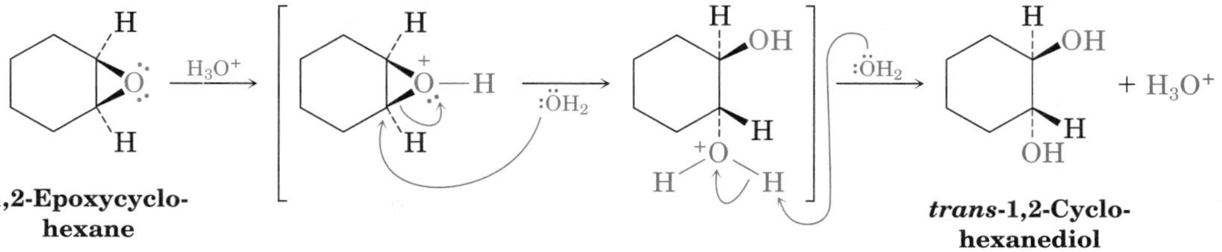

1,2-Epoxycyclohexane $\xrightarrow{H_3O^+}$ **trans-1,2-Cyclohexanediol (86%)** $+ H_3O^+$

Recall the following:

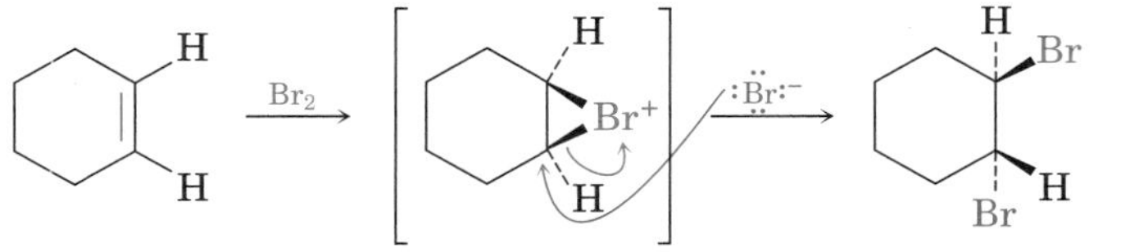

Cyclohexene $\xrightarrow{Br_2}$ **trans-1,2-Dibromocyclohexane**

Epoxide opening is also involved in the mechanism by which the polycyclic aromatic hydrocarbons (PAH's) in chimney soot and cigarette smoke cause cancer. Benzo[a]pyrene, one of the best-studied PAH's, is converted by metabolic oxidation into a diol epoxide. The epoxide ring then reacts with an amino group in cellular DNA to give an altered DNA that is covalently bound to the PAH. With its DNA thus altered, the cell is unable to function normally.

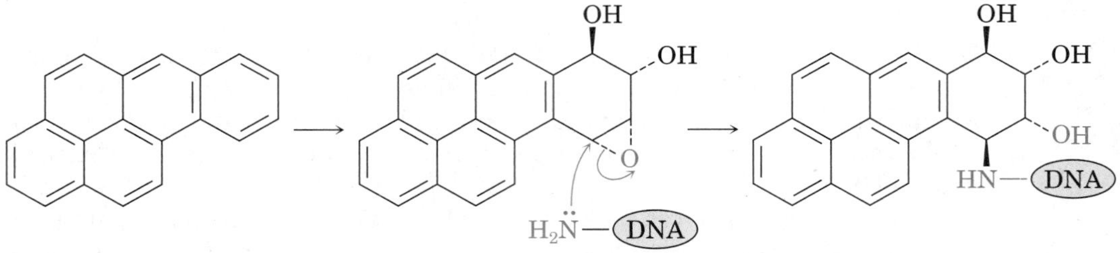

OH, OH, O, $H_2\ddot{N}$—DNA; OH, OH, OH, HN—DNA

Benzo[a]pyrene A diol epoxide

PROBLEM

8.19 Show the steps involved in the reaction of *cis*-2,3-epoxybutane with aqueous acid to yield 2,3-butanediol. What is the stereochemistry of the product if the ring opening takes place by normal back-side S_N2 attack?

PROBLEM

8.20 Answer Problem 8.19 for the reaction of *trans*-2,3-epoxybutane. Is the same product formed?

8.12 Thiols and Sulfides

Sulfur is the element just below oxygen in the periodic table, and many oxygen-containing organic compounds have sulfur analogs. **Thiols, R–SH,** are sulfur analogs of alcohols, and **sulfides, R–S–R′,** are sulfur analogs of ethers. Thiols are named in the same way as alcohols, with the suffix *-thiol* used in place of *-ol.* The –SH group itself is referred to as a **mercapto group.**

CH_3CH_2SH — **Ethanethiol**

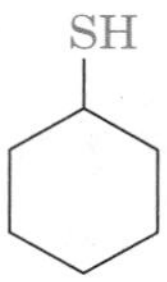

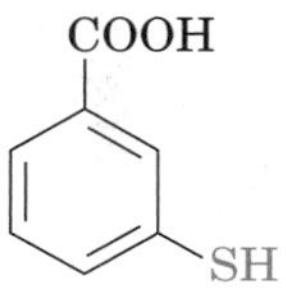

SH (on cyclohexane) — **Cyclohexanethiol**

COOH, SH (on benzene) — ***m*-Mercaptobenzoic acid**

Sulfides are named in the same way as ethers, with *sulfide* used in place of *ether* for simple compounds and with *alkylthio* used in place of *alkoxy* for more complex substances.

$CH_3—S—CH_3$ **Dimethyl sulfide**

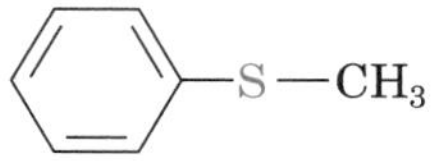

$C_6H_5—S—CH_3$ **Methyl phenyl sulfide**

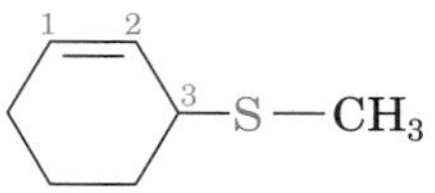

3-cyclohexenyl$—S—CH_3$ **3-(Methylthio)cyclohexene**

Thiols are usually prepared from the corresponding alkyl halide by S_N2 displacement with a sulfur nucleophile such as hydrosulfide anion, SH^-:

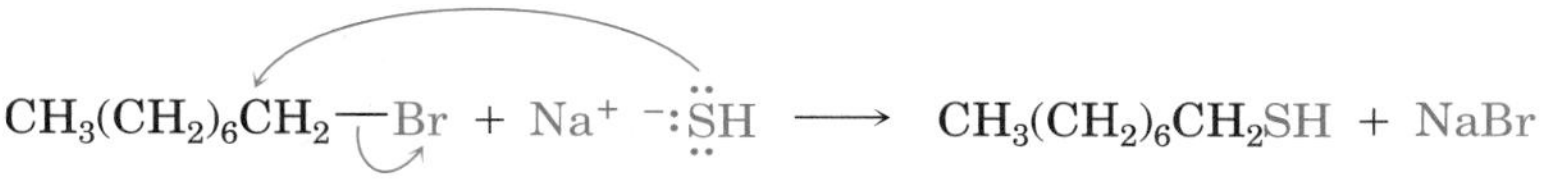

$$CH_3(CH_2)_6CH_2—Br + Na^+ \ ^-{:}\ddot{S}H \longrightarrow CH_3(CH_2)_6CH_2SH + NaBr$$

1-Bromooctane **Sodium hydrosulfide** **1-Octanethiol**

Sulfides are prepared by treating a primary or secondary alkyl halide with a *thiolate ion,* RS^-. Reaction occurs by an S_N2 mechanism that is analogous to the Williamson ether synthesis (Section 8.6). Thiolate anions are among the best nucleophiles known, so these reactions usually work well.

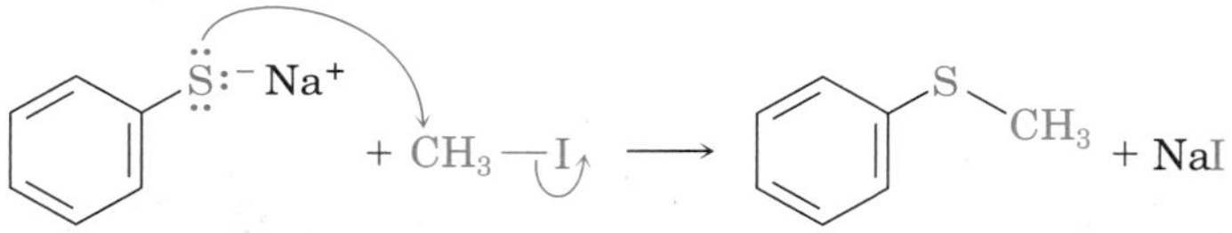

$$C_6H_5\ddot{S}{:}^- Na^+ + CH_3—I \longrightarrow C_6H_5—S—CH_3 + NaI$$

Sodium benzenethiolate **Methyl phenyl sulfide (96%)**

The most unforgettable characteristic of thiols is their appalling odor. Skunk scent, in fact, is due primarily to the simple thiols 3-methyl-1-butanethiol and 2-butene-1-thiol. Thiols can be oxidized by mild reagents such as bromine to yield *disulfides,* R–S–S–R, and disulfides can be reduced back to thiols by treatment with zinc metal and acetic acid:

$$2\,R—SH \underset{Zn,\ H^+}{\overset{Br_2}{\rightleftharpoons}} R—S—S—R + 2\,HBr$$

A thiol A disulfide

We'll see in Section 15.5 that the thiol–disulfide interconversion is extremely important in biochemistry because disulfide "bridges" form cross-links that help stabilize the three-dimensional structure of proteins.

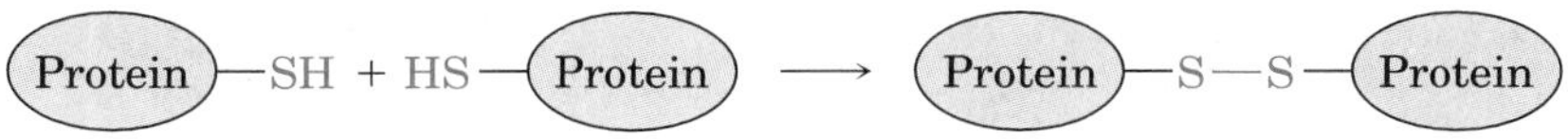

Protein—SH + HS—Protein ⟶ Protein—S—S—Protein

A cross-linked protein

PROBLEM

8.21 Name the following thiols by IUPAC rules:

(a) $CH_3CH_2CH(SH)CH_3$

(b) $CH_3C(CH_3)_2CH_2CH(SH)CH_2CH(CH_3)CH_3$

(c) Cyclopent-2-enyl–SH

PROBLEM

8.22 Name the following compounds by IUPAC rules:

(a) $CH_3CH_2SCH_3$ (b) $CH_3C(CH_3)_2SCH_2CH_3$ (c) 1,2-bis(SCH_3)benzene

PROBLEM

8.23 2-Butene-1-thiol is a component of skunk spray. How would you synthesize this substance from 2-buten-1-ol? From methyl 2-butenoate, $CH_3CH{=}CHCOOCH_3$? More than one step is required in both instances.

INTERLUDE

Ethanol as Chemical, Drug, and Poison

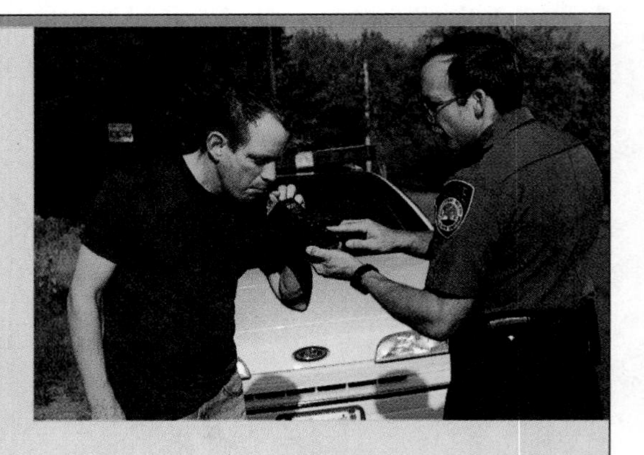

A Breathalyzer test showing a blood alcohol concentration above 0.10% will result in a charge of drunken driving.

The production of ethanol by fermentation of grains and sugars is one of the oldest known organic reactions, going back at least 2500 years. Fermentation is carried out by adding yeast to an aqueous sugar solution, where enzymes break down carbohydrates into ethanol and CO_2:

$$\underset{\text{A carbohydrate}}{C_6H_{12}O_6} \xrightarrow{\text{Yeast}} 2\ CH_3CH_2OH + 2\ CO_2$$

(continued)➤

Nearly 110 million gallons of ethanol are produced each year in the United States, primarily for use as a solvent. Only about 5% of this industrial ethanol comes from fermentation, though; most is obtained by acid-catalyzed hydration of ethylene.

$$H_2C{=}CH_2 + H_2O \xrightarrow[\text{catalyst}]{\text{Acid}} CH_3CH_2OH$$

Ethanol is classified for medical purposes as a central nervous system (CNS) depressant. Its effects (that is, being drunk) resemble the human response to anesthetics. There is an initial excitability and increase in sociable behavior, but this results from depression of inhibition rather than from stimulation. At a blood alcohol concentration of 0.1–0.3%, motor coordination is affected, accompanied by loss of balance, slurred speech, and amnesia. When blood alcohol concentration rises to 0.3–0.4%, nausea and loss of consciousness occur. Above 0.6%, spontaneous respiration and cardiovascular regulation are affected, ultimately leading to death. The LD_{50} (Chapter 1 Interlude) of ethanol is 10.6 g/kg.

The passage of ethanol through the body begins with its absorption in the stomach and small intestine, followed by rapid distribution to all body fluids and organs. In the pituitary gland, ethanol inhibits the production of a hormone that regulates urine flow, causing increased urine production and dehydration. In the stomach, ethanol stimulates production of acid. Throughout the body, ethanol causes blood vessels to dilate, resulting in flushing of the skin and a sensation of warmth as blood moves into capillaries beneath the surface.

The metabolism of ethanol occurs mainly in the liver and proceeds by oxidation in two steps, first to acetaldehyde (CH_3CHO) and then to acetic acid (CH_3COOH). In chronic alcoholics, ethanol and acetaldehyde are toxic, leading to devastating physical and metabolic deterioration. The liver usually suffers the worst damage since it is the major site of alcohol metabolism.

The quick and uniform distribution of ethanol in body fluids, the ease with which it crosses lung membranes, and its ready oxidizability provide the basis for simple tests for blood alcohol concentration. The *Breathalyzer test* measures alcohol concentration in expired air by the color change that occurs when the chromium in the bright orange oxidizing agent potassium dichromate ($K_2Cr_2O_7$) is reduced to blue-green chromium(III). In most states, a blood alcohol level above 0.10% is sufficient for a charge of driving while intoxicated.

Summary and Key Words

alcohol, 251
alkoxide ion, 256
epoxide, 268
ether, 251
hydroquinone, 266
mercapto group, 270
oxirane, 268
phenol, 251
phenoxide ion, 256
quinone, 266
sulfide, 270
thiol, 270
Williamson ether synthesis, 261

Ethers, alcohols, and phenols are organic derivatives of water in which one or both of the water hydrogens have been replaced by organic groups.

Ethers have two organic groups bonded to the same oxygen atom. They are prepared by S_N2 reaction of an alkoxide ion with a primary alkyl halide—the **Williamson synthesis.** Ethers are inert to most reagents but are cleaved by strong acids. **Epoxides**—cyclic ethers with an oxygen atom in a three-membered ring—differ from other ethers in their ease of cleavage. The high reactivity of the strained three-membered ether ring allows epoxides to react with aqueous acid, yielding vicinal glycols (diols).

Alcohols are compounds that have an –OH group bonded to an alkyl residue. They can be prepared in many ways, including hydration of alkenes. The most general method of alcohol synthesis involves reduction of a carbonyl compound. Aldehydes, esters, and carboxylic acids yield primary alcohols on reduction; ketones yield secondary alcohols.

Alcohols are weak acids and can be converted into their **alkoxide anions** on treatment with a strong base or with an alkali metal. Alcohols can also be dehydrated to yield alkenes, transformed into alkyl halides by treatment with PBr_3 or $SOCl_2$, converted into ethers by reaction of their anions with alkyl halides, and oxidized to yield carbonyl compounds. Primary alcohols give either aldehydes or carboxylic acids when oxidized, secondary alcohols yield ketones, and tertiary alcohols are not oxidized.

Phenols are aromatic counterparts of alcohols. Although similar to alcohols in some respects, phenols are more acidic than alcohols because **phenoxide anions** are stabilized by resonance. Phenols undergo electrophilic aromatic substitution and can be oxidized to yield **quinones.**

Sulfides (R–S–R′) and **thiols (R–SH)** are sulfur analogs of ethers and alcohols. Thiols are prepared by S_N2 reaction of an alkyl halide with HS^-, and sulfides are prepared by further alkylation of the thiol with an alkyl halide.

Summary of Reactions

1. Synthesis of alcohols (Section 8.5)
   (a) Reduction of aldehydes to yield primary alcohols

$$\mathrm{R\overset{O}{\overset{\|}{C}}H} \xrightarrow[2.\ H_2O]{1.\ NaBH_4} \mathrm{RCH_2OH}$$

   (b) Reduction of ketones to yield secondary alcohols

$$\mathrm{R\overset{O}{\overset{\|}{C}}R'} \xrightarrow[2.\ H_2O]{1.\ NaBH_4} \mathrm{R\overset{OH}{\overset{|}{C}}HR'}$$

   (c) Reduction of esters to yield primary alcohols

$$\mathrm{R\overset{O}{\overset{\|}{C}}OR'} \xrightarrow[2.\ H_2O]{1.\ LiAlH_4} \mathrm{RCH_2OH}$$

   (d) Reduction of carboxylic acids to yield primary alcohols

$$\mathrm{R\overset{O}{\overset{\|}{C}}OH} \xrightarrow[2.\ H_2O]{1.\ LiAlH_4} \mathrm{RCH_2OH}$$

2. Synthesis of ethers (Section 8.6)

$$\mathrm{RO^-\ Na^+ + R'Br} \xrightarrow[\text{reaction}]{S_N2} \mathrm{ROR'}$$

3. Synthesis of phenols (Section 8.8)

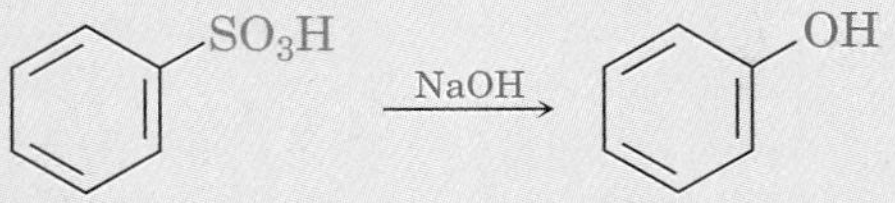

$$\mathrm{C_6H_5SO_3H} \xrightarrow{NaOH} \mathrm{C_6H_5OH}$$

4. Synthesis of epoxides (Section 8.10)

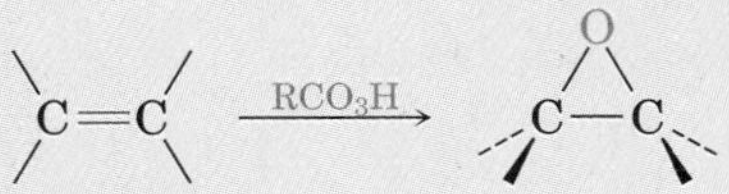

$$\mathrm{C{=}C} \xrightarrow{RCO_3H} \text{epoxide } \mathrm{(C{-}C \text{ bridged by } O)}$$

5. Synthesis of thiols (Section 8.12)

$$\mathrm{Na^+\ {}^-SH + RBr} \xrightarrow[\text{reaction}]{S_N2} \mathrm{RSH}$$

6. Synthesis of sulfides (Section 8.12)

$$\mathrm{RS^-\ Na^+ + R'Br} \xrightarrow[\text{reaction}]{S_N2} \mathrm{RSR'}$$

7. Reactions of alcohols
   (a) Conversion into ethers (Section 8.6)

$$2\,ROH + 2\,Na \longrightarrow 2\,RO^-\;Na^+ + H_2$$

$$RO^-\;Na^+ + R'Br \xrightarrow[\text{reaction}]{S_N2} ROR'$$

   (b) Dehydration to yield alkenes (Section 8.7)

$$-\overset{H}{\underset{|}{\overset{|}{C}}}-\overset{OH}{\underset{|}{\overset{|}{C}}}- \xrightarrow{H_2SO_4} \;\rangle C{=}C\langle\; + H_2O$$

   (c) Oxidation to yield carbonyl compounds (Section 8.7)

$$RCH_2OH \xrightarrow[\text{chlorochromate}]{\text{Pyridinium}} R\overset{O}{\overset{\|}{C}}H \quad \text{An aldehyde}$$

$$RCH_2OH \xrightarrow[H_3O^+]{CrO_3} R\overset{O}{\overset{\|}{C}}OH \quad \text{A carboxylic acid}$$

$$R\overset{OH}{\overset{|}{C}}HR' \xrightarrow[\text{chlorochromate}]{\text{Pyridinium}} R\overset{O}{\overset{\|}{C}}R' \quad \text{A ketone}$$

8. Reactions of ethers; acidic cleavage (Section 8.9)

$$ROR' + HI \longrightarrow ROH + R'I$$

ADDITIONAL PROBLEMS

8.24 Draw structures corresponding to the following IUPAC names:
(a) Ethyl isopropyl ether
(b) 3,4-Dimethoxybenzoic acid
(c) 2-Methyl-2,5-heptanediol
(d) *trans*-3-Ethylcyclohexanol
(e) 4-Allyl-2-methoxyphenol (eugenol, from oil of cloves)

8.25 Name the following compounds according to IUPAC rules:

(a) $HOCH_2CH_2\overset{CH_3}{\overset{|}{C}}HCH_2OH$

(b) $CH_3\underset{HO}{\underset{|}{C}}H\underset{CH_2CH_2CH_3}{\underset{|}{C}}HCH_2CH_3$

(c) Ph, H, OH, H on cyclopentane (wedge/dash stereochemistry)

(d) $(CH_3)_2CH\overset{SH}{\underset{CH_3}{\overset{|}{\underset{|}{C}}}}CH_2CH_2CH_3$

8.26 Draw and name the eight isomeric alcohols with the formula $C_5H_{12}O$.

8.27 Which of the eight alcohols you identified in Problem 8.26 would react with aqueous acidic CrO_3? Show the products you would expect from each reaction.

8.28 Which of the eight alcohols you identified in Problem 8.26 are chiral?

8.29 Draw and name the six ethers that are isomeric with the alcohols you drew in Problem 8.26. Which are chiral?

8.30 Show the HI cleavage products of the ethers you drew in Problem 8.29.

8.31 Predict the likely products of the following cleavage reactions:

(a) $CH_3CH_2OCH(CH_3)CH_3 \xrightarrow{HI,\ H_2O}$ (b) $(CH_3)_3CCH_2OCH_3 \xrightarrow{HI,\ H_2O}$

8.32 What reagents would you use to carry out the following transformations?

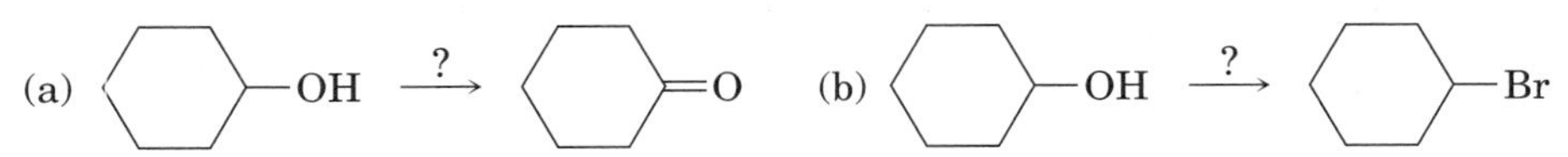

(c) $CH_3CH_2CH_2OH \xrightarrow{?} CH_3CH_2CHO$
(d) $CH_3CH_2CH_2OH \xrightarrow{?} CH_3CH_2COOH$
(e) $CH_3CH_2CH_2OH \xrightarrow{?} CH_3CH_2CH_2O^- Na^+$
(f) $CH_3CH_2CH_2OH \xrightarrow{?} CH_3CH_2CH_2Cl$

8.33 How would you prepare the following compounds from 2-phenylethanol?
(a) Benzoic acid
(b) Ethylbenzene
(c) 2-Bromo-1-phenylethane
(d) Phenylacetic acid ($C_6H_5CH_2COOH$)
(e) Phenylacetaldehyde ($C_6H_5CH_2CHO$)

8.34 Give the structures of the major products you would obtain from reaction of phenol with the following reagents:
(a) Br_2 (1 mol) (b) Br_2 (3 mol)
(c) NaOH, then CH_3I (d) $Na_2Cr_2O_7$, H_3O^+

8.35 What products would you obtain from reaction of 1-butanol with the following reagents?
(a) PBr_3 (b) CrO_3, H_3O^+ (c) Na (d) Pyridinium chlorochromate

8.36 What products would you obtain from reaction of 1-methylcyclohexanol with the following reagents?
(a) HBr (b) H_2SO_4 (c) CrO_3 (d) Na
(e) Product of part (d), then CH_3I

8.37 What alcohols would you oxidize to obtain the following products?

(a) Cyclopentanone (cyclopentane ring with =O) (b) Benzene ring–CHO (c) $CH_3CH(CH_3)COOH$

8.38 Show the alcohols you would obtain by reduction of the following carbonyl compounds:

(a) $CH_3CH(CH_3)CH_2CHO$ (b) Benzene ring with two adjacent COOH groups (c) $CH_3CH_2C(=O)CH_2CH(CH_3)CH_3$

8.39 Reduction of 2-butanone with $NaBH_4$ yields 2-butanol. Is the product chiral? Is it optically active? Explain.

$CH_3C(=O)CH_2CH_3$ **2-Butanone**

8.40 When 4-chloro-1-butanol is treated with a strong base, tetrahydrofuran is produced. Suggest a mechanism for this reaction.

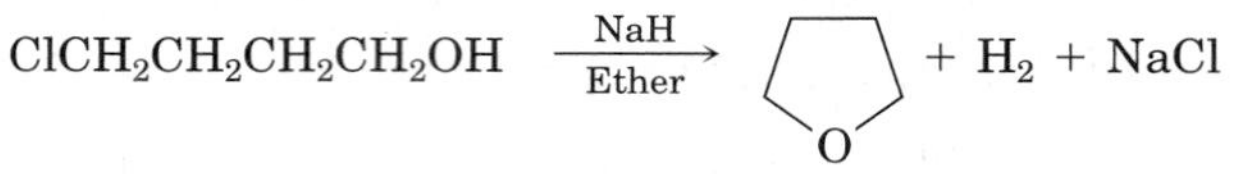

$$ClCH_2CH_2CH_2CH_2OH \xrightarrow[\text{Ether}]{\text{NaH}} \text{(tetrahydrofuran)} + H_2 + NaCl$$

8.41 Which is more acidic, *p*-methylphenol or *p*-bromophenol? Explain.

8.42 Why can't the Williamson ether synthesis be used to prepare diphenyl ether?

8.43 Rank the following substances in order of increasing acidity:

CH_3COCH_3	$CH_3COCH_2COCH_3$	C_6H_5OH	CH_3COOH
Acetone	**2,4-Pentanedione**	**Phenol**	**Acetic acid**
$pK_a = 19$	$pK_a = 9$	$pK_a = 9.9$	$pK_a = 4.7$

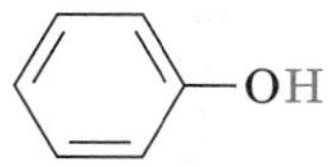

8.44 Which, if any, of the substances in Problem 8.43 are strong enough acids to react substantially with NaOH? (The pK_a of H_2O is 15.7.)

8.45 Is *tert*-butoxide anion a strong enough base to react with water? In other words, does the following reaction take place as written? (The pK_a of *tert*-butyl alcohol is 18.)

$$(CH_3)_3CO^- \, Na^+ + H_2O \xrightarrow{?} (CH_3)_3COH + NaOH$$

8.46 Sodium bicarbonate, $NaHCO_3$, is the sodium salt of carbonic acid (H_2CO_3), $pK_a = 6.4$. Which of the substances shown in Problem 8.43 will react with sodium bicarbonate?

8.47 Assume that you have two unlabeled bottles, one that contains phenol ($pK_a = 9.9$) and one that contains acetic acid ($pK_a = 4.7$). In light of your answer to Problem 8.46, propose a simple way to tell what is in each bottle.

8.48 Starting from benzene, how would you prepare benzyl phenyl ether, $C_6H_5OCH_2C_6H_5$? More than one step is required.

8.49 Since all hamsters look pretty much alike, pairing and mating is governed by chemical means of communication. Investigations have shown that dimethyl disulfide, CH_3SSCH_3, is secreted by female hamsters as a sex attractant for males. How would you synthesize dimethyl disulfide in the laboratory if you wanted to trick your hamster?

8.50 *p*-Nitrophenol ($pK_a = 7.15$) is much more acidic than phenol ($pK_a = 9.9$). Draw as many resonance structures as you can for the *p*-nitrophenoxide anion.

8.51 The herbicide 2,4,5-T (2,4,5-trichlorophenoxyacetic acid) can be prepared by heating a mixture of 2,4,5-trichlorophenol and $ClCH_2COOH$ with NaOH. Show the mechanism of the reaction.

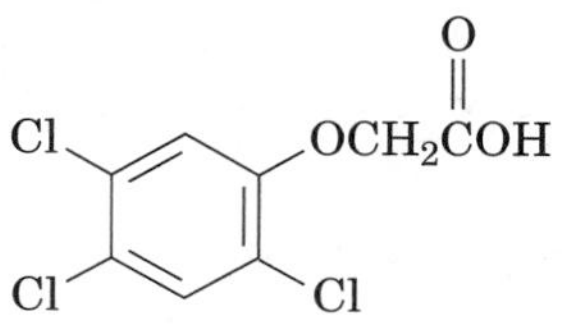

$$2,4,5\text{-}Cl_3C_6H_2OCH_2COOH$$

2,4,5-T

8.52 *tert*-Butyl ethers can be prepared by the reaction of an alcohol with 2-methylpropene in the presence of an acid catalyst. Propose a mechanism for this reaction.

8.53 *tert*-Butyl ethers react with trifluoroacetic acid, CF_3COOH, to yield an alcohol and 2-methylpropene. For example:

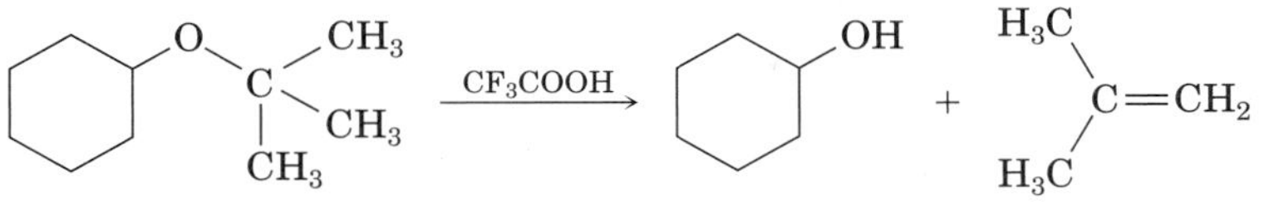

CH_3 CH_3 CH_3 O C $\xrightarrow{CF_3COOH}$ OH + H_3C H_3C $C{=}CH_2$

Tell what kind of reaction is occurring, and propose a mechanism.

8.54 How would you prepare the following ethers?

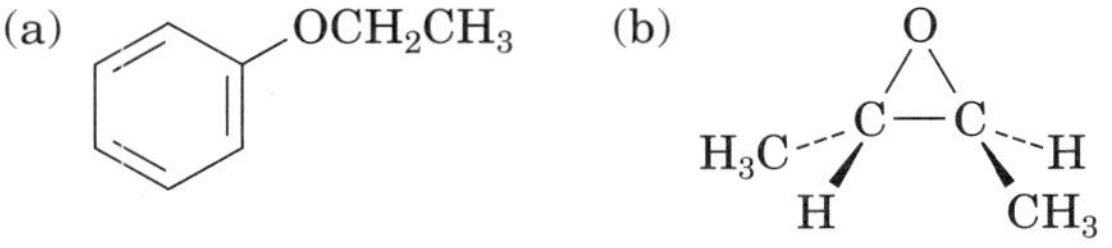

(a) OCH_2CH_3 (b) O H_3C C—C H H CH_3

8.55 Identify the reagents a–d in the following scheme:

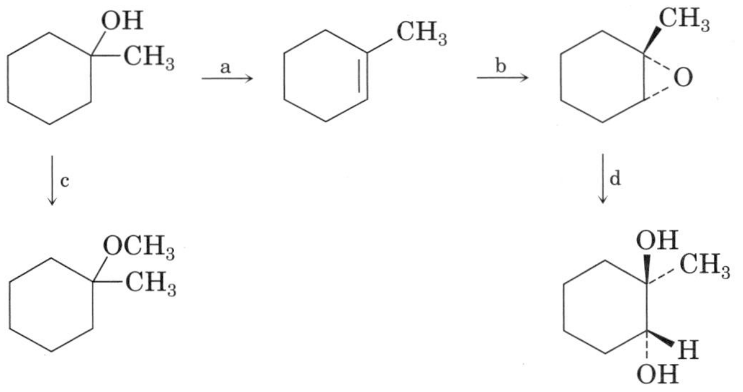

OH CH_3 $\xrightarrow{a}$ CH_3 $\xrightarrow{b}$ CH_3 O

c ↓ d ↓

OCH_3 CH_3 OH CH_3 H OH

8.56 What cleavage product would you expect from reaction of tetrahydrofuran with hot aqueous HI?

O **Tetrahydrofuran**

8.57 Methyl phenyl ether can be cleaved to yield iodomethane and lithium phenoxide when heated with LiI. Propose a mechanism for this reaction.

8.58 The *Zeisel method,* a procedure for determining the number of methoxyl groups (CH_3O–) in a compound, involves heating a weighed amount of the compound with HI. Ether cleavage occurs, and the iodomethane that forms is distilled off and passed into a solution of $AgNO_3$, where it reacts to give AgI. The silver iodide is then weighed, and the number of methoxy groups in the sample is thereby determined. For example, 1.06 g of vanillin, the material responsible for the characteristic odor of vanilla, yields 1.60 g of AgI. If vanillin has a molecular weight of 152, how many methoxyls does it contain?

Visualizing Chemistry

8.59 Give IUPAC names for the following compounds (gray = C, red = O, blue = N, light green = H):

(a) (b)

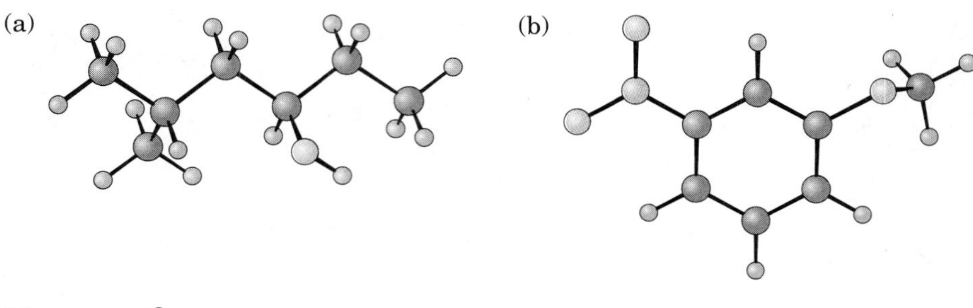

(c)

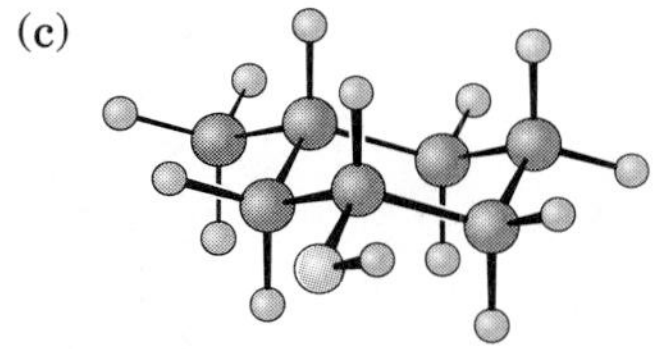

8.60 Predict the product of each of the following reactions (gray = C, red = O, light green = H):

(a)

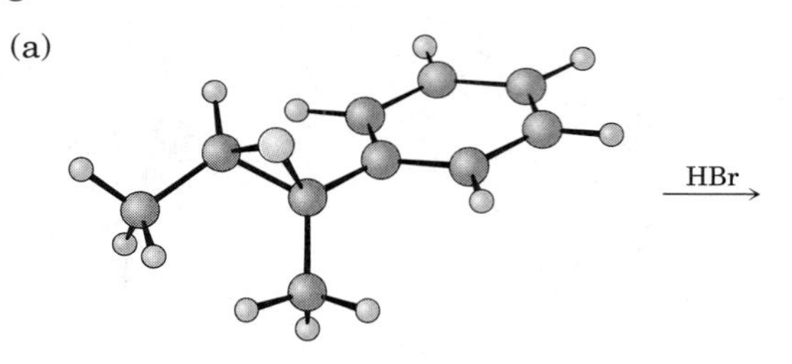

(b)

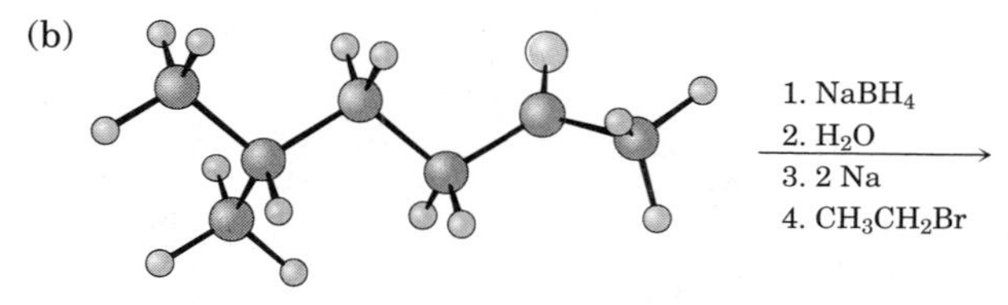

8.61 Name and assign *R* or *S* stereochemistry to the product you would obtain by reaction of the following compound with $NaBH_4$ (gray = C, red = O, light green = H).

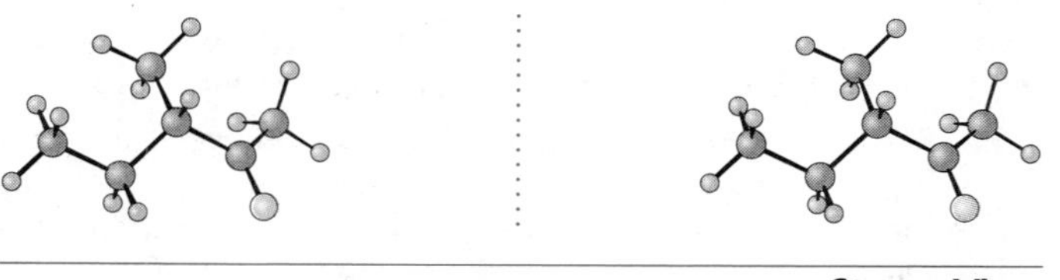

Stereo View

Acetone and formaldehyde are typical carbonyl compounds.

9 Aldehydes and Ketones: Nucleophilic Addition Reactions

In this and the next two chapters, we'll discuss the most important functional group in organic chemistry—the **carbonyl group, C=O.** Carbonyl compounds are everywhere in nature. Most biologically important molecules contain carbonyl groups, as do many pharmaceutical agents and many of the synthetic chemicals that touch our everyday lives. Acetic acid (the chief component of vinegar), acetaminophen (an over-the-counter headache remedy), and Dacron (the polyester material used in clothing) all contain different kinds of carbonyl groups.

Acetic acid (a carboxylic acid)	**Acetaminophen** (an amide)	**Dacron** (a polyester)

9.1 Kinds of Carbonyl Compounds

Table 9.1 shows some of the many kinds of carbonyl compounds. All contain an **acyl group,** $R{-}\overset{\displaystyle O}{\overset{\|}{C}}{-}$, bonded to another part, which may contain a carbon, hydrogen, oxygen, halogen, sulfur, or other atom.

Table 9.1 Some Types of Carbonyl Compounds

Name	*General formula*	*Name ending*	*Name*	*General formula*	*Name ending*
Aldehyde	$R{-}C({=}O){-}H$	*-al*	Ester	$R{-}C({=}O){-}O{-}R'$	*-oate*
Ketone	$R{-}C({=}O){-}R'$	*-one*	Lactone (cyclic ester)	ring: $\text{C}{-}C({=}O){-}O$	None
Carboxylic acid	$R{-}C({=}O){-}O{-}H$	*-oic acid*	Amide	$R{-}C({=}O){-}N<$	*-amide*
Acid halide	$R{-}C({=}O){-}X$ (X = halogen)	*-yl* or *-oyl halide*	Lactam (cyclic amide)	ring: $\text{C}{-}C({=}O){-}N$	None
Acid anhydride	$R{-}C({=}O){-}O{-}C({=}O){-}R'$	*-oic anhydride*			

It's useful to classify carbonyl compounds into two general categories based on the kinds of chemistry they undergo. In one category are aldehydes and ketones; in the other are carboxylic acids and their derivatives:

Aldehydes (RCHO)
Ketones (R_2CO)

The acyl groups in these two families are bonded to substituents (–H and –R, respectively) that *can't stabilize a negative charge and therefore can't act as leaving groups.* Aldehydes and ketones behave similarly and undergo many of the same reactions.

Carboxylic acids (RCOOH) Acid chlorides (RCOCl) Acid anhydrides (RCOOCOR′) Esters (RCOOR′) Amides ($RCONH_2$)	The acyl groups in carboxylic acids and their derivatives are bonded to substituents (oxygen, halogen, nitrogen) that *can stabilize a negative charge and can serve as leaving groups in substitution reactions.* The chemistry of these compounds is therefore similar.

PROBLEM

9.1 Propose structures for molecules that meet the following descriptions:
(a) A ketone, $C_5H_{10}O$
(b) An aldehyde, $C_6H_{10}O$
(c) A keto aldehyde, $C_6H_{10}O_2$
(d) A cyclic ketone, C_5H_8O

9.2 Structure and Properties of Carbonyl Groups

The carbon–oxygen double bond of carbonyl groups is similar in some respects to the carbon–carbon double bond of alkenes (Figure 9.1). The carbonyl carbon atom is sp^2-hybridized and forms three σ bonds. The fourth valence electron remains in a carbon p orbital and forms a π bond to oxygen by overlap with an oxygen p orbital. The oxygen also has two nonbonding pairs of electrons, which occupy its remaining two orbitals. Like alkenes, carbonyl compounds are planar about the double bond and have bond angles of approximately 120°.

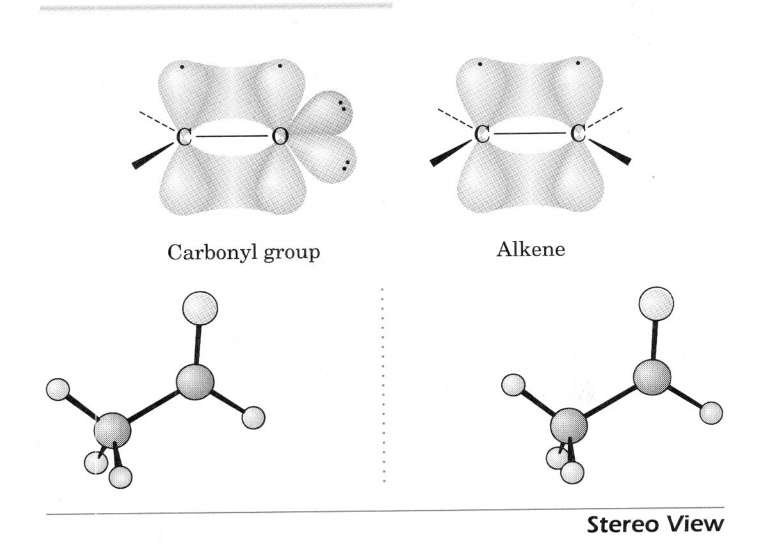

Figure 9.1 Electronic structure of the carbonyl group.

Carbon–oxygen double bonds are polarized because of the high electronegativity of oxygen relative to carbon. Since the carbonyl carbon is positively polarized, it is electrophilic (a Lewis acid) and reacts with nucleophiles. Conversely, the carbonyl oxygen is negatively polarized and nucleophilic (a Lewis base). We'll see in this and the next two chapters that most carbonyl-group reactions are the result of this bond polarization.

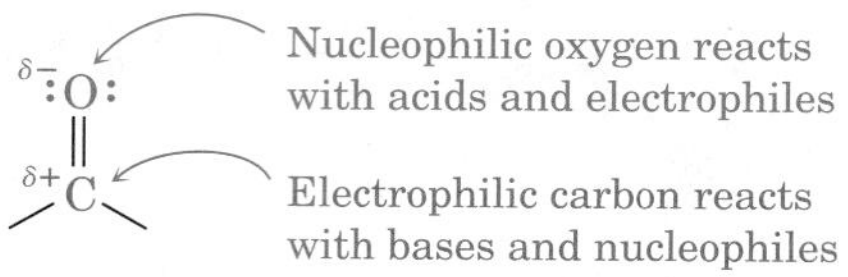

$\delta-$:O: (Nucleophilic oxygen reacts with acids and electrophiles)

$\delta+$ C (Electrophilic carbon reacts with bases and nucleophiles)

9.3 Naming Aldehydes and Ketones

Aldehydes are named by replacing the terminal *-e* of the corresponding alkane name with *-al*. The parent chain must contain the –CHO group, and the –CHO carbon is always numbered as carbon 1. For example:

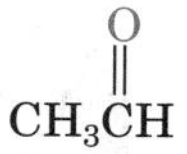

CH_3CHO — **Ethanal (Acetaldehyde)**

CH_3CH_2CHO — **Propanal (Propionaldehyde)**

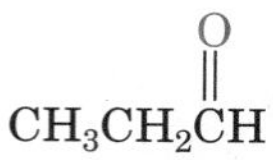

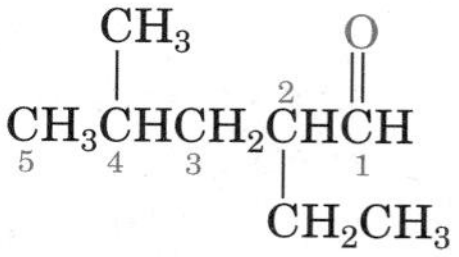

$CH_3CH(CH_3)CH_2CH(CH_2CH_3)CHO$ (C5 4 3 2 1) — **2-Ethyl-4-methylpentanal**

Note that the longest chain in 2-ethyl-4-methylpentanal is a hexane, but this chain does not include the –CHO group and thus is not the parent.

For more complex aldehydes in which the –CHO group is attached to a ring, the suffix *-carbaldehyde* is used:

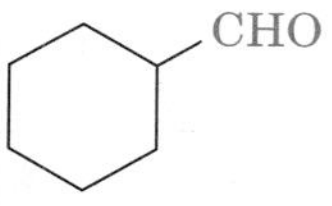

Cyclohexanecarbaldehyde **2-Naphthalenecarbaldehyde**

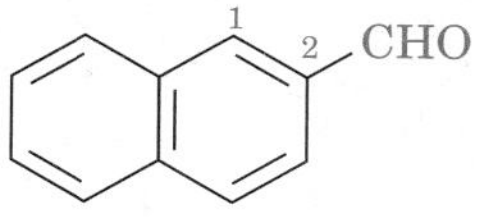

Some simple and well-known aldehydes also have common names, as indicated in Table 9.2.

Ketones are named by replacing the terminal *-e* of the corresponding alkane name with *-one*. The parent chain is the longest one that contains the ketone group, and numbering begins at the end nearer the carbonyl carbon. For example:

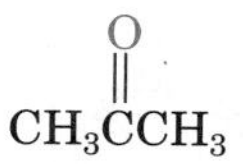

CH_3COCH_3 — **Propanone (Acetone)**

$CH_3CH_2COCH_2CH_2CH_3$ (1 2 3 4 5 6) — **3-Hexanone**

$CH_3CH{=}CHCH_2COCH_3$ (6 5 4 3 2 1) — **4-Hexen-2-one**

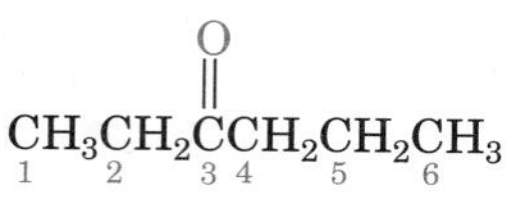

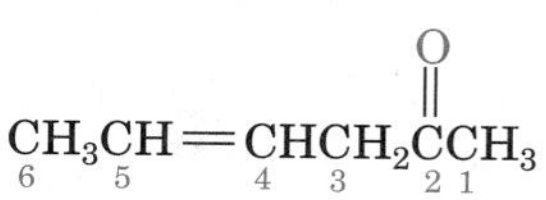

Table 9.2 Common Names of Some Simple Aldehydes

Formula	*Common name*	*Systematic name*
HCHO	Formaldehyde	Methanal
CH_3CHO	Acetaldehyde	Ethanal
CH_3CH_2CHO	Propionaldehyde	Propanal
$CH_3CH_2CH_2CHO$	Butyraldehyde	Butanal
$CH_3CH_2CH_2CH_2CHO$	Valeraldehyde	Pentanal
$H_2C{=}CHCHO$	Acrolein	2-Propenal
C_6H_5–CHO (benzene ring–CHO)	Benzaldehyde	Benzenecarbaldehyde

A few ketones also have common names:

$CH_3\overset{O}{\overset{\|}{C}}CH_3$ — **Acetone**

$C_6H_5\overset{O}{\overset{\|}{C}}CH_3$ — **Acetophenone**

$C_6H_5\overset{O}{\overset{\|}{C}}C_6H_5$ — **Benzophenone**

When it is necessary to refer to the –COR group as a substituent, the general term *acyl* is used. Similarly, $-COCH_3$ is an *acetyl* group, –CHO is a *formyl* group, –COAr is an *aroyl* group, and $-COC_6H_5$ is a *benzoyl* group.

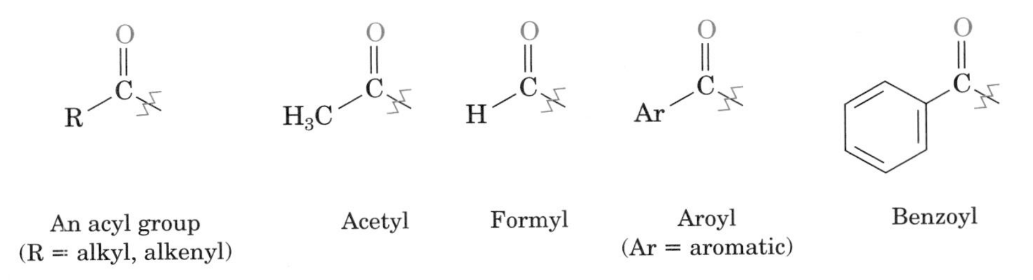

An acyl group (R = alkyl, alkenyl) — Acetyl — Formyl — Aroyl (Ar = aromatic) — Benzoyl

Occasionally, the doubly bonded oxygen must be considered a substituent, and the prefix *oxo-* is used. For example:

$$\underset{6}{CH_3}\underset{5}{CH_2}\underset{4}{CH_2}\underset{3}{\overset{O}{\overset{\|}{C}}}\underset{2}{CH_2}\underset{1}{\overset{O}{\overset{\|}{C}}}OCH_3$$ **Methyl 3-oxohexanoate**

PROBLEM

9.2 Name the following aldehydes and ketones:

(a) $CH_3CH_2\overset{O}{\overset{\|}{C}}CH(CH_3)_2$

(b) C_6H_5–CH_2CH_2CHO (benzene ring–CH_2CH_2CHO)

(c) $CH_3\overset{O}{\overset{\|}{C}}CH_2CH_2CH_2\overset{O}{\overset{\|}{C}}CH_2CH_3$

(d) 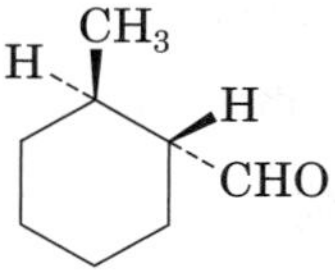 2-methylcyclohexanecarbaldehyde structure (H, CH_3; H, CHO)

(e) $OHCCH_2CH_2CH_2CHO$

(f) 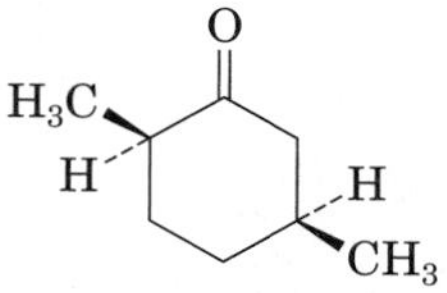 3,5-dimethylcyclohexanone structure (H_3C, H; H, CH_3)

PROBLEM

9.3 Draw structures corresponding to the following IUPAC names:
(a) 3-Methylbutanal
(b) 3-Methyl-3-butenal
(c) 4-Chloro-2-pentanone
(d) Phenylacetaldehyde
(e) 2,2-Dimethylcyclohexanecarbaldehyde
(f) 1,3-Cyclohexanedione

9.4 Synthesis of Aldehydes and Ketones

We've already discussed one of the best methods of preparing aldehydes and ketones: alcohol oxidation (Section 8.7). Primary alcohols are oxidized to give aldehydes, and secondary alcohols are oxidized to give ketones. Pyridinium chlorochromate (PCC) in dichloromethane is usually chosen for making aldehydes, while PCC, CrO_3, and $Na_2Cr_2O_7$ are all effective for making ketones:

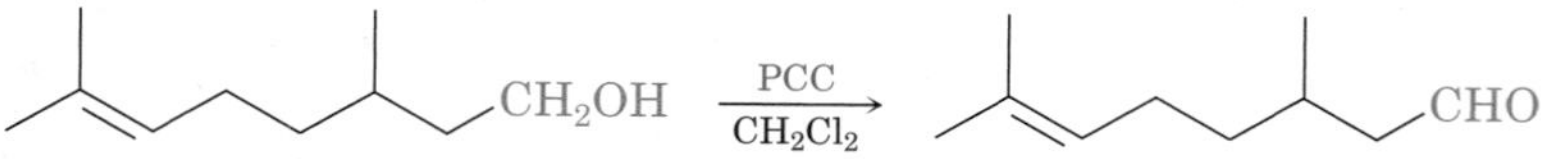

$$\text{Citronellol } (\ldots CH_2OH) \xrightarrow[CH_2Cl_2]{PCC} \text{Citronellal } (\ldots CHO)$$

Citronellol → **Citronellal (82%)**

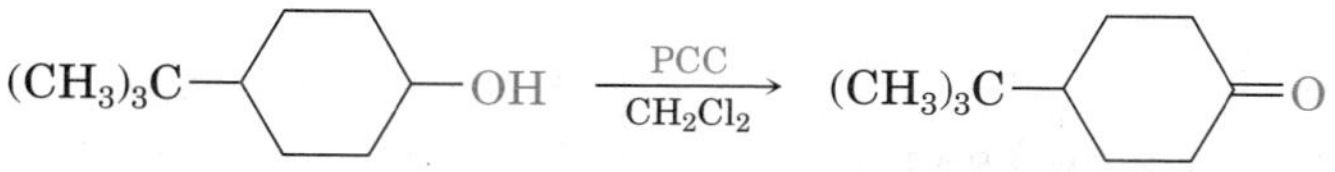

$$(CH_3)_3C\text{-}C_6H_{10}\text{-}OH \xrightarrow[CH_2Cl_2]{PCC} (CH_3)_3C\text{-}C_6H_{10}{=}O$$

4-*tert*-Butylcyclohexanol → **4-*tert*-Butylcyclohexanone (90%)**

Other methods for preparing ketones include the hydration of a terminal alkyne to yield a methyl ketone (Section 4.15) and the Friedel–Crafts acylation of an aromatic ring to yield an alkyl aryl ketone (Section 5.8).

$$CH_3(CH_2)_3C{\equiv}CH \xrightarrow[Hg(OAc)_2]{H_3O^+} CH_3(CH_2)_3\overset{O}{\overset{\|}{C}}{-}CH_3$$

1-Hexyne → **2-Hexanone (78%)**

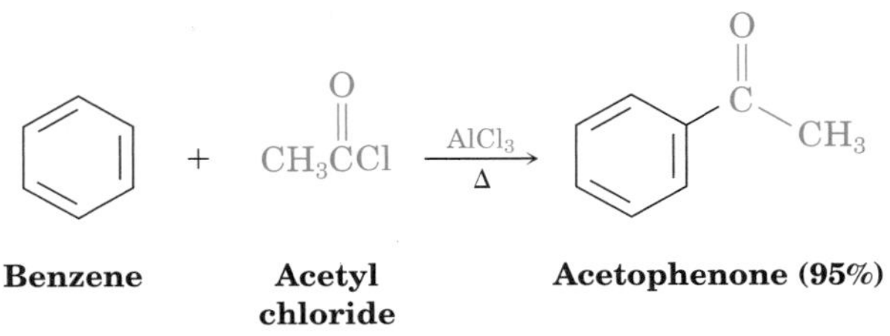

$$\text{Benzene} + CH_3\overset{O}{\overset{\|}{C}}Cl \xrightarrow[\Delta]{AlCl_3} C_6H_5\overset{O}{\overset{\|}{C}}CH_3$$

Benzene **Acetyl chloride** **Acetophenone (95%)**

PROBLEM

9.4 How could you prepare pentanal from the following starting materials?
(a) 1-Pentanol
(b) $CH_3CH_2CH_2CH_2COOH$
(c) 5-Decene

PROBLEM

9.5 How could you prepare 2-hexanone from the following starting materials?
(a) 2-Hexanol
(b) 1-Hexyne
(c) 2-Methyl-1-hexene

PROBLEM

9.6 How would you carry out the following transformations? More than one step may be required.
(a) 3-Hexene ⟶ 3-Hexanone
(b) Benzene ⟶ 1-Phenylethanol

9.5 Oxidation of Aldehydes

Aldehydes are easily oxidized to yield carboxylic acids, RCHO → RCOOH, but ketones are unreactive toward oxidation. This reactivity difference is a consequence of structure: Aldehydes have a –CHO proton that can be removed during oxidation, but ketones do not.

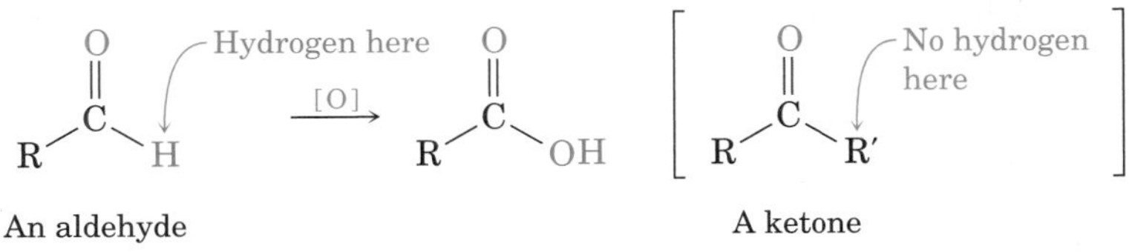

$$R-\overset{O}{\overset{\|}{C}}-H \xrightarrow{[O]} R-\overset{O}{\overset{\|}{C}}-OH \qquad \left[R-\overset{O}{\overset{\|}{C}}-R' \right]$$

Hydrogen here; No hydrogen here

An aldehyde A ketone

One of the simplest methods for oxidizing an aldehyde is to use silver ion, Ag^+, in dilute aqueous ammonia (**Tollens' reagent**). As the oxidation proceeds, a shiny mirror of silver metal is deposited on the walls of the reaction flask, forming the basis of a simple test to detect the presence of an aldehyde functional group in a sample of unknown structure. A small amount of the unknown is dissolved in ethanol in a test tube, and a few drops of Tollens' reagent are added. If the test tube becomes silvery, the unknown is an aldehyde.

$$C_6H_5CHO \xrightarrow[NH_4OH]{AgNO_3} C_6H_5COOH + Ag$$

Benzaldehyde **Benzoic acid**

PRACTICE PROBLEM 9.1

What product would you obtain from the reaction of 3-methylbutanal with Tollens' reagent?

Solution Write the structure of the aldehyde, and then replace the –H bonded to the carbonyl group by –OH:

$$CH_3CH(CH_3)CH_2C(=O)—H \xrightarrow[\text{reagent}]{\text{Tollens'}} CH_3CH(CH_3)CH_2C(=O)—OH$$

3-Methylbutanal **3-Methylbutanoic acid**

PROBLEM

9.7 Predict the products of the reaction of the following substances with Tollens' reagent:
(a) Pentanal (b) 2,2-Dimethylhexanal (c) Cyclohexanone

9.6 Reactions of Aldehydes and Ketones: Nucleophilic Additions

The most common reaction of aldehydes and ketones is the **nucleophilic addition reaction,** in which a nucleophile (:Nu or $:Nu^-$) adds to the electrophilic carbon of the carbonyl group. Hydroxide ion (OH^-), hydride ion (H^-), carbon anions (**carbanions, R_3C^-**), water (H_2O), ammonia (NH_3), and alcohols (ROH) are several of many possibilities. In fact, the reduction of a ketone or aldehyde by reaction with $NaBH_4$ (Section 8.5) is simply a nucleophilic addition reaction of $H:^-$ ion.

$$>C=O \xrightarrow[2.\ H_3O^+]{1.\ Nu:^-} >C(OH)Nu$$

The mechanism of a generalized nucleophilic addition is shown in Figure 9.2. As indicated, the nucleophile uses a pair of its electrons to form a bond to the carbon atom of the C=O group. The C=O double bond breaks, the carbon atom rehybridizes from sp^2 to sp^3, and two electrons from the C=O bond move to the oxygen atom giving an alkoxide ion. Addition of H^+ to the alkoxide ion then yields a neutral alcohol product.

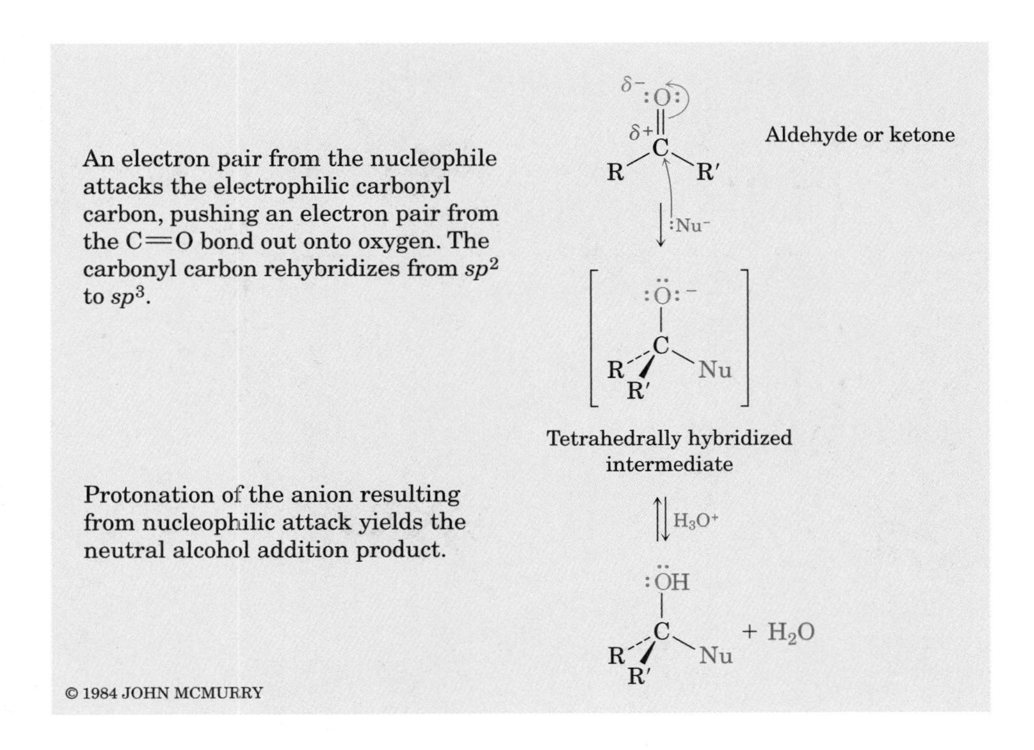

An electron pair from the nucleophile attacks the electrophilic carbonyl carbon, pushing an electron pair from the C=O bond out onto oxygen. The carbonyl carbon rehybridizes from sp^2 to sp^3.
Protonation of the anion resulting from nucleophilic attack yields the neutral alcohol addition product.
Aldehyde or ketone
δ−
δ+
:Nu⁻
Tetrahedrally hybridized intermediate
H_3O^+
+ H_2O
© 1984 JOHN MCMURRY

Figure 9.2 General mechanism of a nucleophilic addition reaction.

Aldehydes are generally more reactive than ketones in nucleophilic additions for steric reasons. The presence of two relatively large substituents in ketones versus one large substituent in aldehydes means that attacking nucleophiles are able to approach aldehydes more readily (Figure 9.3, p.290).

PRACTICE PROBLEM 9.2

What product would you expect from nucleophilic addition of aqueous hydroxide ion to acetaldehyde?

Solution Hydroxide ion adds to the C=O carbon atom, giving an alkoxide ion intermediate that is protonated to yield a 1,1-dialcohol:

CH_3CHO $\xrightarrow{:\ddot{O}H^-}$ $[CH_3CH(O^-)OH]$ $\xrightarrow{H_3O^+}$ $CH_3CH(OH)OH$

Acetaldehyde

PROBLEM

9.8 What product would you expect if the nucleophile cyanide ion, CN^-, were to add to acetone and the intermediate were to be protonated?

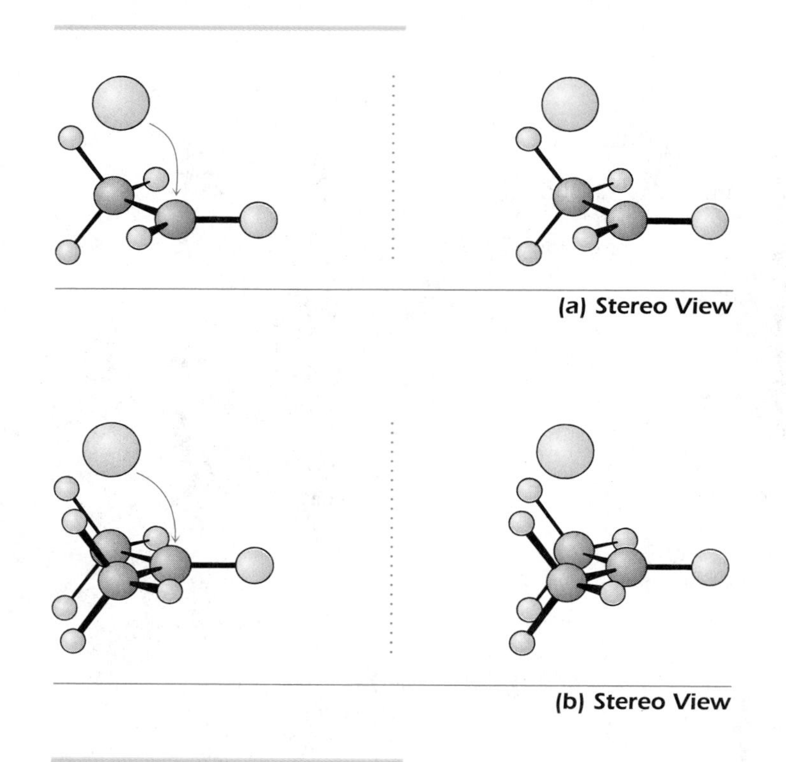

Figure 9.3 Nucleophilic attack on an aldehyde (a) is relatively unhindered, but attack on a ketone (b) is sterically hindered because of the two relatively large substituents attached to the carbonyl-group carbon.

PROBLEM

9.9 The reduction of a ketone to a secondary alcohol on treatment with $NaBH_4$ (Section 8.5) is a nucleophilic addition reaction in which the nucleophile hydride ion ($:H^-$) adds to the carbonyl group, and the alkoxide ion intermediate is then protonated. Show the mechanism of this reduction.

PROBLEM

9.10 Which would you expect to be more reactive toward nucleophilic additions, propanal or 2,2-dimethylpropanal? Explain.

9.7 Nucleophilic Addition of Water: Hydration

Aldehydes and ketones undergo a nucleophilic addition reaction with water to yield 1,1-diols, or **geminal (gem) diols.** The reaction is reversible, and a gem diol can eliminate water to regenerate a ketone or aldehyde:

$$H_3C{-}C({=}O){-}CH_3 + H_2O \rightleftharpoons (H_3C)_2C(OH)_2$$

Acetone **Acetone hydrate (a gem diol)**

The position of the equilibrium between gem diols and aldehydes/ketones depends on the structure of the carbonyl compound. Although the equilibrium strongly favors the carbonyl compound in most cases, the gem diol is favored for a few simple aldehydes. For example, an aqueous solution of acetone consists of about 0.1% gem diol and 99.9% ketone, whereas an aqueous solution of formaldehyde consists of 99.9% gem diol and 0.1% aldehyde.

The nucleophilic addition of water to aldehydes and ketones is slow in pure water but is catalyzed by both acid and base. As is always the case, these catalysts don't change the *position* of the equilibrium; they affect only the rate at which the hydration reaction occurs.

The base-catalyzed reaction takes place in several steps, as shown in Figure 9.4. The attacking nucleophile is the negatively charged hydroxide ion.

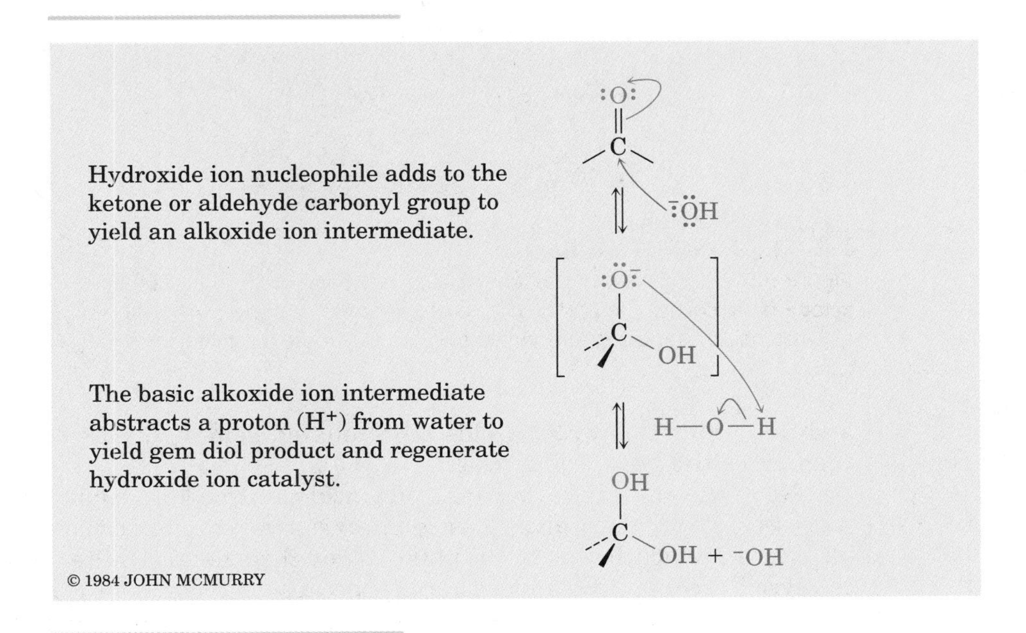

Hydroxide ion nucleophile adds to the ketone or aldehyde carbonyl group to yield an alkoxide ion intermediate.
The basic alkoxide ion intermediate abstracts a proton (H+) from water to yield gem diol product and regenerate hydroxide ion catalyst.
:O:
C
:ÖH
OH
H—O—H
OH + ⁻OH
© 1984 JOHN MCMURRY

Figure 9.4 Mechanism of the base-catalyzed hydration reaction of a ketone or aldehyde. Hydroxide ion is a more reactive nucleophile than neutral water.

The acid-catalyzed hydration reaction also takes place in several steps (Figure 9.5, p. 292). The acid catalyst first protonates the Lewis-basic oxygen atom of the carbonyl group, and subsequent nucleophilic

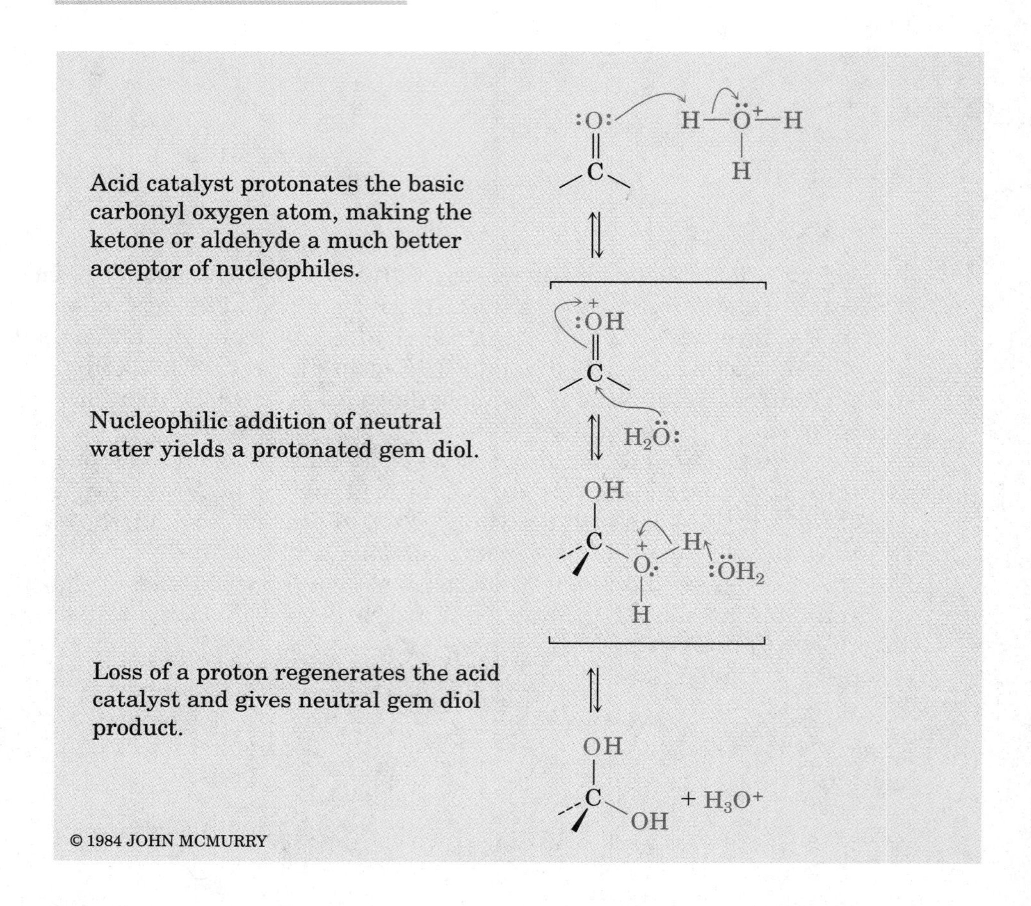
Acid catalyst protonates the basic carbonyl oxygen atom, making the ketone or aldehyde a much better acceptor of nucleophiles.
Nucleophilic addition of neutral water yields a protonated gem diol.
Loss of a proton regenerates the acid catalyst and gives neutral gem diol product.
:O:
C
H—O+—H
H
:OH
H2O:
OH
O
H
:OH2
OH
OH
+ H3O+
© 1984 JOHN MCMURRY

Figure 9.5 Mechanism of the acid-catalyzed hydration reaction of a ketone or aldehyde. The acid catalyst protonates the carbonyl starting material, thus making it more electrophilic and more reactive.

addition of neutral water yields a protonated gem diol. Loss of a proton then gives the gem diol product.

Note the difference between the acid-catalyzed and base-catalyzed processes. The *base*-catalyzed reaction takes place rapidly because hydroxide ion is a much better nucleophilic *donor* than neutral water. The *acid*-catalyzed reaction takes place rapidly because the carbonyl compound is converted by protonation into a much better electrophilic *acceptor.* Most nucleophilic addition reactions can be similarly catalyzed by either acid or base.

PROBLEM

9.11 When dissolved in water, trichloroacetaldehyde (chloral, CCl_3CHO) exists primarily as the gem diol, chloral hydrate (better known as "knockout drops"). Show the structure of chloral hydrate.

PROBLEM

9.12 The oxygen in water is primarily (99.8%) ^{16}O, but water enriched with the heavy isotope ^{18}O is also available. When a ketone or aldehyde is dissolved in $H_2{}^{18}O$, the isotopic label becomes incorporated into the carbonyl group: $R_2C{=}O + H_2O^* \rightarrow R_2C{=}O^* + H_2O$ (where $O^* = {}^{18}O$). Explain.

9.8 Nucleophilic Addition of Alcohols: Acetal Formation

Aldehydes and ketones react with alcohols in the presence of an acid catalyst to yield **acetals, $R_2C(OR')_2$,** compounds that have two ether-like –OR groups bonded to the same carbon:

$$R_2C{=}O + 2\,R'OH \xrightleftharpoons[\text{catalyst}]{\text{Acid}} R_2C(OR')_2 + H_2O$$

Ketone/aldehyde → An acetal

Acetal formation involves the acid-catalyzed nucleophilic addition of an alcohol to the carbonyl group in a manner similar to that of the acid-catalyzed hydration. The initial nucleophilic addition step yields a hydroxy ether called a **hemiacetal,** which reacts further with a second equivalent of alcohol to yield the acetal plus water. For example, reaction of cyclohexanone with methanol yields the dimethyl acetal.

$$\text{A ketone} \xrightarrow[\text{H}^+\text{ catalyst}]{CH_3OH} [\text{A hemiacetal (OH, } OCH_3)] \xrightarrow[\text{H}^+\text{ catalyst}]{CH_3OH} \text{An acetal } (OCH_3, OCH_3) + H_2O$$

A ketone — A hemiacetal — An acetal

As with hydration (Section 9.7), all the steps during acetal formation are reversible, and the reaction can be made to go either forward (from carbonyl compound to acetal) or backward (from acetal to carbonyl compound), depending on the reaction conditions. The forward reaction is favored by conditions that remove water from the medium and thus drive the equilibrium to the right. The backward reaction is favored in the presence of a large excess of water.

Acetals are valuable to organic chemists because they can serve as **protecting groups** for aldehydes and ketones. To see what this means, imagine that you are faced with having to reduce the keto ester methyl 4-oxopentanoate to obtain the keto alcohol 5-hydroxy-2-pentanone. This reaction can't be done in a single step because of the presence in the molecule of the ketone carbonyl group. If you were to treat methyl 4-oxopentanoate with $LiAlH_4$, both ester and ketone groups would be reduced.

$$CH_3\overset{O}{\overset{\|}{C}}CH_2CH_2\overset{O}{\overset{\|}{C}}OCH_3 \xrightarrow{?} CH_3\overset{O}{\overset{\|}{C}}CH_2CH_2CH_2OH$$

Methyl 4-oxopentanoate → **5-Hydroxy-2-pentanone**

This situation isn't unusual. It often happens that one functional group in a complex molecule interferes with intended chemistry on another functional group elsewhere in the molecule. In such situations, it's often possible to circumvent the problem by *protecting* the interfering functional group to render it unreactive, carrying out the desired reaction, and then removing the protecting group.

Aldehydes and ketones can be protected by converting them into acetals. Acetals, like other ethers, are stable to bases, reducing agents, and various nucleophiles, but they are acid-sensitive. Thus, you can selectively reduce the ester group in methyl 4-oxopentanoate by converting the keto group into an acetal, treating the compound with $LiAlH_4$ in ether, and then removing the acetal protecting group by treatment with aqueous acid.

$$CH_3\overset{O}{\overset{\|}{C}}CH_2CH_2\overset{O}{\overset{\|}{C}}OCH_3 \xrightarrow[H^+\ catalyst]{2\ CH_3OH} CH_3\underset{OCH_3}{\overset{OCH_3}{C}}CH_2CH_2\overset{O}{\overset{\|}{C}}OCH_3$$

Methyl 4-oxopentanoate

$$\xrightarrow[2.\ H_2O]{1.\ LiAlH_4} CH_3\underset{OCH_3}{\overset{OCH_3}{C}}CH_2CH_2CH_2OH \xrightarrow{H_3O^+} 2\ CH_3OH + CH_3\overset{O}{\overset{\|}{C}}CH_2CH_2CH_2OH$$

5-Hydroxy-2-pentanone

PRACTICE PROBLEM 9.3

What product would you obtain from the acid-catalyzed reaction of 2-methylcyclopentanone with methanol?

Solution Replace the oxygen of the ketone with two $-OCH_3$ groups from the alcohol:

$$\text{2-methylcyclopentanone} \xrightarrow[H^+\ catalyst]{CH_3OH} \text{1,1-dimethoxy-2-methylcyclopentane} + H_2O$$

PROBLEM

9.13 What product would you expect from the acid-catalyzed reaction of cyclohexanone and ethanol?

PROBLEM

9.14 Show the mechanism of the acid-catalyzed formation of a *cyclic* acetal from ethylene glycol and acetone.

$$H_3C-C(=O)-CH_3 \xrightarrow[\text{H}^+ \text{ catalyst}]{HOCH_2CH_2OH} \text{(cyclic acetal: } (H_3C)(CH_3)C \text{ in O–CH}_2\text{CH}_2\text{–O ring)} + H_2O$$

PROBLEM

9.15 Show how you might carry out the following transformation. (A protection step is needed.)

$$HC(=O)CH_2CH_2C(=O)OCH_3 \longrightarrow HC(=O)CH_2CH_2CH_2OH$$

9.9 Nucleophilic Addition of Amines: Imine Formation

Ammonia and primary amines, $R'NH_2$, add to aldehydes and ketones to yield **imines, $R_2C{=}NR'$.** Imines are formed by addition to the carbonyl group of the nucleophilic amine, followed by loss of water from the amino alcohol addition product.

$$C{=}O \xrightarrow{:NH_2R} \left[C(OH)(NHR) \right] \longrightarrow C{=}N{-}R + H_2O$$

A ketone or aldehyde — Amino alcohol intermediate — An imine

Imine derivatives, such as *oximes* and *2,4-dinitrophenylhydrazones* (abbreviated as 2,4-DNP's), are also easily prepared by reaction of a ketone or aldehyde with the appropriate $H_2N–Y$ compound. Such imines, which are usually crystalline, easy-to-handle materials, are often prepared as a means of converting liquid aldehydes or ketones into solid derivatives.

Oxime

$$C_6H_{10}{=}O + NH_2OH \longrightarrow C_6H_{10}{=}N{-}OH + H_2O$$

Cyclohexanone — **Hydroxylamine** — **Cyclohexanone oxime (mp 90°C)**

2,4-Dinitrophenylhydrazone

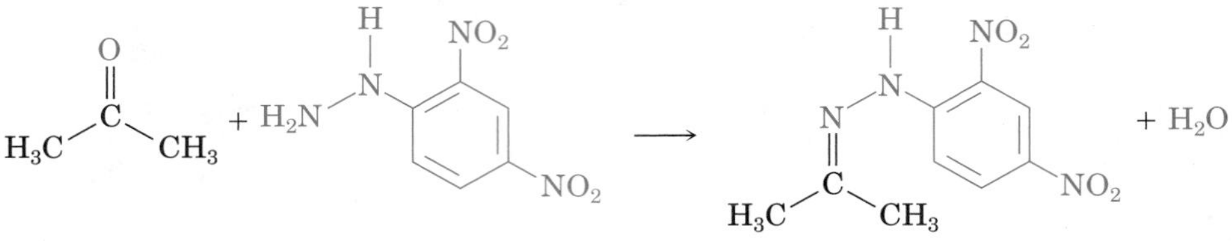

$$H_3C\text{–}C(=O)\text{–}CH_3 + H_2N\text{–}NH\text{–}C_6H_3(NO_2)_2 \longrightarrow (H_3C)(CH_3)C{=}N\text{–}NH\text{–}C_6H_3(NO_2)_2 + H_2O$$

Acetone **2,4-Dinitrophenylhydrazine** **Acetone 2,4-dinitrophenylhydrazone (mp 126°C)**

PRACTICE PROBLEM 9.4

What product do you expect from the reaction of 2-butanone with hydroxylamine, NH_2OH?

Solution Take oxygen from the ketone and two hydrogens from the amine to form water, and then join the fragments that remain:

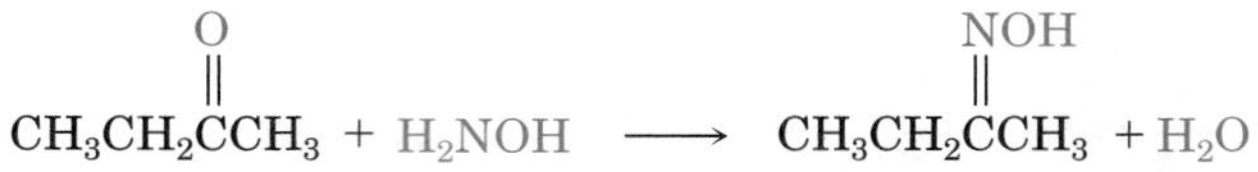

$$CH_3CH_2\overset{\overset{O}{\|}}{C}CH_3 + H_2NOH \longrightarrow CH_3CH_2\overset{\overset{NOH}{\|}}{C}CH_3 + H_2O$$

PROBLEM

9.16 Write the products you would obtain from treatment of cyclohexanone with the following reagents:
(a) NH_2OH (b) 2,4-Dinitrophenylhydrazine (c) $NaBH_4$

9.10 Nucleophilic Addition of Grignard Reagents: Alcohol Formation

Grignard reagents, RMgX, react with aldehydes and ketones to yield alcohols in the same way that $NaBH_4$ does (Section 8.5):

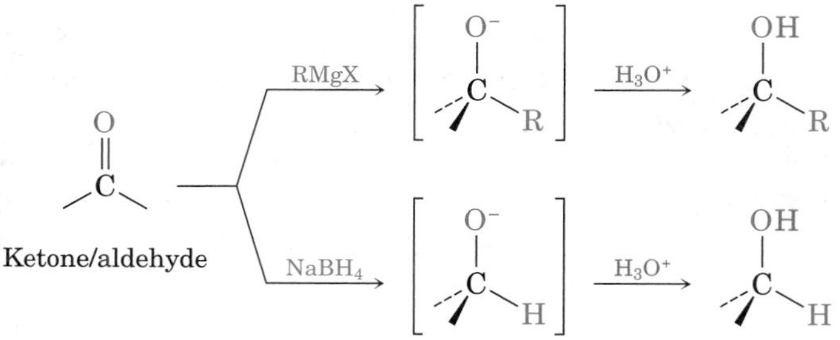

$$\text{Ketone/aldehyde } \left(\gt C{=}O\right) \xrightarrow{RMgX} \left[\gt C(O^-)R\right] \xrightarrow{H_3O^+} \gt C(OH)R$$

$$\text{Ketone/aldehyde } \left(\gt C{=}O\right) \xrightarrow{NaBH_4} \left[\gt C(O^-)H\right] \xrightarrow{H_3O^+} \gt C(OH)H$$

As we saw in Section 7.4, Grignard reagents are prepared by reaction of alkyl, aryl, or vinylic halides with magnesium metal in ether solvent:

$$R\text{—}X \xrightarrow[\text{Ether}]{Mg} \overset{\delta-}{R}\text{—}\overset{\delta+}{MgX}$$

An organohalide A Grignard reagent

The carbon–magnesium bond of a Grignard reagent is polarized so that the carbon atom is both nucleophilic and basic. Grignard reagents therefore react as though they were carbanions, $:R^-$, and they undergo nucleophilic addition to aldehydes and ketones just as water and alcohols do. Unlike the addition of water and alcohols, though, the nucleophilic addition of a Grignard reagent is irreversible. The reaction first produces a tetrahedrally hybridized magnesium alkoxide intermediate, which is then protonated to yield the neutral alcohol on treatment with aqueous acid.

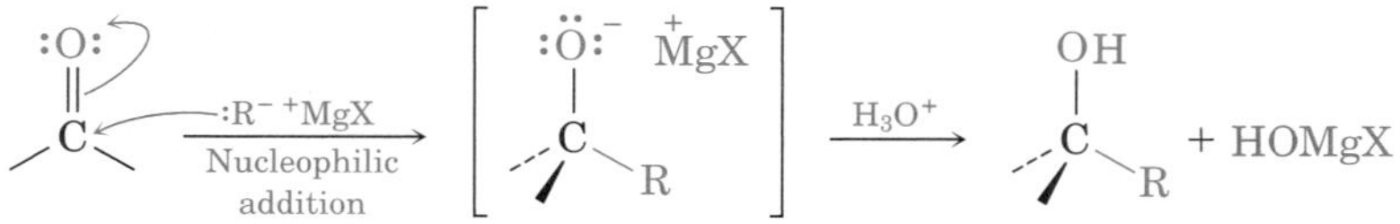

$:R^- \; ^+MgX$ (Nucleophilic addition); H_3O^+; + HOMgX

Carbonyl — Tetrahedral intermediate — Alcohol

A great many alcohols can be obtained from Grignard reactions, depending on the reagents used. For example, formaldehyde, CH_2O, reacts with Grignard reagents to give primary alcohols, RCH_2OH; aldehydes react with Grignard reagents to give secondary alcohols; and ketones react similarly to yield tertiary alcohols:

Formaldehyde reaction

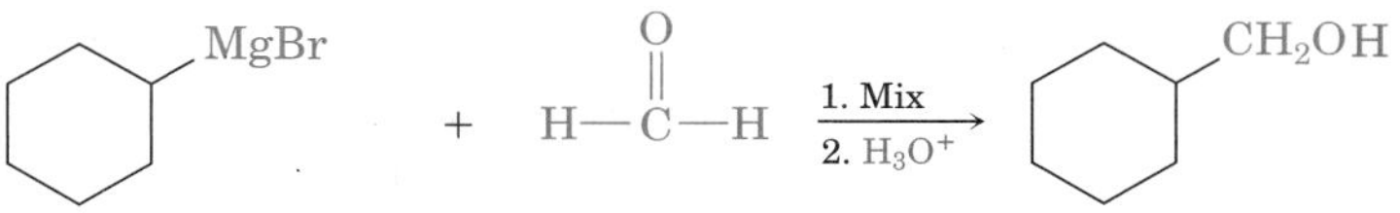

1. Mix 2. H_3O^+

Cyclohexylmagnesium bromide + **Formaldehyde** → **Cyclohexylmethanol (65%) (a 1° alcohol)**

Aldehyde reaction

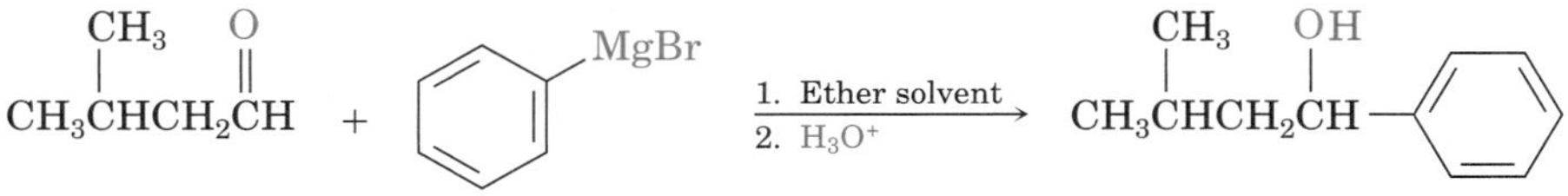

$CH_3CH(CH_3)CH_2CHO$ + PhMgBr; 1. Ether solvent 2. H_3O^+ → $CH_3CH(CH_3)CH_2CH(OH)Ph$

3-Methylbutanal + **Phenylmagnesium bromide** → **3-Methyl-1-phenyl-1-butanol (73%) (a 2° alcohol)**

Ketone reaction

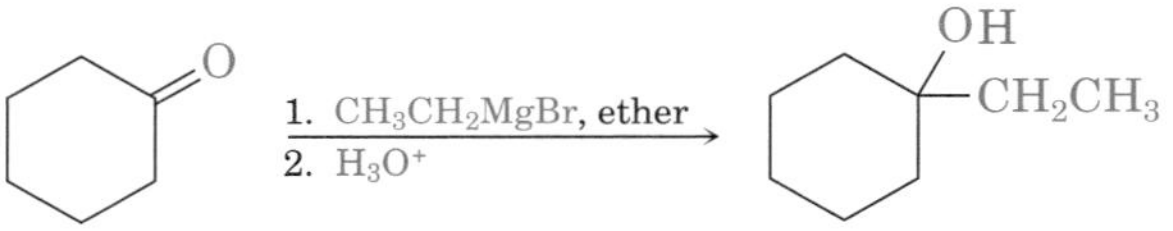

1. CH_3CH_2MgBr, ether 2. H_3O^+

Cyclohexanone → **1-Ethylcyclohexanol (89%) (a 3° alcohol)**

Although useful, the Grignard reaction also has limitations. For example, Grignard reagents can't be prepared from organohalides if there are other reactive functional groups in the same molecule. A compound that is both an alkyl halide and a ketone won't form a Grignard reagent—it reacts with itself instead. Similarly, a compound that is both an alkyl halide and a carboxylic acid, alcohol, or amine can't form a Grignard reagent because the acidic RCOO**H**, RO**H**, or RN**H**$_2$ protons in the molecule simply react with the basic Grignard reagent as it's formed. In general, Grignard reagents can't be prepared from compounds that have the following functional groups in the molecule:

Br—(Molecule)—FG

where FG = —OH, —NH, —SH, —COOH — The Grignard reagent is protonated by these groups.

FG = —CH(=O), —CR(=O), —CNR$_2$(=O), —C≡N, —NO$_2$, —SO$_2$R — The Grignard reagent adds to these groups.

PRACTICE PROBLEM 9.5

How can you use the addition of a Grignard reagent to a ketone to synthesize 2-phenyl-2-propanol?

Solution Draw the structure of the product, and identify the groups bonded to the alcohol carbon atom. In this instance, there are two methyl groups (–CH_3) and one phenyl (–C_6H_5). One of the three must come from a Grignard reagent, and the remaining two must come from a ketone. Thus, the possibilities are addition of CH_3MgBr to acetophenone and addition of C_6H_5MgBr to acetone:

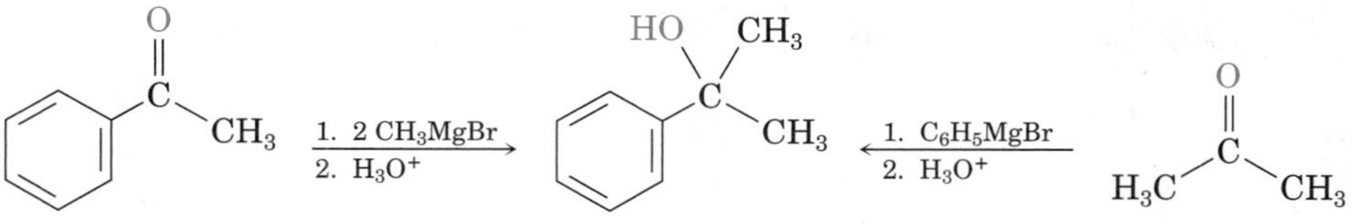

Acetophenone —(1. 2 CH_3MgBr; 2. H_3O^+)→ 2-Phenyl-2-propanol ←(1. C_6H_5MgBr; 2. H_3O^+)— Acetone

Acetophenone **2-Phenyl-2-propanol** **Acetone**

PROBLEM

9.17 Show the products obtained from addition of CH_3MgBr to the following compounds:
(a) Cyclopentanone (b) Benzophenone (diphenyl ketone)
(c) 3-Hexanone

PROBLEM

9.18 How might you use a Grignard addition reaction to prepare the following alcohols?
(a) 2-Methyl-2-propanol (b) 1-Methylcyclohexanol
(c) 3-Methyl-3-pentanol

9.11 Some Biological Nucleophilic Addition Reactions

We'll see in Chapter 17 that living organisms synthesize the molecules of life using many of the same reactions that chemists use in the laboratory. This is particularly true of carbonyl-group reactions, where nucleophilic addition steps are an important part of the biological synthesis of many vital molecules. For example, one of the pathways by which amino acids are made involves a nucleophilic addition reaction of ammonia to α-keto acids. To choose a specific case, the bacterium *Bacillus subtilis* synthesizes alanine from pyruvic acid and ammonia. The key step is the nucleophilic addition of ammonia to the ketone carbonyl group of pyruvic acid to give an imine that is further reduced by enzymes.

$$CH_3C(=O)COOH + :NH_3 \longrightarrow \left[CH_3C(=NH)COOH \right] \xrightarrow[\text{enzyme}]{\text{Reducing}} CH_3CH(NH_2)COOH$$

Pyruvic acid — An imine — **Alanine**

Other examples of nucleophilic carbonyl additions occur frequently in carbohydrate chemistry. For example, the six-carbon sugar glucose reacts as if it were an aldehyde. It can, for example, be oxidized to yield a carboxylic acid. Spectroscopic examination of glucose shows, however, that no aldehyde group is present; instead, glucose exists as a *cyclic hemiacetal.* The hydroxyl group at carbon 5 adds to the aldehyde at carbon 1 in an internal nucleophilic addition step.

Glucose (open form) ⇌ **Glucose** (hemiacetal form)

Further reaction between molecules of glucose leads to the carbohydrate polymer *cellulose,* which constitutes the major building block of plant cell walls. Cellulose consists simply of glucose units joined by acetal bonds between carbon 1 of one glucose and the –OH group at carbon 4 of another glucose, as shown at the top of the next page.

Glucose $\Longrightarrow$ **Cellulose** (Acetal)

We'll study this and other reactions of carbohydrates in Chapter 14.

INTERLUDE

Insect Antifeedants

Approximately 15% of the world's crops are lost each year to insects.

It has been estimated by the World Health Organization that approximately 15% of the world's crops are lost to insects each year and that more than 2 billion dollars is spent each year on crop insecticides. Though remarkably effective, the powerful, broad-spectrum insecticides in current use are far from an ideal solution for insect control because of their potential toxicity to animals and their lack of selectivity. Beneficial as well as harmful insects are killed indiscriminately.

One new approach to insect control, which promises to be more selective and less ecologically harmful than present-day insecticides, is the use of *insect antifeedants,* substances that prevent an insect from eating but do not kill it directly. The idea of using antifeedants came from the observations of chemical ecologists that many plants have evolved elaborate and sophisticated chemical defenses against insects. Among these defenses, some plants contain chemicals that appear to block the ability or desire of an insect to feed. The insect often remains nearby, where it dies of starvation.

Most naturally occurring antifeedants are relatively complex molecules. Polygodial, for instance, is a dialdehyde active against African army worms, and ajugarin I shows activity against locusts. Both polygodial and

(continued)➤

ajugarin I are probably too complex to be manufactured economically, but recent research has discovered a number of substances that are structurally simpler than naturally occurring antifeedants yet retain potent activity. Among these substances synthesized in the laboratory is the acetal shown below.

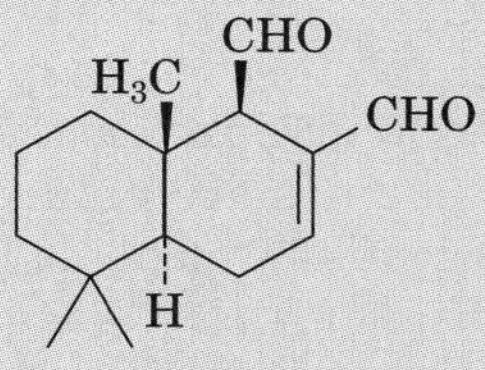

Polygodial

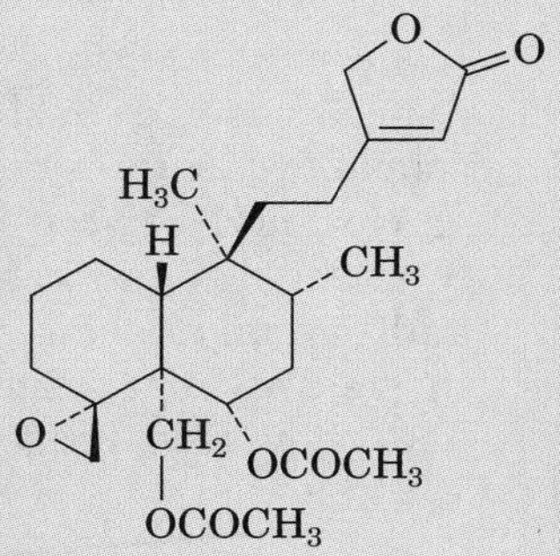

Ajugarin I

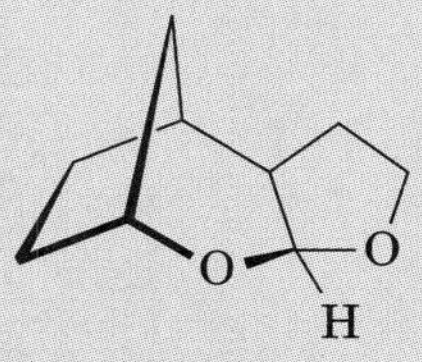

Synthetic compound

Summary and Key Words

acetal, 293
acyl group, 282
carbanion, 288
carbonyl group, 281
geminal, 290
gem diol, 290
hemiacetal, 293
imine, 295
nucleophilic addition reaction, 288
protecting group, 293
Tollens' reagent, 287

Carbonyl compounds can be classified into two general categories:

RCHO
R_2CO

Aldehydes and **ketones** are similar in their reactivity and are distinguished by the fact that the substituents on the **acyl** carbon can't act as leaving groups.

RCOOH
RCOCl
RCOOCOR′
RCOOR′
$RCONH_2$

Carboxylic acids and their derivatives—**acid chlorides, acid anhydrides, esters,** and **amides**—are distinguished by the fact that the substituents on the acyl carbon *can* act as leaving groups.

A carbon–oxygen double bond is structurally similar to a carbon–carbon double bond. The carbonyl carbon atom is sp^2-hybridized and forms both an sp^2 σ bond and a p π bond to oxygen. Carbonyl groups are strongly polarized because of the electronegativity of oxygen.

Aldehydes are usually prepared by oxidation of primary alcohols. Ketones are similarly prepared by oxidation of secondary alcohols.

Aldehydes and ketones behave similarly in much of their chemistry, though aldehydes are generally more reactive than ketones. Both undergo **nucleophilic addition reactions,** and a variety of products can be prepared. For example, aldehydes and ketones undergo a reversible addition reaction with water to yield 1,1-dialcohols, or **gem diols.** Similarly, they react with alcohols to yield **acetals, $R_2C(OR')_2$,** which are valuable as carbonyl **protecting groups.** Primary amines add to aldehydes and ketones to give **imines, $R_2C{=}NR'$.** In addition, aldehydes and ketones are reduced by $NaBH_4$ to yield primary and secondary alcohols, respectively, and they also react with Grignard reagents to give alcohols.

Summary of Reactions

1. Reaction of ketones and aldehydes with alcohols to yield acetals (Section 9.8)

$$R_2C{=}O + ROH \underset{\text{catalyst}}{\overset{H^+}{\rightleftharpoons}} \left[R_2C(OH)(OR) \right] \rightleftharpoons R_2C(OR)(OR) + H_2O$$

2. Reaction of ketones and aldehydes with amines to yield imines (Section 9.9)

$$R_2C{=}O + RNH_2 \longrightarrow R_2C{=}NR + H_2O$$

3. Reaction of ketones and aldehydes with Grignard reagents to yield alcohols (Section 9.10)

$$R_2C{=}O \xrightarrow[\text{2. } H_3O^+]{\text{1. RMgX}} R_2C(OH)R$$

ADDITIONAL PROBLEMS

9.19 Identify the different kinds of carbonyl groups in the following molecules:

(a) CO_2H O OCCH$_3$

Aspirin

(b) N CH_3 CO_2CH_3 H $OCOC_6H_5$ H

Cocaine

(c) CH_2OH CHOH O O O OH

Ascorbic acid (vitamin C)

9.20 What is the structural difference between an aldehyde and a ketone?

9.21 Draw structures corresponding to the following names:

(a) Bromoacetone
(b) 3-Methyl-2-butanone
(c) 3,5-Dinitrobenzaldehyde
(d) 3,5-Dimethylcyclohexanone
(e) 2,2,4,4-Tetramethyl-3-pentanone
(f) Butanedial
(g) (*S*)-2-Hydroxypropanal
(h) 3-Phenyl-2-propenal

9.22 Draw and name the seven aldehydes and ketones with the formula $C_5H_{10}O$.

9.23 Which of the compounds you identified in Problem 9.22 are chiral?

9.24 Draw structures of molecules that meet the following descriptions:

(a) A cyclic ketone, C_6H_8O
(b) A diketone, $C_6H_{10}O_2$
(c) An aryl ketone, $C_9H_{10}O$
(d) A 2-bromoaldehyde, C_5H_9BrO

9.25 Give IUPAC names for the following structures:

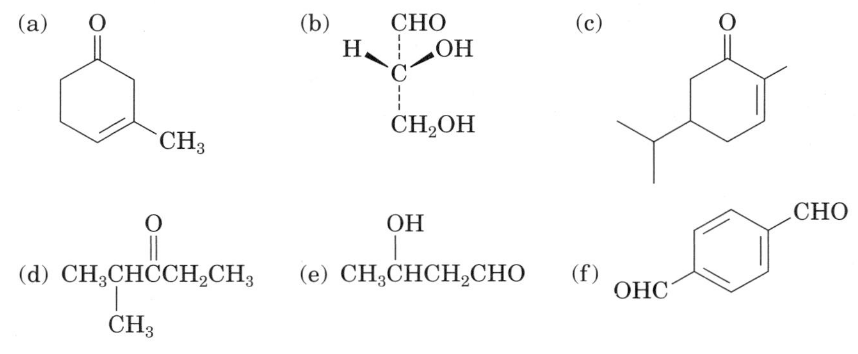

9.26 Give an example of each of the following:

(a) An acetal
(b) A gem diol
(c) An oxime
(d) An imine
(e) A hemiacetal

9.27 Predict the products of the reaction of phenylacetaldehyde, $C_6H_5CH_2CHO$, with the following reagents:

(a) $NaBH_4$, then H_3O^+ (b) Tollens' reagent (c) NH_2OH
(d) CH_3MgBr, then H_3O^+ (e) CH_3OH, H^+ catalyst

9.28 Answer Problem 9.27 for the reaction of acetophenone, $C_6H_5COCH_3$, with the same reagents.

9.29 Identify the nucleophile that has added to acetone to give the following products:

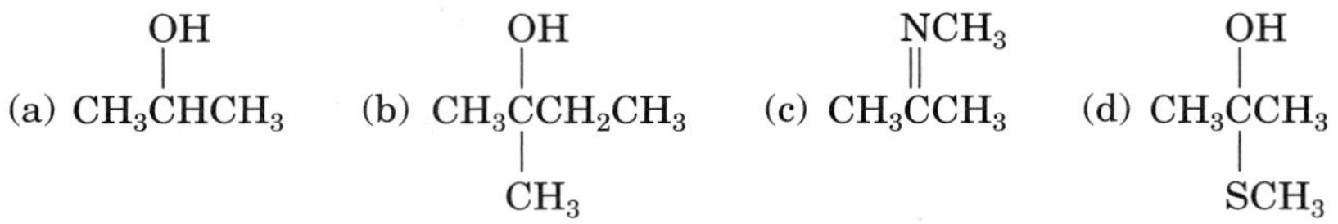

(a) $CH_3CH(OH)CH_3$ (b) $CH_3C(OH)(CH_3)CH_2CH_3$ (c) $CH_3C(=NCH_3)CH_3$ (d) $CH_3C(OH)(SCH_3)CH_3$

9.30 Reaction of 2-butanone with HCN yields a *cyanohydrin* product [$R_2C(OH)CN$] having a new stereocenter. What stereochemistry would you expect the product to have? (Review Section 6.13.)

9.31 In light of your answer to Problem 9.30, what stereochemistry would you expect the product from the reaction of phenylmagnesium bromide with 2-butanone to have?

9.32 Starting from 2-cyclohexenone and any other reagents needed, how would you prepare the following substances? (More than one step may be required.)

(a) 1,3-Cyclohexadiene (b) 1-Methylcyclohexanol
(c) Cyclohexanol (d) 1-Phenyl-2-cyclohexen-1-ol

9.33 How can you explain the observation that the S_N2 reaction of (dibromomethyl)-benzene, $C_6H_5CHBr_2$, with NaOH yields benzaldehyde rather than (dihydroxy-methyl)benzene, $C_6H_5CH(OH)_2$?

9.34 Use a Grignard reaction on an aldehyde or ketone to synthesize the following compounds:

(a) 2-Pentanol (b) 2-Phenyl-2-butanol
(c) 1-Ethylcyclohexanol (d) Diphenylmethanol

9.35 Show the products that result from the reaction of phenylmagnesium bromide with the following reagents:

(a) CH_2O (b) Benzophenone ($C_6H_5COC_6H_5$) (c) 3-Pentanone

9.36 Show how you could make the following alcohols using a Grignard reaction:

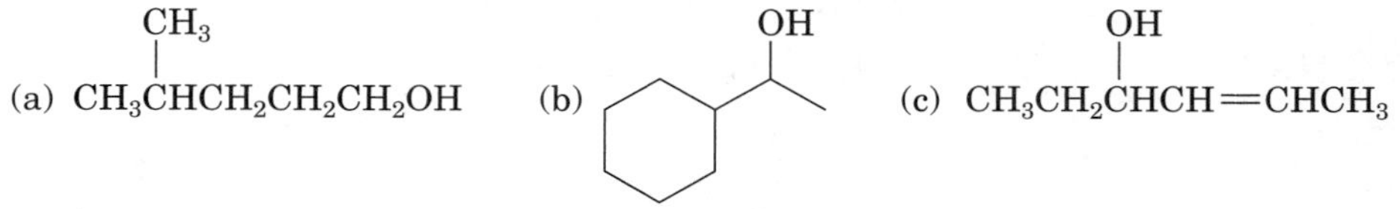

(a) $CH_3CH(CH_3)CH_2CH_2CH_2OH$ (b) [cyclohexyl–CH(OH)–CH$_3$ structure] (c) $CH_3CH_2CH(OH)CH{=}CHCH_3$

9.37 How could you convert bromobenzene into benzoic acid, C_6H_5COOH? (More than one step is required.)

9.38 Show the structures of the intermediate hemiacetals and the final acetals that result from the following reactions:

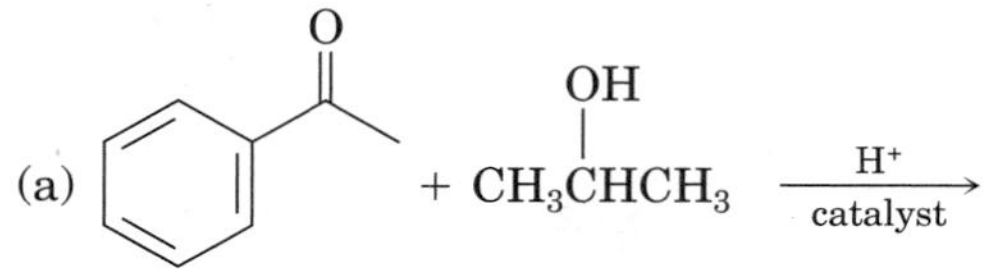

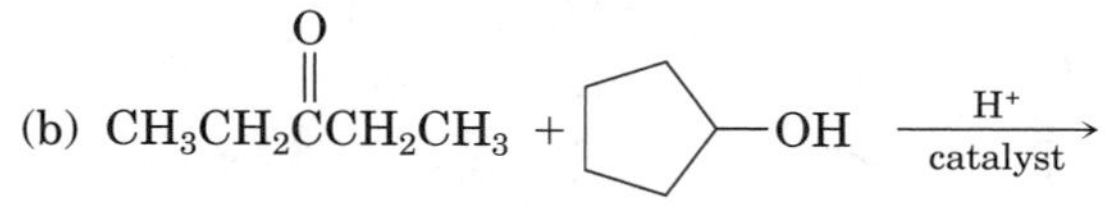

(a) [acetophenone, $C_6H_5COCH_3$] + $CH_3CH(OH)CH_3$ $\xrightarrow[\text{catalyst}]{H^+}$

(b) $CH_3CH_2C(=O)CH_2CH_3$ + [cyclopentyl]–OH $\xrightarrow[\text{catalyst}]{H^+}$

9.39 Show the structures of the alcohols and aldehydes or ketones you would use to make the following acetals:

(a) $CH_3CH_2CH(CH_3)CH_2CH(OCH_3)OCH_3$ (b) CH_3CH_2O OCH_2CH_3 (c)

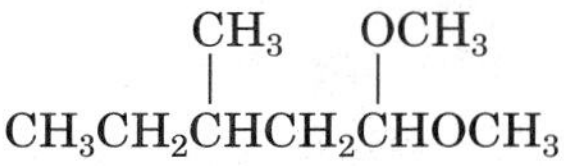

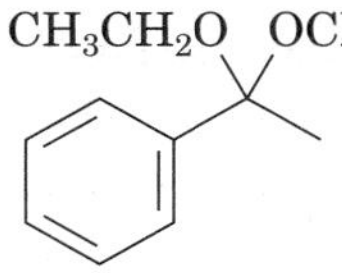

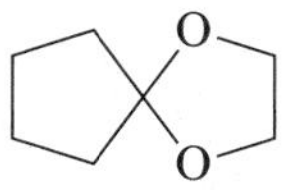

9.40 Show the products from the reaction of 2-pentanone with the following reagents:

(a) NH_2OH (b) $NHNH_2$, O_2N, NO_2 (c) CH_3CH_2OH, H^+

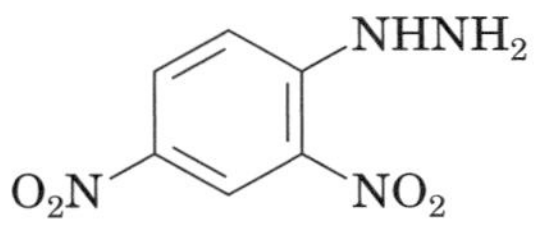

9.41 How would you synthesize the following compounds from cyclohexanone?
(a) 1-Methylcyclohexene (b) *cis*-1,2-Cyclohexanediol
(c) 1-Bromo-1-methylcyclohexane (d) 1-Cyclohexylcyclohexanol

9.42 How can you explain the observation that treatment of 4-hydroxycyclohexanone with 1 equiv of CH_3MgBr yields none of the expected addition product, whereas treatment with an excess of the Grignard reagent leads to a good yield of 1-methyl-1,4-cyclohexanediol?

9.43 Carvone is the major constituent of spearmint oil. What products would you expect from the reaction of carvone with the following reagents?
(a) $LiAlH_4$, then H_3O^+ (b) C_6H_5MgBr, then H_3O^+
(c) H_2, Pd catalyst (d) CH_3OH, H^+

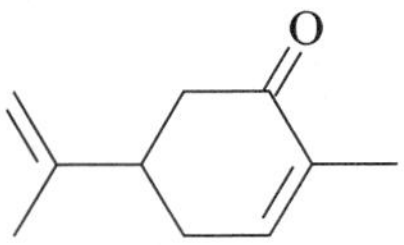

O

Carvone

9.44 When 4-hydroxybutanal is treated with methanol in the presence of an acid catalyst, 2-methoxytetrahydrofuran is obtained. Propose a mechanism to account for this result.

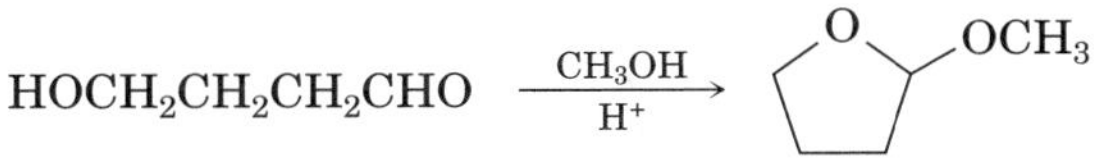

$$HOCH_2CH_2CH_2CHO \xrightarrow[H^+]{CH_3OH}$$ O OCH_3

9.45 Using your knowledge of the reactivity differences between aldehydes and ketones, show how the following two selective reductions might be carried out. One of the schemes requires a protection step.

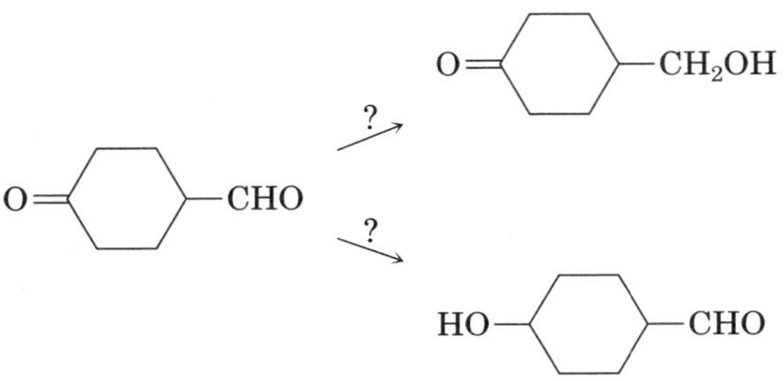

O= CH_2OH

? ?

O= CHO

HO CHO

9.46 Treatment of a ketone or aldehyde with a thiol in the presence of an acid catalyst yields a *thioacetal,* $R_2C(SR')_2$. To what other reaction is this thioacetal formation analogous? Propose a mechanism for the reaction.

9.47 Treatment of a ketone or aldehyde with hydrazine, H_2NNH_2, yields an *azine,* $R_2C{=}N{-}N{=}CR_2$. Propose a mechanism for the reaction.

9.48 When glucose is treated with $NaBH_4$, reaction occurs to yield *sorbitol,* a commonly used food additive. Show how this reduction occurs

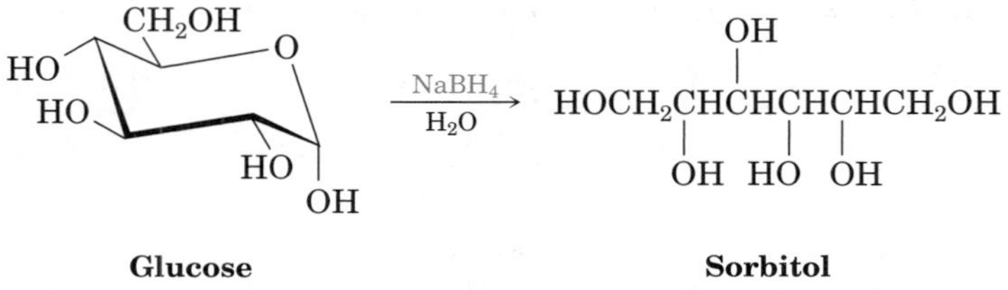

$$\text{Glucose} \xrightarrow[H_2O]{NaBH_4} HOCH_2CH(OH)CH(OH)CH(OH)CH(OH)CH_2OH \ (\text{Sorbitol})$$

Glucose **Sorbitol**

9.49 Ketones react with dimethylsulfonium methylide to yield epoxides by a mechanism that involves an initial nucleophilic addition followed by an intramolecular S_N2 substitution. Show the mechanism.

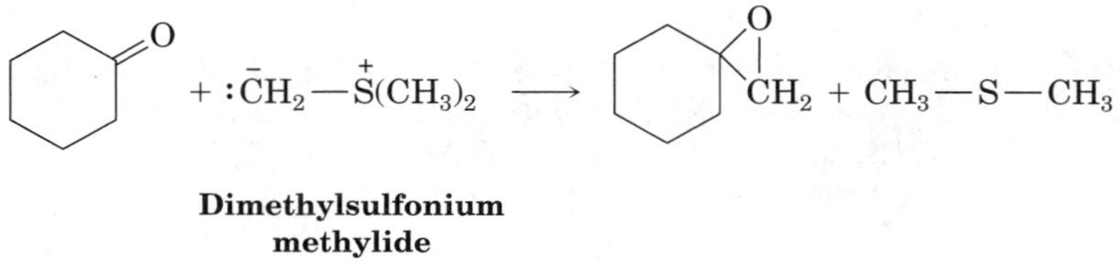

$$\text{Cyclohexanone} + :\overset{-}{C}H_2{-}\overset{+}{S}(CH_3)_2 \longrightarrow \text{epoxide } (CH_2) + CH_3{-}S{-}CH_3$$

Dimethylsulfonium methylide

9.50 How would you prepare tamoxifen, a drug used in the treatment of breast cancer, from benzene, the following ketone, and any other reagents needed?

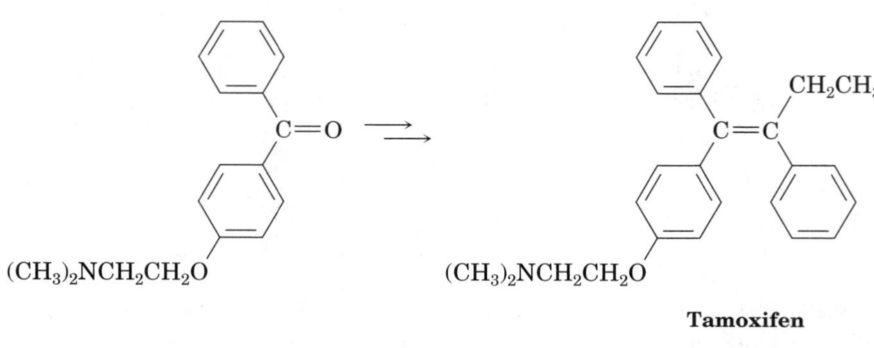

$(CH_3)_2NCH_2CH_2O$–C₆H₄–C(=O)–C₆H₅ $\xrightarrow{?}$ $(CH_3)_2NCH_2CH_2O$–C₆H₄–C(C₆H₅)=C(C₆H₅)CH_2CH_3

Tamoxifen

9.51 Identify the reagents a–d in the following scheme:

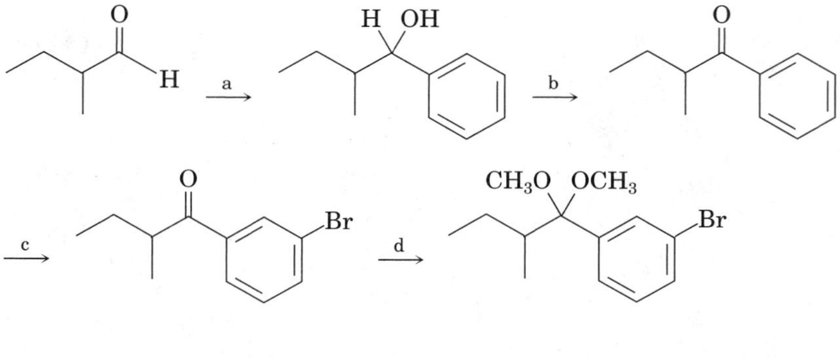

2-methylbutanal $\xrightarrow{a}$ H OH (secondary alcohol with phenyl) $\xrightarrow{b}$ ketone (phenyl) $\xrightarrow{c}$ ketone (m-Br phenyl) $\xrightarrow{d}$ CH_3O OCH_3 acetal (m-Br phenyl)

Visualizing Chemistry

9.52 Identify the kinds of carbonyl groups in the following molecules (gray = C, red = O, blue = N, light green = H).

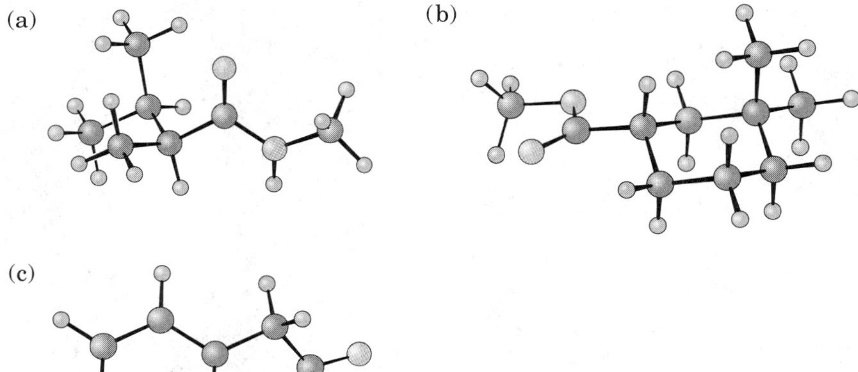

9.53 Identify the reactants from which the following molecules were prepared. If an acetal, identify the carbonyl compound and the alcohol; if an imine, identify the carbonyl compound and the amine; and if an alcohol, identify the carbonyl compound and the Grignard reagent (gray = C, red = O, blue = N, light green = H).

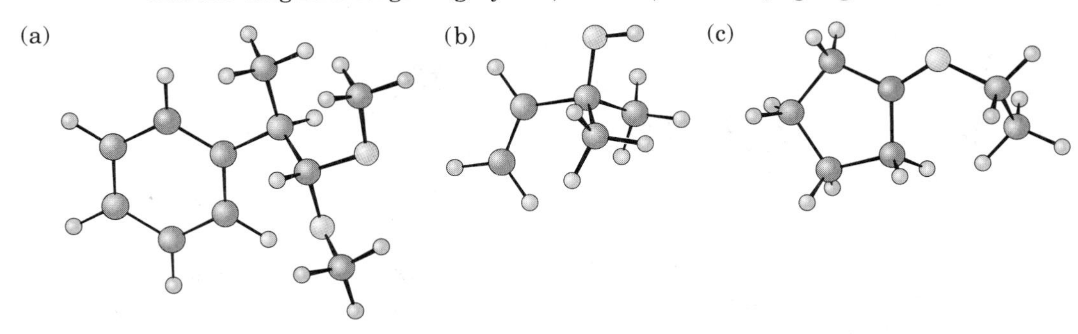

9.54 Compounds called *cyanohydrins* result from the nucleophilic addition of HCN to an aldehyde or ketone. Draw and name the carbonyl compound that the following cyanohydrin was prepared from (gray = C, red = O, blue = N, light green = H).

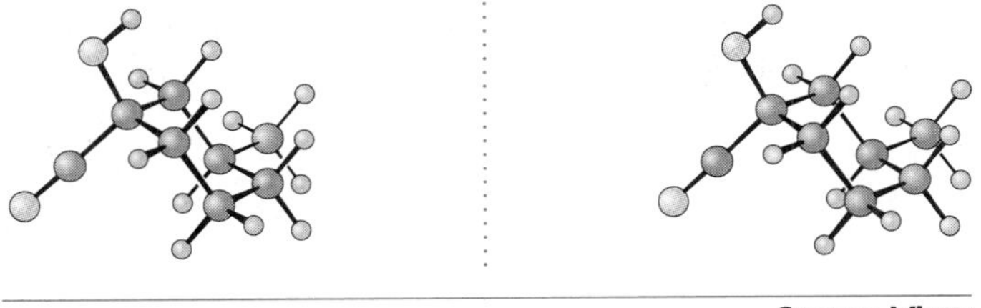

Stereo View

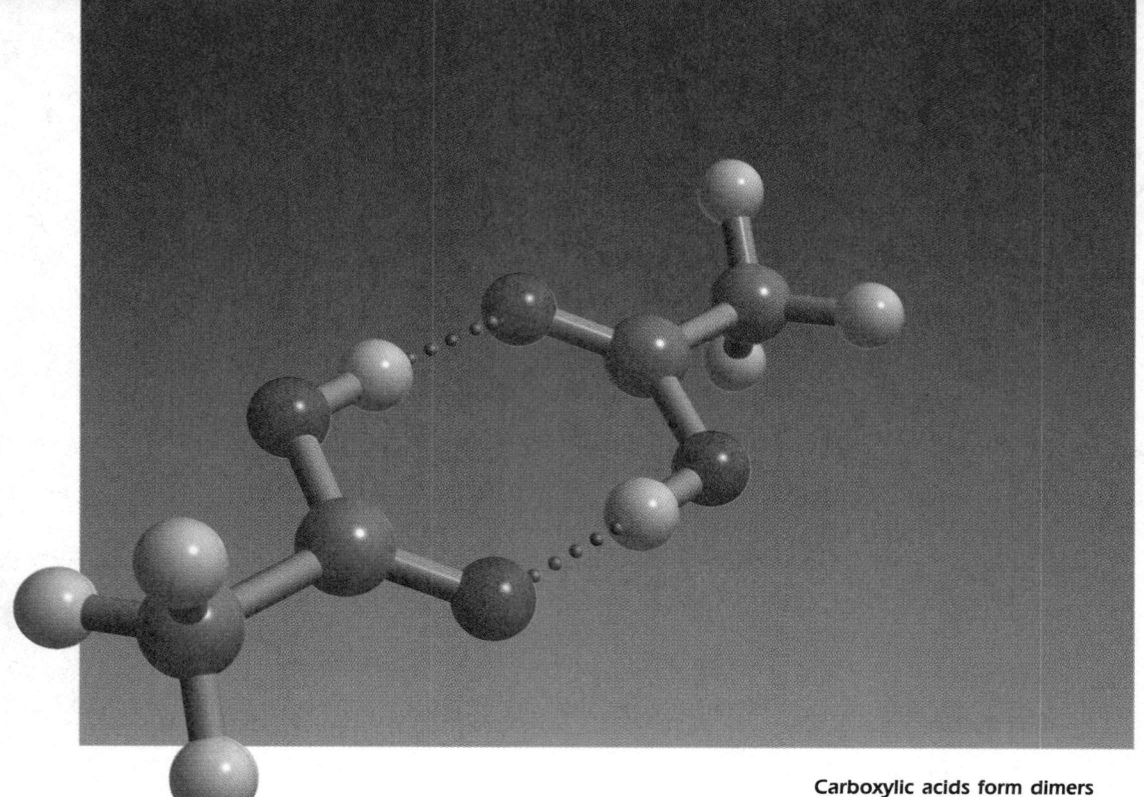

Carboxylic acids form dimers that are held together by hydrogen bonds.

10 Carboxylic Acids and Derivatives

Carboxylic acids and their derivatives are carbonyl compounds in which the acyl group is bonded to an electronegative atom such as oxygen, halogen, nitrogen, or sulfur. Although there are many different kinds of carboxylic acid derivatives, we'll be concerned only with four of the most common types in addition to the acids themselves: acid halides, acid anhydrides, esters, and amides. In contrast to aldehydes and ketones, these compounds contain an acyl group bonded to a substituent that can act as a leaving group in substitution reactions. Also in this chapter, we'll discuss *nitriles,* a class of compounds closely related to carboxylic acids.

$R{-}C({=}O){-}OH$

Carboxylic acid

$R{-}C({=}O){-}X$

Acid halide (X = F, Cl, Br, I)

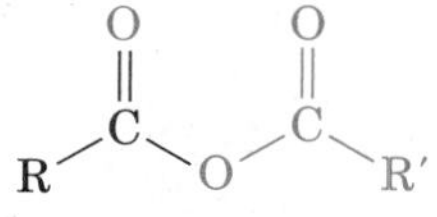

Acid anhydride

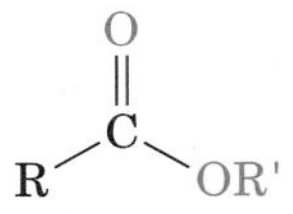

$RCOOR'$ (R–C(=O)–OR')

Ester

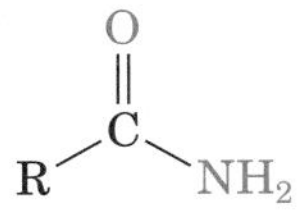

$RCONH_2$ (R–C(=O)–NH$_2$)

Amide

$R{-}C{\equiv}N$

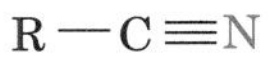

Nitrile

10.1 Naming Carboxylic Acids and Derivatives

Carboxylic Acids: RCOOH

Simple open-chain carboxylic acids are named by replacing the terminal *-e* of the alkane name with *-oic acid.* The –COOH carbon (the **carboxyl** carbon) is always numbered C1.

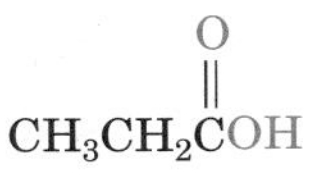

CH_3CH_2COH (C=O)

Propanoic acid

$CH_3CH(CH_3)CH_2CH_2COH$ (C=O), numbered 5 4 3 2 1

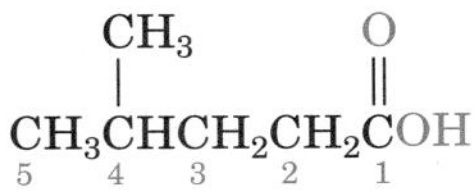

4-Methylpentanoic acid

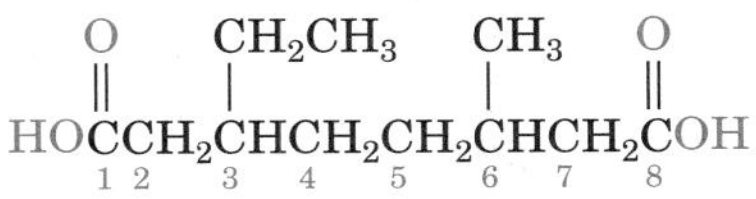

$HOC(=O)CH_2CH(CH_2CH_3)CH_2CH_2CH(CH_3)CH_2C(=O)OH$, numbered 1 2 3 4 5 6 7 8

3-Ethyl-6-methyloctanedioic acid

Alternatively, compounds that have a –COOH group bonded to a ring are named by using the suffix *-carboxylic acid.* In this alternate system, the carboxylic acid carbon is *attached to* C1 on the ring but is not itself numbered.

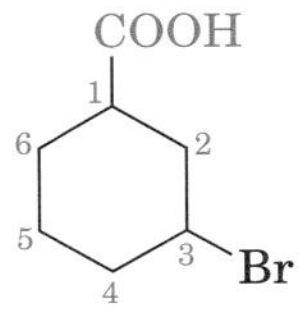

COOH on C1 of a cyclohexane ring (numbered 1–6), Br on C3

3-Bromocyclohexanecarboxylic acid

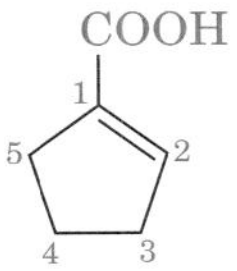

COOH on C1 of a cyclopentene ring (numbered 1–5), C1=C2 double bond

1-Cyclopentenecarboxylic acid

Because many carboxylic acids were among the first organic compounds to be isolated and purified, there are a large number of acids with common names (Table 10.1, p. 310). We'll use systematic names in this book, with the exception of formic (methanoic) acid, HCOOH, and acetic (ethanoic) acid, CH_3COOH, whose names are so well known that it makes little sense to refer to them in any other way.

PROBLEM

10.1 Give IUPAC names for compounds (a)–(e):

(a) 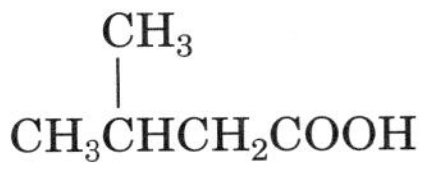 $CH_3CH(CH_3)CH_2COOH$

(b)  $CH_3CH(Br)CH_2CH_2COOH$

Table 10.1 Some Common Names of Carboxylic Acids and Acyl Groups

Carboxylic acid		*Acyl group*	
Structure	*Name*	*Name*	*Structure*
HCOOH	Formic	Formyl	HCO–
CH_3COOH	Acetic	Acetyl	CH_3CO–
CH_3CH_2COOH	Propionic	Propionyl	CH_3CH_2CO–
$CH_3CH_2CH_2COOH$	Butyric	Butyryl	$CH_3(CH_2)_2CO$–
HOOCCOOH	Oxalic	Oxalyl	–OCCO–
$HOOCCH_2COOH$	Malonic	Malonyl	$–OCCH_2CO–$
$HOOCCH_2CH_2COOH$	Succinic	Succinyl	$–OC(CH_2)_2CO–$
$H_2C{=}CHCOOH$	Acrylic	Acryloyl	$H_2C{=}CHCO$–
C_6H_5–COOH (benzene ring with COOH)	Benzoic	Benzoyl	C_6H_5–C(=O)– (benzene ring with C(=O)–)

(c) $CH_3CH{=}CHCH_2CH_2COOH$ (d) $CH_3CH_2\overset{\displaystyle COOH}{\overset{|}{C}}HCH_2CH_2CH_3$

(e)

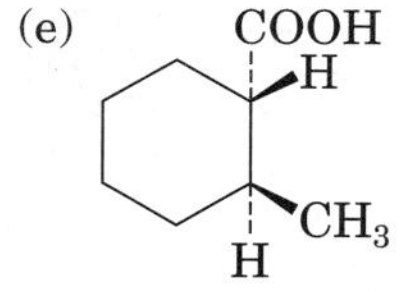

PROBLEM

10.2 Draw structures corresponding to the following names:
(a) 2,3-Dimethylhexanoic acid (b) 4-Methylpentanoic acid
(c) *o*-Hydroxybenzoic acid (d) *trans*-1,2-Cyclobutanedicarboxylic acid

Acid Halides: RCOX

Acid halides are named by identifying first the acyl group and then the halide. The acyl group name is derived from the acid name by replacing the *-ic acid* ending with *-yl,* or the *-carboxylic acid* ending with *-carbonyl.* For example

$CH_3\overset{O}{\overset{\|}{C}}Cl$ — **Acetyl chloride (from acetic acid)**

C_6H_5–C(=O)–Br — **Benzoyl bromide (from benzoic acid)**

C_6H_{11}–C(=O)–Cl — **Cyclohexanecarbonyl chloride (from cyclohexanecarboxylic acid)**

Acid Anhydrides: RCO_2COR'

Anhydrides from simple carboxylic acids and cyclic anhydrides from dicarboxylic acids are named by replacing the word *acid* with *anhydride:*

Acetic anhydride **Benzoic anhydride** **Succinic anhydride**

Amides: $RCONH_2$

Amides with an unsubstituted $-NH_2$ group are named by replacing the *-oic acid* or *-ic acid* ending with *-amide,* or by replacing the *-carboxylic acid* ending with *-carboxamide:*

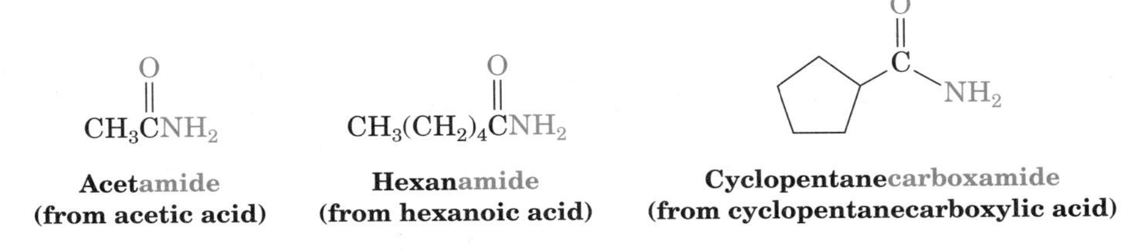

Acetamide (from acetic acid) **Hexanamide** (from hexanoic acid) **Cyclopentanecarboxamide** (from cyclopentanecarboxylic acid)

If the nitrogen atom is substituted, the amide is named by first identifying the substituent group and then the parent. The substituents are preceded by the letter *N* to identify them as being directly attached to nitrogen.

$CH_3CH_2\overset{O}{\overset{\|}{C}}NHCH_3$

***N*-Methylpropanamide** ***N,N*-Diethylcyclohexanecarboxamide**

Esters: RCO_2R'

Systematic names for esters are derived by first giving the name of the alkyl group attached to oxygen and then identifying the carboxylic acid. In so doing, the *-ic acid* ending is replaced by *-ate:*

$CH_3\overset{O}{\overset{\|}{C}}OCH_2CH_3$ $CH_3O\overset{O}{\overset{\|}{C}}CH_2\overset{O}{\overset{\|}{C}}OCH_3$

Ethyl acetate (the ethyl ester of acetic acid) **Dimethyl malonate** (the dimethyl ester of malonic acid) ***tert*-Butylcyclohexanecarboxylate** (the *tert*-butyl ester of cyclohexanecarboxylic acid)

Nitriles: R–C≡N

Compounds containing the –C≡N functional group are called **nitriles.** Simple acyclic nitriles are named by adding *-nitrile* as a suffix to the alkane name, with the nitrile carbon numbered C1.

$$\underset{5}{CH_3}\underset{4}{\overset{\overset{\displaystyle CH_3}{|}}{C}H}\underset{3}{CH_2}\underset{2}{CH_2}\underset{1}{CN}$$

4-Methylpentanenitrile

More complex nitriles are named as derivatives of carboxylic acids by replacing the *-ic acid* or *-oic acid* ending with *-onitrile,* or by replacing the *-carboxylic acid* ending with *-carbonitrile.* In this system, the nitrile carbon atom is attached to C1 but is not itself numbered:

$CH_3C{\equiv}N$

Acetonitrile (from acetic acid)

Benzonitrile (from benzoic acid)

2,2-Dimethylcyclohexanecarbonitrile (from 2,2-dimethylcyclohexanecarboxylic acid)

PROBLEM

10.3 Give IUPAC names for the following structures:

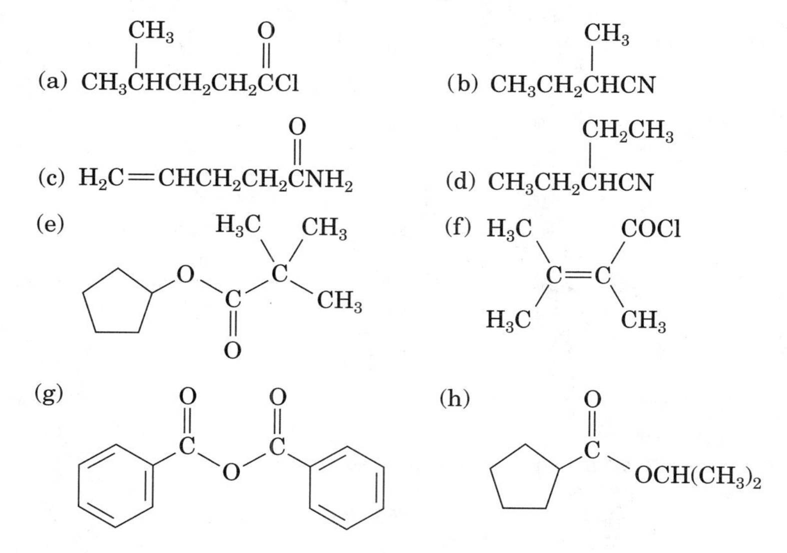

PROBLEM

10.4 Draw structures corresponding to the following names:

(a) 2,2-Dimethylpropanoyl chloride
(b) *N*-Methylbenzamide
(c) 5,5-Dimethylhexanenitrile
(d) *tert*-Butyl butanoate

(e) *trans*-2-Methylcyclohexanecarboxamide
(f) *p*-Methylbenzoic anhydride
(g) *cis*-3-Methylcyclohexanecarbonyl bromide
(h) *p*-Bromobenzonitrile

10.2 Occurrence, Structure, and Properties of Carboxylic Acids

Carboxylic acids occupy a central place among carbonyl compounds, both in nature and in the laboratory. Vinegar, for example, is a dilute solution of acetic acid, CH_3COOH; butanoic acid, $CH_3CH_2CH_2COOH$, is responsible for the rancid odor of sour butter; and hexanoic acid (caproic acid), $CH_3(CH_2)_4COOH$, is partially responsible for the unmistakable aroma of goats (Latin *caper*, "goat").

Since the carboxylic acid functional group, –COOH, is structurally similar to both ketones and alcohols, we might expect to see similar properties. As in ketones, the carbonyl carbon is sp^2-hybridized. Carboxylic acid groups are therefore planar, with C–C–O and O–C–O bond angles of approximately 120°.

Like alcohols, carboxylic acids form strong intermolecular hydrogen bonds. Most carboxylic acids, in fact, exist as *dimers* held together by two hydrogen bonds:

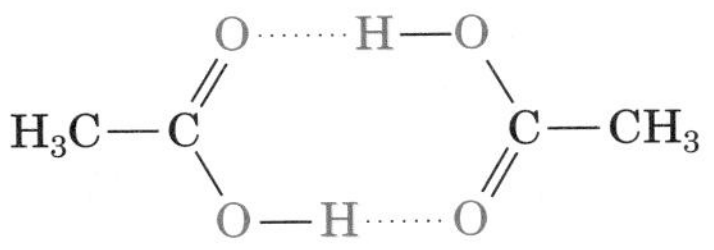

```
         O·······H—O
        //          \
H3C—C                C—CH3
        \           //
         O—H·······O
```

Acetic acid dimer

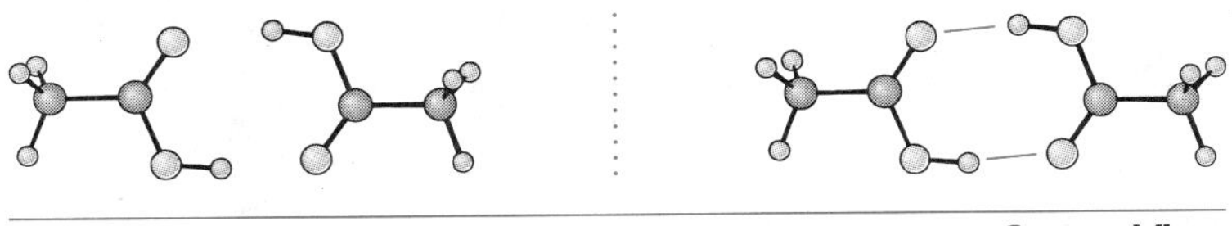

Stereo View

This strong hydrogen bonding has a noticeable effect on boiling points, and carboxylic acids normally boil at much higher temperatures than alkanes or alkyl halides of similar molecular weight. Acetic acid, for example, boils at 118°C, whereas chloropropane boils at 46.6°C.

10.3 Acidity of Carboxylic Acids

As their name implies, carboxylic acids are *acidic*. They therefore react with bases such as NaOH to give metal carboxylate salts, $RCO_2^-\ Na^+$. Although carboxylic acids with more than six carbon atoms are only slightly soluble in water, alkali metal salts of carboxylic acids are generally quite water-soluble because they are ionic. It's often possible to take advantage of this solubility to purify acids by extracting their salts into aqueous base, then reacidifying and extracting the pure acid back into an organic solvent.

$$RCOOH + NaOH \xrightarrow{H_2O} RCOO^- Na^+ + H_2O$$

A carboxylic acid (water-insoluble) — A carboxylic acid salt (water-soluble)

For most carboxylic acids, the acidity constant K_a (Section 1.12) is near 10^{-5}. Acetic acid, for example, has $K_a = 1.76 \times 10^{-5}$, which corresponds to a pK_a of 4.75. In practical terms, a K_a value near 10^{-5} means that only about 1% of the molecules in a 0.1 M aqueous solution are dissociated, as opposed to the 100% dissociation observed for strong mineral acids like HCl and H_2SO_4.

As indicated by the list of K_a values in Table 10.2, there is a considerable range in the strengths of various carboxylic acids. Trichloroacetic acid ($K_a = 0.23$), for example, is more than 12,000 times as strong as acetic acid ($K_a = 1.76 \times 10^{-5}$). How can we account for such differences?

Table 10.2 Acid Strengths of Some Carboxylic Acids

Name	K_a	pK_a	
HCl (hydrochloric acid)[a]	(10^7)	(−7)	Stronger acid
CCl_3COOH	0.23	0.64	↑
$CHCl_2COOH$	3.3×10^{-2}	1.48	
$CH_2ClCOOH$	1.4×10^{-3}	2.85	
HCOOH	1.77×10^{-4}	3.75	
C_6H_5COOH	6.46×10^{-5}	4.19	
$H_2C{=}CHCOOH$	5.6×10^{-5}	4.25	
CH_3COOH	1.76×10^{-5}	4.75	
CH_3CH_2OH (ethanol)[a]	(10^{-16})	(16)	Weaker acid

[a]Values for HCl and ethanol are shown for reference.

Because the dissociation of a carboxylic acid is an equilibrium process, anything that stabilizes the carboxylate anion favors increased dissociation and increases acidity. Thus, introducing an electron-withdrawing chlorine atom spreads out the negative charge on the anion and makes chloroacetic acid stronger than acetic acid by a factor of 75. Introducing two electronegative chlorine atoms makes dichloroacetic acid some 3000 times as strong as acetic acid, and introducing three makes trichloroacetic acid more than 12,000 times as strong.

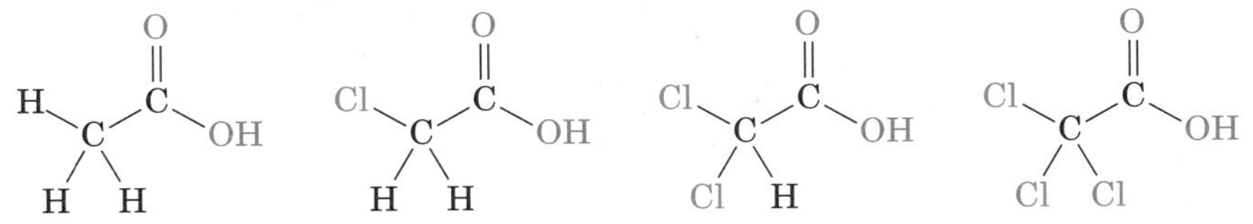

$$\underset{pK_a = 4.75}{H_3C{-}COOH} \qquad \underset{pK_a = 2.85}{ClCH_2{-}COOH} \qquad \underset{pK_a = 1.48}{Cl_2CH{-}COOH} \qquad \underset{pK_a = 0.64}{Cl_3C{-}COOH}$$

Weaker acid → Stronger acid

Although weaker than mineral acids, carboxylic acids are nevertheless much stronger acids than alcohols. Ethanol, for example, has a K_a of approximately 10^{-16}, making it a weaker acid than acetic acid by a factor of 10^{11}.

Why are carboxylic acids so much more acidic than alcohols even though both contain O–H groups? The easiest way to answer this question is to look at the relative stability of a carboxylate anion versus an alkoxide anion. In an alkoxide ion, the negative charge is confined to the one oxygen atom. In a carboxylate ion, however, the negative charge is *delocalized,* or spread out over both oxygen atoms. In other words, a carboxylate anion is a stabilized resonance hybrid (Section 4.12) of two equivalent structures.

$$\underset{\textbf{Alcohol}}{CH_3CH_2\ddot{O}H} + H_2O \rightleftharpoons \underset{\text{Unstabilized alkoxide ion}}{CH_3CH_2\ddot{O}\!:^-} + H_3O^+$$

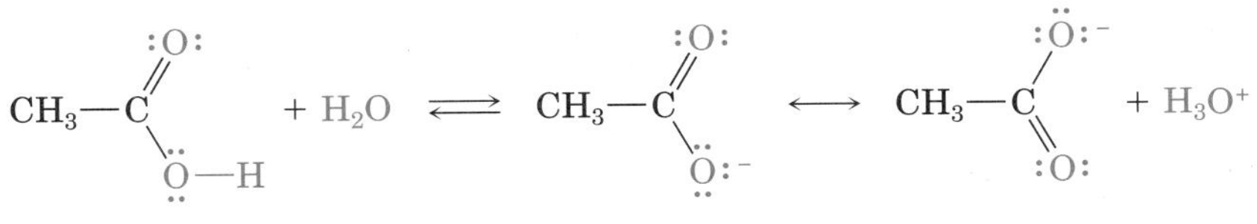

$$\underset{\textbf{Carboxylic acid}}{CH_3{-}C(=\!O){-}\ddot{O}{-}H} + H_2O \rightleftharpoons CH_3{-}C(=\!O){-}\ddot{O}\!:^- \longleftrightarrow CH_3{-}C(\ddot{O}\!:^-)\!=\!O + H_3O^+$$

Resonance-stabilized carboxylate ion (two equivalent resonance forms)

Since a carboxylate ion is more stable than an alkoxide ion, it is lower in energy and is present in greater amount at equilibrium, as shown in the reaction energy diagram in Figure 10.1 (p. 316). Put another way, dissociation of a carboxylic acid has a smaller ΔH than dissociation of an alcohol, leading to a larger equilibrium constant, K_a.

PRACTICE PROBLEM 10.1

Which would you expect to be the stronger acid, benzoic acid or *p*-nitrobenzoic acid?

Solution We know from its effect on aromatic substitution (Section 5.9) that a nitro group is electron-withdrawing and can stabilize a negative charge. Thus, a

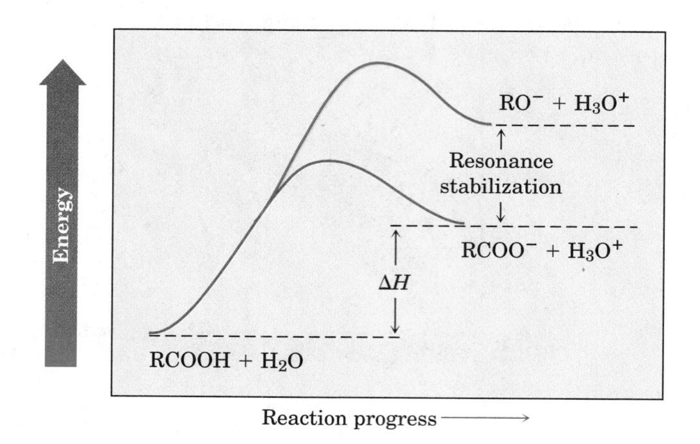

Figure 10.1 A reaction energy diagram for the dissociation of an alcohol (green curve) and a carboxylic acid (red curve). Resonance stabilization of the carboxylate anion lowers ΔH for dissociation of the acid, leading to a larger K_a. (The starting energy levels of alcohol and acid are shown at the same point for ease of comparison.)

p-nitrobenzoate ion is more stable than a benzoate ion, and *p*-nitrobenzoic acid is stronger than benzoic acid.

O, C, $^{-}$O, NO_2

Nitro group withdraws electrons from ring and stabilizes negative charge.

PROBLEM

10.5 Draw structures for the products of the following reactions:
(a) Benzoic acid + $NaOCH_3$ $\longrightarrow$ (b) $(CH_3)_3CCOOH$ + KOH $\longrightarrow$

PROBLEM

10.6 Rank the following compounds in order of increasing acidity: sulfuric acid, methanol, phenol, *p*-nitrophenol, acetic acid.

PROBLEM

10.7 Rank the following compounds in order of increasing acidity:
(a) CH_3CH_2COOH, $BrCH_2COOH$, $BrCH_2CH_2COOH$
(b) Benzoic acid, ethanol, *p*-cyanobenzoic acid

10.4 Synthesis of Carboxylic Acids

Let's review briefly the methods for preparing carboxylic acids that we've already seen:

1. A substituted alkylbenzene can be oxidized with $KMnO_4$ to give a substituted benzoic acid (Section 5.11):

$$O_2N{-}C_6H_4{-}CH_3 \xrightarrow[H_2O,\ 95°C]{KMnO_4} O_2N{-}C_6H_4{-}\overset{O}{\overset{\|}{C}}OH$$

p-Nitrotoluene → **p-Nitrobenzoic acid (88%)**

2. Primary alcohols and aldehydes can be oxidized to give carboxylic acids (Sections 8.7 and 9.5). A primary alcohol is often oxidized with CrO_3 or $Na_2Cr_2O_7$; an aldehyde is oxidized with Tollens' reagent ($AgNO_3$ in NH_4OH).

$$CH_3(CH_2)_8CH_2OH \xrightarrow[H_2O,\ H_2SO_4]{CrO_3} CH_3(CH_2)_8\overset{O}{\overset{\|}{C}}OH$$

1-Decanol → **Decanoic acid (93%)**

$$CH_3CH_2CH_2CH_2CH_2\overset{O}{\overset{\|}{C}}H \xrightarrow[NH_4OH]{AgNO_3} CH_3CH_2CH_2CH_2CH_2\overset{O}{\overset{\|}{C}}OH$$

Hexanal → **Hexanoic acid (85%)**

Hydrolysis of Nitriles

Carboxylic acids can be prepared from nitriles, R–C≡N, by reaction with aqueous acid or base (*hydrolysis*). Since nitriles themselves are usually prepared by an S_N2 reaction between an alkyl halide and cyanide ion, CN^-, the two-step sequence of cyanide ion displacement followed by nitrile hydrolysis is an excellent method for converting an alkyl halide into a carboxylic acid (RBr → RC≡N → RCOOH). As with all S_N2 reactions, the method works best with primary alkyl halides, although secondary alkyl halides can sometimes be used (Section 7.7).

A good example of the reaction occurs in the commercial synthesis of the antiarthritis drug, fenoprofen, a nonsteroidal anti-inflammatory agent (see Chapter 5 Interlude) marketed under the name Mylan.

$$C_6H_5{-}O{-}C_6H_4{-}\underset{\underset{Br}{|}}{C}HCH_3 \xrightarrow[\text{2. }^-OH/H_2O\text{; 3. }H_3O^+]{\text{1. NaCN}} C_6H_5{-}O{-}C_6H_4{-}\underset{\underset{CH_3}{|}}{C}HCOOH$$

Fenoprofen
(an antiarthritic agent)

Carboxylation of Grignard Reagents

Yet another method for preparing carboxylic acids is by reaction of a Grignard reagent, RMgX, with carbon dioxide. This **carboxylation** reaction is carried out either by pouring a solution of the Grignard reagent over dry ice (solid CO_2) or by bubbling a stream of CO_2 gas through the Grignard reagent solution.

$$\text{1-Bromo-2,4,6-trimethylbenzene} \xrightarrow[\text{Ether}]{\text{Mg}} \text{2,4,6-trimethylphenylmagnesium bromide} \xrightarrow[\text{2. } H_3O^+]{\text{1. } CO_2\text{, ether}} \text{2,4,6-Trimethylbenzoic acid (87\%)}$$

1-Bromo-2,4,6-trimethylbenzene **2,4,6-Trimethylbenzoic acid (87%)**

PRACTICE PROBLEM 10.2

How would you convert 2-chloro-2-methylpropane into 2,2-dimethylpropanoic acid?

Solution Since 2-chloro-2-methylpropane is a tertiary alkyl halide, it won't undergo S_N2 substitution with cyanide ion. Thus, the only way you could carry out the desired reaction is to convert the alkyl halide into a Grignard reagent and then add CO_2.

$$(CH_3)_3CCl + Mg \longrightarrow (CH_3)_3CMgCl \xrightarrow[\text{2. } H_3O^+]{\text{1. } CO_2} (CH_3)_3C\overset{O}{\overset{\|}{C}}OH$$

PROBLEM

10.8 Predict the products of the following reactions:

(a) Bromobenzene $\xrightarrow[\text{2. } CO_2 \text{; 3. } H_3O^+]{\text{1. Mg, ether}}$?

(b) $CH_3\underset{}{\overset{CH_3}{\overset{|}{C}}}HCH_2CH_2Br \xrightarrow[\text{2. NaOH, } H_2O\text{; 3. } H_3O^+]{\text{1. NaCN}}$?

PROBLEM

10.9 Show the steps in the conversion of iodomethane to acetic acid by the nitrile hydrolysis route. Would a similar route work for the conversion of iodobenzene to benzoic acid? Explain.

PROBLEM

10.10 Show all the steps in the conversion of iodobenzene to benzoic acid by the Grignard carboxylation route. Would a similar route work for the conversion of iodomethane to acetic acid?

10.5 Nucleophilic Acyl Substitution Reactions

We saw in Chapter 9 that the addition of nucleophiles to the polar C=O bond is a general feature of aldehyde and ketone chemistry. Carboxylic acids and their derivatives also react with nucleophiles, but the initially formed intermediate expels a substituent originally bonded to the carbonyl carbon, leading to the formation of a new carbonyl compound by a **nucleophilic acyl substitution reaction** (Figure 10.2).

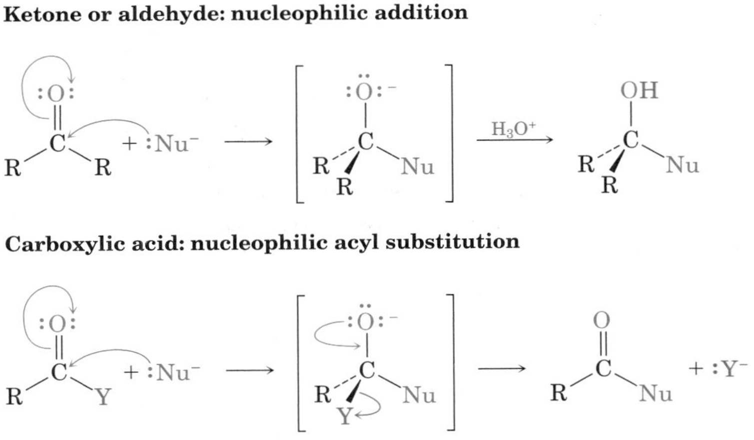

Ketone or aldehyde: nucleophilic addition

Carboxylic acid: nucleophilic acyl substitution

Figure 10.2 The general mechanisms of nucleophilic addition and nucleophilic acyl substitution reactions. Both reactions begin with the addition of a nucleophile to a polar C=O bond to give a tetrahedral, alkoxide ion intermediate. The intermediate formed from an aldehyde or ketone is then protonated to give an alcohol, while the intermediate formed from a carboxylic acid derivative expels a leaving group to give a new carbonyl compound.

The different behavior toward nucleophiles of aldehydes/ketones and carboxylic acid derivatives is a consequence of structure. Carboxylic acid derivatives have an acyl function bonded to a group –Y that can leave as a stable anion. As soon as addition of a nucleophile occurs, the group –Y leaves and a new carbonyl compound forms. Ketones and aldehydes have no such leaving group, however, and therefore don't undergo substitution.

Note that the overall nucleophilic substitution reaction of an acyl derivative is superficially similar to what occurs during S_N2 reaction of an alkyl halide (Section 7.7) in that a leaving group is replaced by an incoming nucleophile. The *mechanisms* of the two reactions are very different, however: S_N2 reactions occur in a single step by back-side displacement of a leaving group, whereas nucleophilic acyl substitutions take place in two steps through a tetrahedrally hybridized, alkoxide ion intermediate.

In comparing the reactivity of different acyl derivatives, the more highly polar a compound is, the more reactive it is. Thus, acid chlorides are the most reactive compounds because the electronegative chlorine atom strongly polarizes the carbonyl group, whereas amides are the least reactive compounds:

$RCONH_2$ < $RCOOR'$ < $RCOOCOR$ < $RCOCl$

Amide < **Ester** < **Acid anhydride** < **Acid chloride**

Less reactive → **Reactivity** → More reactive

An important consequence of these reactivity differences is that it's usually possible to convert a more reactive acid derivative into a less reactive one. Acid chlorides, for example, can be converted into esters and amides, but amides and esters can't be converted into acid chlorides. Remembering the reactivity order is therefore a useful way to keep track of a large number of reactions (Figure 10.3).

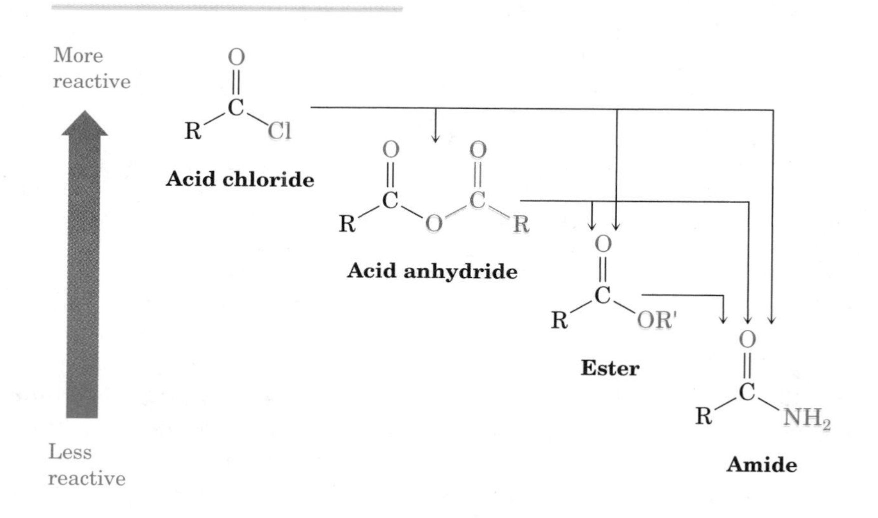

Figure 10.3 Interconversions of carboxylic acid derivatives. More reactive compounds can be converted into less reactive ones, but not vice versa.

PRACTICE PROBLEM 10.3

Which is more reactive in a nucleophilic acyl substitution reaction with hydroxide ion, CH_3CONH_2 or CH_3COCl?

Solution Since Cl is more electronegative than N, the carbonyl group of an acid chloride is more polar than the carbonyl group of an amide, and acid chlorides are more reactive than amides.

PROBLEM

10.11 Which compound in each of the following sets is more reactive in nucleophilic acyl substitution reactions?

(a) CH_3COCl or CH_3COOCH_3 (b) $(CH_3)_2CHCONH_2$ or $CH_3CH_2COOCH_3$

(c) CH_3COOCH_3 or $CH_3COOCOCH_3$ (d) CH_3COOCH_3 or CH_3CHO

PROBLEM

10.12 How can you account for the fact that methyl trifluoroacetate, CF_3COOCH_3, is more reactive than methyl acetate, CH_3COOCH_3, in nucleophilic acyl substitution reactions?

10.6 Reactions of Carboxylic Acids

Reduction: Conversion of Acids into Alcohols ($RCO_2H \longrightarrow RCH_2OH$)

We saw in Section 8.5 that carboxylic acids are reduced by lithium aluminum hydride ($LiAlH_4$) to yield primary alcohols:

$$CH_3(CH_2)_7CH{=}CH(CH_2)_7\overset{O}{\overset{\|}{C}}OH \xrightarrow[2.\ H_3O^+]{1.\ LiAlH_4} CH_3(CH_2)_7CH{=}CH(CH_2)_7CH_2OH$$

Oleic acid **_cis_-9-Octadecen-1-ol (87%)**

Conversion of Acids into Acid Chlorides ($RCO_2H \longrightarrow RCOCl$)

The most useful reactions of carboxylic acids are those that convert the –COOH group into another acyl function by a nucleophilic substitution. Acid chlorides, anhydrides, and esters can all be prepared from carboxylic acids.

Acid chlorides are usually prepared by treatment of carboxylic acids with thionyl chloride, $SOCl_2$. The net effect is substitution of the –OH group by –Cl. For example:

$$\text{2,4,6-}(CH_3)_3C_6H_2COOH \xrightarrow[CHCl_3]{SOCl_2} \text{2,4,6-}(CH_3)_3C_6H_2COCl + HCl + SO_2$$

2,4,6-Trimethylbenzoic acid **2,4,6-Trimethylbenzoyl chloride (90%)**

Conversion of Acids into Acid Anhydrides ($RCO_2H \longrightarrow RCO_2COR$)

Acid anhydrides are formally derived from two molecules of carboxylic acid by removing one molecule of water. In practice, however, anhydrides are difficult to prepare directly from the corresponding acids, and only acetic anhydride is commonly used.

$$H_3C-C(=O)-O-C(=O)-CH_3$$

Acetic anhydride

Conversion of Acids into Esters ($RCO_2H \longrightarrow RCO_2R'$)

Perhaps the most useful reaction of carboxylic acids is their conversion into esters. Among the many methods for accomplishing this transformation is the S_N2 reaction between a carboxylate anion nucleophile and a primary alkyl halide (Section 7.7):

$$CH_3CH_2CH_2\overset{O}{\overset{\|}{C}}\ddot{\underset{\cdot\cdot}{O}}{:}^-\ Na^+ + CH_3{-}I \xrightarrow[\text{reaction}]{S_N2} CH_3CH_2CH_2\overset{O}{\overset{\|}{C}}OCH_3 + NaI$$

Sodium butanoate → **Methyl butanoate, an ester (97%)**

Alternatively, esters can be synthesized by a nucleophilic acyl substitution reaction of a carboxylic acid with an alcohol. Called the **Fischer esterification reaction,** this method involves heating the carboxylic acid with an acid catalyst in an alcohol solvent.

$$C_6H_5-C(=O)-OH + CH_3CH_2OH \xrightarrow[\text{catalyst}]{H_2SO_4} C_6H_5-C(=O)-OCH_2CH_3 + H_2O$$

Benzoic acid → **Ethyl benzoate (91%)**

The Fischer esterification reaction, whose mechanism is shown in Figure 10.4, is a nucleophilic acyl substitution of –OR for –OH. The acid catalyst first protonates an oxygen atom of the –COOH group, which gives the carboxylic acid a positive charge and makes it more reactive toward nucleophiles. An alcohol molecule then adds to the protonated carboxylic acid, and subsequent loss of water yields the ester product.

All steps in the Fischer esterification reaction are reversible, and the position of the equilibrium can be driven either forward or backward depending on the reaction conditions. Ester formation is favored when alcohol is used as solvent, but carboxylic acid is favored when water is used as solvent.

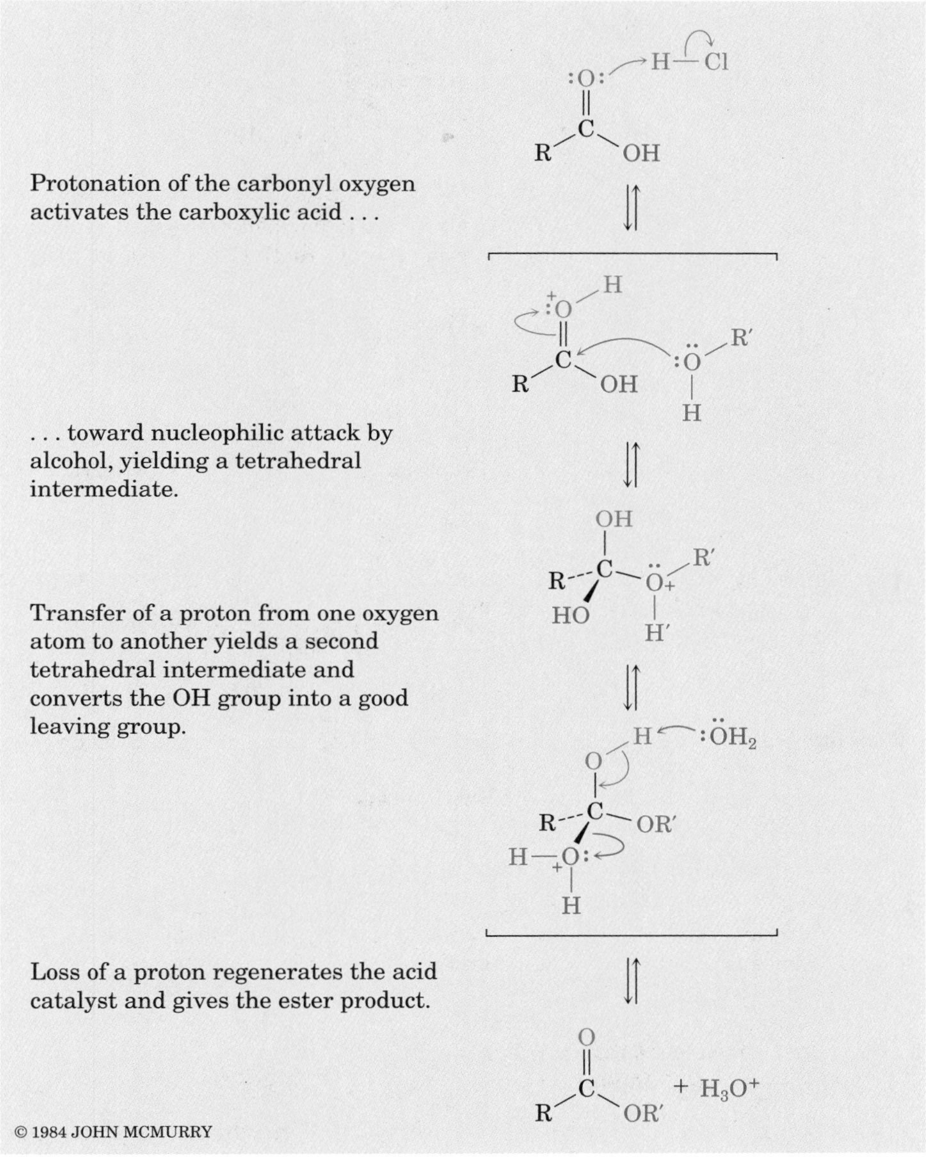

Protonation of the carbonyl oxygen activates the carboxylic acid . . .

. . . toward nucleophilic attack by alcohol, yielding a tetrahedral intermediate.

Transfer of a proton from one oxygen atom to another yields a second tetrahedral intermediate and converts the OH group into a good leaving group.

Loss of a proton regenerates the acid catalyst and gives the ester product.

Figure 10.4 Mechanism of the Fischer esterification reaction of a carboxylic acid to yield an ester. The reaction is an acid-catalyzed nucleophilic acyl substitution.

PRACTICE PROBLEM 10.4

How might you prepare the following ester using a Fischer esterification reaction?

$$C_6H_5-C(=O)-OCH_2CH_2CH_3$$

Solution The trick is to identify the two parts of the ester. The target molecule is propyl benzoate, so it can be prepared by treating benzoic acid with 1-propanol.

$$\underset{\textbf{Benzoic acid}}{C_6H_5-C(=O)-OH} + \underset{\textbf{1-Propanol}}{HOCH_2CH_2CH_3} \xrightarrow[\text{catalyst}]{H^+} \underset{\textbf{Propyl benzoate}}{C_6H_5-C(=O)-OCH_2CH_2CH_3}$$

PROBLEM

10.13 What products would you obtain by treating benzoic acid with the following reagents? Formulate the reactions.
(a) $SOCl_2$ (b) CH_3OH, HCl (c) $LiAlH_4$ (d) NaOH

PROBLEM

10.14 Show how you might prepare the following esters using a Fischer esterification reaction:
(a) Butyl acetate (b) Methyl butanoate

PROBLEM

10.15 If 5-hydroxypentanoic acid is treated with an acid catalyst, an intramolecular esterification reaction occurs. What is the structure of the product? (*Intramolecular* means within the same molecule.)

Conversion of Acids into Amides ($RCO_2H \longrightarrow RCONH_2$)

Amides are carboxylic acid derivatives in which the acid –OH group has been replaced by a nitrogen substituent, $-NH_2$, –NHR, or $-NR_2$. Amides are difficult to prepare directly from acids because amines are bases, which convert acidic carboxyl groups into their carboxylate anions. Because the carboxylate anion has a negative charge, it no longer undergoes attack by nucleophiles except at high temperatures.

$$R-C(=O)-OH + :NH_3 \rightleftharpoons R-C(=O)-O^- \; NH_4^+$$

We'll see a method for converting an acid directly into an amide in Section 15.8 in connection with the synthesis of proteins from amino acids.

10.7 Chemistry of Acid Halides

Acid chlorides are prepared from carboxylic acids by reaction with thionyl chloride, $SOCl_2$, as we saw in the previous section:

$$RC(=O)OH \xrightarrow{SOCl_2} RC(=O)Cl$$

Acid halides are among the most reactive of the various carboxylic acid derivatives and can be converted into many other kinds of substances. As illustrated in Figure 10.5, the halogen can be replaced by –OH to yield an acid, by –OR to yield an ester, or by $-NH_2$ to yield an amide. Although Figure 10.5 illustrates these reactions only for acid chlorides, similar processes take place with other acid halides.

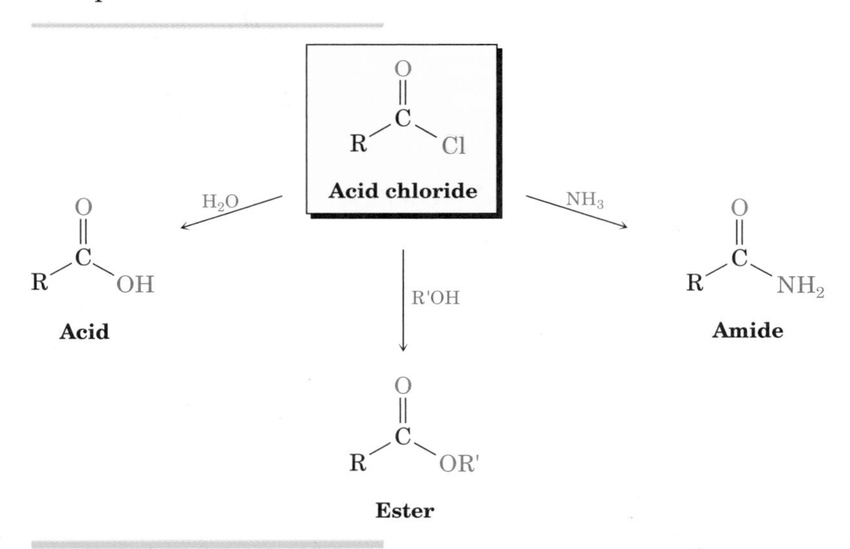

Figure 10.5 Some nucleophilic acyl substitution reactions of acid chlorides.

Hydrolysis: Conversion of Acid Chlorides into Acids ($RCOCl \longrightarrow RCO_2H$)

Acid chlorides react with water to yield carboxylic acids. This hydrolysis reaction is a typical nucleophilic acyl substitution process and is initiated by attack of the nucleophile water on the acid chloride carbonyl group. The initially formed tetrahedral intermediate undergoes loss of HCl to yield the product:

$$RC(=O)Cl + :\ddot{O}H_2 \longrightarrow \left[R-C(O^-)(Cl)-\overset{+}{O}H_2 \longrightarrow RC(=O)-\overset{+}{O}H_2 \right] \xrightarrow{:Base} RC(=O)OH$$

An acid chloride — A carboxylic acid

Alcoholysis: Conversion of Acid Chlorides into Esters ($RCOCl \longrightarrow RCO_2R'$)

Acid chlorides react with alcohols to yield esters in a reaction analogous to their reaction with water to yield acids:

$$H_3C\text{–}C(=O)\text{–}Cl + CH_3CH_2CH_2CH_2OH \xrightarrow{\text{Pyridine}} H_3C\text{–}C(=O)\text{–}OCH_2CH_2CH_2CH_3 + HCl$$

Acetyl chloride **1-Butanol** **Butyl acetate (90%)**

Since HCl is generated as a by-product of alcoholysis, the reaction is usually carried out in the presence of an amine base such as pyridine (see Section 12.6), which reacts with the HCl as it's formed and prevents it from causing side reactions.

PRACTICE PROBLEM 10.5

Show how you could prepare ethyl benzoate by reaction of an acid chloride with an alcohol.

Solution As its name implies, ethyl benzoate can be made by reaction of *ethyl* alcohol with the acid chloride of *benzoic* acid:

$$C_6H_5\text{–}C(=O)\text{–}Cl + CH_3CH_2OH \xrightarrow[\text{solvent}]{\text{Pyridine}} C_6H_5\text{–}C(=O)\text{–}OCH_2CH_3$$

Benzoyl chloride **Ethanol** **Ethyl benzoate**

PROBLEM

10.16 How could you prepare the following esters using the reaction of an acid chloride with an alcohol?
(a) $CH_3CH_2COOCH_3$ (b) $CH_3COOCH_2CH_3$ (c) Cyclohexyl acetate

Aminolysis: Conversion of Acid Chlorides into Amides ($RCOCl \longrightarrow RCONH_2$)

Acid chlorides react rapidly with ammonia and with amines to give amides. Both mono- and disubstituted amines can be used. For example, 2-methylpropanamide is prepared by reaction of 2-methylpropanoyl chloride with ammonia. Note that one extra equivalent of ammonia is added to react with the HCl generated.

$$(CH_3)_2CHC(=O)Cl + 2\ {:}NH_3 \longrightarrow (CH_3)_2CHC(=O)NH_2 + \overset{+}{N}H_4\overset{-}{Cl}$$

2-Methylpropanoyl chloride **2-Methylpropanamide (83%)**

PRACTICE PROBLEM 10.6

Show how you would prepare *N*-methylpropanamide by reaction of an acid chloride with an amine.

Solution Reaction of *methyl*amine with *propanoyl* chloride gives *N*-methylpropanamide:

$$CH_3CH_2\overset{O}{\overset{\|}{C}}Cl + 2\ CH_3NH_2 \longrightarrow CH_3CH_2\overset{O}{\overset{\|}{C}}NHCH_3 + CH_3NH_3^+\ Cl^-$$

Propanoyl chloride **Methylamine** ***N*-Methylpropanamide**

PROBLEM

10.17 Write the steps in the mechanism of the reaction between ammonia and 2-methylpropanoyl chloride to yield 2-methylpropanamide.

PROBLEM

10.18 What amines would react with what acid chlorides to give the following amide products?
(a) $CH_3CH_2CONH_2$
(b) $(CH_3)_2CHCH_2CONHCH_3$
(c) *N,N*-Dimethylpropanamide
(d) *N,N*-Diethylbenzamide

10.8 Chemistry of Acid Anhydrides

The best method of preparing acid anhydrides is by a nucleophilic acyl substitution reaction of an acid chloride with a carboxylic acid anion. Both symmetrical and unsymmetrical acid anhydrides can be prepared in this way.

$$R\overset{O}{\overset{\|}{C}}O^-\ Na^+ + Cl\overset{O}{\overset{\|}{C}}CH_3 \xrightarrow[25°C]{Ether} H\overset{O}{\overset{\|}{C}}O\overset{O}{\overset{\|}{C}}CH_3$$

Sodium formate **Acetyl chloride** **Acetic formic anhydride (64%)**

The chemistry of acid anhydrides is similar to that of acid chlorides. Although anhydrides react more slowly than acid chlorides, the kinds of reactions the two functional groups undergo are the same. Thus, acid anhydrides react with water to form acids, with alcohols to form esters, and with amines to form amides (Figure 10.6, p. 328).

Acetic anhydride is often used to prepare acetate esters of complex alcohols and to prepare substituted acetamides from amines. For example,

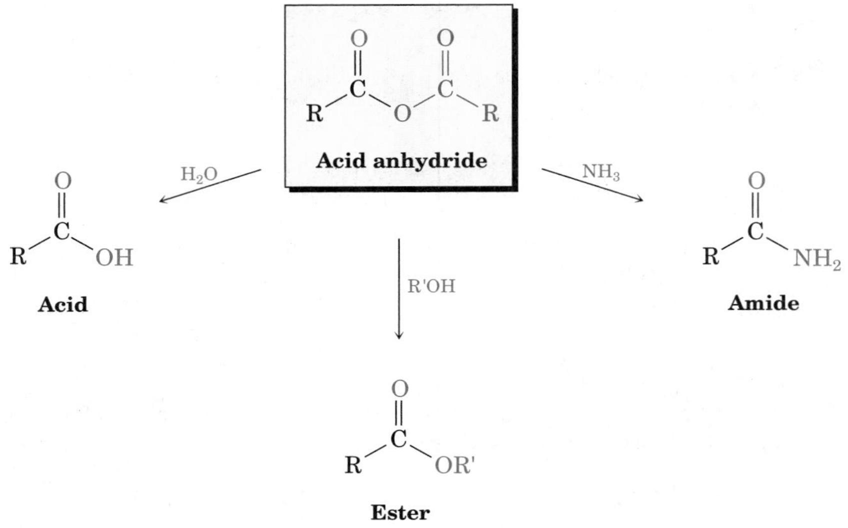

Figure 10.6 Some reactions of acid anhydrides.

aspirin (an ester) is prepared by the reaction of acetic anhydride with *o*-hydroxybenzoic acid. Similarly, acetaminophen (an amide) is prepared by reaction of acetic anhydride with *p*-hydroxyaniline.

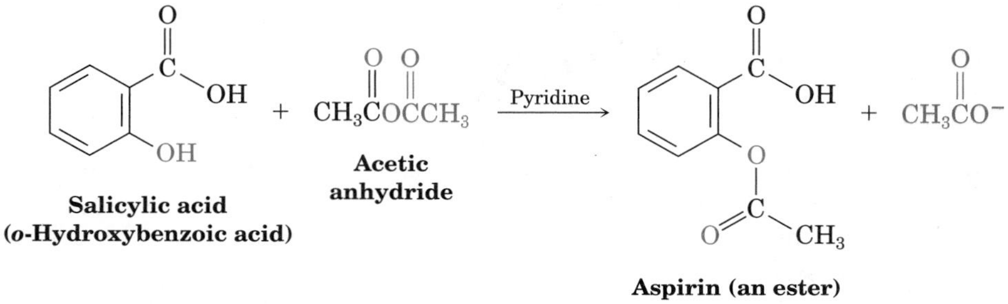

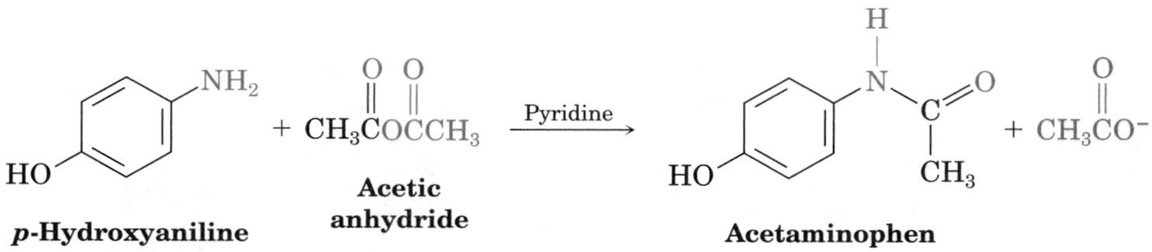

$$\text{Salicylic acid } (o\text{-Hydroxybenzoic acid}) + CH_3COCCH_3 \text{ (Acetic anhydride)} \xrightarrow{\text{Pyridine}} \text{Aspirin (an ester)} + CH_3CO^-$$

$$p\text{-Hydroxyaniline} + CH_3COCCH_3 \text{ (Acetic anhydride)} \xrightarrow{\text{Pyridine}} \text{Acetaminophen} + CH_3CO^-$$

Notice in both of these examples that only "half" of the anhydride molecule is used; the other half acts as the leaving group during the nucleophilic acyl substitution step and produces carboxylate anion as a by-product. Thus, anhydrides are inefficient to use, and acid chlorides are normally preferred for introducing acyl substituents other than acetyl groups.

PRACTICE PROBLEM 10.7

What is the product of the following reaction?

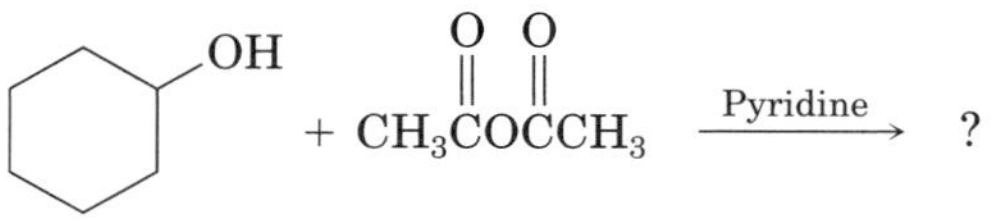

$$\text{Cyclohexanol (OH)} + CH_3\overset{O}{\overset{\|}{C}}O\overset{O}{\overset{\|}{C}}CH_3 \xrightarrow{\text{Pyridine}} ?$$

Solution Reaction of cyclohexanol with acetic anhydride yields cyclohexyl acetate by nucleophilic acyl substitution of the $-OCOCH_3$ group of the anhydride by the $-OR$ group of the alcohol:

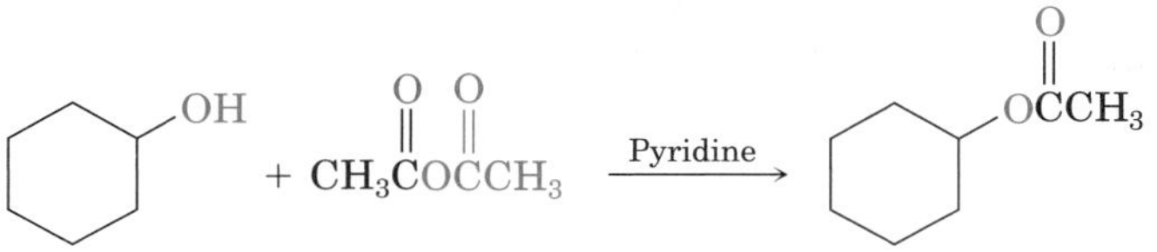

$$\text{(OH)} + CH_3\overset{O}{\overset{\|}{C}}O\overset{O}{\overset{\|}{C}}CH_3 \xrightarrow{\text{Pyridine}} \text{(}O\overset{O}{\overset{\|}{C}}CH_3\text{)}$$

Cyclohexanol **Cyclohexyl acetate**

PROBLEM

10.19 Write the steps in the mechanism of the reaction between *p*-hydroxyaniline and acetic anhydride to prepare acetaminophen.

PROBLEM

10.20 What product would you expect to obtain from the reaction of 1 equivalent of methanol with a cyclic anhydride such as phthalic anhydride?

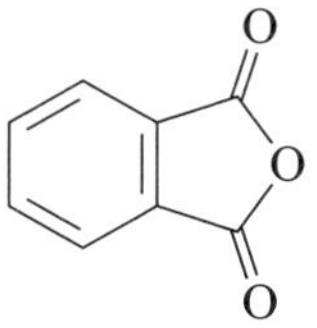

Phthalic anhydride

10.9 Chemistry of Esters

Esters are among the most widespread of naturally occurring compounds. Many simple esters are pleasant-smelling liquids that are responsible for the fragrant odors of fruits and flowers. Methyl butanoate, for example, has been isolated from pineapple oil, and isopentyl acetate has been found in banana oil. The ester linkage is also present in animal fats and other biologically important molecules.

$CH_3CH_2CH_2\overset{\overset{O}{\|}}{C}OCH_3$

Methyl butanoate (from pineapples)

$CH_3\overset{\overset{O}{\|}}{C}OCH_2CH_2\overset{\overset{CH_3}{|}}{C}HCH_3$

Isopentyl acetate (from bananas)

$$\begin{array}{l} CH_2O\overset{\overset{O}{\|}}{C}R \\ | \\ CHO\overset{\overset{O}{\|}}{C}R \\ | \\ CH_2O\overset{\overset{O}{\|}}{C}R \end{array}$$

A fat (R = C_{11-17} chains)

Esters are usually prepared either from acids or acid chlorides by the methods already discussed. Thus, carboxylic acids are converted directly into esters either by S_N2 reaction of a carboxylate ion with a primary alkyl halide or by reaction of the acid with an alcohol (Section 10.6). Acid chlorides are converted into esters by reaction with an alcohol in the presence of pyridine (Section 10.7).

$$RCOOH \xrightarrow{NaOH} RCOO^- Na^+ \xrightarrow[(S_N2\ reaction)]{R'X} RCOOR'$$

$$RCOOH \xrightarrow[H^+\ catalyst]{R'OH} RCOOR'$$

$$RCOOH \xrightarrow{SOCl_2} RCOCl \xrightarrow[Pyridine]{R'OH} RCOOR'$$

Esters show the same kinds of chemistry we've seen for other acyl derivatives, but they're less reactive toward nucleophiles than acid chlorides or anhydrides. Figure 10.7 shows some general reactions of esters.

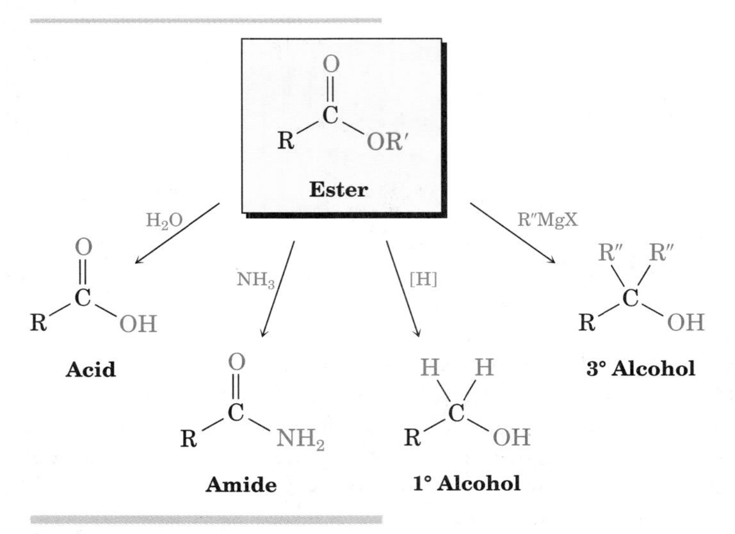

Figure 10.7 Some general reactions of esters.

Hydrolysis: Conversion of Esters into Acids ($RCO_2R' \longrightarrow RCO_2H$)

Esters are hydrolyzed either by aqueous base or by aqueous acid to yield a carboxylic acid plus an alcohol:

$$\underset{\textbf{Ester}}{R-\overset{\displaystyle O}{\overset{\|}{C}}-OR'} \xrightarrow[\text{NaOH or } H_3O^+]{H_2O} \underset{\textbf{Acid}}{R-\overset{\displaystyle O}{\overset{\|}{C}}-OH} + R'OH$$

Hydrolysis in basic solution is called **saponification** (Latin *sapo,* "soap"). (As we'll see in Section 16.3, soap is made by the base-induced ester hydrolysis of animal fat.) Ester hydrolysis occurs by a typical nucleophilic acyl substitution pathway in which OH^- nucleophile adds to the ester carbonyl group, yielding a tetrahedral intermediate. Loss of OR^- then gives a carboxylic acid, which is deprotonated to give the acid salt:

$$\underset{\textbf{Ester}}{R-\overset{\displaystyle O}{\overset{\|}{C}}-OR'} + :\ddot{O}H^- \longrightarrow \underset{\text{Tetrahedral intermediate}}{\left[R-\overset{\displaystyle :\ddot{O}:^-}{\overset{|}{\underset{\displaystyle R'O}{C}}}-OH\right]} \longrightarrow \underset{\textbf{Acid}}{R-\overset{\displaystyle O}{\overset{\|}{C}}-OH} + R'O^- \longrightarrow \underset{\textbf{Acid salt}}{R-\overset{\displaystyle O}{\overset{\|}{C}}-O^-} + R'OH$$

PRACTICE PROBLEM 10.8

Write the products of the following saponification reaction:

$$\underset{\textbf{Ethyl 3-methylbutanoate}}{CH_3\overset{\displaystyle CH_3}{\overset{|}{C}}HCH_2\overset{\displaystyle O}{\overset{\|}{C}}OCH_2CH_3} \xrightarrow[2.\ H_3O^+]{1.\ NaOH,\ H_2O} ?$$

Solution Esters are cleaved by aqueous base into their acid and alcohol components by breaking the bond between the carbonyl carbon and the alcohol oxygen:

$$\underset{\textbf{Ethyl 3-methylbutanoate}}{CH_3\overset{\displaystyle CH_3}{\overset{|}{C}}HCH_2\overset{\displaystyle O}{\overset{\|}{C}}OCH_2CH_3} \xrightarrow[2.\ H_3O^+]{1.\ NaOH,\ H_2O} \underset{\textbf{3-Methylbutanoic acid}}{CH_3\overset{\displaystyle CH_3}{\overset{|}{C}}HCH_2\overset{\displaystyle O}{\overset{\|}{C}}OH} + \underset{\textbf{Ethanol}}{CH_3CH_2OH}$$

PROBLEM

10.21 Show the products of hydrolysis of the following esters:
(a) Isopropyl acetate (b) Methyl cyclohexanecarboxylate

PROBLEM

10.22 Why do you suppose saponification of esters is not reversible? In other words, why doesn't treatment of a carboxylic acid with an alkoxide ion give an ester?

Aminolysis: Conversion of Esters into Amides ($RCO_2R' \longrightarrow RCONH_2$)

Esters react with ammonia and amines to yield amides. The reaction is not often used, however, because it is usually simpler to start from the acid chloride rather than from the ester.

$$C_6H_5COOCH_3 \xrightarrow[\text{Ether}]{NH_3} C_6H_5CONH_2 + CH_3OH$$

Methyl benzoate → **Benzamide**

Reduction: Conversion of Esters into Alcohols ($RCO_2R' \longrightarrow RCH_2OH$)

Esters are reduced by treatment with $LiAlH_4$ to yield primary alcohols (Section 8.5):

$$CH_3CH_2CH{=}CHCOCH_2CH_3 \xrightarrow[\text{2. } H_3O^+]{\text{1. } LiAlH_4\text{, ether}} CH_3CH_2CH{=}CHCH_2OH + CH_3CH_2OH$$

Ethyl 2-pentenoate → **2-Penten-1-ol (91%)**

Hydride ion first adds to the carbonyl group, followed by elimination of an alkoxide ion to yield an aldehyde intermediate. Further reduction of the aldehyde gives the primary alcohol.

$$RC(=O)OR' + :H^- \xrightarrow[\text{Ether}]{LiAlH_4} \left[R\text{–}C(O^-)(H)(OR') \longrightarrow RC(=O)H + R'O^- \right]$$

$$\left[RC(=O)H + :H^- \xrightarrow[\text{Ether}]{LiAlH_4} R\text{–}C(O^-)(H)H \right] \xrightarrow{H_3O^+} R\text{–}C(OH)(H)H$$

PRACTICE PROBLEM 10.9

What products would you obtain by reduction of propyl benzoate with $LiAlH_4$?

Solution Reduction of an ester with $LiAlH_4$ yields two molecules of alcohol product, one from the acyl part of the ester and one from the alkoxy part. Thus, reduction of propyl benzoate yields benzyl alcohol (from the acyl group) and 1-propanol (from the alkoxyl group).

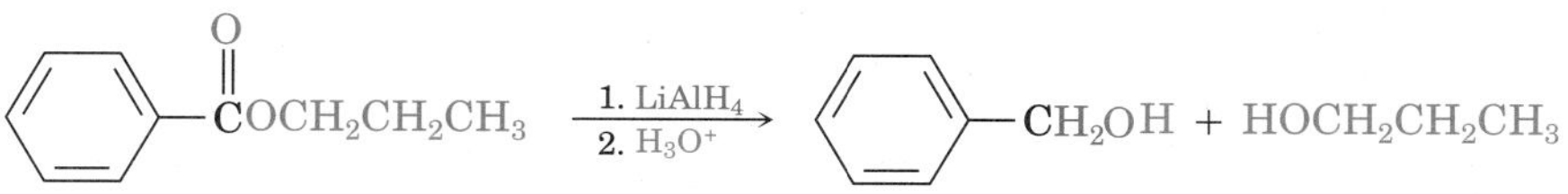

$$C_6H_5COOCH_2CH_2CH_3 \xrightarrow[2.\ H_3O^+]{1.\ LiAlH_4} C_6H_5CH_2OH + HOCH_2CH_2CH_3$$

Propyl benzoate **Benzyl alcohol** **1-Propanol**

PROBLEM

10.23 Show the products you would obtain by reduction of the following esters with $LiAlH_4$:

(a) $CH_3CH_2CH_2CH(CH_3)COOCH_3$ (b) $C_6H_5COOC_6H_5$

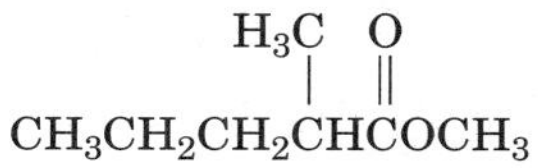

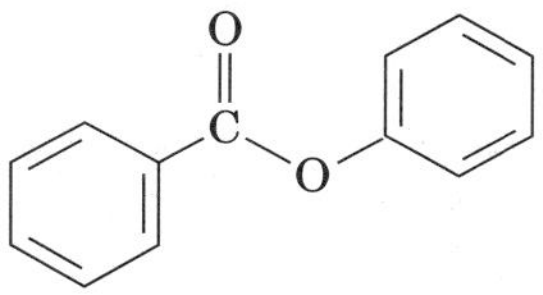

PROBLEM

10.24 What product would you expect from the reaction of a cyclic ester such as butyrolactone with $LiAlH_4$?

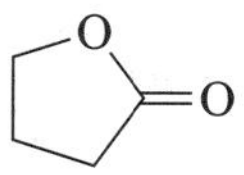

Butyrolactone

Reaction of Esters with Grignard Reagents

Grignard reagents react with esters to yield tertiary alcohols in which two of the substituents on the hydroxyl-bearing carbon are identical. For example, methyl benzoate reacts with 2 equiv of CH_3MgBr to yield 2-phenyl-2-propanol. The reaction occurs by addition of a Grignard reagent to the ester, elimination of alkoxide ion to give an intermediate ketone, and further addition to the ketone to yield the tertiary alcohol (Figure 10.8, p. 334).

PRACTICE PROBLEM 10.10

How could you use the reaction of a Grignard reagent with an ester to prepare 1,1-diphenyl-1-propanol?

Solution The product of the reaction between a Grignard reagent and an ester is a tertiary alcohol in which the alcohol carbon and one of the attached groups have come from the ester, and the remaining two groups bonded to the alcohol carbon have come from the Grignard reagent. Since 1,1-diphenyl-1-propanol has two phenyl

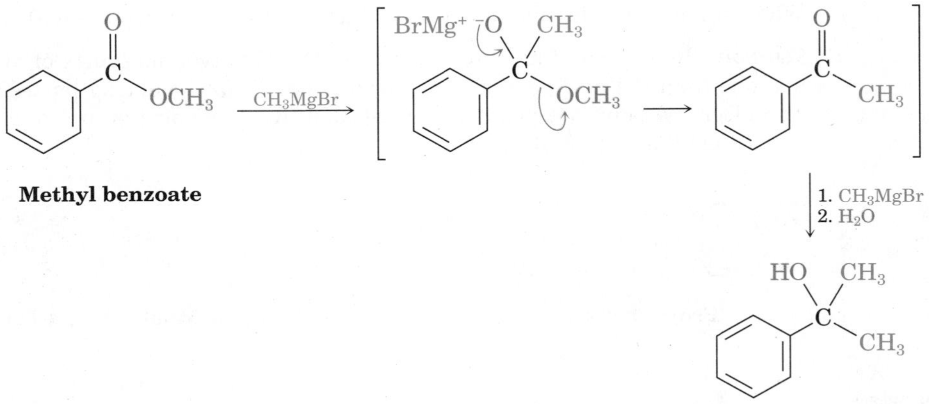

Methyl benzoate

CH_3MgBr

1. CH_3MgBr
2. H_2O

2-Phenyl-2-propanol (95%)

Figure 10.8 Mechanism of the reaction of a Grignard reagent with an ester to yield a tertiary alcohol. A ketone intermediate is involved.

groups and one ethyl group bonded to the alcohol carbon, it must be prepared from reaction of a phenylmagnesium halide with an ester of propanoic acid:

$$2\ C_6H_5MgBr + CH_3CH_2\overset{\overset{\displaystyle O}{\|}}{C}OCH_3 \longrightarrow CH_3CH_2{-}\underset{\underset{\displaystyle C_6H_5}{|}}{\overset{\overset{\displaystyle OH}{|}}{C}}{-}C_6H_5$$

1,1-Diphenyl-1-propanol

PROBLEM

10.25 What ester and what Grignard reagent might you use to prepare the following alcohols?

(a) 2-Phenyl-2-propanol
(b) 1,1-Diphenylethanol
(c) 3-Ethyl-3-heptanol

10.10 Chemistry of Amides

Amides are usually prepared by reaction of an acid chloride with an amine, as we saw in Section 10.7. Ammonia, monosubstituted amines, and disubstituted amines all undergo this reaction.

$$\text{RCOCl} \xrightarrow{NH_3} \text{RCONH}_2 \qquad \text{RCOCl} \xrightarrow{R'NH_2} \text{RCONHR}' \qquad \text{RCOCl} \xrightarrow{R'_2NH} \text{RCONR}'_2$$

Amides are much less reactive than acid chlorides, acid anhydrides, and esters. We'll see in Chapter 15, for example, that the amide linkage is stable enough to serve as the basic unit from which proteins are made.

$$H_2N{-}CH(R){-}C(=O){-}OH \Longrightarrow {-}NH{-}CH(R){-}C(=O){-}NH{-}CH(R'){-}C(=O){-}NH{-}CH(R''){-}C(=O){-}$$

Amino acids — A protein (polyamide)

Hydrolysis: Conversion of Amides into Acids ($RCONH_2 \longrightarrow RCOOH$)

Amides undergo hydrolysis to yield carboxylic acids plus amine on heating in either aqueous acid or base. Although the reaction is slow and requires prolonged heating, the overall transformation is a typical nucleophilic acyl substitution of –OH for $-NH_2$.

$$\underset{\textbf{Amide}}{RC(=O)NH_2} \xrightarrow[\text{Heat}]{H_3O^+ \text{ or } HO^-,\ H_2O} \underset{\textbf{Acid}}{RC(=O)OH} + NH_3$$

Reduction: Conversion of Amides into Amines ($RCONH_2 \longrightarrow RCH_2NH_2$)

Like other carboxylic acid derivatives, amides are reduced by $LiAlH_4$. The product of this reduction, however, is an *amine* rather than an alcohol:

$$\underset{\textbf{Benzamide}}{C_6H_5C(=O)NH_2} \xrightarrow[2.\ H_2O]{1.\ LiAlH_4,\ \text{ether}} \underset{\textbf{Benzylamine (93\%)}}{C_6H_5CH_2NH_2}$$

The effect of amide reduction is to convert the amide carbonyl group into a methylene group (C=O → CH_2). This kind of reaction is specific for amides and does not occur with other carboxylic acid derivatives.

PRACTICE PROBLEM 10.11

How could you prepare *N*-ethylaniline by reduction of an amide with $LiAlH_4$?

Solution Since reduction of an amide with $LiAlH_4$ yields an amine, the starting material for synthesis of *N*-ethylaniline must be *N*-phenylacetamide.

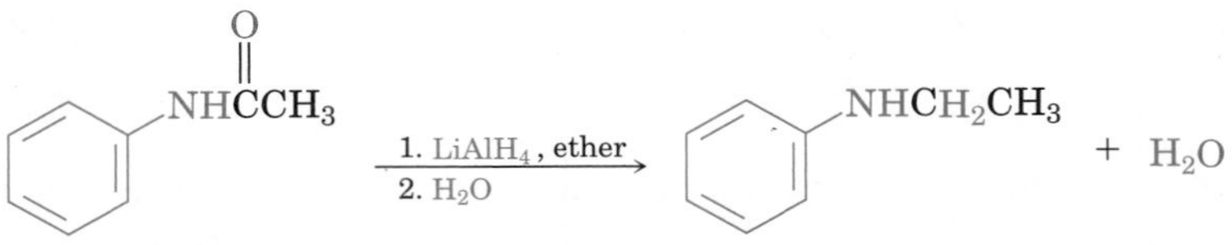

$$C_6H_5NHCOCH_3 \xrightarrow[2.\ H_2O]{1.\ LiAlH_4,\ ether} C_6H_5NHCH_2CH_3 + H_2O$$

***N*-Phenylacetamide** → ***N*-Ethylaniline**

PROBLEM

10.26 How would you convert *N*-ethylbenzamide into the following products?
(a) Benzoic acid (b) Benzyl alcohol
(c) *N*-Ethylbenzylamine, $C_6H_5CH_2NHCH_2CH_3$

PROBLEM

10.27 The reduction of an amide with $LiAlH_4$ to yield an amine is effective with both acyclic and cyclic amides (*lactams*). What product would you obtain from reduction of 5,5-dimethyl-2-pyrrolidone with $LiAlH_4$?

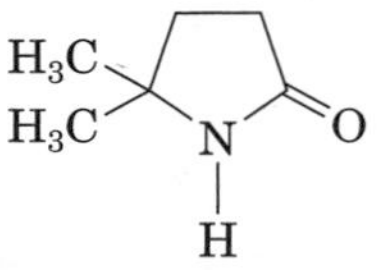

5,5-Dimethyl-2-pyrrolidone
(a lactam)

10.11 Chemistry of Nitriles

Nitriles, R–C≡N, are not related to carboxylic acids in the same sense that acyl derivatives are, but the chemistries of nitriles and carboxylic acids are so similar that the two classes of compounds can be considered together. Both functional groups have a carbon atom with three bonds to an electronegative atom, and both contain a π bond.

$$R-C\equiv N \qquad\qquad R-C(=O)OH$$

A nitrile—three bonds to nitrogen

An acid—three bonds to two oxygens

The simplest method of preparing nitriles is by the S_N2 reaction of cyanide ion with a primary alkyl halide (Section 7.7):

$$RCH_2Br + Na^+ \, CN^- \xrightarrow[\text{reaction}]{S_N2} RCH_2CN + NaBr$$

Reactions of Nitriles

The chemistry of nitriles is similar in many respects to the chemistry of carbonyl compounds. Thus, nitriles undergo many of the same kinds of reactions as carboxylic acid derivatives (Figure 10.9).

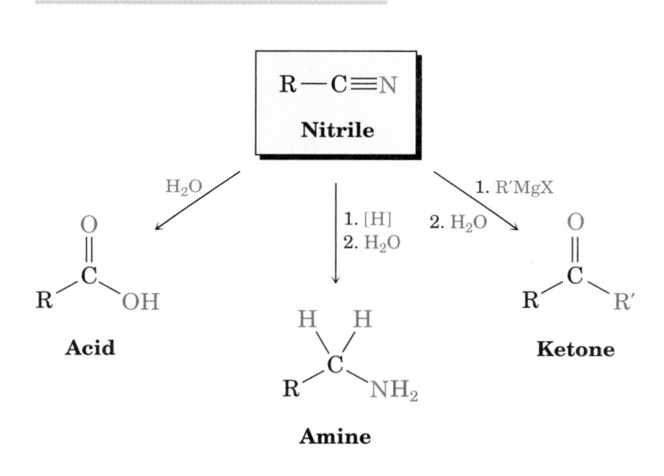

Figure 10.9 Some reactions of nitriles.

Like carbonyl groups, nitriles are strongly polarized. The nitrile carbon atom is electrophilic and undergoes attack by nucleophiles to yield an sp^2-hybridized intermediate imine anion that is analogous to the sp^3-hybridized intermediate alkoxide anion formed by addition of a nucleophile to a carbonyl group. Once formed, the intermediate imine anion can then go on to yield further products.

Carbonyl compound $\xrightarrow{:Nu^-}$ [Alkoxide ion] $\longrightarrow$ Products

Alkoxide ion

Nitrile $R-C\equiv N$ ($\delta+$, $\delta-$) $\xrightarrow{:Nu^-}$ [Imine anion] $\longrightarrow$ Products

Imine anion

Hydrolysis: Conversion of Nitriles into Carboxylic Acids (RCN ⟶ RCO_2H)

Nitriles are hydrolyzed in either acidic or basic solution to yield carboxylic acids and ammonia (or an amine).

$$R{-}C{\equiv}N \xrightarrow[\text{or NaOH, } H_2O]{H_3O^+} R{-}\overset{\overset{\large O}{\|}}{C}{-}OH + NH_3$$

Reduction: Conversion of Nitriles into Amines (RCN ⟶ RCH_2NH_2)

Reduction of a nitrile with $LiAlH_4$ gives a primary amine, just as reduction of an ester gives a primary alcohol.

$$o\text{-}CH_3C_6H_4{-}C{\equiv}N \xrightarrow[\text{2. } H_2O]{\text{1. } LiAlH_4\text{, ether}} o\text{-}CH_3C_6H_4{-}CH_2NH_2$$

o-Methylbenzonitrile → **o-Methylbenzylamine (88%)**

Reaction of Nitriles with Grignard Reagents

Grignard reagents, RMgX, add to nitriles to give intermediate imine anions that can be hydrolyzed to yield ketones:

$$R{-}C{\equiv}N: + \ddot{R}'^{-}\ \overset{+}{Mg}X \longrightarrow \left[R{-}\overset{\overset{:\ddot{N}^{-}\ \overset{+}{Mg}X}{\|}}{C}{-}R' \right] \xrightarrow{H_3O^+} R{-}\overset{\overset{O}{\|}}{C}{-}R' + NH_3$$

Nitrile — Imine anion — **Ketone**

For example, benzonitrile reacts with ethylmagnesium bromide to give propiophenone in high yield:

$$C_6H_5{-}C{\equiv}N \xrightarrow[\text{2. } H_3O^+]{\text{1. } CH_3CH_2MgBr\text{, ether}} C_6H_5{-}\overset{\overset{O}{\|}}{C}{-}CH_2CH_3$$

Benzonitrile → **Propiophenone (89%)**

PRACTICE PROBLEM 10.12 ....

Show how you could prepare 2-hexanone by reaction of a Grignard reagent with a nitrile.

Solution There are two ways to prepare a ketone from a nitrile by Grignard addition:

$$\left.\begin{array}{l} CH_3C{\equiv}N + CH_3CH_2CH_2CH_2MgBr \\ CH_3MgBr + N{\equiv}CCH_2CH_2CH_2CH_3 \end{array}\right\} \longrightarrow CH_3\overset{O}{\overset{\|}{C}}CH_2CH_2CH_2CH_3$$

PROBLEM

10.28 What nitrile would you react with what Grignard reagent to prepare the following ketones?

(a) $CH_3CH_2COCH_2CH_3$
(b) $CH_3CH_2COCH(CH_3)_2$
(c) Acetophenone (methyl phenyl ketone)
(d) [dicyclohexyl ketone structure: two cyclohexane rings joined by C=O]

PROBLEM

10.29 How would you prepare the following substances from the indicated starting materials? More than one step is required in each case.

(a) $(CH_3)_2CHCH_2CH_2NH_2$ from $(CH_3)_2CHCH_2I$
(b) 1-Phenyl-2-butanone from benzyl bromide, $C_6H_5CH_2Br$

10.12 Nylons and Polyesters: Step-Growth Polymers

There are two main classes of synthetic polymers: chain-growth polymers and step-growth polymers. **Chain-growth polymers,** such as polyethylene and the other alkene polymers we saw in Section 4.8, are prepared by chain-reaction processes in which an initiator first adds to the double bond of an alkene monomer to produce a reactive intermediate. This intermediate adds to a second alkene monomer unit, and the polymer chain lengthens as more monomer units add successively to the end of the growing chain.

Step-growth polymers are produced by polymerization reactions between two difunctional molecules. Each new bond is formed in a discrete step, independent of all other bonds in the polymer, and chain reactions are not involved. The key bond-forming step is often a nucleophilic acyl substitution of a carbonyl compound. Some commercially important step-growth polymers are shown in Table 10.3 (p. 340).

Nylons

The best-known step-growth polymers are the polyamides, or **nylons,** first prepared by Wallace Carothers at the Du Pont Company. Like other amides, nylons are usually prepared by reaction between a diamine and a diacid. For example, nylon 66 is prepared by heating the six-carbon adipic acid

Table 10.3 Some Important Step-Growth Polymers and Their Uses

Monomer name	*Formula*	*Trade or common name of polymer*	*Uses*
Hexamethylenediamine	$H_2N(CH_2)_6NH_2$	Nylon 66	Fibers, clothing, tire cord, bearings
Adipic acid	$HOOC(CH_2)_4COOH$		
Ethylene glycol	$HOCH_2CH_2OH$	Dacron, Terylene, Mylar	Fibers, clothing tire cord, film
Dimethyl terephthalate	$CH_3OOC-C_6H_4-COOCH_3$ (para)		
Caprolactam	seven-membered cyclic amide ring (C=O, N—H)	Nylon 6, Perlon	Fibers, large cast articles

(hexanedioic acid) with the six-carbon hexamethylenediamine (1,6-hexanediamine) at 280°C:

$$\underset{\textbf{Adipic acid}}{HO\overset{O}{\overset{\|}{C}}CH_2CH_2CH_2CH_2\overset{O}{\overset{\|}{C}}OH} + \underset{\textbf{Hexamethylenediamine}}{H_2NCH_2CH_2CH_2CH_2CH_2CH_2NH_2} \xrightarrow{\text{Heat}}$$

$$\underset{\textbf{Nylon 66}}{\left(\overset{O}{\overset{\|}{C}}CH_2CH_2CH_2CH_2\overset{O}{\overset{\|}{C}}-NHCH_2CH_2CH_2CH_2CH_2CH_2NH \right)_n} + 2n\ H_2O$$

Nylons are used both in engineering applications and in making fibers. A combination of high-impact strength and abrasion resistance makes nylon an excellent metal substitute for bearings and gears. As fibers, nylon is used in a variety of applications, from clothing to tire cord to mountaineering ropes.

Polyesters

Just as polyamides are made by reaction between diacids and diamines, **polyesters** are step-growth polymers made by reaction between diacids and dialcohols. The most generally useful polyester is made by a nucleophilic acyl substitution reaction between dimethyl terephthalate (dimethyl 1,4-benzenedicarboxylate) and ethylene glycol. The product is used under the trade name Dacron to make clothing fiber and tire cord, and under the name Mylar to make plastic film and recording tape. The tensile strength of polyester film is nearly equal to that of steel.

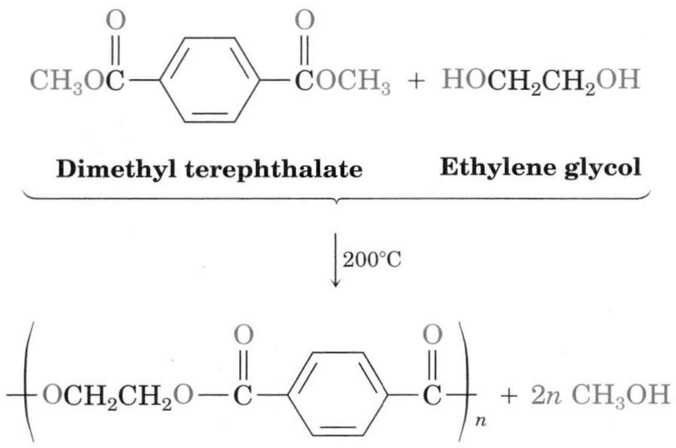

$$CH_3OC(=O){-}C_6H_4{-}C(=O)OCH_3 + HOCH_2CH_2OH$$

Dimethyl terephthalate **Ethylene glycol**

$$\xrightarrow{200^\circ C}$$

$$\left(OCH_2CH_2O{-}C(=O){-}C_6H_4{-}C(=O) \right)_n + 2n\ CH_3OH$$

Polyester, Dacron, Mylar

PRACTICE PROBLEM 10.13

Draw the structure of Qiana, a polyamide made by high-temperature reaction of hexanedioic acid with 1,4-cyclohexanediamine.

Solution

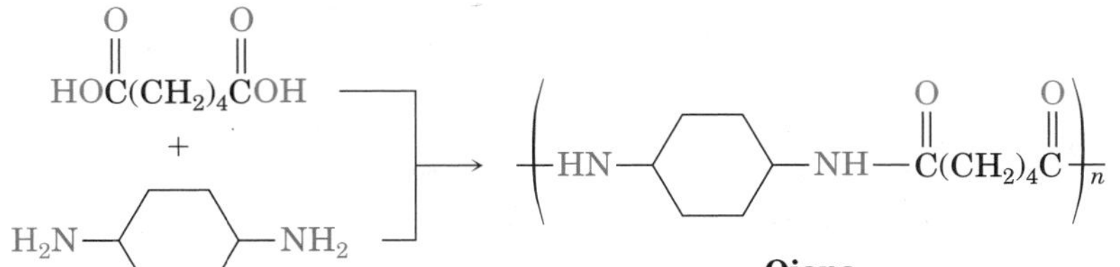

$$HOC(=O)(CH_2)_4C(=O)OH + H_2N{-}C_6H_{10}{-}NH_2 \longrightarrow \left(HN{-}C_6H_{10}{-}NH{-}C(=O)(CH_2)_4C(=O) \right)_n$$

Qiana

PROBLEM

10.30 Kevlar, a nylon polymer used in bulletproof vests, is made by reaction of 1,4-benzenedicarboxylic acid with 1,4-benzenediamine. Show the structure of Kevlar.

INTERLUDE

β-Lactam Antibiotics

Penicillium mold growing in a Petri dish.

The value of hard work and logical thinking shouldn't be underestimated, but sheer luck also plays a role in most real scientific breakthroughs. What has been called "the supreme example [of luck] in all scientific history" occurred in the late summer of 1928 when the Scottish bacteriologist Alexander Fleming went on vacation, leaving in his lab a culture plate recently inoculated with the bacterium *Staphylococcus aureus.*

While Fleming was away, an extraordinary chain of events occurred. First, a 9 day cold spell lowered the laboratory temperature to a point where the *Staphylococcus* on the plate could not grow. During this time, spores from a colony of the mold *Penicillium notatum* being grown on the floor below wafted up into Fleming's lab and landed in the culture plate. The temperature then rose, and both *Staphylococcus* and *Penicillium* began to grow. On returning from vacation, Fleming discarded the plate into a tray of antiseptic, intending to sterilize it. Evidently, though, the plate did not sink deeply enough into the antiseptic, because when Fleming happened to glance at it a few days later, what he saw changed the course of human history. He noticed that the growing *Penicillium* mold appeared to dissolve the colonies of staphylococci.

Fleming realized that the mold must be producing a chemical that killed bacteria, and he spent several years trying to isolate the substance. Finally, in 1939, the Australian pathologist Howard Florey and the German refugee Ernst Chain managed to isolate the active substance, called *penicillin.* The dramatic ability of penicillin to cure infections in mice was soon demonstrated, and successful tests in humans followed shortly thereafter. By 1943, penicillin was being produced on a large scale for military use, and by 1944 it was being used on civilians. Fleming, Florey, and Chain shared the 1945 Nobel Prize in Medicine.

Now called *benzylpenicillin,* or penicillin G, the substance discovered by Fleming is just one member of a large class of so-called *β-lactam antibiotics,* compounds with a four-membered lactam (cyclic amide) ring. The four-membered lactam ring is fused to a five-membered, sulfur-containing ring, and the carbon atom next to the lactam carbonyl group is bonded to an acylamino substituent, RCONH–. This acylamino side chain can be varied in the laboratory to provide literally hundreds of penicillin

(*continued*)➤

analogs with different biological activity profiles. Ampicillin, for instance, has an α-aminophenylacetamido substituent, $PhCH(NH_2)CONH–$.

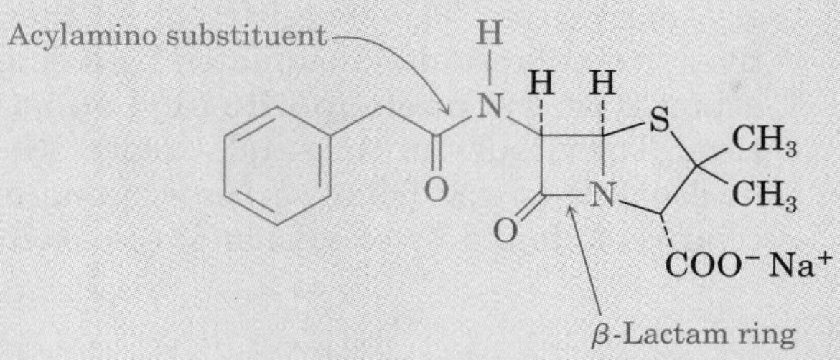

Acylamino substituent

β-Lactam ring

Benzylpenicillin (Penicillin G)

Closely related to the penicillins are the *cephalosporins,* a group of β-lactam antibiotics that contain an unsaturated six-membered, sulfur-containing ring. Cephalexin, marketed under the trade name Keflex, is an example. Cephalosporins generally have much greater antibacterial activity than penicillins, particularly against resistant strains of bacteria.

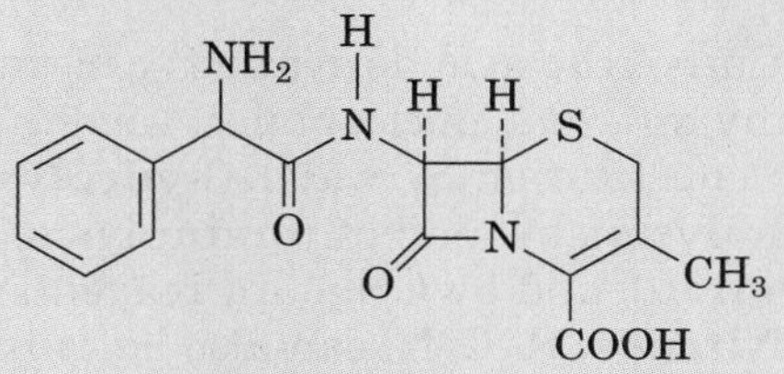

Cephalexin (a cephalosporin)

The biological activity of penicillins and cephalosporins is due to the presence of the strained β-lactam ring, which reacts with and deactivates the transpeptidase enzyme needed to synthesize and repair bacterial cell walls. With its wall either incomplete or weakened, the bacterial cell ruptures and dies.

Summary and Key Words

carboxyl group, 309
carboxylation, 317
chain-growth polymer, 339
Fischer esterification reaction, 322
nitrile, 312
nucleophilic acyl substitution reaction, 319
nylon, 339

Carboxylic acids are useful building blocks for synthesizing other molecules, both in nature and in the chemical laboratory. The distinguishing characteristic of carboxylic acids is their acidity. Although weaker than mineral acids like HCl, carboxylic acids are much more acidic than alcohols because carboxylate ions are stabilized by resonance. Most carboxylic acids have pK_a values near 5, but the exact acidity of an acid depends on its structure. Carboxylic acids substituted by an electron-withdrawing group are more acidic (have a lower pK_a) because their carboxylate ions are more stable.

Carboxylic acids can be transformed into a variety of **acyl derivatives** in which the acid –OH group has been replaced by another substituent. **Acid chlorides, acid anhydrides, esters,** and **amides** are the most common acyl derivatives. The chemistry of all these acyl derivatives is similar and is dominated by a single general reaction type: the **nucleophilic acyl substitution reaction.** These substitutions take place by addition of a nucleophile to the polar carbonyl group of the acid derivative, followed by expulsion of the leaving group.

$$\text{RC(=O)Y} + :\text{Nu}^- \longrightarrow \left[\text{R}-\text{C}(\text{O}^-)(\text{Y})-\text{Nu}\right] \longrightarrow \text{RC(=O)Nu} + :\text{Y}^-$$

where Y = Cl, Br, I (acid halide); OR (ester); OCOR (anhydride); or NH_2 (amide)

Carboxylic acid derivatives can undergo reaction with many different nucleophiles. Among the most important are substitution by water (**hydrolysis**), by alcohols (**alcoholysis**), by amines (**aminolysis**), by hydride ion (**reduction**), and by Grignard reagents.

Nitriles, R–C≡N, can also be considered as carboxylic acid derivatives because they undergo nucleophilic additions to the polar C≡N bond in the same way carbonyl compounds do. The most important reactions of nitriles are their hydrolysis to yield carboxylic acids, their reduction to yield primary amines, and their reaction with Grignard reagents to yield ketones.

Summary of Reactions

1. Reactions of carboxylic acids (Section 10.6)
   (a) Conversion into acid chlorides

$$\text{RCOOH} \xrightarrow{SOCl_2} \text{RCOCl}$$

   (b) Conversion into esters (Fischer esterification)

$$\text{RCOOH} + \text{R'OH} \xrightarrow[\text{catalyst}]{H^+} \text{RCOOR'}$$

2. Reactions of acid halides (Section 10.7)
   (a) Conversion into carboxylic acids

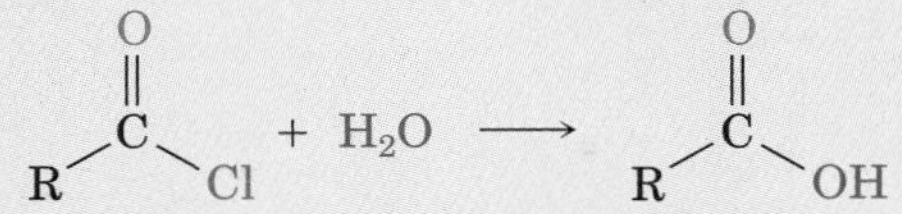

$$\mathrm{RCOCl} + \mathrm{H_2O} \longrightarrow \mathrm{RCOOH}$$

   (b) Conversion into esters

$$\mathrm{RCOCl} + \mathrm{R'OH} \xrightarrow{\text{Pyridine}} \mathrm{RCOOR'}$$

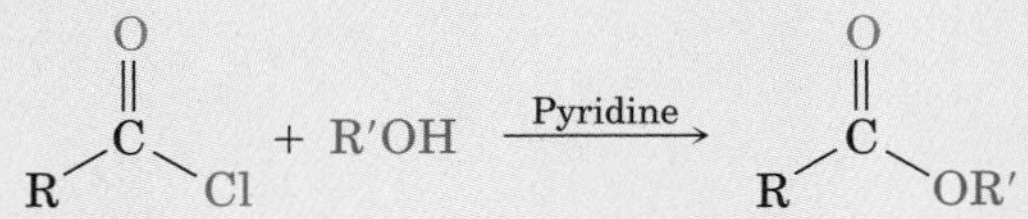

   (c) Conversion into amides

$$\mathrm{RCOCl} + \mathrm{NH_3} \longrightarrow \mathrm{RCONH_2}$$

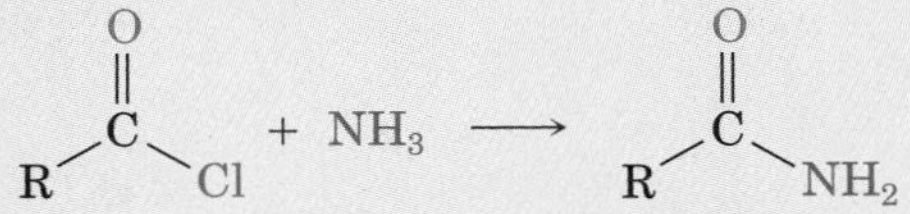

3. Reactions of acid anhydrides (Section 10.8)
   (a) Conversion into esters

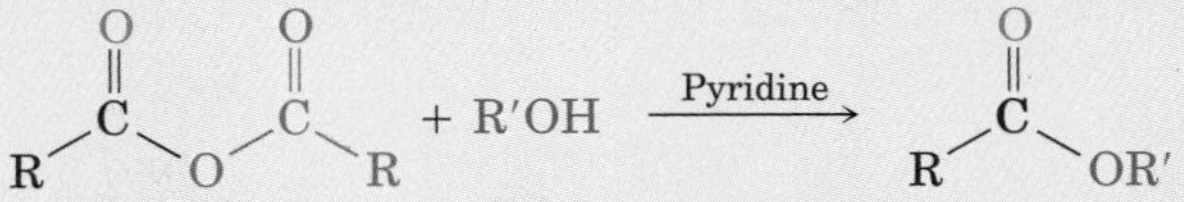

$$\mathrm{RCOOCOR} + \mathrm{R'OH} \xrightarrow{\text{Pyridine}} \mathrm{RCOOR'}$$

   (b) Conversion into amides

$$\mathrm{RCOOCOR} + \mathrm{NH_3} \xrightarrow{\text{Pyridine}} \mathrm{RCONH_2}$$

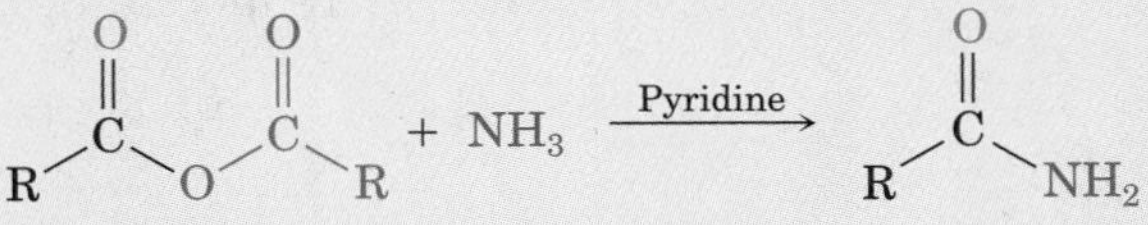

4. Reactions of esters (Section 10.9)
   (a) Conversion into acids

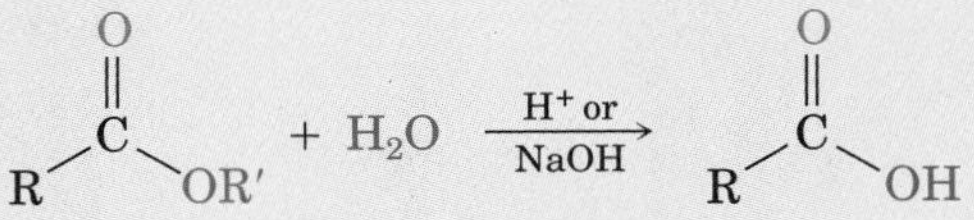

$$\mathrm{RCOOR'} + \mathrm{H_2O} \xrightarrow[\text{NaOH}]{\mathrm{H^+}\text{ or}} \mathrm{RCOOH}$$

   (b) Conversion into amides

$$\mathrm{RCOOR'} + \mathrm{NH_3} \longrightarrow \mathrm{RCONH_2}$$

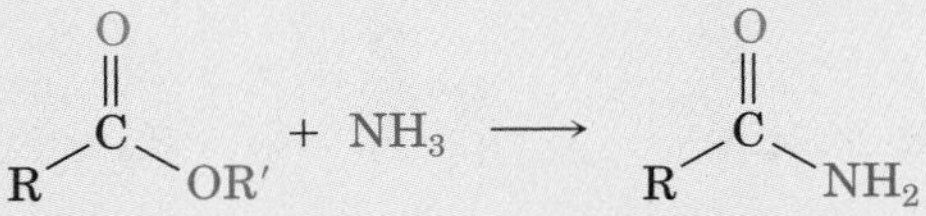

   (c) Conversion into primary alcohols by reduction

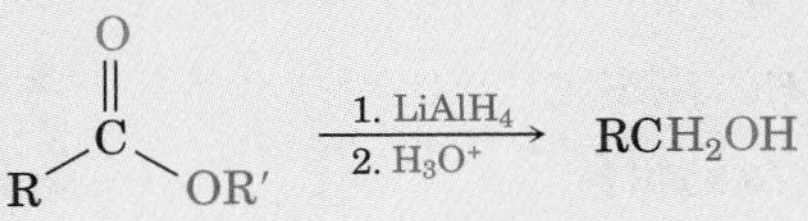

$$\mathrm{RCOOR'} \xrightarrow[\text{2. } \mathrm{H_3O^+}]{\text{1. } \mathrm{LiAlH_4}} \mathrm{RCH_2OH}$$

(d) Conversion into tertiary alcohols by Grignard reaction

$$\mathrm{R{-}C({=}O){-}OR'} \xrightarrow[\text{2. } H_3O^+]{\text{1. } R''MgX} \mathrm{R{-}C(OH)(R''){-}R''}$$

5. Reactions of amides (Section 10.10)

(a) Conversion into carboxylic acids

$$\mathrm{R{-}C({=}O){-}NH_2} + H_2O \xrightarrow[\text{NaOH}]{H^+ \text{ or}} \mathrm{R{-}C({=}O){-}OH}$$

(b) Conversion into amines by reduction

$$\mathrm{R{-}C({=}O){-}NH_2} \xrightarrow[\text{2. } H_2O]{\text{1. } LiAlH_4} RCH_2NH_2$$

6. Reactions of nitriles (Section 10.11)

(a) Conversion into carboxylic acids

$$\mathrm{R{-}C{\equiv}N} + H_2O \xrightarrow[\text{NaOH}]{H^+ \text{ or}} \mathrm{R{-}C({=}O){-}OH}$$

(b) Conversion into amines by reduction

$$\mathrm{R{-}C{\equiv}N} \xrightarrow[\text{2. } H_2O]{\text{1. } LiAlH_4} RCH_2NH_2$$

(c) Conversion into ketones by Grignard reaction

$$\mathrm{R{-}C{\equiv}N} \xrightarrow[\text{2. } H_2O]{\text{1. } R'MgX} \mathrm{R{-}C({=}O){-}R'}$$

ADDITIONAL PROBLEMS

10.31 Give IUPAC names for the following carboxylic acids:

(a) $CH_3CH(COOH)CH_2CH_2CH(COOH)CH_3$

(b) $(CH_3)_3CCOOH$

(c) 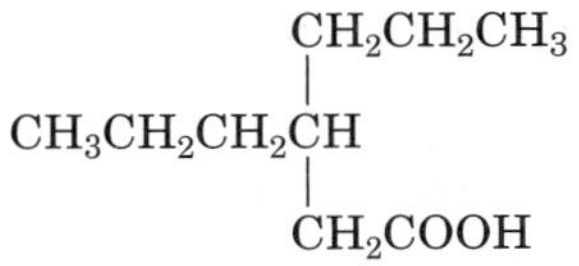 $CH_3CH_2CH_2CH(CH_2CH_2CH_3)CH_2COOH$

$$\begin{array}{l} \qquad\qquad\quad CH_2CH_2CH_3 \\ \qquad\qquad\quad | \\ CH_3CH_2CH_2CH \\ \qquad\qquad\quad | \\ \qquad\qquad\quad CH_2COOH \end{array}$$

(d) 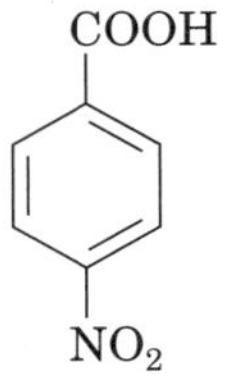 [benzene ring with COOH and para NO_2]

(e) 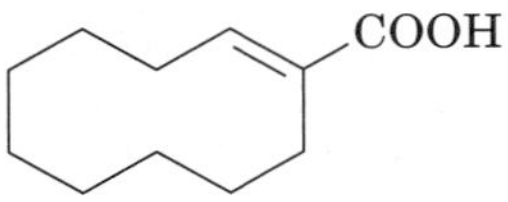 [cyclodecene ring with COOH on the double-bond carbon]

(f) $BrCH_2CH(Br)CH_2CH_2COOH$

10.32 Give IUPAC names for the following carboxylic acid derivatives:

(a) 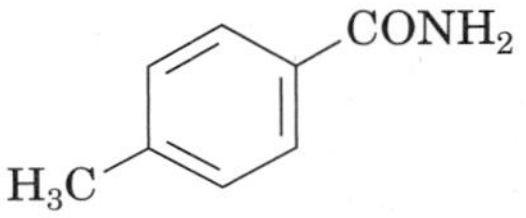 [benzene ring with $CONH_2$ and H_3C substituents, meta]

(b) $(CH_3CH_2)_2CHCH{=}CHCN$

(c) $CH_3O_2CCH_2CH_2CO_2CH_3$

(d) 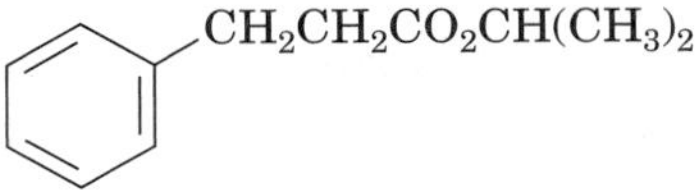 [benzene ring]$-CH_2CH_2CO_2CH(CH_3)_2$

(e) 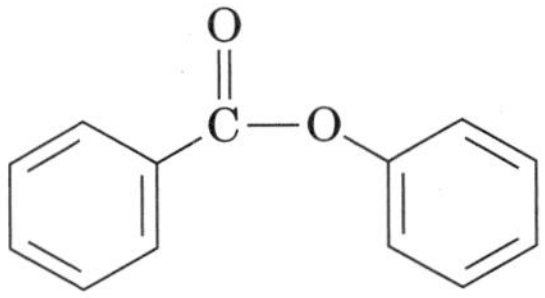 [benzene ring]$-\overset{O}{\overset{\|}{C}}-O-$[benzene ring]

(f) $CH_3CH(Br)CH_2CONHCH_3$

(g) 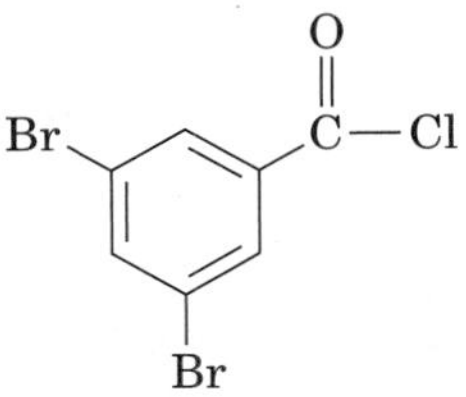 [benzene ring with $\overset{O}{\overset{\|}{C}}-Cl$ and two Br substituents, 3,5-positions]

(h) 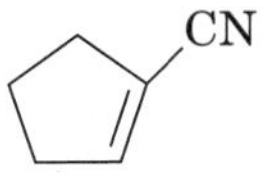 [cyclopentene ring with CN on the double-bond carbon]

10.33 Draw structures corresponding to the following IUPAC names:

(a) 4,5-Dimethylheptanoic acid
(b) *cis*-1,2-Cyclohexanedicarboxylic acid
(c) Heptanedioic acid
(d) Triphenylacetic acid
(e) 2,2-Dimethylhexanamide
(f) Phenylacetamide
(g) 2-Cyclobutenecarbonitrile
(h) Ethyl cyclohexanecarboxylate

10.34 Acetic acid boils at 118°C, but its ethyl ester boils at 77°C. Why is the boiling point of the acid so much higher, even though it has a lower molecular weight?

10.35 Draw and name the eight carboxylic acids with formula $C_6H_{12}O_2$. Which are chiral?

10.36 Draw and name compounds that meet the following descriptions:

(a) Three acid chlorides, C_6H_9ClO
(b) Three amides, $C_7H_{11}NO$
(c) Three nitriles, C_5H_7N
(d) Three esters, $C_5H_8O_2$

10.37 The following reactivity order has been found for the saponification of alkyl acetates by aqueous NaOH:

$$CH_3COOCH_3 > CH_3COOCH_2CH_3 > CH_3COOCH(CH_3)_2 > CH_3COOC(CH_3)_3$$

How can you explain this reactivity order?

10.38 Citric acid has pK_a = 3.14, and tartaric acid has pK_a = 2.98. Which acid is stronger?

10.39 Order the compounds in each of the following sets with respect to increasing acidity:
(a) Acetic acid, chloroacetic acid, trifluoroacetic acid
(b) Benzoic acid, *p*-bromobenzoic acid, *p*-nitrobenzoic acid
(c) Acetic acid, phenol, cyclohexanol

10.40 How can you explain the fact that 2-chlorobutanoic acid has pK_a = 2.86, 3-chlorobutanoic acid has pK_a = 4.05, 4-chlorobutanoic acid has pK_a = 4.52, and butanoic acid itself has pK_a = 4.82?

10.41 Rank the following compounds in order of their reactivity toward nucleophilic acyl substitution:
(a) CH_3COOCH_3 (b) CH_3COCl
(c) CH_3CONH_2 (d) $CH_3CO_2COCH_3$

10.42 How can you prepare acetophenone (methyl phenyl ketone) from the following starting materials? (More than one step may be required.)
(a) Benzonitrile (b) Bromobenzene
(c) Methyl benzoate (d) Benzene

10.43 How might you prepare the following products from butanoic acid? (More than one step may be required.)
(a) 1-Butanol (b) Butanal
(c) 1-Bromobutane (d) Pentanenitrile
(e) 1-Butene (f) Butylamine, $CH_3CH_2CH_2CH_2NH_2$

10.44 Predict the product of the reaction of *p*-methylbenzoic acid with each of the following reagents:
(a) $LiAlH_4$ (b) CH_3OH, HCl (c) $SOCl_2$ (d) NaOH, then CH_3I

10.45 A chemist in need of 2,2-dimethylpentanoic acid decided to synthesize some by reaction of 2-chloro-2-methylpentane with NaCN, followed by hydrolysis of the product. After carrying out the reaction sequence, however, none of the desired product could be found. What do you suppose went wrong?

10.46 Which method of carboxylic acid synthesis, Grignard carboxylation or nitrile hydrolysis, would you use for each of the following reactions? Explain.

(a) o-HOC$_6$H$_4$CH$_2$Br (benzene ring with CH_2Br and OH) $\longrightarrow$ o-HOC$_6$H$_4$CH$_2$COOH (benzene ring with CH_2COOH and OH)

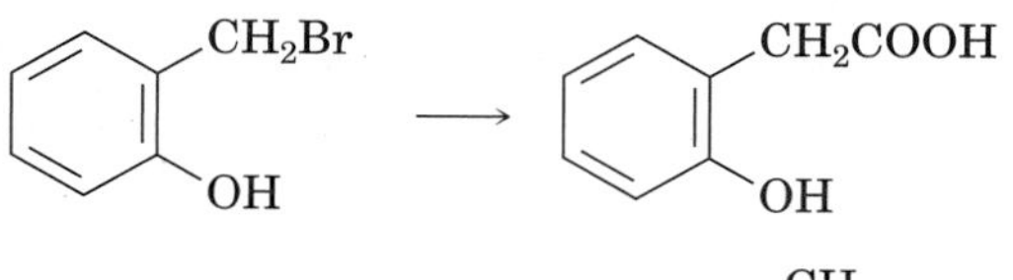

(b) $CH_3CH_2CH(Br)CH_3 \longrightarrow CH_3CH_2CH(CH_3)COOH$

(c) $CH_3C(=O)CH_2CH_2CH_2I \longrightarrow CH_3C(=O)CH_2CH_2CH_2COOH$

(d) $HOCH_2CH_2CH_2Br \longrightarrow HOCH_2CH_2CH_2COOH$

10.47 How can you explain the observation that an attempted Fischer esterification of 2,4,6-trimethylbenzoic acid with methanol/HCl is unsuccessful? No ester is obtained, and the starting acid is recovered unchanged.

10.48 Acid chlorides undergo reduction with $LiAlH_4$ in the same way that esters do to yield primary alcohols. What are the products of the following reactions?

(a) $CH_3CH(CH_3)CH_2CH_2COCl \xrightarrow[2.\ H_2O]{1.\ LiAlH_4}$? (b) 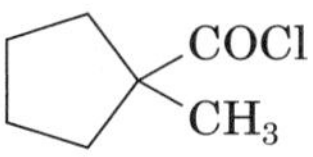 1-methylcyclopentanecarbonyl chloride (cyclopentane ring bearing COCl and CH_3) $\xrightarrow[2.\ H_2O]{1.\ LiAlH_4}$?

10.49 The reaction of an acid chloride with $LiAlH_4$ to yield a primary alcohol (Problem 10.48) takes place in two steps. The first step is a nucleophilic acyl substitution of H^- for Cl^- to yield an aldehyde, and the second step is nucleophilic addition of H^- to the aldehyde to yield an alcohol. Write the mechanism of the reduction of CH_3COCl.

10.50 Acid chlorides undergo reaction with Grignard reagents at −78°C to yield ketones. Propose a mechanism for the reaction.

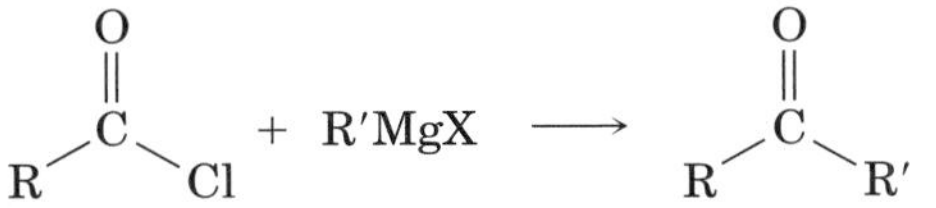

$$RCOCl + R'MgX \longrightarrow RCOR'$$

10.51 If the reaction of an acid chloride with a Grignard reagent (Problem 10.50) is carried out at room temperature, a tertiary alcohol is formed.
(a) Propose a mechanism for this reaction.
(b) What are the products of the reaction of CH_3MgBr with the acid chlorides given in Problem 10.48?

10.52 When dimethyl carbonate, $CH_3OCO_2CH_3$, is treated with phenylmagnesium bromide, triphenylmethanol is formed. Explain.

10.53 Predict the product, if any, of reaction between propanoyl chloride and the following reagents. (See Problems 10.49 and 10.50.)
(a) Excess CH_3MgBr in ether (b) NaOH in H_2O (c) Methylamine
(d) $LiAlH_4$ (e) Cyclohexanol (f) Sodium acetate

10.54 Answer Problem 10.53 for reaction between methyl propanoate and the listed reagents.

10.55 What esters and what Grignard reagents would you use to make the following alcohols?

(a) $CH_3CH_2CH_2C(OH)(CH_3)CH_3$ (b) 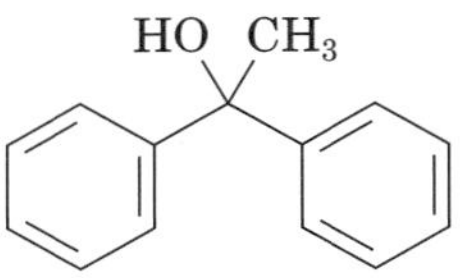 $(C_6H_5)_2C(OH)CH_3$

10.56 Show two ways to make the following esters:

(a) $CH_3CH(CH_3)CH_2CH_2COCH_2CH_3$ (C=O ester) (b) 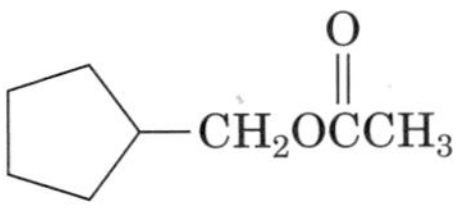 cyclopentyl–CH_2OCCH_3 (C=O ester)

10.57 What products would you obtain on saponification of the following esters?

(a) 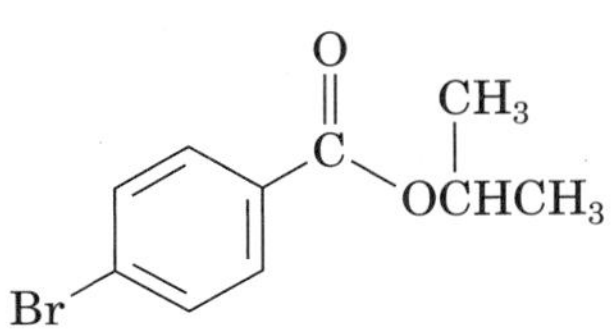 $p\text{-}BrC_6H_4COOCH(CH_3)CH_3$ (b) Cyclohexyl propanoate

10.58 When *methyl* acetate is heated in pure ethanol containing a small amount of HCl catalyst, *ethyl* acetate results. Propose a mechanism for this reaction.

10.59 *tert*-Butoxycarbonyl azide, an important reagent used in protein synthesis, is prepared by treating *tert*-butoxycarbonyl chloride with sodium azide. Propose a mechanism for this reaction.

$$\begin{array}{c}H_3C\ \ O\\ |\ \ \ \ \|\\ CH_3COCCl\\ |\\ H_3C\end{array} + NaN_3 \longrightarrow \begin{array}{c}H_3C\ \ O\\ |\ \ \ \ \|\\ CH_3COCN_3\\ |\\ H_3C\end{array} + NaCl$$

10.60 What product would you expect to obtain on treatment of the cyclic ester butyrolactone with excess phenylmagnesium bromide?

Butyrolactone

10.61 *N,N*-Diethyl-*m*-toluamide (DEET) is the active ingredient in many insect repellents. How might you synthesize DEET from *m*-bromotoluene?

_N,N_-Diethyl-_m_-toluamide

10.62 In the *iodoform reaction,* a triiodomethyl ketone reacts with aqueous NaOH to yield a carboxylate ion and iodoform (triiodomethane). Propose a mechanism for this reaction.

$$R-\overset{O}{\overset{\|}{C}}-CI_3 \xrightarrow{NaOH,\ H_2O} R-\overset{O}{\overset{\|}{C}}-O^- + CHI_3$$

10.63 The K_a for bromoacetic acid is approximately 1×10^{-3}. What percentage of the acid is dissociated in a 0.10 M aqueous solution?

10.64 As indicated in Table 10.3, the step-growth polymer called nylon 6 is prepared from caprolactam. The reaction involves initial reaction of caprolactam with water to give an intermediate amino acid, followed by heating to form the polymer. Propose mechanisms for both steps, and show the structure of nylon 6.

Visualizing Chemistry

10.65 Name the following compounds (gray = C, red = O, blue = N, light green = H):

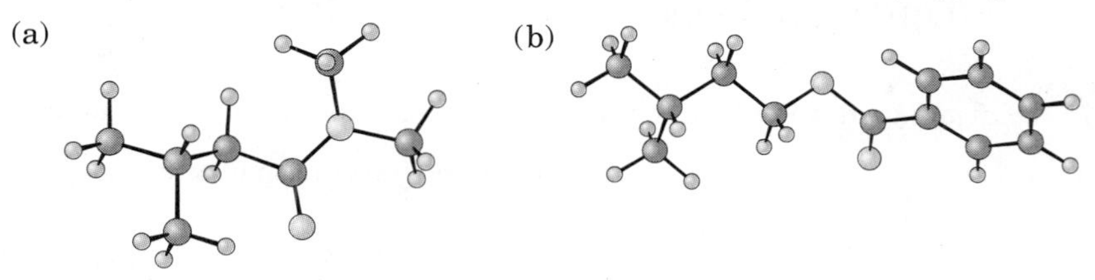

10.66 Show how you could prepare each of the following compounds starting with an appropriate carboxylic acid and any other reagents needed (gray = C, red = O, blue = N, yellow-green = Cl, light green = H):

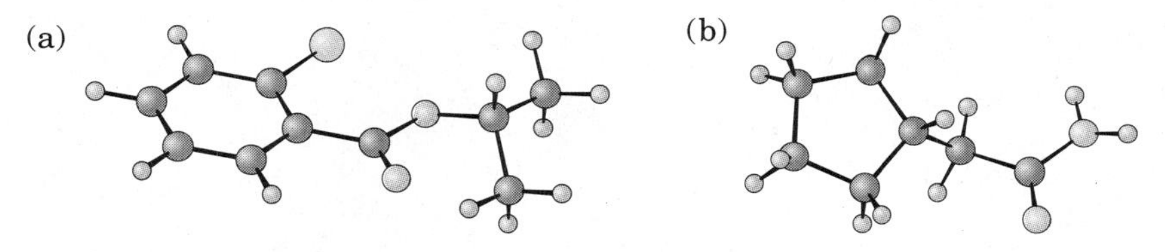

10.67 The following structure represents a tetrahedral alkoxide ion intermediate formed by addition of a nucleophile to a carboxylic acid derivative. Identify the nucleophile, the leaving group, the reactant, and the product (gray = C, red = O, blue = N, yellow-green = Cl, light green = H).

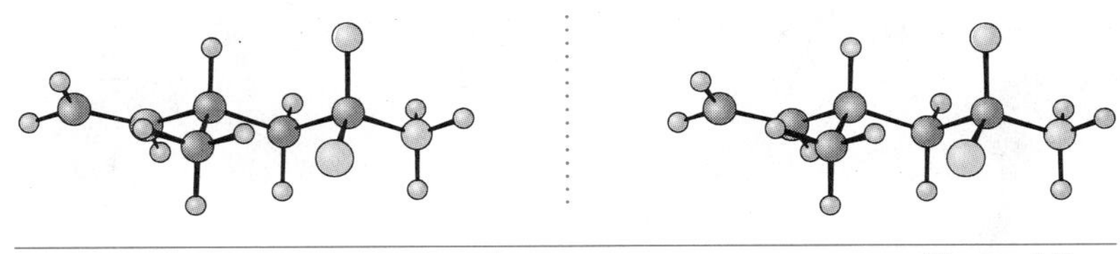

Stereo View

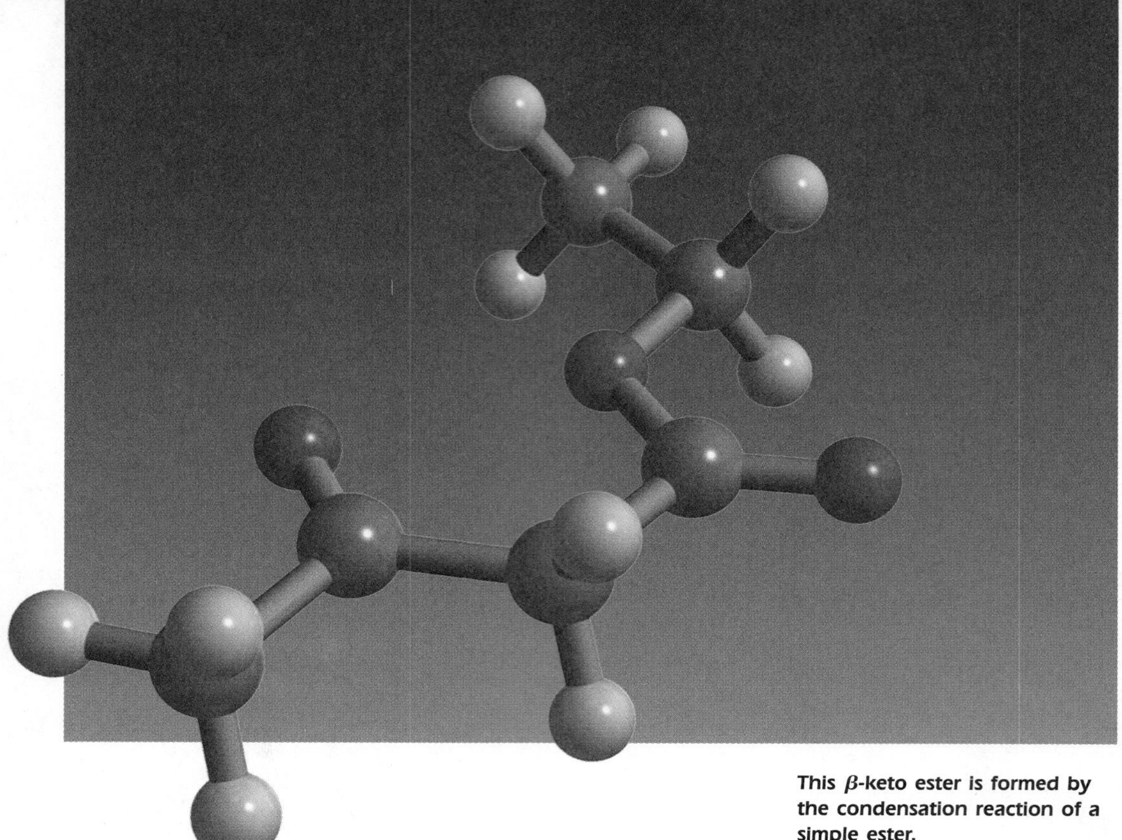

This β-keto ester is formed by the condensation reaction of a simple ester.

11 Carbonyl Alpha-Substitution Reactions and Condensation Reactions

Much of the chemistry of carbonyl compounds can be explained by just four fundamental reactions. We've already looked in detail at two of the four: the nucleophilic addition reaction (Chapter 9) and the nucleophilic acyl substitution reaction (Chapter 10). In this chapter, we'll look at the remaining two: the *alpha-substitution reaction* and the *carbonyl condensation reaction.*

Alpha-substitution reactions occur at the position *next to* the carbonyl group—the **alpha (α) position**—and result in the substitution of an α hydrogen atom by some other group:

$$\text{—C(=O)—C}_\alpha\text{—H} \xrightarrow{\text{Sub}^+} \text{—C(=O)—C—Sub} + \text{H}^+$$

Carbonyl condensation reactions take place when *two* carbonyl compounds react with each other in such a way that the α carbon of one partner becomes bonded to the carbonyl carbon of the second partner:

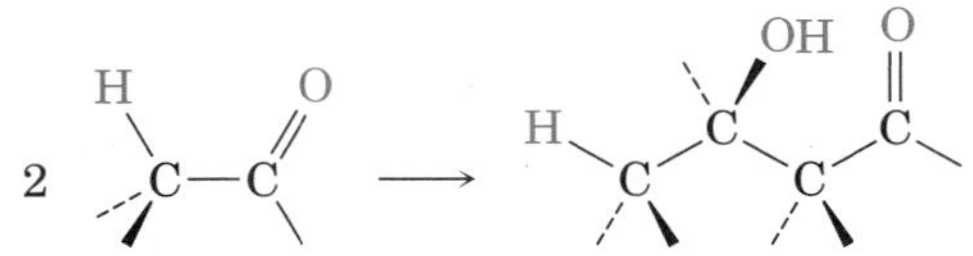

The key feature of both α-substitution reactions and carbonyl condensation reactions is that they take place through the formation of either *enol* or *enolate ion* intermediates. Let's begin our study by learning more about these two species.

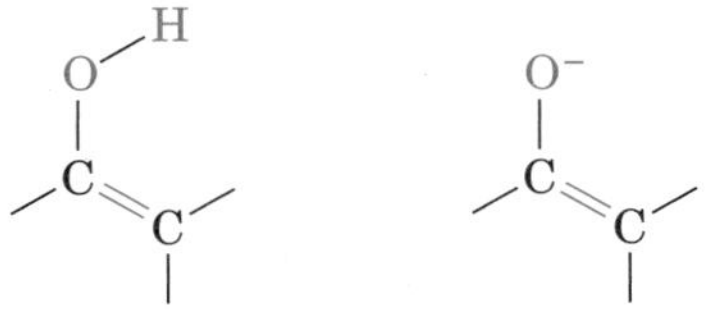

An enol An enolate ion

11.1 Keto–Enol Tautomerism

A carbonyl compound that has a hydrogen atom on its α carbon rapidly interconverts with its corresponding **enol** (*ene* + *ol;* unsaturated alcohol) isomer. As we saw in Section 4.15, this interconversion between keto and enol forms is a special kind of isomerism called *tautomerism* (Greek *tauto,* "the same," and *meros,* "part"). The individual isomers are called **tautomers.**

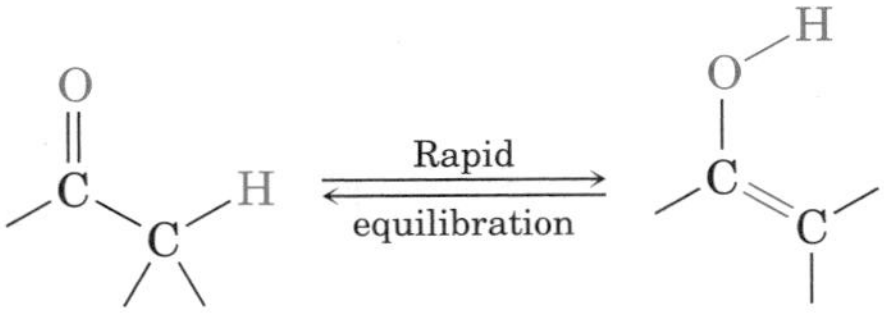

Rapid equilibration

Keto tautomer Enol tautomer

Note that two isomers must interconvert *rapidly* to be considered tautomers. Thus, keto and enol isomers of carbonyl compounds *are* tautomers, but two alkene isomers such as 1-butene and 2-butene are not, because they don't interconvert rapidly.

Most carbonyl compounds exist almost entirely in the keto form at equilibrium, and it's usually difficult to isolate the pure enol. Cyclohexanone, for example, contains only about 0.000 1% of its enol tautomer at room temperature, and acetone contains only about 0.000 001% enol. The amount of enol tautomer is even less for carboxylic acids, esters, and amides. Even though enols are difficult to isolate and are present to only a small extent at equilibrium, they are nevertheless critically important intermediates in the chemistry of carbonyl compounds.

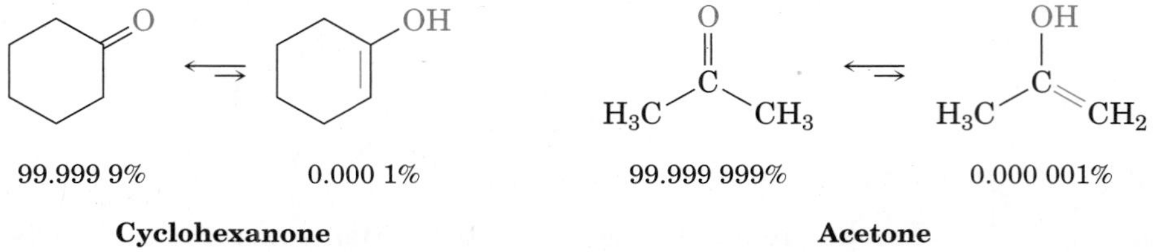

99.999 9%	0.000 1%	99.999 999%	0.000 001%
Cyclohexanone		Acetone	

Keto–enol tautomerism of carbonyl compounds is catalyzed by both acids and bases. Acid catalysis involves protonation of the carbonyl oxygen atom (a Lewis base) to give an intermediate cation that then loses H^+ from the α carbon to yield the enol (Figure 11.1). This proton loss from the positively charged intermediate is analogous to what occurs during an E1 reaction when a carbocation loses H^+ from the neighboring carbon to form an alkene (Section 7.10).

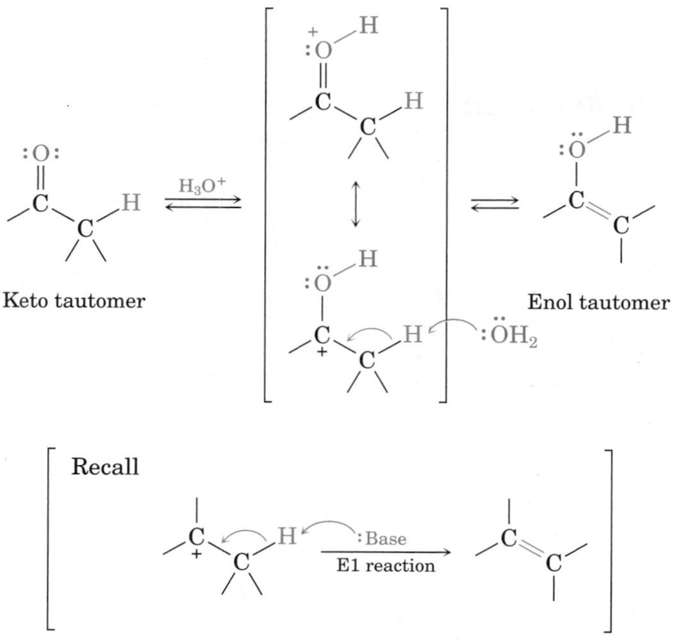

H_3O^+

Keto tautomer

Enol tautomer

$:\ddot{O}H_2$

Recall

:Base

E1 reaction

Figure 11.1 Mechanism of acid-catalyzed enol formation.

Base-catalyzed enol formation occurs because the presence of a carbonyl group makes the hydrogens on the α carbon weakly acidic. Thus, a carbonyl compound can act as an acid and lose one of its α hydrogens to the base. The resultant resonance-stabilized anion, an **enolate ion,** is then protonated to yield a neutral compound. If protonation of the enolate ion takes place on the α carbon, the keto tautomer is regenerated, and no net change occurs. If, however, protonation takes place on the oxygen atom, then an enol tautomer is formed (Figure 11.2).

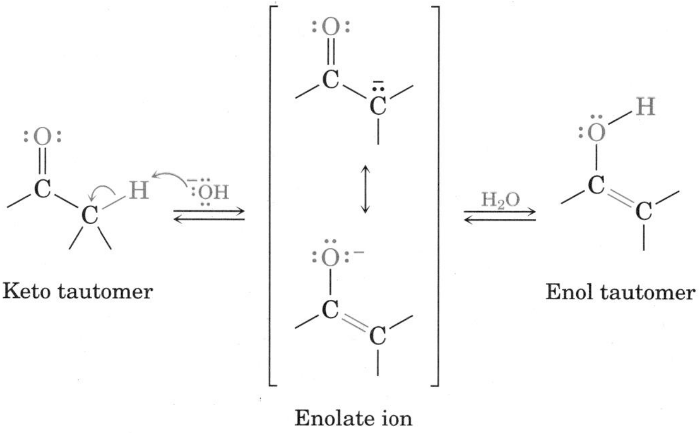

$$\text{Keto tautomer} \underset{}{\overset{:\ddot{O}H^-}{\rightleftharpoons}} \left[\text{Enolate ion}\right] \underset{}{\overset{H_2O}{\rightleftharpoons}} \text{Enol tautomer}$$

Keto tautomer

Enolate ion

Enol tautomer

Figure 11.2 Mechanism of base-catalyzed enol formation. The intermediate enolate anion, a resonance hybrid of two forms, can be protonated either on carbon to regenerate the starting ketone or on oxygen to give an enol.

Note that only the protons on the α position of carbonyl compounds are acidic. The protons at beta (β), gamma (γ), delta (δ), and other positions aren't acidic, because the resulting anions can't be resonance-stabilized by the carbonyl group.

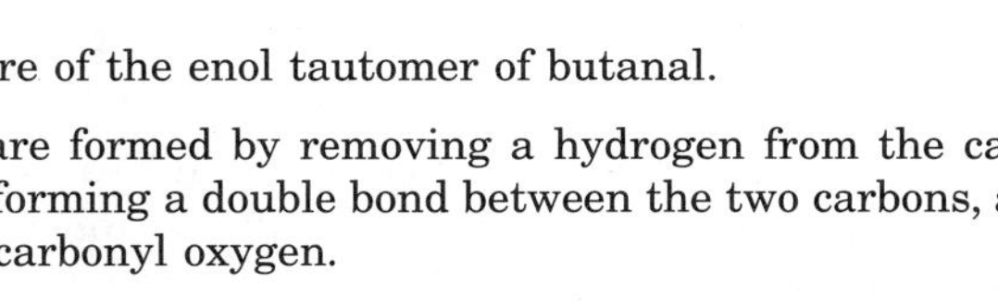

Acidic — Not acidic

$$-\overset{O}{\overset{\|}{C}}-\underset{\alpha}{\overset{H}{\overset{|}{C}}}-\underset{\beta}{\overset{H}{\overset{|}{C}}}-\underset{\gamma}{\overset{H}{\overset{|}{C}}}-\underset{\delta}{\overset{H}{\overset{|}{C}}}-$$

PRACTICE PROBLEM 11.1

Show the structure of the enol tautomer of butanal.

Solution Enols are formed by removing a hydrogen from the carbon next to the carbonyl carbon, forming a double bond between the two carbons, and replacing the hydrogen on the carbonyl oxygen.

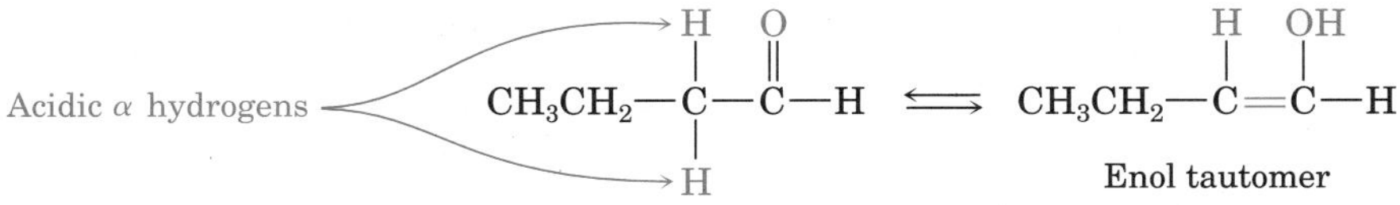

Acidic α hydrogens

$$CH_3CH_2-\underset{H}{\overset{H}{C}}-\overset{O}{\overset{\|}{C}}-H \rightleftharpoons CH_3CH_2-\overset{H}{C}=\overset{OH}{C}-H$$

Enol tautomer

PROBLEM

11.1 Draw structures for the enol tautomers of the following compounds:
(a) Cyclopentanone (b) Acetyl chloride (c) Ethyl acetate
(d) Acetic acid (e) Acetophenone (methyl phenyl ketone)

PROBLEM

11.2 How many acidic hydrogens does each of the molecules listed in Problem 11.1 have? Identify them.

PROBLEM

11.3 Account for the fact that 2-methylcyclohexanone can form two enol tautomers. Show the structures of both.

11.2 Reactivity of Enols: The Mechanism of Alpha-Substitution Reactions

What kind of chemistry might we expect of enols? Since their double bonds are electron-rich, enols behave as nucleophiles and react with electrophiles in much the same way as alkenes (Section 4.1). Because of electron donation from the oxygen lone-pair electrons, however, enols are even more reactive than alkenes.

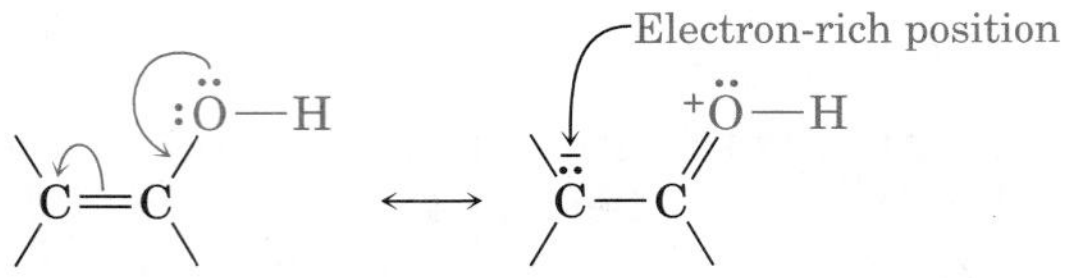

Electron-rich position

When an *alkene* reacts with an electrophile, such as Br_2, addition of Br^+ occurs to give an intermediate carbocation that reacts with Br^- to give the addition product. When an *enol* reacts with an electrophile, however, the addition step is the same but the intermediate cation loses the –OH proton to regenerate a carbonyl compound. The net result of the reaction is α substitution by the mechanism shown in Figure 11.3.

11.3 Alpha Halogenation of Aldehydes and Ketones

Aldehydes and ketones are halogenated at their α positions by reaction with Cl_2, Br_2, or I_2 in acidic solution. Bromine is most often used, and acetic acid is often employed as solvent. The reaction is a typical α-substitution process that proceeds through an enol intermediate.

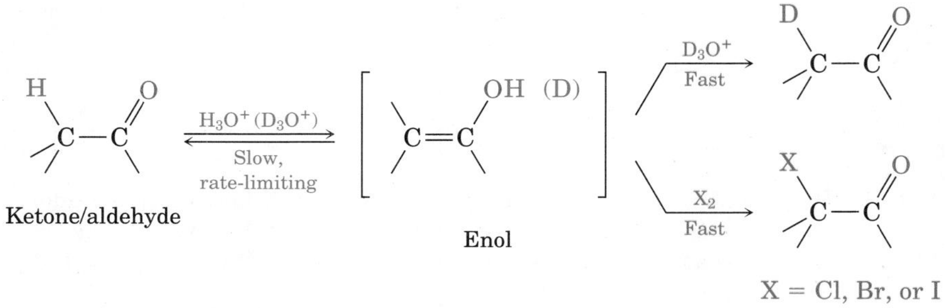

H_3O^+ (D_3O^+)
Slow, rate-limiting

Ketone/aldehyde

Enol

D_3O^+
Fast

X_2
Fast

X = Cl, Br, or I

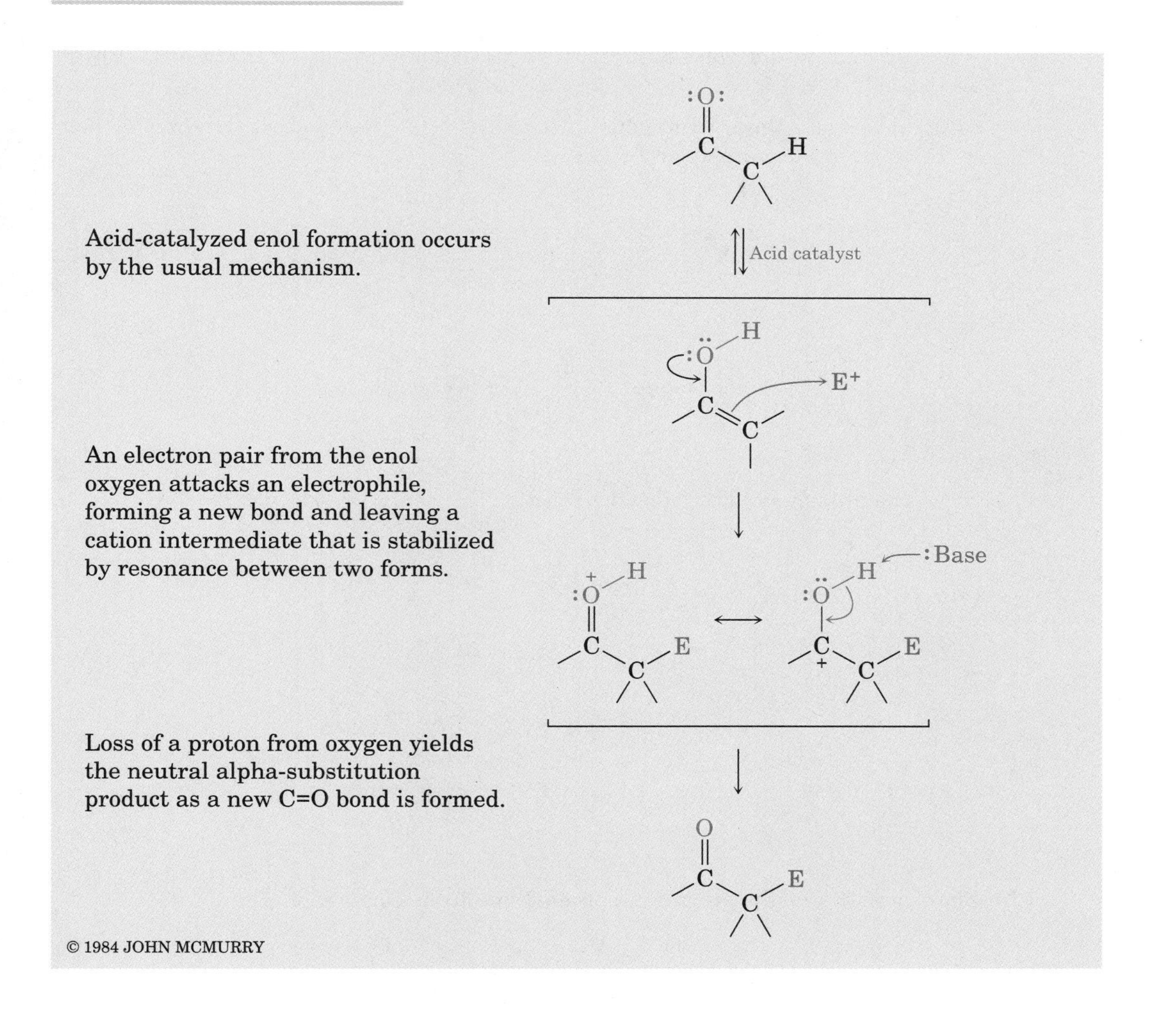

Acid-catalyzed enol formation occurs by the usual mechanism.
Acid catalyst
An electron pair from the enol oxygen attacks an electrophile, forming a new bond and leaving a cation intermediate that is stabilized by resonance between two forms.
E+
:Base
Loss of a proton from oxygen yields the neutral alpha-substitution product as a new C=O bond is formed.
© 1984 JOHN MCMURRY

Figure 11.3 The general mechanism of a carbonyl α-substitution reaction with an electrophile, E^+.

α-Bromo ketones are useful because they undergo elimination of HBr on treatment with base to yield α,β-unsaturated ketones. For example, 2-bromo-2-methylcyclohexanone gives 2-methyl-2-cyclohexenone when heated in the organic base pyridine. The reaction takes place by an E2 elimination pathway (Section 7.9) and is an excellent way of introducing C=C bonds into molecules.

$$\text{2-Methylcyclohexanone} \xrightarrow[\text{CH}_3\text{COOH}]{\text{Br}_2} \text{2-Bromo-2-methylcyclohexanone} \xrightarrow[\Delta]{\text{Pyridine}} \text{2-Methyl-2-cyclohexenone (62\%)} + \text{HBr}$$

2-Methylcyclo-hexanone **2-Bromo-2-methyl-cyclohexanone** **2-Methyl-2-cyclo-hexenone (62%)**

PRACTICE PROBLEM 11.2

What product would you obtain from the reaction of cyclopentanone with Br_2 in acetic acid?

Solution Locate the α hydrogens in the starting ketone and replace one of them by Br to carry out an α-substitution reaction:

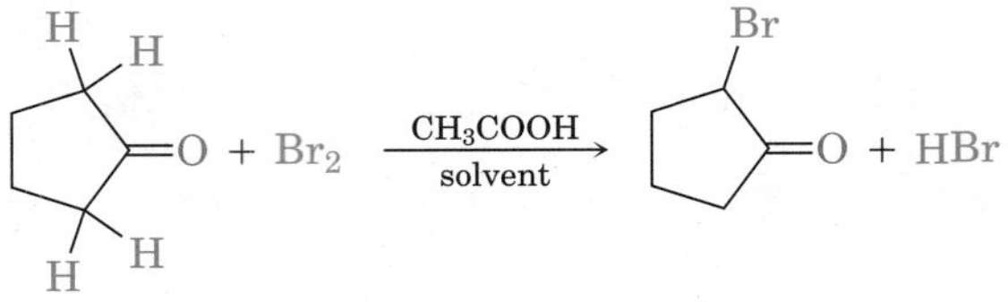

$$\text{Cyclopentanone} + Br_2 \xrightarrow[\text{solvent}]{CH_3COOH} \text{2-Bromocyclopentanone} + HBr$$

Cyclopentanone **2-Bromocyclopentanone**

PROBLEM

11.4 Show the products of the following reactions:

(a) $CH_3CH(CH_3)C(=O)CH(CH_3)CH_3 + Cl_2 \xrightarrow[\text{solvent}]{CH_3COOH}$?

(b) 2,2-Dimethylcyclohexanone $+ Br_2 \xrightarrow[\text{solvent}]{CH_3COOH}$?

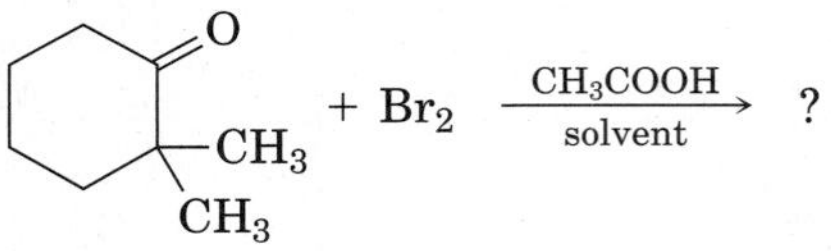

PROBLEM

11.5 Show how you might prepare 1-penten-3-one from 3-pentanone:

$$CH_3CH_2C(=O)CH_2CH_3 \longrightarrow CH_3CH_2C(=O)CH{=}CH_2$$

PROBLEM

11.6 When optically active (R)-3-phenyl-2-butanone is exposed to aqueous acid, a loss of optical activity occurs, and racemic 3-phenyl-2-butanone is produced. Explain. (Review Section 6.13.)

11.4 Acidity of Alpha Hydrogen Atoms: Enolate Ion Formation

During the discussion of base-catalyzed enol formation in Section 11.1, we said that carbonyl compounds are weak acids. Strong bases can abstract an acidic α proton from a carbonyl compound to form a resonance-stabilized enolate ion:

:Base

An enolate anion

Why are carbonyl compounds weakly acidic? If we compare acetone, pK_a = 19.3, with ethane, p$K_a \approx 60$, we find that the presence of the carbonyl group increases the acidity of the neighboring C–H by a factor of 10^{40}.

Acetone
(pK_a = 19.3)

Ethane
(p$K_a \approx 60$)

The easiest way to understand the acidity of carbonyl compounds is to look at an orbital picture of an enolate ion (Figure 11.4). Proton abstraction from a carbonyl compound occurs when the α C–H bond is oriented parallel to the p orbitals of the carbonyl group. The α carbon of the enolate ion is sp^2-hybridized and has a p orbital that overlaps the carbonyl p orbitals. Thus, the negative charge is shared by the electronegative oxygen atom, and the enolate ion is stabilized by resonance between two forms.

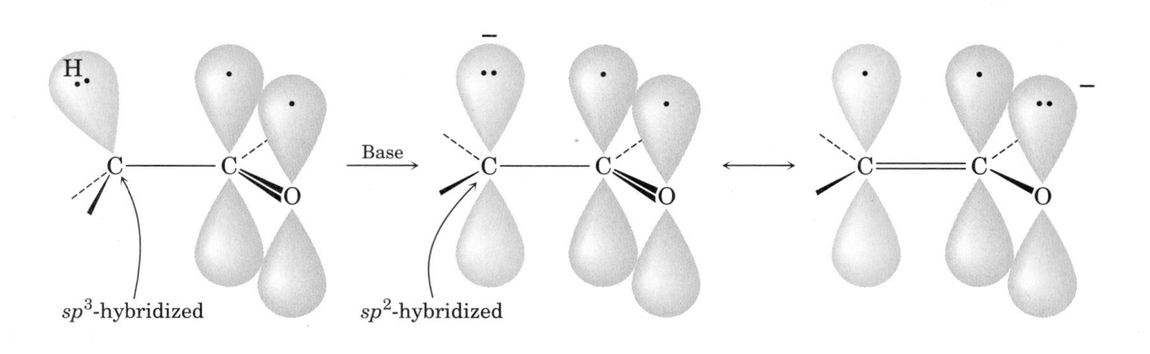

Figure 11.4 Mechanism of enolate ion formation by abstraction of an acidic α hydrogen from a carbonyl compound.

Carbonyl compounds are more acidic than alkanes for the same reason that carboxylic acids are more acidic than alcohols (Section 10.3). In both cases, the anions are stabilized by resonance. Enolate ions differ from carboxylate ions, though, because their two resonance forms aren't equivalent. The resonance form with the negative charge on the enolate oxygen atom is lower in energy than the form with the charge on carbon. Nevertheless, the principle behind resonance stabilization is the same in both cases.

CH_3CH_3 versus

Ethane
(pK_a ≈ 60)

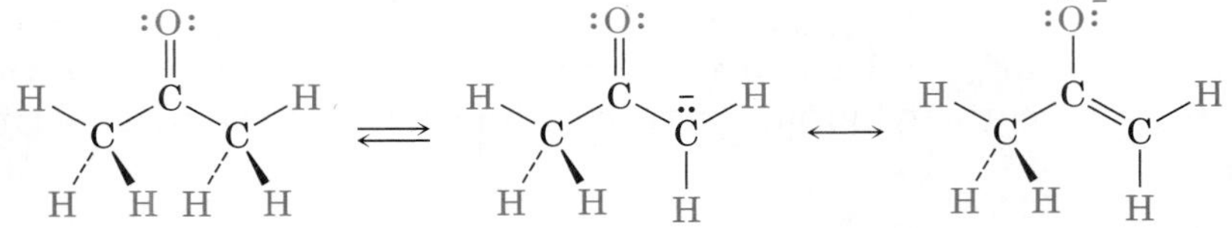

Acetone
(pK_a = 19.3)

Nonequivalent resonance forms

CH_3OH versus

Methanol
(pK_a = 15.5)

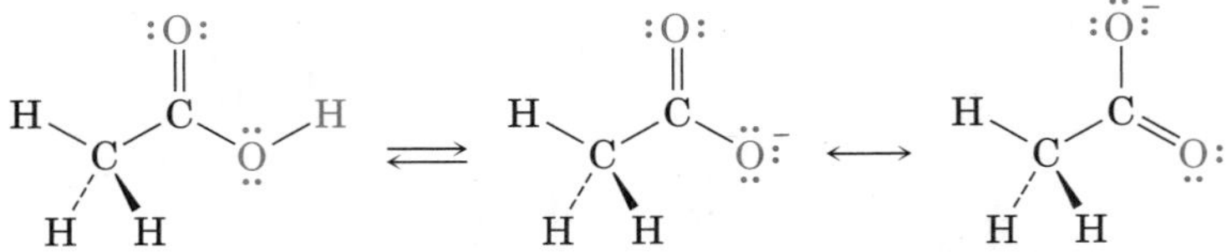

Acetic acid
(pK_a = 4.7)

Equivalent resonance forms

Because α hydrogen atoms of carbonyl compounds are only weakly acidic, strong bases are needed to form enolate ions. If an alkoxide ion such as sodium ethoxide is used, ionization of acetone takes place only to the extent of about 0.1% because ethanol (pK_a = 16) is a stronger acid than acetone (pK_a = 19.3). If, however, a more powerful base such as sodium amide ($NaNH_2$, the sodium salt of ammonia) or sodium hydride (NaH, the sodium salt of H_2) is used, then a carbonyl compound is completely converted into its enolate ion.

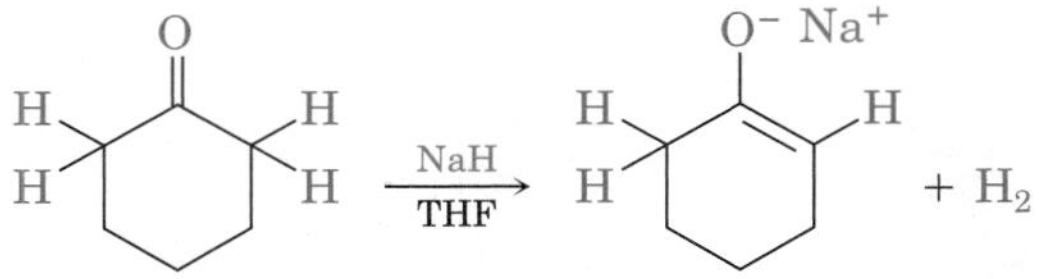

$\xrightarrow[\text{THF}]{\text{NaH}}$ + H_2

Cyclohexanone **Cyclohexanone enolate (100%)**

Table 11.1 lists the approximate pK_a values of various kinds of carbonyl compounds and shows how these values compare with other common acids.

When a C–H bond is flanked by *two* carbonyl groups, its acidity is enhanced even more. Thus, Table 11.1 shows that 1,3-diketones (called β-diketones), 3-keto esters (β-keto esters), and 1,3-diesters are more acidic than water. The enolate ions derived from these β-dicarbonyl compounds are stabilized by delocalization of the negative charge onto both of the neighboring carbonyl oxygens. For example, there are three resonance forms for the enolate ion from 2,4-pentanedione:

Table 11.1 Acidity Constants for Some Organic Compounds

Compound type	*Compound*	pK_a	
Carboxylic acid	CH_3COOH	5	Stronger acid
1,3-Diketone	$CH_2(COCH_3)_2$	9	
3-Keto ester	$CH_3COCH_2CO_2CH_3$	11	
1,3-Diester	$CH_2(CO_2CH_3)_2$	13	
Water	HOH	15.74	
Primary alcohol	CH_3CH_2OH	16	
Acid chloride	CH_3COCl	16	
Aldehyde	CH_3CHO	17	
Ketone	CH_3COCH_3	19	
Ester	$CH_3CO_2CH_3$	25	
Nitrile	CH_3CN	25	
Dialkylamide	$CH_3CON(CH_3)_2$	30	
Ammonia	NH_3	35	Weaker acid

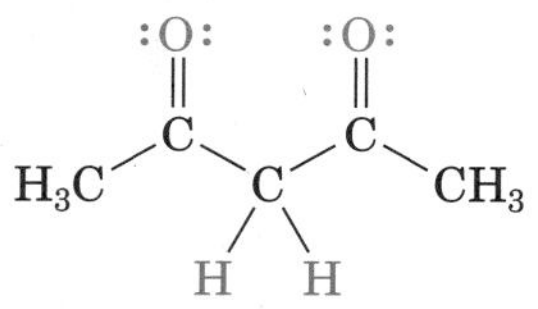

2,4-Pentanedione (pK_a = 9)

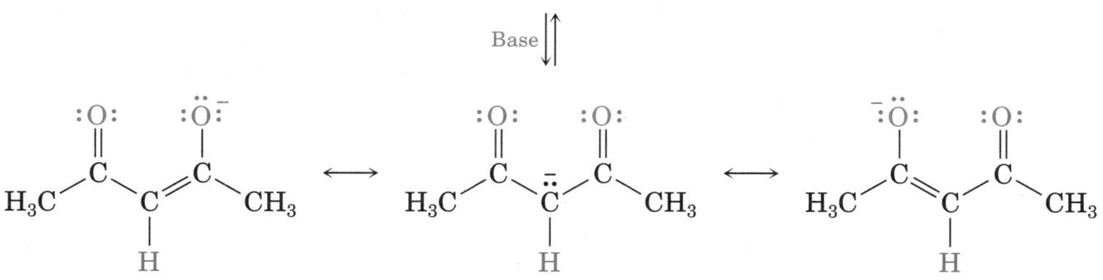

PRACTICE PROBLEM 11.3

Draw structures of the two enolate ions you could obtain by deprotonation of 3-methylcyclohexanone.

Solution Locate the acidic hydrogens and then remove them one at a time to generate the possible enolate ions. In this case, 3-methylcyclohexanone can be deprotonated either at C2 or at C6.

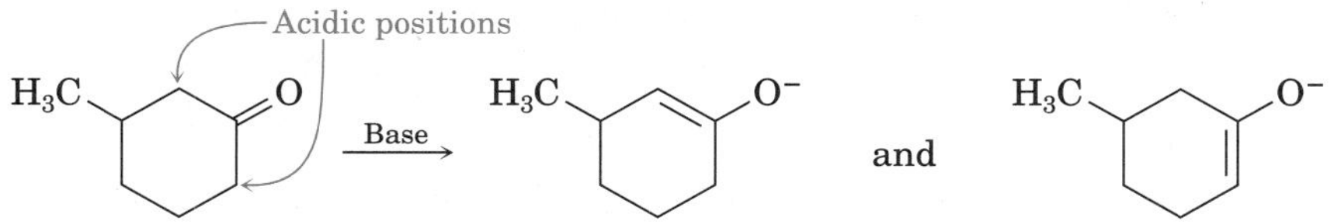

PROBLEM

11.7 Identify all acidic hydrogens in the following molecules:
(a) CH_3CH_2CHO (b) $(CH_3)_3CCOCH_3$ (c) CH_3COOH
(d) $CH_3CH_2CH_2C{\equiv}N$ (e) 1,3-Cyclohexanedione

PROBLEM

11.8 Show the enolate ions you would obtain by deprotonation of the following carbonyl compounds:
(a) Butanal (b) 2-Butanone (c) 2-Methylcyclohexanone

PROBLEM

11.9 Draw three resonance forms for the most stable enolate ion you would obtain by deprotonation of methyl 3-oxobutanoate.

$$CH_3\overset{O}{\overset{\|}{C}}CH_2\overset{O}{\overset{\|}{C}}OCH_3$$ **Methyl 3-oxobutanoate**

11.5 Reactivity of Enolate Ions

Enolate ions are more useful than enols for two reasons. First, pure enols can't normally be isolated. Instead, enols are usually generated only as transient intermediates in low concentration. By contrast, stable solutions of pure enolate ions are easily prepared from many carbonyl compounds by treatment with a strong base. Second, enolate ions are more reactive than enols. Whereas enols are neutral, enolate ions have a negative charge that makes them much better nucleophiles. Thus, the α position of enolate ions is highly reactive toward electrophiles.

$$\text{C}{=}\text{C}{-}\ddot{\text{O}}\text{H}$$

Enol: neutral, moderately reactive, very difficult to isolate

$$\text{C}{=}\text{C}{-}\ddot{\text{O}}{:}^{-} \longleftrightarrow {}^{-}{:}\text{C}{-}\text{C}{=}\text{O}$$

Enolate: negatively charged, very reactive, easily prepared

As resonance hybrids of two nonequivalent forms, enolate ions can be thought of either as α-keto carbanions ($^{-}$C–C=O) or as vinylic alkoxides (C=C–O^{-}). Thus, enolate ions can react with electrophiles either on oxygen or on carbon. Reaction on oxygen yields an enol derivative, whereas reaction on carbon yields an α-substituted carbonyl compound (Figure 11.5). Both kinds of reactivity are known, but reaction on carbon is more common.

11.6 Alkylation of Enolate Ions

Perhaps the most useful of all reactions of enolate ions is their **alkylation** by treatment with an alkyl halide. The alkylation reaction forms a

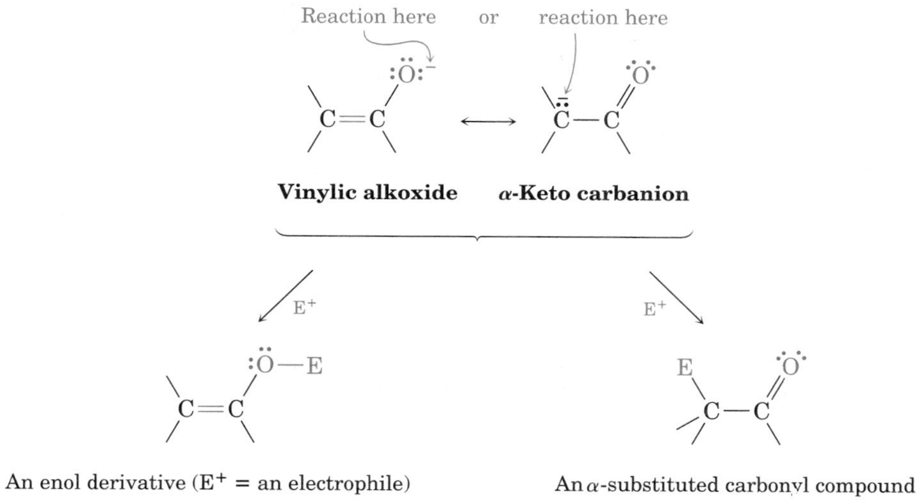

Reaction here or reaction here

Vinylic alkoxide ⟷ **α-Keto carbanion**

E^+ E^+

An enol derivative (E^+ = an electrophile) An α-substituted carbonyl compound

Figure 11.5 Two modes of enolate ion reactivity. Reaction on carbon to yield an α-substituted carbonyl product is more common.

new C–C bond, thereby joining two smaller pieces into one larger molecule. Alkylation occurs when the nucleophilic enolate ion reacts with the electrophilic alkyl halide in an S_N2 reaction, displacing the halide ion by back-side attack:

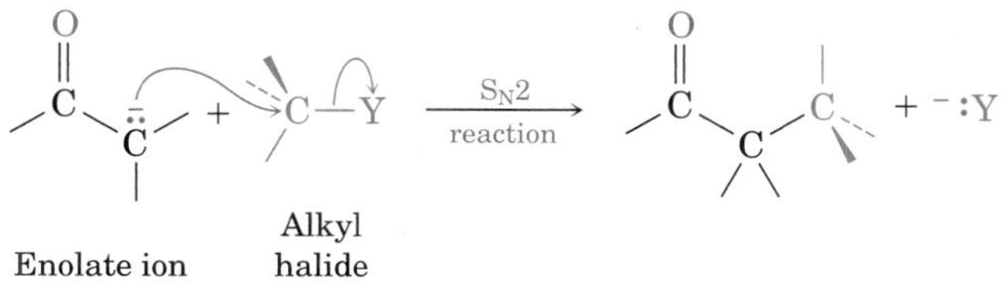

Enolate ion + Alkyl halide $\xrightarrow[\text{reaction}]{S_N2}$ + $^-$:Y

Like all S_N2 reactions (Section 7.7), alkylations are successful only when a primary alkyl halide (RCH_2X) or methyl halide (CH_3X) is used, because a competing E2 elimination occurs if a secondary or tertiary halide is used. The leaving group X can be chloride, bromide, or iodide.

The **malonic ester synthesis,** one of the best-known carbonyl alkylation reactions, is an excellent method for preparing a substituted acetic acid from an alkyl halide:

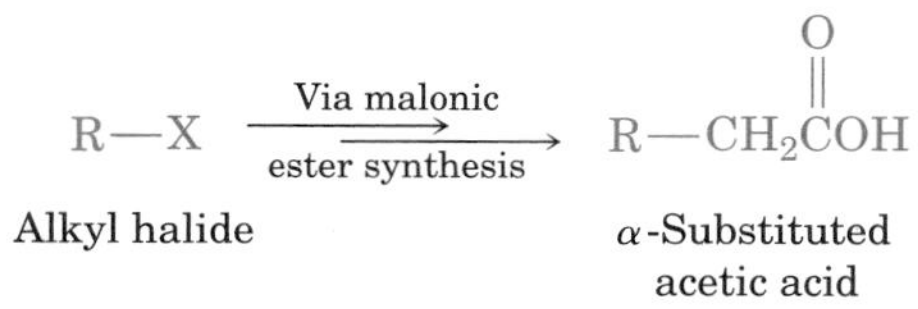

$$\underset{\text{Alkyl halide}}{R{-}X} \xrightarrow[\text{ester synthesis}]{\text{Via malonic}} \underset{\alpha\text{-Substituted acetic acid}}{R{-}CH_2COOH}$$

Diethyl propanedioate, commonly called diethyl malonate or *malonic ester,* is relatively acidic ($pK_a = 13$) because its α hydrogen atoms are flanked by two carbonyl groups. Thus, malonic ester is easily converted into its enolate ion by reaction with sodium ethoxide in ethanol. The enolate ion, in turn, is readily alkylated by treatment with an alkyl halide, yielding an α-substituted malonic ester.

$$\underset{\textbf{Malonic ester}}{H{-}\overset{CO_2CH_2CH_3}{\underset{H}{C}}{-}CO_2CH_2CH_3} \xrightarrow[CH_3CH_2OH]{Na^+\ ^-OCH_2CH_3} \underset{\textbf{Sodiomalonic ester}}{Na^+\ ^-{:}\overset{CO_2CH_2CH_3}{\underset{H}{C}}{-}CO_2CH_2CH_3} \xrightarrow{RX} \underset{\text{An alkylated malonic ester}}{R{-}\overset{CO_2CH_2CH_3}{\underset{H}{C}}{-}CO_2CH_2CH_3} + NaX$$

The product of a malonic ester alkylation has one acidic α hydrogen left, and the alkylation process can therefore be repeated a second time to yield a dialkylated malonic ester:

$$R{-}\overset{CO_2CH_2CH_3}{\underset{H}{C}}{-}CO_2CH_2CH_3 \xrightarrow[2.\ R'X]{1.\ Na^+\ ^-OCH_2CH_3} \underset{\text{A dialkylated malonic ester}}{R{-}\overset{CO_2CH_2CH_3}{\underset{R'}{C}}{-}CO_2CH_2CH_3} + NaX$$

Once formed, an alkylated malonic ester can be hydrolyzed and *decarboxylated* (lose CO_2) when heated with aqueous HCl, yielding a substituted carboxylic acid. Note that decarboxylation is not a general reaction of carboxylic acids but is a unique feature of compounds like malonic acids that have a second carbonyl group two atoms away from the –COOH.

$$R{-}\overset{CO_2CH_2CH_3}{\underset{R'}{C}}{-}CO_2CH_2CH_3 \xrightarrow[\Delta]{H_3O^+} R{-}\overset{R'}{\underset{H}{C}}{-}COOH + CO_2 + 2\ CH_3CH_2OH$$

The overall result of the malonic ester synthesis is to convert an alkyl halide into a carboxylic acid and to lengthen the carbon chain by two atoms ($RX \rightarrow RCH_2COOH$). Note in the following example that the abbreviation “Et” is used as a space-saving way of indicating an ethyl group.

$$\underset{\textbf{1-Bromobutane}}{CH_3CH_2CH_2CH_2Br} + \underset{\textbf{Sodio diethylmalonate}}{Na^+\ ^-{:}CH(CO_2Et)_2} \longrightarrow \underset{\textbf{(84\%)}}{CH_3CH_2CH_2CH_2CH(CO_2Et)_2}$$

$$\downarrow H_3O^+,\ \Delta$$

$$\underset{\textbf{Hexanoic acid (75\%)}}{CH_3CH_2CH_2CH_2CH_2\overset{O}{\overset{\|}{C}}OH}$$

PRACTICE PROBLEM 11.4

How would you prepare heptanoic acid by a malonic ester synthesis?

Solution The malonic ester synthesis converts an alkyl halide into a carboxylic acid having two more carbon atoms. Thus, a seven-carbon acid chain must be derived from a five-carbon alkyl halide such as 1-bromopentane.

$$CH_3CH_2CH_2CH_2CH_2Br + CH_2(CO_2Et)_2 \xrightarrow[\text{2. } H_3O^+\text{, heat}]{\text{1. } Na^+ \; ^-OEt} CH_3CH_2CH_2CH_2CH_2CH_2\overset{O}{\overset{\|}{C}}OH$$

PROBLEM

11.10 What alkyl halide would you use to prepare the following compounds by a malonic ester synthesis?
(a) Butanoic acid (b) 3-Phenylpropanoic acid (c) 5-Methylhexanoic acid

PROBLEM

11.11 Show how you could use a malonic ester synthesis to prepare the following compounds:
(a) 4-Methylpentanoic acid (b) 2-Methylpentanoic acid

11.7 Carbonyl Condensation Reactions

We've seen by now that carbonyl compounds can behave either as electrophiles or as nucleophiles. In a nucleophilic addition reaction or a nucleophilic acyl substitution reaction, the carbonyl group behaves as an electrophile by accepting electrons from an attacking nucleophile. In an α-substitution reaction, however, the carbonyl compound behaves as a nucleophile when it is converted into an enol or enolate ion.

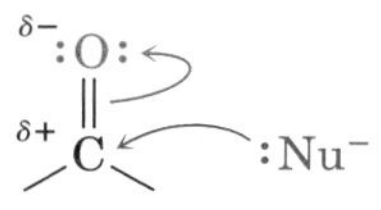

Electrophilic carbonyl group is attacked by nucleophiles

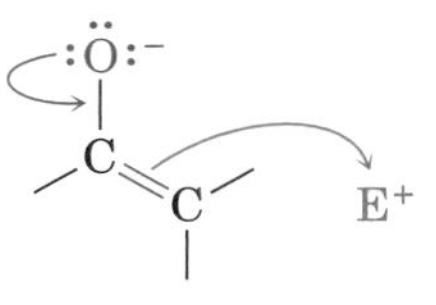

Nucleophilic enolate ion attacks electrophiles

Carbonyl condensation reactions, the fourth and last general category of carbonyl-group reactions we'll study, involve both kinds of reactivity. These reactions take place between two carbonyl partners and involve a combination of nucleophilic addition and α-substitution steps. One partner (the nucleophilic *donor*) is converted into an enolate ion and undergoes an α-substitution reaction, while the other partner (the electrophilic *acceptor*) undergoes a nucleophilic addition reaction. There are numerous variations of carbonyl condensation reactions, depending on the two carbonyl partners, but the general mechanism remains the same (Figure 11.6).

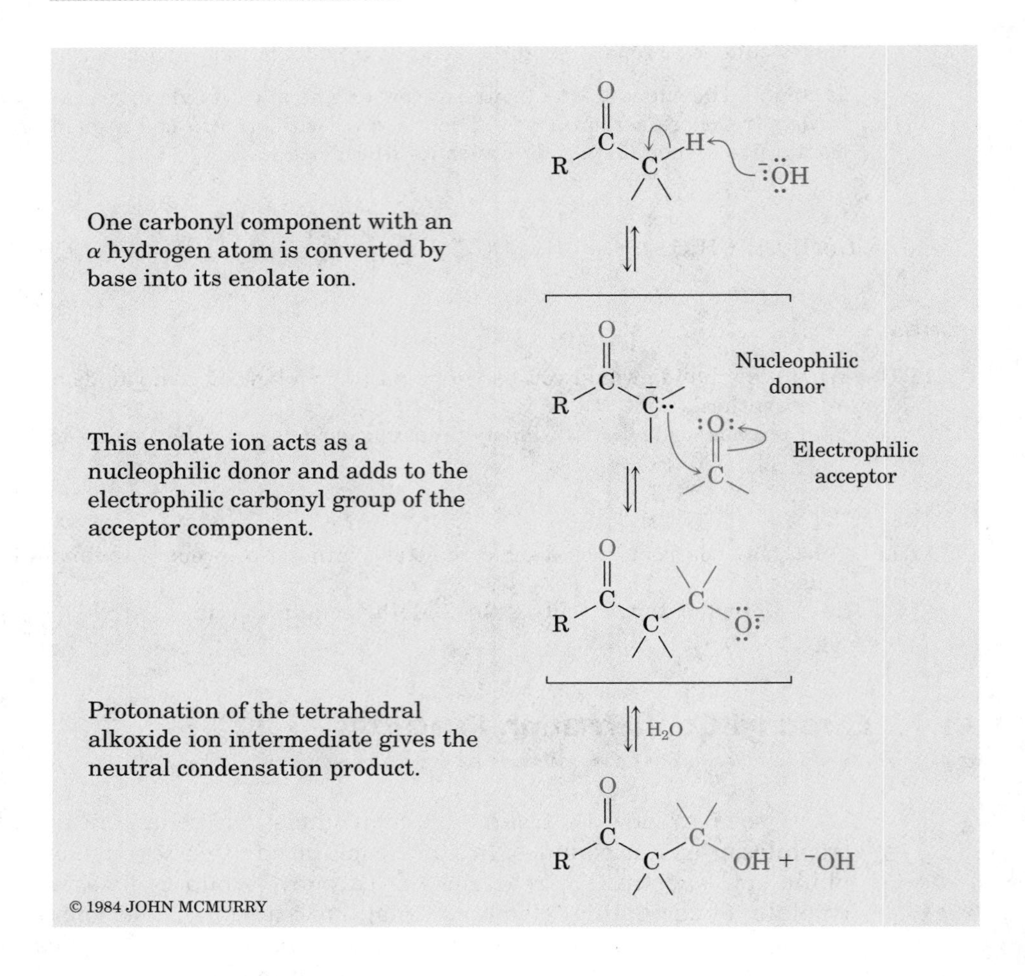

One carbonyl component with an α hydrogen atom is converted by base into its enolate ion.
This enolate ion acts as a nucleophilic donor and adds to the electrophilic carbonyl group of the acceptor component.
Nucleophilic donor
Electrophilic acceptor
Protonation of the tetrahedral alkoxide ion intermediate gives the neutral condensation product.
H2O
OH + ⁻OH
© 1984 JOHN MCMURRY

Figure 11.6 The general mechanism of a carbonyl condensation reaction. One partner (the donor) acts as a nucleophile, while the other partner (the acceptor) acts as an electrophile.

11.8 Condensations of Aldehydes and Ketones: The Aldol Reaction

When acetaldehyde is dissolved in an alcohol solvent and treated with a base, such as sodium hydroxide or sodium ethoxide, a rapid and reversible condensation reaction occurs. The product is a β-hydroxy aldehyde product known commonly as *aldol* (*ald*ehyde + alcoh*ol*). Called the **aldol reaction,** base-catalyzed condensation is a general reaction of all aldehydes and ketones with α hydrogen atoms. If the ketone or aldehyde does not have an α hydrogen atom, aldol condensation can't occur.

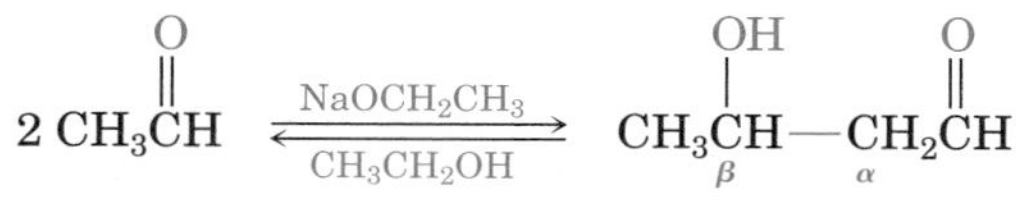

$$2\ \mathrm{CH_3\overset{O}{\overset{\|}{C}}H} \underset{\mathrm{CH_3CH_2OH}}{\overset{\mathrm{NaOCH_2CH_3}}{\rightleftharpoons}} \mathrm{CH_3\underset{\beta}{\overset{OH}{\overset{|}{C}}H}{-}\underset{\alpha}{C}H_2\overset{O}{\overset{\|}{C}}H}$$

Acetaldehyde **Aldol (a β-hydroxy aldehyde)**

The exact position of the aldol equilibrium depends both on reaction conditions and on substrate structure. As the following examples indicate, the aldol equilibrium generally favors the condensation product for monosubstituted acetaldehydes (RCH_2CHO) but favors the starting material for disubstituted acetaldehydes (R_2CHCHO) and for ketones.

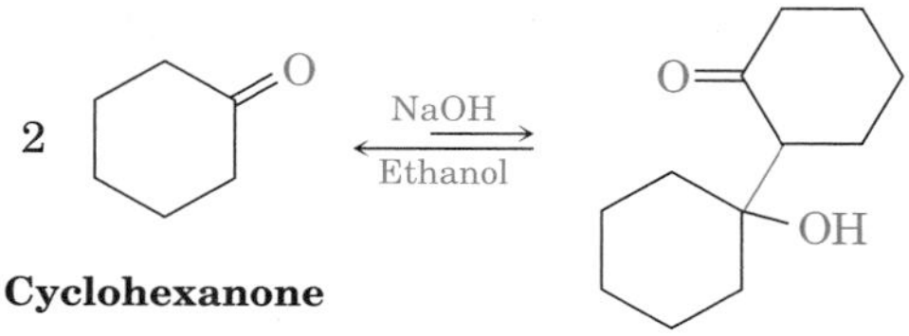

2 Cyclohexanone $\underset{\text{Ethanol}}{\overset{\text{NaOH}}{\rightleftharpoons}}$ 2-(1-hydroxycyclohexyl)cyclohexanone (O=, OH)

Cyclohexanone

2 Phenylacetaldehyde ($C_6H_5CH_2CHO$) $\underset{\text{Ethanol}}{\overset{\text{NaOH}}{\rightleftharpoons}}$ $C_6H_5CH_2CH(OH)CH(C_6H_5)CHO$

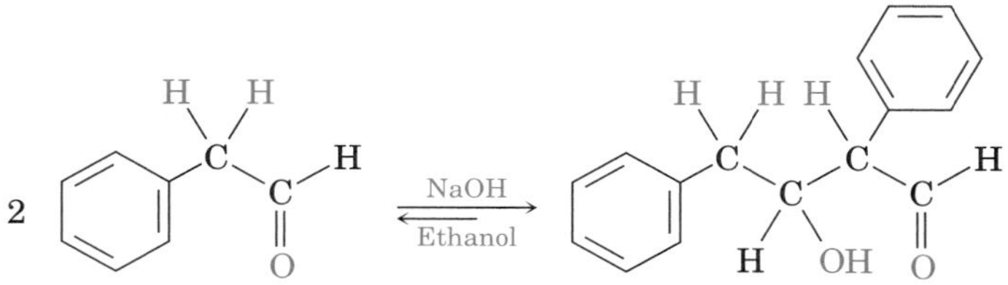

Phenylacetaldehyde 90%

PRACTICE PROBLEM 11.5

What is the structure of the aldol product derived from propanal?

Solution An aldol reaction combines two molecules of starting material, forming a bond between the α carbon of one partner and the carbonyl carbon of the second partner:

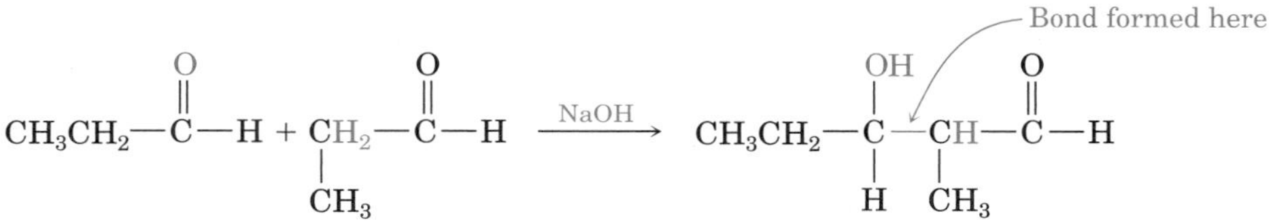

$$\mathrm{CH_3CH_2{-}\overset{O}{\overset{\|}{C}}{-}H} + \mathrm{\underset{CH_3}{\underset{|}{CH_2}}{-}\overset{O}{\overset{\|}{C}}{-}H} \xrightarrow{\mathrm{NaOH}} \mathrm{CH_3CH_2{-}\underset{H}{\underset{|}{\overset{OH}{\overset{|}{C}}}}{-}\underset{CH_3}{\underset{|}{CH}}{-}\overset{O}{\overset{\|}{C}}{-}H}$$

Bond formed here

PROBLEM

11.12 Which of the following compounds can undergo the aldol reaction, and which cannot? Explain.

(a) Cyclohexanone
(b) Benzaldehyde
(c) 2,2,6,6-Tetramethylcyclohexanone
(d) Formaldehyde

PROBLEM

11.13 Show the product of the aldol reaction of the following compounds:
(a) Butanal (b) Cyclopentanone (c) Acetophenone

11.9 Dehydration of Aldol Products: Synthesis of Enones

The β-hydroxy ketones and β-hydroxy aldehydes formed in aldol reactions are easily dehydrated to yield conjugated **enones** (*ene* + *one*). In fact, it's this loss of water that gives the aldol *condensation* its name, since water condenses out of the reaction.

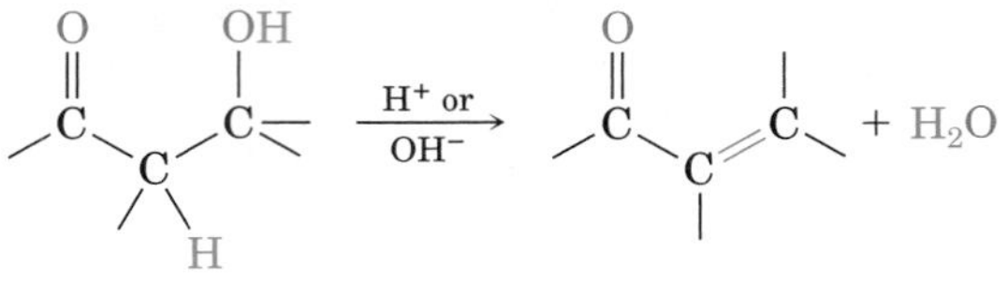

$\xrightarrow[\text{OH}^-]{\text{H}^+ \text{ or}}$ + H_2O

A β-hydroxy ketone or aldehyde

A conjugated enone

Most alcohols are resistant to dehydration by dilute acid or base (Section 8.7), but –OH groups that are two carbons away from a carbonyl group are special. Under basic conditions, an α hydrogen is abstracted, and the resultant enolate ion expels the nearby OH^- leaving group. Under acidic conditions, the –OH group is protonated and H_2O is then expelled as the leaving group.

Base-catalyzed

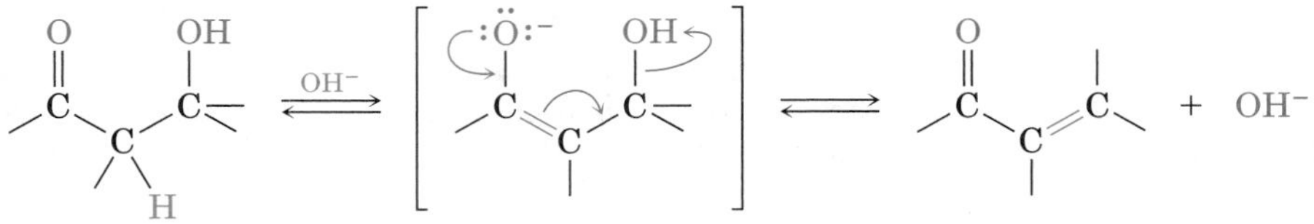

$\xrightleftharpoons{\text{OH}^-}$ [ ] $\rightleftharpoons$ + OH^-

Enolate ion

Acid-catalyzed

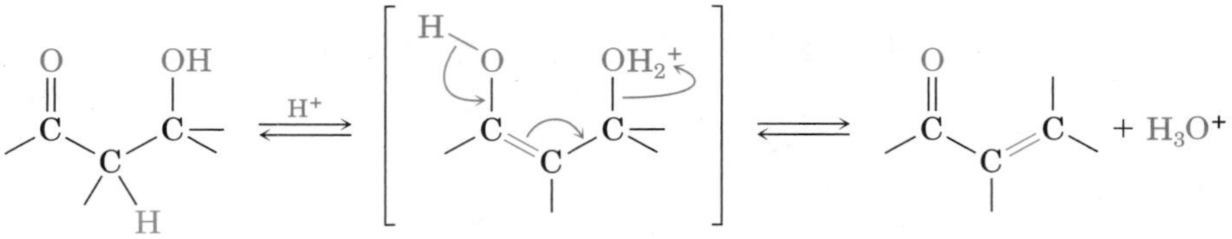

$\xrightleftharpoons{\text{H}^+}$ [ ] $\rightleftharpoons$ + H_3O^+

Enol

The reaction conditions needed for aldol dehydration are often only a bit more vigorous (slightly higher temperature, for instance) than the conditions needed for the aldol condensation itself. As a result, conjugated enones are often obtained directly from aldol reactions without ever isolating the intermediate β-hydroxy carbonyl compounds.

Conjugated enones are more stable than nonconjugated enones for the same reasons that conjugated dienes are more stable than nonconjugated dienes (Section 4.10). Interaction between the π electrons of the C=C bond and the π electrons of the C=O group allows delocalization of the π electrons over all four atomic centers.

Conjugated enone (more stable)

Nonconjugated enone (less stable)

PRACTICE PROBLEM 11.6

What is the structure of the enone obtained from aldol condensation of acetaldehyde?

Solution In the aldol reaction, H_2O is eliminated by removing two hydrogens from the acidic α position of one partner and the oxygen from the second partner:

$$H_3C-CH{=}O + H_2CH-CHO \xrightarrow{\text{NaOH}} H_3C-CH{=}CH-CHO + H_2O$$

2-Butenal

PROBLEM

11.14 Write the structures of the enone products you would obtain from aldol condensation of the following compounds:
(a) Acetone
(b) Cyclopentanone
(c) Acetophenone
(d) Propanal

PROBLEM

11.15 Aldol condensation of 2-butanone leads to a mixture of two enones (ignoring double-bond stereochemistry). Draw them.

11.10 Condensations of Esters: The Claisen Condensation Reaction

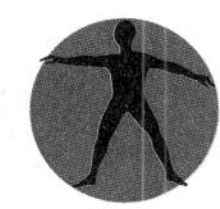

Esters, like aldehydes and ketones, are weakly acidic. When an ester with an α hydrogen is treated with 1 equivalent of a base such as sodium ethoxide, a reversible condensation reaction occurs to yield a β-keto ester product. For example, ethyl acetate yields ethyl acetoacetate on treatment with base. This reaction between two ester components is known as the **Claisen condensation reaction.**

$$2\ CH_3\overset{\overset{O}{\|}}{C}OCH_2CH_3 \xrightarrow[2.\ H_3O^+]{1.\ Na^+\ ^-OEt,\ ethanol} CH_3\underset{\beta}{\overset{\overset{O}{\|}}{C}}-\underset{\alpha}{C}H_2\overset{\overset{O}{\|}}{C}OCH_2CH_3 + CH_3CH_2OH$$

Ethyl acetate **Ethyl acetoacetate**
a β-keto ester (75%)

The mechanism of the Claisen reaction is similar to that of the aldol reaction. As shown in Figure 11.7, the reaction involves the nucleophilic addition of an ester enolate ion to the carbonyl group of a second ester molecule. From the point of view of the enolate ion partner, the Claisen condensation is simply an α-substitution reaction. From the point of view of the other partner, the Claisen condensation is a nucleophilic acyl substitution reaction.

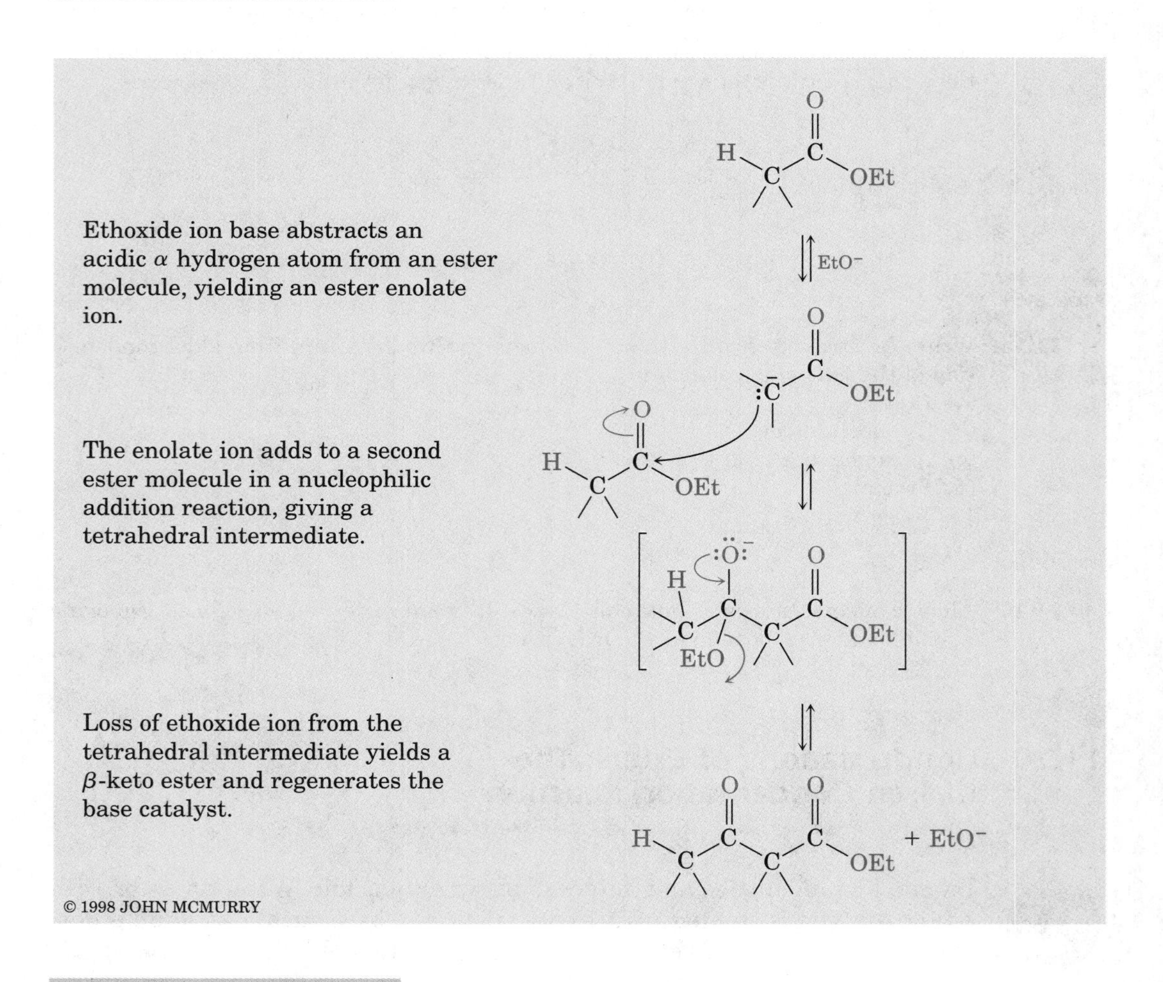

Ethoxide ion base abstracts an acidic α hydrogen atom from an ester molecule, yielding an ester enolate ion.
The enolate ion adds to a second ester molecule in a nucleophilic addition reaction, giving a tetrahedral intermediate.
Loss of ethoxide ion from the tetrahedral intermediate yields a β-keto ester and regenerates the base catalyst.
H
C
O
OEt
EtO⁻
EtO
+ EtO⁻
© 1998 JOHN MCMURRY

Figure 11.7 The mechanism of the Claisen condensation reaction.

The only difference between an aldol condensation and a Claisen condensation involves the fate of the initially formed tetrahedral intermediate. The tetrahedral intermediate in the aldol reaction is protonated to give a stable alcohol—exactly the behavior previously seen for aldehydes and ketones (Section 9.6). The tetrahedral intermediate in the Claisen reaction, however, expels a leaving group to yield an acyl substitution product—exactly the behavior previously seen for esters (Section 10.5).

PRACTICE PROBLEM 11.7

What product would you obtain from Claisen condensation of methyl propanoate?

Solution The Claisen condensation of an ester results in the loss of one molecule of alcohol and the formation of a product in which an acyl group of one partner bonds to the α carbon of the second partner:

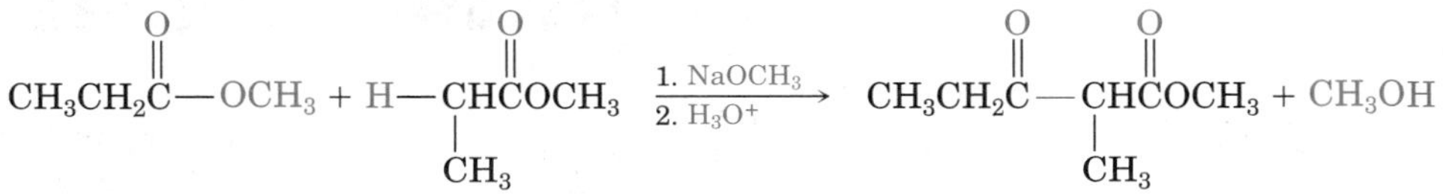

$$CH_3CH_2\overset{\overset{O}{\|}}{C}—OCH_3 + H—\underset{\underset{CH_3}{|}}{C}H\overset{\overset{O}{\|}}{C}OCH_3 \xrightarrow[2.\ H_3O^+]{1.\ NaOCH_3} CH_3CH_2\overset{\overset{O}{\|}}{C}—\underset{\underset{CH_3}{|}}{C}H\overset{\overset{O}{\|}}{C}OCH_3 + CH_3OH$$

2 Methyl propanoate **Methyl 2-methyl-3-oxopentanoate**

PROBLEM

11.16 Which of the following esters can't undergo Claisen condensation? Explain.
(a) Methyl formate
(b) Methyl propenoate
(c) Methyl propanoate

PROBLEM

11.17 Show the products you would obtain by Claisen condensation of the following esters:
(a) $(CH_3)_2CHCH_2COOCH_3$
(b) Methyl phenylacetate
(c) Methyl cyclohexylacetate

INTERLUDE

Carbonyl Compounds in Metabolism

You are what you eat. The major classes of food molecules are metabolized using the carbonyl-group reactions discussed in the last three chapters.

Biochemistry *is* carbonyl chemistry. Almost all metabolic processes used by all living organisms involve one or more of the four fundamental carbonyl-group reactions we've seen in the last three chapters. The digestion and metabolic breakdown of all the major classes of food molecules—fats, carbohydrates, and proteins—take place by nucleophilic addition reactions, nucleophilic acyl substitutions, α substitutions, and carbonyl condensations. Similarly, hormones and other crucial biological molecules are built up from smaller precursors by these same carbonyl-group reactions.

Take *glycolysis,* for example, the metabolic pathway by which organisms convert glucose to pyruvate as the first step in extracting energy from carbohydrates:

$$\text{Glucose} \xrightarrow{\text{Glycolysis}} 2\ CH_3-\overset{\overset{O}{\|}}{C}-\overset{\overset{O}{\|}}{C}-O^-$$

Glucose **Pyruvate**

Glycolysis is a ten-step process that begins with conversion of glucose from its cyclic hemiacetal form to its open-chain aldehyde form—a reverse nucleophilic addition reaction. The aldehyde then undergoes tautomerization to yield an enol, which undergoes yet another tautomerization to give the ketone fructose.

(*continued*)➤

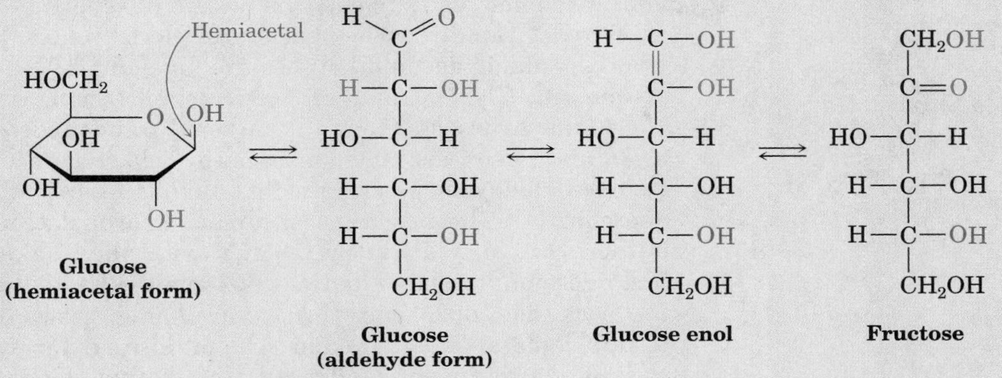

Hemiacetal

Glucose (hemiacetal form) ⇌ **Glucose (aldehyde form)** ⇌ **Glucose enol** ⇌ **Fructose**

Fructose, a β-hydroxy ketone, is then cleaved into two three-carbon molecules—one ketone and one aldehyde—by a reverse aldol reaction. Still further carbonyl-group reactions then occur until pyruvate results.

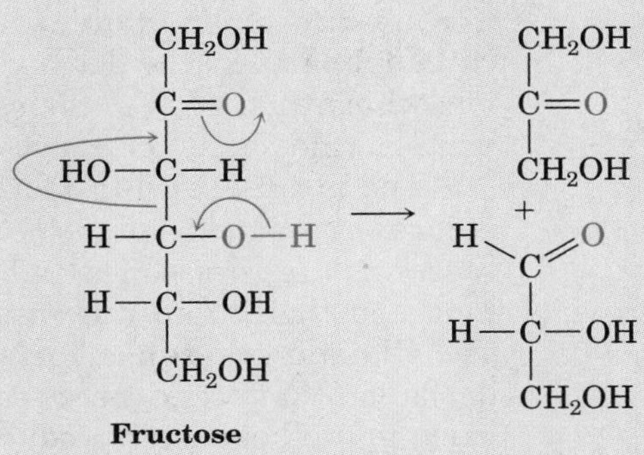

Fructose

The few examples just given are only an introduction; we'll look at several of the major metabolic pathways in more detail in Chapter 17. You haven't seen the end of carbonyl-group chemistry, though. A good grasp of carbonyl-group reactions is crucial to an understanding of biochemistry.

Summary and Key Words

aldol reaction, 366
alkylation, 362
alpha (α) position, 352
alpha-substitution reaction, 352
carbonyl condensation reaction, 352

Alpha substitutions and **carbonyl condensations** are two of the four fundamental reaction types in carbonyl-group chemistry. Alpha-substitution reactions, which take place via **enol** or **enolate ion** intermediates, result in the replacement of an α hydrogen atom by another substituent.

Carbonyl compounds are in rapid equilibrium with their enols, a process known as *tautomerism*. Enol

tautomers are normally present to only a small extent, and pure enols usually can't be isolated. Nevertheless, enols react rapidly with a variety of electrophiles. For example, aldehydes and ketones are halogenated by reaction with Cl_2, Br_2, or I_2 in acetic acid solution.

Alpha hydrogen atoms in carbonyl compounds are acidic and can be abstracted by bases to yield enolate ions. Ketones, aldehydes, esters, amides, and nitriles can all be deprotonated. The most important reaction of enolate ions is their S_N2 **alkylation** by reaction with alkyl halides. The nucleophilic enolate ion attacks an alkyl halide from the back side, displacing the leaving halide group and yielding an α-alkylated product. The **malonic ester synthesis,** which involves alkylation of diethyl malonate with an alkyl halide, is a good method for preparing a monoalkylated or dialkylated acetic acid.

A carbonyl condensation reaction takes place between two carbonyl components and involves a combination of nucleophilic addition and α-substitution steps. One carbonyl partner (the donor) is converted into its enolate ion, which then adds to the carbonyl group of the second partner (the acceptor).

The **aldol reaction** is a carbonyl condensation that occurs between two ketone or aldehyde components. Aldol reactions are reversible, leading first to a β-hydroxy ketone and then to an α,β-unsaturated ketone, or **enone.** The **Claisen condensation reaction** is a carbonyl condensation that occurs between two ester components and leads to a β-keto ester product.

Summary of Reactions

1. Halogenation of ketones and aldehydes (Section 11.3)

$$\underset{\diagup\,\diagdown}{-\overset{O}{\overset{\|}{C}}-\overset{}{C}}-H + X_2 \xrightarrow{CH_3COOH} \underset{\diagup\,\diagdown}{-\overset{O}{\overset{\|}{C}}-\overset{}{C}}-X \qquad \text{where X = Cl, Br, or I}$$

2. Malonic ester synthesis (Section 11.6)

$$H-\overset{H}{\underset{CO_2R}{C}}-CO_2R \xrightarrow[2.\ R'X]{1.\ Base} R'-\overset{H}{\underset{CO_2R}{C}}-CO_2R \xrightarrow{H_3O^+} R'CH_2COOH$$

3. Aldol reaction of ketones and aldehydes (Section 11.8)

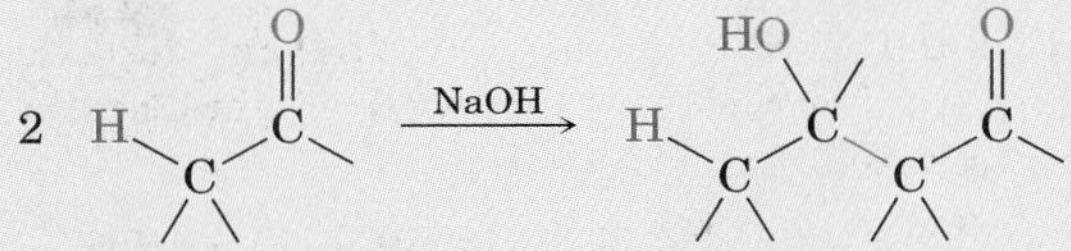

$$2\ \mathrm{H{-}C{-}C(=O){-}} \xrightarrow{\text{NaOH}} \mathrm{H{-}C{-}C(HO){-}C{-}C(=O){-}}$$

4. Claisen condensation reaction of esters (Section 11.10)

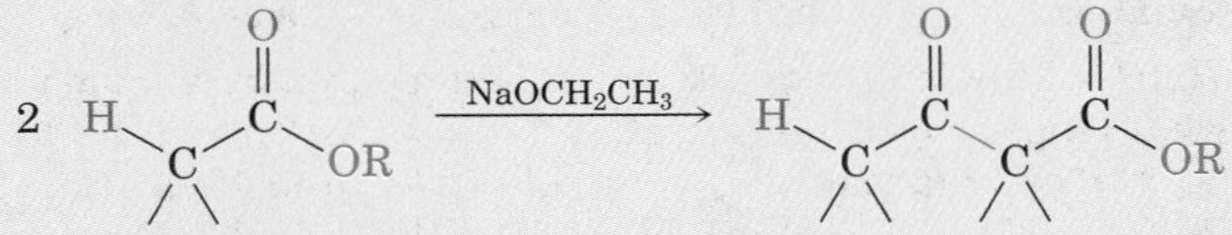

$$2\ \mathrm{H{-}C{-}C(=O)OR} \xrightarrow{\text{NaOCH}_2\text{CH}_3} \mathrm{H{-}C{-}C(=O){-}C{-}C(=O)OR}$$

ADDITIONAL PROBLEMS

11.18 Indicate all acidic hydrogen atoms in the following molecules:

(a) $HOCH_2\overset{O}{\overset{\|}{C}}CH_3$ (b) $HOCH_2CH_2\overset{O}{\overset{\|}{C}}C(CH_3)_3$ (c) 1,3-Cyclopentanedione
(d) $CH_3CH{=}CHCHO$

11.19 Draw structures for the possible monoenol tautomers of 1,3-cyclohexanedione. How many enol forms are possible, and which would you expect to be most stable? Explain.

11.20 Rank the following compounds in order of increasing acidity:

$$CH_3CH_2COOH,\quad CH_3COCH_3,\quad CH_3CH_2OH,\quad CH_3COCH_2COCH_3$$

11.21 Why do you suppose 2,4-pentanedione is 76% enolized at equilibrium although acetone is enolized only to the extent of about 0.0001%?

11.22 Write resonance structures for the following anions:

(a) $CH_3\overset{O}{\overset{\|}{C}}\overset{\ddot{-}}{C}H\overset{O}{\overset{\|}{C}}CH_3$ (b) $:\bar{C}H_2C{\equiv}N$ (c) $CH_3CH{=}CH\overset{\ddot{-}}{C}H\overset{O}{\overset{\|}{C}}CH_3$

(d) $N{\equiv}C\overset{\ddot{-}}{C}HCO_2C_2H_5$

11.23 When acetone is treated with acid in deuterated water, D_2O, deuterium becomes incorporated into the molecule. Propose a mechanism.

$$CH_3\overset{O}{\overset{\|}{C}}CH_3 + D_2O \xrightleftharpoons{\text{DCl}} CH_3\overset{O}{\overset{\|}{C}}CH_2D$$

11.24 Why is an enolate ion generally more reactive than an enol?

11.25 How do the mechanisms of base-catalyzed enolization and acid-catalyzed enolization differ?

11.26 When optically active (*R*)-2-methylcyclohexanone is treated with aqueous HCl or NaOH, racemic 2-methylcyclohexanone is produced. Explain.

11.27 When optically active (*R*)-3-methylcyclohexanone is treated with aqueous HCl or NaOH, no racemization occurs. Instead, the optically active ketone is recovered unchanged. How can you reconcile this observation with your answer to Problem 11.26?

11.28 Monoalkylated acetic acids (RCH_2COOH) and dialkylated acetic acids ($R_2CHCOOH$) can be prepared by malonic ester synthesis, but trialkylated acetic acids (R_3CCOOH) can't be prepared. Explain.

11.29 Which of the following compounds would you expect to undergo aldol condensation? Draw the product in each case.
(a) 2,2-Dimethylpropanal (b) Cyclobutanone
(c) Benzophenone (diphenyl ketone) (d) Decanal

11.30 Which of the following esters can be prepared by a malonic ester synthesis? Show what reagents you would use.
(a) Ethyl pentanoate (b) Ethyl 3-methylbutanoate
(c) Ethyl 2-methylbutanoate (d) Ethyl 2,2-dimethylpropanoate

11.31 The aldol condensation reaction can be carried out intramolecularly by treatment of a diketone with base. What diketone would you start with to prepare 3-methyl-2-cyclohexenone? Show the reaction.

11.32 Nonconjugated β,γ-unsaturated ketones such as 3-cyclohexenone are in an acid-catalyzed equilibrium with their conjugated α,β-unsaturated isomers. Propose a mechanism for the acid-catalyzed interconversion.

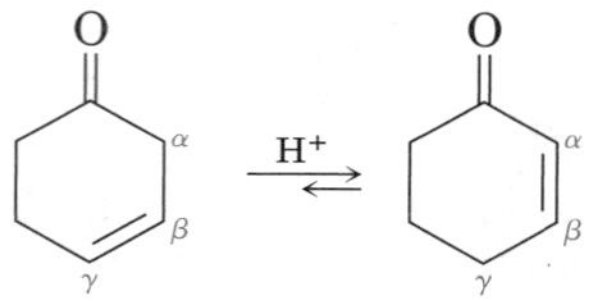

O α β γ $\xrightleftharpoons{H^+}$ O α β γ

11.33 The α,β to β,γ interconversion of unsaturated ketones (see Problem 11.32) is catalyzed by base as well as by acid. Propose a mechanism.

11.34 One consequence of the base-catalyzed α,β to β,γ isomerization of unsaturated ketones (see Problem 11.33) is that C5-substituted 2-cyclopentenones can be interconverted with C2-substituted 2-cyclopentenones. Propose a mechanism for this isomerization.

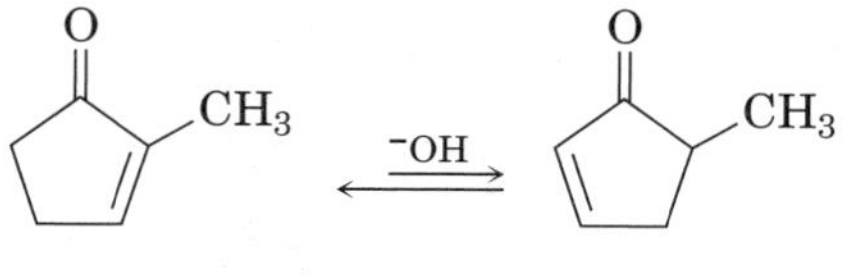

O CH_3 $\xrightleftharpoons{^-OH}$ O CH_3

11.35 If a 1:1 mixture of ethyl acetate and ethyl propanoate is treated with base under Claisen condensation conditions, a mixture of four β-keto ester products is obtained. Show their structures.

11.36 If a mixture of ethyl acetate and ethyl benzoate is treated with base, a mixture of two Claisen condensation products is obtained. Explain.

11.37 Cinnamaldehyde, the aromatic constituent of cinnamon oil, can be synthesized by a mixed aldol-like reaction between benzaldehyde and acetaldehyde. Formulate the reaction. What other product would you expect to obtain?

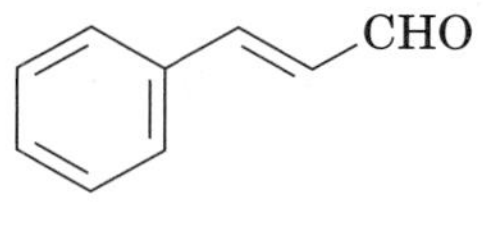

CHO

Cinnamaldehyde

11.38 How might you prepare the following compounds using an aldol condensation reaction?
(a) $C_6H_5C(CH_3)=CHCOC_6H_5$
(b) 4-Methyl-3-penten-2-one

11.39 1-Butanol is synthesized commercially from acetaldehyde by a three-step route that involves an aldol reaction followed by two reductions. How might you carry out this transformation?

11.40 By starting with a *dihalide,* cyclic compounds can be prepared using the malonic ester synthesis. What product would you expect to obtain from the reaction of diethyl malonate, 1,4-dibromobutane, and 2 equivalents of base?

11.41 What product would you expect to obtain from aldol condensation of hexanedial, $OHCCH_2CH_2CH_2CH_2CHO$? (See Problem 11.31.)

11.42 How can you account for the fact that *cis-* and *trans*-4-*tert*-butyl-2-methylcyclohexanone are interconverted by base treatment? Which of the two isomers is more stable, and why? (See Section 2.11.)

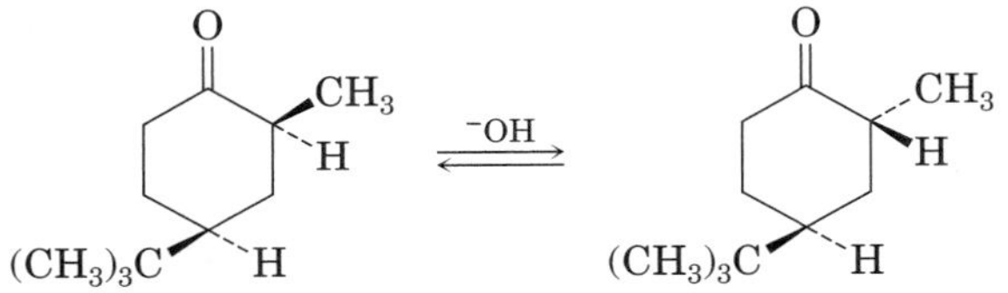

O, CH_3, H, $(CH_3)_3C$, H $\underset{}{\overset{^{-}OH}{\rightleftharpoons}}$ O, CH_3, H, $(CH_3)_3C$, H

11.43 Show how you might convert geraniol, the chief constituent of rose oil, into ethyl geranylacetate.

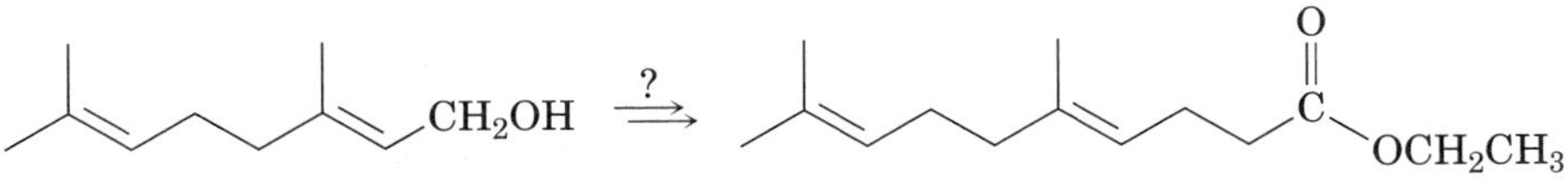

CH_2OH $\overset{?}{\Longrightarrow}$ O, C, OCH_2CH_3

Geraniol **Ethyl geranylacetate**

11.44 The *acetoacetic ester synthesis* is closely related to the malonic ester synthesis, but involves alkylation with the anion of ethyl acetoacetate rather than diethyl malonate. Treatment of the ethyl acetoacetate anion with an alkyl halide, followed by decarboxylation, yields a ketone product:

$$CH_3\overset{O}{\overset{\|}{C}}CH_2\overset{O}{\overset{\|}{C}}OCH_2CH_3 \xrightarrow[\substack{2.\ RX \\ 3.\ H_3O^+}]{1.\ Na^+\ ^{-}OCH_2CH_3} CH_3\overset{O}{\overset{\|}{C}}CH_2{-}R + CO_2 + HOCH_2CH_3$$

How would you prepare the following compounds using an acetoacetic ester synthesis?
(a) 4-Phenyl-2-butanone
(b) 5-Methyl-2-hexanone
(c) 3-Methyl-2-hexanone

11.45 Which of the following compounds can't be prepared by an acetoacetic ester synthesis (see Problem 11.44)? Explain.
(a) 2-Butanone
(b) Phenylacetone
(c) Acetophenone
(d) 3,3-Dimethyl-2-butanone

11.46 The Claisen condensation is reversible. That is, a β-keto ester can be cleaved by base into two fragments. Show the mechanism by which the following cleavage occurs:

$$C_6H_5\overset{O}{\overset{\|}{C}}CH_2\overset{O}{\overset{\|}{C}}OCH_2CH_3 \xrightarrow[H_2O]{NaOH} C_6H_5\overset{O}{\overset{\|}{C}}O^-\ Na^+ + CH_3\overset{O}{\overset{\|}{C}}OCH_2CH_3$$

11.47 Treatment of an α,β-unsaturated carbonyl compound with base yields an anion by removal of H^+ from the γ carbon. Why are hydrogens on the γ carbon atom acidic?

$$C_6H_5\overset{O}{\overset{\|}{C}}{-}CH{=}CHCH_3 \xrightarrow{Base} C_6H_5\overset{O}{\overset{\|}{C}}{-}CH{=}CH\ddot{C}H_2^{\,-}$$

11.48 We'll see in Chapter 15 that amino acids can be prepared by reaction of alkyl halides with diethyl acetamidomalonate, followed by heating the initial alkylation product with aqueous HCl. Show how you would prepare alanine, $CH_3CH(NH_2)COOH$, one of the 20 amino acids found in proteins.

$$CH_3\overset{O}{\overset{\|}{C}}NH\underset{CO_2CH_2CH_3}{\underset{|}{C}H}\overset{O}{\overset{\|}{C}}OCH_2CH_3$$ **Diethyl acetamidomalonate**

Visualizing Chemistry

11.49 What ketones or aldehydes might the following enones have been prepared from by aldol reaction? (Gray = C, red = O, light green = H.)

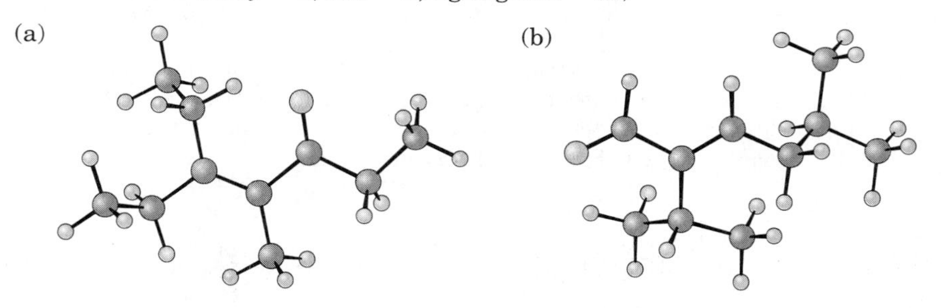

11.50 The following structure represents an intermediate formed by addition of an ester enolate ion to a second ester molecule. Identify the reactant, the leaving group, and the product (gray = C, red = O, light green = H).

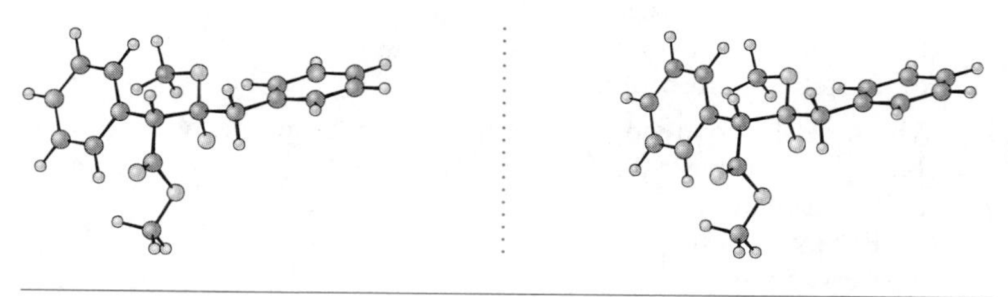

Stereo View

11.51 For a given α hydrogen atom to be acidic, the C–H bond must be parallel to the p orbitals of the C=O double bond (that is, perpendicular to the plane of the adjacent carbonyl group). Identify the most acidic hydrogen atom in the following structure. Is it axial or equatorial? (Gray = C, red = O, light green = H.)

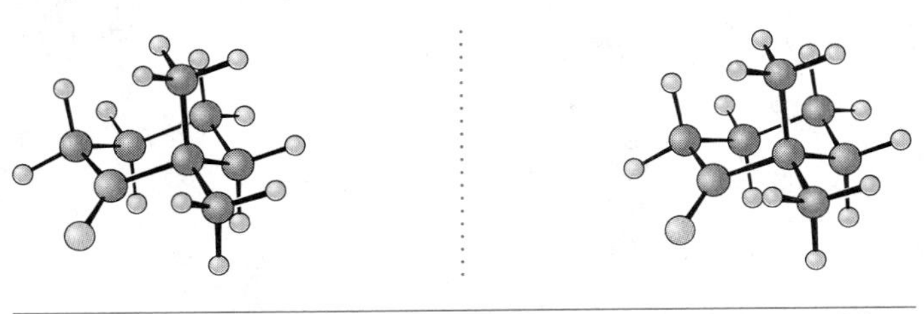

Amino acids, such as isoleucine, contain a basic amine functional group.

12 Amines

Amines are organic derivatives of ammonia in the same way that alcohols and ethers are organic derivatives of water. Like ammonia, amines contain a nitrogen atom with a lone pair of electrons, making amines both basic and nucleophilic.

Amines are classified either as **primary** (RNH_2), **secondary** (R_2NH), or **tertiary** (R_3N), depending on the number of organic substituents attached to nitrogen. For example, methylamine (CH_3NH_2) is a primary amine and trimethylamine [$(CH_3)_3N$] is a tertiary amine. Note that this usage of the terms *primary, secondary,* and *tertiary* is different from our previous usage. When we speak of a tertiary alcohol or alkyl halide, we refer to the degree of substitution at the alkyl *carbon* atom, but when we speak of a tertiary amine, we refer to the degree of substitution at the *nitrogen* atom.

$$H_3C-\overset{\overset{\large CH_3}{|}}{\underset{\underset{\large CH_3}{|}}{C}}-OH \qquad H_3C-\overset{\overset{\large CH_3}{|}}{\underset{\underset{\large CH_3}{|}}{N}} \qquad H_3C-\overset{\overset{\large CH_3}{|}}{\underset{\underset{\large CH_3}{|}}{C}}-NH_2$$

***tert*-Butyl alcohol** **(a tertiary alcohol)** — **Trimethylamine** **(a tertiary amine)** — ***tert*-Butylamine** **(a primary amine)**

Compounds with four groups attached to nitrogen are also known, but the nitrogen atom must carry a positive charge. Such compounds are called **quaternary ammonium salts.**

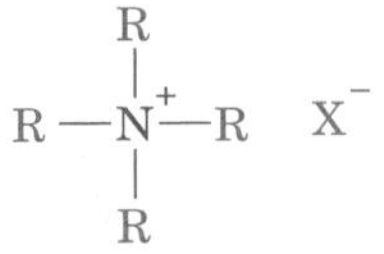

$$R_4N^+ \; X^-$$ A quaternary ammonium salt

Amines can be either alkyl-substituted (**alkylamines**) or aryl-substituted (**arylamines**). Although much of the chemistry of the two classes is similar, we'll soon see that there are also important differences.

$CH_3CH_2\ddot{N}H_2$ — **Ethylamine (an alkylamine)**

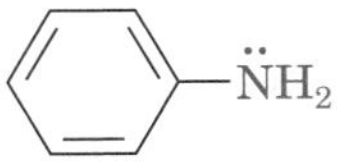

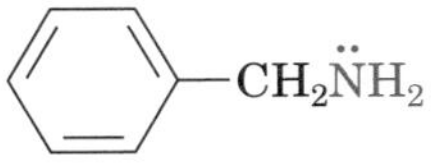

$C_6H_5\ddot{N}H_2$ — **Aniline (an arylamine)**

$C_6H_5CH_2\ddot{N}H_2$ — **Benzylamine (an alkylamine)**

PRACTICE PROBLEM 12.1

Classify the following amines as primary, secondary, or tertiary:

(a) 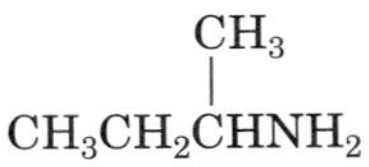 $CH_3CH_2CH(CH_3)NH_2$ (b) 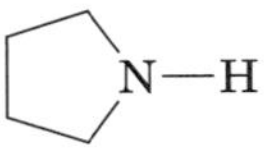 pyrrolidine (ring N—H) (c) 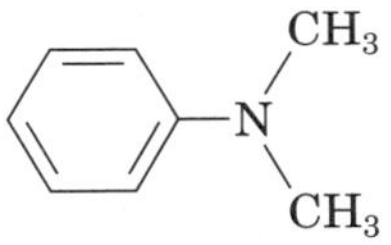 $C_6H_5N(CH_3)_2$

Solution Amine (a) is primary, (b) is secondary, and (c) is tertiary.

PROBLEM

12.1 Classify each of the following compounds as either a primary, secondary, or tertiary amine, or as a quaternary ammonium salt:

(a) $(CH_3)_2CHNH_2$ (b) $(CH_3CH_2)_2NH$

(c) 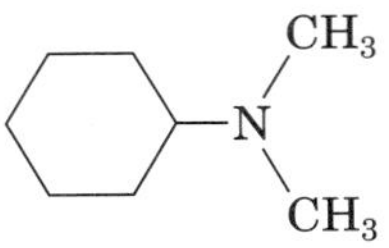 cyclohexyl–$N(CH_3)_2$ (d) 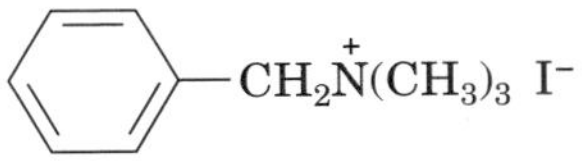 $C_6H_5CH_2\overset{+}{N}(CH_3)_3\ I^-$

PROBLEM

12.2 Draw structures of compounds that meet the following descriptions:
(a) A secondary amine with one isopropyl group
(b) A tertiary amine with one phenyl group and one ethyl group
(c) A quaternary ammonium salt with four different groups bonded to nitrogen

12.1 Naming Amines

Primary amines, RNH_2, are named in the IUPAC system by adding the suffix *-amine* to the name of the organic substituent:

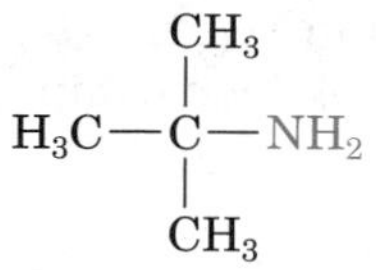

$$H_3C-C(CH_3)_2-NH_2$$

***tert*-Butylamine**

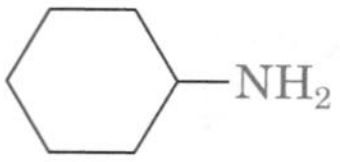

Cyclohexylamine

$$H_2NCH_2CH_2CH_2CH_2NH_2$$

1,4-Butanediamine

Amines that have additional functional groups are named by considering the $-NH_2$ as an *amino* substituent on the parent molecule:

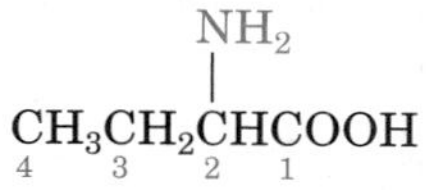

$$\underset{4}{CH_3}\underset{3}{CH_2}\underset{2}{CH(NH_2)}\underset{1}{COOH}$$

2-Aminobutanoic acid

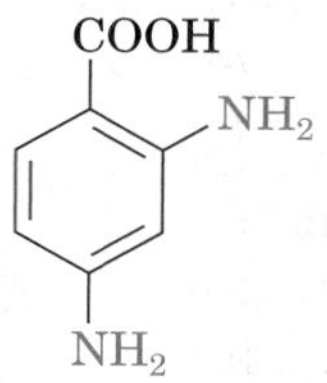

2,4-Diaminobenzoic acid

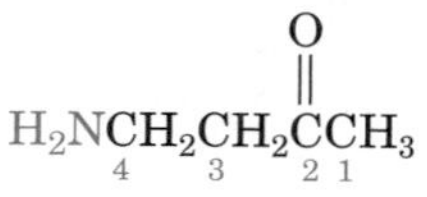

$$H_2N\underset{4}{CH_2}\underset{3}{CH_2}\underset{2}{C(=O)}\underset{1}{CH_3}$$

4-Amino-2-butanone

Symmetrical secondary and tertiary amines are named by adding the prefix *di-* or *tri-* to the alkyl group:

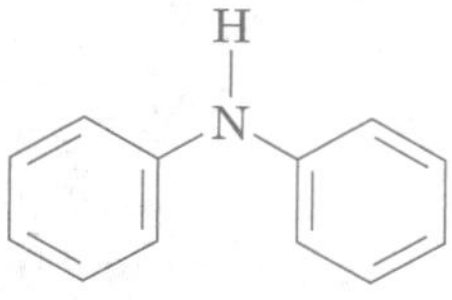

Diphenylamine

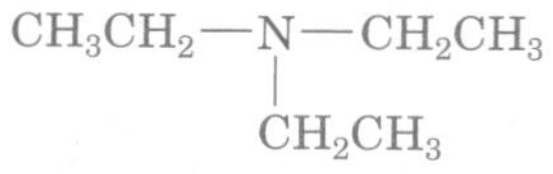

$$CH_3CH_2-N(CH_2CH_3)-CH_2CH_3$$

Triethylamine

Unsymmetrically substituted secondary and tertiary amines are named as *N*-substituted primary amines. The largest organic group is chosen as the parent, and the other groups are considered as *N*-substituents on the parent (*N* because they're attached to nitrogen).

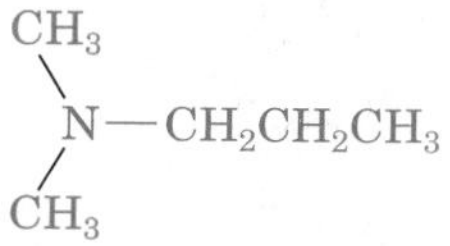

$$(CH_3)_2N-CH_2CH_2CH_3$$

***N,N*-Dimethylpropylamine**
(propylamine is the parent name; the two methyl groups are substituents on nitrogen)

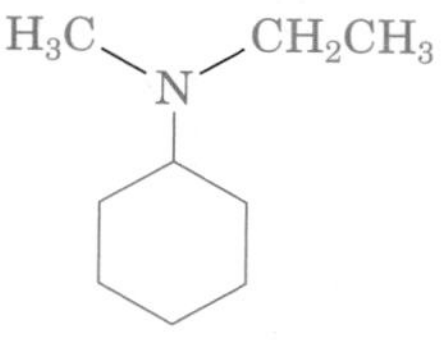

***N*-Ethyl-*N*-methylcyclohexylamine**
(cyclohexylamine is the parent name; methyl and ethyl are *N*-substituents)

There are few common names for simple amines, although phenylamine is usually called *aniline.*

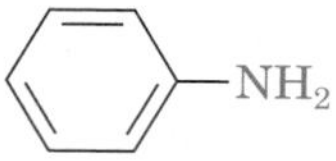

Aniline

Heterocyclic amines—compounds in which the nitrogen atom occurs as part of a ring—are also common, and each different heterocyclic ring system has its own parent name. In all cases, the nitrogen atom is numbered as position 1.

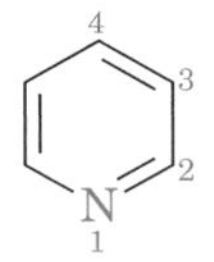

Pyridine 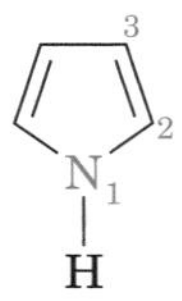 **Pyrrole** 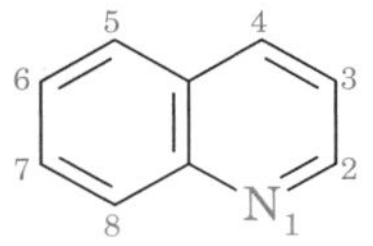 **Quinoline** 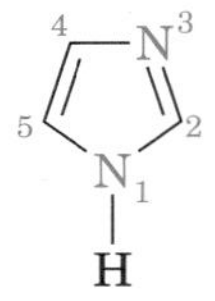 **Imidazole** 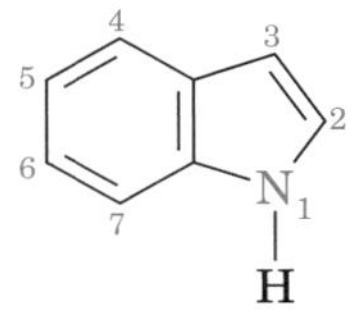 **Indole** 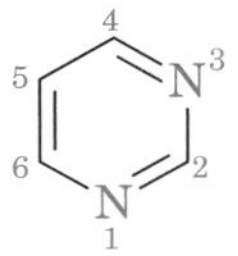 **Pyrimidine**

PROBLEM

12.3 Name the following compounds by IUPAC rules:

(a) $CH_3NHCH_2CH_3$ (b) 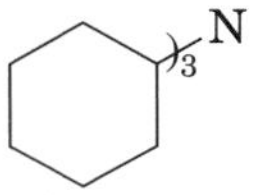 (c)

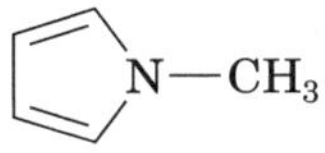

(d) 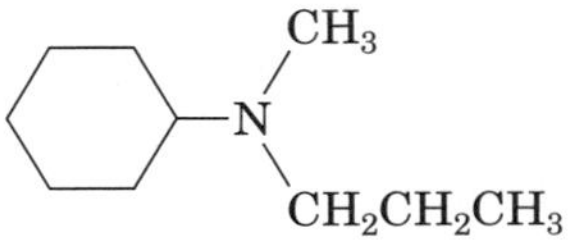 (e)

CH_3, $CH_2CH_2CH_3$ $H_2NCH_2CH_2CH(CH_3)NH_2$

PROBLEM

12.4 Draw structures corresponding to the following IUPAC names:

(a) Triethylamine (b) *N*-Methylaniline (c) Tetraethylammonium bromide
(d) *p*-Bromoaniline (e) *N*-Ethyl-*N*-methylcyclopentylamine

12.2 Structure and Properties of Amines

The bonding in amines is similar to the bonding in ammonia. The nitrogen atom is sp^3-hybridized, with the three substituents occupying three corners of a regular tetrahedron and the lone pair of electrons occupying the fourth corner. As expected, the C–N–C bond angles are very close to the 109° tetrahedral value. For trimethylamine, the C–N–C angle is 108°, and the C–N bond length is 1.47 Å.

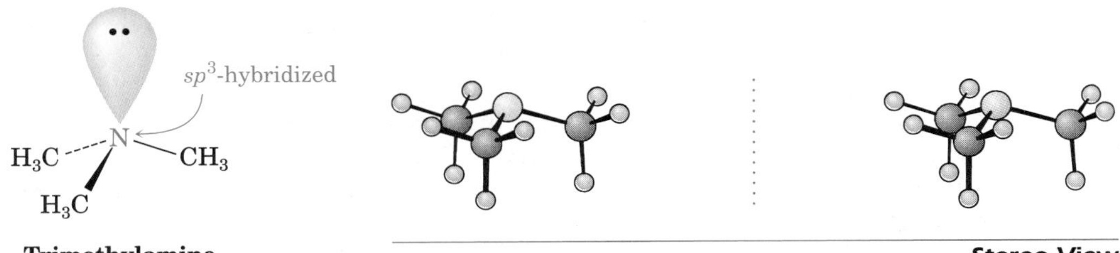

Trimethylamine **Stereo View**

Like alcohols, amines are highly polar, and those with fewer than five carbon atoms are generally water-soluble. Also like alcohols, primary and secondary amines form hydrogen bonds and therefore have higher boiling points than alkanes of similar molecular weight.

One other characteristic property of amines is their *odor.* Low-molecular-weight amines such as trimethylamine have a distinctive fishlike aroma, while diamines such as putrescine (1,4-butanediamine) have odors that are as putrid as their names suggest.

12.3 Amine Basicity

The chemistry of amines is dominated by the lone pair of electrons on nitrogen. Because of this lone pair, amines are both basic and nucleophilic. They react with acids to form acid–base salts, and they react with electrophiles in many of the polar reactions seen in previous chapters.

$$R_3N: + H{-}A \rightleftharpoons R_3\overset{+}{N}{-}H + :A^-$$

An amine (a Lewis base) — An acid — A salt

Amines are much stronger bases than alcohols, ethers, or water. When an amine is dissolved in water, an equilibrium is established in which water acts as an acid and donates H^+ to the amine. By finding the equilibrium constant for the reaction, we can define a **basicity constant, K_b,** that measures the ability of an amine to accept a proton, and we can thereby establish a relative order of base strength.

For the reaction: $RNH_2 + H_2O \rightleftharpoons RNH_3^+ + OH^-$

$$K_b = \frac{[RNH_3^+][OH^-]}{[RNH_2]}$$

$$pK_b = -\log K_b$$

The larger the K_b (and the smaller the pK_b), the more favorable the equilibrium and the stronger the base; the smaller the K_b (and the larger the pK_b), the weaker the base.

Table 12.1 gives the pK_b values of some common amines and indicates that substitution has relatively little effect on alkylamine basicity. Most simple alkylamines have pK_b's in the narrow range 3–4, regardless of their exact structure.

The most important conclusion from Table 12.1 is that arylamines, such as aniline, are weaker bases than alkylamines by a factor of about 10^6. The

Table 12.1 Basicity of Some Common Amines

Name	*Structure*	pK_b	
Triethylamine	$(CH_3CH_2)_3N$	2.99	More basic
Ethylamine	$CH_3CH_2NH_2$	3.19	
Dimethylamine	$(CH_3)_2NH$	3.27	
Methylamine	CH_3NH_2	3.34	
Diethylamine	$(CH_3CH_2)_2NH$	3.51	
Trimethylamine	$(CH_3)_3N$	4.19	
Ammonia	NH_3	4.74	
Pyridine	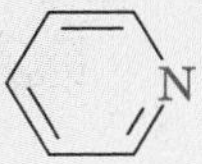 N	8.75	
Aniline	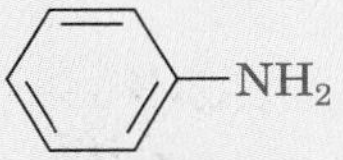 $-NH_2$	9.37	Less basic

nitrogen lone-pair electrons of arylamines are delocalized by orbital overlap with the π orbitals of the aromatic ring, and are therefore less available for bonding to an acid. In resonance terms, arylamines are more stable than alkylamines because of their five resonance structures:

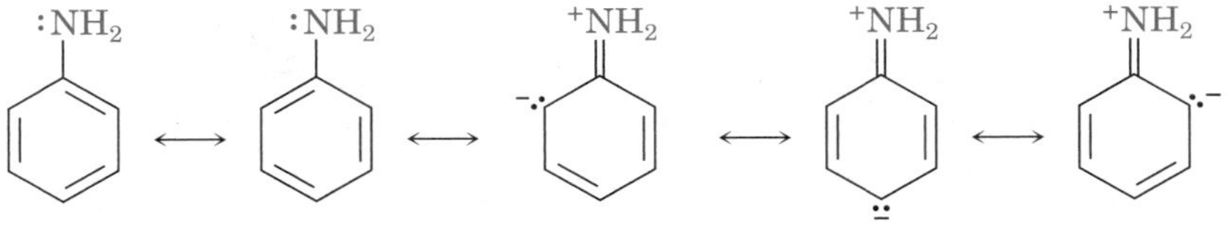

$:NH_2$ ⟷ $:NH_2$ ⟷ $^+NH_2$ ⟷ $^+NH_2$ ⟷ $^+NH_2$

In contrast to amines, *amides* ($RCONH_2$) are completely nonbasic. Amides don't react with acids, and their aqueous solutions are neutral. The main reason for the decreased basicity of amides relative to amines is that the nitrogen lone-pair electrons are delocalized by orbital overlap with the neighboring carbonyl-group π orbital. The electrons are therefore much less available for bonding to an acid. In resonance terms, amides are more stable and less reactive than amines because they are hybrids of two resonance forms:

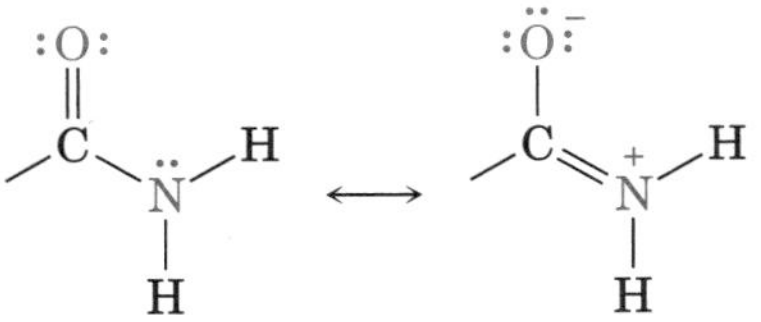

:O: C H N H ⟷ :O:⁻ C H N⁺ H

It's often possible to take advantage of the basicity of amines to purify them. For example, if a mixture of an amine (basic) and a ketone (neutral) is dissolved in an organic solvent and aqueous HCl is added, the basic amine dissolves in the acidic water as its ammonium ion, while the ketone remains in the organic solvent. Separation of the water layer and neutralization of the ammonium ion by addition of NaOH then provides the pure amine (Figure 12.1).

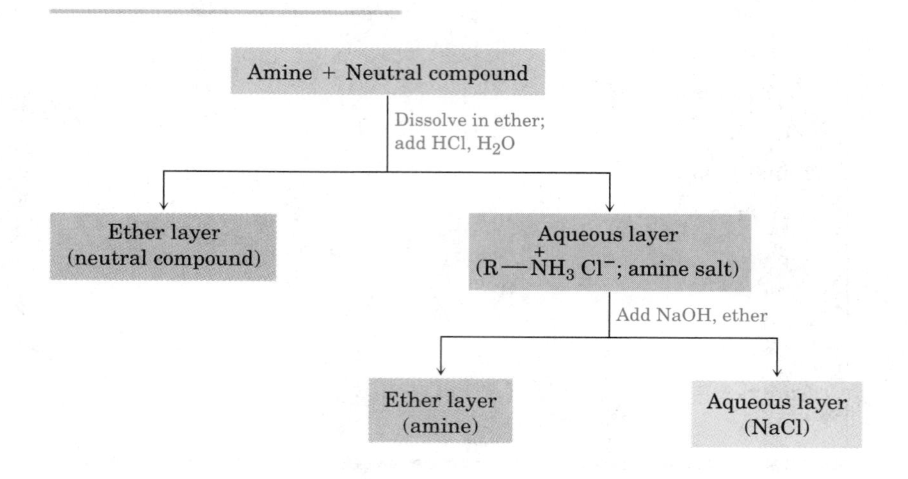

Figure 12.1 Separation and purification of a basic amine from a mixture.

PRACTICE PROBLEM 12.2

Which would you expect to be the stronger base, aniline or *p*-nitroaniline?

Solution Since a nitro group on a benzene ring is strongly electron-withdrawing (Section 5.9), it pulls electrons from the $-NH_2$ group, making them less available for donation to acids and making *p*-nitroaniline a weaker base than aniline.

PRACTICE PROBLEM 12.3

Predict the product of the following reaction:

$$CH_3CH_2NHCH_3 + HCl \longrightarrow ?$$

Solution Amines are protonated by acids to yield ammonium salts:

$$CH_3CH_2NHCH_3 + HCl \longrightarrow CH_3CH_2\overset{+}{N}H_2CH_3\ Cl^-$$

PROBLEM

12.5 Which would you expect to be the stronger base, aniline or *p*-methylaniline? Explain.

PROBLEM

12.6 Predict the product of the following reaction:

CH_3

N + HBr ⟶ ?

H

PROBLEM

12.7 Which compound in each of the following pairs is more basic?
(a) $CH_3CH_2NH_2$ or $CH_3CH_2CONH_2$ (b) NaOH or $C_6H_5NH_2$
(c) CH_3NHCH_3 or $CH_3NHC_6H_5$ (d) CH_3OCH_3 or $(CH_3)_3N$

12.4 Synthesis of Amines

S_N2 Reactions of Alkyl Halides

Alkylamines are excellent nucleophiles in S_N2 reactions (Section 7.7). As a result, the simplest method of amine synthesis is by S_N2 reaction of ammonia or an alkylamine with an alkyl halide. If ammonia is used, a primary amine results; if a primary amine is used, a secondary amine results; and so on. Even tertiary amines react with alkyl halides to yield quaternary ammonium salts, $R_4N^+ X^-$.

	S_N2 reaction				
Ammonia	$\ddot{N}H_3 + R{-}X$	$\longrightarrow$ $RNH_3^+ X^-$	$\xrightarrow{NaOH}$	RNH_2	Primary
Primary	$R\ddot{N}H_2 + R{-}X$	$\longrightarrow$ $R_2NH_2^+ X^-$	$\xrightarrow{NaOH}$	R_2NH	Secondary
Secondary	$R_2\ddot{N}H + R{-}X$	$\longrightarrow$ $R_2NH^+ X^-$	$\xrightarrow{NaOH}$	R_3N	Tertiary
Tertiary	$R_3\ddot{N} + R{-}X$	$\longrightarrow$ $R_4N^+ X^-$	Quaternary ammonium salt		

Unfortunately, these reactions don't stop cleanly after a single alkylation has occurred. Because primary, secondary, and tertiary amines all have similar reactivity, the initially formed monoalkylated amine often undergoes further reaction to yield a mixture of products. For example, treatment of 1-bromooctane with a twofold excess of ammonia leads to a mixture containing only a 45% yield of octylamine. A nearly equal amount of dioctylamine is produced by double alkylation, along with smaller amounts of trioctylamine and tetraoctylammonium bromide.

$$CH_3(CH_2)_6CH_2Br + :NH_3 \longrightarrow CH_3(CH_2)_6CH_2\ddot{N}H_2 + [CH_3(CH_2)_6CH_2]_2\ddot{N}H$$

1-Bromooctane **Octylamine (45%)** **Dioctylamine (43%)**

$$+ [CH_3(CH_2)_6CH_2]_3N: + [CH_3(CH_2)_6CH_2]_4\overset{+}{N}\ \overset{-}{Br}$$

Trace Trace

PRACTICE PROBLEM 12.4

How could you prepare diethylamine from ammonia?

Solution Look at the starting material (NH_3) and the product $(CH_3CH_2)_2NH$, and note the difference. Since two ethyl groups have become bonded to the nitrogen atom, the reaction must involve ammonia and 2 equivalents of an ethyl halide:

$$2\ CH_3CH_2Br + NH_3 \longrightarrow (CH_3CH_2)_2NH$$

PROBLEM

12.8 How could you prepare the following amines from ammonia?
(a) Triethylamine
(b) Tetramethylammonium bromide

Reduction of Nitriles and Amides

We've already seen how amines can be prepared by reduction of amides (Section 10.10) and nitriles (Section 10.11) with $LiAlH_4$. The two-step sequence of S_N2 reaction of an alkyl halide with cyanide ion, followed by reduction, is a good method for converting an alkyl halide into a primary amine having one more carbon atom than the original halide. Amide reduction provides a method for converting a carboxylic acid into an amine having the same number of carbon atoms.

$$\underset{\textbf{Alkyl halide}}{RX} \xrightarrow{NaCN} RCN \xrightarrow[2.\ H_2O]{1.\ LiAlH_4,\ ether} \underset{\textbf{1° amine}}{RCH_2NH_2}$$

$$\underset{\textbf{Carboxylic acid}}{R{-}\overset{\overset{\displaystyle O}{\|}}{C}{-}OH} \xrightarrow[2.\ NH_3]{1.\ SOCl_2} R{-}\overset{\overset{\displaystyle O}{\|}}{C}{-}NH_2 \xrightarrow[2.\ H_2O]{1.\ LiAlH_4,\ ether} \underset{\textbf{1° amine}}{RCH_2NH_2}$$

PRACTICE PROBLEM 12.5

What amide would you use to prepare *N*-ethylcyclohexylamine?

Solution Reduction of an amide with $LiAlH_4$ yields an amine in which the amide carbonyl group has been replaced by a methylene ($-CH_2-$) unit, $RCONR_2 \rightarrow RCH_2NR_2$. Since *N*-ethylcyclohexylamine has only one $-CH_2-$ carbon attached to nitrogen (the ethyl group), the product must come from reduction of *N*-cyclohexylacetamide:

$$\underset{\textbf{N-Cyclohexylacetamide}}{C_6H_{11}{-}NH{-}\overset{\overset{\displaystyle O}{\|}}{C}{-}CH_3} \xrightarrow[2.\ H_2O]{1.\ LiAlH_4} \underset{\textbf{N-Ethylcyclohexylamine}}{C_6H_{11}{-}NHCH_2CH_3}$$

PRACTICE PROBLEM 12.6

What nitrile would yield butylamine on reaction with $LiAlH_4$?

Solution Reduction of a nitrile with $LiAlH_4$ yields a primary amine whose $-CH_2NH_2$ part comes from the $-C{\equiv}N$ group. Thus, butylamine must have come from butanenitrile:

$$\underset{\textbf{Butanenitrile}}{CH_3CH_2CH_2C{\equiv}N} \xrightarrow[2.\ H_2O]{1.\ LiAlH_4} \underset{\textbf{Butylamine}}{CH_3CH_2CH_2CH_2NH_2}$$

PROBLEM ....

12.9 Propose structures for amides that might be precursors of the following amines:
(a) Propylamine (b) Dipropylamine (c) Benzylamine, $C_6H_5CH_2NH_2$

PROBLEM ....

12.10 Propose structures for nitriles that might be precursors of the following amines:

(a) $CH_3CH(CH_3)CH_2CH_2NH_2$ (b) Benzylamine, $C_6H_5CH_2NH_2$

....

Reduction of Nitrobenzenes

Arylamines are prepared by nitration of an aromatic starting material, followed by reduction of the nitro group. The reduction step can be carried out in different ways, depending on the circumstances. Catalytic hydrogenation over platinum works well, but is sometimes incompatible with the presence elsewhere in the molecule of other reducible groups, such as C=C bonds. Iron, zinc, tin, and stannous chloride ($SnCl_2$) in aqueous acid are also effective.

$$(CH_3)_3C\text{-}C_6H_4\text{-}NO_2 \xrightarrow[\text{or 1. Fe, } H_3O^+ \text{; 2. NaOH}]{H_2\text{, Pt catalyst}} (CH_3)_3C\text{-}C_6H_4\text{-}NH_2$$

***p-tert*-Butylnitrobenzene** → ***p-tert*-Butylaniline**

PRACTICE PROBLEM 12.7 ....

How could you synthesize *p*-methylaniline from benzene? More than one step is required.

Solution A methyl group is introduced onto a benzene ring by a Friedel–Crafts reaction with $CH_3Cl/AlCl_3$ (Section 5.8), and an amino group is introduced onto a ring by nitration and reduction. The overall sequence is

$$C_6H_6 \xrightarrow[AlCl_3]{CH_3Cl} C_6H_5CH_3 \xrightarrow[H_2SO_4]{HNO_3} p\text{-}CH_3C_6H_4NO_2 \xrightarrow[\text{2. NaOH}]{\text{1. Fe, } H_3O^+} p\text{-}CH_3C_6H_4NH_2$$

Benzene → **Toluene** → ***p*-Nitrotoluene** → ***p*-Methylaniline**

PROBLEM ....

12.11 How could you synthesize the following amines from benzene? More than one step is required in each case.
(a) *m*-Aminobenzoic acid (b) 2,4,6-Tribromoaniline

....

12.5 Reactions of Amines

We've already seen the two most important reactions of alkylamines: alkylation and acylation. As we saw in the previous section, primary, secondary, and tertiary amines can be alkylated by reaction with alkyl halides. Primary and secondary (but not tertiary) amines can also be acylated by nucleophilic acyl substitution reactions with acid chlorides or acid anhydrides (Sections 10.7 and 10.8). The products are amides.

$$RCOCl + NH_3 \xrightarrow[\text{solvent}]{\text{Pyridine}} RCONH_2 + HCl$$

$$RCOCl + R'NH_2 \xrightarrow[\text{solvent}]{\text{Pyridine}} RCONHR' + HCl$$

$$RCOCl + R'_2NH \xrightarrow[\text{solvent}]{\text{Pyridine}} RCONR'_2 + HCl$$

PROBLEM

12.12 Write an equation for the reaction of diethylamine with acetyl chloride to yield *N,N*-diethylacetamide.

Diazonium Salts: The Sandmeyer Reaction

Primary amines react with nitrous acid, HNO_2, in a *diazotization* reaction to yield *diazonium salts,* $R{-}\overset{+}{N}{\equiv}N\ X^-$. Although alkyl diazonium salts are too reactive to be isolated, aryl diazonium salts, $Ar{-}\overset{+}{N}{\equiv}N\ X^-$, are more stable.

$$C_6H_5NH_2 + HNO_2 + H_2SO_4 \longrightarrow C_6H_5\overset{+}{N}{\equiv}N\ HSO_4^- + 2\ H_2O$$

Aryl diazonium salts are extremely useful compounds, because the diazonio group (N_2^+) can be replaced by nucleophiles in a substitution reaction:

$$C_6H_5\overset{+}{N}{\equiv}N\ HSO_4^- + :Nu^- \longrightarrow C_6H_5Nu + N_2$$

Many different nucleophiles react with aryl diazonium salts, yielding many different kinds of substituted benzenes. The overall sequence of (1) nitration, (2) reduction, (3) diazotization, and (4) nucleophilic replacement, is probably the single most versatile method for preparing substituted aromatic rings (Figure 12.2).

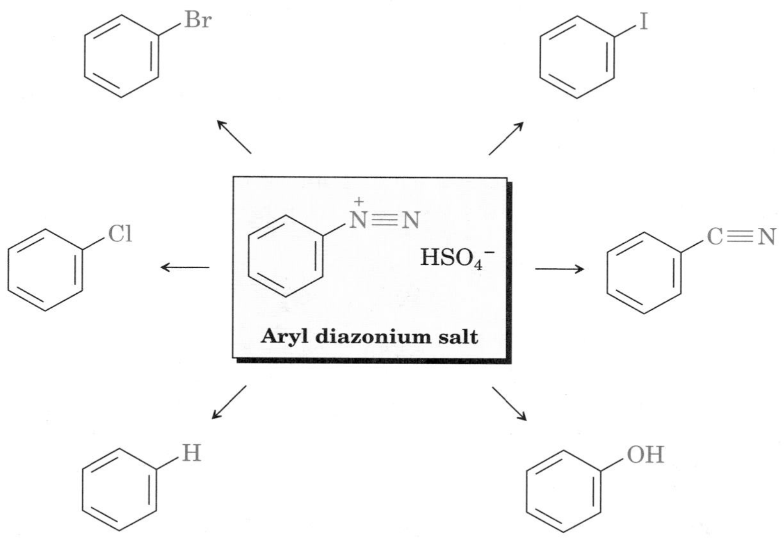
Br
I
Cl
N≡N
HSO4−
C≡N
Aryl diazonium salt
H
OH

Figure 12.2 Preparation of substituted aromatic compounds by diazonio replacement reactions.

Aryl chlorides and bromides are prepared by reaction of an aryl diazonium salt with HX in the presence of a small amount of cuprous halide (CuX) catalyst, a process called the **Sandmeyer reaction.** Aryl iodides are prepared by reaction with sodium iodide.

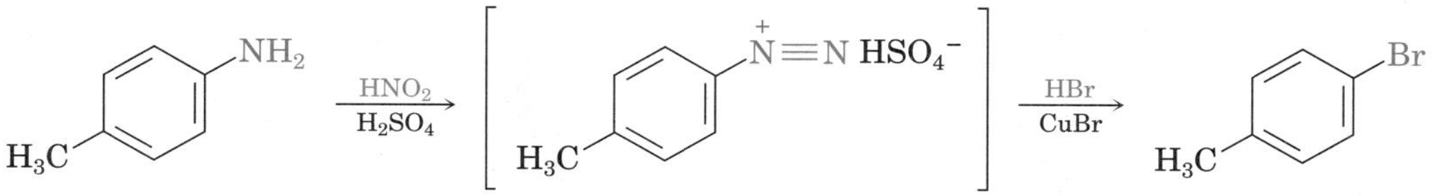
NH2
H3C
HNO2
H2SO4
N≡N HSO4−
H3C
HBr
CuBr
Br
H3C

***p*-Methylaniline** **_p_-Bromotoluene (73%)**

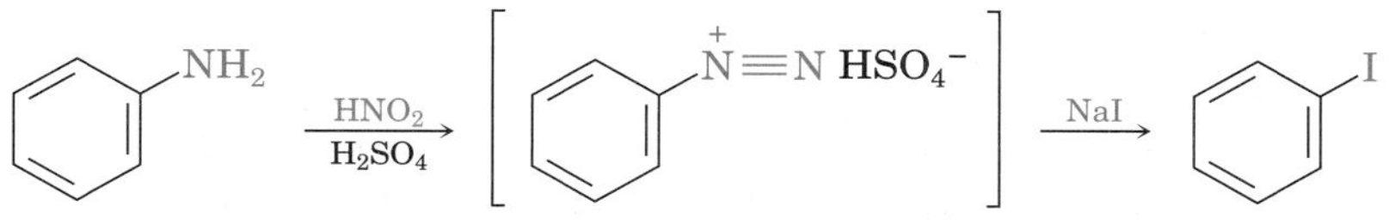
NH2
HNO2
H2SO4
N≡N HSO4−
NaI
I

Aniline **Iodobenzene (67%)**

Treatment of an aryl diazonium salt with KCN and a small amount of cuprous cyanide, CuCN, yields a nitrile, ArCN. This reaction is particularly useful because it allows the replacement of a nitrogen substituent ($-NH_2$)

by a carbon substituent (–CN). The nitrile can then be elaborated into other functional groups such as –COOH or $-CH_2NH_2$. For example, hydrolysis of *o*-methylbenzonitrile, produced by Sandmeyer reaction of *o*-methylbenzenediazonium bisulfate with cuprous cyanide, yields *o*-methylbenzoic acid:

$$\text{o-Methylaniline} \xrightarrow[H_2SO_4]{HNO_2} \text{o-Methylbenzenediazonium bisulfate} \xrightarrow[CuCN]{KCN} \text{o-Methylbenzonitrile} \xrightarrow{H_3O^+} \text{o-Methylbenzoic acid}$$

o-Methylaniline **o-Methylbenzene-diazonium bisulfate** **o-Methylbenzonitrile** **o-Methylbenzoic acid**

The diazonio group can also be replaced by –OH or by –H. Phenols are prepared by addition of the aryl diazonium salt to hot aqueous acid. For example:

$$\text{m-Nitroaniline} \xrightarrow[H_2SO_4]{HNO_2} m\text{-}O_2NC_6H_4\overset{+}{N}{\equiv}N\ HSO_4^- \xrightarrow{H_3O^+} \text{m-Nitrophenol (86\%)}$$

m-Nitroaniline **m-Nitrophenol (86%)**

Replacement of the diazonio group by –H is accomplished by reaction of the diazonium salt with hypophosphorous acid, H_3PO_2. For example, *p*-methylaniline can be converted into 3,5-dibromotoluene by a sequence involving bromination, diazotization, and hypophosphorous acid treatment:

$$\text{p-Methylaniline} \xrightarrow{2\ Br_2} \text{2,6-dibromo-4-methylaniline} \xrightarrow[H_2SO_4]{HNO_2} \text{2,6-dibromo-4-methylbenzenediazonium } HSO_4^- \xrightarrow{H_3PO_2} \text{3,5-Dibromotoluene}$$

p-Methylaniline **3,5-Dibromotoluene**

PRACTICE PROBLEM 12.8

How would you prepare *p*-methylphenol from benzene using a diazonio replacement reaction?

Solution Working backward, the immediate precursor of the target molecule might be *p*-methylbenzenediazonium ion, which could be prepared from *p*-nitrotoluene. *p*-Nitrotoluene, in turn, could be prepared by nitration of toluene, which could be prepared by Friedel–Crafts methylation of benzene.

$$C_6H_6 \xrightarrow[AlCl_3]{CH_3Cl} C_6H_5CH_3 \xrightarrow[H_2SO_4]{HNO_3} O_2N\text{-}C_6H_4\text{-}CH_3$$

$$O_2N-C_6H_4-CH_3 \xrightarrow{\text{Fe, HCl}} H_2N-C_6H_4-CH_3 \xrightarrow[\text{2. } H_3O^+]{\text{1. } HNO_2} HO-C_6H_4-CH_3$$

PROBLEM ....

12.13 How would you prepare *p*-bromobenzonitrile from bromobenzene using a diazonio replacement reaction?

PROBLEM ....

12.14 How would you prepare the following compounds from benzene?
(a) *m*-Bromobenzoic acid
(b) *m*-Bromochlorobenzene

....

Diazonium Coupling Reactions

In addition to their reactivity in nucleophilic substitution reactions, aryl diazonium salts undergo a coupling reaction with activated aromatic rings, such as phenols and substituted anilines, to yield brightly colored **azo compounds, Ar–N=N–Ar′,** which are widely used as dyes. *p*-(Dimethylamino)azobenzene, for example, is a bright yellow dye that was once used as a coloring agent in margarine.

$$C_6H_5-\overset{+}{N}{\equiv}N\ HSO_4^- + C_6H_5-\ddot{N}(CH_3)_2 \longrightarrow C_6H_5-N{=}N-C_6H_4-N(CH_3)_2$$

Benzenediazonium bisulfate | ***N,N*-Dimethylaniline** | ***p*-(Dimethylamino)azobenzene (yellow crystals, mp 127°C)**

Diazonium coupling reactions are typical electrophilic aromatic substitutions (Sections 5.5–5.8) in which the positively charged diazonium ion is the electrophile that reacts with the electron-rich ring of a phenol or an arylamine. Reaction almost always occurs at the para position, although ortho attack can take place if the para position is blocked.

PROBLEM ....

12.15 Propose a synthesis of *p*-(dimethylamino)azobenzene from benzene.

PROBLEM ....

12.16 Show the mechanism of the azo coupling reaction between phenol and benzenediazonium bisulfate.

....

12.6 Heterocyclic Amines

Cyclic organic compounds are classified either as *carbocycles* or as *heterocycles*. **Carbocycles** contain only carbon atoms in their rings, while **heterocycles** contain one or more different atoms in addition to carbon. Heterocyclic amines are particularly common in organic chemistry, and many have important biological properties. For example, the antiulcer agent cimetidine and the sedative phenobarbital are heterocyclic amines:

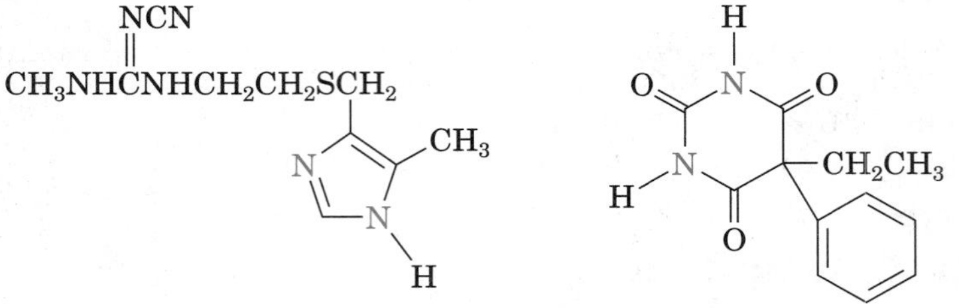

NCN
||
$CH_3NHCNHCH_2CH_2SCH_2$ — CH_3, N, N, H

H, O, N, O, H, N, CH_2CH_3, O

Cimetidine (an antiulcer agent) **Phenobarbital (a sedative)**

For the most part, heterocyclic amines have the same chemistry as their open-chain counterparts. In certain cases, though, particularly when the ring is unsaturated, heterocycles have unique and interesting properties. Let's look at several examples.

Pyrrole, a Five-Membered Aromatic Heterocycle

Pyrrole, a five-membered heterocyclic amine, has two double bonds and one nitrogen. Although pyrrole is both an amine and a conjugated diene, its chemistry is not consistent with either of these structural features. Unlike most amines, pyrrole isn't basic; unlike most conjugated dienes, pyrrole doesn't undergo electrophilic addition reactions. How can we explain these observations?

In fact, pyrrole is *aromatic*. Even though it has a five-membered ring, pyrrole has six π electrons in a cyclic, conjugated π orbital system, just as benzene does (Section 5.3). Each of the four carbon atoms contributes one π electron, and the sp^2-hybridized nitrogen atom contributes two more (its lone pair). The six π electrons occupy *p* orbitals with lobes above and below the plane of the flat ring, as shown in Figure 12.3.

Like benzene, pyrrole undergoes substitution of a ring hydrogen atom on reaction with an electrophile. Substitution normally occurs at the position next to nitrogen, as the following nitration shows:

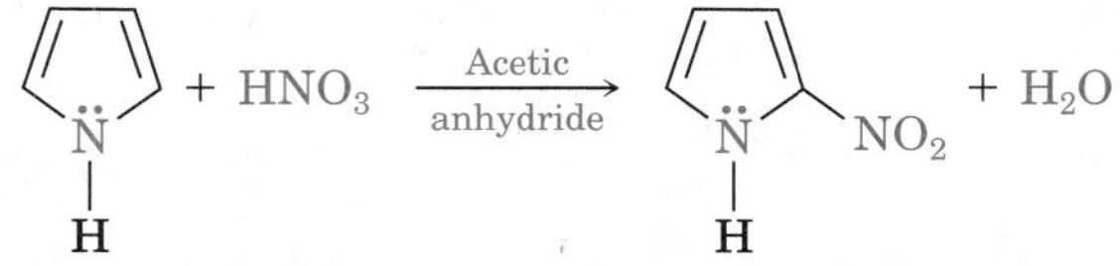

Pyrrole (N–H) + HNO_3 $\xrightarrow{\text{Acetic anhydride}}$ 2-nitropyrrole (N–H, NO_2) + H_2O

Pyrrole **2-Nitropyrrole (83)%**

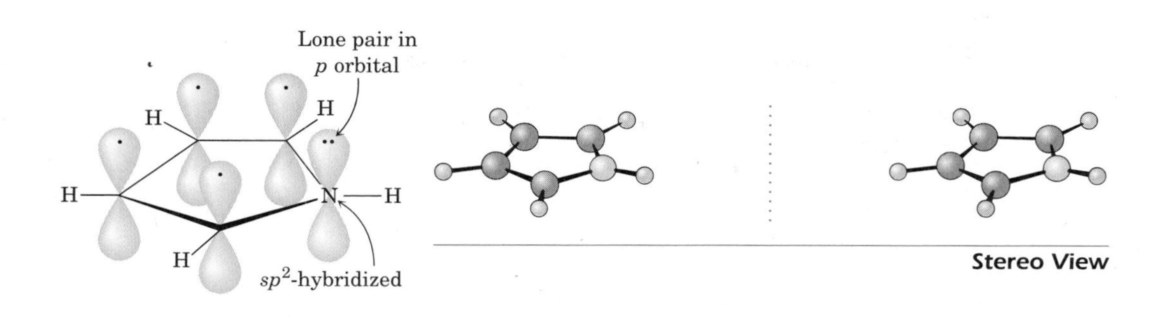

Six π electrons

Figure 12.3 Pyrrole, an aromatic heterocycle, has a π electron structure similar to that of benzene.

Substituted pyrrole rings form the basic building blocks from which many important plant and animal pigments are constructed. Among these is *heme,* an iron-containing tetrapyrrole found in blood.

H_3C $CH{=}CH_2$ H_3C CH_3 N N—Fe—N N $HOOCCH_2CH_2$ $CH{=}CH_2$ $HOOCCH_2CH_2$ CH_3

Heme

PROBLEM

12.17 Pyrrole undergoes other typical electrophilic substitution reactions in addition to nitration. What products would you expect to obtain from reaction of *N*-methylpyrrole with the following reagents?
(a) Br_2 (b) CH_3Cl, $AlCl_3$ (c) CH_3COCl, $AlCl_3$

PROBLEM

12.18 Review the mechanism of the bromination of benzene (Section 5.6) and then propose a mechanism for the nitration of pyrrole.

Pyridine, a Six-Membered Aromatic Heterocycle

Pyridine is a nitrogen-containing heterocyclic analog of benzene. Like benzene, pyridine is a flat molecule with bond angles of approximately 120° and with C–C bond lengths of 1.39 Å, intermediate between normal single and double bonds. Also like benzene, pyridine is aromatic with six π electrons in a cyclic, conjugated π orbital system. The sp^2-hybridized nitrogen atom and the five carbon atoms each contribute one π electron to the cyclic, conjugated p orbitals of the ring. Unlike the situation in pyrrole, however, the lone-pair electrons on the pyridine nitrogen atom are not part of the π orbital system but instead occupy an sp^2 orbital in the plane of the ring (Figure 12.4).

sp^2-hybridized

Lone pair in sp^2 orbital

Six π electrons

Figure 12.4 Electronic structure of pyridine, a nitrogen-containing analog of benzene.

Pyridine, $pK_b = 8.75$, is less basic than typical alkylamines, but is nevertheless used in a wide variety of organic reactions when a base catalyst is required. The reaction of an acid chloride with an alcohol to yield an ester, for example, is commonly done in the presence of pyridine (Section 10.7).

Substituted pyridines, such as the B_6 complex vitamins pyridoxal and pyridoxine, are important biologically. Present in yeast, cereal, and other foodstuffs, the B_6 vitamins are necessary for the synthesis of some amino acids.

Pyridoxal **Pyridoxine**

PROBLEM

12.19 The five-membered heterocycle imidazole contains two nitrogen atoms, one "pyrrole-like" and one "pyridine-like." Draw an orbital picture of imidazole, and indicate the orbital in which each nitrogen has its electron lone pair.

Imidazole

Fused-Ring Aromatic Heterocycles

Quinoline, isoquinoline, and indole are *fused-ring* heterocycles that contain a benzene ring and a heterocyclic ring sharing a common bond. All three fused-ring systems occur widely in nature, and many members of the class have useful biological activity. Thus, quinine, a quinoline derivative found in the bark of the South American cinchona tree, is an important antimalarial drug. *N,N*-Dimethyltryptamine, which contains an indole ring, is a powerful hallucinogen.

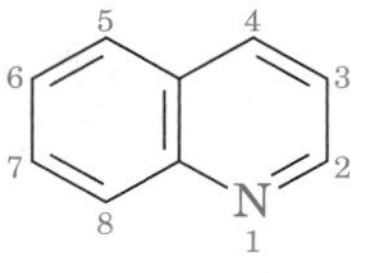

Quinoline 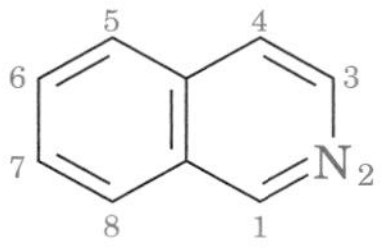 **Isoquinoline** 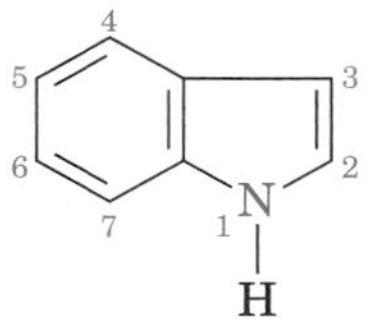 **Indole**

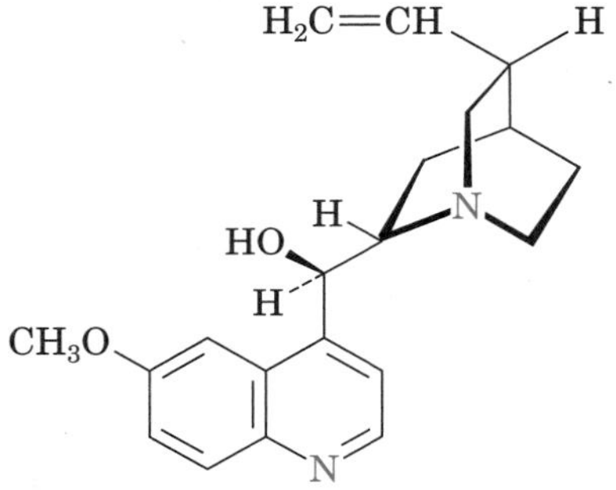

Quinine, an antimalarial drug
(a quinoline alkaloid)

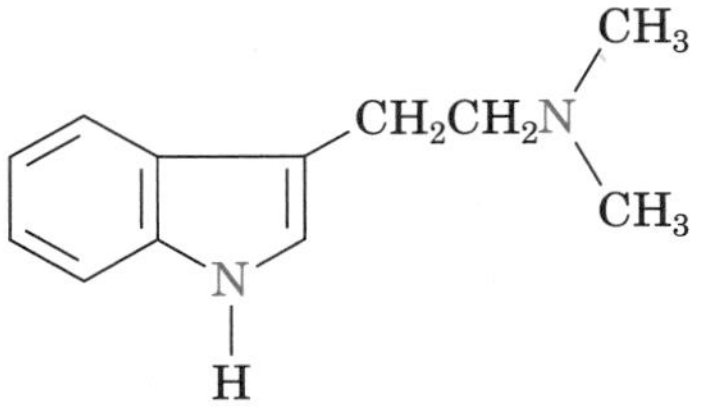

***N,N*-Dimethyltryptamine, a hallucinogen**
(an indole alkaloid)

INTERLUDE

Naturally Occurring Amines: Morphine Alkaloids

Morphine and several of its relatives are isolated from the opium poppy, *Papaver somniferum.*

Naturally occurring amines derived from plant sources were once known as "vegetable alkali" because their aqueous solutions are slightly basic, but they are now referred to as *alkaloids.* The study of alkaloids provided much of the impetus for the growth of organic chemistry in the nineteenth century and remains today a fascinating area of research. Let's look briefly at one particular group, the morphine alkaloids.

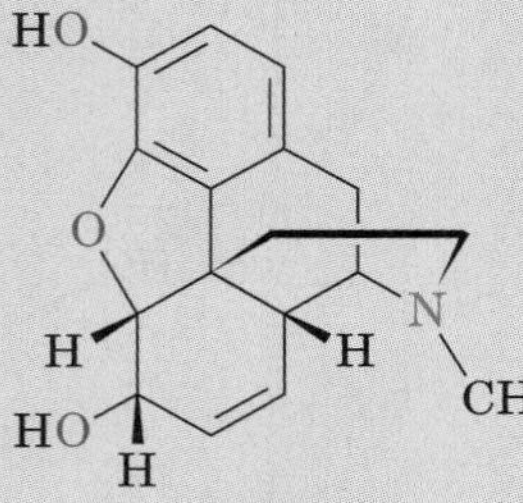

Morphine (an analgesic)

Stereo View

The medical uses of morphine alkaloids have been known at least since the seventeenth century, when crude extracts of the opium poppy, *Papaver somniferum,* were used for the relief of pain. Morphine was the first pure alkaloid to be isolated from the poppy, but its close relative, codeine, also occurs naturally. Codeine, which is simply the methyl ether of morphine, is used in prescription cough medicines and as an analgesic. Heroin, another close relative of morphine, does not occur naturally but is synthesized by diacetylation of morphine.

(continued)➤

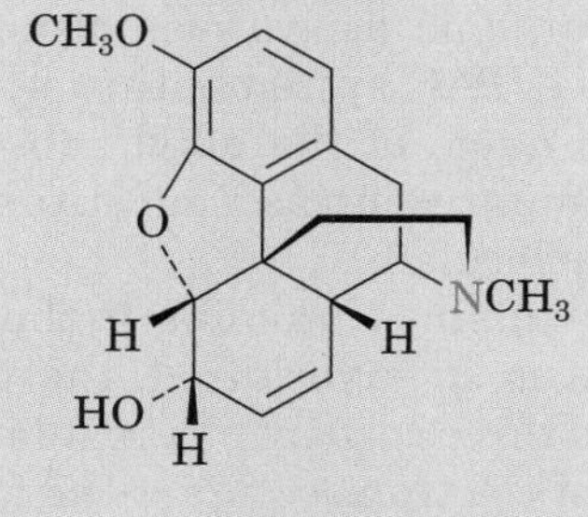

Codeine

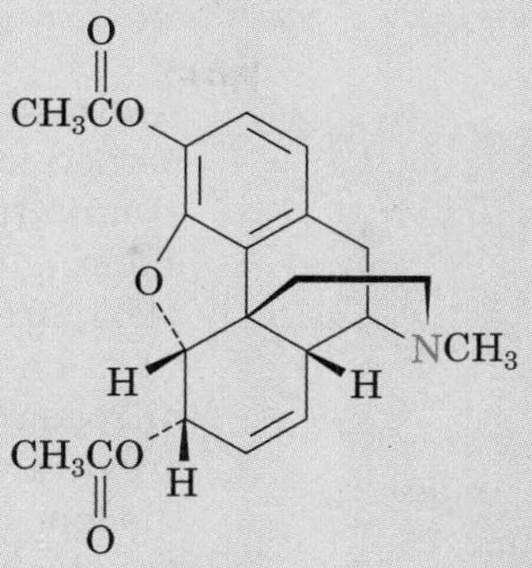

Heroin

Chemical investigations into the structure of morphine occupied some of the finest chemical minds of the nineteenth and early twentieth centuries, and it was not until 1924 that the puzzle was finally solved by Robert Robinson.

Morphine and its relatives are extremely useful pharmaceutical agents, yet they also pose an enormous social problem because of their addictive properties. Much effort has therefore gone into understanding how morphine works and into developing modified morphine analogs that retain the analgesic activity but don't cause physical dependence. Our present understanding is that morphine binds to opiate receptor sites in the brain. It doesn't interfere with the transmission of a pain signal to the brain but rather changes the brain's reception of the signal.

Hundreds of morphine-like molecules have been synthesized and tested for their analgesic properties. Research has shown that only part of the complex framework of morphine is necessary for biological activity. According to the "morphine rule," biological activity requires: (1) an aromatic ring attached to (2) a quaternary carbon atom and (3) a tertiary amine situated (4) two carbon atoms farther away. Meperidine (Demerol), a widely used analgesic, and methadone, a substance used in the treatment of heroin addiction, are two compounds that fit the morphine rule.

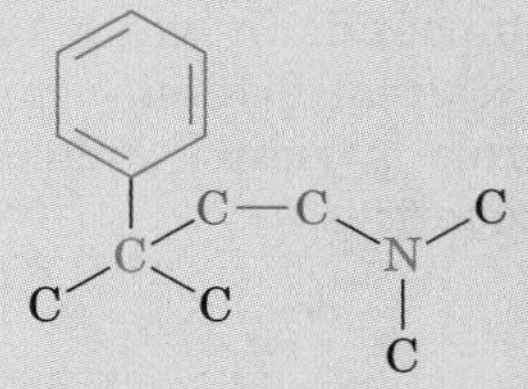

The morphine rule: an aromatic ring, attached to a quaternary carbon, attached to two more carbons, attached to a tertiary amine

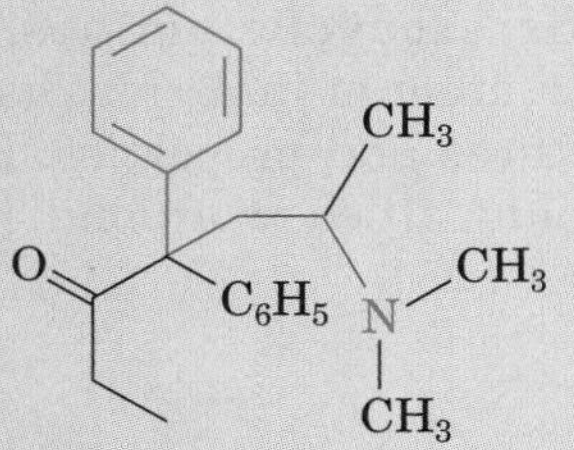

Methadone

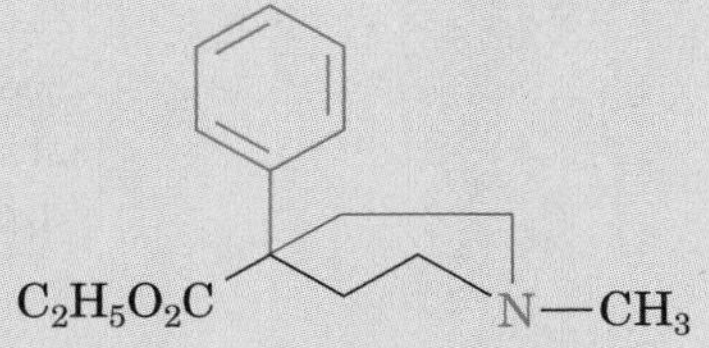

Meperidine

Summary and Key Words

alkylamine, 381
amine, 380
arylamine, 381
azo compound, 393
basicity constant, K_b, 384
carbocycle, 394
heterocycle, 394
heterocyclic amine, 383
primary amine, 380
quaternary ammonium salt, 381
Sandmeyer reaction, 391
secondary amine, 380
tertiary amine, 380

Amines are organic derivatives of ammonia. They are named in the IUPAC system either by adding the suffix *-amine* to the name of the alkyl substituent or by considering the amino group as a substituent on a more complex parent molecule.

Bonding in amines is similar to that in ammonia. The nitrogen atom is sp^3-hybridized, the three substituents are directed to three corners of a regular tetrahedron, and the lone pair of electrons occupies the fourth corner of the tetrahedron.

The chemistry of amines is dominated by the presence of the lone-pair electrons on nitrogen, which make amines both basic and nucleophilic. **Arylamines** are generally weaker bases than **alkylamines** because their lone-pair electrons are delocalized by orbital overlap with the aromatic π electron system.

The simplest method of amine synthesis involves S_N2 reaction of ammonia or an amine with an alkyl halide. Alkylation of ammonia yields a primary amine; alkylation of a primary amine yields a secondary amine; and so on. Amines can also be prepared from amides and nitriles by reduction with $LiAlH_4$. Arylamines are prepared by nitration of an aromatic ring followed by reduction of the nitro group.

Many of the reactions that amines undergo are familiar from previous chapters. Thus, amines react with alkyl halides in S_N2 reactions and with acid chlorides in nucleophilic acyl substitution reactions. Arylamines are converted by treatment with nitrous acid into aryl diazonium salts, $Ar{-}N_2^+ X^-$. The diazonio group can then be replaced by nucleophiles in the **Sandmeyer reaction** to give a variety of substituted aromatic compounds. Aryl chlorides, bromides, iodides, nitriles, and phenols can be prepared.

Heterocyclic amines, compounds in which the nitrogen atom is in a ring, have a great diversity in their structures and properties. Pyrrole, pyridine, indole, and quinoline all show aromatic properties.

Summary of Reactions

1. Synthesis of amines (Section 12.4)
   (a) Alkylamines by S_N2 reaction

$$NH_3 + RX \longrightarrow RNH_2$$
$$RNH_2 + RX \longrightarrow R_2NH$$
$$R_2NH + RX \longrightarrow R_3N$$
$$R_3N + RX \longrightarrow R_4N^+ X^-$$

   (b) Arylamines by reduction of nitroarenes

$$C_6H_5NO_2 \xrightarrow{Fe,\ H_3O^+} C_6H_5NH_2$$

2. Reactions of amines (Section 12.5)
   (a) Formation and reactions of aryl diazonium salts

$$C_6H_5NH_2 \xrightarrow[H_2SO_4]{HNO_2} C_6H_5\overset{+}{N}{\equiv}N\ HSO_4^-$$

The diazonium salt then reacts with:
- HCl, CuCl → C_6H_5Cl
- HBr, CuBr → C_6H_5Br
- NaI → C_6H_5I
- KCN, CuCN → C_6H_5CN
- H_3O^+ → C_6H_5OH
- H_3PO_2 → C_6H_5H

   (b) Diazo coupling

$$C_6H_5\overset{+}{N}{\equiv}N\ HSO_4^- + C_6H_5OH \text{ (or } -NR_2) \longrightarrow C_6H_5N{=}N{-}C_6H_4{-}OH \text{ (or } -NR_2)$$

ADDITIONAL PROBLEMS

12.20 Classify each of the amine (not amide) nitrogen atoms in the following substances as primary, secondary, or tertiary:

(a) $(C_2H_5)_2N-C(=O)-$ ... $N-CH_3$ ... $N-H$

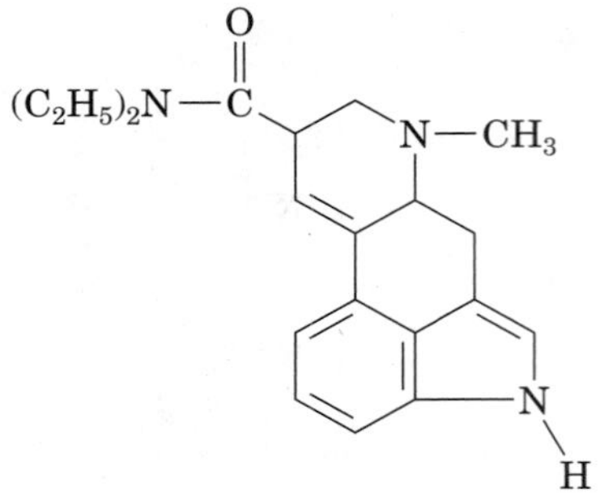

Lysergic acid diethylamide

(b) **Caffeine**

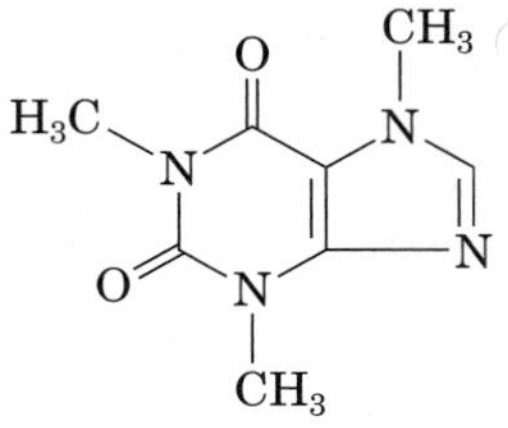

12.21 Draw structures corresponding to the following IUPAC names:

(a) *N,N*-Dimethylaniline
(b) *N*-Methylcyclohexylamine
(c) (Cyclohexylmethyl)amine
(d) (2-Methylcyclohexyl)amine
(e) 3-(*N,N*-Dimethylamino)propanoic acid

12.22 Name the following compounds according to IUPAC rules:

(a) NH_2, Br, Br (benzene ring)
(b) cyclopentyl$-CH_2CH_2NH_2$
(c) cyclopentyl$-NHCH_2CH_3$
(d) cyclopentyl$-N(CH_3)_2$
(e) pyrrolidine $N-CH_2CH_2CH_3$
(f) $H_2NCH_2CH_2CH_2CN$

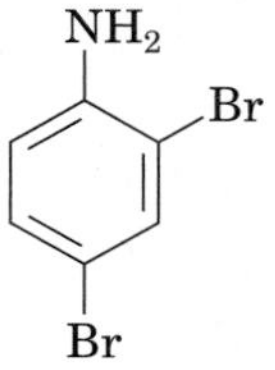

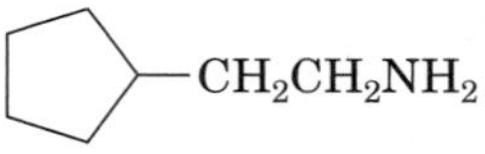

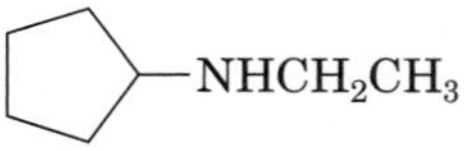

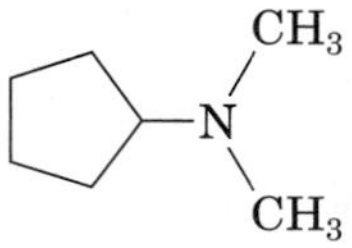

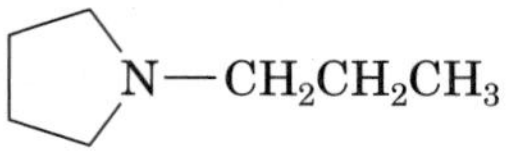

12.23 How can you explain the fact that trimethylamine (bp 3°C) has a lower boiling point than dimethylamine (bp 7°C)?

12.24 Mescaline, a powerful hallucinogen derived from the peyote cactus, has the systematic name 2-(3,4,5-trimethoxyphenyl)ethylamine. Draw its structure.

12.25 There are eight isomeric amines with the formula $C_4H_{11}N$. Draw them, name them, and classify each as primary, secondary, or tertiary.

12.26 Propose structures for amines that fit the following descriptions:

(a) A secondary arylamine
(b) A 1,3,5-trisubstituted arylamine
(c) An achiral quaternary ammonium salt
(d) A five-membered heterocyclic amine

12.27 Show the products of the following reactions:

(a) $CH_3CH_2CH_2NH_2 + CH_3Br \longrightarrow$
(b) Cyclohexylamine + HBr $\longrightarrow$
(c) $CH_3CH_2CONH_2 + LiAlH_4 \longrightarrow$
(d) Benzonitrile + $LiAlH_4 \longrightarrow$

12.28 How might you prepare the following amines from ammonia and any alkyl halides needed?
(a) Hexylamine
(b) Benzylamine
(c) Tetramethylammonium iodide
(d) *N*-Methylcyclohexylamine

12.29 How might you prepare the following amines from 1-bromobutane?
(a) Butylamine (b) Dibutylamine (c) Pentylamine

12.30 How might you prepare each of the amines in Problem 12.29 from 1-butanol?

12.31 How would you prepare benzylamine, $C_6H_5CH_2NH_2$, from each of the following starting materials?
(a) Benzamide (b) Benzoic acid (c) Nitrobenzene (d) Chlorobenzene

12.32 Write equations for the reaction of *p*-bromobenzenediazonium bisulfate with the following reagents:
(a) H_3O^+ (b) HBr, CuBr (c) H_3PO_2 (d) KCN, CuCN

12.33 Show how you might prepare benzoic acid from aniline. A diazonio replacement reaction is needed.

12.34 How might you prepare pentylamine from the following starting materials?
(a) Pentanamide
(b) Pentanenitrile
(c) Pentanoic acid

12.35 Which compound is more basic, $CH_3CH_2NH_2$ or $CF_3CH_2NH_2$? Explain.

12.36 Which compound is more basic, *p*-aminobenzaldehyde or aniline?

12.37 Which compound is more basic, triethylamine or aniline? Does the following reaction proceed as written?

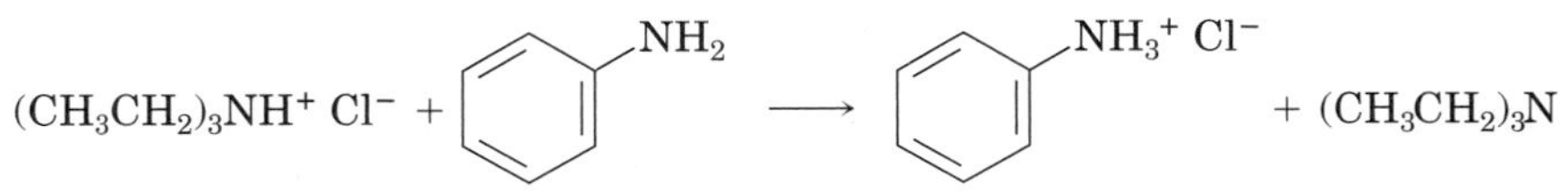

$$(CH_3CH_2)_3NH^+\ Cl^- + C_6H_5NH_2 \longrightarrow C_6H_5NH_3^+\ Cl^- + (CH_3CH_2)_3N$$

12.38 1,6-Hexanediamine, one of the starting materials used for the manufacture of nylon 66, can be synthesized by a route that begins with the addition of Cl_2 to 1,3-butadiene (Section 4.10). How would you carry out the complete synthesis?

12.39 Another method for making 1,6-hexanediamine (see Problem 12.38) starts from adipic acid (hexanedioic acid). How would you carry out the synthesis?

12.40 Give the structures of the major organic products you would obtain from the reaction of *m*-methylaniline with the following reagents:
(a) Br_2 (1 mol) (b) CH_3I (excess) (c) CH_3COCl, pyridine

12.41 Suppose that you were given a mixture of toluene, aniline, and phenol. Describe how you would separate the mixture into its three pure components.

12.42 Would you expect diphenylamine to be more basic or less basic than aniline? Explain.

12.43 Draw structures for the following amines:
(a) 2-Ethylpyrrole (b) 2,3-Dimethylaniline (c) 3-Methylindole

12.44 Furan, the oxygen-containing analog of pyrrole, is aromatic in the same way that pyrrole is. Draw an orbital picture of furan, and show how it has six electrons in its cyclic conjugated π orbitals.

12.45 By analogy with the chemistry of pyrrole, what product would you expect from the reaction of furan with Br_2 (see Problem 12.44)?

12.46 How would you synthesize 1,3,5-tribromobenzene from benzene? (A diazonio replacement reaction is needed.)

12.47 We've seen that amines are basic and amides are neutral. *Imides,* compounds with two carbonyl groups flanking an N–H, are actually acidic. Show by drawing resonance structures of the anion why imides are acidic.

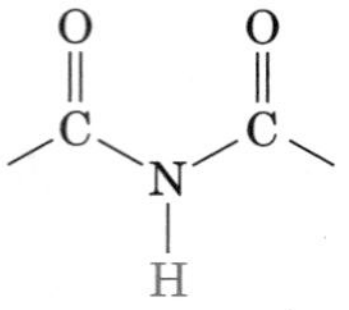

An imide

12.48 Tyramine is an alkaloid found, among other places, in mistletoe and in ripe cheese. How would you prepare tyramine from toluene?

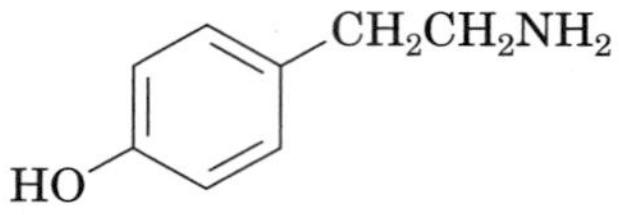

$CH_2CH_2NH_2$

HO

Tyramine

12.49 Atropine, $C_{17}H_{23}NO_3$, is a poisonous alkaloid isolated from the leaves and roots of the deadly nightshade, *Atropa belladonna.* In low doses, atropine acts as a muscle relaxant: 0.5 ng (1 nanogram = 10^{-9} g) is sufficient to cause pupil dilation. On reaction with aqueous NaOH, atropine yields tropic acid, $C_6H_5CH(CH_2OH)COOH$, and tropine, $C_8H_{15}NO$. Tropine, an optically inactive alcohol, yields tropidene on dehydration. Propose a structure for atropine.

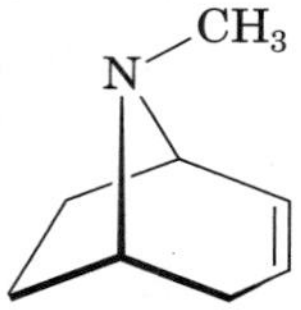

Tropidene

12.50 Choline, a component of the phospholipids in cell membranes, can be prepared by S_N2 reaction of trimethylamine with ethylene oxide. Show the structure of choline, and propose a mechanism for the reaction.

12.51 Methyl orange is an azo dye that is widely used as a pH indicator. How would you synthesize methyl orange from benzene?

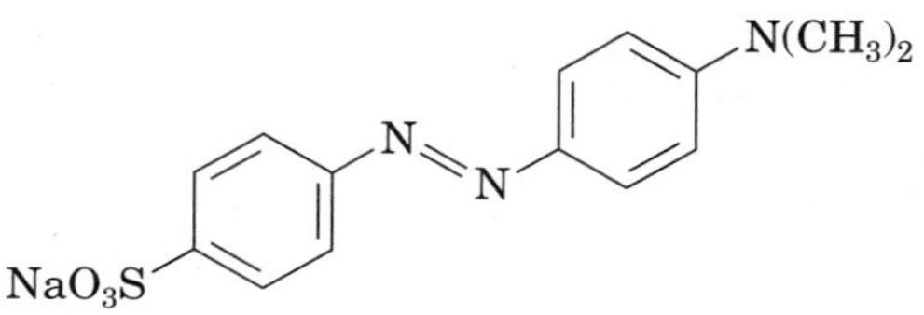

$N(CH_3)_2$

NaO_3S

Methyl orange

12.52 Fill in the missing reagents a–f in the following scheme:

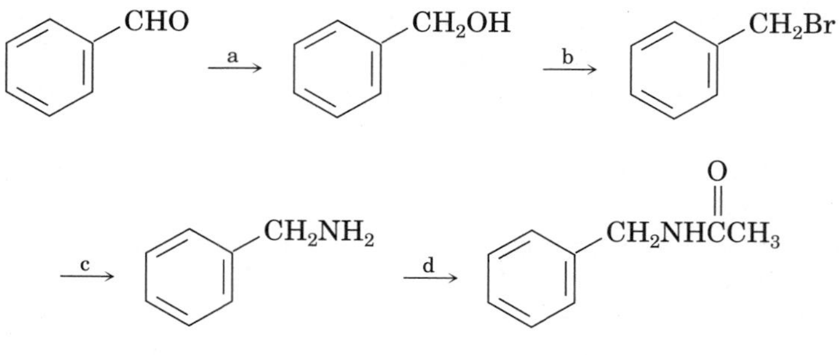

CHO $\xrightarrow{a}$ CH_2OH $\xrightarrow{b}$ CH_2Br $\xrightarrow{c}$ CH_2NH_2 $\xrightarrow{d}$ CH_2NHCCH_3 (O)

$$\xrightarrow{e} \quad O_2N\text{-}C_6H_4\text{-}CH_2NH\overset{O}{\overset{\|}{C}}CH_3 \quad \xrightarrow{f} \quad H_2N\text{-}C_6H_4\text{-}CH_2NH\overset{O}{\overset{\|}{C}}CH_3$$

Visualizing Chemistry

12.53 Name the following amines, and identify each as primary, secondary, or tertiary (gray = C, blue = N, light green = H):

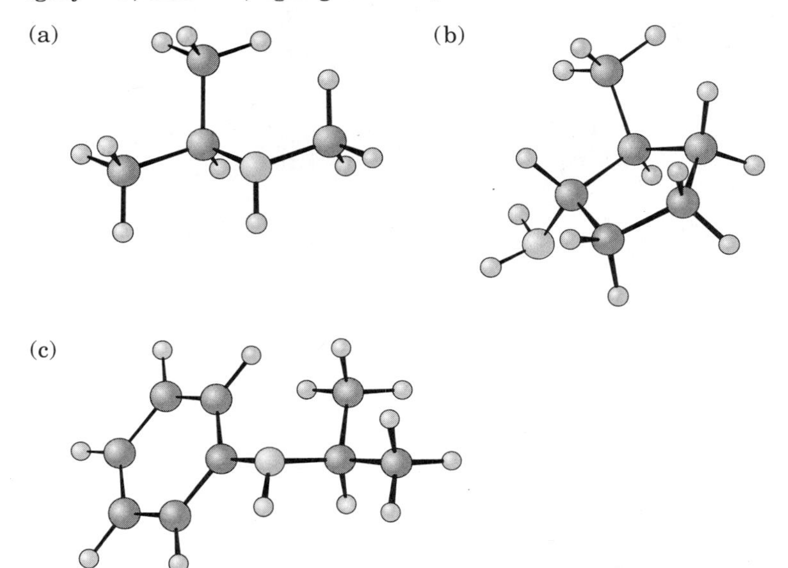

12.54 The following amine contains three nitrogen atoms. Rank them in order of increasing basicity (gray = C, blue = N, red = O, light green = H).

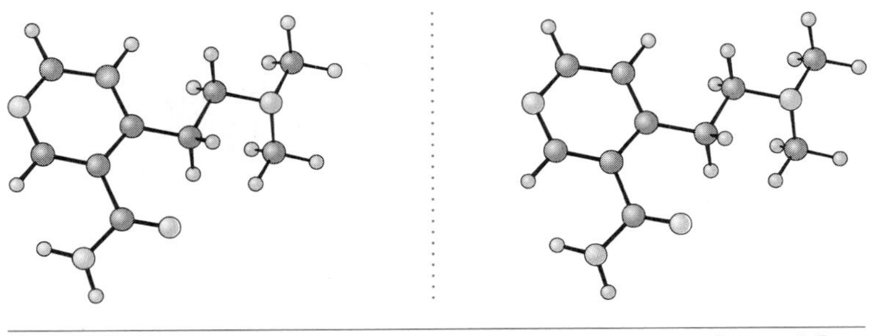

Stereo View

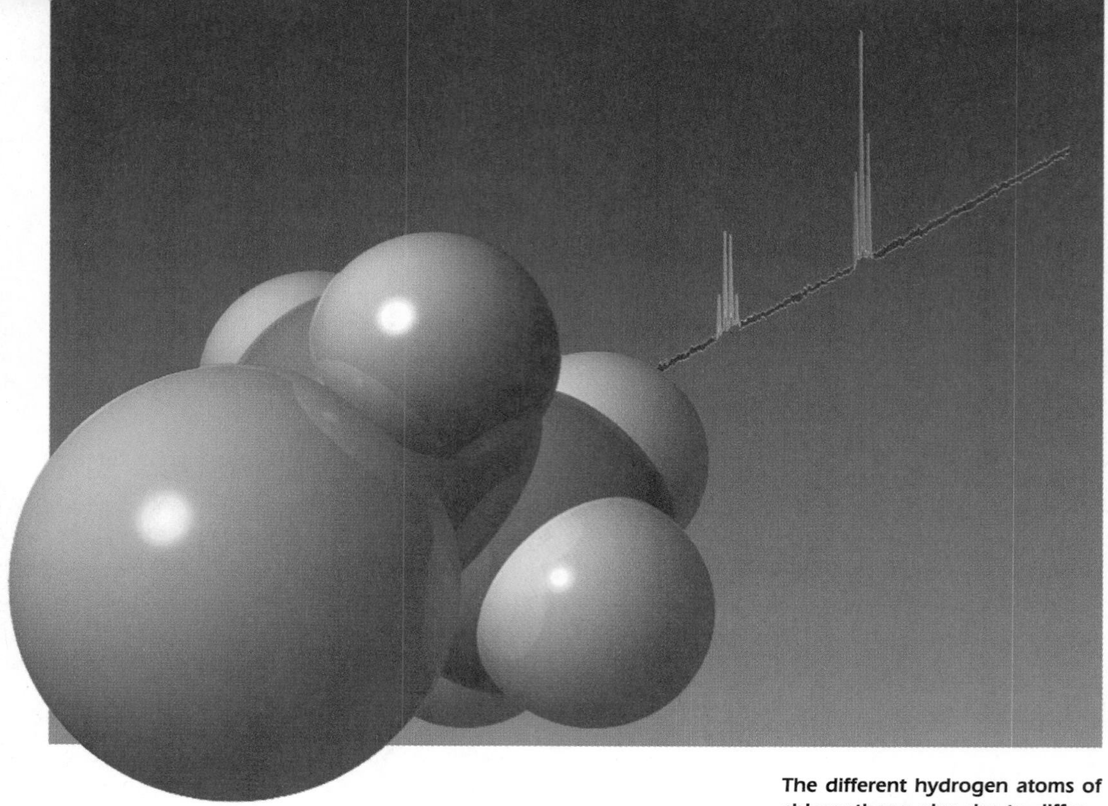

The different hydrogen atoms of chloroethane give rise to different absorption peaks in NMR spectroscopy.

13 Structure Determination

Every time a reaction is run, the products must be isolated, purified, and identified. In the nineteenth and early twentieth centuries, determining the structure of an organic molecule was a time-consuming process requiring skill and patience. In the past few decades, though, extraordinary advances have been made in chemical instrumentation. Sophisticated instruments are now available that greatly simplify structure determination.

What are the instruments for determining structures, and how are they used? We'll answer these questions by looking at three of the most useful methods of structure determination—infrared spectroscopy (IR), ultraviolet spectroscopy (UV), and nuclear magnetic resonance spectroscopy (NMR)—each of which yields a different kind of structural information.

Infrared spectroscopy	What functional groups are present?
Ultraviolet spectroscopy	Is a conjugated π electron system present?
Nuclear magnetic resonance spectroscopy	What carbon–hydrogen framework is present?

13.1 Infrared Spectroscopy and the Electromagnetic Spectrum

Infrared (IR) spectroscopy is a method of structure determination that depends on the interaction of molecules with infrared radiant energy. Before beginning a study of infrared spectroscopy, however, we need to look into the nature of radiant energy and the electromagnetic spectrum.

Visible light, X rays, microwaves, radio waves, and so forth, are all different kinds of **electromagnetic radiation.** Collectively, they make up the **electromagnetic spectrum,** shown in Figure 13.1. The electromagnetic spectrum is loosely divided into regions, with the familiar visible region accounting for only a small portion of the overall spectrum (from 3.8×10^{-5} to 7.8×10^{-5} cm in wavelength). The visible region is flanked by the infrared and ultraviolet regions.

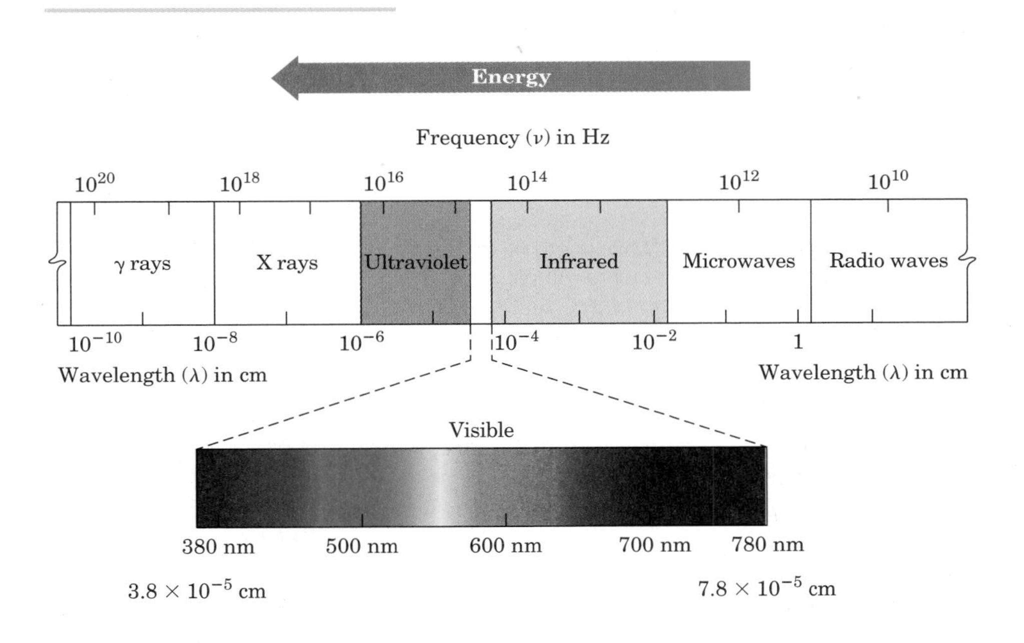

Figure 13.1 The electromagnetic spectrum.

Electromagnetic radiation has dual behavior. In some respects it has the properties of particles (called *photons*), yet in other respects it behaves as a wave of energy traveling at the speed of light. Like all waves, electromagnetic radiation is characterized by a *frequency,* a *wavelength,* and an *amplitude* (Figure 13.2, p. 408). The **frequency, ν** (Greek nu), is the number of peaks that pass by a fixed point per unit time, usually given in reciprocal seconds (s^{-1}), or **hertz, Hz** (1 Hz = 1 s^{-1}). The **wavelength, λ** (Greek lambda), is the distance from one wave maximum to the next. The **amplitude** is the height of the wave, measured from the midpoint between peak and trough to the maximum. (The intensity of radiant energy, whether

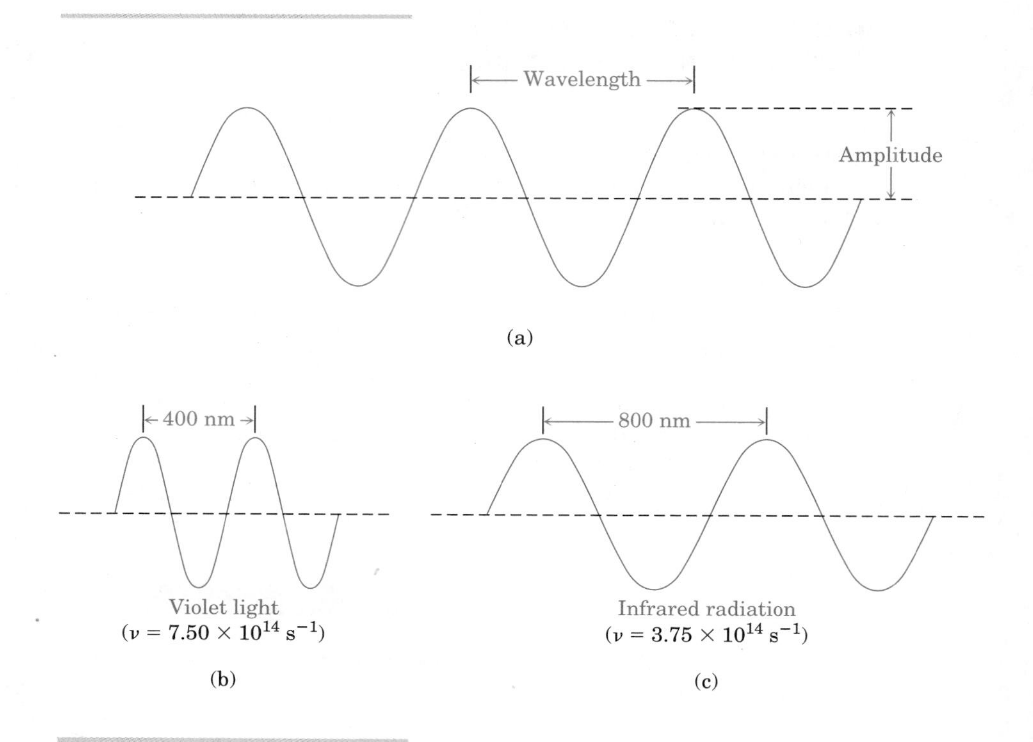

Figure 13.2 (a) Wavelength (λ) is the distance between two successive wave maxima. Amplitude is the height of the wave measured from the center. (b)–(c) What we perceive as different kinds of electromagnetic radiation are simply waves with different wavelengths and frequencies.

a feeble beam or a blinding glare, is proportional to the square of the wave's amplitude.)

Multiplying the length of a wave in centimeters (cm) by its frequency in reciprocal seconds (s^{-1}) gives the speed of the wave in centimeters per second (cm/s). The rate of travel of all electromagnetic radiation in a vacuum is a constant value, commonly called the "speed of light" and abbreviated *c*. It is one of the most accurately known of all physical constants, with a numerical value of $2.997\ 924\ 58 \times 10^{10}$ cm/s, usually rounded off to 3.00×10^{10} cm/s.

$$\text{Wavelength} \times \text{Frequency} = \text{Speed}$$

$$\lambda\ (\text{cm}) \times \nu\ (\text{s}^{-1}) = c\ (\text{cm/s})$$

which can be rewritten as: $\lambda = \dfrac{c}{\nu}$ or $\nu = \dfrac{c}{\lambda}$

Electromagnetic energy is transmitted only in discrete amounts, called *quanta*. The amount of energy ε corresponding to 1 quantum of energy (or 1 photon) with a given frequency ν is expressed by the equation

$$\varepsilon = h\nu = \frac{hc}{\lambda}$$

where ε = Energy of 1 photon (1 quantum)
h = Planck's constant (6.62×10^{-34} J · s = 1.58×10^{-34} cal · s)
ν = Frequency (s^{-1})
λ = Wavelength (cm)
c = Speed of light (3.00×10^{10} cm/s)

This equation says that the energy of a given photon varies *directly* with its frequency ν but *inversely* with its wavelength λ. High frequencies and short wavelengths correspond to high-energy radiation, such as gamma rays; low frequencies and long wavelengths correspond to low-energy radiation, such as radio waves.

When an organic compound is exposed to electromagnetic radiation, it absorbs energy of certain wavelengths and passes, or transmits, energy of other wavelengths. Thus, if we irradiate an organic compound with energy of many wavelengths and determine which are absorbed and which are transmitted, we can determine the **absorption spectrum** of the compound. The results are displayed on a graph that plots wavelength versus the amount of radiation transmitted.

The spectrum of ethanol irradiated with infrared radiation is shown in Figure 13.3. The horizontal axis shows the wavelength in micrometers (μm), and the vertical axis shows the intensity of the various energy absorptions in percent transmittance. The *baseline* corresponding to 0% absorption (or 100% transmittance) runs along the top of the chart, and a downward spike means that energy absorption has occurred at that wavelength.

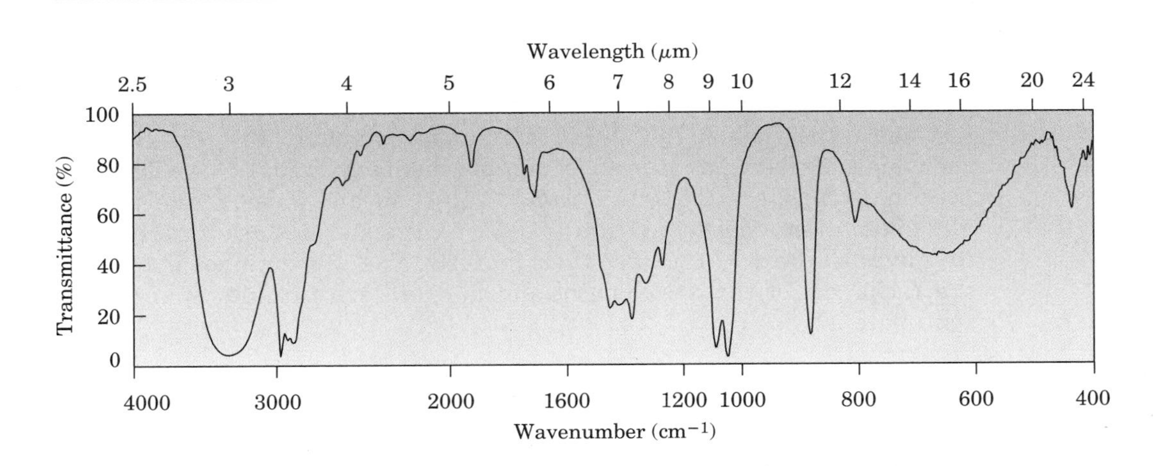

Figure 13.3 The infrared spectrum of ethanol, CH_3CH_2OH. A transmittance of 100% means that all the energy is passing through the sample. A lower transmittance means that some energy is being absorbed. Thus, each downward spike corresponds to an energy absorption.

PRACTICE PROBLEM 13.1

Which is higher in energy, FM radio waves with a frequency of 1.015×10^8 Hz (101.5 MHz) or visible light with a frequency of 5×10^{14} Hz?

Solution The equation $\varepsilon = h\nu$ says that energy increases as frequency increases. Thus, visible light is higher in energy than radio waves.

PRACTICE PROBLEM 13.2

What is the wavelength of visible light with a frequency of 4.5×10^{14} Hz?

Solution Frequency and wavelength are related by the equation $\lambda = c/\nu$, where c is the speed of light (3.0×10^{10} cm/s):

$$\lambda = \frac{3.0 \times 10^{10} \text{ cm/s}}{4.5 \times 10^{14} \text{ s}^{-1}} = 6.7 \times 10^{-5} \text{ cm}$$

PROBLEM

13.1 How does the energy of infrared radiation with $\lambda = 1 \times 10^{-4}$ cm compare with that of an X ray having $\lambda = 3 \times 10^{-7}$ cm?

PROBLEM

13.2 Which is higher in energy, radiation with $\nu = 4 \times 10^9$ Hz or radiation with $\lambda = 9 \times 10^{-4}$ cm?

13.2 Infrared Spectroscopy of Organic Molecules

The infrared region of the electromagnetic spectrum covers the range from just above the visible (7.8×10^{-5} cm) to approximately 10^{-2} cm, but only the middle of the region is used by organic chemists (Figure 13.4). This mid-portion extends from 2.5×10^{-3} to 2.5×10^{-4} cm, and wavelengths are usually given in *micrometers* (1 μm = 10^{-6} m). Frequencies are usually given in **wavenumbers ($\tilde{\nu}$)**, rather than in hertz. The wavenumber is equal to the reciprocal of the wavelength, and is expressed in units of reciprocal centimeters (cm^{-1}):

$$\text{Wavenumber:} \quad \tilde{\nu}\ (\text{cm}^{-1}) = \frac{1}{\lambda\ (\text{cm})}$$

Why does a molecule absorb some wavelengths of infrared energy but not others? All molecules have a certain amount of energy, which causes bonds to stretch and bend and other molecular vibrations to occur. The amount of energy a molecule contains is not continuously variable, though, but is *quantized*. That is, a molecule can vibrate only at specific frequencies corresponding to specific energy levels.

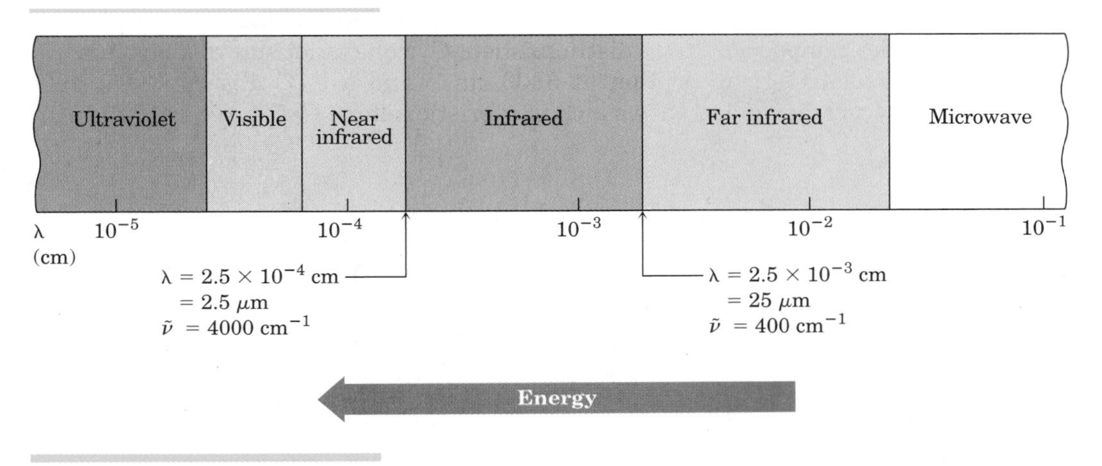

Figure 13.4 The infrared (IR) region of the electromagnetic spectrum.

Take bond stretching, for example. Although we usually speak of bond lengths as if they were fixed, the numbers given are actually averages. In reality, bonds are constantly stretching and bending, lengthening and contracting. Thus, a typical C–H bond with an average bond length of 1.10 Å is actually vibrating at a specific frequency, alternately stretching and compressing as if there were a spring connecting the two atoms.

When the molecule is irradiated with electromagnetic radiation, energy is absorbed if the frequency of the radiation matches the vibrational frequency. The result of energy absorption is an increased amplitude for the vibration; in other words, the "spring" connecting the two atoms stretches and compresses a bit further. Since each frequency corresponds to a specific molecular motion, we can find what kinds of motions a molecule has by measuring its IR spectrum. By then interpreting those motions, we can find out what kinds of bonds (functional groups) are present in the molecule.

IR spectrum ⟶ **What molecular motions?** ⟶ **What functional groups?**

The full interpretation of an IR spectrum is difficult because most organic molecules have dozens of different bond stretching and bending motions. Thus, an IR spectrum usually contains dozens of absorptions. Fortunately, we don't need to interpret an IR spectrum fully to get useful information, because *functional groups have characteristic infrared absorptions that don't change from one compound to another.* The C=O absorption of a ketone is almost always in the range 1680–1750 cm^{-1}; the O–H absorption of an alcohol is almost always in the range 3200–3650 cm^{-1}; and the C=C absorption of an alkene is almost always in the range 1640–1680 cm^{-1}. By learning to recognize where characteristic functional-group absorptions occur, it's possible to interpret infrared spectra.

Look at the IR spectra of cyclohexanol and cyclohexanone in Figure 13.5 to see how IR spectra can be used. Although both spectra contain many peaks, the characteristic absorptions of the different functional groups allow the compounds to be distinguished. Cyclohexanol shows a characteristic alcohol O–H absorption at 3300 cm^{-1} and a C–O absorption at 1060 cm^{-1}; cyclohexanone shows a characteristic ketone C=O peak at 1715 cm^{-1}.

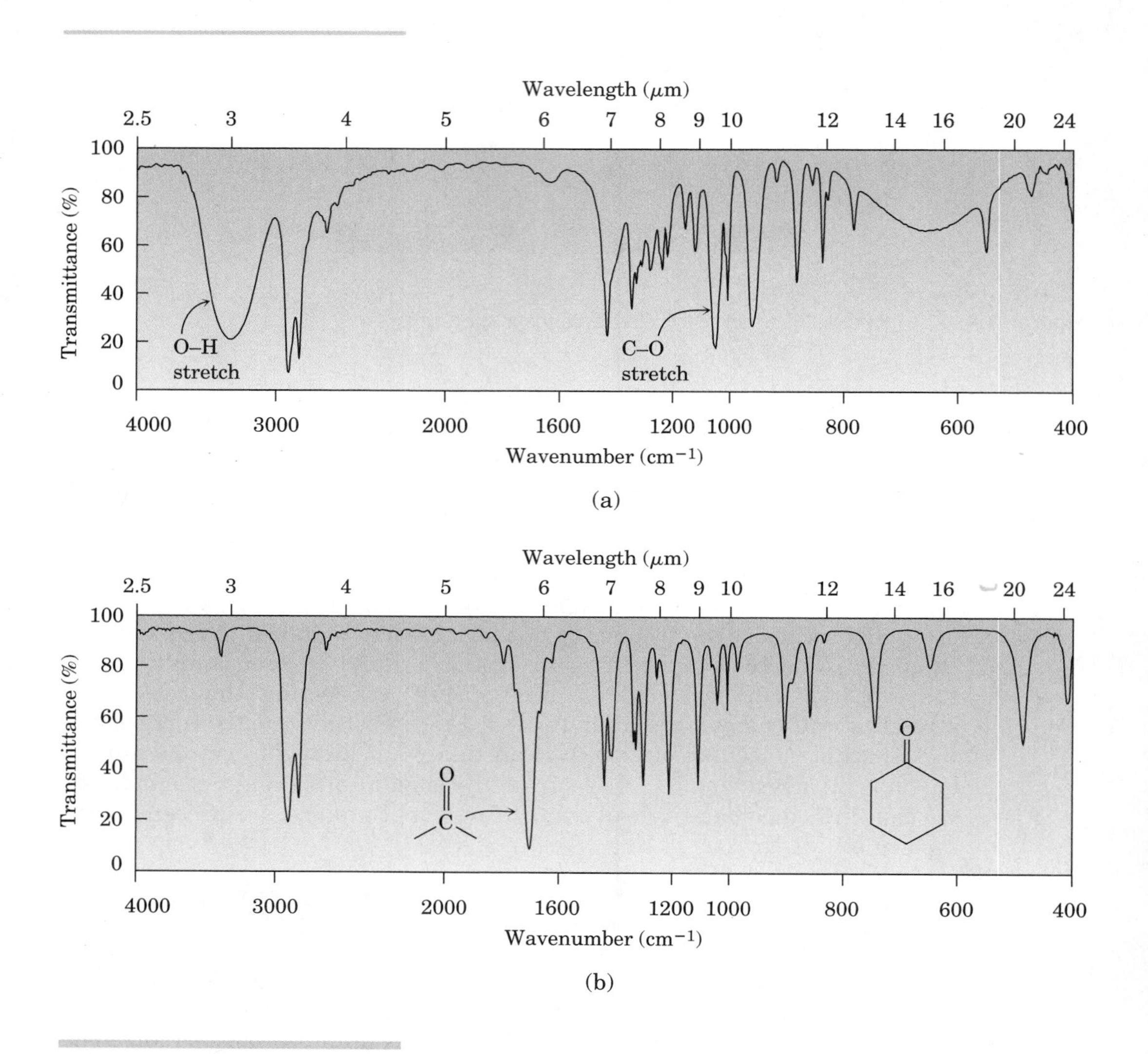

Figure 13.5 Infrared spectra of (a) cyclohexanol and (b) cyclohexanone. Such spectra are easily obtained in minutes with milligram amounts of material.

One further point about infrared spectroscopy: It's also possible to obtain structural information from an IR spectrum by noticing which absorptions are *not* present. If the spectrum of an unknown has no absorption near 3400 cm^{-1}, the unknown is not an alcohol; if the spectrum has no absorption near 1715 cm^{-1}, the unknown is not a ketone; and so on. Table 13.1 lists characteristic IR absorption frequencies of some common functional groups.

Table 13.1 Characteristic Infrared Absorptions of Some Functional Groups

Functional group class	*Band position* (cm^{-1})	*Intensity of absorption*
Alkanes; alkyl groups		
C—H	2850–2960	Medium to strong
Alkenes		
=C—H	3020–3100	Medium
C=C	1640–1680	Medium
Alkynes		
≡C—H	3300	Strong
—C≡C—	2100–2260	Medium
Alkyl halides		
C—Cl	600–800	Strong
C—Br	500–600	Strong
C—I	500	Strong
Alcohols		
O—H	3200–3650	Strong, broad
C—O	1050–1150	Strong
Aromatics		
(aromatic ring) C—H	3030	Medium
(benzene ring C=C)	1600, 1500	Strong
Amines		
N—H	3300–3500	Medium
C—N	1030, 1230	Medium
Carbonyl compounds[a]		
C=O	1680–1750	Strong
Carboxylic acids		
O—H	2500–3100	Strong, very broad
Nitriles		
C≡N	2210–2260	Medium
Nitro compounds		
NO_2	1540	Strong

[a]Acids, esters, aldehydes, and ketones.

It helps in remembering the positions of various IR absorptions to divide the infrared range from 4000 to 200 cm^{-1} into four regions, as shown in Figure 13.6 (p. 414).

1. The region from 4000 to 2500 cm^{-1} corresponds to N–H, C–H, and O–H bond stretching motions. Both N–H and O–H bonds absorb in the 3300–3600 cm^{-1} range, whereas C–H bond stretching occurs near

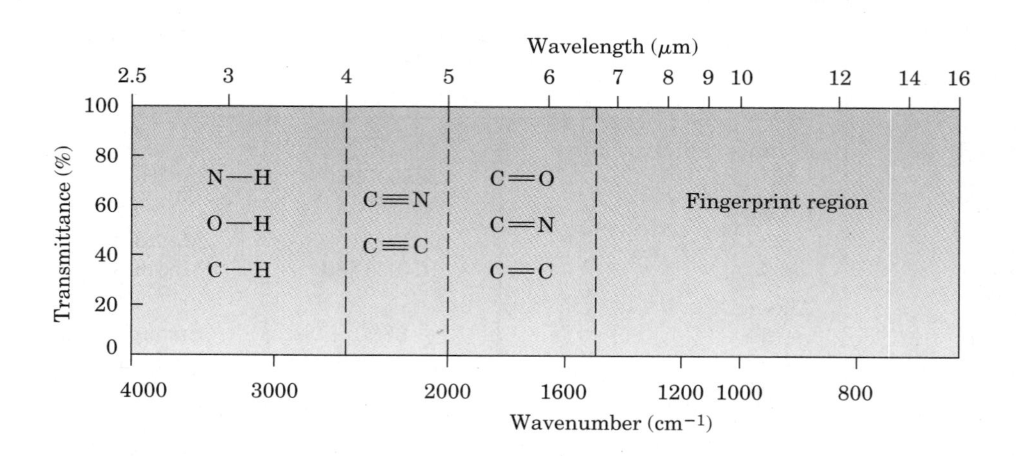

Figure 13.6 *Regions in the infrared spectrum.*

3000 cm^{-1}. Since almost all organic compounds have C–H bonds, almost all IR spectra have an intense absorption in this region.

2. The region from 2500 to 2000 cm^{-1} is where triple-bond stretching occurs. Both nitriles (RC≡N) and alkynes (RC≡CR) absorb here.
3. The region from 2000 to 1500 cm^{-1} is where C=O, C=N, and C=C double bonds absorb. Carbonyl groups generally absorb from 1680 to 1750 cm^{-1}, and alkene stretching normally occurs in the narrow range from 1640 to 1680 cm^{-1}. The exact position of a C=O absorption is often diagnostic of the exact kind of carbonyl group in the molecule. Esters usually absorb at 1735 cm^{-1}, aldehydes at 1725 cm^{-1}, and open-chain ketones at 1715 cm^{-1}.
4. The region below 1500 cm^{-1} is the so-called *fingerprint region.* A large number of absorptions due to various C–O, C–C, and C–N single-bond vibrations occur here, forming a unique pattern that acts as an identifying "fingerprint" of each organic molecule.

PRACTICE PROBLEM 13.3

Refer to Table 13.1 and make educated guesses about the functional groups that cause the following IR absorptions:
(a) 1735 cm^{-1} (b) 3500 cm^{-1}

Solution

(a) An absorption at 1735 cm^{-1} is in the carbonyl-group region of the IR spectrum, probably an ester.
(b) An absorption at 3500 cm^{-1} is in the –OH (alcohol) region.

PRACTICE PROBLEM 13.4

Acetone and 2-propen-1-ol (H_2C=$CHCH_2OH$) are isomers. How could you distinguish them by IR spectroscopy?

Solution Acetone has a strong C=O absorption at 1715 cm^{-1}. 2-Propen-1-ol has an –OH absorption at 3500 cm^{-1} and a C=C absorption at 1660 cm^{-1}.

PROBLEM

13.3 What functional groups might molecules contain if they show IR absorptions at the following frequencies?
(a) 1715 cm^{-1} (b) 1540 cm^{-1} (c) 2210 cm^{-1}
(d) 1720 and 2500–3100 cm^{-1} (e) 3500 and 1735 cm^{-1}

PROBLEM

13.4 How might you use IR spectroscopy to help distinguish between the following pairs of isomers?
(a) Ethanol and dimethyl ether (b) Cyclohexane and 1-hexene
(c) Propanoic acid and 3-hydroxypropanal

13.3 Ultraviolet Spectroscopy

The **ultraviolet (UV)** region of the electromagnetic spectrum extends from the low-wavelength end of the visible region (4×10^{-5} cm) to 10^{-6} cm. The portion of greatest interest to organic chemists, though, is the narrow range from 2×10^{-5} cm to 4×10^{-5} cm. Absorptions in this region are measured in *nanometers* (nm), where 1 nm = 10^{-9} m = 10^{-7} cm. Thus, the ultraviolet range of interest is from 200 to 400 nm (Figure 13.7).

We saw in Section 13.1 that an organic molecule either absorbs or transmits electromagnetic energy when irradiated, depending on the radiation's

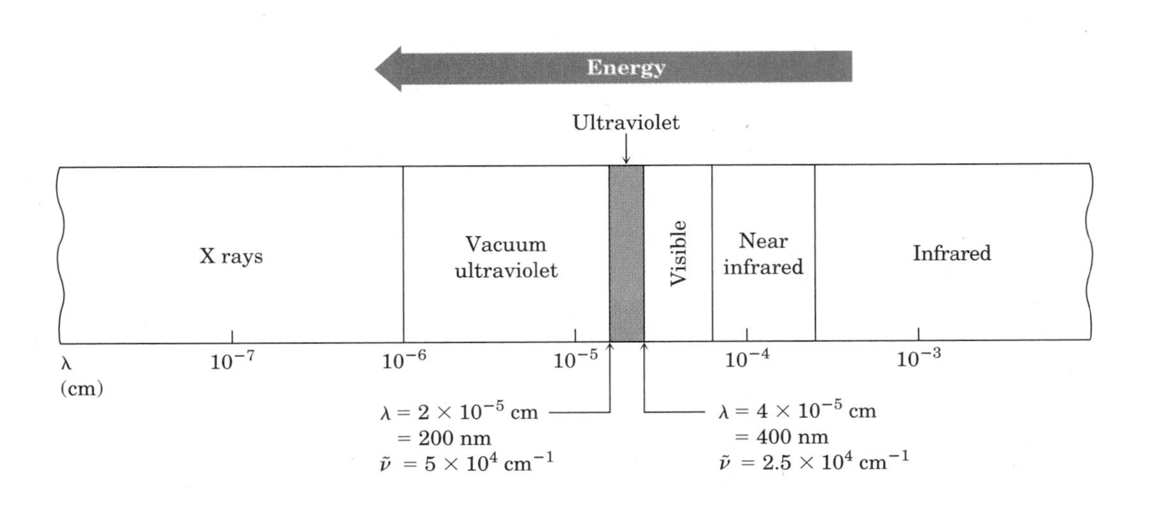

Figure 13.7 The ultraviolet (UV) region of the electromagnetic spectrum.

energy level. With IR radiation, the energy absorbed corresponds to the amount necessary to increase molecular bending or stretching vibrations. With UV radiation, the energy absorbed corresponds to the amount necessary to raise the energy level of a π electron in an unsaturated molecule.

Ultraviolet spectra are recorded by irradiating a sample with UV light of continuously changing wavelength. When the wavelength of light corresponds to the amount of energy required to excite a π electron to a higher level, energy is absorbed. The absorption is detected and displayed on a chart that plots wavelength versus percent radiation absorbed.

A typical UV spectrum—that of 1,3-butadiene—is shown in Figure 13.8. Unlike IR spectra, which generally have many peaks, UV spectra are usually quite simple. Often, there is only a single broad peak, which is identified by noting the wavelength at the very top (λ_{max}). For 1,3-butadiene, λ_{max} = 217 nm.

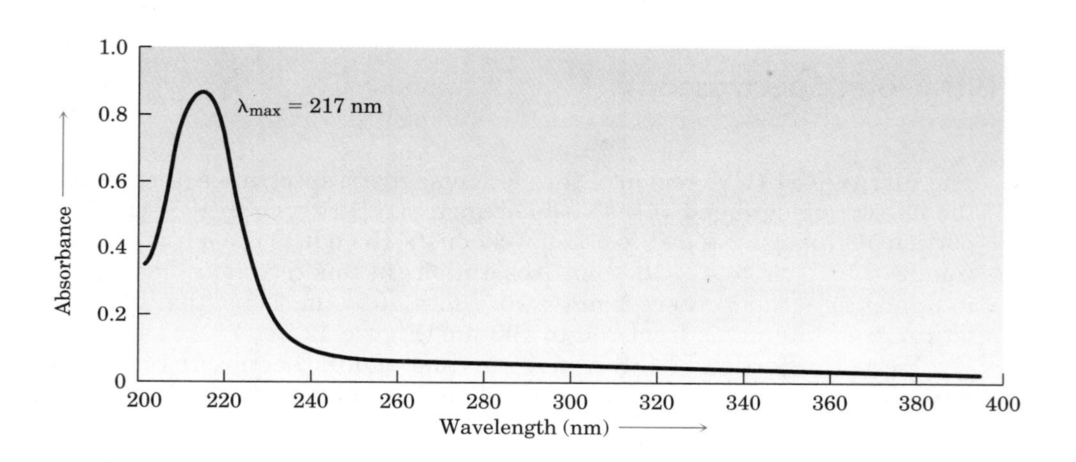

Figure 13.8 Ultraviolet spectrum of 1,3-butadiene.

13.4 Interpreting Ultraviolet Spectra: The Effect of Conjugation

The wavelength of radiation necessary to cause an electronic excitation in an unsaturated molecule depends on the nature of the π electron system in the molecule. Working backward, it's possible to obtain information about the π electron system by measuring the molecule's UV spectrum.

One of the most important factors affecting the wavelength of a UV absorption is the extent of conjugation (Section 4.10). It turns out that the energy required for an electronic transition decreases as the extent of conjugation increases. Thus, 1,3-buta*di*ene shows an absorption at λ_{max} = 217 nm, 1,3,5-hexa*tri*ene absorbs at λ_{max} = 258 nm, and 1,3,5,7-octa*tetra*ene has λ_{max} = 290 nm. (*Remember:* Longer wavelength means lower energy.)

Other kinds of conjugated π electron systems besides dienes and polyenes also show ultraviolet absorptions. Conjugated enones, such as 3-buten-2-one, and aromatic molecules, such as benzene, also have characteristic UV absorptions that aid in structure determination. The UV absorption maxima of some representative conjugated molecules are given in Table 13.2.

Table 13.2 Ultraviolet Absorption Maxima of Some Conjugated Molecules

Name	*Structure*	λ_{max} (nm)
Ethylene	$H_2C{=}CH_2$	171
2-Methyl-1,3-butadiene	$H_2C{=}C(CH_3){-}CH{=}CH_2$	220
1,3-Cyclohexadiene	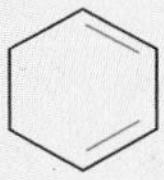 (ring structure)	256
1,3,5-Hexatriene	$H_2C{=}CH{-}CH{=}CH{-}CH{=}CH_2$	258
3-Buten-2-one	$H_2C{=}CH{-}C(CH_3){=}O$	219
Benzene	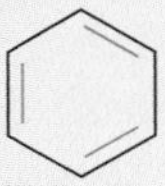 (ring structure)	254

PRACTICE PROBLEM 13.5

1,5-Hexadiene and 1,3-hexadiene are isomers. How can you distinguish them by UV spectroscopy?

Solution 1,3-Hexadiene is a conjugated diene, but 1,5-hexadiene is nonconjugated. Only the conjugated isomer shows a UV absorption above 200 nm.

PROBLEM

13.5 Which of the following compounds show UV absorptions in the range 200–400 nm?

(a) 1,3-Cyclohexadiene (b) 1,4-Cyclohexadiene
(c) Methyl propenoate (d) *p*-Bromotoluene
(e) 2-Methylcyclohexanone (f) 2-Methyl-2-cyclohexenone

PROBLEM

13.6 How can you distinguish between 1,3-hexadiene and 1,3,5-hexatriene by UV spectroscopy?

13.5 Nuclear Magnetic Resonance Spectroscopy

Nuclear magnetic resonance (NMR) spectroscopy is the most valuable spectroscopic technique available. It's the method that organic chemists turn to first for structural information.

We've seen that IR spectroscopy provides information about a molecule's functional groups and that UV spectroscopy provides information about a molecule's conjugated π electron system. NMR spectroscopy doesn't replace or duplicate either of these techniques; rather, it complements them by providing a "map" of the carbon–hydrogen framework in an organic molecule. Taken together, IR, UV, and NMR spectroscopies often make it possible to find the structures of even very complex molecules.

IR spectroscopy	Functional groups
UV spectroscopy	π electron system
NMR spectroscopy	Map of carbon–hydrogen framework

How does NMR spectroscopy work? Many kinds of nuclei, including ^{1}H and ^{13}C, behave as if they were a child's top spinning about an axis. Since they're positively charged, these spinning nuclei act like tiny magnets and interact with an external magnetic field (denoted $\boldsymbol{H}_0$). In the absence of an external magnetic field, the nuclear spins of magnetic nuclei are oriented randomly. When a sample containing these nuclei is placed between the poles of a strong magnet, however, the nuclei adopt specific orientations, much as a compass needle orients itself in the earth's magnetic field.

A spinning ^{1}H or ^{13}C nucleus can orient so that its own tiny magnetic field is aligned either with (*parallel* to) or against (*antiparallel* to) the external field. The two orientations don't have the same energy and therefore aren't equally likely. The parallel orientation is slightly lower in energy, making this spin state slightly favored over the antiparallel orientation (Figure 13.9).

If the oriented nuclei are now irradiated with electromagnetic radiation of the right frequency, energy absorption occurs and the lower-energy state "spin-flips" to the higher-energy state. When this spin-flip occurs, the nuclei are said to be in resonance with the applied radiation—hence the name, *nuclear magnetic resonance.*

The exact frequency necessary for resonance depends both on the strength of the external magnetic field and on the identity of the nucleus. If a very strong external field is applied, the energy difference between the two spin states is large, and higher-energy (higher-frequency) radiation is required. If a weaker magnetic field is applied, less energy is required to effect the transition between nuclear spin states.

In practice, superconducting magnets that produce enormously powerful fields up to 14.1 tesla (T) are sometimes used, but field strengths in the range 1.41–4.7 T are more common.[1] At a magnetic field strength of 1.41 T,

[1]The SI unit of magnetic flux density is the *tesla* (T), which has replaced the older unit *gauss* (G); $1\ T = 10^4\ G = 1\ J/(A \cdot m^2)$.

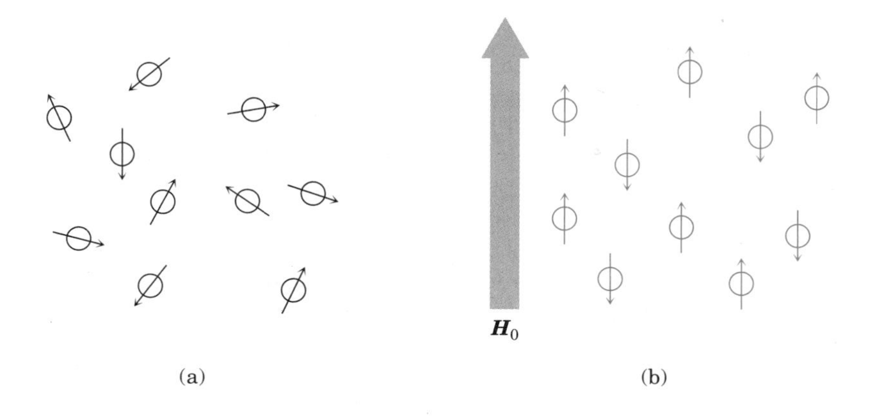

Figure 13.9 Nuclear spins are oriented randomly in the absence of an external magnetic field (a), but have a specific orientation in the presence of an external field H_0 (b). Note that some of the spins (red) are aligned parallel to the external field, and others (blue) are antiparallel. The parallel spin state is lower in energy.

so-called *radiofrequency* (rf) energy in the 60 MHz range (1 MHz = 10^6 Hz) is required to bring a ^1H nucleus into resonance, and rf energy of 15 MHz is required to bring a ^{13}C nucleus into resonance.

PROBLEM

13.7 NMR spectroscopy uses electromagnetic radiation with a frequency of 6×10^7 Hz. Is this a greater or lesser amount of energy than that used by IR spectroscopy?

13.6 The Nature of NMR Absorptions

From the description thus far, you might expect all protons (^1H nuclei) in a molecule to absorb energy at the same frequency and all ^{13}C nuclei to absorb at the same frequency.[2] If this were true, we would observe only a single NMR absorption in the ^1H or ^{13}C spectrum of a molecule, a situation that would be of little use for structure determination. In fact, the absorption frequency is not the same for all ^1H or all ^{13}C nuclei.

All nuclei are surrounded by electron clouds. When an external magnetic field is applied to a molecule, the electron clouds around nuclei set up tiny local magnetic fields of their own. These local fields act in opposition to the applied field, so that the *effective* field actually felt by the nucleus is a bit weaker than the applied field:

$$\boldsymbol{H}_{\text{effective}} = \boldsymbol{H}_{\text{applied}} - \boldsymbol{H}_{\text{local}}$$

[2]In speaking about NMR, the words *proton* and *hydrogen* are often used interchangeably.

In describing this effect of local fields, we say that the carbon and hydrogen nuclei are **shielded** from the full effect of the applied field by their surrounding electrons. Since each specific ^{1}H or ^{13}C nucleus in a molecule is in a slightly different electronic environment, each specific nucleus is shielded to a slightly different extent, and the effective magnetic field felt by each is not the same. These slight differences can be detected, and we therefore see different NMR signals for each chemically distinct ^{1}H or ^{13}C nucleus.

Figure 13.10 shows both the ^{1}H and the ^{13}C NMR spectra of methyl acetate, $CH_3CO_2CH_3$. In both spectra, the horizontal axis tells the effective field strength felt by the nuclei, and the vertical axis indicates the intensity of absorption of rf energy. Each peak in the NMR spectrum corresponds to a chemically distinct hydrogen or carbon in the molecule. Note, though, that ^{1}H and ^{13}C spectra can't be observed at the same time on the same

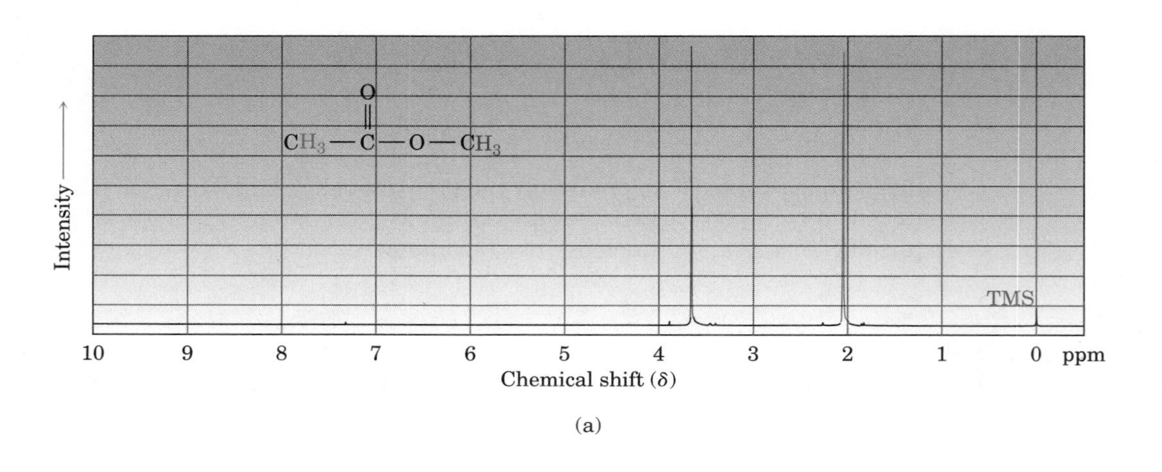

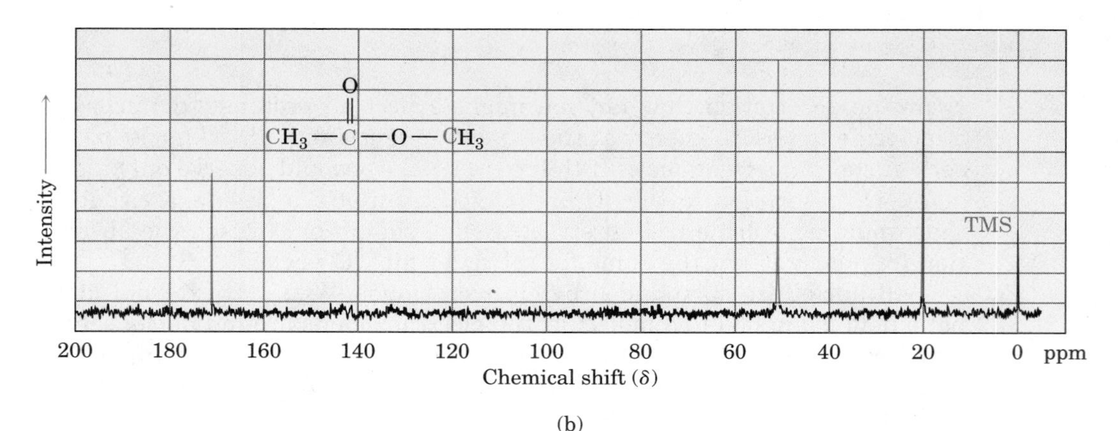

Figure 13.10 (a) The ^{1}H NMR spectrum and (b) the ^{13}C NMR spectrum of methyl acetate, $CH_3CO_2CH_3$. The peak in each spectrum marked TMS is explained in the next section.

spectrometer, because different amounts of energy are required to spin-flip the different kinds of nuclei. The two spectra must be recorded separately.

The ^{13}C spectrum of methyl acetate shown in Figure 13.10b has three peaks, one for each of the three chemically distinct carbons in the molecule. The ^{1}H spectrum shows only *two* peaks, however, even though methyl acetate has *six* hydrogens. One peak is due to the CH_3CO hydrogens and the other to the OCH_3 hydrogens. Because the three hydrogens in each methyl group have the same chemical (and magnetic) environment, they are shielded to the same extent and are said to be *equivalent. Chemically equivalent nuclei show a single absorption.* The two methyl groups themselves, however, are nonequivalent, so the two sets of hydrogens absorb at different positions.

PRACTICE PROBLEM 13.6

How many signals would you expect *p*-dimethylbenzene to show in its ^{1}H and ^{13}C NMR spectra?

Solution Because of the molecule's symmetry, the two methyl groups in *p*-dimethylbenzene are equivalent, and all four ring hydrogens are equivalent. Thus, there are only two absorptions in the ^{1}H NMR spectrum. Also because of symmetry, there are only three absorptions in the ^{13}C NMR spectrum: one for the two equivalent methyl-group carbons, one for the four equivalent C–H ring carbons, and one for the two equivalent ring carbons bonded to the methyl groups.

H H
H_3C CH_3
H H
Two kinds of hydrogens
Three kinds of carbons

***p*-Dimethylbenzene**

PROBLEM

13.8 How many absorptions would you expect each of the following compounds to show in its ^{1}H and ^{13}C NMR spectra?

(a) Methane (b) Ethane (c) Propane (d) Cyclohexane
(e) Dimethyl ether (f) Benzene (g) $(CH_3)_3COH$ (h) Chloroethane
(i) $(CH_3)_2C{=}C(CH_3)_2$

PROBLEM

13.9 How can you explain the fact that 2-chloropropene shows signals for three kinds of hydrogens in its ^{1}H NMR spectrum?

13.7 Chemical Shifts

NMR spectra are displayed on charts that show the applied field strength increasing from left to right (Figure 13.10). Thus, the left side of the chart is the low-field (or *downfield*) side, and the right side is the high-field (or *upfield*) side. To define the position of an absorption, the NMR chart is calibrated and a reference point is used. In practice, a small amount of tetramethylsilane [TMS, $(CH_3)_4Si$] is added to the sample so that a reference absorption line is produced when the spectrum is run. TMS is used as a reference for both ^1H and ^{13}C spectra, because in both kinds of spectra it produces a single peak that occurs upfield (farther right on the chart) of other absorptions normally found in organic molecules.

The place on the chart at which a nucleus absorbs is called its **chemical shift.** By convention, the chemical shift of TMS is called zero, and other peaks normally occur downfield (to the left on the chart). NMR charts are calibrated in units of frequency using an arbitrary scale called the **delta scale,** where 1 delta unit (δ) is equal to 1 part per million (ppm) of the spectrometer operating frequency. For example, if we were using a 100 MHz instrument to measure the ^1H NMR spectrum of a substance, 1 δ would be 1 ppm of 100,000,000 Hz, or 100 Hz. Similarly, if we were measuring the spectrum with a 300 MHz instrument, then 1 $\delta = 300$ Hz.

Although this method of calibrating NMR charts may seem complex, there's a good reason for it. There are many different kinds of NMR spectrometers operating at many different frequencies and magnetic field strengths. By using a system of measurement in which NMR absorptions are expressed in *relative* terms (parts per million relative to spectrometer frequency) rather than in absolute terms (Hz), comparisons of spectra obtained on different instruments are possible. *The chemical shift of an NMR absorption in δ units is constant, regardless of the operating frequency of the instrument.* A ^1H nucleus that absorbs at 2.0 δ on a 100 MHz instrument (2.0 ppm $\times$ 100 MHz = 200 Hz to the left of TMS) also absorbs at 2.0 δ on a 300 MHz instrument (2.0 ppm $\times$ 300 MHz = 600 Hz to the left of TMS).

PRACTICE PROBLEM 13.7

Cyclohexane shows an absorption at 1.43 δ in its ^1H NMR spectrum. How many hertz away from TMS is this on a spectrometer operating at 60 MHz? On a spectrometer operating at 220 MHz?

Solution On a 60 MHz spectrometer, 1 $\delta = 60$ Hz. Thus, 1.43 $\delta = 86$ Hz away from the TMS reference peak. On a 220 MHz spectrometer, 1 $\delta = 220$ Hz and 1.43 $\delta =$ 315 Hz.

PROBLEM

13.10 When the ^1H NMR spectrum of acetone is recorded on a 60 MHz instrument, a single sharp resonance line at 2.1 δ is observed.
(a) How far away from TMS (in hertz) does the acetone absorption occur?
(b) What is the position of the acetone absorption in δ units on a 100 MHz instrument?
(c) How many hertz away from TMS does the absorption in the 100 MHz spectrum correspond to?

PROBLEM

13.11 The following ^1H NMR resonances were recorded on a spectrometer operating at 60 MHz. Convert each into δ units.
(a) $CHCl_3$, 436 Hz (b) CH_3Cl, 183 Hz
(c) CH_3OH, 208 Hz (d) CH_2Cl_2, 318 Hz

13.8 Chemical Shifts in ^{1}H NMR Spectra

Everything we've said thus far about NMR spectroscopy applies to both ^{1}H and ^{13}C spectra. Now, though, let's focus only on ^{1}H NMR spectroscopy to see how it can be used in organic structure determination. Most ^{1}H NMR absorptions occur in the range 0–10 δ, which can be divided into the five regions shown in Figure 13.11. By remembering the positions of these regions, it's possible to tell at a glance what kinds of protons a molecule contains.

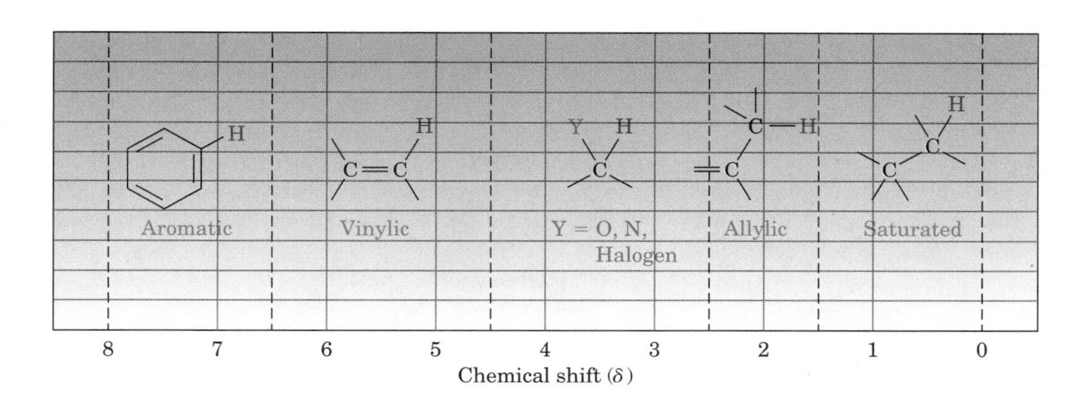

Figure 13.11 Regions of the ^{1}H NMR spectrum.

Table 13.3 (p. 424) shows the correlation of ^{1}H chemical shift with electronic environment in more detail. In general, protons bonded to sp^3 carbons absorb at higher fields (right-hand side of the spectrum), whereas protons bonded to sp^2 carbons absorb at lower fields (left-hand side of the spectrum). Protons on carbons that are bonded to electronegative atoms such as N, O, or halogen also absorb at lower fields.

PRACTICE PROBLEM 13.8

Methyl 2,2-dimethylpropanoate $(CH_3)_3CCO_2CH_3$ has two peaks in its ^{1}H NMR spectrum. At what approximate chemical shifts do they come?

Solution The CH_3O– protons absorb around 3.5–4.0 δ because they are on carbon bonded to oxygen. The $(CH_3)_3C$– protons absorb around 1.0 δ because they are typical alkane protons. (See Figure 13.12, p. 425.)

PROBLEM

13.12 Each of the following compounds exhibits a single ^{1}H NMR peak. Approximately where would you expect each to absorb?
(a) Ethane (b) Acetone (c) Benzene (d) Trimethylamine

Table 13.3 Correlation of ^1H Chemical Shift with Environment

Type of hydrogen		*Chemical shift (δ)*	*Type of hydrogen*		*Chemical shift (δ)*
Reference	$(CH_3)_4Si$	0	Alcohol, ether	O–C–H	3.3–4.0
Saturated primary	$-CH_3$	0.7–1.3	Alkynyl	C≡C—H	2.5–2.7
Saturated secondary	$-CH_2-$	1.2–1.4	Vinylic	C=C–H	5.0–6.5
Saturated tertiary	C–H	1.4–1.7	Aromatic	Ar—H	6.5–8.0
Allylic	C=C–C–H	1.6–2.2	Aldehyde	O=C–H	9.7–10.0
Methyl ketone	O=C–CH_3	2.1–2.4	Carboxylic acid	O=C–O–H	11.0–12.0
Aromatic methyl	Ar—CH_3	2.5–2.7	Alcohol	C–O–H	2.5–5.0 (Variable)
Alkyl halide X = Cl, Br, I	X–C–H	2.5–4.0			

13.9 Integration of ^1H NMR Spectra: Proton Counting

Look at the ^1H NMR spectrum of methyl 2,2-dimethylpropanoate in Figure 13.12. There are two peaks, corresponding to the two kinds of protons, but the peaks aren't the same size. The peak at 1.2 δ, due to the $(CH_3)_3C$– protons, is larger than the peak at 3.7 δ, due to the $-OCH_3$ protons.

The area under each peak is proportional to the number of protons causing that peak. By electronically measuring, or **integrating,** the area under each peak, it's possible to measure the relative number of each kind of proton in a molecule. Integrated peak areas are superimposed over the spectrum in a "stair-step" manner, with the height of each step proportional to the area of the peak, and therefore proportional to the relative number of protons causing the peak. To compare the size of one peak to another, simply take a ruler and measure the heights of the various steps. For

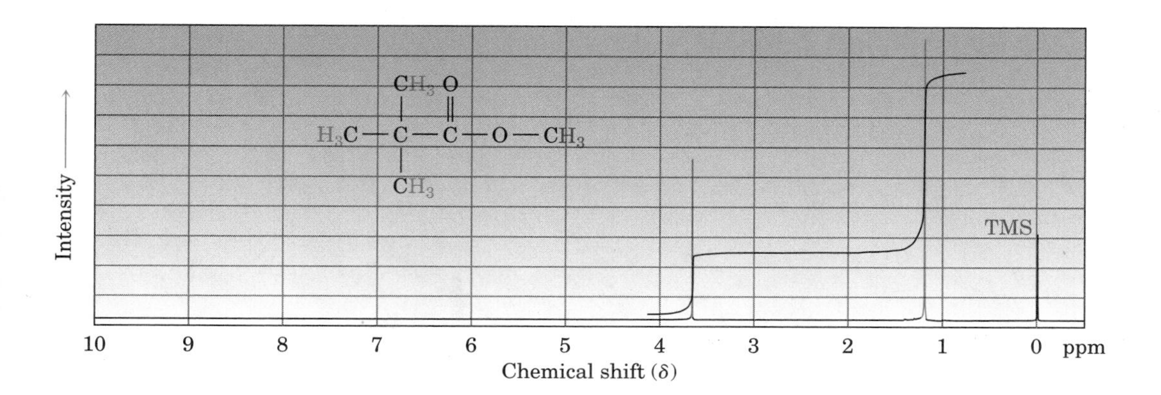

Figure 13.12 The ¹H NMR spectrum of methyl 2,2-dimethylpropanoate. Integrating the peaks in a "stair-step" manner shows that they have a 1:3 ratio, corresponding to the ratio of the numbers of protons (3:9) responsible for each peak.

example, the two peaks in methyl 2,2-dimethylpropanoate are found to have a 1:3 (or 3:9) ratio when integrated—exactly what we expect because the three $-OCH_3$ protons are equivalent and the nine $(CH_3)_3C-$ protons are equivalent.

PROBLEM

13.13 How many peaks would you expect in the ¹H NMR spectrum of *p*-dimethylbenzene (*p*-xylene)? What ratio of peak areas would you expect to find on integration of the spectrum? Refer to Table 13.3 for approximate chemical shift values, and sketch what the spectrum might look like.

13.10 Spin–Spin Splitting in ¹H NMR Spectra

In the ¹H NMR spectra we've seen thus far, each chemically distinct proton in a molecule has given rise to a single peak. It often happens, though, that a given absorption splits into *multiple* peaks. For example, the ¹H NMR spectrum of bromoethane in Figure 13.13 (p. 426) indicates that the $-CH_2Br$ protons appear as four peaks (a *quartet*) at 3.42 δ, and the $-CH_3$ protons appear as a *triplet* at 1.68 δ.

Called **spin–spin splitting,** the phenomenon of multiple absorptions is caused by the interaction, or **coupling,** of the nuclear spins of neighboring atoms. In other words, the tiny magnetic field of one nucleus affects the magnetic field felt by a neighboring nucleus.

To understand the reasons for spin–spin splitting, look at the $-CH_3$ protons in bromoethane. The three equivalent $-CH_3$ protons are neighbored by

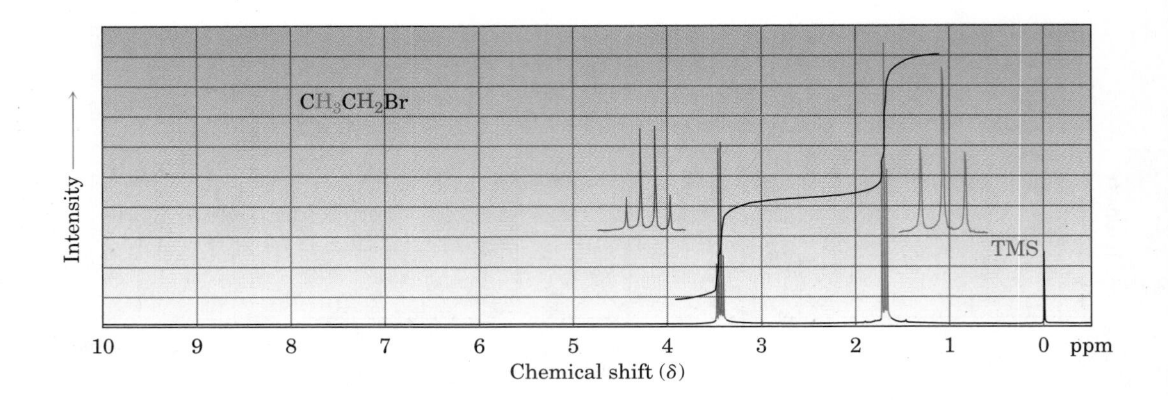

Figure 13.13 The ¹H NMR spectrum of bromoethane, CH_3CH_2Br. The $-CH_2Br$ protons appear as a quartet at 3.42 δ, and the $-CH_3$ protons appear as a triplet at 1.68 δ.

two other magnetic nuclei, the $-CH_2Br$ protons. Each of the $-CH_2Br$ protons has its own nuclear spin, which can align either with or against the applied magnetic field, producing a tiny effect that is felt by the neighboring $-CH_3$ protons.

There are three ways in which the two $-CH_2Br$ protons can align, as shown in Figure 13.14. If both protons align *with* the applied magnetic field,

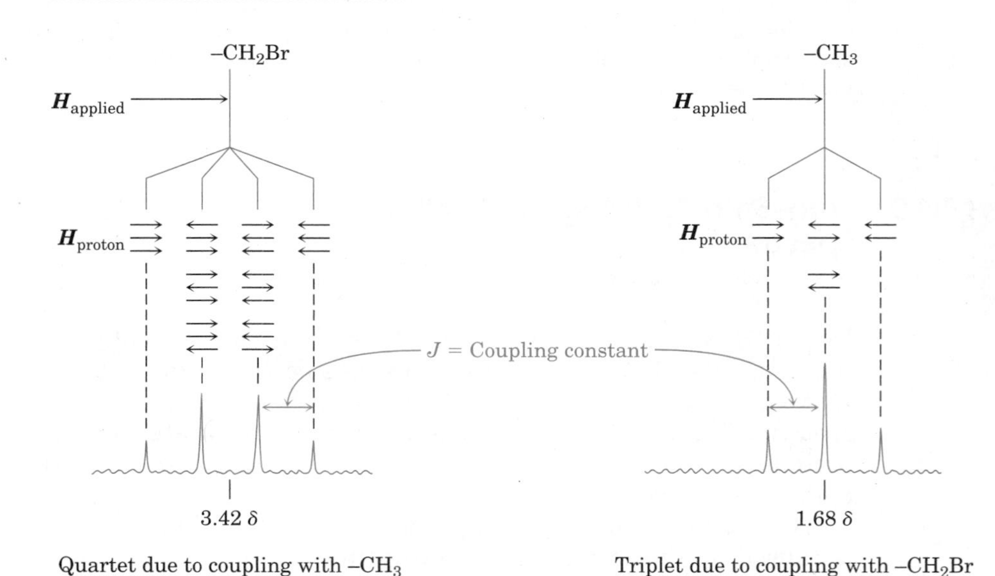

Figure 13.14 The origin of spin–spin splitting in bromoethane. The nuclear spins of neighboring protons, indicated by horizontal arrows, align either with or against the applied field, causing the splitting of absorptions into multiplets.

the total effective field felt by the neighboring $-CH_3$ protons is slightly larger than it would otherwise be. Consequently, the applied field necessary to cause resonance is slightly reduced. Alternatively, if one $-CH_2Br$ proton aligns *with* and one aligns *against* the applied field (two possible ways), there is no effect on the neighboring $-CH_3$ protons. Finally, if both $-CH_2Br$ protons align *against* the applied field, the effective field felt by the $-CH_3$ protons is slightly smaller than it would otherwise be, and the applied field needed for resonance must be slightly increased.

Any given molecule can adopt only one of the three possible alignments of $-CH_2Br$ spins, but in a large collection of molecules, all three spin states will be represented in a 1:2:1 statistical ratio. We therefore find that the neighboring $-CH_3$ protons come into resonance at three slightly different values of the applied field, and we see a 1:2:1 triplet in the NMR spectrum. One resonance is a little above where it would be without coupling, one is at the same place it would be without coupling, and the third resonance is a little below where it would be without coupling.

In the same way that the $-CH_3$ protons of bromoethane are split into a triplet in the NMR spectrum, the $-CH_2Br$ protons are split into a quartet. The three spins of the neighboring $-CH_3$ protons align in four combinations: all three with the applied field, two with and one against (three possibilities), one with and two against (three possibilities), or all three against. Thus, four peaks are produced for the $-CH_2Br$ protons in a 1:3:3:1 ratio.

As a general rule (called the ***n* + 1 rule**), protons that have n equivalent neighboring protons split into $n + 1$ peaks in their NMR absorption. For example, the spectrum of 2-bromopropane in Figure 13.15 shows a doublet at 1.71 δ and a seven-line multiplet, or *septet,* at 4.28 δ. The septet is caused by splitting of the –CHBr– proton signal by six equivalent neighboring protons on the two methyl groups ($n + 1 = 7$ when $n = 6$). The doublet is due to splitting of the six equivalent $-CH_3$ protons by the single –CHBr– proton.

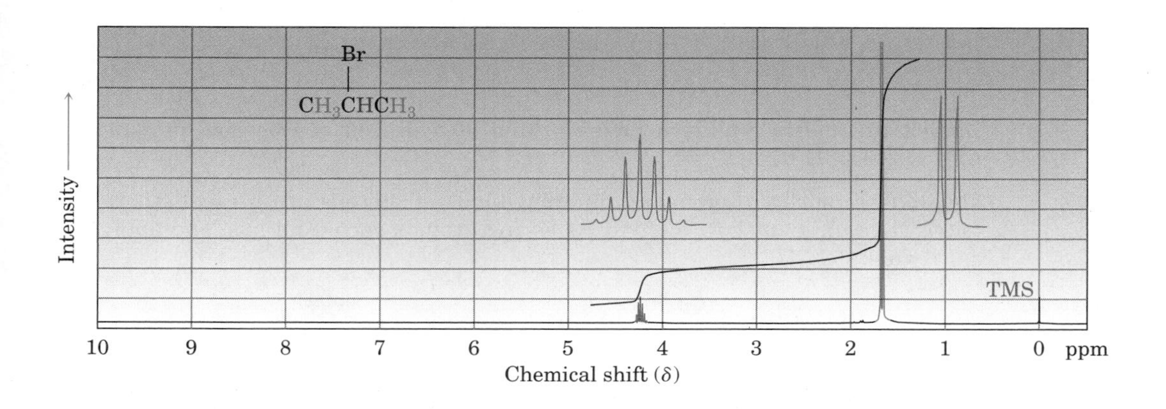

Figure 13.15 The ¹H NMR spectrum of 2-bromopropane. The $-CH_3$ proton signal at 1.71 δ is split into a doublet, and the –CHBr– proton signal at 4.28 δ is split into a septet.

The distance between peaks in a multiplet is called the **coupling constant, *J*.** Coupling constants are measured in hertz and generally fall in the range 0–18 Hz. Though the exact value of J depends on the geometry of the molecule, a typical value for an open-chain alkane is J = 6–8 Hz. Note that the same coupling constant is shared by both groups of hydrogens whose spins are coupled. In bromoethane, for instance, the $-CH_2Br$ protons are coupled to the $-CH_3$ protons with coupling constant J = 7 Hz. The $-CH_3$ protons are similarly coupled to the $-CH_2Br$ protons with the same J = 7 Hz coupling constant.

Three important rules about spin–spin splitting are illustrated by the spectra of bromoethane in Figure 13.13 and 2-bromopropane in Figure 13.15:

1. *Chemically equivalent protons don't show spin–spin splitting.* The equivalent protons can be on the same carbon or on different carbons, but their signals still appear as singlets and don't split.

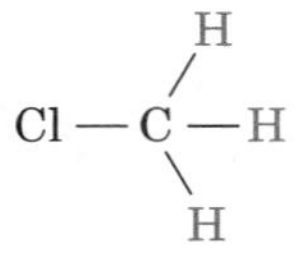

Three C–H protons are chemically equivalent; no splitting occurs.

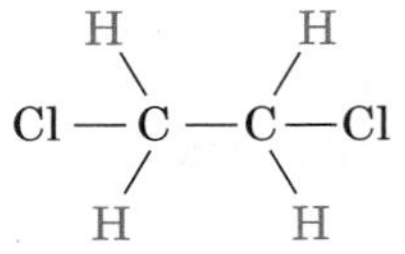

Four C–H protons are chemically equivalent; no splitting occurs.

2. *The signal of a proton with n equivalent neighboring protons is split into a multiplet of n + 1 peaks with coupling constant J.* Protons that are more than two carbon atoms apart usually don't split each other's signals.

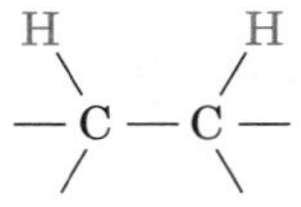

Splitting observed

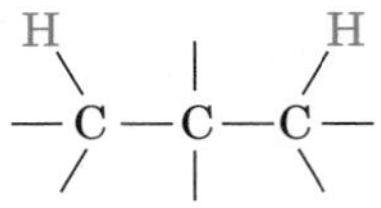

Splitting not usually observed

3. *Two groups of protons coupled to each other have the same coupling constant J.*

PRACTICE PROBLEM 13.9

Predict the splitting pattern for each kind of hydrogen in isopropyl propanoate, $CH_3CH_2CO_2CH(CH_3)_2$.

Solution First, find how many different kinds of protons are present (there are four). Then, find out how many neighboring protons each kind has, and apply the $n + 1$ rule:

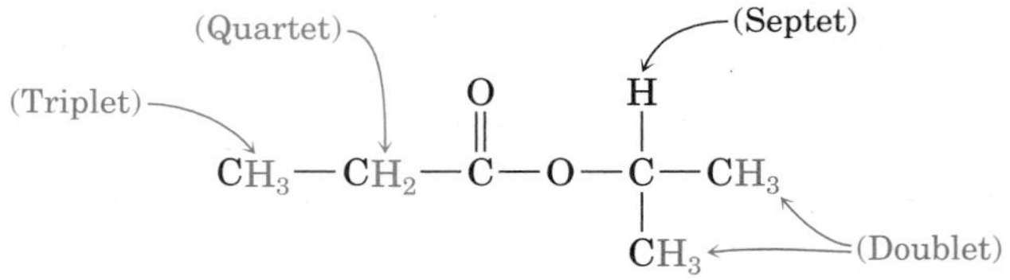

(Quartet) (Septet) (Triplet) (Doublet)

Isopropyl propanoate

PROBLEM

13.14 Predict the splitting patterns for each proton in the following molecules:
(a) $(CH_3)_3CH$ (b) CH_3CHBr_2 (c) $CH_3OCH_2CH_2Br$
(d) $CH_3CH_2COOCH_3$ (e) $ClCH_2CH_2CH_2Cl$ (f) $(CH_3)_2CHCOOCH_3$

PROBLEM

13.15 Propose structures for compounds that show the following ^1H NMR spectra:
(a) C_2H_6O; one singlet (b) $C_3H_6O_2$; two singlets
(c) C_3H_7Cl; one doublet and one septet

13.11 Uses of ^1H NMR Spectra

^1H NMR spectroscopy is used to help identify the products of nearly every reaction run in the laboratory. For example, we said in Section 4.4 that acid-catalyzed addition of H_2O to an alkene occurs with Markovnikov orientation; that is, the more highly substituted alcohol is formed. With the help of ^1H NMR, we can now prove this statement.

Does addition of H_2O to 1-methylcyclohexene yield 1-methylcyclohexanol or 2-methylcyclohexanol?

CH_3 $\xrightarrow[H_2SO_4]{H_2O}$ CH_3 / OH or CH_3 / OH

1-Methylcyclohexene **1-Methylcyclohexanol** **2-Methylcyclohexanol**

The ^1H NMR spectrum of the reaction product is shown in Figure 13.16. Although many of the ring protons overlap into a broad, poorly de-

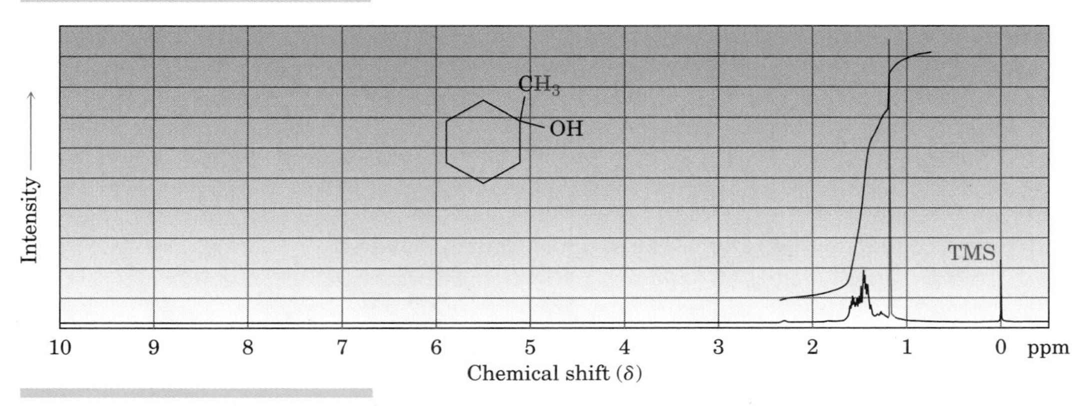

Figure 13.16 The ^1H NMR spectrum of the reaction product from H_2O and 1-methylcyclohexene. The presence of the $-CH_3$ absorption at 1.2 δ and the absence of any absorptions near 4 δ identify the product as 1-methylcyclohexanol.

fined multiplet centered around 1.6 δ, the spectrum also shows a large singlet absorption in the saturated methyl region at 1.2 δ, indicating that the product has a methyl group with no neighboring hydrogens ($R_3C–CH_3$). Furthermore, the spectrum shows no absorptions around 4 δ, where we would expect the signal of an R_2CHOH proton to occur. Thus, the reaction product must be 1-methylcyclohexanol.

13.12 ^{13}C NMR Spectroscopy

In some ways, it's surprising that carbon NMR is even possible. After all, ^{12}C, the most abundant carbon isotope, has no nuclear spin and is not observable by NMR. The only naturally occurring carbon isotope with a magnetic moment is ^{13}C, but its natural abundance is only 1.1%. Thus, only about 1 of every 100 carbon atoms in an organic molecule is observable by NMR. Fortunately, the technical problems caused by this low abundance have been overcome by improved electronics and computer techniques, and ^{13}C NMR has now become a routine structural tool.

At its simplest, ^{13}C NMR makes it possible to count the number of carbon atoms in a molecule. In addition, it's possible to get information about the environment of each carbon by observing its chemical shift. As illustrated by the ^{13}C NMR spectrum of methyl acetate shown earlier (Figure 13.10b), we normally observe a single, sharp resonance line for each kind of carbon atom in a molecule. Thus, methyl acetate has three nonequivalent carbon atoms and three peaks in its ^{13}C NMR spectrum. (Coupling between adjacent carbon atoms isn't seen, because the low natural abundance of ^{13}C makes it unlikely that two such nuclei will be adjacent in a molecule.)

Most ^{13}C resonances are between 0 and 220 δ, with the exact chemical shift dependent on a carbon's environment in the molecule. Figure 13.17 shows how environment and chemical shift are correlated.

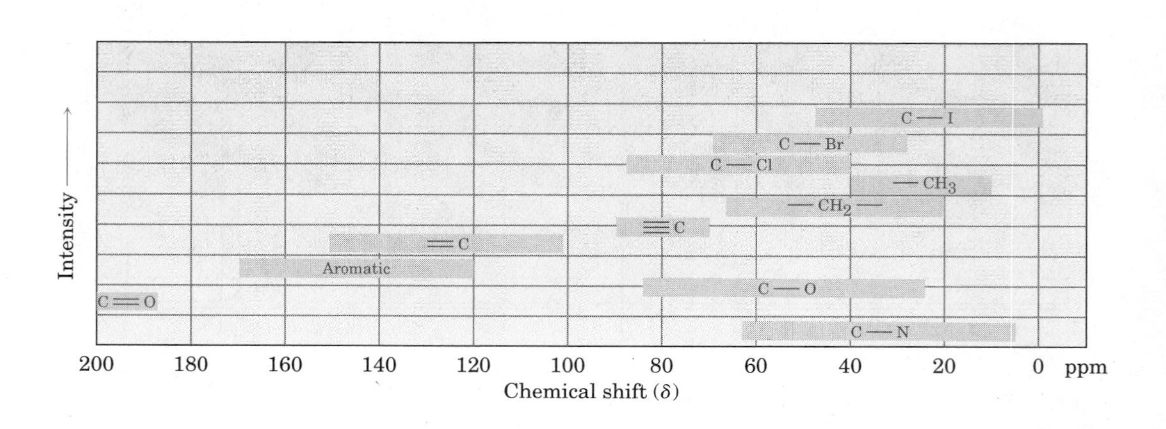

Figure 13.17 Chemical shift correlations for ^{13}C NMR.

The factors that determine chemical shifts are complex, but it's possible to make some generalizations. One trend is that a carbon bonded to an electronegative atom like oxygen, nitrogen, or halogen absorbs downfield of a typical alkane carbon. Another trend is that sp^3-hybridized carbons absorb in the range 0–100 δ, and sp^2 carbons absorb in the range 100–220 δ. Carbonyl carbons (C=O) are particularly distinct in the ^{13}C NMR spectrum and are easily observed at the extreme low-field side of the chart, in the range 170–220 δ. For example, the ^{13}C NMR spectrum of *p*-bromoacetophenone in Figure 13.18 shows an absorption for the carbonyl carbon at 197 δ.

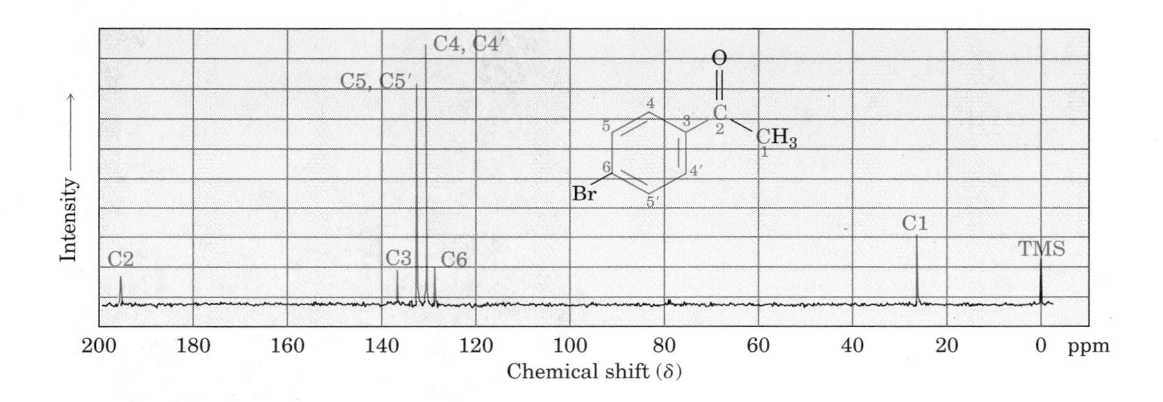

Figure 13.18 The ^{13}C NMR spectrum of *p*-bromoacetophenone, $BrC_6H_4COCH_3$.

The ^{13}C NMR spectrum of *p*-bromoacetophenone is interesting for another reason as well. Note that only six absorptions are observed even though the molecule has eight carbons. *p*-Bromoacetophenone has a symmetry plane that makes carbons 4 and 4′, and carbons 5 and 5′, equivalent. Thus, the six ring carbons show only four absorptions in the range 128–137 δ. In addition, the $-CH_3$ carbon absorbs at 26 δ.

PRACTICE PROBLEM 13.10

How many peaks would you expect in the ^{13}C NMR spectrum of methylcyclopentane?

Solution Methylcyclopentane has a symmetry plane. Thus, it has only four kinds of carbons and only four peaks in its ^{13}C NMR spectrum.

PROBLEM

13.16 How many peaks would you expect in the ^{13}C NMR spectra of the following compounds?

(a) Cyclopentane (b) 1,3-Dimethylbenzene
(c) 1,2-Dimethylbenzene (d) 1-Methylcyclohexene

PROBLEM

13.17 Propose structures for compounds whose ^{13}C NMR spectra fit the following descriptions:
(a) A hydrocarbon with seven peaks in its spectrum
(b) A six-carbon compound with only five peaks in its spectrum
(c) A four-carbon compound with three peaks in its spectrum

INTERLUDE

Magnetic Resonance Imaging (MRI)

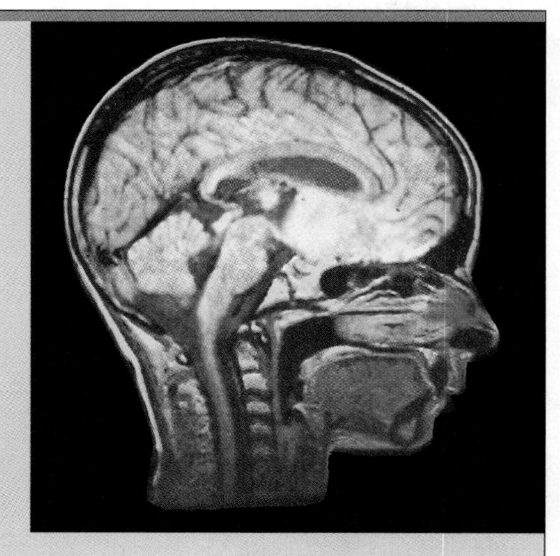

This NMR image through the head of a healthy child shows clear details of the brain and spinal cord.

As practiced by organic chemists, NMR spectroscopy is a valued method of structure determination. A small amount of sample, typically 10 mg or less, is dissolved in 1 mL or so of a suitable solvent, the solution is placed in a thin glass tube, and the tube is placed into the narrow (1–2 cm) gap between the poles of a strong magnet. Imagine, though, that a much larger NMR instrument were available. Instead of a few milligrams, the sample size could be tens of kilograms; instead of a narrow gap between magnet poles, the gap could be large enough for a person to climb into so that an NMR spectrum of body parts could be obtained. What you've just imagined is an instrument for *magnetic resonance imaging (MRI),* a diagnostic technique that has created enormous excitement in the medical community because of its advantages over X-ray or radioactive imaging methods. (You don't need to be dissolved in solvent or sit in a thin glass tube while undergoing MRI.)

Like NMR spectroscopy, MRI takes advantage of the magnetic properties of certain nuclei, typically hydrogen, and of the signals emitted when those nuclei are stimulated by radiofrequency energy. Unlike what happens in NMR spectroscopy, though, MRI instruments use powerful computers and data manipulation techniques to look at the three-dimensional *location* of magnetic nuclei in the body rather than at the chemical nature of

(*continued*)➤

the nuclei. As noted, most MRI instruments currently look at hydrogen, present in abundance wherever there is water or fat in the body.

The signals produced vary with the density of hydrogen atoms and with the nature of their surroundings, allowing the identification of different types of tissue and even the visualization of motion. For example, the volume of blood leaving the heart in a single stroke can be measured, allowing observation of the heart in motion. Soft tissues that do not show up well on X rays can also be seen clearly, allowing diagnosis of brain tumors, strokes, and other conditions. The technique is also valuable in diagnosing knee damage, because it is a painless alternative to arthroscopy, in which an endoscope is physically introduced into the knee joint.

Several elements in addition to hydrogen can be detected by MRI, and the applications of images based on ^{31}P atoms are being explored. The technique holds great promise for studies of metabolism.

Summary and Key Words

absorption spectrum, 409
amplitude, 407
chemical shift, 422
coupling, 425
coupling constant (J), 428
delta scale, 422
electromagnetic radiation, 407
electromagnetic spectrum, 407
frequency (ν), 407
hertz (Hz), 407
infrared (IR) spectroscopy, 407
integration, 424
$n + 1$ rule, 427
nuclear magnetic resonance (NMR) spectroscopy, 418
shielding, 420
spin–spin splitting, 425
ultraviolet (UV) spectroscopy, 415
wavelength (λ), 407
wavenumber ($\tilde{\nu} = 1/\lambda$), 410

Three main spectroscopic methods are used to determine the structures of organic molecules. Each of the three gives a different kind of information:

Infrared spectroscopy	What functional groups are present?
Ultraviolet spectroscopy	Is a conjugated π electron system present?
Nuclear magnetic resonance spectroscopy	What carbon–hydrogen framework is present?

When an organic molecule is irradiated with **infrared (IR)** energy, frequencies of light corresponding to the energy levels of molecular bending and stretching motions are absorbed. Each kind of functional group has a characteristic set of IR absorptions that allows the group to be identified. For example, an alkene C=C bond absorbs in the range 1640–1680 cm^{-1}, a saturated ketone absorbs near 1715 cm^{-1}, and a nitrile absorbs near 2230 cm^{-1}. By observing which frequencies of IR radiation are absorbed by a molecule *and which are not,* the functional groups in a molecule can be identified.

Ultraviolet (UV) spectroscopy is applicable to conjugated π electron systems. When a conjugated molecule is irradiated with ultraviolet light, energy absorption occurs, leading to excitation of π electrons to higher energy levels. The greater the extent of conjugation, the longer the wavelength needed for excitation.

Nuclear magnetic resonance (NMR) spectroscopy is the most valuable of the common spectroscopic techniques. When ^{1}H and ^{13}C nuclei are placed in a magnetic field, their spins orient either with or against the field. On irradiation with radiofrequency (rf) waves, energy is absorbed and the nuclear spins flip from the lower-energy state to the higher-energy state. This absorption of energy is detected, amplified, and displayed as an NMR spectrum. NMR spectra display four general features:

1. **Number of peaks.** Each nonequivalent kind of ^{1}H or ^{13}C nucleus in a molecule gives rise to a different peak.
2. **Chemical shift.** The exact position of each peak is called its chemical shift and is correlated to the chemical environment of each ^{1}H or ^{13}C nucleus. Most ^{1}H absorptions fall in the range 0–10 δ downfield from the TMS reference signal.
3. **Integration.** The area under each peak can be electronically integrated to determine the relative number of protons responsible for each peak.
4. **Spin–spin splitting.** Neighboring nuclear spins can **couple,** splitting an NMR absorption into multiplets. The NMR signal of a ^{1}H nucleus neighbored by n adjacent protons splits into $n + 1$ peaks with coupling constant J.

ADDITIONAL PROBLEMS

13.18 What kinds of functional groups might compounds contain if they show the following IR absorptions?
(a) 1670 cm^{-1} (b) 1735 cm^{-1} (c) 1540 cm^{-1}
(d) 1715 cm^{-1} and 2500–3100 cm^{-1} (broad)

13.19 At what approximate positions might the following compounds show IR absorptions?
(a) Benzoic acid
(b) Methyl benzoate
(c) *p*-Hydroxybenzonitrile
(d) 3-Cyclohexenone
(e) Methyl 4-oxopentanoate

13.20 The following ^{1}H NMR absorptions, determined on a spectrometer operating at 60 MHz, are given in hertz downfield from the TMS standard. Convert the absorptions to δ units.
(a) 131 Hz (b) 287 Hz (c) 451 Hz

13.21 At what positions, in hertz downfield from TMS standard, would the NMR absorptions in Problem 13.20 appear on a spectrometer operating at 100 MHz?

13.22 The following NMR absorptions, given in δ units, were obtained on a spectrometer operating at 80 MHz. Convert the chemical shifts from δ units into hertz downfield from TMS.
(a) 2.1 δ (b) 3.45 δ (c) 6.30 δ

13.23 If C–O single-bond stretching occurs at 1000 cm^{-1} and C=O double-bond stretching occurs at 1700 cm^{-1}, which of the two requires more energy? How does your answer correlate with the relative strengths of single and double bonds?

13.24 Tell what is meant by each of the following terms:
(a) Chemical shift
(b) Coupling constant
(c) λ_{max}
(d) Spin–spin splitting
(e) Wavenumber
(f) Applied magnetic field

13.25 When measured on a spectrometer operating at 60 MHz, chloroform ($CHCl_3$) shows a single sharp absorption at 7.3 δ.
(a) How many parts per million downfield from TMS does chloroform absorb?
(b) How many hertz downfield from TMS does chloroform absorb if the measurement is carried out on a spectrometer operating at 360 MHz?
(c) What is the position of the chloroform absorption in δ units measured on a 360 MHz spectrometer?

13.26 How many absorptions would you expect in the ^{13}C NMR spectra of the following compounds?
(a) 1,1-Dimethylcyclohexane
(b) Ethyl methyl ether
(c) Cyclohexanone
(d) 2-Methyl-2-butene
(e) *cis*-2-Pentene
(f) *trans*-2-Pentene

13.27 How many types of nonequivalent protons are there in each of the molecules listed in Problem 13.26?

13.28 Describe the ^{1}H NMR spectra you would expect for the following compounds:
(a) CH_3CHCl_2
(b) $CH_3CO_2CH_2CH_3$
(c) $(CH_3)_3CCH_2CH_3$

13.29 The following compounds all show a single peak in their ^{1}H NMR spectra. List them in order of expected increasing chemical shift: CH_4, CH_2Cl_2, cyclohexane, CH_3COCH_3, $H_2C{=}CH_2$, benzene

13.30 Propose structures for compounds that meet the following descriptions:
(a) C_5H_8, with IR absorptions at 3300 and 2150 cm^{-1}
(b) C_4H_8O, with a strong IR absorption at 3400 cm^{-1}
(c) C_4H_8O, with a strong IR absorption at 1715 cm^{-1}
(d) C_8H_{10}, with IR absorptions at 1600 and 1500 cm^{-1}

13.31 How would you use IR spectroscopy to distinguish between the following pairs of isomers?
(a) $(CH_3)_3N$ and $CH_3CH_2NHCH_3$
(b) CH_3COCH_3 and $CH_2{=}CHCH_2OH$
(c) CH_3COCH_3 and CH_3CH_2CHO

13.32 How would you use ^{1}H NMR spectroscopy to distinguish between the isomer pairs shown in Problem 13.31?

13.33 How could you use ^{13}C NMR spectroscopy to distinguish between the isomer pairs shown in Problem 13.31?

13.34 Assume that you're carrying out the dehydration of 1-methylcyclohexanol to yield 1-methylcyclohexene. How could you use IR spectroscopy to determine when the reaction is complete? What characteristic absorptions would you expect for both starting material and product?

13.35 The IR spectrum of a compound with the formula C_7H_6O is shown. Propose a likely structure.

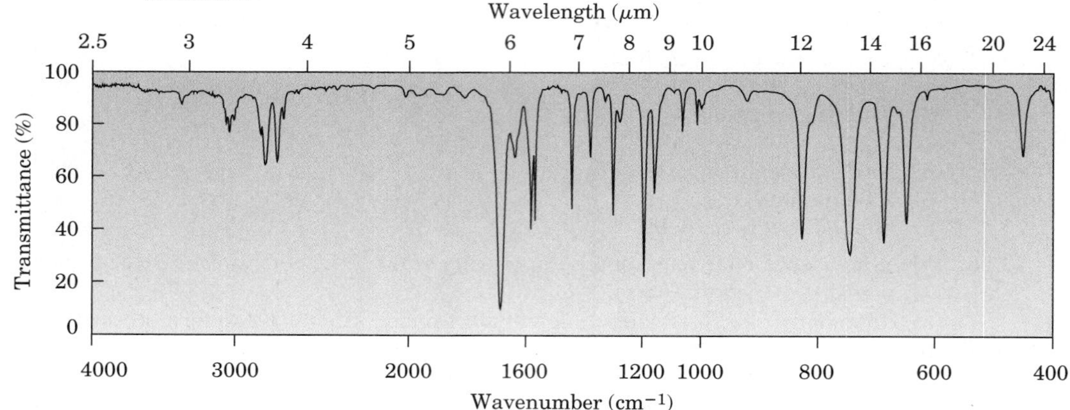

13.36 Dehydration of 1-methylcyclohexanol might lead to either of two isomeric alkenes, 1-methylcyclohexene or methylenecyclohexane. How could you use NMR spectroscopy (both ^1H and ^{13}C) to determine the structure of the product?

$=CH_2$ **Methylenecyclohexane**

13.37 3,4-Dibromohexane can undergo base-induced double dehydrobromination to yield either 3-hexyne or 2,4-hexadiene. How could you use UV spectroscopy to help identify the product? How could you use ^1H NMR spectroscopy?

13.38 Describe the ^1H and ^{13}C NMR spectra you expect for the following compounds:
(a) $ClCH_2CH_2CH_2Cl$
(b) $CH_3COCH_2CH_2Cl$

13.39 Propose structures for compounds with the following formulas that show only one peak in their ^1H NMR spectra:
(a) C_5H_{12}
(b) C_5H_{10}
(c) $C_4H_8O_2$

13.40 Assume that you have a compound with formula C_3H_6O.
(a) Propose as many structures as you can that fit the molecular formula (there are seven).
(b) If your compound has an IR absorption at 1715 cm^{-1}, what can you conclude?
(c) If your compound has a single ^1H NMR absorption at 2.1 δ, what is its structure?

13.41 Propose structures for compounds that fit the following ^1H NMR data:

(a) $C_5H_{10}O$
6 H doublet at 0.95 δ, $J = 7$ Hz
3 H singlet at 2.10 δ
1 H multiplet at 2.43 δ

(b) C_3H_5Br
3 H singlet at 2.32 δ
1 H singlet at 5.25 δ
1 H singlet at 5.54 δ

13.42 How can you use ^1H and ^{13}C NMR to help distinguish among the following four isomers?

CH_2-CH_2 / CH_2-CH_2 (cyclobutane) $\quad H_2C=CHCH_2CH_3 \quad CH_3CH=CHCH_3 \quad CH_3C(CH_3)=CH_2$

13.43 How can you use ^1H NMR to help distinguish between the two isomers shown at the top of the next page?

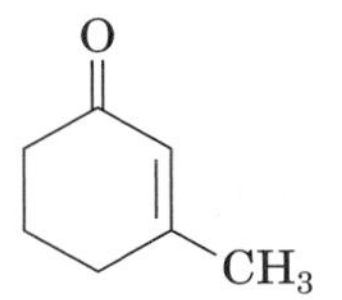

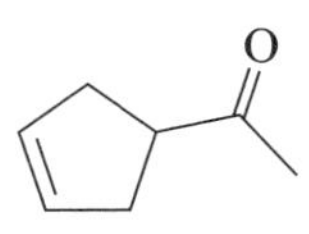

3-Methyl-2-cyclohexenone **4-Cyclopentenyl methyl ketone**

13.44 How can you use ^{13}C NMR to help distinguish between the isomers in Problem 13.43?

13.45 How can you use UV spectroscopy to help distinguish between the isomers in Problem 13.43?

13.46 The energy E of electromagnetic radiation, expressed in units of kilojoules per mole (kJ/mol), can be determined by the following formula:

$$E = \frac{1.20 \times 10^{-2}}{\lambda \text{ (in cm)}} \text{ kJ/mol}$$

What is the energy of infrared radiation of wavelength 10^{-4} cm?

13.47 Using the formula given in Problem 13.46, calculate the energy required to effect the electronic excitation of 1,3-butadiene ($\lambda_{max} = 217$ nm).

13.48 Using the equation given in Problem 13.46, calculate the amount of energy required to spin-flip a proton in a spectrometer operating at 100 MHz. Does increasing the spectrometer frequency from 60 MHz to 100 MHz increase or decrease the amount of energy necessary for resonance?

13.49 The ^1H NMR spectrum of compound A, $C_3H_6Br_2$, is shown. Propose a structure for A, and explain how the spectrum fits your structure.

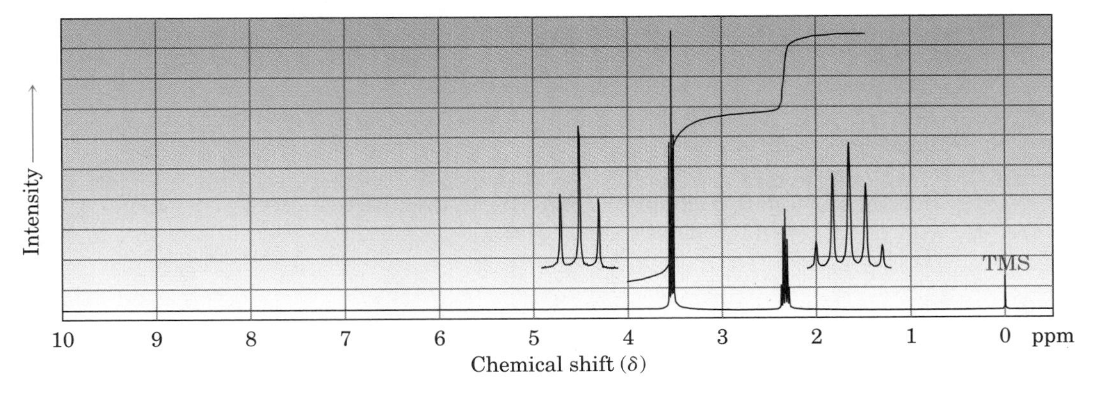

13.50 The compound whose ^1H NMR spectrum is shown has the formula $C_4H_7O_2Cl$ and has an IR absorption peak at 1740 cm^{-1}. Propose a structure.

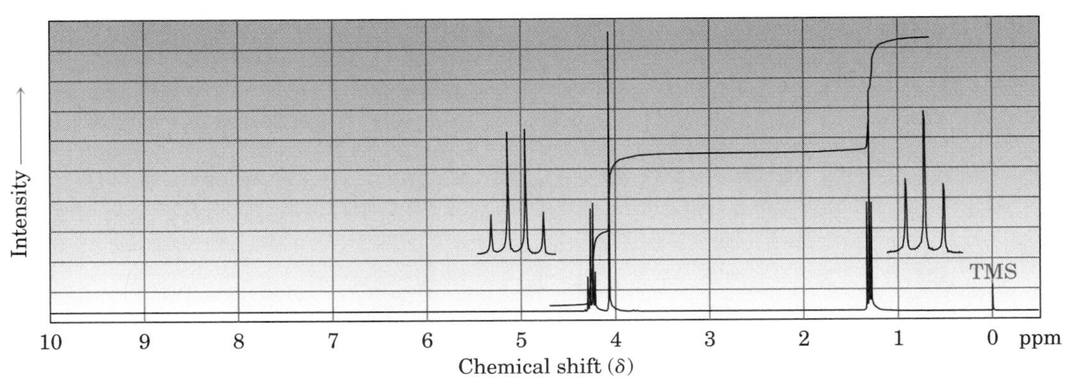

13.51 Propose a structure for a compound with formula C_4H_9Br that has the following ^1H NMR spectrum:

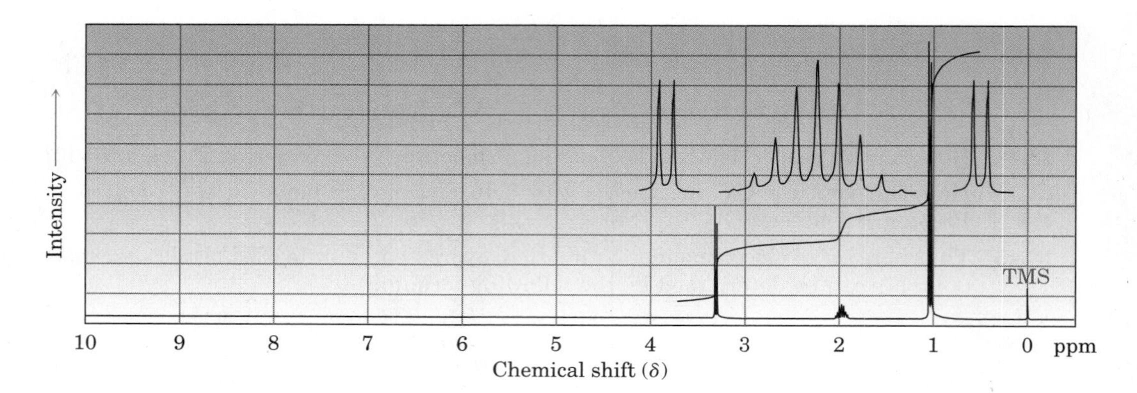

13.52 Propose structures for compounds that fit the following ^1H NMR data:

(a) $C_4H_6Cl_2$
3 H singlet at 2.18 δ
2 H doublet at 4.16 δ, $J = 7$ Hz
1 H triplet at 5.71 δ, $J = 7$ Hz

(b) $C_{10}H_{14}$
9 H singlet at 1.30 δ
5 H singlet at 7.30 δ

13.53 Nitriles (RC≡N) react with Grignard reagents (RMgBr). The reaction product from 2-methylpropanenitrile with methylmagnesium bromide has the following spectroscopic properties. Propose a structure.

$$(CH_3)_2CHC{\equiv}N \xrightarrow[2.\ H_3O^+]{1.\ CH_3MgBr} ?$$

2-Methylpropanenitrile

Molecular weight = 86
IR: 1715 cm^{-1}
^1H NMR: 1.05 δ (6 H doublet, $J = 7$ Hz); 2.12 δ (3 H singlet); 2.67 δ (1 H septet, $J = 7$ Hz)
^{13}C NMR: 18.2, 27.2, 41.6, 211.2 δ

Visualizing Chemistry

13.54 Into how many peaks would you expect the ^1H NMR signal of the indicated protons to be split? (Gray = C, red = O, yellow-green = Cl, light green = H.)

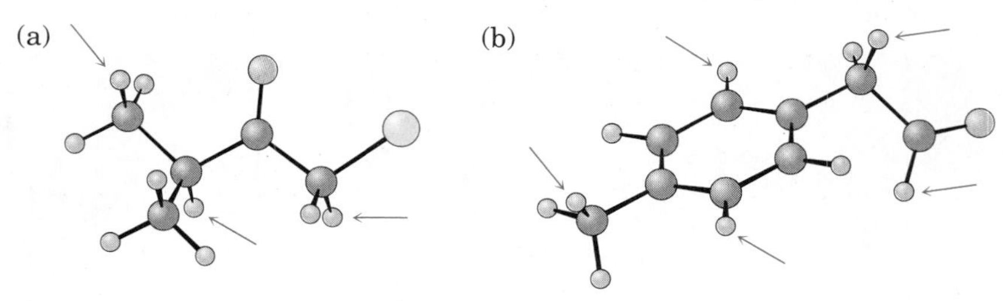

13.55 Where in the infrared spectrum would you expect each of the following compounds to absorb? (Gray = C, red = O, blue = N, light green = H.)

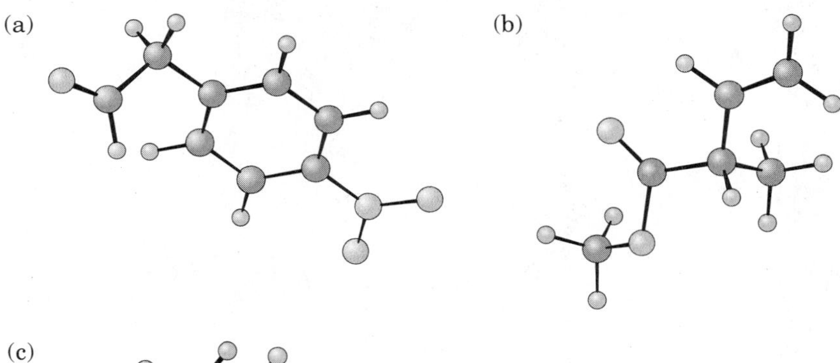

13.56 How many absorptions would you expect the following compound to have in its ^{13}C NMR spectrum? (Gray = C, light green = H.)

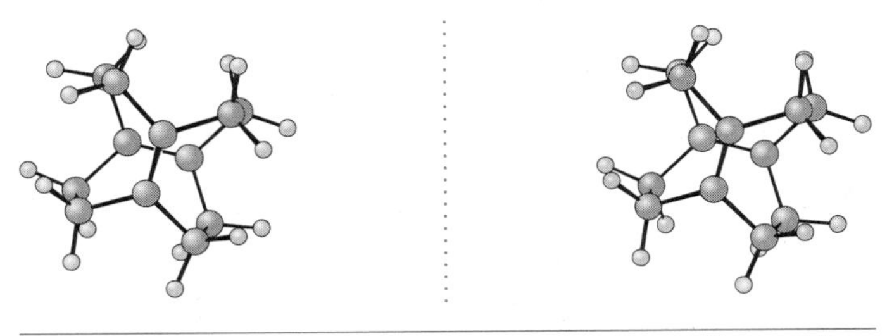

Stereo View

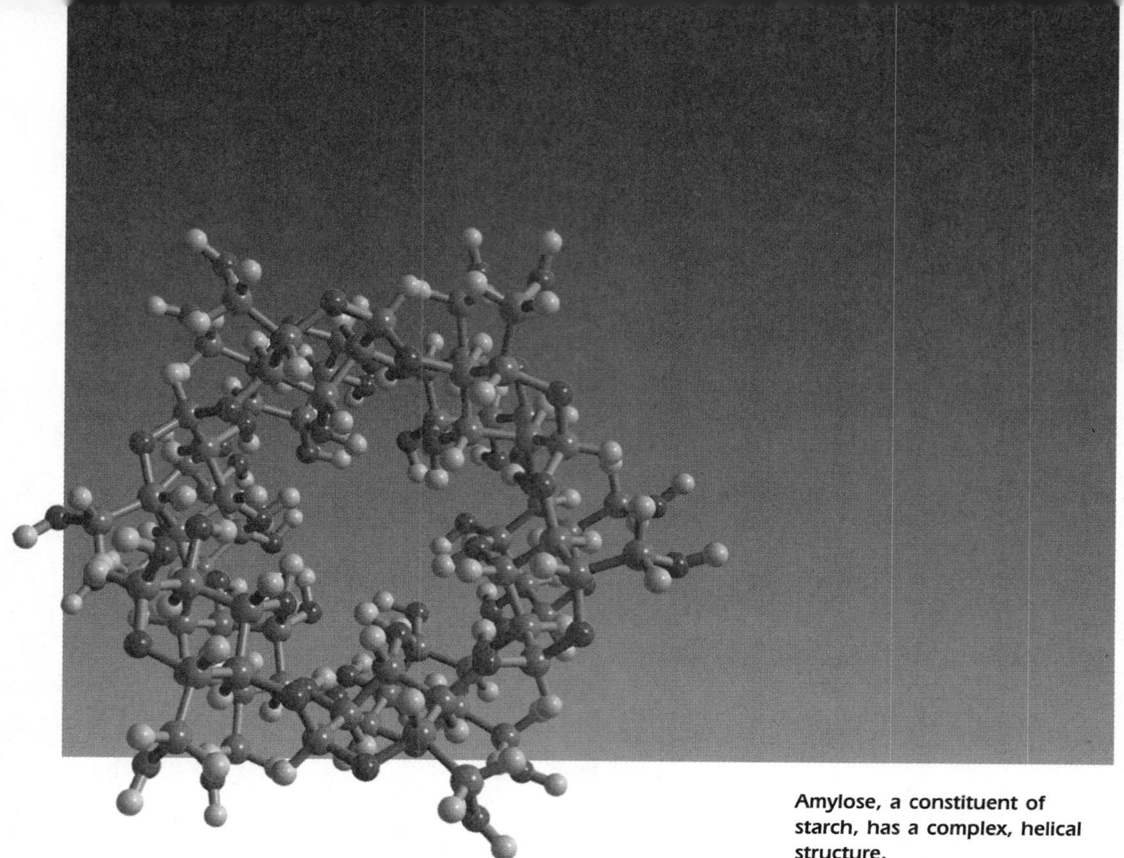

Amylose, a constituent of starch, has a complex, helical structure.

14 Biomolecules: Carbohydrates

Carbohydrates are everywhere; they occur in every living organism and are essential to life. The sugar and starch in food, and the cellulose in wood, paper, and cotton are nearly pure carbohydrate. Modified carbohydrates form part of the coating around living cells, other carbohydrates are found in the DNA that carries genetic information, and still others are used as medicines.

The word **carbohydrate** derives historically from the fact that glucose, the first simple carbohydrate to be obtained pure, has the molecular formula $C_6H_{12}O_6$ and was originally thought to be a "hydrate of carbon," $C_6(H_2O)_6$. This view was soon abandoned, but the name persisted. Today, the term *carbohydrate* is used to refer loosely to the broad class of polyhydroxylated aldehydes and ketones commonly called *sugars.*

$$
\begin{array}{c}
\mathrm{H} \diagdown \quad \diagup\!\!\!\!= \mathrm{O} \\
\mathrm{C} \\
| \\
\mathrm{H-C-OH} \\
| \\
\mathrm{HO-C-H} \\
| \\
\mathrm{H-C-OH} \\
| \\
\mathrm{H-C-OH} \\
| \\
\mathrm{CH_2OH}
\end{array}
$$

Glucose (also called dextrose), a pentahydroxyhexanal

Carbohydrates are made by green plants during photosynthesis, a complex process in which sunlight provides the energy to convert CO_2 into glucose. Many molecules of glucose are then chemically linked for storage by the plant in the form of either cellulose or starch. It has been estimated that more than 50% of the dry weight of the earth's biomass—all plants and animals—consists of glucose polymers. When eaten and then metabolized, carbohydrates provide the major source of energy required by organisms. Thus, carbohydrates act as the chemical intermediaries by which solar energy is stored and used to support life.

$$6\,CO_2 + 6\,H_2O \xrightarrow{\text{Sunlight}} 6\,O_2 + \underset{\textbf{Glucose}}{C_6H_{12}O_6} \longrightarrow \text{Cellulose, starch}$$

14.1 Classification of Carbohydrates

Carbohydrates are generally classed into two groups, *simple* and *complex.* **Simple sugars,** or **monosaccharides,** are carbohydrates like glucose and fructose that can't be hydrolyzed into smaller molecules. **Complex carbohydrates** are composed of two or more simple sugars linked together. Sucrose (table sugar), for example, is a **disaccharide** made up of one glucose molecule linked to one fructose molecule. Similarly, cellulose is a **polysaccharide** made up of several thousand glucose molecules linked together. Hydrolysis of polysaccharides breaks them down into their constituent monosaccharide units.

$$1\ \text{Sucrose} \xrightarrow{H_3O^+} 1\ \text{Glucose} + 1\ \text{Fructose}$$

$$\text{Cellulose} \xrightarrow{H_3O^+} \sim 3000\ \text{Glucose}$$

Monosaccharides are further classified as either **aldoses** or **ketoses.** The *-ose* suffix is used as the family name ending for carbohydrates, and the *aldo-* and *keto-* prefixes identify the nature of the carbonyl group, whether aldehyde or ketone. The number of carbon atoms in the monosaccharide is indicated by using *tri-*, *tetr-*, *pent-*, *hex-*, and so forth, in the name. For example, glucose is an *aldohexose,* a six-carbon aldehydo sugar; fructose is a *ketohexose,* a six-carbon keto sugar; and ribose is an *aldopentose,* a five-carbon aldehydo sugar. Most of the commonly occurring simple sugars are either aldopentoses or aldohexoses.

```
   H\  //O            CH2OH              H\  //O
     C                  |                  C
     |                 C=O                 |
  H—C—OH                |               H—C—OH
     |             HO—C—H                  |
 HO—C—H                 |               H—C—OH
     |              H—C—OH                 |
  H—C—OH                |               H—C—OH
     |              H—C—OH                 |
  H—C—OH                |                CH2OH
     |                CH2OH
   CH2OH
```

Glucose (an aldohexose) **Fructose (a ketohexose)** **Ribose (an aldopentose)**

PRACTICE PROBLEM 14.1

Classify the following monosaccharide:

```
   H\  //O
     C
     |
  H—C—OH
     |
  H—C—OH
     |
  H—C—OH        Allose
     |
  H—C—OH
     |
   CH2OH
```

Solution Since allose has six carbons and an aldehyde carbonyl group, it is an aldohexose.

PROBLEM

14.1 Classify each of the following monosaccharides:

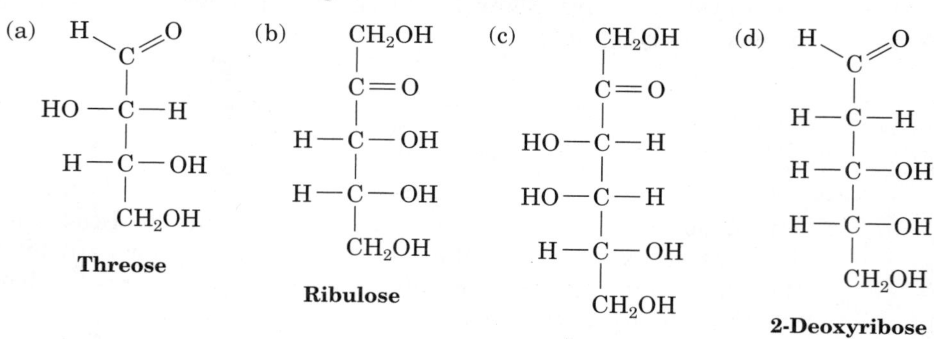

14.2 Configurations of Monosaccharides: Fischer Projections

Since all carbohydrates have stereocenters, it was recognized long ago that a standard method of representation is needed to describe carbohydrate stereoisomers. In 1891, Emil Fischer suggested a method based on the projection of a tetrahedral carbon atom onto a flat surface. These **Fischer projections** were soon adopted and are now a standard means of depicting stereochemistry at stereocenters.

A tetrahedral carbon atom is represented in a Fischer projection by two perpendicular lines. The horizontal lines indicate bonds coming out of the page, and the vertical lines indicate bonds going into the page:

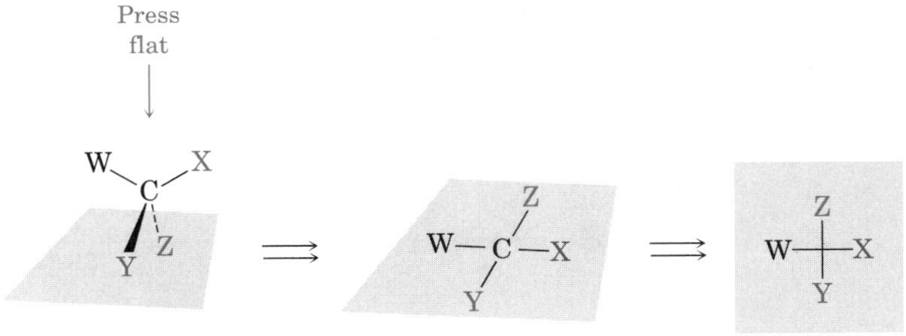

Press flat

Fischer projection

By convention, the carbonyl carbon is placed at or near the top in Fischer projections. Thus, (*R*)-glyceraldehyde, the simplest monosaccharide, is represented as shown in Figure 14.1 (p. 444).

Carbohydrates with more than one stereocenter are shown by stacking the atoms, one on top of the other. Once again, the carbonyl carbon is at or near the top of the Fischer projection. Glucose, for example, has four stereocenters stacked on top of one another in a Fischer projection:

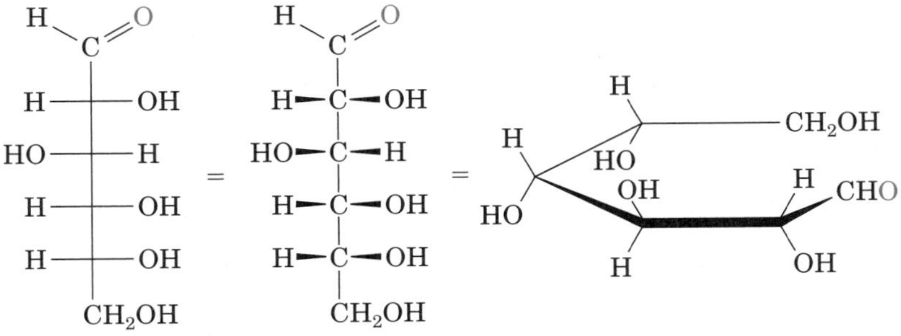

Glucose (carbonyl group at top)

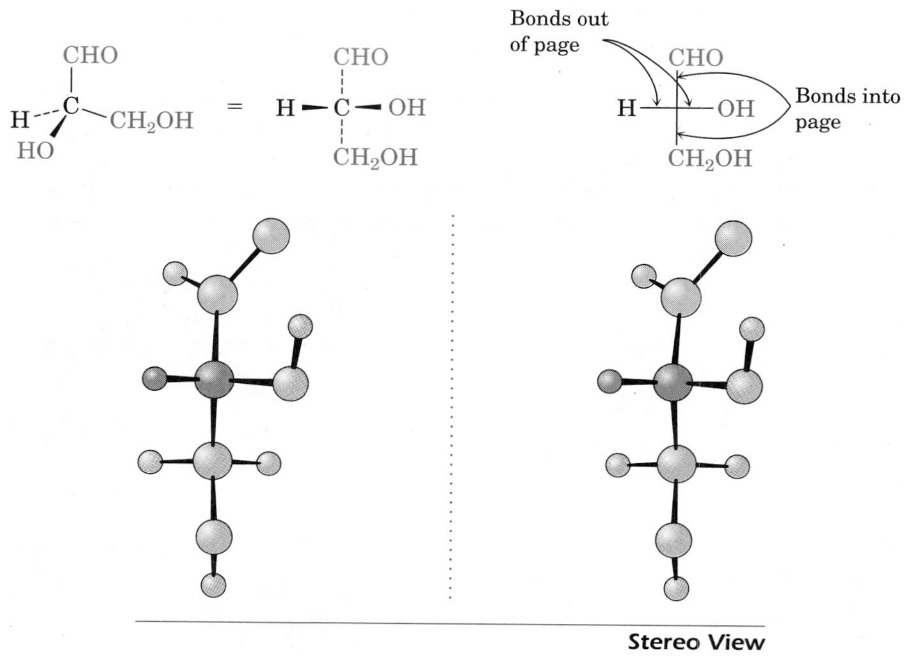

CHO, H, CH_2OH, HO = CHO, H—C—OH, CH_2OH

Bonds out of page — CHO, H, OH, CH_2OH — Bonds into page

Stereo View

Figure 14.1 Fischer projection of (*R*)-glyceraldehyde.

PRACTICE PROBLEM 14.2

Convert the following tetrahedral representation of (*R*)-2-butanol into a Fischer projection:

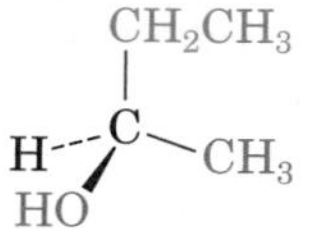

CH_2CH_3, H, C, CH_3, HO **(*R*)-2-Butanol**

Solution Orient the molecule so that two horizontal bonds are facing you and two vertical bonds are receding away from you. Then press the molecule flat into the paper, indicating the stereocenter as the intersection of two crossed lines:

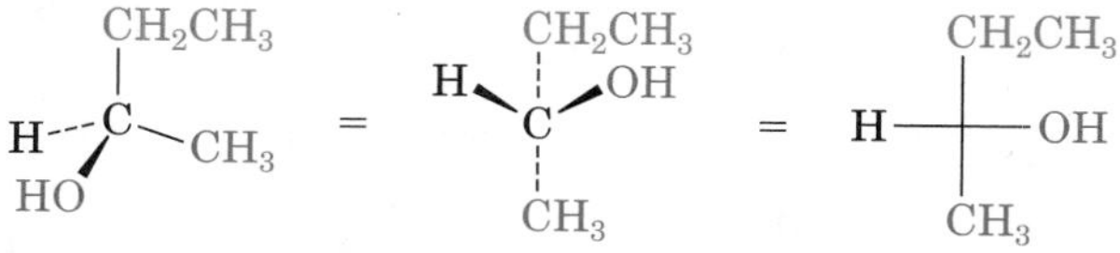

CH_2CH_3, H, C, CH_3, HO = CH_2CH_3, H, OH, C, CH_3 = CH_2CH_3, H—OH, CH_3

(*R*)-2-Butanol

PRACTICE PROBLEM 14.3

Convert the Fischer projection of lactic acid shown at the top of the next page into a tetrahedral representation, and indicate whether the molecule is (*R*) or (*S*).

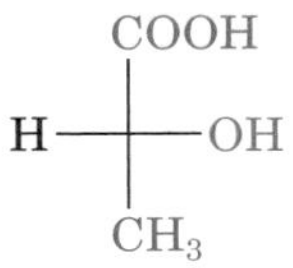

COOH
H—C—OH Lactic acid
CH$_3$

Solution After placing a carbon atom at the intersection of the two crossed lines, imagine that the two horizontal bonds are coming toward you and the two vertical bonds are receding away from you. The projection represents (R)-lactic acid.

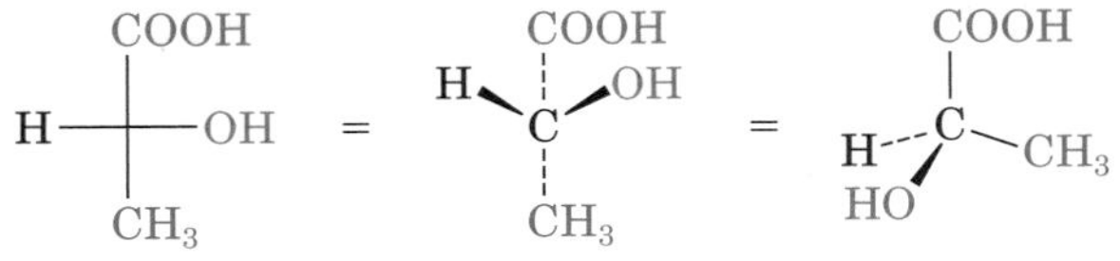

(R)-Lactic acid

PROBLEM

14.2 Convert the following tetrahedral representation of (S)-glyceraldehyde into a Fischer projection:

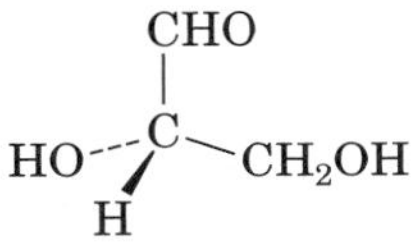

(S)-Glyceraldehyde

PROBLEM

14.3 Draw Fischer projections of both (R)-2-chlorobutane and (S)-2-chlorobutane.

PROBLEM

14.4 Convert the following Fischer projections into tetrahedral representations, and assign R or S stereochemistry to each:

(a) COOH; H$_2$N—C—H; CH$_3$

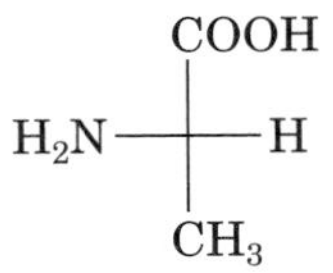

(b) CHO; H—C—OH; CH$_3$

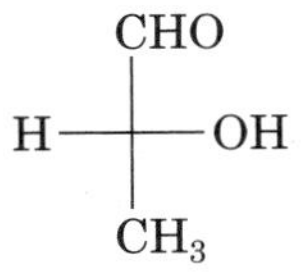

(c) CH$_3$; H—C—CHO; CH$_2$CH$_3$

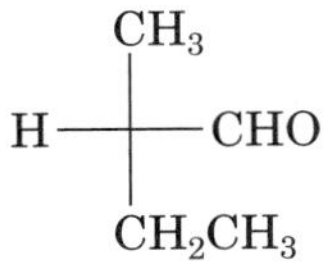

14.3 D,L Sugars

Glyceraldehyde has only one stereocenter and therefore has two enantiomeric (mirror-image) forms. Only the dextrorotatory enantiomer occurs naturally, however. That is, a sample of naturally occurring glyceraldehyde placed in a polarimeter (Section 6.4) rotates plane-polarized light in a clockwise direction, denoted (+). Since (+)-glyceraldehyde is known to have the R configuration at C2, it can be represented as in Figure 14.2. For historical reasons dating from long before the adoption of the R,S system, (R)-(+)-glyceraldehyde is also referred to as D-glyceraldehyde because it is dextrorotatory. The other enantiomer, (*S*)-(−)-glyceraldehyde, is known as *L-glyceraldehyde* because it is levorotatory.

Because of the way that monosaccharides are synthesized in nature, glucose, fructose, ribose, and most other naturally occurring monosaccharides have the same stereochemical configuration as D-glyceraldehyde at the stereocenter farthest from the carbonyl group. In Fischer projections, therefore, most naturally occurring sugars have the –OH group at the lowest stereocenter pointing to the *right* (Figure 14.2). Such compounds are referred to as **D sugars.**

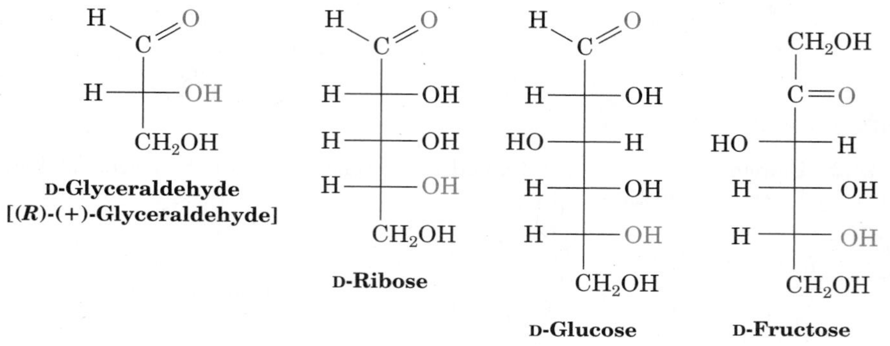

D-Glyceraldehyde [(*R*)-(+)-Glyceraldehyde]

D-Ribose

D-Glucose

D-Fructose

Figure 14.2 Some naturally occurring D sugars. The –OH at the stereocenter farthest from the carbonyl group is on the right in Fischer projections.

In contrast to the D sugars, all **L sugars** have the –OH group at the stereocenter farthest from the carbonyl group on the *left* in Fischer projections. Thus, L sugars are mirror images (enantiomers) of D sugars. Note that the D and L notations have no relation to the direction in which a given sugar rotates plane-polarized light. A D sugar may be either dextrorotatory or levorotatory. The prefix D indicates only that the stereochemistry of the bottommost stereocenter is the same as that of D-glyceraldehyde and is to the right in Fischer projection when the molecule is drawn in the standard way with the carbonyl group at or near the top.

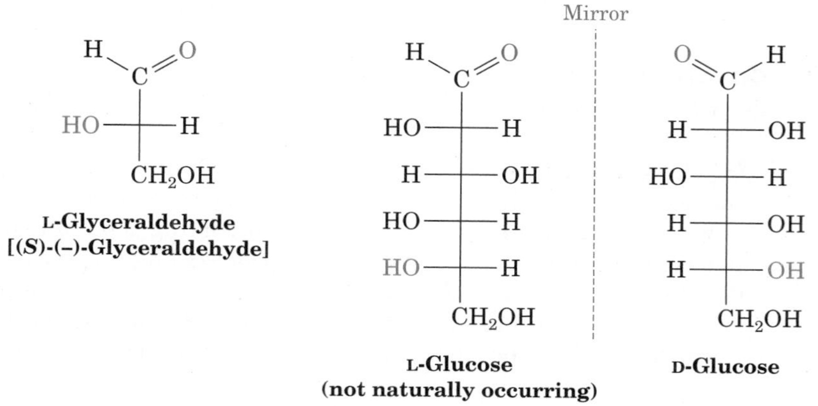

Mirror

L-Glyceraldehyde [(S)-(–)-Glyceraldehyde]

L-Glucose (not naturally occurring)

D-Glucose

PRACTICE PROBLEM 14.4

Look at the Fischer projection of D-fructose in Figure 14.2, and then draw a Fischer projection of L-fructose.

Solution Since L-fructose is the enantiomer (mirror image) of D-fructose, take the structure of D-fructose and reverse the configuration at each stereocenter:

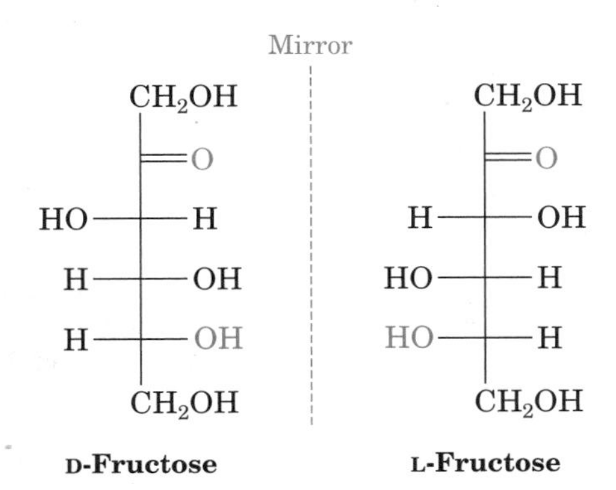

PROBLEM

14.5 Which of the following are L sugars and which are D sugars?

(a) CHO; HO—H; HO—H; CH_2OH

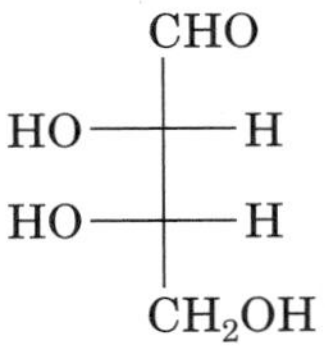

(b) CHO; H—OH; HO—H; H—OH; CH_2OH

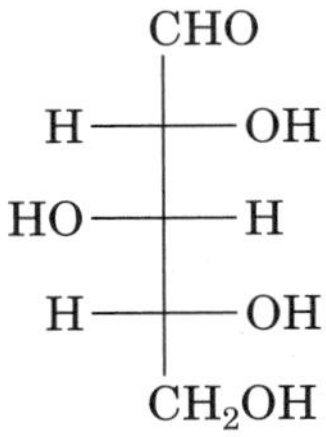

(c) CH_2OH; C=O; HO—H; H—OH; CH_2OH

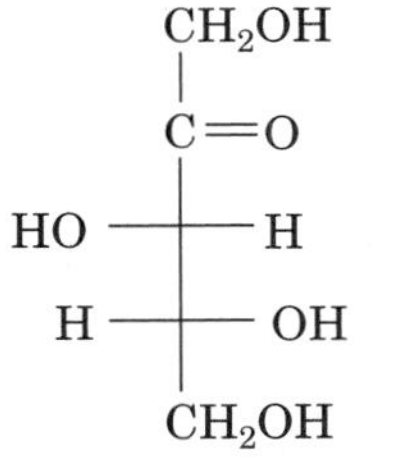

PROBLEM

14.6 Draw the enantiomers (mirror images) of the carbohydrates shown in Problem 14.5, and identify each as a D sugar or an L sugar.

14.4 Configurations of Aldoses

Aldotetroses are four-carbon sugars with two stereocenters. Thus, there are $2^2 = 4$ possible stereoisomeric aldotetroses, or two D,L pairs of enantiomers, named *erythrose* and *threose.*

Aldopentoses have three stereocenters and a total of $2^3 = 8$ possible stereoisomers, or four D,L pairs of enantiomers. These four pairs are named *ribose, arabinose, xylose,* and *lyxose.* All D aldopentoses except lyxose occur widely in nature. D-Ribose is an important part of RNA (ribonucleic acid), L-arabinose is found in many plants, and D-xylose is found in wood.

Aldohexoses have four stereocenters, for a total of $2^4 = 16$ possible stereoisomers, or eight D,L pairs of enantiomers. The names of the eight are *allose, altrose, glucose, mannose, gulose, idose, galactose,* and *talose.* Only D-glucose, from starch and cellulose, and D-galactose, from gums and fruit pectins, are widely distributed in nature. D-Mannose and D-talose also occur naturally, but in lesser abundance.

Fischer projections of the four-, five-, and six-carbon aldoses are shown in Figure 14.3 for the D series. Starting from D-glyceraldehyde, we can imagine constructing the two D aldotetroses by inserting a new stereocenter just below the aldehyde carbon. Each of the two D aldotetroses then leads to two D aldopentoses (four total), and each of the four D aldopentoses leads to two D aldohexoses (eight total).

PROBLEM

14.7 Write Fischer projections for the following L sugars. Remember that an L sugar is the mirror image of the corresponding D sugar shown in Figure 14.3.
(a) L-Arabinose (b) L-Threose (c) L-Galactose

PROBLEM

14.8 How many aldoheptoses are possible? How many of them are D sugars, and how many are L sugars?

PROBLEM

14.9 Draw Fischer projections for the two D aldoheptoses (Problem 14.8) whose stereochemistry at C3, C4, C5, and C6 is the same as that of glucose at C2, C3, C4, and C5.

14.5 Cyclic Structures of Monosaccharides: Hemiacetal Formation

We said during the discussion of carbonyl-group chemistry in Section 9.8 that alcohols undergo a rapid and reversible nucleophilic addition reaction with ketones and aldehydes to form hemiacetals:

$$RCHO + R'OH \underset{}{\overset{H^+ \text{ catalyst}}{\rightleftharpoons}} RCH(OH)OR'$$

An aldehyde — A hemiacetal

If both the hydroxyl and the carbonyl group are in the same molecule, an *intramolecular* nucleophilic addition can take place, leading to the formation of a *cyclic* hemiacetal. Five- and six-membered cyclic hemiacetals form particularly easily, and many carbohydrates therefore exist in an equilibrium between open-chain and cyclic forms. For example, glucose exists in aqueous solution primarily in the six-membered, **pyranose** form resulting from intramolecular nucleophilic addition of the –OH group at C5 to the C1

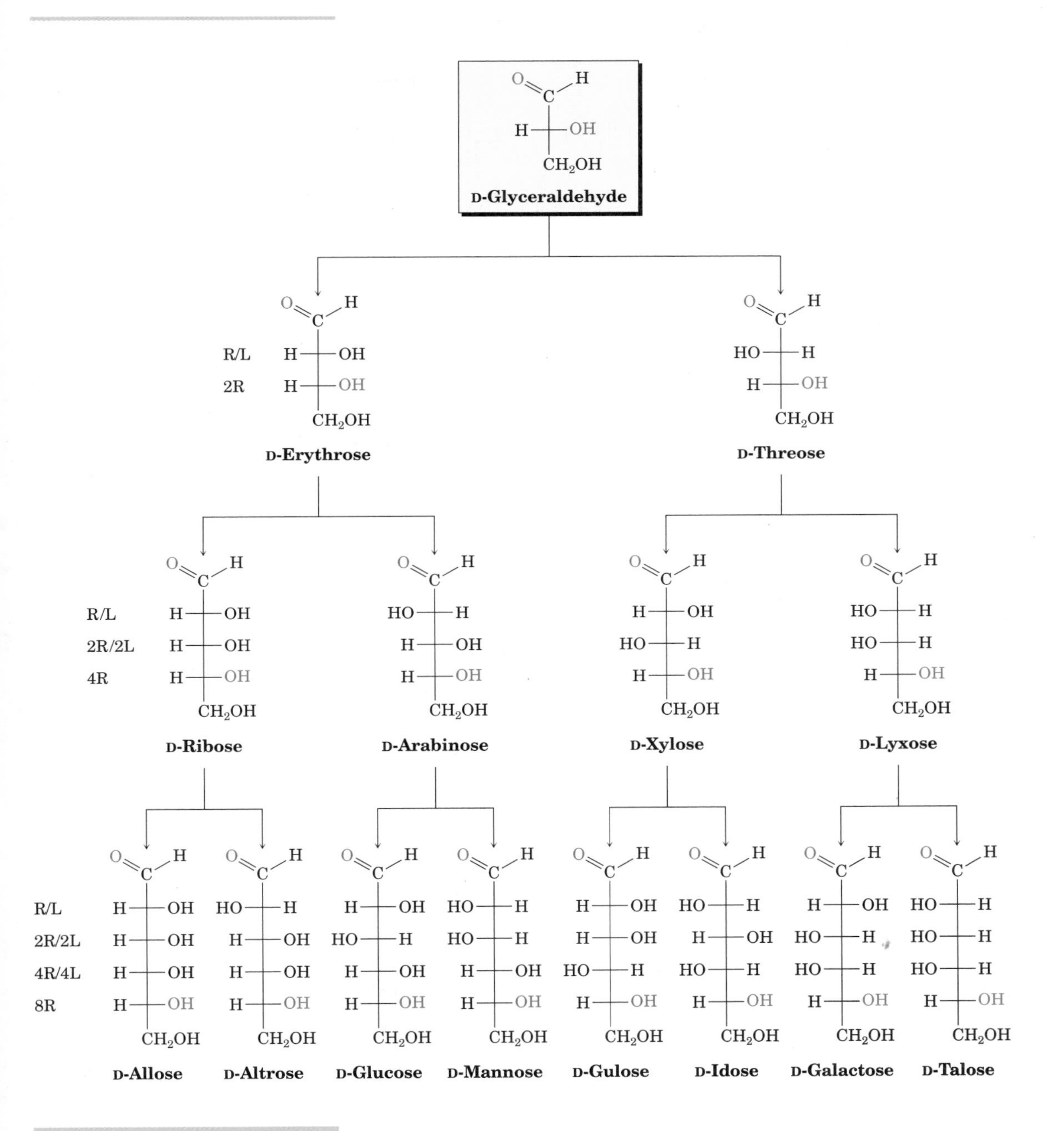

Figure 14.3 Configurations of D aldoses. The structures are arranged from left to right so that the –OH groups on C2 alternate right/left (R/L) in going across a series. Similarly, the –OH groups at C3 alternate two right/two left (2R/2L); the –OH groups at C4 alternate 4R/4L; and the –OH groups at C5 are to the right in all eight (8R).

aldehyde group. Fructose, on the other hand, exists to the extent of about 80% in the pyranose form and about 20% in the five-membered, **furanose** form resulting from addition of the –OH group at C5 to the C2 ketone. The terms *pyranose* for a six-membered ring and *furanose* for a five-membered ring are derived from the names of the simple cyclic ethers pyran and furan. The cyclic forms of glucose and fructose are shown in Figure 14.4.

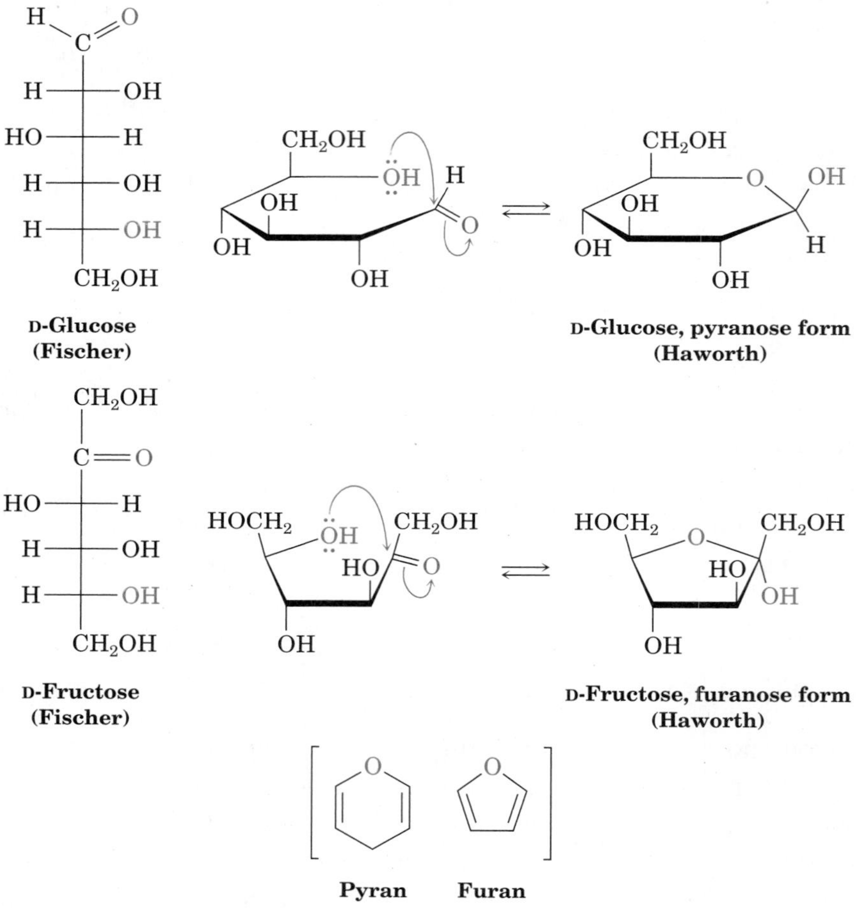

D-Glucose (Fischer)

D-Glucose, pyranose form (Haworth)

D-Fructose (Fischer)

D-Fructose, furanose form (Haworth)

Pyran Furan

Figure 14.4 Glucose and fructose in their cyclic pyranose and furanose forms.

Pyranose and furanose rings are often represented using **Haworth projections** rather than Fischer projections, as shown in Figure 14.4. In a Haworth projection, the hemiacetal ring is drawn as if it were flat and is viewed edge-on, with the oxygen atom at the upper right. Although convenient, this view isn't really accurate because pyranose rings are actually chair-shaped like cyclohexane (Section 2.9) rather than flat. Nevertheless, Haworth projections are widely used because they make it possible to see at a glance the cis–trans relationships among –OH groups on the ring.

When converting from one kind of projection to the other, it's helpful to remember that an –OH on the *right* in a Fischer projection is *down* in a Haworth projection and an –OH on the *left* in a Fischer projection is *up* in a Haworth projection. For D sugars, the $-CH_2OH$ group is always up in a Haworth projection, whereas for L sugars, the $-CH_2OH$ group is down. Figure 14.5 illustrates the conversion for glucose.

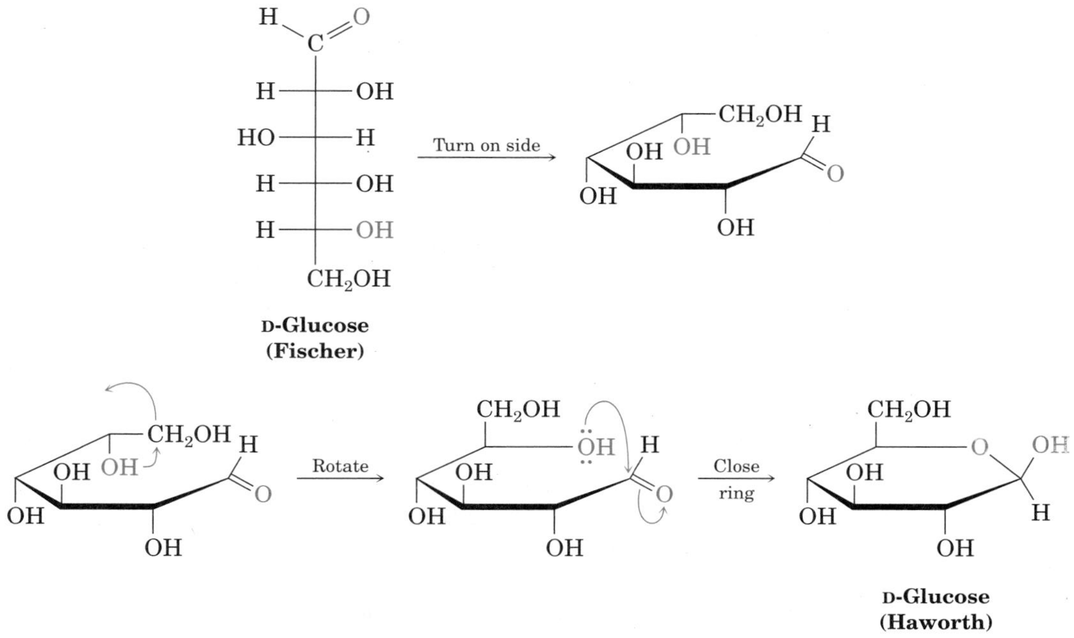

Turn on side

D-Glucose (Fischer)

Rotate

Close ring

D-Glucose (Haworth)

Figure 14.5 Interconversion of Fischer and Haworth projections of D-glucose.

PRACTICE PROBLEM 14.5

D-Mannose differs from D-glucose in its stereochemistry at C2. Draw a Haworth projection of D-mannose in its pyranose form.

Solution First draw a Fischer projection of D-mannose. Then lay it on its side and curl it around so that the –CHO group (C1) is toward the front and the $-CH_2OH$ (C6) is toward the rear. Now connect the –OH at C5 to the C1 carbonyl group to form a pyranose ring.

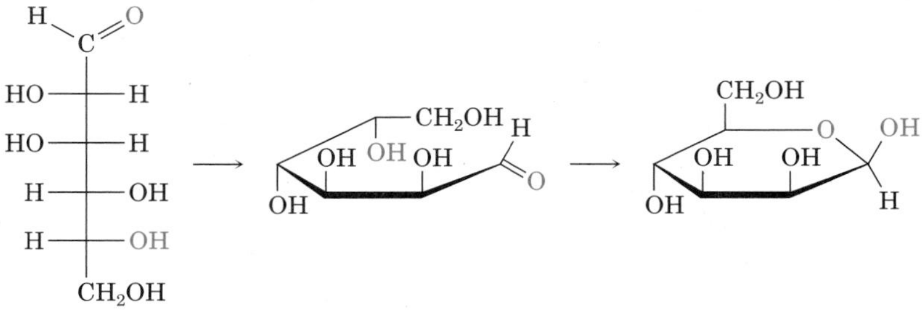

PROBLEM

14.10 D-Galactose differs from D-glucose in its stereochemistry at C4. Draw a Haworth projection of D-galactose in its pyranose form.

PROBLEM

14.11 Ribose exists largely in a furanose form, produced by addition of the C4 –OH group to the C1 aldehyde. Draw a Haworth projection of D-ribose in its furanose form.

14.6 Monosaccharide Anomers: Mutarotation

When an open-chain monosaccharide cyclizes to a furanose or pyranose form, a new stereocenter is formed at the former carbonyl carbon. Two diastereomers, called **anomers,** are produced, with the hemiacetal carbon referred to as the **anomeric center.** For example, glucose cyclizes reversibly in aqueous solution to yield a 36:64 mixture of two anomers. The minor anomer, with the C1 –OH group trans to the $-CH_2OH$ substituent at C5 (and therefore *down* in a Haworth projection), is called the **alpha (α) anomer;** its complete name is α-D-glucopyranose. The major anomer, with the C1 –OH group cis to the $-CH_2OH$ substituent at C5 (and therefore *up* in a Haworth projection), is called the **beta (β) anomer;** its complete name is β-D-glucopyranose.

CH₂OH OH H OH OH OH =O ⇌ CH₂OH O OH OH OH OH Trans + CH₂OH O OH OH OH OH Cis

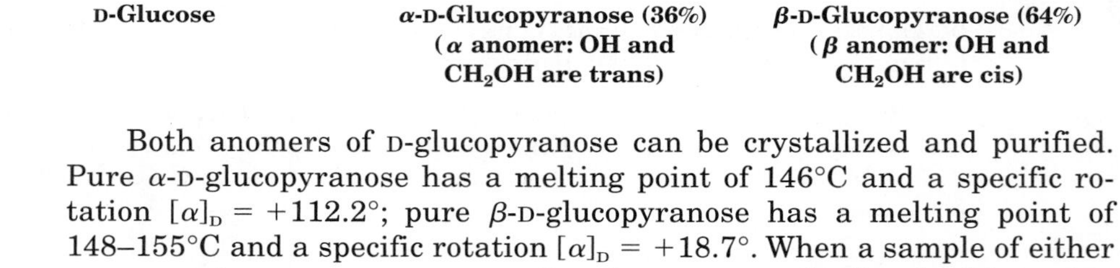

D-Glucose

α-D-Glucopyranose (36%)
(α anomer: OH and CH_2OH are trans)

β-D-Glucopyranose (64%)
(β anomer: OH and CH_2OH are cis)

Both anomers of D-glucopyranose can be crystallized and purified. Pure α-D-glucopyranose has a melting point of 146°C and a specific rotation $[\alpha]_D = +112.2°$; pure β-D-glucopyranose has a melting point of 148–155°C and a specific rotation $[\alpha]_D = +18.7°$. When a sample of either pure α-D-glucopyranose or pure β-D-glucopyranose is dissolved in water, however, the optical rotation slowly changes and ultimately reaches a constant value of +52.6°. The specific rotation of the α anomer solution decreases from +112.2° to +52.6°, and the specific rotation of the β anomer solution increases from +18.7° to +52.6°. Known as **mutarotation,** this spontaneous change in optical rotation is due to the slow conversion of the pure α and β enantiomers into the 36:64 equilibrium mixture.

Mutarotation occurs by a reversible ring opening of each anomer to the open-chain aldehyde form, followed by reclosure. Although equilibration is slow at neutral pH, it is catalyzed by both acid and base.

D-Glucose

α-D-Glucopyranose (36%)
$[\alpha]_D = +112.2°$

β-D-Glucopyranose (64%)
$[\alpha]_D = +18.7°$

PRACTICE PROBLEM 14.6

Draw Haworth projections of the two pyranose anomers of D-galactose, and identify each as α or β.

Solution The α anomer has the –OH group at C1 pointing down, and the β anomer has the –OH group at C1 pointing up.

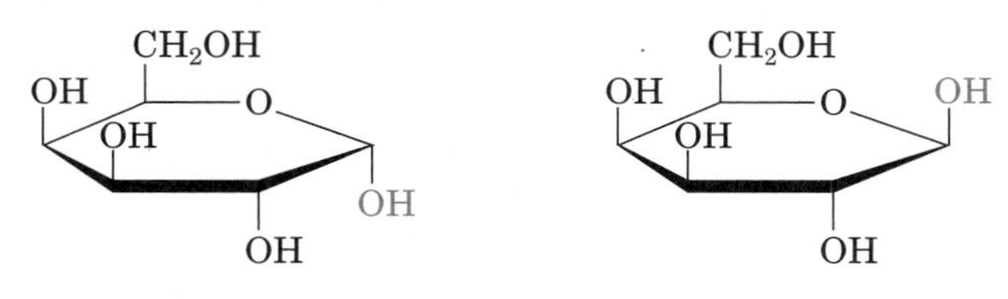

α-D-Galactopyranose **β-D-Galactopyranose**

PROBLEM

14.12 Draw the two anomers of D-fructose in their furanose forms.

PROBLEM

14.13 If the specific rotation of pure α-D-glucopyranose is $+112.2°$ and the specific rotation of pure β-D-glucopyranose is $+18.7°$, calculate the equilibrium percentages of α and β anomers from the equilibrium specific rotation of $+52.6°$.

PROBLEM

14.14 Many other sugars besides glucose exhibit mutarotation. For example, α-D-galactopyranose has $[\alpha]_D = +150.7°$, and β-D-galactopyranose has $[\alpha]_D = +52.8°$. If either anomer is dissolved in water and allowed to reach equilibrium, the specific rotation of the solution is $+80.2°$. What are the percentages of each anomer at equilibrium?

14.7 Conformations of Monosaccharides

Haworth projections are easy to draw, but they don't give an accurate three-dimensional picture of a molecule. Pyranose rings, like cyclohexane rings (Section 2.10), have a chairlike geometry with axial and equatorial substituents. Any substituent that is up in a Haworth projection is also up in a chair conformation, and any substituent that is down in a Haworth projection is also down in a chair conformation. Haworth projections can be converted into chair representations by following three steps:

Step 1 Draw the Haworth projection with the ring oxygen atom at the upper right.

Step 2 Raise the leftmost carbon atom (C4) *above* the ring plane.

Step 3 Lower the anomeric carbon atom (C1) *below* the ring plane.

Figure 14.6 shows how this is done for α-D-glucopyranose and β-D-glucopyranose.

Note that in β-D-glucopyranose, all the substituents on the ring are equatorial. Thus, β-D-glucopyranose is the least sterically crowded and most stable of the eight D aldohexoses.

PROBLEM

14.15 Draw β-D-galactopyranose in its chair conformation. Label all the ring substituents as axial or equatorial.

PROBLEM

14.16 Draw β-D-mannopyranose in its chair conformation, and label all substituents as axial or equatorial. Which would you expect to be more stable, mannose or galactose (Problem 14.15)?

14.8 Reactions of Monosaccharides

Since monosaccharides contain only two kinds of functional groups, carbonyls and hydroxyls, most of the chemistry of monosaccharides is the now familiar chemistry of these two groups.

Ester and Ether Formation

Monosaccharides behave as simple alcohols in much of their chemistry. For example, carbohydrate –OH groups can be converted into esters and ethers, which are often much easier to work with than the free sugars. Because of their many –OH groups, monosaccharides are usually soluble in water but insoluble in organic solvents. Ester and ether derivatives, however, are soluble in organic solvents and are easily crystallized.

Esterification is carried out by treating the carbohydrate with an acid chloride or acid anhydride in the presence of a base (Sections 10.7 and 10.8).

Stereo View

Oxygen at upper right

α-D-Glucopyranose

Stereo View

β-D-Glucopyranose

Figure 14.6 Chair representations of α-D-glucopyranose and β-D-glucopyranose.

All the –OH groups react, including the anomeric one. For example, β-D-glucopyranose is converted into its pentaacetate by treatment with acetic anhydride in pyridine solution:

$(CH_3CO)_2O$, Pyridine, 0°C

β-D-Glucopyranose → **Penta-*O*-acetyl-β-D-glucopyranose (91%)**

Carbohydrates are converted into ethers by treatment with an alkyl halide in the presence of base (the Williamson ether synthesis; Section 8.6). Silver oxide is a particularly mild and useful base for this reaction, since hydroxide and alkoxide bases tend to degrade the sensitive sugar molecules. For example, α-D-glucopyranose is converted into its pentamethyl ether in 85% yield on reaction with iodomethane and silver oxide:

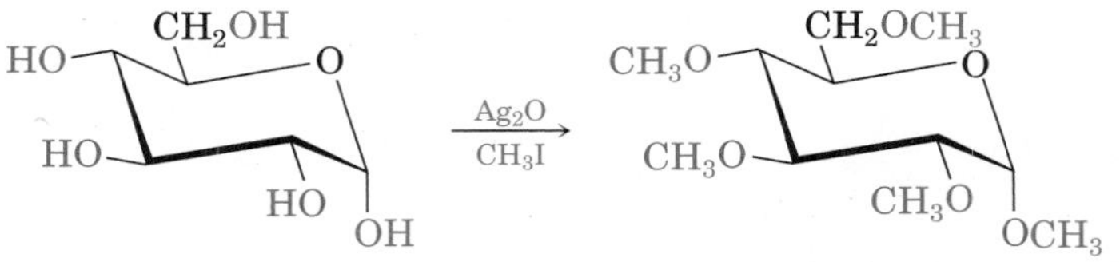

$\xrightarrow[CH_3I]{Ag_2O}$

α-D-Glucopyranose **α-D-Glucopyranose pentamethyl ether (85%)**

PROBLEM

14.17 Draw the products you would obtain by reaction of β-D-ribofuranose with:
(a) CH_3I, Ag_2O (b) $(CH_3CO)_2O$, pyridine

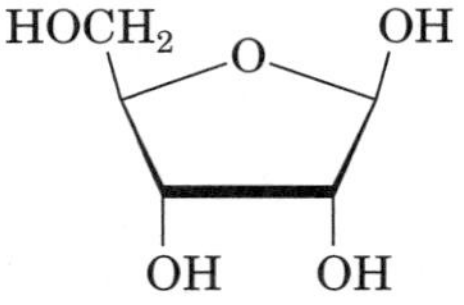

β-D-Ribofuranose

Glycoside Formation

We saw in Section 9.8 that treatment of a hemiacetal with an alcohol and an acid catalyst yields an acetal:

$$\text{C(OH)(OR)} + \text{ROH} \xrightleftharpoons{\text{HCl}} \text{C(OR)(OR)} + H_2O$$

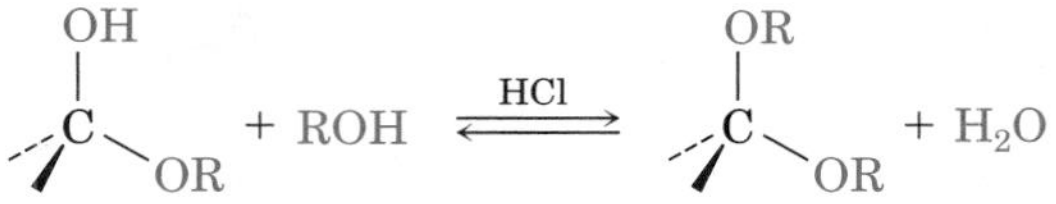

In the same way, treatment of a monosaccharide hemiacetal with an alcohol and an acid catalyst yields an acetal in which the anomeric –OH group has been replaced by an –OR group. For example, reaction of glucose with methanol gives methyl β-D-glucopyranoside:

$\xrightleftharpoons[H_3O^+]{CH_3OH,\ HCl}$ $+ H_2O$

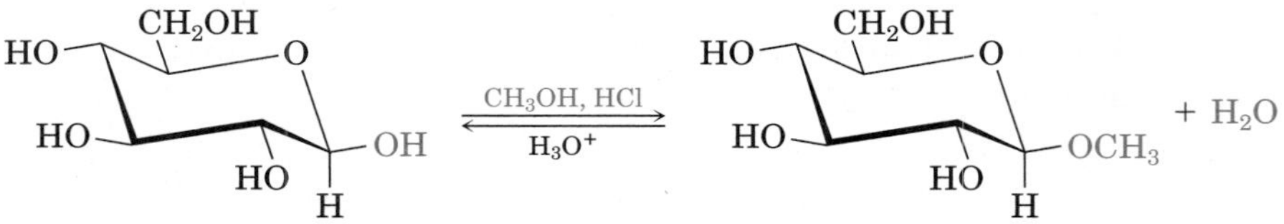

β-D-Glucopyranose (a hemiacetal) **Methyl β-D-glucopyranoside (an acetal)**

Called **glycosides,** carbohydrate acetals are named by first citing the alkyl group and then replacing the *-ose* ending of the sugar with *-oside.* Like all acetals, glycosides are stable to water. They aren't in equilibrium with an open-chain form, and they don't show mutarotation. They can, however, be converted back to the free monosaccharide by hydrolysis with aqueous acid.

Glycosides are common in nature, and a great many biologically active molecules contain glycosidic linkages. For example, digitoxin, the active component of the digitalis preparations used for treatment of heart disease, is a glycoside consisting of a complex steroid alcohol linked to a trisaccharide (Figure 14.7). Note that the three sugars are also linked by glycoside bonds.

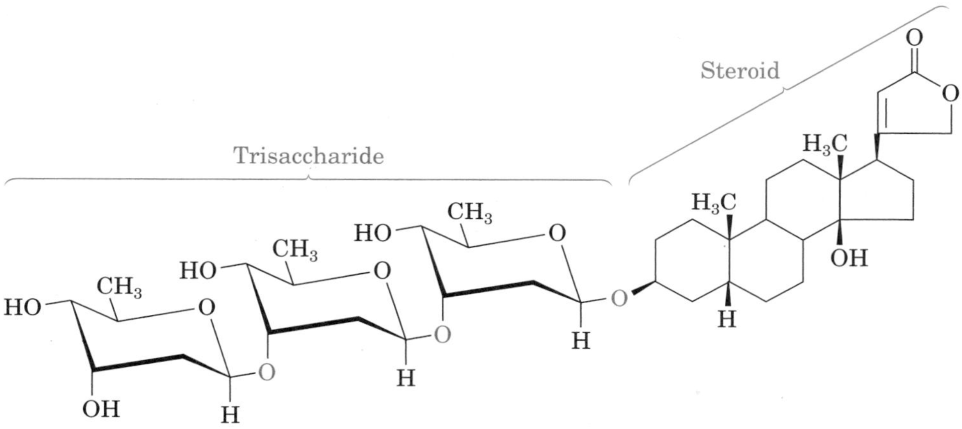

Steroid

Trisaccharide

Figure 14.7 The structure of digitoxin, a complex glycoside.

PRACTICE PROBLEM 14.7

What product would you expect from the acid-catalyzed reaction of β-D-ribofuranose with methanol?

Solution The acid-catalyzed reaction of a monosaccharide with an alcohol yields a glycoside in which the anomeric –OH group is replaced by the –OR group of the alcohol:

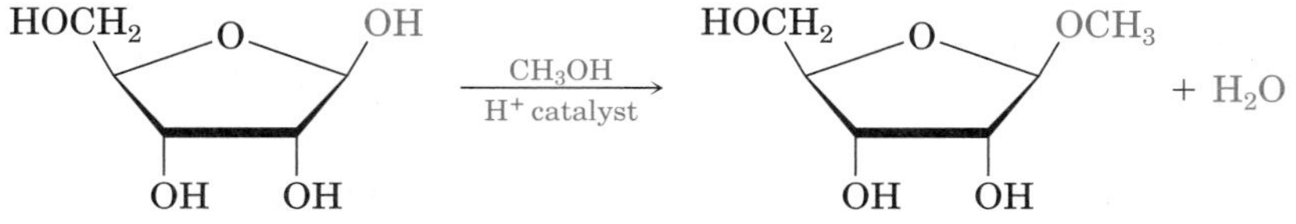

$$\text{β-D-Ribofuranose} \xrightarrow[\text{H}^+ \text{ catalyst}]{CH_3OH} \text{Methyl β-D-ribofuranoside} + H_2O$$

β-D-Ribofuranose **Methyl β-D-ribofuranoside**

PROBLEM

14.18 Draw the product you would obtain from the acid-catalyzed reaction of β-D-galactopyranose with ethanol.

Reduction of Monosaccharides

Treatment of an aldose or a ketose with $NaBH_4$ reduces it to a polyalcohol called an **alditol.** The reaction occurs by reaction of the open-chain form present in the aldehyde ⇌ hemiacetal equilibrium.

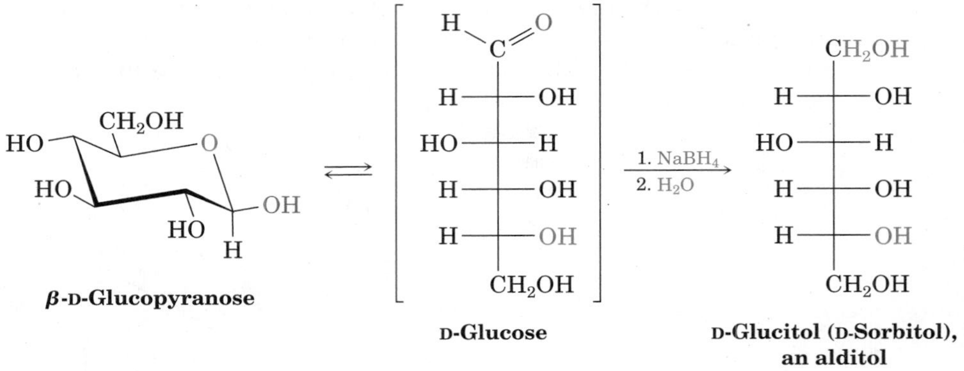

1. $NaBH_4$
2. H_2O

***β*-D-Glucopyranose**

D-Glucose

D-Glucitol (D-Sorbitol), an alditol

D-Glucitol, the alditol produced on reduction of D-glucose, is itself a naturally occurring substance that has been isolated from many fruits and berries. It is used under the name D-sorbitol as a sweetener and sugar substitute in many foods.

PRACTICE PROBLEM 14.8

Show the structure of the alditol you would obtain from reduction of D-galactose.

Solution First, draw D-galactose in its open-chain form. Then convert the –CHO group at C1 into a $–CH_2OH$ group.

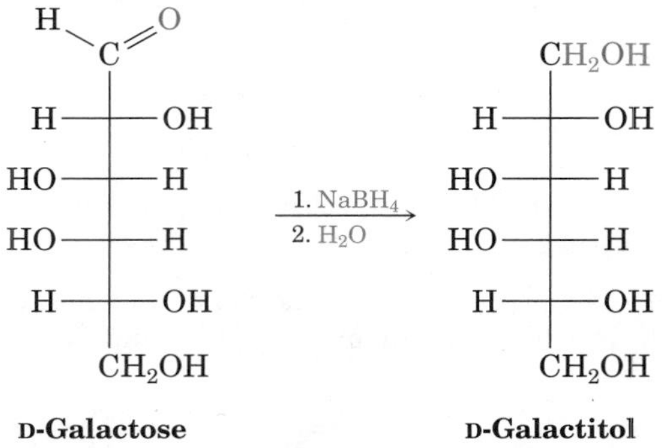

1. $NaBH_4$
2. H_2O

D-Galactose

D-Galactitol

PROBLEM

14.19 How can you account for the fact that reduction of D-glucose leads to an optically active alditol (D-glucitol) whereas reduction of D-galactose leads to an optically inactive alditol (see Section 6.8)?

PROBLEM

14.20 Reduction of L-gulose with $NaBH_4$ leads to the same alditol (D-glucitol) as reduction of D-glucose. Explain.

Oxidation of Monosaccharides

Like other aldehydes, aldoses are easily oxidized to yield the corresponding carboxylic acids, called **aldonic acids.** Aldoses react with Tollens' reagent (Ag^{+} in aqueous ammonia), Fehling's reagent (Cu^{2+} with aqueous sodium tartrate), and Benedict's reagent (Cu^{2+} with aqueous sodium citrate) to yield the oxidized sugar and a reduced metallic species. All three reactions serve as simple chemical tests for what are called **reducing sugars** (*reducing,* because the sugar reduces the metallic oxidizing agent).

β-D-Galactose ⇌ [open-chain form] —Tollens' reagent or Fehling's reagent→ **D-Galactonic acid (an aldonic acid)**

If Tollens' reagent is used, metallic silver is produced as a shiny mirror on the walls of the reaction flask or test tube (Section 9.5). If Fehling's or Benedict's reagent is used, a reddish precipitate of Cu_2O signals a positive result. Some diabetes self-test kits sold in drugstores for home use employ Benedict's test. As little as 0.1% glucose in urine gives a positive test.

All aldoses are reducing sugars because they contain aldehyde carbonyl groups, but glycosides are nonreducing. Glycosides don't react with Tollens' or Fehling's reagents because the acetal group can't open to an aldehyde under basic conditions.

If warm dilute HNO_3 (nitric acid) is used as the oxidizing agent, aldoses are oxidized to dicarboxylic acids called **aldaric acids.** Both the aldehyde carbonyl and the terminal –CH_2OH group are oxidized in this reaction.

β-D-Glucose ⇌ [open-chain form] —HNO_3, Δ→ **D-Glucaric acid (an aldaric acid)**

PROBLEM

14.21 D-Glucose yields an optically active aldaric acid on treatment with nitric acid, but D-allose yields an optically inactive aldaric acid. Explain.

PROBLEM

14.22 Which of the other six D aldohexoses yield optically active aldaric acids, and which yield optically inactive aldaric acids? (See Problem 14.21.)

14.9 Disaccharides

We saw in the previous section that reaction of a monosaccharide hemiacetal yields a glycoside in which the anomeric –OH group is replaced by an –OR substituent. If the alcohol is another sugar, the glycoside product is a disaccharide.

Cellobiose and Maltose

Disaccharides can contain a glycosidic acetal bond between the anomeric carbon of one sugar and an –OH group at *any* position on the other sugar. A glycosidic link between C1 of the first sugar and C4 of the second sugar, called a **1,4′ link,** is particularly common. (The prime indicates that the 4′ position is on a sugar other than the nonprime 1 position.)

A glycosidic bond to the anomeric carbon can be either α or β. *Maltose,* the disaccharide obtained by partial hydrolysis of starch, consists of two D-glucopyranoses joined by a 1,4′-α-glycoside bond. *Cellobiose,* the disaccharide obtained by partial hydrolysis of cellulose, consists of two D-glucopyranoses joined by a 1,4′-β-glycoside bond.

Stereo View

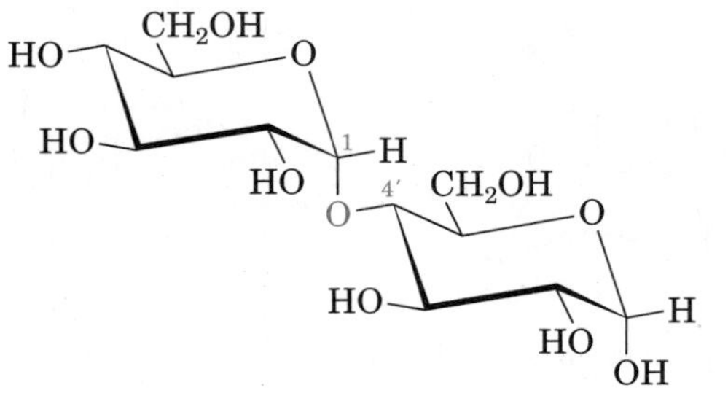

HO, CH_2OH, O, HO, HO, 1, H, O, 4′, CH_2OH, O, HO, HO, H, OH

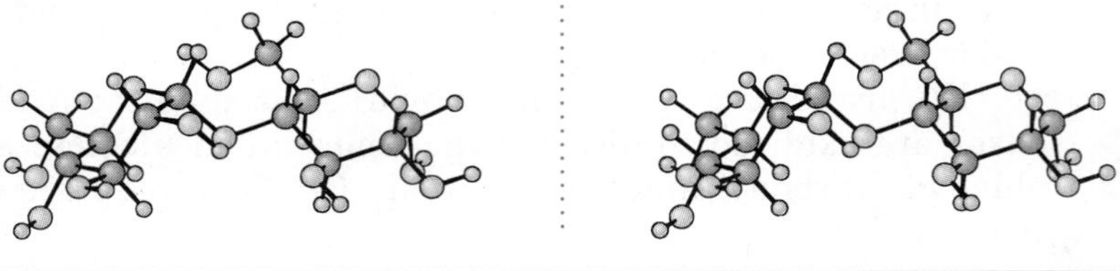

Maltose, a 1,4′-α-glycoside
[4-*O*-(α-D-Glucopyranosyl)-α-D-glucopyranose]

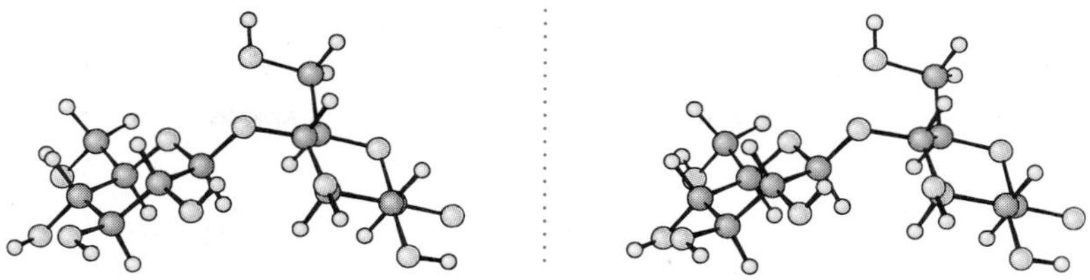

Stereo View

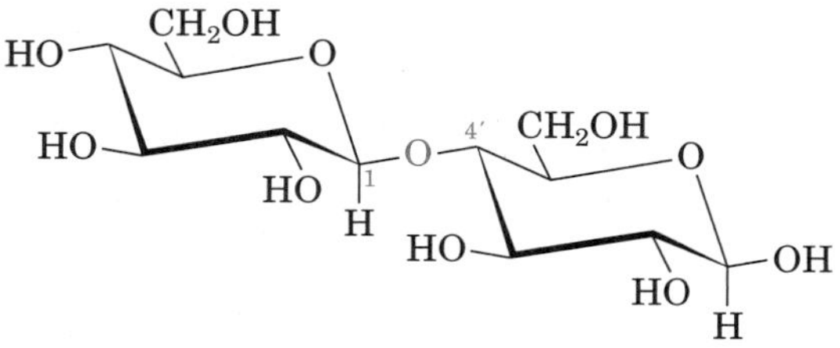

Cellobiose, a 1,4′-β-glycoside
[4-*O*-(β-D-Glucopyranosyl)-β-D-glucopyranose]

Both maltose and cellobiose are reducing sugars because the right-hand saccharide unit in each has a hemiacetal group. Both are therefore in equilibrium with aldehyde forms, which can reduce Tollens' or Fehling's reagent. For a similar reason, both maltose and cellobiose exhibit mutarotation.

Despite the similarities of their structures, maltose and cellobiose are dramatically different biologically. Cellobiose can't be digested by humans and can't be fermented by yeast. Maltose, however, is digested without difficulty and is readily fermented.

PROBLEM

14.23 Draw the structures of the products obtained from reaction of cellobiose with the following reagents:
(a) $NaBH_4$ (b) $AgNO_3$, H_2O, NH_3

Sucrose

Sucrose, or ordinary table sugar, is probably the most abundant pure organic chemical in the world. Whether from sugar cane (20% by weight) or from sugar beets (15% by weight), and whether raw or refined, all table sugar is sucrose.

Sucrose is a disaccharide that yields 1 equivalent of glucose and 1 equivalent of fructose on hydrolysis. This 1:1 mixture of glucose and fructose is often referred to as *invert sugar* because the sign of optical rotation changes (inverts) during the hydrolysis from sucrose, $[\alpha]_D = +66.5°$, to a glucose/fructose mixture, $[\alpha]_D = -22°$. Insects such as honeybees have enzymes called *invertases* that catalyze the hydrolysis of sucrose to glucose + fructose. Honey, in fact, is primarily a mixture of glucose, fructose, and sucrose.

Unlike most other disaccharides, sucrose is not a reducing sugar and does not exhibit mutarotation. These observations imply that sucrose has no hemiacetal group and suggest that the glucose and fructose units must *both* be glycosides. This can happen only if the two sugars are joined by a glycoside link between their anomeric carbons, C1 of glucose and C2 of fructose.

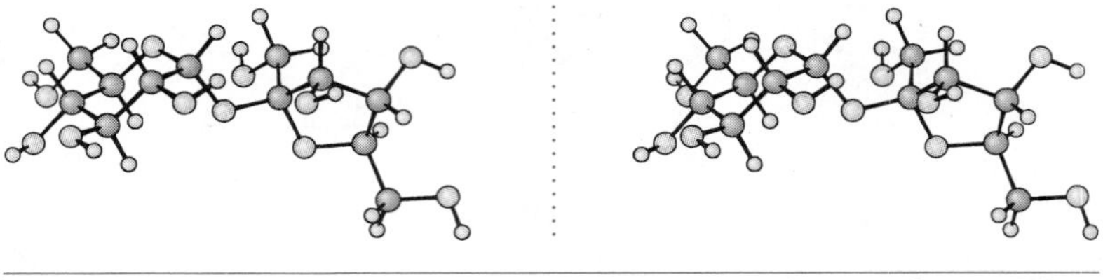

Stereo View

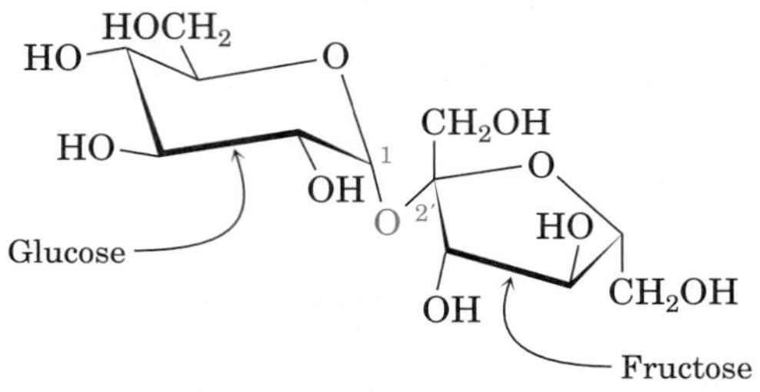

Sucrose, a 1,2′-glycoside
[2-*O*-(α-D-Glucopyranosyl)-β-D-fructofuranoside]

14.10 Polysaccharides

Polysaccharides are carbohydrates in which tens, hundreds, or even thousands of simple sugars are linked by glycoside bonds. Since these compounds have no free anomeric –OH groups (except for one at the end of the chain), they aren't reducing sugars and don't show mutarotation. Cellulose and starch are the two most widely occurring polysaccharides.

Cellulose

Cellulose consists of D-glucose units linked by the 1,4′-β-glycoside bonds we saw in cellobiose. Several thousand glucose units are linked to form one large molecule, and different molecules can then interact to form a large aggregate structure held together by hydrogen bonds:

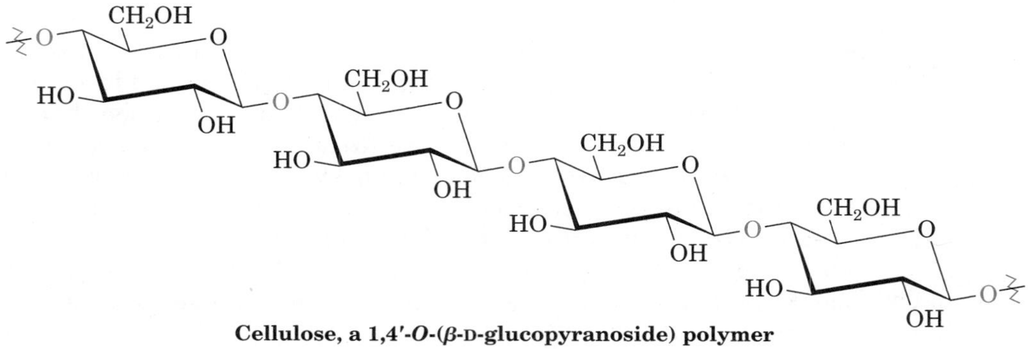

Cellulose, a 1,4′-*O*-(β-D-glucopyranoside) polymer

Nature uses cellulose primarily as a structural material to impart strength and rigidity to plants. Wood, leaves, grasses, and cotton are primarily cellulose. Cellulose also serves as a raw material for the manufacture of cellulose acetate, known commercially as acetate rayon:

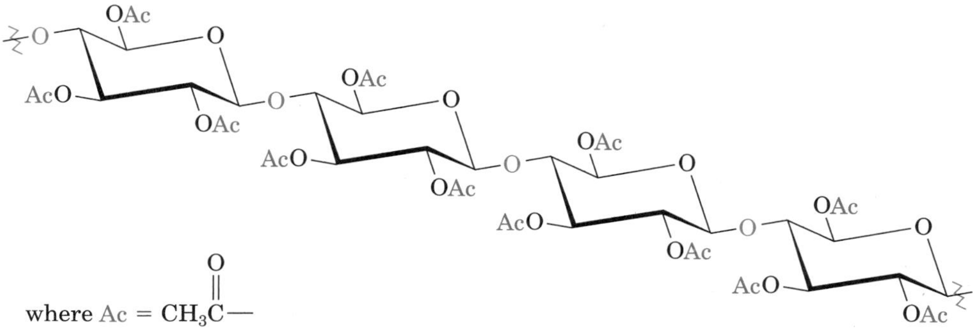

where Ac = $CH_3C(=O)—$

A segment of cellulose acetate (acetate rayon)

Starch and Glycogen

Starch, a glucose polymer whose monosaccharide units are linked by the 1,4′-α-glycoside bonds we saw in maltose, can be separated into two fractions called *amylopectin* and *amylose.* Amylose, which accounts for about 20% by weight of starch, consists of several hundred glucose molecules linked by 1,4′-α-glycoside bonds:

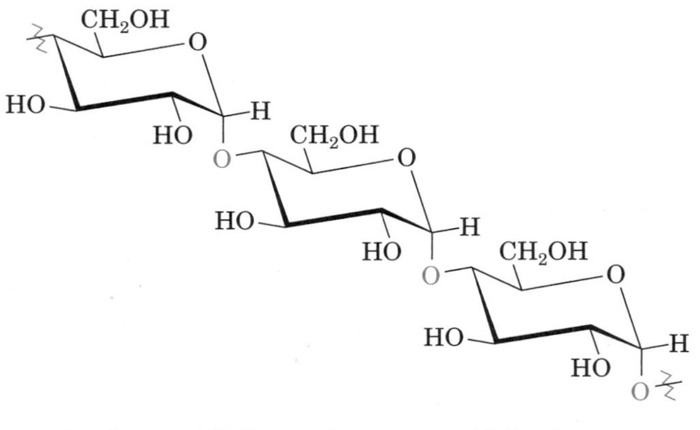

Amylose, a 1,4′-*O*-(α-D-glucopyranoside) polymer

Amylopectin, which accounts for the remaining 80% of starch, is more complex in structure than amylose. Unlike cellulose or amylose, which are linear polymers, amylopectin contains 1,6′-α-glycoside *branches* approximately every 25 glucose units. As a result, amylopectin has an exceedingly complex three-dimensional structure (Figure 14.8, p. 464).

Nature uses starch as the medium by which plants store energy for later use; potatoes, corn, and cereal grains all contain large amounts of starch. When eaten, starch is digested in the mouth and stomach by enzymes called *glycosidases,* which catalyze the hydrolysis of glycoside bonds

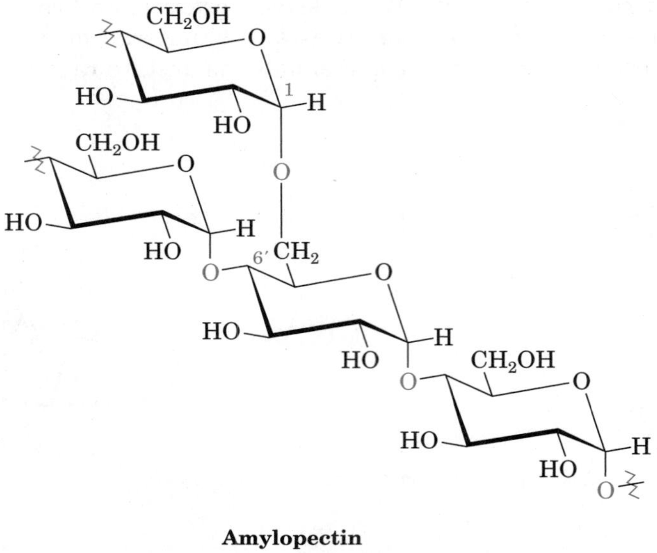

Amylopectin

Figure 14.8 A 1,6′-α-glycoside branch in amylopectin.

and release individual molecules of glucose. Like most enzymes, glycosidases are highly selective in their action. They hydrolyze only the α-glycoside links in starch and leave the β-glycoside links in cellulose untouched. Thus, humans can eat potatoes and grains but not grass.

Glycogen is a polysaccharide that serves the same energy-storage function in animals that starch serves in plants. Dietary carbohydrate not needed for immediate energy is converted by the body to glycogen for long-term storage. Like the amylopectin found in starch, glycogen has a complex three-dimensional structure with both 1,4′ and 1,6′ links (Figure 14.9). Glycogen molecules are larger than those of amylopectin—up to 100,000 glucose units—and contain even more branches.

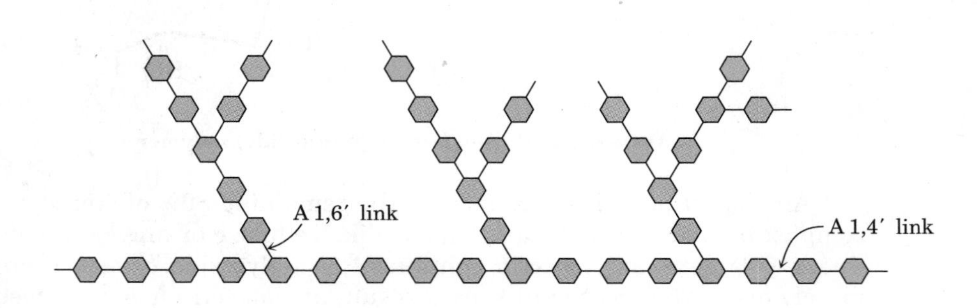

Figure 14.9 A representation of the structure of glycogen. The hexagons represent glucose units linked by 1,4′ and 1,6′ acetal bonds.

14.11 Other Important Carbohydrates

In addition to the common carbohydrates mentioned in previous sections, there are a variety of important carbohydrate-derived materials whose structures have been chemically modified. Their structural resemblance to sugars is clear, but they aren't simple aldoses or ketoses.

Deoxy sugars differ from other sugars by having one of their oxygen atoms "missing." In other words, an –OH group is replaced by an –H. The most common deoxy sugar is 2-deoxyribose, a monosaccharide found in DNA (deoxyribonucleic acid). Note that 2-deoxyribose adopts a furanose (five-membered) form:

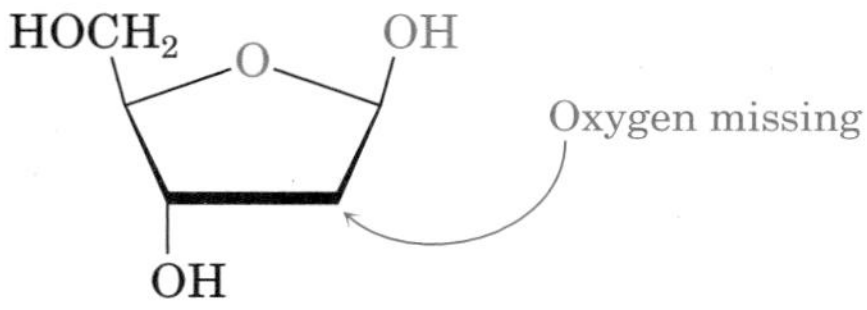

Oxygen missing

2-Deoxyribose

Amino sugars, such as D-glucosamine, have one of their –OH groups replaced by an $-NH_2$. The *N*-acetyl amide derived from D-glucosamine is the monosaccharide unit from which *chitin,* the hard crust that protects insects and shellfish, is made. Still other amino sugars are found in antibiotics such as streptomycin and gentamicin.

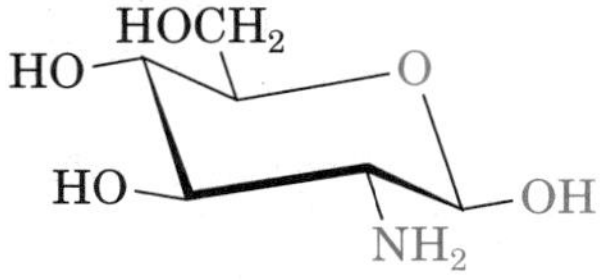

β-D-Glucosamine
(an amino sugar)

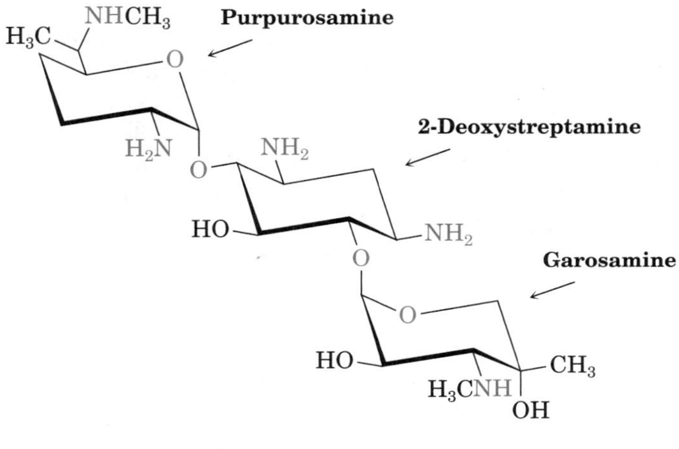

Purpurosamine

2-Deoxystreptamine

Garosamine

Gentamicin
(an antibiotic)

14.12 Cell-Surface Carbohydrates

Carbohydrates were once thought to be uninteresting compounds whose only biological purposes were to serve as structural materials and as energy sources. Although carbohydrates do indeed fill these two roles, it's now known that they have many other important biochemical functions as well. Polysaccharides are centrally involved in the critical process by which one cell type recognizes another. Small polysaccharide chains, covalently bound by glycoside links to –OH groups on proteins (*glycoproteins*), act as biochemical labels on cell surfaces, as illustrated by the human blood-group antigens.

It has been known for a century that human blood can be classified into four blood-group types—A, B, AB, and O—and that blood from a donor of one type can't be transfused into a recipient with another type unless the two types are compatible (Table 14.1). Should an incompatible mix be made, the red blood cells clump together, or *agglutinate.*

Table 14.1 Human Blood-Group Compatibilities

Donor blood type	*Acceptor blood type*			
	A	B	AB	O
A	o	×	o	×
B	×	o	o	×
AB	×	×	o	×
O	o	o	o	o

o = Compatible; × = Incompatible.

The agglutination of incompatible red blood cells, which indicates that the recipient's immune system has recognized the presence of foreign cells and has formed antibodies to them, results from the presence of polysaccharide markers on the surface of the cells. Type A, B, and O red blood cells each have characteristic markers called *antigenic determinants*; type AB cells have both type A and type B markers. The structures of all three blood-group determinants are shown in Figure 14.10.

Note that all three antigenic determinants contain *N*-acetylamino sugars as well as the unusual monosaccharide L-fucose. The antigenic determinant of blood group O is a trisaccharide, whereas those of blood groups A and B are tetrasaccharides. Type A and B determinants differ only in the substitution of an acetylamino group (–$NHCOCH_3$) for an –OH in the terminal galactose unit.

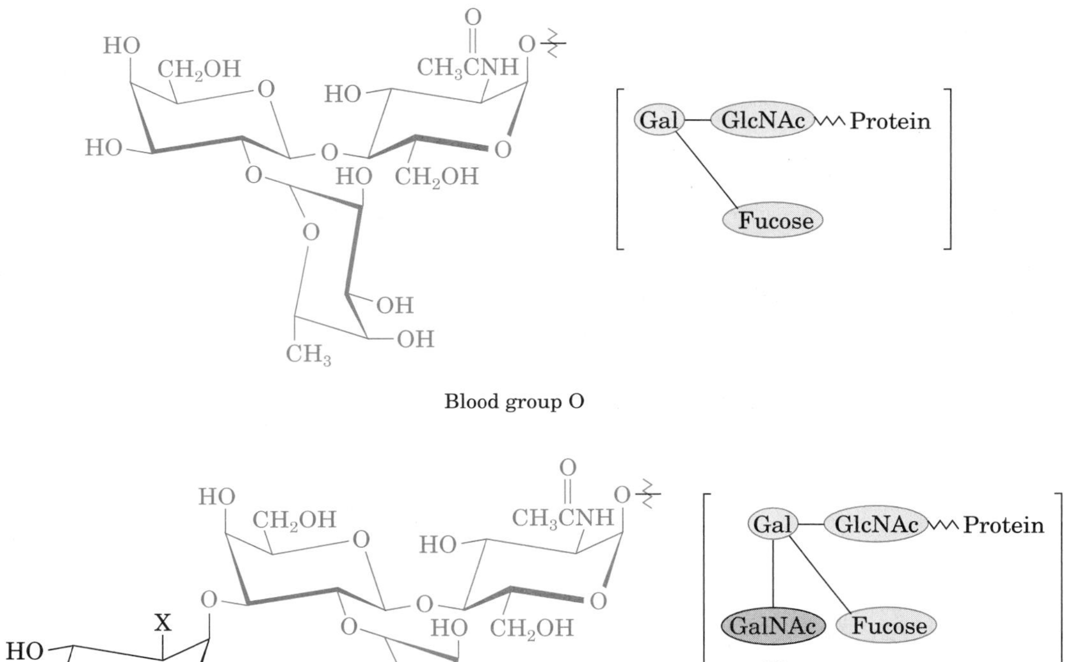

Figure 14.10 Structures of the A, B, and O blood-group antigenic determinants. (Gal is D-galactose; GlcNAc is *N*-acetylglucosamine; GalNAc is *N*-acetylgalactosamine.)

INTERLUDE

Sweetness

Dietary disaster!

Say the word *sugar* and most people immediately think of sweet-tasting candies, desserts, and such. In fact most of the simple carbohydrates *do* taste sweet, but the degree of sweetness varies greatly from one sugar to another. With sucrose (table sugar) as a reference point, fructose is nearly twice as sweet, but lactose is only about one-sixth as sweet. Comparisons are difficult, though, because sweetness is a matter of taste, and the ranking of sugars is a matter of personal opinion. Nevertheless, the ordering in Table 14.2 is generally accepted.

Table 14.2 Sweetness of Some Sugars and Sugar Substitutes

Name	*Type*	*Sweetness*
Lactose	Disaccharide	0.16
Glucose	Monosaccharide	0.75
Sucrose	Disaccharide	1.00
Fructose	Monosaccharide	1.75
Cyclamate	Synthetic	300
Aspartame	Synthetic	1500
Saccharin	Synthetic	3500

The desire of many people to cut their caloric intake has led to the development of synthetic sweeteners such as aspartame, saccharin, and cyclamate. All are far sweeter than natural sugars, but doubts have been raised as to their long-term safety. Cyclamate was banned briefly in the United States (but not in Canada), and saccharin has been banned in

(*continued*)➤

Canada (but not in the United States). None of the three has any structural resemblance to carbohydrates.

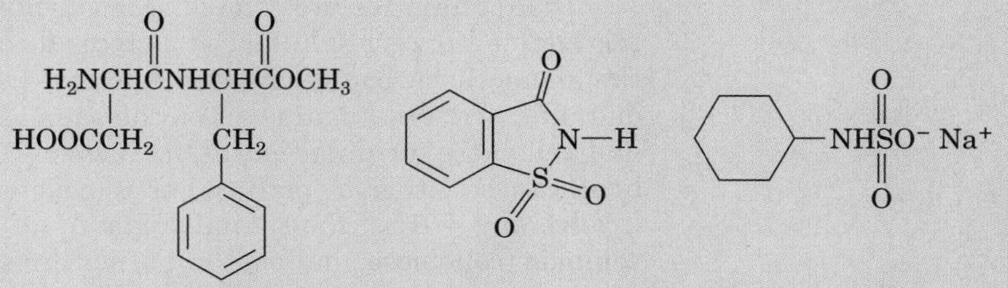

Aspartame **Saccharin** **Sodium cyclamate**

Summary and Key Words

aldaric acid, 459
alditol, 458
aldonic acid, 459
aldose, 441
alpha (α) anomer, 452
amino sugar, 465
anomer, 452
anomeric center, 452
beta (β) anomer, 452
carbohydrate, 440
complex carbohydrate; 441
deoxy sugar, 465
disaccharide, 441
Fischer projection, 443
furanose, 450
glycoside, 456
Haworth projection, 450
ketose, 441
1,4′ link, 460
monosaccharide, 441
mutarotation, 452
polysaccharide, 441
pyranose, 448
reducing sugar, 459
simple sugar, 441
D sugar, 446
L sugar, 446

Carbohydrates are polyhydroxy aldehydes and ketones. They are classified according to the number of carbon atoms and the kind of carbonyl group they contain. Thus, glucose is an *aldohexose,* a six-carbon aldehydo sugar. **Monosaccharides** are further classified as either **D** or **L sugars,** depending on the stereochemistry of the stereocenter farthest from the carbonyl group. Most naturally occurring sugars are in the D series.

Monosaccharides normally exist as cyclic hemiacetals rather than as open-chain aldehydes or ketones. The hemiacetal linkage results from reaction of the carbonyl group with an –OH group three or four carbon atoms away. A five-membered ring hemiacetal is a **furanose,** and a six-membered ring hemiacetal is a **pyranose.** Cyclization leads to the formation of a new stereocenter (the **anomeric center**) and to production of two diastereomeric hemiacetals called **alpha (α)** and **beta (β) anomers.**

Stereochemical relationships among monosaccharides are portrayed in several ways. **Fischer projections** display stereocenters as a pair of crossed lines; **Haworth projections** provide a more accurate view. Any group to the right in a Fischer projection is down in a Haworth projection.

Much of the chemistry of monosaccharides is the familiar chemistry of alcohol and carbonyl functional groups. Thus, the –OH groups of carbohydrates form esters and ethers in the normal way. The carbonyl group of a monosaccharide can be reduced with $NaBH_4$ to yield an **alditol,** can be oxidized with Tollens' or Fehling's reagent to yield an **aldonic acid,** can be oxidized with warm HNO_3 to yield an **aldaric acid,** and can be treated with an alcohol in the presence of acid catalyst to yield a **glycoside.**

Disaccharides are complex carbohydrates in which two simple sugars are linked by a glycoside bond between the anomeric carbon of one unit and an –OH of the second unit. The two sugars can be the same, as in maltose and cellobiose, or different, as in sucrose. The glycoside bond can be either α (maltose) or β (cellobiose) and can involve any –OH of the second sugar. A 1,4′ link is most common (cellobiose, maltose), but other links, such as 1,2′ (sucrose), also occur. **Polysaccharides,** such as cellulose, starch, and glycogen, are used in nature both as structural materials and for long-term energy storage.

ADDITIONAL PROBLEMS

14.24 Classify the following sugars by type (for example, glucose is an aldohexose):

(a) CH_2OH–C=O–CH_2OH

CH_2OH
C=O
CH_2OH

(b)

CH_2OH
H—C—OH
C=O
H—C—OH
CH_2OH

(c)

CHO
H—C—OH
HO—C—H
H—C—OH
HO—C—H
H—C—OH
CH_2OH

14.25 Write open-chain structures for a ketotetrose and a ketopentose.

14.26 Write an open-chain structure for a deoxyaldohexose.

14.27 Write an open-chain structure for a five-carbon amino sugar.

14.28 The structure of ascorbic acid (vitamin C) is shown. Does ascorbic acid have a D or L configuration?

OH
HO–C=C–C=O (ring closed through O)
H—C—O
HO—C—H
CH_2OH

Ascorbic acid

14.29 Draw a Haworth projection of ascorbic acid (see Problem 14.28).

14.30 Define the following terms, and give an example of each:

(a) Monosaccharide (b) Anomeric center (c) Haworth projection
(d) Fischer projection (e) Glycoside (f) Reducing sugar
(g) Pyranose form (h) 1,4′ Link (i) D-Sugar

14.31 The following cyclic structure is that of gulose. Is this a furanose or pyranose form? Is it an α or β anomer? Is it a D sugar or L sugar?

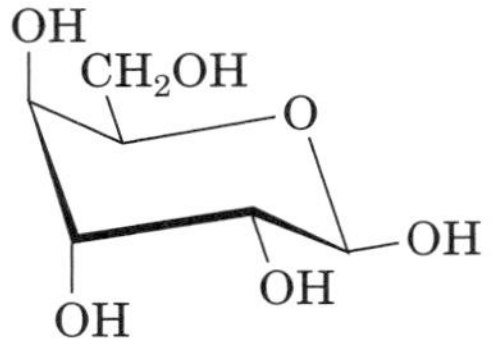

OH
CH_2OH
O
OH
OH
OH

Gulose

14.32 Uncoil gulose (see Problem 14.31), and write it in its open-chain form.

14.33 Draw D-ribulose in its five-membered cyclic β hemiacetal form.

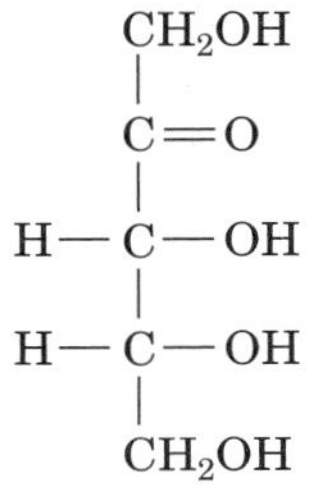

CH_2OH
|
C=O
|
H—C—OH
|
H—C—OH
|
CH_2OH

Ribulose

14.34 Look up the structure of D-talose in Figure 14.3, and draw the β anomer in its pyranose form. Identify the ring substituents as axial or equatorial.

14.35 Draw structures for the products you would expect to obtain from the reaction of β-D-talopyranose (see Problem 14.34) with each of the following reagents:
(a) $NaBH_4$
(b) Warm dilute HNO_3
(c) $AgNO_3$, NH_3, H_2O
(d) CH_3CH_2OH, H^+
(e) CH_3I, Ag_2O
(f) $(CH_3CO)_2O$, pyridine

14.36 What is the stereochemical relationship of D-allose to L-allose? What generalizations can you make about the following properties of the two sugars?
(a) Melting point (b) Solubility in water
(c) Specific rotation (d) Density

14.37 What is the stereochemical relationship of D-ribose to L-xylose? What generalizations can you make about the following properties of the two sugars?
(a) Melting point (b) Solubility in water
(c) Specific rotation (d) Density

14.38 How many D-2-ketohexoses are there? Draw them.

14.39 One of the D-2-ketohexoses (see Problem 14.38) is called *sorbose.* On treatment with $NaBH_4$, sorbose yields a mixture of gulitol and iditol. What is the structure of D-sorbose? (Gulitol and iditol are the alditols obtained by reduction of gulose and idose.)

14.40 Another D-2-ketohexose, *psicose,* yields a mixture of allitol and altritol when reduced with $NaBH_4$ (see Problems 14.38 and 14.39). What is the structure of psicose?

14.41 Fructose exists at equilibrium as an approximately 2:1 mixture of β-D-fructopyranose and β-D-fructofuranose. Draw both forms in Haworth projection.

14.42 Draw Fischer projections of the following substances:
(a) (*R*)-2-Methylbutanoic acid
(b) (*S*)-3-Methyl-2-pentanone

14.43 Convert the following Fischer projections into tetrahedral representations:

(a) Br at top; H—(C)—OCH_3; CH_3 at bottom

(b) CH_3 at top; H—(C)—NH_2; CH_2CH_3 at bottom

14.44 Which of the eight D aldohexoses yield optically inactive (meso) alditols on reduction with $NaBH_4$?

14.45 What other D aldohexose gives the same alditol as D-talose? (See Problem 14.44.)

14.46 Which of the eight D aldohexoses give the same aldaric acids as their L enantiomers?

14.47 Which of the other three D aldopentoses gives the same aldaric acid as D-lyxose?

14.48 The *Ruff degradation* is a method used to shorten an aldose chain by one carbon atom. The original C1 carbon atom is cut off, and the original C2 carbon atom becomes the aldehyde of the chain-shortened aldose. For example, D-glucose, an aldohexose, is converted by Ruff degradation into D-arabinose, an aldopentose. What other D aldohexose would also yield D-arabinose on Ruff degradation?

14.49 D-Galactose and D-talose yield the same aldopentose on Ruff degradation (Problem 14.48). What does this tell you about the stereochemistry of galactose and talose? Which D aldopentose is obtained?

14.50 The aldaric acid obtained by nitric acid oxidation of D-erythrose, one of the D aldotetroses, is optically inactive. The aldaric acid obtained from oxidation of the other D aldotetrose, D-threose, however, is optically active. How does this information allow you to assign structures to the two D aldotetroses?

14.51 Gentiobiose is a rare disaccharide found in saffron and gentian. It is a reducing sugar and forms only glucose on hydrolysis with aqueous acid. If gentiobiose contains a 1,6′-β-glycoside link, what is its structure?

14.52 The *cyclitols* are a group of carbocyclic sugar derivatives with the general formula 1,2,3,4,5,6-cyclohexanehexaol—that is, a cyclohexane ring with one –OH on each carbon. Draw the structures of the nine stereoisomeric cyclitols in Haworth projections.

14.53 Raffinose, a trisaccharide found in sugar beets, is formed by a 1,6′ α linkage of D-galactose to the glucose unit of sucrose. Draw the structure of raffinose.

14.54 Is raffinose (see Problem 14.53) a reducing sugar? Explain.

14.55 Glucose and fructose can be interconverted by treatment with dilute aqueous NaOH. Propose a mechanism (see Section 11.1).

Visualizing Chemistry

14.56 Identify the following aldoses (see Figure 14.3), and tell whether each is a D or L sugar (gray = C, red = O, light green = H).

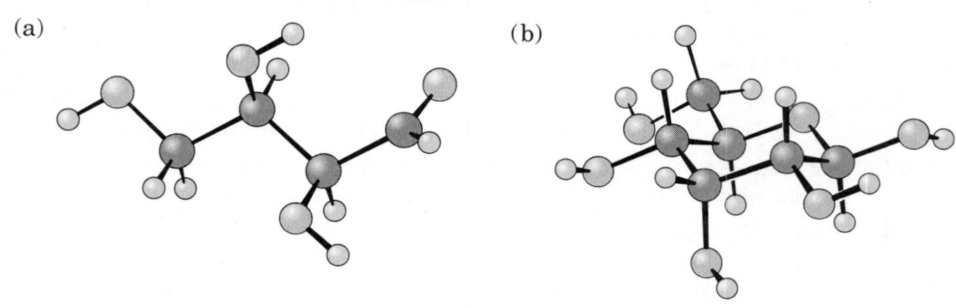

14.57 Draw Fischer projections of the following molecules, placing the carbonyl group at the top in the usual way. Identify each as a D or L sugar (gray = C, red = O, light green = H).

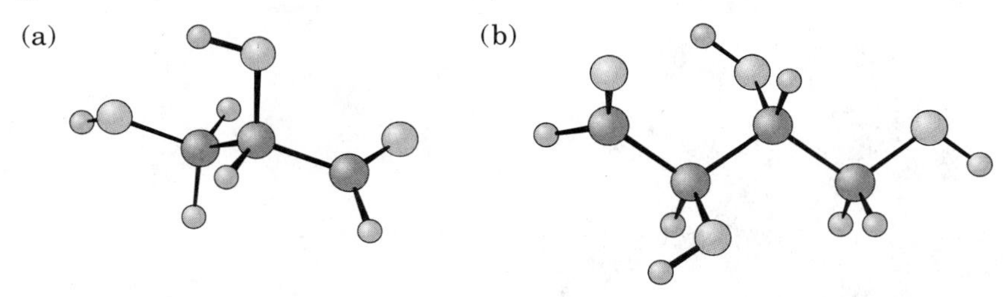

14.58 The following structure is that of an L aldohexose in its pyranose form. Identify it (see Figure 14.3). (Gray = C, red = O, light green = H.)

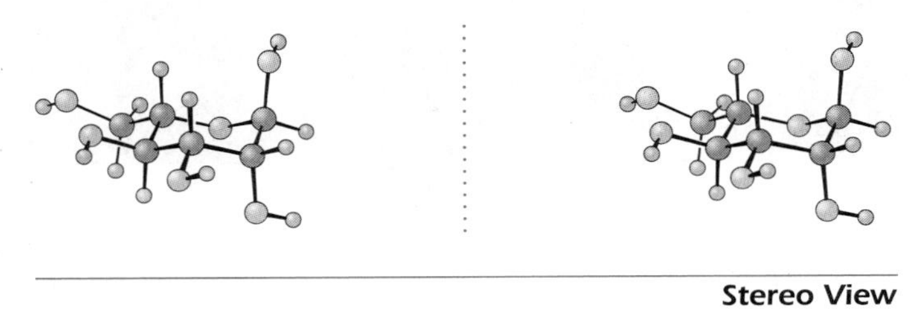

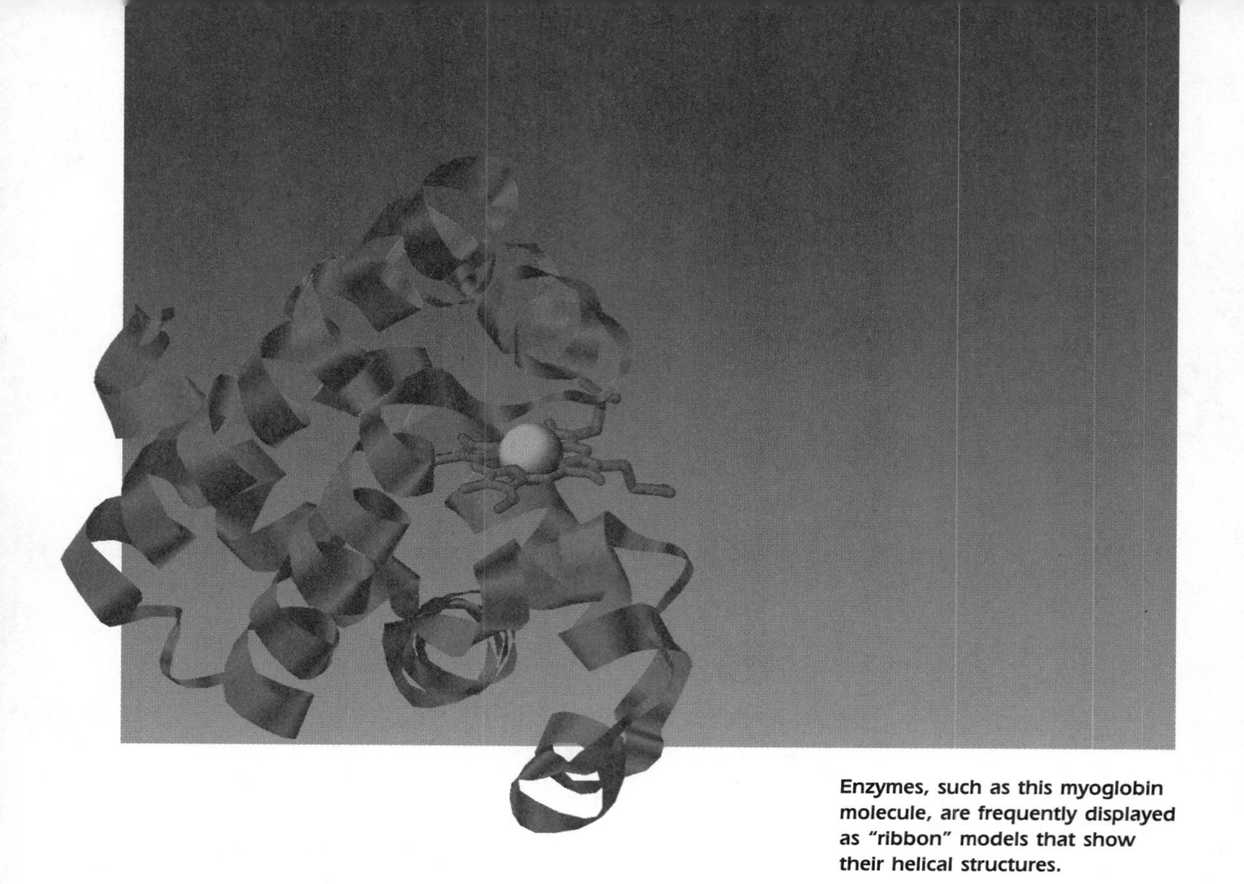

Enzymes, such as this myoglobin molecule, are frequently displayed as "ribbon" models that show their helical structures.

15 Biomolecules: Amino Acids, Peptides, and Proteins

Proteins are large biomolecules that occur in every living organism. They are of many types and have many biological functions. The keratin of skin and fingernails, the insulin that regulates glucose metabolism in the body, and the DNA polymerase that catalyzes the synthesis of DNA in cells are all proteins. Regardless of their appearance or function, all proteins are chemically similar. All are made up of many *amino acid* units linked together by amide bonds in a long chain.

Amino acids, as their name implies, are difunctional. They contain both a basic amino group and an acidic carboxyl group:

H_3C H

C

H_2N C=O

OH

Alanine, an amino acid

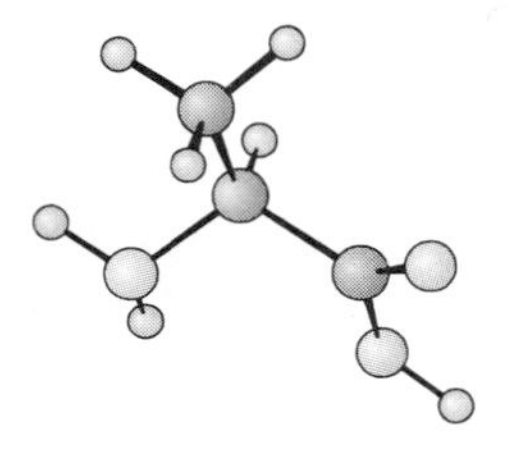

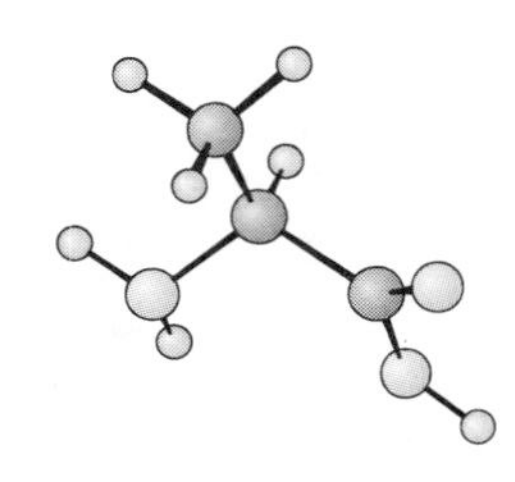

Stereo View

Their value as building blocks for proteins stems from the fact that large numbers of amino acids can link together by forming amide bonds between the $-NH_2$ of one amino acid and the $-COOH$ of another. For classification purposes, chains with fewer than 50 amino acids are usually called **peptides,** while the term *protein* is used for longer chains.

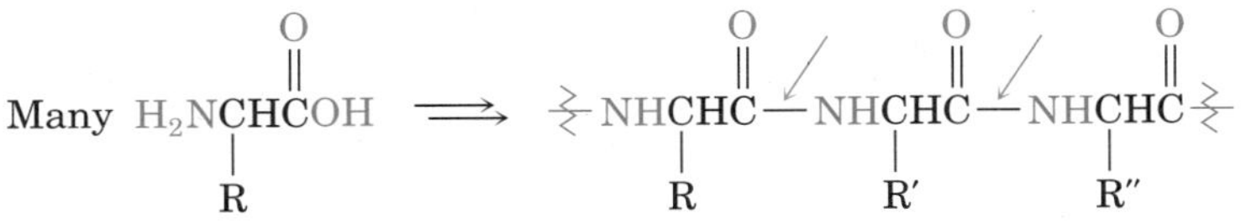

Many $H_2NCH(R)COOH$ ⟹ ~NHCH(R)C(=O)—NHCH(R′)C(=O)—NHCH(R″)C(=O)~

A polypeptide (many amide bonds)

15.1 Structures of Amino Acids

The structures of the 20 amino acids commonly found in proteins are shown in Table 15.1 (p. 476). All 20 are ***α*-amino acids,** meaning that the amino group in each is attached to the α carbon atom—the one next to the carbonyl group. Note that 19 of the 20 amino acids are primary amines, RNH_2, and differ only in the nature of their side chains. Proline, however, is a secondary amine whose nitrogen and α carbon atom are part of a five-membered ring.

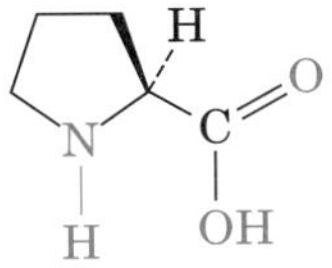

Proline, a secondary *α*-amino acid

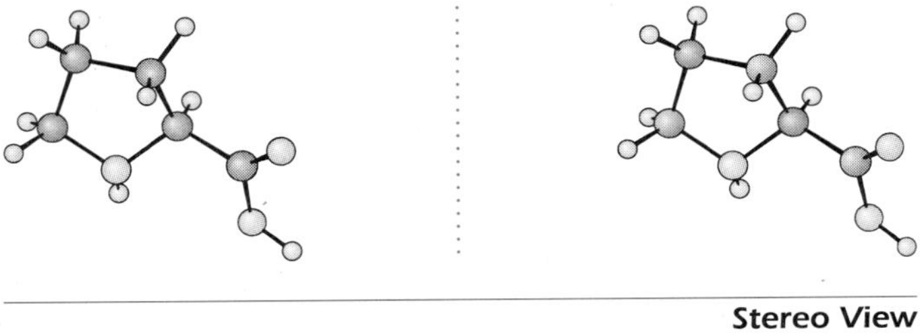

Stereo View

Note also that each of the amino acids in Table 15.1 is referred to by a three-letter shorthand code: Ala for alanine, Gly for glycine, and so on. In addition, a one-letter code, shown in parentheses in the table, is frequently used.

With the exception of glycine, H_2NCH_2COOH, the α carbons of the amino acids are stereocenters. Two enantiomeric forms of each amino acid are therefore possible, but nature uses only a single enantiomer to build proteins. In Fischer projections, naturally occurring amino acids are represented by placing the carboxyl group at the top as if drawing a carbohydrate (Section 14.2) and then placing the amino group on the left. Because of their stereochemical similarity to L sugars (Section 14.3), the naturally occurring α-amino acids are often referred to as L-amino acids.

Table 15.1 Structures of the 20 Common Amino Acids Found in Proteins (Names of the amino acids essential to the human diet are shown in red.)

Name	*Abbreviations*		*Molecular weight*	*Structure*	*Isoelectric point*	pK_{a1} α-COOH	pK_{a2} α-NH_3^+
Neutral Amino Acids							
Alanine	Ala	(A)	89	$CH_3CH(NH_2)COOH$	6.00	2.34	9.69
Asparagine	Asn	(N)	132	$H_2NC(=O)CH_2CH(NH_2)COOH$	5.41	2.02	8.80
Cysteine	Cys	(C)	121	$HSCH_2CH(NH_2)COOH$	5.07	1.96	10.28
Glutamine	Gln	(Q)	146	$H_2NC(=O)CH_2CH_2CH(NH_2)COOH$	5.65	2.17	9.13
Glycine	Gly	(G)	75	$CH_2(NH_2)COOH$	5.97	2.34	9.60
Isoleucine	Ile	(I)	131	$CH_3CH_2CH(CH_3)CH(NH_2)COOH$	6.02	2.36	9.60
Leucine	Leu	(L)	131	$CH_3CH(CH_3)CH_2CH(NH_2)COOH$	5.98	2.36	9.60
Methionine	Met	(M)	149	$CH_3SCH_2CH_2CH(NH_2)COOH$	5.74	2.28	9.21
Phenylalanine	Phe	(F)	165	C_6H_5–$CH_2CH(NH_2)COOH$	5.48	1.83	9.13
Proline	Pro	(P)	115	pyrrolidine ring (N–H) bearing C(=O)OH	6.30	1.99	10.60
Serine	Ser	(S)	105	$HOCH_2CH(NH_2)COOH$	5.68	2.21	9.15

Table 15.1 (continued)

Name	*Abbreviations*		*Molecular weight*	*Structure*	*Isoelectric point*	pK_{a1} α-COOH	pK_{a2} α-NH_3^+
Threonine	Thr	(T)	119	$CH_3CH(OH)CH(NH_2)COOH$	5.60	2.09	9.10
Tryptophan	Trp	(W)	204	indol-3-yl–$CH_2CH(NH_2)COOH$ (indole N–H)	5.89	2.83	9.39
Tyrosine	Tyr	(Y)	181	HO–C_6H_4–$CH_2CH(NH_2)COOH$	5.66	2.20	9.11
Valine	Val	(V)	117	$CH_3CH(CH_3)CH(NH_2)COOH$	5.96	2.32	9.62
Acidic Amino Acids							
Aspartic acid	Asp	(D)	133	$HOOCCH_2CH(NH_2)COOH$	2.77	1.88	9.60
Glutamic acid	Glu	(E)	147	$HOOCCH_2CH_2CH(NH_2)COOH$	3.22	2.19	9.67
Basic Amino Acids							
Arginine	Arg	(R)	174	$H_2NC(=NH)NHCH_2CH_2CH_2CH(NH_2)COOH$	10.76	2.17	9.04
Histidine	His	(H)	155	imidazol-4-yl–$CH_2CH(NH_2)COOH$ (ring N–H)	7.59	1.82	9.17
Lysine	Lys	(K)	146	$H_2NCH_2CH_2CH_2CH_2CH(NH_2)COOH$	9.74	2.18	8.95

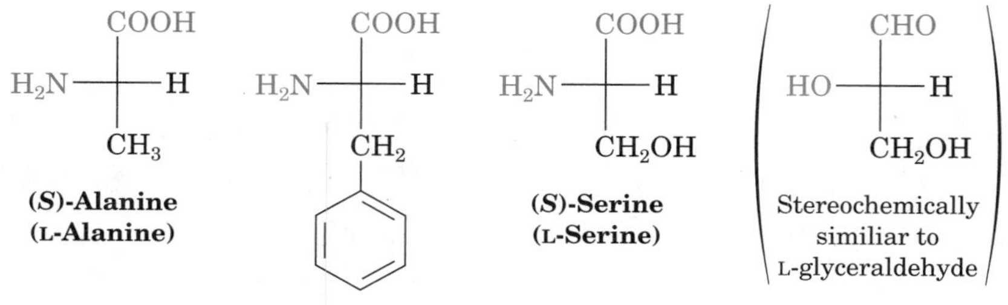

COOH / H_2N—C—H / CH_3
(*S*)-Alanine
(L-Alanine)

COOH / H_2N—C—H / CH_2—C₆H₅
(*S*)-Phenylalanine
(L-Phenylalanine)

COOH / H_2N—C—H / CH_2OH
(*S*)-Serine
(L-Serine)

CHO / HO—C—H / CH_2OH
Stereochemically similiar to L-glyceraldehyde

The 20 common amino acids can be further classified as either neutral, acidic, or basic, depending on the structure of their side chains. Fifteen of the 20 have neutral side chains, but 2 (aspartic acid and glutamic acid) have an extra carboxylic acid function in their side chain, and 3 (lysine, arginine, and histidine) have a basic amino group in their side chain.

All 20 of the amino acids are necessary for protein synthesis, but humans can synthesize only 10 of the 20. The other 10 are called *essential amino acids* because they must be obtained from food. Failure to include an adequate dietary supply of any of these essential amino acids leads to poor growth and general failure to thrive.

PROBLEM

15.1 Look carefully at the 20 amino acids in Table 15.1. How many contain aromatic rings? How many contain sulfur? How many are alcohols? How many have hydrocarbon side chains?

PROBLEM

15.2 Eighteen of the 19 L-amino acids have the *S* configuration at the α carbon. Cysteine is the only L-amino acid that has an *R* configuration. Explain.

PROBLEM

15.3 Draw L-alanine in the standard three-dimensional format using solid, wedged, and dashed lines.

15.2 Dipolar Structure of Amino Acids

Since amino acids contain both acidic and basic groups in the same molecule, they undergo an internal acid–base reaction and exist primarily in the form of a dipolar ion, or **zwitterion** (German *zwitter,* "hybrid").

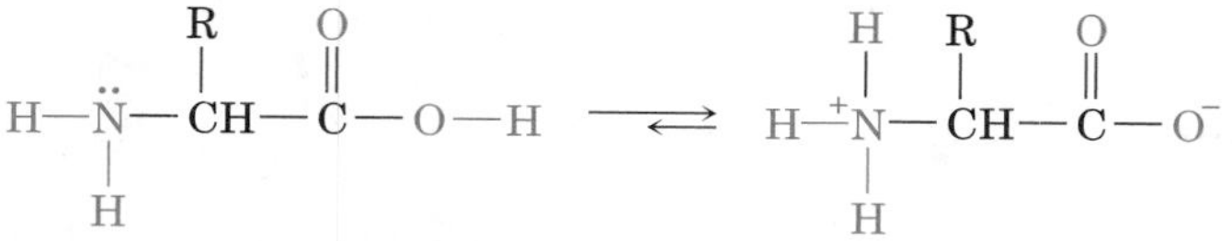

$$\mathrm{H{-}\ddot{N}H{-}CH(R){-}C({=}O){-}O{-}H \rightleftharpoons H{-}\overset{+}{N}H_2{-}CH(R){-}C({=}O){-}O^-}$$

A zwitterion

Amino acid zwitterions are a kind of internal salt and have many of the physical properties associated with inorganic salts. Thus, amino acids are crystalline, have high melting points, and are soluble in water but insoluble in hydrocarbons. In addition, amino acids are *amphoteric*: They can react either as acids or as bases, depending on the circumstances. In aqueous acid solution, an amino acid zwitterion *accepts* a proton to yield a cation; in aqueous basic solution, the zwitterion *loses* a proton to form an anion.

In acid solution

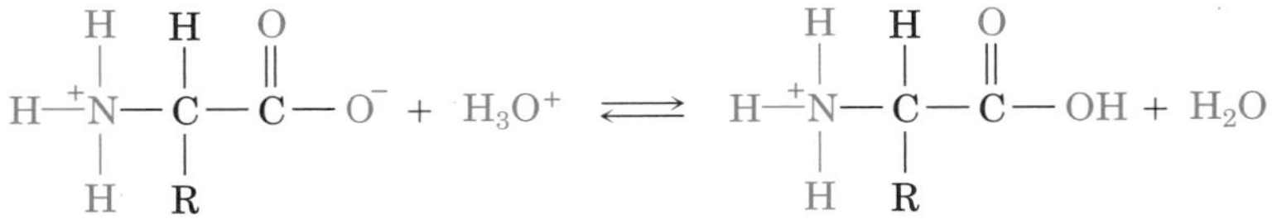

$$H_3\overset{+}{N}-CHR-\overset{O}{\overset{\|}{C}}-O^- + H_3O^+ \rightleftharpoons H_3\overset{+}{N}-CHR-\overset{O}{\overset{\|}{C}}-OH + H_2O$$

In base solution

$$H_3\overset{+}{N}-CHR-\overset{O}{\overset{\|}{C}}-O^- + {}^-OH \rightleftharpoons H_2N-CHR-\overset{O}{\overset{\|}{C}}-O^- + H_2O$$

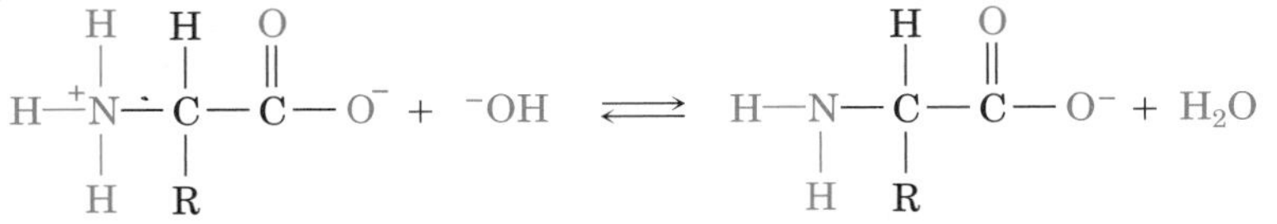

Note that it is the carboxylate anion, $-COO^-$, rather than the $-NH_2$ group that acts as the base and accepts the proton in acid solution. Similarly, it is the ammonium cation, $-NH_3^+$, rather than the $-COOH$ group that acts as the acid and donates a proton in base solution.

PRACTICE PROBLEM 15.1

Write an equation for the reaction of glycine hydrochloride with:
(a) 1 equiv NaOH (b) 2 equiv NaOH

Solution Glycine hydrochloride has the structure

$$Cl^- \; H_3\overset{+}{N}CH_2\overset{O}{\overset{\|}{C}}OH$$

(a) Reaction with the first equivalent of NaOH removes the acidic $-COOH$ proton:

$$Cl^- \; H_3\overset{+}{N}CH_2\overset{O}{\overset{\|}{C}}OH + NaOH \longrightarrow H_3\overset{+}{N}CH_2\overset{O}{\overset{\|}{C}}O^- + H_2O + NaCl$$

(b) Reaction with a second equivalent of NaOH removes the $-NH_3^+$ proton:

$$H_3\overset{+}{N}CH_2\overset{O}{\overset{\|}{C}}O^- + NaOH \longrightarrow H_2NCH_2\overset{O}{\overset{\|}{C}}O^- \; Na^+ + H_2O$$

PROBLEM

15.4 Draw phenylalanine in its zwitterionic form.

PROBLEM

15.5 Write structural formulas for the following equations:
(a) Phenylalanine + 1 equiv NaOH → ? (b) Product of (a) + 1 equiv HCl → ?
(c) Product of (a) + 2 equiv HCl → ?

15.3 Isoelectric Points

In acid solution (low pH), an amino acid is protonated and exists primarily as a cation. In base solution (high pH), an amino acid is deprotonated and exists primarily as an anion. Thus, at some intermediate pH, the amino acid must be exactly balanced between anionic and cationic forms and exist primarily as the neutral, dipolar zwitterion. This pH is called the amino acid's **isoelectric point.**

$$H_3\overset{+}{N}CH(R)COOH \underset{H_3O^+}{\rightleftharpoons} H_3\overset{+}{N}CH(R)CO_2^- \underset{^-OH}{\rightleftharpoons} H_2NCH(R)CO_2^-$$

Low pH (protonated) → High pH (deprotonated)

Isoelectric point (neutral zwitterion)

The isoelectric point of an amino acid depends on its structure, with values for the 20 common amino acids given in Table 15.1. The 15 amino acids with neutral side chains have isoelectric points near neutrality, in the pH range 5.0–6.5. (These values aren't exactly at neutral pH = 7 because carboxyl groups are stronger acids in aqueous solution than amino groups are bases.) The 2 amino acids with acidic side chains have isoelectric points at lower (more acidic) pH, which suppresses dissociation of the extra –COOH group, and the 3 amino acids with basic side chains have isoelectric points at higher (more basic) pH, which suppresses protonation of the extra amino group. For example, aspartic acid has its isoelectric point at pH = 2.77, and lysine has its isoelectric point at pH = 9.74.

We can take advantage of the differences in isoelectric points to separate a mixture of amino acids (or a mixture of proteins) into its pure constituents. In a technique known as **electrophoresis,** a solution of amino acids is placed near the center of a strip of paper or gel. The paper or gel is moistened with an aqueous buffer of a given pH, and electrodes are connected to the ends of the strip. When an electric potential is applied, the amino acids with negative charges (those that are deprotonated because their isoelectric points are below the pH of the buffer) migrate slowly toward the positive electrode. At the same time, the amino acids with positive charges (those that are protonated because their isoelectric points are above the pH of the buffer) migrate toward the negative electrode.

Different amino acids migrate at different rates, depending on their isoelectric points and on the pH of the buffer. Thus, the mixture of amino acids can be separated. Figure 15.1 illustrates this separation for a mixture of lysine (basic), glycine (neutral), and aspartic acid (acidic).

PRACTICE PROBLEM 15.2

Draw structures of the predominant forms of glycine at pH 3.0, pH 6.0, and pH 9.0.

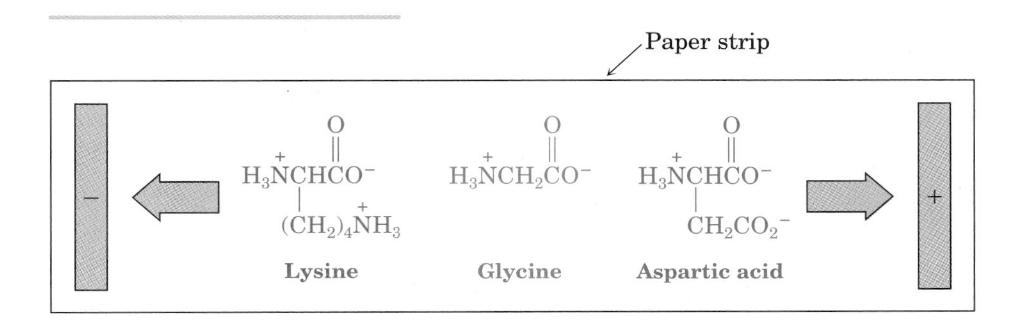

Figure 15.1 Separation of an amino acid mixture by electrophoresis. At pH 5.97, glycine molecules are primarily neutral and do not migrate; lysine molecules are largely protonated and migrate toward the negative electrode; aspartic acid molecules are largely deprotonated and migrate toward the positive electrode. (Lysine has its isoelectric point at pH 9.74, glycine at 5.97, and aspartic acid at 2.77.)

Solution According to Table 15.1, the isoelectric point of glycine is 5.97. At a pH substantially lower than 6.0, glycine is protonated; at pH 6.0, glycine is zwitterionic; and at a pH substantially higher than 6.0, glycine is deprotonated.

$$H_3\overset{+}{N}CH_2\overset{O}{\overset{\|}{C}}OH \qquad H_3\overset{+}{N}CH_2\overset{O}{\overset{\|}{C}}O^- \qquad H_2NCH_2\overset{O}{\overset{\|}{C}}O^-$$

At pH 3.0 At pH 6.0 At pH 9.0

PROBLEM

15.6 Draw the structure of the predominant form of each of the following amino acids:
(a) Lysine at pH 2.0 (b) Aspartic acid at pH 6.0
(c) Lysine at pH 11.0 (d) Alanine at pH 3.0

PROBLEM

15.7 For the mixtures of amino acids indicated, predict the direction of migration of each component (toward the positive or negative electrode) and the relative rate of migration during electrophoresis.
(a) Valine, glutamic acid, and histidine at pH 7.6
(b) Glycine, phenylalanine, and serine at pH 5.7
(c) Glycine, phenylalanine, and serine at pH 6.0

15.4 Peptides and Proteins

Peptides and proteins are amino acid polymers in which the amino acid units, or **residues,** are linked together by amide bonds. The amino group of one residue forms an amide bond with the carboxyl of a second residue, the amino group of the second residue forms an amide bond with the carboxyl of a third, and so on. For example, alanylserine is the *dipeptide* formed when an amide bond is formed between the alanine carboxyl and the serine amino group:

Alanine (Ala) + **Serine (Ser)** ⟶ **Alanylserine (Ala-Ser)**

Stereo View

Note that two dipeptides can result from reaction between alanine and serine, depending on which carboxyl group reacts with which amino group. If the alanine amino group reacts with the serine carboxyl, serylalanine results:

Serine (Ser) + **Alanine (Ala)** ⟶ **Serylalanine (Ser-Ala)**

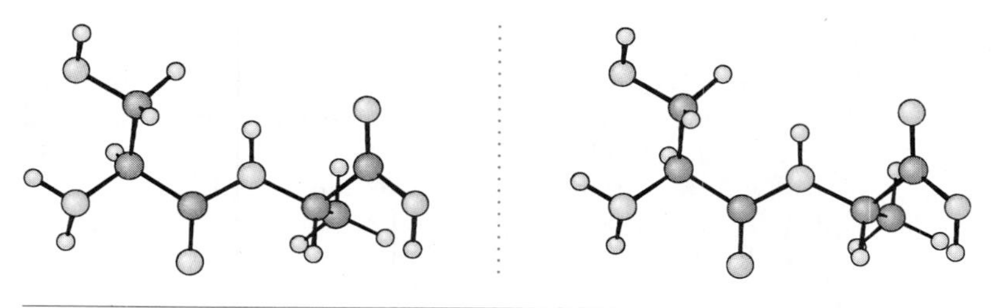

Stereo View

By convention, peptides are written with the **N-terminal amino acid** (the one with the free –NH_2 group) on the left and the **C-terminal amino acid** (the one with the free –COOH group) on the right. The name of the peptide is indicated using the three-letter abbreviations listed in Table 15.1. Thus, serylalanine is abbreviated Ser-Ala, and alanylserine is abbreviated Ala-Ser.

The number of possible isomeric peptides increases rapidly as the number of amino acid units increases. There are six ways in which three amino acids can be joined and more than 40,000 ways in which the eight amino acids in the hormone angiotensin II can be joined (Figure 15.2).

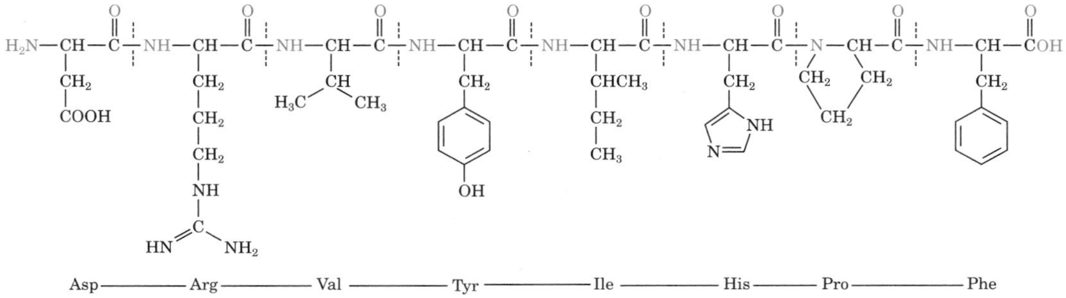

Asp — Arg — Val — Tyr — Ile — His — Pro — Phe

Figure 15.2 *The structure of angiotensin II, a hormone in blood plasma that regulates blood pressure.*

PRACTICE PROBLEM 15.3

Draw the structure of Ala-Val.

Solution By convention, the N-terminal amino acid is written on the left and the C-terminal amino acid on the right. Thus, alanine is N-terminal, valine is C-terminal, and the amide bond is formed between the alanine –COOH and the valine $-NH_2$:

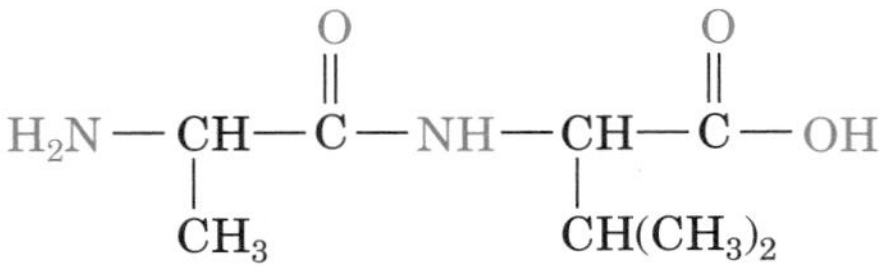

$$H_2N-\underset{\displaystyle CH_3}{\underset{|}{CH}}-\overset{\displaystyle O}{\overset{\|}{C}}-NH-\underset{\displaystyle CH(CH_3)_2}{\underset{|}{CH}}-\overset{\displaystyle O}{\overset{\|}{C}}-OH \qquad \textbf{Ala-Val}$$

PRACTICE PROBLEM 15.4

Name the six tripeptides that contain methionine, lysine, and isoleucine.

Solution Met-Lys-Ile Lys-Met-Ile Ile-Met-Lys

Met-Ile-Lys Lys-Ile-Met Ile-Lys-Met

PROBLEM

15.8 Draw structures of the two dipeptides made from leucine and cysteine.

PROBLEM

15.9 Using the three-letter shorthand notations for each amino acid, name the six possible isomeric tripeptides that contain valine, tyrosine, and glycine.

PROBLEM

15.10 Draw the structure of Met-Pro-Val-Gly, and indicate where the amide bonds are.

15.5 Covalent Bonding in Peptides

In addition to the amide bonds that link amino acid residues in peptides, a second kind of covalent bonding occurs when a disulfide linkage, RS–SR, is formed between two cysteine residues. The linkage is sometimes indicated by writing CyS, with a capital "S" (for sulfur), and then drawing a line from one CyS to the other: CyS CyS. As we saw in Section 8.12, disulfides are formed from thiols (RSH) by mild oxidation and are converted back to thiols by mild reduction:

$$\underbrace{\text{NH–CH(C=O)–}CH_2\text{–SH} \quad \text{HS–}CH_2\text{–CH(C=O)–NH}}_{\textbf{Two cysteines (thiols)}} \underset{\text{Reduction}}{\overset{\text{Oxidation}}{\rightleftharpoons}} \underbrace{\text{NH–CH(C=O)–}CH_2\text{–S–S–}CH_2\text{–CH(C=O)–NH}}_{\textbf{A disulfide}}$$

Disulfide bonds between cysteine residues in two different peptide chains link the otherwise separate chains together. A disulfide bond between cysteine residues in the same chain creates a loop in the chain. Such is the case with the nonapeptide vasopressin, an antidiuretic hormone involved in controlling water balance in the body. Note that the C-terminal end of vasopressin occurs as the primary amide, $–CONH_2$, rather than as the free acid.

Disulfide bridge

CyS-Tyr-Phe-Glu-Asn-CyS-Pro-Arg-Gly-NH_2

Vasopressin

15.6 Peptide Structure Determination: Amino Acid Analysis

Determining the structure of a peptide or protein requires answering three questions: What amino acids are present? How much of each is present? In what sequence do the amino acids occur in the peptide chain? The answers to the first two questions are provided by an instrument called an *amino acid analyzer.*

An amino acid analyzer is an automated instrument based on techniques worked out in the 1950s by W. Stein and S. Moore at the Rockefeller University in New York. In preparation for analysis, the peptide is broken into its constituent amino acids by reducing all disulfide bonds and hydrolyzing all amide bonds with aqueous HCl. The resultant amino acid mixture is then analyzed by placing it at the top of a glass column filled with a special adsorbent material and pumping a series of aqueous buffers through the column. The various amino acids migrate down the column at different rates depending on their structures and are thus separated as they exit (*elute* from) the end of the column.

As each amino acid elutes from the glass column, it mixes with a solution of *ninhydrin,* a reagent that forms a purple color on reaction with an α-amino acid. The purple color is detected by a spectrometer, which measures its intensity and charts it as a function of time.

$$\text{Ninhydrin} + H_2NCH(R)COOH \xrightarrow[H_2O]{^-OH} \text{(Purple color)} + RCHO + CO_2$$

Ninhydrin

An α-amino acid

(Purple color)

Because the time required for a given amino acid to elute from a standard column is reproducible, the identities of all amino acids in a peptide are determined simply by noting the various elution times. The amount of each amino acid in the sample is determined by measuring the intensity of the purple color resulting from its reaction with ninhydrin. Thus, the identity and percentage composition of each amino acid in a peptide can be found. Figure 15.3 shows the results of amino acid analysis of a standard equimolar mixture of 17 α-amino acids.

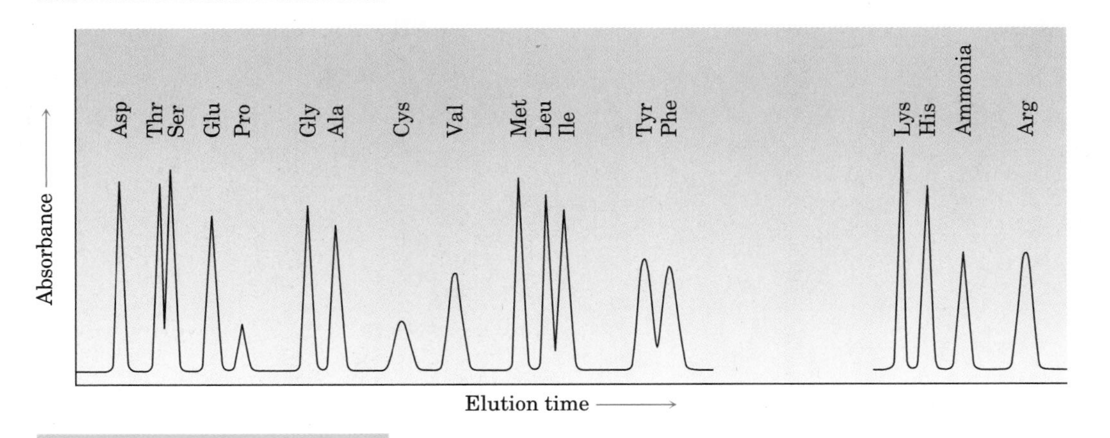

Figure 15.3 Amino acid analysis of an equimolar mixture of 17 amino acids.

PROBLEM

15.11 Write an equation for the reaction of valine with ninhydrin.

15.7 Peptide Sequencing: The Edman Degradation

After the identities and amounts of the amino acids in a peptide are known, the peptide is *sequenced* to find the order in which the amino acids are linked. The general idea of peptide sequencing is to cleave one residue at a time from the end of the peptide chain (either C terminus or N terminus). That terminal amino acid is then separated and identified, and the cleavage reaction is repeated on the chain-shortened peptide until the entire sequence is known. Most peptide sequencing is now done by **Edman degradation,** an efficient method of N-terminal analysis. Automated protein sequenators are available that allow a series of 50 or so repetitive sequencing steps to be carried out.

Edman degradation involves treatment of a peptide with phenyl isothiocyanate, $C_6H_5{-}N{=}C{=}S$, followed by mild acid hydrolysis, as shown in Figure 15.4. The first step attaches a marker to the $-NH_2$ group of the N-terminal amino acid, and the second step splits the N-terminal residue from the chain, yielding a *phenylthiohydantoin* derivative plus the chain-shortened peptide. The phenylthiohydantoin is identified by comparison with known derivatives of the common amino acids, and the chain-shortened peptide is resubmitted to another round of Edman degradation.

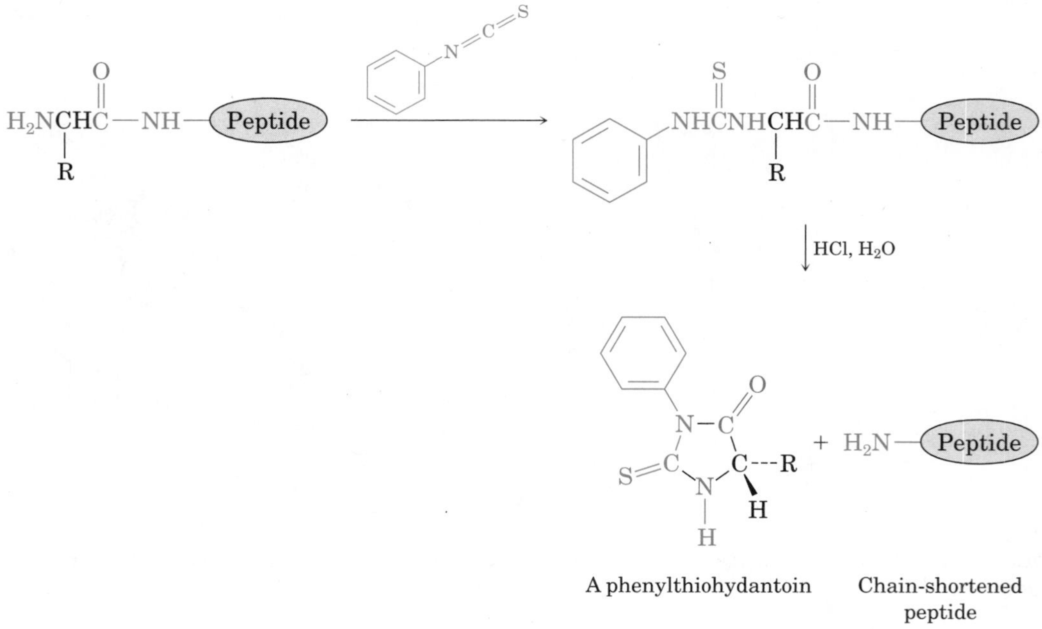

$H_2NCH(R)C(=O){-}NH{-}$Peptide → $C_6H_5NHC(=S)NHCH(R)C(=O){-}NH{-}$Peptide

HCl, H_2O

A phenylthiohydantoin + $H_2N{-}$Peptide (Chain-shortened peptide)

Figure 15.4 Edman degradation of a peptide chain.

Complete sequencing of large proteins by Edman degradation is impractical since the method is limited to about 50 cycles due to buildup of unwanted by-products. Instead, a large protein chain is first cleaved by partial hydrolysis into a number of smaller fragments, the sequence of each fragment is determined, and the individual pieces are then fitted together.

Partial hydrolysis of a protein can be carried out either chemically with aqueous acid or enzymatically with enzymes such as trypsin and chymotrypsin. Acid hydrolysis is unselective and leads to a more or less random mixture of small fragments. Enzymatic hydrolysis, however, is quite specific. Trypsin catalyzes hydrolysis only at the carboxyl side of the basic amino acids arginine and lysine; chymotrypsin cleaves only at the carboxyl side of the aryl-substituted amino acids phenylalanine, tyrosine, and tryptophan. For example:

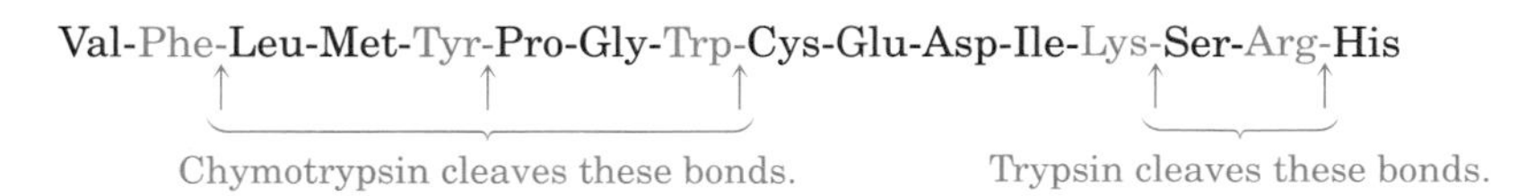

As an example of peptide sequencing, look at a hypothetical structure determination of angiotensin II (Figure 15.2), a hormonal octapeptide involved in controlling hypertension by regulating the sodium–potassium salt balance in the body.

1. Amino acid analysis of angiotensin II shows the composition: Arg, Asp, His, Ile, Phe, Pro, Tyr, Val.
2. N-terminal analysis by the Edman method shows that angiotensin II has an aspartic acid residue at its N terminus.
3. Partial hydrolysis of angiotensin II with dilute HCl might yield the following fragments, whose sequences can be determined by Edman degradation:
   (a) Asp-Arg-Val (b) Ile-His-Pro (c) Arg-Val-Tyr
   (d) Pro-Phe (e) Val-Tyr-Ile
4. Matching the overlapping regions of the various fragments provides the full sequence of angiotensin II:

```
Asp-Arg-Val
    :   :
    Arg-Val-Tyr
        :   :
        Val-Tyr-Ile
                :
                Ile-His-Pro
                        :
                        Pro-Phe
Asp-Arg-Val-Tyr-Ile-His-Pro-Phe
```

Angiotensin II

The structure of angiotensin II is simple—the entire sequence could be done easily by a protein sequenator—but the methods and logic used here are the same as those used to solve far more complex structures. Indeed, protein chains with more than 400 amino acids have been sequenced by these methods.

PRACTICE PROBLEM 15.5

A hexapeptide with the composition Arg, Gly, Leu, Pro_3 has proline at both C-terminal and N-terminal positions. What is the structure of the hexapeptide if partial hydrolysis gives Gly-Pro-Arg, Arg-Pro, and Pro-Leu-Gly?

Solution Line up the overlapping fragments:

Pro-Leu-Gly
:
Gly-Pro-Arg
:
Arg-Pro

The final sequence is Pro-Leu-Gly-Pro-Arg-Pro.

PROBLEM

15.12 What fragments would result if angiotensin II were cleaved with trypsin? With chymotrypsin?

PROBLEM

15.13 Give the amino acid sequence of a hexapeptide containing Arg, Gly, Ile, Leu, Pro, and Val that produces the following fragments on partial acid hydrolysis: Pro-Leu-Gly, Arg-Pro, Gly-Ile-Val.

PROBLEM

15.14 Propose two structures for a tripeptide that gives Leu, Ala, and Phe on hydrolysis but doesn't react with phenyl isothiocyanate.

15.8 Peptide Synthesis

After a peptide's structure has been determined, synthesis is often the next goal. Although simple amides are usually formed by reaction between amines and acid chlorides (Section 10.7), peptide synthesis is much more complex because of the requirement for specificity. Many different amide links must be formed in a specific order rather than at random.

The solution to the specificity problem is *protection*. We can force a reaction to take only the desired course by protecting all the $-NH_2$ and $-COOH$ functional groups except for those we want to react. For example, if we wanted to couple alanine with leucine to synthesize Ala-Leu, we could protect the $-NH_2$ group of alanine and the $-COOH$ group of leucine to render them unreactive. With only the alanine carboxyl and the leucine amine available, we could then form the proper amide bond and remove the protecting groups:

$$\underset{\textbf{Alanine}}{H_2NCH(CH_3)COOH} \xrightarrow[-NH_2]{\text{Protect}} (H_2N)\text{—}CH(CH_3)COOH$$

$$\underset{\textbf{Leucine}}{H_2NCH(CH_2CH(CH_3)_2)COOH} \xrightarrow[-COOH]{\text{Protect}} H_2NCH(CH_2CH(CH_3)_2)\text{—}(COOH)$$

$$\xrightarrow[\text{2. Deprotect}]{\text{1. Form amide}} \underset{\textbf{Ala-Leu}}{H_2NCH(CH_3)C(=O)\text{—}NHCH(CH_2CH(CH_3)_2)COOH}$$

Carboxyl groups are often protected simply by converting them into methyl esters. Ester groups are easily made from carboxylic acids and are easily hydrolyzed by mild treatment with aqueous NaOH:

$$\underset{\textbf{Leucine}}{H_2NCH(CH_2CH(CH_3)_2)COOH} \xrightarrow[HCl]{CH_3OH} \underset{\textbf{Methyl leucinate}}{H_2NCH(CH_2CH(CH_3)_2)COOCH_3} \xrightarrow[\text{2. } H_3O^+]{\text{1. NaOH, } H_2O} \underset{\textbf{Leucine}}{H_2NCH(CH_2CH(CH_3)_2)COOH}$$

Amino groups are often protected as their *tert*-butoxycarbonyl amide (BOC) derivatives. The BOC protecting group is easily introduced by reaction of the amino acid with di-*tert*-butyl dicarbonate and is removed by brief treatment with a strong acid such as trifluoroacetic acid, CF_3COOH.

$$\underset{\textbf{Alanine}}{H_2NCH(CH_3)COOH} + \underset{\textbf{Di-}\textbf{\textit{tert}}\textbf{-butyl dicarbonate}}{(CH_3)_3COC(=O)OC(=O)OC(CH_3)_3} \xrightarrow{(CH_3CH_2)_3N} \underset{\textbf{BOC-Ala}}{(CH_3)_3COC(=O)\text{—}NHCH(CH_3)COOH}$$

A peptide bond is formed by treating a mixture of the protected acid and amine with dicyclohexylcarbodiimide (DCC). Although its mechanism of action is complex, DCC functions by first converting the acid into a reactive intermediate, which then undergoes further nucleophilic acyl substitution reaction with the amine.

$$\underset{\text{An acid}}{R{-}\overset{\overset{O}{\|}}{C}{-}OH} + \underset{\text{An amine}}{R'NH_2} \xrightarrow{DCC} \underset{\text{An amide}}{R{-}\overset{\overset{O}{\|}}{C}{-}NHR'} + \underset{\textbf{Dicyclohexylurea}}{C_6H_{11}{-}\overset{H}{\overset{|}{N}}{-}\overset{\overset{O}{\|}}{C}{-}\overset{H}{\overset{|}{N}}{-}C_6H_{11}}$$

where $DCC = C_6H_{11}{-}N{=}C{=}N{-}C_6H_{11}$

The five steps needed to synthesize Ala-Leu are summarized below:

Steps 1–2 The amino group of alanine is protected as the BOC derivative, and the carboxyl group of leucine is protected as the methyl ester.

$$\text{Ala} + (t\text{-BuO}\overset{\overset{O}{\|}}{C})_2O \longrightarrow \text{BOC–Ala} \qquad \text{Leu} + CH_3OH \xrightarrow{H^+\ \text{cat.}} \text{Leu–OCH}_3$$

Step 3 The two protected amino acids are coupled using DCC.

$$\text{BOC–Ala} + \text{Leu–OCH}_3 \xrightarrow{DCC} \text{BOC–Ala-Leu–OCH}_3$$

Step 4 The BOC protecting group is removed by acid treatment.

$$\text{BOC–Ala-Leu–OCH}_3 \xrightarrow{CF_3COOH} \text{Ala-Leu–OCH}_3$$

Step 5 The methyl ester is removed by basic hydrolysis.

$$\text{Ala-Leu–OCH}_3 \xrightarrow[H_2O]{NaOH} \text{Ala-Leu}$$

These steps can be repeated to add one amino acid at a time to a growing chain or to link two peptide chains together. Many remarkable achievements in peptide synthesis have been reported, including a complete synthesis of human insulin. Insulin, whose structure is shown in Figure 15.5, is composed of two chains totaling 51 amino acids and linked by two cysteine disulfide bridges. Its structure was determined by Frederick Sanger, who received the 1958 Nobel Prize for his work.

PRACTICE PROBLEM 15.6

Write equations for the reaction of methionine with:
(a) CH_3OH, HCl (b) Di-*tert*-butyl dicarbonate

Solution

$$\text{(a)}\ H_2N\underset{\underset{CH_2CH_2SCH_3}{|}}{CH}\overset{\overset{O}{\|}}{C}OH + CH_3OH \xrightarrow{HCl} H_2N\underset{\underset{CH_2CH_2SCH_3}{|}}{CH}\overset{\overset{O}{\|}}{C}OCH_3 + H_2O$$

```
                    Gly
                    Ile
A chain (21 units)  Val
                    Glu
                    Gln-CyS-CyS-Thr-Ser-Ile-CyS-Ser-Leu-Tyr-Gln-Leu-Glu-Asn-Tyr-CyS-Asn

                    His-Leu-CyS-Gly-Ser-His-Leu-Val-Glu-Ala-Leu-Tyr-Leu-Val-CyS
                    Glu                                                      Gly
B chain (30 units)  Asn                                                      Glu
                    Val                                                      Arg
                    Phe                            Thr-Lys-Pro-Thr-Tyr-Phe-Phe-Gly
```

Figure 15.5 Structure of human insulin. The two separate chains totaling 51 amino acids are linked by two disulfide bridges.

(b) $H_2NCH(CH_2CH_2SCH_3)C(=O)OH + (CH_3)_3COC(=O)OC(=O)OC(CH_3)_3 \longrightarrow (CH_3)_3COC(=O)NHCH(CH_2CH_2SCH_3)C(=O)OH$

PROBLEM

15.15 Write the structures of the intermediates in the five-step synthesis of Leu-Ala from alanine and leucine.

PROBLEM

15.16 Show all the steps involved in the synthesis of the tripeptide Val-Phe-Gly.

15.9 Classification of Proteins

Proteins are classified into two major types according to their composition. **Simple proteins,** such as blood serum albumin, are those that yield only amino acids and no other compounds on hydrolysis. **Conjugated proteins,** which are much more common than simple proteins, yield other compounds in addition to amino acids on hydrolysis. As shown in Table 15.2 (p. 492), conjugated proteins can be further classified according to the chemical nature of the non-amino acid portion.

Another way to classify proteins is as either *fibrous* or *globular,* according to their three-dimensional shape. **Fibrous proteins,** such as collagen and keratin, consist of polypeptide chains arranged side by side in long filaments. Because these proteins are tough and insoluble in water, they are used in nature for structural materials such as tendons, hooves, horns, and muscles. **Globular proteins,** by contrast, are usually coiled into compact, nearly spherical shapes. These proteins are generally soluble in

Table 15.2 Some Conjugated Proteins

Name	*Composition*
Glycoproteins	Proteins bonded to a carbohydrate; cell membranes have a glycoprotein coating
Lipoproteins	Proteins bonded to fats and oils (lipids); these proteins transport cholesterol and other fats through the body
Metalloproteins	Proteins bonded to a metal ion; the enzyme cytochrome oxidase, necessary for biological energy production, is an example
Nucleoproteins	Proteins bonded to RNA (ribonucleic acid); these are found in cell ribosomes
Phosphoproteins	Proteins bonded to a phosphate group; milk casein, which stores nutrients for growing embryos, is an example

water and are mobile within cells. Most of the 2000 or so known enzymes are globular. Table 15.3 lists some common examples of both kinds.

Table 15.3 Some Common Fibrous and Globular Proteins

Name	*Occurrence and use*
Fibrous proteins (insoluble)	
Collagens	Animal hide, tendons, connective tissues
Elastins	Blood vessels, ligaments
Fibrinogen	Necessary for blood clotting
Keratins	Skin, wool, feathers, hooves, silk, fingernails
Myosins	Muscle tissue
Globular proteins (soluble)	
Hemoglobin	Involved in oxygen transport
Immunoglobulins	Involved in immune response
Insulin	Hormone for controlling glucose metabolism
Ribonuclease	Enzyme controlling RNA synthesis

Yet a third way to classify proteins is according to function. As shown in Table 15.4, there is an extraordinary diversity to the biological roles of proteins.

Table 15.4 Some Biological Functions of Proteins

Type	*Function and example*
Enzymes	Proteins such as chymotrypsin that act as biological catalysts
Hormones	Proteins such as insulin that regulate body processes
Protective proteins	Proteins such as antibodies that fight infection
Storage proteins	Proteins such as casein that store nutrients
Structural proteins	Proteins such as keratin, elastin, and collagen that form the structure of an organism
Transport proteins	Proteins such as hemoglobin that transport oxygen and other substances through the body

15.10 Protein Structure

Proteins are so large that the word *structure* takes on a broader meaning than it does with other organic compounds. In fact, chemists speak of four different levels of structure when describing proteins. At its simplest, protein structure is the sequence in which amino acid residues are bound together. Called the **primary structure** of a protein, this is the most fundamental structural level.

There is much more to protein structure than amino acid sequence. The chemical properties of a protein are also dependent on higher levels of structure—on exactly how the peptide backbone folds to give the molecule a specific three-dimensional shape. Thus, the term **secondary structure** refers to the way in which *segments* of the peptide backbone orient into a regular pattern; **tertiary structure** refers to the way in which the *entire* protein molecule coils into an overall three-dimensional shape; and **quaternary structure** refers to the way in which several protein molecules come together to yield a large aggregate structure. Let's look at three examples—α-keratin (fibrous), fibroin (fibrous), and myoglobin (globular)—to see how higher structure affects a protein's properties.

α-Keratin

α-Keratin is the fibrous structural protein found in wool, hair, nails, and feathers. Studies show that α-keratin is coiled into a right-handed helical secondary structure like that of a telephone cord. Illustrated in Figure 15.6, this so-called **α-helix** is stabilized by hydrogen bonding between amide N–H groups and other amide C=O groups four residues away. Although the strength of a single hydrogen bond (about 20 kJ/mol; 5 kcal/mol) is only about 5% the strength of a C–C or C–H covalent bond, the large number of hydrogen bonds made possible by helical winding imparts a great deal of stability to the α-helical structure. Each coil of the helix (the *repeat distance*) contains 3.6 amino acid residues; the distance between coils is 5.4 Å.

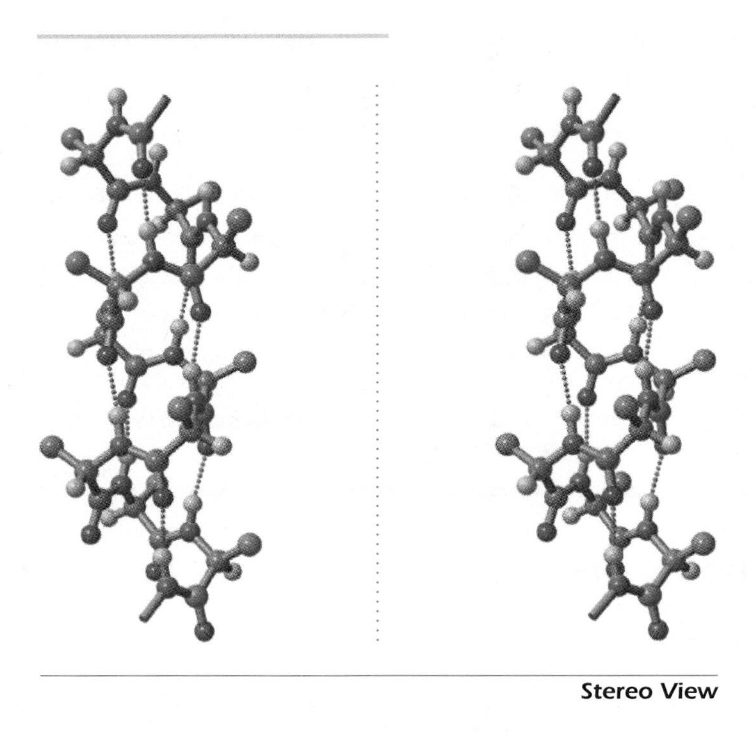

Figure 15.6 The helical secondary structure of α-keratin.

Further evidence shows that the α-keratins of wool and hair also have a quaternary structure. The individual helical strands are themselves coiled about one another in stiff bundles to form a *superhelix* that accounts for the threadlike properties and strength of these proteins. Although α-keratin is the best example of an almost entirely helical protein, most globular proteins contain α-helical *segments.* Both hemoglobin and myoglobin, for example, contain many short helical sections in their chains.

Fibroin

Fibroin, the fibrous protein found in silk, has a secondary structure known as a **β-pleated sheet** in which polypeptide chains line up in a parallel arrangement held together by hydrogen bonds between chains (Figure 15.7). Although not as common as the α-helix, small β-pleated-sheet regions are often found in proteins where sections of peptide chains double back on themselves.

Myoglobin

Myoglobin is a small globular protein containing 153 amino acid residues in a single chain. A relative of hemoglobin, myoglobin is found in the skeletal muscles of sea mammals, where it stores oxygen needed to sustain the animals during long dives. Myoglobin consists of eight helical segments connected by bends to form a compact, nearly spherical, tertiary structure (Figure 15.8).

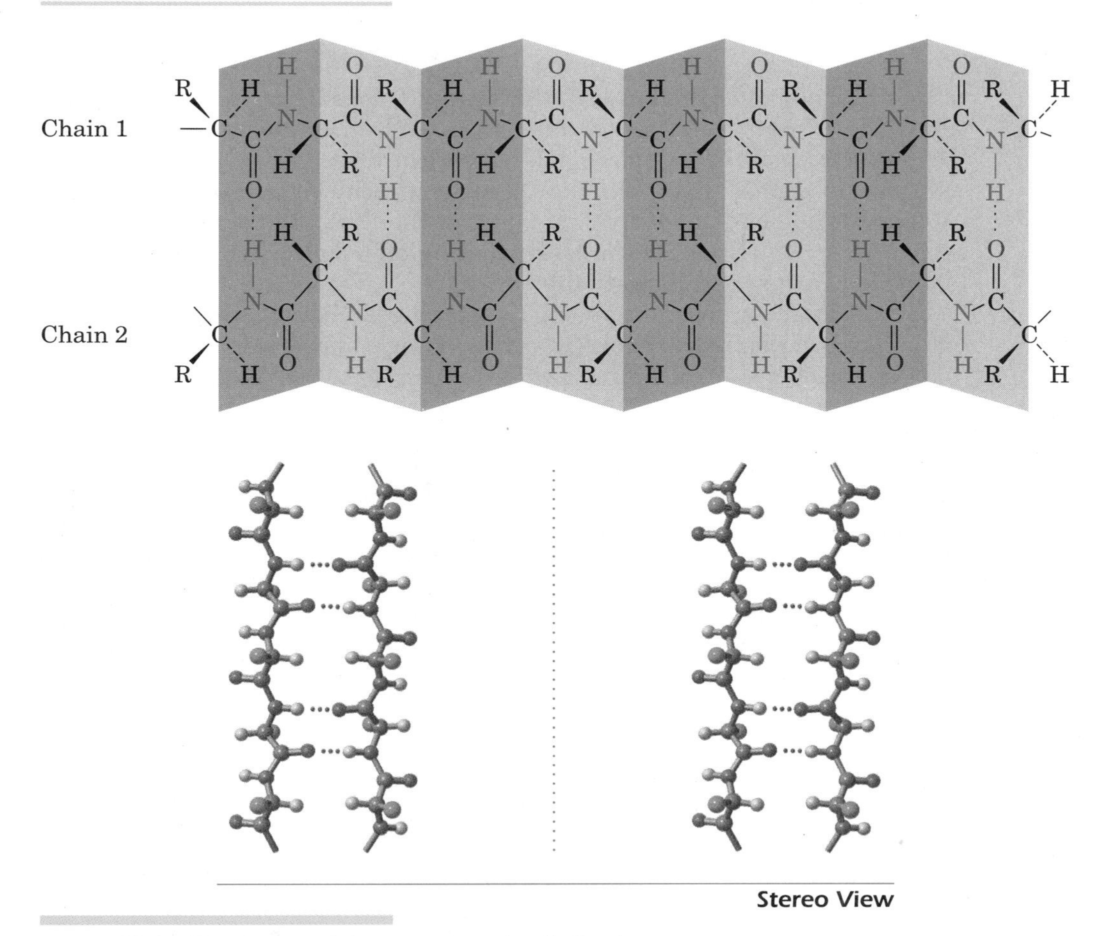

Figure 15.7 The β-pleated-sheet structure in silk fibroin.

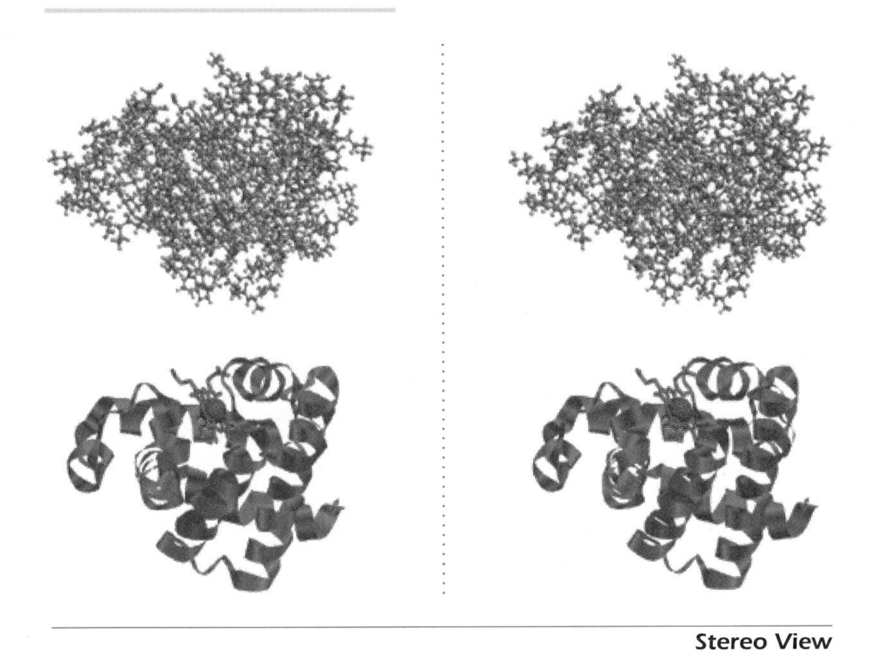

Figure 15.8 Secondary and tertiary structure of myoglobin.

Why does myoglobin adopt the shape it does? The forces that determine the tertiary structure of myoglobin and other globular proteins are the same forces that act on all molecules, regardless of size. By bending and twisting in exactly the right way, myoglobin achieves maximum stability. Although the bends appear irregular and the three-dimensional structure appears random, this isn't the case. All myoglobin molecules adopt this same shape because it is the most stable.

Particularly important among the forces stabilizing a protein's tertiary structure are the hydrophobic (water-repelling) interactions of hydrocarbon side chains on neutral amino acids. The amino acids with neutral, nonpolar side chains have a strong tendency to congregate on the interior of a protein molecule, away from the aqueous medium. The acidic or basic amino acids with charged side chains, by contrast, tend to congregate on the exterior of the protein where they can be solvated by water.

Also important for stabilizing a protein's tertiary structure are the formation of disulfide bridges between cysteine residues, the formation of hydrogen bonds between nearby amino acids, and the formation of ionic attractions, called *salt bridges,* between positively and negatively charged sites on the protein. The various kinds of stabilizing forces are summarized in Figure 15.9.

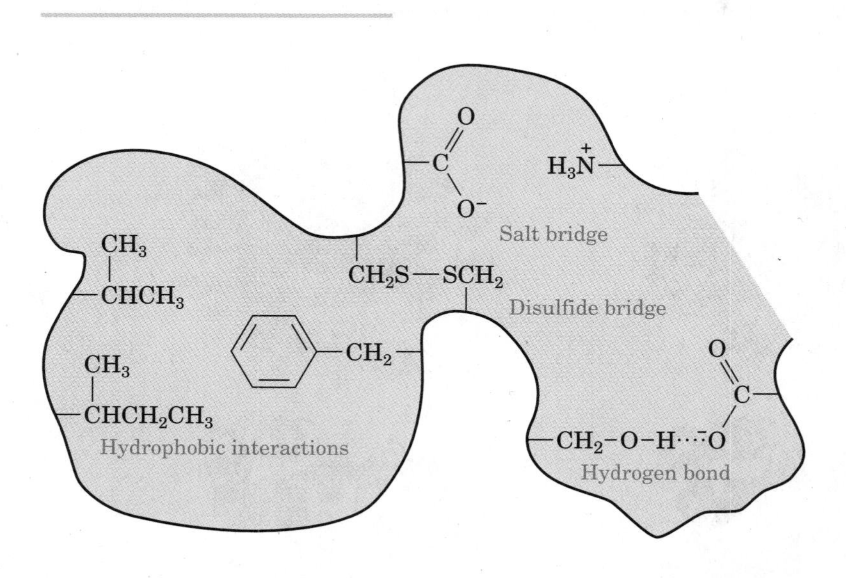

Figure 15.9 Interactions among amino acid side chains that stabilize a protein's tertiary structure.

15.11 Enzymes

Enzymes are large proteins that act as catalysts for biological reactions. Unlike many of the simple catalysts that chemists use in the laboratory, enzymes are usually specific in their action. Often, in fact, an enzyme can catalyze only a single reaction of a single compound, called the enzyme's *substrate.* For example, the enzyme amylase found in the human digestive tract catalyzes only the hydrolysis of starch to yield glucose; cellulose and other polysaccharides are untouched by amylase.

Different enzymes have different specificities. Some, such as amylase, are specific for a single substrate, but others operate on a range of substrates. Papain, for instance, a globular protein of 212 amino acids isolated from papaya fruit, catalyzes the hydrolysis of many kinds of peptide bonds. In fact, it's this ability to hydrolyze peptide bonds that makes papain useful as a meat tenderizer and a cleaner for contact lenses.

$$\text{-(NHCH(R)C(=O)—NHCH(R')C(=O)—NHCH(R'')C(=O))-} \xrightarrow[\text{H}_2\text{O}]{\text{Papain}} \text{H}_2\text{NCH(R)COOH} + \text{H}_2\text{NCH(R')COOH} + \text{H}_2\text{NCH(R'')COOH}$$

Like all catalysts, enzymes don't affect the equilibrium constant of a reaction and can't bring about chemical changes that are otherwise unfavorable. Enzymes act only to lower the activation energy for a reaction, thereby making the reaction take place faster or at a lower temperature. Starch and water, for example, react very slowly in the absence of a catalyst because the activation energy is too high. But when amylase is present, the energy barrier is lowered and the hydrolysis reaction occurs rapidly.

15.12 Structure and Classification of Enzymes

In addition to their protein part, most enzymes also have a small, nonprotein part called a **cofactor.** The protein part in such enzymes is called an **apoenzyme,** and the combination of apoenzyme plus cofactor is called a **holoenzyme.** Only holoenzymes have biological activity; the cofactor or the apoenzyme alone can't catalyze reactions.

$$\text{Holoenzyme} = \text{Cofactor} + \text{Apoenzyme}$$

Cofactors can be either inorganic ions, such as Zn^{2+}, or small organic molecules called **coenzymes.** The requirement of many enzymes for inorganic cofactors is the main reason for our dietary need of trace minerals. Iron, zinc, copper, manganese, and numerous other metal ions are all essential minerals that act as enzyme cofactors, though the exact biological role is not known in all cases.

A variety of organic molecules act as coenzymes. Many, though not all, coenzymes are **vitamins,** small organic molecules that must be obtained in the diet and are required in trace amounts for proper growth. Table 15.5 lists the 13 known vitamins required in the human diet and their enzyme functions.

Table 15.5 Vitamins and Their Enzyme Functions

Vitamin	*Enzyme function*	*Deficiency symptom*
Water-soluble vitamins		
Ascorbic acid (vitamin C)	Hydrolases	Bleeding gums, bruising
Thiamin (vitamin B_1)	Reductases	Fatigue, depression
Riboflavin (vitamin B_2)	Reductases	Cracked lips, scaly skin
Pyridoxine (vitamin B_6)	Transaminases	Anemia, irritability
Niacin	Reductases	Dermatitis, dementia
Folic acid (vitamin M)	Methyltransferases	Megaloblastic anemia
Vitamin B_{12}	Isomerases	Megaloblastic anemia, neurodegeneration
Pantothenic acid	Acyltransferases	Weight loss, irritability
Biotin (vitamin H)	Carboxylases	Dermatitis, anorexia, depression
Fat-soluble vitamins		
Vitamin A	Visual system	Night blindness, dry skin
Vitamin D	Calcium metabolism	Rickets, osteomalacia
Vitamin E	Antioxidant	Hemolysis of red blood cells
Vitamin K	Blood clotting	Hemorrhage, delayed blood clotting

Enzymes are grouped into six main classes according to the kind of reaction they catalyze (Table 15.6). *Hydrolases* catalyze hydrolysis reactions; *isomerases* catalyze isomerizations; *ligases* catalyze the bonding together of two molecules with participation of adenosine triphosphate (ATP); *lyases* catalyze the breaking away of a small molecule such as H_2O from a substrate; *oxidoreductases* catalyze oxidations and reductions; and *transferases* catalyze the transfer of a group from one substrate to another.

Although some enzymes, like papain and trypsin, have uninformative common names, the systematic name of an enzyme has two parts, ending with *-ase*. The first part identifies the enzyme's substrate, and the second part identifies its class. For example, *hexose kinase* is an enzyme that catalyzes the transfer of a phosphate group from adenosine triphosphate to glucose.

Table 15.6 Classification of Enzymes

Main class	*Some subclasses*	*Type of reaction catalyzed*
Hydrolases	Lipases	Hydrolysis of an ester group
	Nucleases	Hydrolysis of a phosphate group
	Proteases	Hydrolysis of an amide group
Isomerases	Epimerases	Isomerization of stereocenter
Ligases	Carboxylases	Addition of CO_2
	Synthetases	Formation of new bond
Lyases	Decarboxylases	Loss of CO_2
	Dehydrases	Loss of H_2O
Oxidoreductases	Dehydrogenases	Introduction of double bond by removal of H_2
	Oxidases	Oxidation
	Reductases	Reduction
Transferases	Kinases	Tranfer of a phosphate group
	Transaminases	Transfer of an amino group

PROBLEM

15.17 To what classes do the following enzymes belong?
(a) Pyruvate decarboxylase
(b) Chymotrypsin
(c) Alcohol dehydrogenase

INTERLUDE

Protein and Nutrition

Every body needs protein, especially these bodies.

Everyone needs protein. Children need large amounts of protein for proper growth, and adults need protein to replace what is lost each day by the body's normal biochemical reactions. Dietary protein is necessary because our bodies can synthesize only 10 of the 20 common amino acids from simple precursor molecules; the other 10 amino acids must be obtained from food by digestion of edible proteins. Table 15.7 shows the estimated essential amino acid requirements of an infant and an adult.

Table 15.7 Estimated Essential Amino Acid Requirements

	Daily requirement (mg/kg body weight)	
Amino acid	*Infant*	*Adult*
Arginine	?	?
Histidine	33	?
Isoleucine	83	12
Leucine	35	16
Lysine	99	12
Methionine (+ Cysteine)	49	10
Phenylalanine (+ Tyrosine)	141	16
Threonine	68	8
Tryptophan	21	3
Valine	92	14

Not all foods provide sufficient amounts of the 10 essential amino acids to meet our minimum daily needs. Most meat and dairy products are satisfactory, but many vegetable sources, such as wheat and corn, are *incomplete*; that is, many vegetable proteins have such a limited amount of one or more essential amino acids that they can't sustain the growth of laboratory animals. Wheat is limited in lysine, for example, and corn is limited in both lysine and tryptophan.

(*continued*)➤

Using an incomplete food as the sole source of protein can cause nutritional deficiencies, particularly in children. Vegetarians must therefore be careful to adopt a varied diet that provides proteins from several sources. Legumes and nuts, for example, are useful for overcoming the deficiencies of wheat and grains. Some of the limiting amino acids in various foods are listed in Table 15.8.

Table 15.8 Limiting Amino Acids in Some Foods

Food	*Limiting, amino acid*
Wheat, grains	Lysine, threonine
Peas, beans, legumes	Methionine, tryptophan
Nuts, seeds	Lysine
Leafy green vegetables	Methionine

Summary and Key Words

amino acid, 474
α-amino acid, 475
apoenzyme, 497
C-terminal amino acid, 482
coenzyme, 497
cofactor, 497
conjugated protein, 491
Edman degradation, 486
electrophoresis, 480
enzyme, 497
fibrous protein, 491
globular protein, 491
α-helix, 493
holoenzyme, 497
isoelectric point, 480
N-terminal amino acid, 482
peptide, 475
β-pleated sheet, 494
primary structure, 493
protein, 474
quaternary structure, 493

Large **proteins** and small **peptides** are biomolecules made of **α-amino acid residues** linked together by amide bonds. Twenty amino acids are commonly found in proteins; all are α-amino acids, and all except glycine have stereochemistry similar to that of L sugars.

Determining the structure of a peptide or protein requires several steps. The identity and amount of each amino acid present in a peptide can be determined by *amino acid analysis.* The peptide is first hydrolyzed to its constituent α-amino acids, which are then separated and identified. Next, the peptide is *sequenced.* **Edman degradation** by treatment with phenyl isothiocyanate cleaves one residue from the **N terminus** of the peptide and forms an easily identifiable derivative of that residue. A series of Edman degradations can sequence peptide chains up to 50 residues in length.

Peptide synthesis involves the use of *protecting* groups. An N-protected amino acid with a free –COOH group is coupled using DCC to an O-protected amino acid with a free $-NH_2$ group. Amide formation occurs, the protecting groups are removed, and the sequence is repeated. Amines are usually protected as their *tert*-butoxycarbonyl (BOC) derivatives; acids are usually protected as esters.

Proteins are classified as either **globular** or **fibrous,** depending on their **secondary** and **tertiary structures.**

Fibrous proteins such as α-keratin are tough and water-insoluble; globular proteins such as myoglobin are water-soluble and mobile within cells. Most of the 2000 or so known enzymes are globular proteins.

Enzymes are globular proteins that act as biological catalysts. They are classified into six groups according to the kind of reaction they catalyze: *oxidoreductases* catalyze oxidations and reductions; *transferases* catalyze transfers of groups; *hydrolases* catalyze hydrolysis; *isomerases* catalyze isomerizations; *lyases* catalyze bond breakages; and *ligases* catalyze bond formations.

In addition to their protein part, many enzymes contain **cofactors,** which can be either metal ions or small organic molecules. If the cofactor is an organic molecule, it is called a **coenzyme.** The combination of protein **(apoenzyme)** plus coenzyme is called a **holoenzyme.** Often, the coenzyme is a **vitamin,** a small molecule that must be obtained in the diet and is required in trace amounts for proper growth and functioning.

ADDITIONAL PROBLEMS

15.18 What does the prefix "α" mean when referring to an α-amino acid?

15.19 What amino acids do the following abbreviations stand for?
(a) Ser (b) Thr (c) Pro (d) Phe (e) Glu

15.20 What kinds of molecules are found in the following conjugated proteins in addition to the protein part?
(a) Nucleoproteins (b) Glycoproteins (c) Lipoproteins

15.21 Why is cysteine such an important amino acid for determining the tertiary structure of a protein?

15.22 The *endorphins* are a group of naturally occurring compounds in the brain that act to control pain. The active part of an endorphin is a pentapeptide called an *enkephalin,* which has the structure Tyr-Gly-Gly-Phe-Met. Draw the structure of this enkephalin.

15.23 What kinds of reactions do the following classes of enzymes catalyze?
(a) Hydrolases (b) Lyases (c) Transferases

15.24 What kind of reaction does each of the following enzymes catalyze?
(a) A protease (b) A kinase (c) A carboxylase

15.25 Although only *S* amino acids occur in proteins, several *R* amino acids are found elsewhere in nature. For example, (*R*)-serine is found in earthworms and (*R*)-alanine is found in insect larvae. Draw Fischer projections of (*R*)-serine and (*R*)-alanine.

15.26 Draw a Fischer projection of (*S*)-proline, the only secondary amino acid.

15.27 Define the following terms:
(a) Amphoteric (b) Isoelectric point (c) Peptide
(d) N terminus (e) C terminus (f) Zwitterion

15.28 Using the three-letter code names for each amino acid, write the structures of all the peptides containing the following amino acids:
(a) Val, Leu, Ser (b) Ser, Leu_2, Pro

15.29 Draw the following amino acids in their zwitterionic forms:
(a) Serine (b) Tyrosine (c) Threonine

15.30 Draw structures of the predominant forms of lysine and aspartic acid at pH 3.0 and pH 9.7.

15.31 At what pH would you carry out an electrophoresis experiment if you wanted to separate a mixture of histidine, serine, and glutamic acid? Explain.

15.32 Which of the following amino acids are more likely to be found on the outside of a globular protein, and which on the inside? Explain.
(a) Valine (b) Aspartic acid (c) Isoleucine (d) Lysine

15.33 Predict the product of the reaction of valine with the following reagents:
(a) CH_3CH_2OH, H^+ (b) NaOH, H_2O (c) Di-*tert*-butyl dicarbonate

15.34 Write full structures for the following peptides, and indicate the positions of the amide bonds:
(a) Val-Phe-Cys (b) Glu-Pro-Ile-Leu

15.35 The amino acid threonine, (2*S*,3*R*)-2-amino-3-hydroxybutanoic acid, has two stereocenters and a stereochemistry similar to that of the four-carbon sugar D-threose. Draw a Fischer projection of threonine.

15.36 Draw the Fischer projection of a diastereomer of threonine (see Problem 15.35).

15.37 The amino acid analysis data in Figure 15.3 indicate that proline is not easily detected by reaction with ninhydrin. Suggest a reason.

15.38 *Cytochrome c,* an enzyme found in the cells of all aerobic organisms, plays a role in respiration. Elemental analysis of cytochrome *c* reveals it to contain 0.43% iron. What is the minimum molecular weight of this enzyme?

15.39 Draw the structure of the phenylthiohydantoin product you would expect to obtain from Edman degradation of the following peptides:
(a) Val-Leu-Gly (b) Ala-Pro-Phe

15.40 Arginine, which contains a *guanidino* group in its side chain, is the most basic of the 20 common amino acids. How can you account for this basicity? (*Hint*: Use resonance structures to see how the protonated guanidino group is stabilized.)

$$H_2N{-}C({=}NH){-}NHCH_2CH_2CH_2CH(NH_2)COOH$$

Guanidino group

Arginine

15.41 Show the steps involved in a synthesis of Phe-Ala-Val.

15.42 When an unprotected α-amino acid is treated with dicyclohexylcarbodiimide (DCC), a 2,5-diketopiperazine results. Explain.

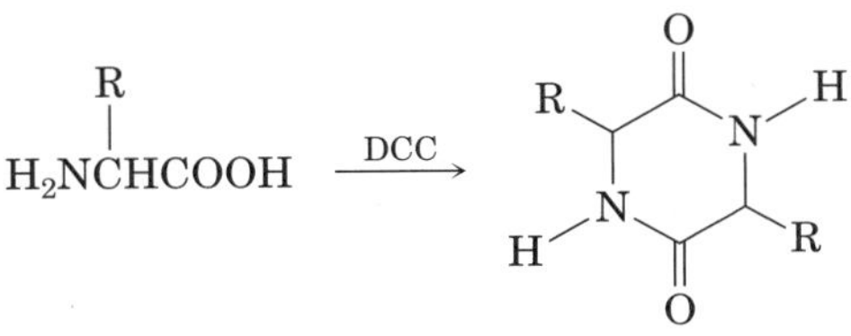

$$H_2NCH(R)COOH \xrightarrow{DCC}$$

A 2,5-diketopiperazine

15.43 Which amide bonds in the following polypeptide are cleaved by trypsin? By chymotrypsin?

Phe-Leu-Met-Lys-Tyr-Asp-Gly-Gly-Arg-Val-Ile-Pro-Tyr

15.44 Look up the structure of human insulin (Figure 15.5), and indicate where in each chain the molecule is cleaved by trypsin and by chymotrypsin.

15.45 A heptapeptide shows the composition Asp, Gly, Leu, Phe, Pro_2, Val on amino acid analysis. Edman degradation shows glycine to be the N-terminal group. Acidic hydrolysis gives the following fragments:

Val-Pro-Leu, Gly, Gly-Asp-Phe-Pro, Phe-Pro-Val

Propose a structure for the starting heptapeptide.

15.46 Give the amino acid sequence of hexapeptides that produce the following fragments on partial acid hydrolysis:

(a) Arg, Gly, Ile, Leu, Pro, Val gives Pro-Leu-Gly, Arg-Pro, Gly-Ile-Val
(b) Asp, Leu, Met, Trp, Val_2 gives Val-Leu, Val-Met-Trp, Trp-Asp-Val

15.47 How can you account for the fact that proline is never encountered in a protein α-helix? The α-helical segments of myoglobin and other proteins stop when a proline residue is encountered in the chain.

15.48 Draw as many resonance forms as you can for the purple anion obtained by reaction of ninhydrin with an amino acid:

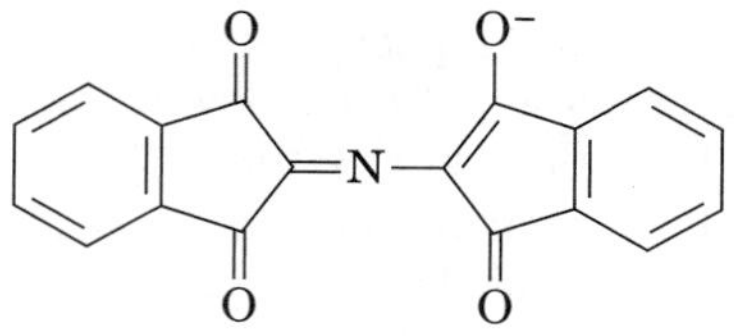

15.49 A nonapeptide gives the following fragments when cleaved by chymotrypsin and by trypsin:

Trypsin cleavage: Val-Val-Pro-Tyr-Leu-Arg and Ser-Ile-Arg

Chymotrypsin cleavage: Leu-Arg and Ser-Ile-Arg-Val-Val-Pro-Tyr

What is the structure of the nonapeptide?

15.50 *Oxytocin,* a nonapeptide hormone secreted by the pituitary gland, stimulates uterine contraction and lactation during childbirth. Its sequence was determined from the following evidence:

1. Oxytocin is a cyclic peptide containing a disulfide bridge between two cysteine residues.
2. When the disulfide bridge is reduced, oxytocin has the constitution Asn, Cys_2, Gln, Gly, Ile, Leu, Pro, Tyr.
3. Partial hydrolysis of reduced oxytocin yields seven fragments:

   Asp-Cys Ile-Glu Cys-Tyr Leu-Gly Tyr-Ile-Glu
   Glu-Asp-Cys Cys-Pro-Leu

4. Gly is the C-terminal group.
5. Both Glu and Asp are present as their side-chain amides (Gln and Asn) rather than as free side-chain acids.

What is the amino acid sequence of reduced oxytocin? What is the structure of oxytocin?

15.51 *Aspartame,* a nonnutritive sweetener marketed under the trade name NutraSweet, is the methyl ester of a simple dipeptide, Asp-Phe-OCH_3.

(a) Draw the full structure of aspartame.

(b) The isoelectric point of aspartame is 5.9. Draw the principal structure present in aqueous solution at this pH.

(c) Draw the principal form of aspartame present at physiological pH 7.6.

(d) Show the products of hydrolysis on treatment of aspartame with H_3O^+.

Visualizing Chemistry

15.52 Identify the following amino acids (gray = C, red = O, blue = N, light green = H):

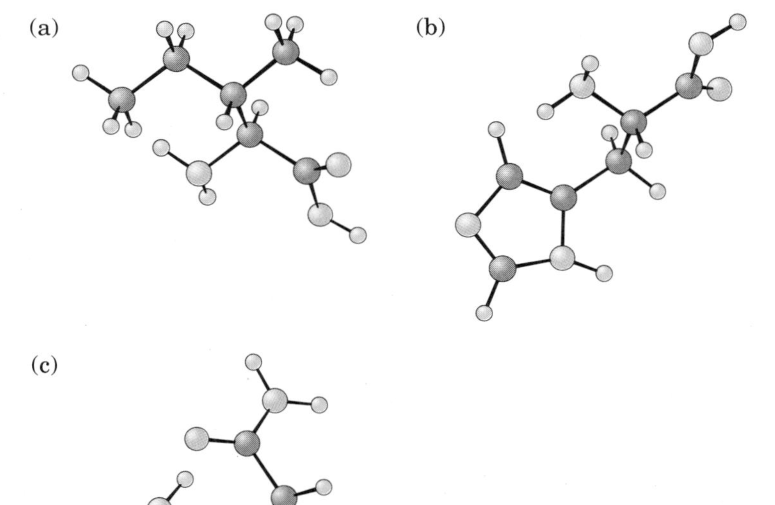

15.53 Give the sequence of the following tetrapeptide (gray = C, red = O, blue = N, yellow = S, light green = H):

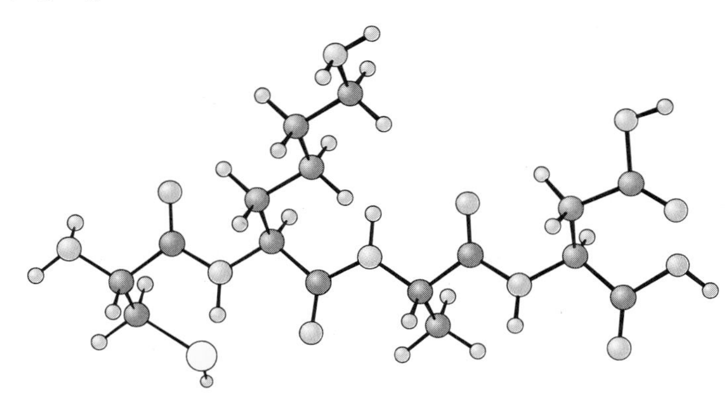

15.54 Isoleucine and threonine (Problem 15.35) are the only two amino acids with two stereocenters. Assign *R* or *S* configuration to the methyl-bearing carbon atom of isoleucine (gray = C, red = O, blue = N, light green = H):

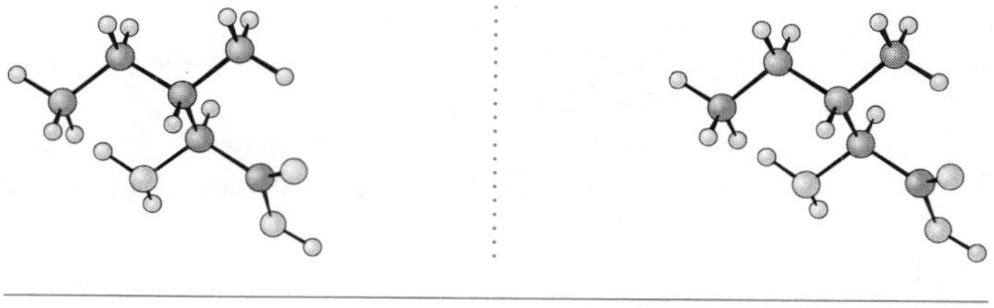

Stereo View

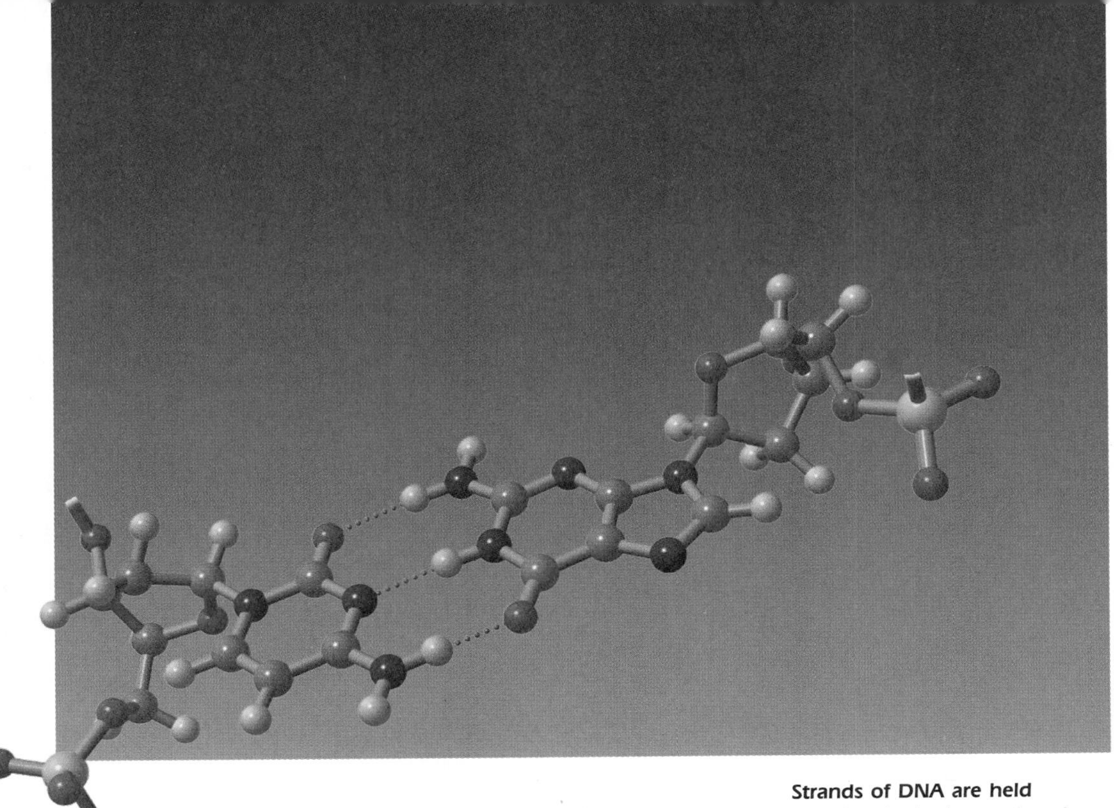

Strands of DNA are held together by hydrogen bonds between nucleotides, as shown by this interaction between cytosine and guanine.

16 Biomolecules: Lipids and Nucleic Acids

In the previous two chapters, we've discussed the organic chemistry of carbohydrates and proteins, two of the four major classes of biomolecules. Let's now look at the two remaining classes: *lipids* and *nucleic acids.* Though chemically quite different from one another, all four types of biomolecules are essential for life.

16.1 Lipids

Lipids are the naturally occurring organic substances that can be isolated from cells and tissues by extraction with a nonpolar organic solvent. Note that this definition differs from those used for carbohydrates and proteins in that lipids are defined by a physical property (solubility) rather than by structure.

Lipids are classified into two general types: those like fats and waxes, which contain ester linkages and can be hydrolyzed, and those like cholesterol and other steroids, which don't have ester linkages and can't be hydrolyzed.

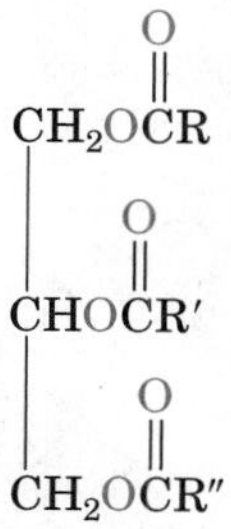

$$\begin{array}{l} CH_2OCR \\ | \quad \ \ \overset{\|}{\ } \text{(C=O)} \\ CHOCR' \\ | \\ CH_2OCR'' \end{array}$$

Animal fat, an ester
(R, R′, R″ = C_{11}–C_{19} chains)

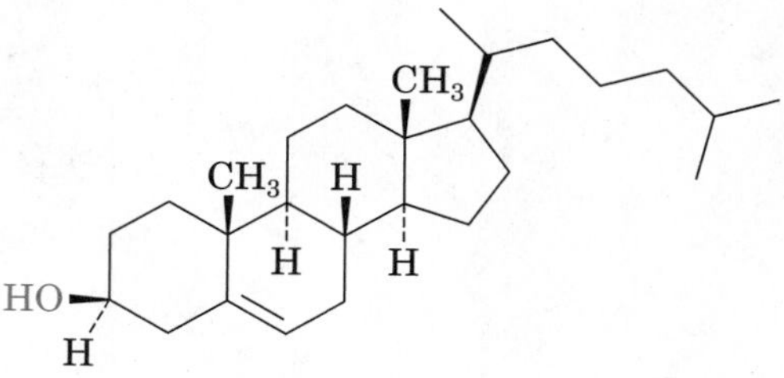

Cholesterol

PROBLEM

16.1 Beeswax contains, among other things, a lipid with the structure $CH_3(CH_2)_{20}COO(CH_2)_{27}CH_3$. What products would you obtain by reaction of this lipid with aqueous NaOH followed by acidification?

16.2 Fats and Oils

Animal fats and vegetable oils are the most widely occurring lipids. Although they appear different—animal fats like butter and lard are solids, whereas vegetable oils like corn oil and peanut oil are liquids—their structures are closely related. Chemically, fats and oil are **triacylglycerols** (also called *TAG's,* or *triglycerides*), triesters of glycerol with three long-chain carboxylic acids. Hydrolysis of a fat or oil with aqueous NaOH yields glycerol and three long-chain **fatty acids**:

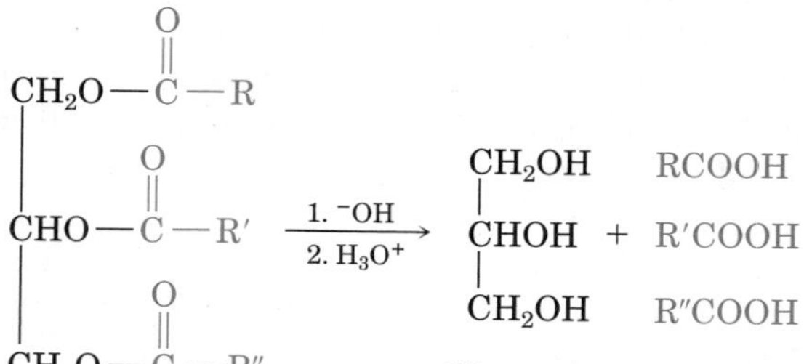

$$\begin{array}{l} CH_2O-C(=O)-R \\ | \\ CHO-C(=O)-R' \\ | \\ CH_2O-C(=O)-R'' \end{array} \xrightarrow[2.\ H_3O^+]{1.\ ^-OH} \begin{array}{l} CH_2OH \\ | \\ CHOH \\ | \\ CH_2OH \end{array} + \begin{array}{l} RCOOH \\ R'COOH \\ R''COOH \end{array}$$

A fat — **Glycerol** — Fatty acids

The fatty acids obtained by hydrolysis of triacylglycerols are generally unbranched and contain an even number of carbon atoms between 12 and 20. If double bonds are present, they usually have *Z* (cis) geometry. The three fatty acids of a specific molecule need not be the same, and a fat or oil from a given source is likely to be a complex mixture of many different triacylglycerols. Table 16.1 lists some commonly occurring fatty acids, and Table 16.2 lists the approximate composition of fats and oils from different sources.

Table 16.1 Structures of Some Common Fatty Acids

Name	*Carbons*	*Structure*	*Melting point (°C)*
Saturated			
Lauric	12	$CH_3(CH_2)_{10}COOH$	44
Myristic	14	$CH_3(CH_2)_{12}COOH$	58
Palmitic	16	$CH_3(CH_2)_{14}COOH$	63
Stearic	18	$CH_3(CH_2)_{16}COOH$	70
Arachidic	20	$CH_3(CH_2)_{18}COOH$	75
Unsaturated			
Palmitoleic	16	$CH_3(CH_2)_5CH{=}CH(CH_2)_7COOH$ (cis)	32
Oleic	18	$CH_3(CH_2)_7CH{=}CH(CH_2)_7COOH$ (cis)	16
Ricinoleic	18	$CH_3(CH_2)_5CH(OH)CH_2CH{=}CH(CH_2)_7COOH$ (cis)	5
Linoleic	18	$CH_3(CH_2)_4CH{=}CHCH_2CH{=}CH(CH_2)_7COOH$ (cis, cis)	−5
Arachidonic	20	$CH_3(CH_2)_4(CH{=}CHCH_2)_4CH_2CH_2COOH$ (all cis)	−50

Table 16.2 Approximate Fatty Acid Composition of Some Common Fats and Oils

	Saturated fatty acids (%)				*Unsaturated fatty acids (%)*		
Source	C_{12} *Lauric*	C_{14} *Myristic*	C_{16} *Palmitic*	C_{18} *Stearic*	C_{18} *Oleic*	C_{18} *Ricinoleic*	C_{18} *Linoleic*
Animal fat							
Lard	—	1	25	15	50	—	6
Butter	2	10	25	10	25	—	5
Human fat	1	3	25	8	46	—	10
Whale blubber	—	8	12	3	35	—	10
Vegetable oil							
Coconut	50	18	8	2	6	—	1
Corn	—	1	10	4	35	—	45
Olive	—	1	5	5	80	—	7
Peanut	—	—	7	5	60	—	20
Linseed	—	—	5	3	20	—	20
Castor bean	—	—	—	1	8	85	4

About 40 different fatty acids occur naturally. Palmitic acid (C_{16}) and stearic acid (C_{18}) are the most abundant saturated fatty acids; oleic and linoleic acids (both C_{18}) are the most abundant unsaturated ones. Oleic acid is monounsaturated since it has only one double bond, whereas linoleic, linolenic, and arachidonic acids are **polyunsaturated fatty acids,** or **PUFA's,** because they have more than one double bond. Linoleic and linolenic acids are essential in the human diet; infants grow poorly and develop skin lesions if fed a diet of nonfat milk for prolonged periods.

$$CH_3CH_2CH_2CH_2CH_2CH_2CH_2CH_2CH_2CH_2CH_2CH_2CH_2CH_2CH_2CH_2CH_2\overset{\overset{\displaystyle O}{\|}}{C}OH$$

Stearic acid

$$CH_3CH_2CH\overset{\downarrow}{=}CHCH_2CH\overset{\downarrow}{=}CHCH_2CH\overset{\downarrow}{=}CHCH_2CH_2CH_2CH_2CH_2CH_2CH_2\overset{\overset{\displaystyle O}{\|}}{C}OH$$

Linolenic acid, a polyunsaturated fatty acid (PUFA)

The data in Table 16.1 show that unsaturated fatty acids generally have lower melting points than their saturated counterparts, a trend that also holds for triacylglycerols. Since vegetable oils generally have a higher proportion of unsaturated to saturated fatty acids than animal fats do (Table 16.2), they have lower melting points. The difference is a consequence of structure. Saturated fats have a uniform shape that allows them to pack together easily in a crystal. Unsaturated vegetable oils, however, have C=C bonds, which introduce bends and kinks into the hydrocarbon chains and make crystal formation difficult. The more double bonds there are, the harder it is for the molecule to crystallize, and the lower the melting point.

The C=C bonds in vegetable oils can be reduced by catalytic hydrogenation (Section 4.6) to produce saturated solid or semisolid fats. Margarine and solid cooking fats such as Crisco are produced by hydrogenating soybean, peanut, or cottonseed oil until the preferred consistency is obtained.

PRACTICE PROBLEM 16.1

Draw the structure of glyceryl tripalmitate, a typical fat molecule.

Solution Glyceryl tripalmitate is the triester of glycerol with three molecules of palmitic acid, $CH_3(CH_2)_{14}COOH$:

$$\begin{array}{l} CH_2OC(=O)CH_2CH_2CH_2CH_2CH_2CH_2CH_2CH_2CH_2CH_2CH_2CH_2CH_2CH_2CH_3 \\ | \\ CHOC(=O)CH_2CH_2CH_2CH_2CH_2CH_2CH_2CH_2CH_2CH_2CH_2CH_2CH_2CH_2CH_3 \\ | \\ CH_2OC(=O)CH_2CH_2CH_2CH_2CH_2CH_2CH_2CH_2CH_2CH_2CH_2CH_2CH_2CH_2CH_3 \end{array}$$

Glyceryl tripalmitate

PROBLEM

16.2 Draw structures of the following compounds. Which would you expect to have a higher melting point?
(a) Glyceryl trioleate
(b) Glyceryl monooleate distearate

PROBLEM

16.3 Fats and oils can be either optically active or optically inactive, depending on their structures. Draw the structure of an optically active fat that gives 2 equivalents of palmitic acid and 1 equivalent of stearic acid on hydrolysis. Draw the structure of an optically inactive fat that gives the same products on hydrolysis.

16.3 Soaps

Soap has been known since at least 600 BC, when the Phoenicians prepared a curdy material by boiling goat fat with extracts of wood ash. The cleansing properties of soap weren't generally recognized, however, and the use of soap didn't become widespread until the eighteenth century. Chemically, soap is a mixture of the sodium or potassium salts of long-chain fatty acids produced by hydrolysis (*saponification*) of animal fat with alkali:

$$\begin{array}{l} \text{CH}_2\text{OCR} \ (\text{C}=\text{O}) \\ | \\ \text{CHOCR} \ (\text{C}=\text{O}) \\ | \\ \text{CH}_2\text{OCR} \ (\text{C}=\text{O}) \end{array} \xrightarrow[\text{H}_2\text{O}]{\text{NaOH}} 3\ \text{RCO}^-\ \text{Na}^+ \ (\text{C}=\text{O}) + \begin{array}{l} \text{CH}_2\text{OH} \\ | \\ \text{CHOH} \\ | \\ \text{CH}_2\text{OH} \end{array}$$

A fat
($R = C_{11}$–C_{19} aliphatic chains)

Soap

Glycerol

Crude soap curds contain glycerol and excess alkali as well as soap but can be purified by boiling with water and adding NaCl to precipitate the pure sodium carboxylate salts. The smooth soap that results is dried, perfumed, and pressed into bars. Dyes are added for colored soaps, antiseptics are added for medicated soaps, pumice is added for scouring soaps, and air is blown in for soaps that float.

Soaps act as cleansers because the two ends of a soap molecule are so different. The carboxylate end of the long-chain molecule is ionic and therefore *hydrophilic* (water-loving). As a result, it tries to dissolve in water. The long aliphatic chain portion of the molecule, however, is nonpolar and *hydrophobic* (water-fearing). It tries to avoid water and to dissolve in grease. The net effect of these two opposing tendencies is that soaps are attracted to both grease and water and are therefore useful as cleansers.

When soaps are dispersed in water, the long hydrocarbon tails cluster together into a hydrophobic ball, while the ionic heads on the surface of the

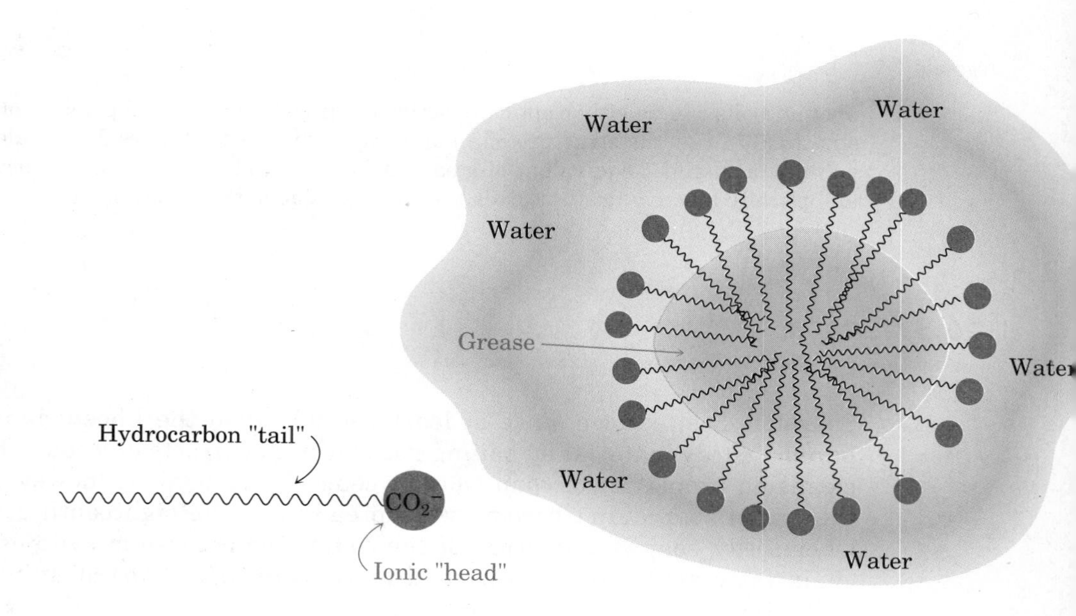

Figure 16.1 A soap micelle solubilizing a grease particle in water.

cluster stick out into the water layer. These spherical clusters, called **micelles,** are shown schematically in Figure 16.1. Grease and oil droplets are solubilized in water when they become coated by the hydrophobic nonpolar tails of soap molecules in the center of micelles. Once solubilized, the grease and dirt can be rinsed away.

Soaps make life much more pleasant than it would otherwise be, but they also have drawbacks. In hard water, which contains metal ions, soluble sodium carboxylates are converted into insoluble calcium and magnesium salts, leaving the familiar ring of scum around bathtubs and the gray tinge on clothes. Chemists have circumvented these problems by synthesizing a class of synthetic detergents based on salts of long-chain alkylbenzenesulfonic acids. The principle of synthetic detergents is identical to that of soaps: The alkylbenzene end of the molecule is attracted to grease, and the ionic sulfonate end is attracted to water. Unlike soaps, though, sulfonate detergents don't form insoluble metal salts in hard water and don't leave an unpleasant scum.

$$R-C_6H_4-S(=O)_2-O^- Na^+$$

A synthetic detergent
(R = a mixture of C_{12} aliphatic chains)

PROBLEM

16.4 Draw the structure of magnesium oleate, one of the components of bathtub scum.

PROBLEM

16.5 Write the saponification reaction of glyceryl monopalmitate dioleate with aqueous NaOH.

16.4 Phospholipids

Just as waxes, fats, and oils are esters of carboxylic acids, **phospholipids** are esters of phosphoric acid, H_3PO_4. There are two main kinds of phospholipids: *phosphoglycerides* and *sphingolipids.*

Phosphoglycerides are closely related to fats and oils in that they contain a glycerol backbone linked by ester bonds to two fatty acids and one phosphoric acid. Although the fatty acid residues can be any of the C_{12}–C_{20} units normally present in fats, the acyl group at C1 is usually saturated, and that at C2 is usually unsaturated. The phosphate group at C3 is also bonded by a separate ester link to an amino alcohol such as choline, $HOCH_2CH_2\overset{+}{N}(CH_3)_3$, or ethanolamine, $HOCH_2CH_2NH_2$.

The most important phosphoglycerides are the *lecithins* and the *cephalins.* Note that these compounds are chiral and that they have the L, or *R*, configuration at C2.

L configuration

$$\begin{array}{l} \phantom{R'-\overset{O}{\overset{\|}{C}}-O-}CH_2O-\overset{O}{\overset{\|}{C}}-R \\ R'-\overset{O}{\overset{\|}{C}}-O-\overset{|}{\underset{|}{C}}-H \\ \phantom{R'-\overset{O}{\overset{\|}{C}}-O-}CH_2O-\overset{O}{\overset{\|}{\underset{O^-}{\underset{|}{P}}}}-O-CH_2CH_2\overset{+}{N}(CH_3)_3 \end{array}$$

Phosphatidylcholine, a lecithin

$$\begin{array}{l} \phantom{R'-\overset{O}{\overset{\|}{C}}-O-}CH_2O-\overset{O}{\overset{\|}{C}}-R \\ R'-\overset{O}{\overset{\|}{C}}-O-\overset{|}{\underset{|}{C}}-H \\ \phantom{R'-\overset{O}{\overset{\|}{C}}-O-}CH_2O-\overset{O}{\overset{\|}{\underset{O^-}{\underset{|}{P}}}}-O-CH_2CH_2\overset{+}{N}H_3 \end{array}$$

Phosphatidylethanolamine, a cephalin

where R is saturated and R′ is unsaturated

Found widely in plant and animal tissues, phosphoglycerides are the major lipid component of cell membranes (approximately 40%). Like soaps, phosphoglycerides have a long, nonpolar hydrocarbon tail bonded to a polar ionic head (the phosphate group). Cell membranes are composed mostly of phosphoglycerides oriented into a **lipid bilayer** about 50 Å thick. As shown in Figure 16.2, the hydrophobic tails aggregate in the center of the bilayer in much the same way that soap tails aggregate in the center of a micelle (Figure 16.1). The bilayer thus forms an effective barrier to the passage of ions and other polar components into and out of the cell.

The second major group of phospholipids is the **sphingolipids.** These substances, which have *sphingosine* or a related dihydroxyamine as their

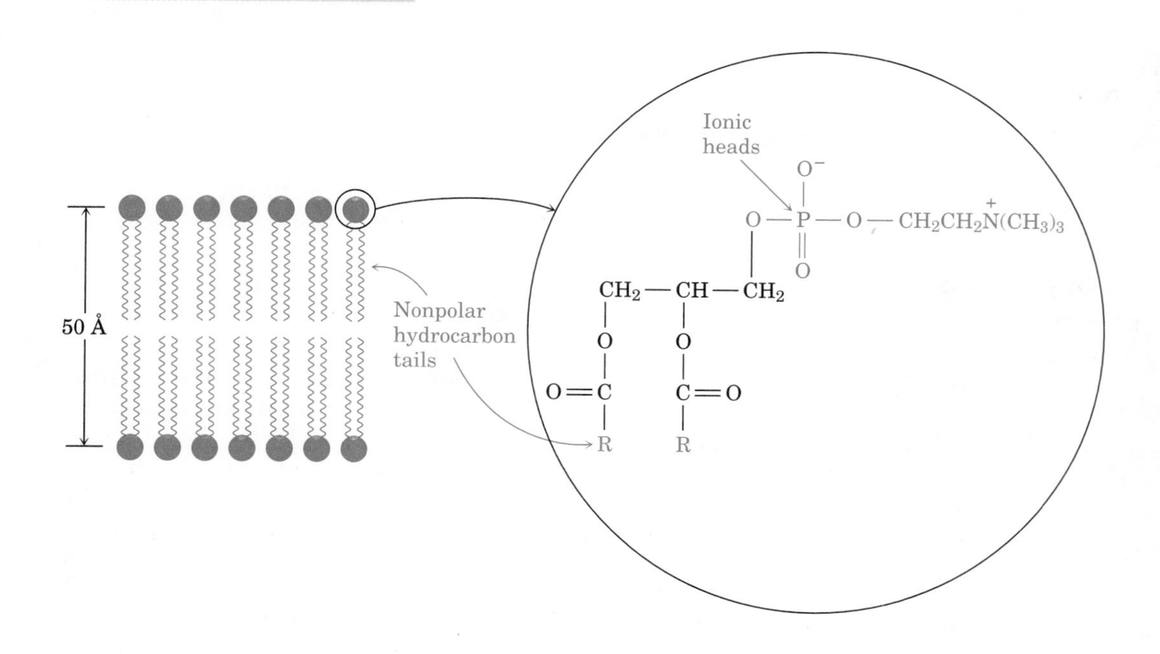

Figure 16.2 Aggregation of phosphoglycerides into the lipid bilayer that composes cell membranes.

backbones, are constituents of plant and animal cell membranes. They are particularly abundant in brain and nerve tissue, where compounds called *sphingomyelins* are a major constituent of the coating around nerve fibers.

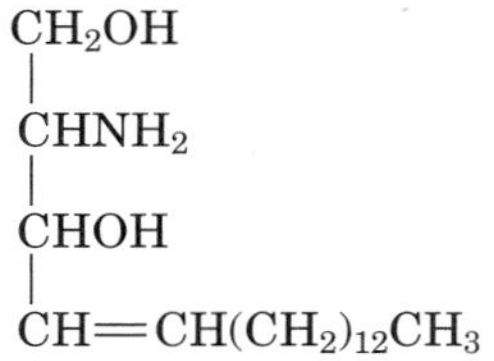

$$\begin{array}{l} CH_2OH \\ | \\ CHNH_2 \\ | \\ CHOH \\ | \\ CH{=}CH(CH_2)_{12}CH_3 \end{array}$$

Sphingosine

$$\begin{array}{l} CH_2O-\overset{\overset{\displaystyle O}{\|}}{\underset{\underset{\displaystyle O^-}{|}}{P}}-OCH_2CH_2\overset{+}{N}(CH_3)_3 \\ | \\ CHNHCO(CH_2)_{16\text{–}24}CH_3 \\ | \\ CHOH \\ | \\ CH{=}CH(CH_2)_{12}CH_3 \end{array}$$

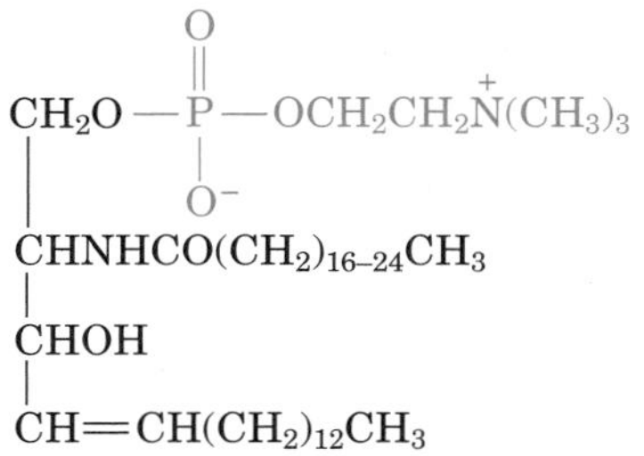

Sphingomyelin, a sphingolipid

16.5 Steroids

In addition to fats and phospholipids, the lipid extracts of plants and animals also contain **steroids,** molecules whose structures are based on the tetracyclic ring system shown below. The four rings are designated A, B, C, and D, beginning at the lower left, and the carbon atoms are numbered beginning in the A ring. The three six-membered rings (A, B, and C) adopt minimum-energy chair conformations, but are constrained by their rigid conformations from undergoing the usual cyclohexane ring-flip (Section 2.11).

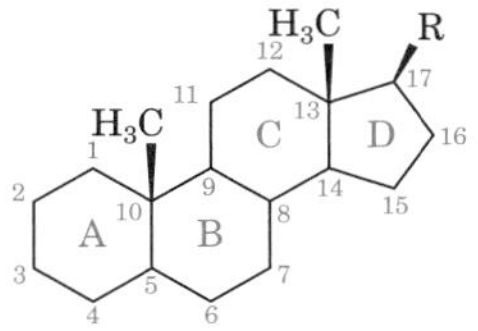

A steroid
(R = various side chains)

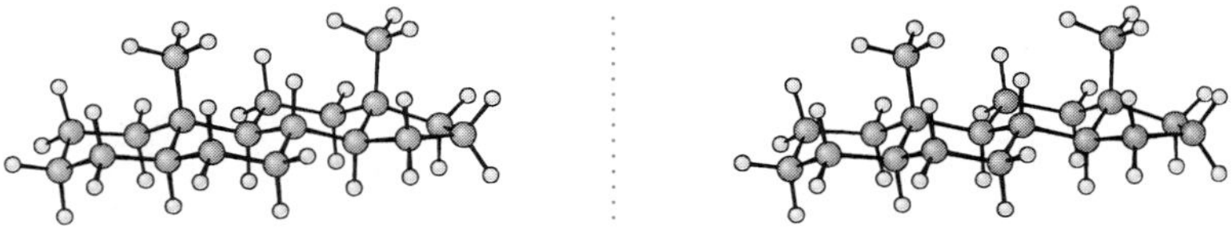

Stereo View

In humans, most steroids function as **hormones,** chemical messengers that are secreted by glands and carried through the bloodstream to target tissues. There are two main classes of steroid hormones: the *sex hormones,* which control maturation and reproduction, and the *adrenocortical hormones,* which regulate a variety of metabolic processes.

Sex Hormones

Testosterone and androsterone are the two most important male sex hormones, or **androgens.** Androgens are responsible for the development of male secondary sex characteristics during puberty and for promoting tissue and muscle growth. Both are synthesized in the testes from cholesterol.

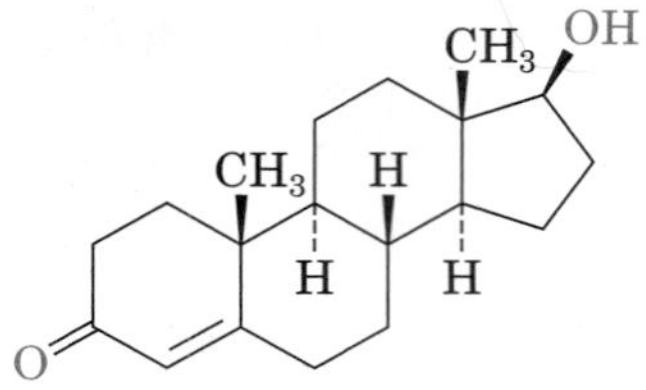

Testosterone 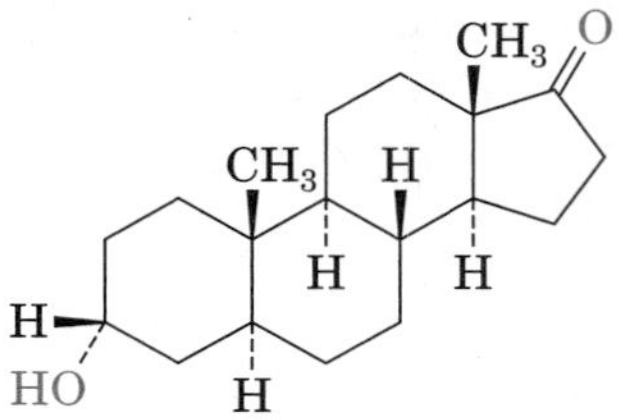 **Androsterone**

(Androgens)

Estrone and estradiol are the two most important female sex hormones, or **estrogens.** Synthesized in the ovaries from testosterone, estrogenic hormones are responsible for the development of female secondary sex characteristics and for regulation of the menstrual cycle. Note that both have a benzene-like aromatic A ring. In addition, another kind of sex hormone called a *progestin* is essential for preparing the uterus for implantation of a fertilized ovum during pregnancy. *Progesterone* is the most important progestin.

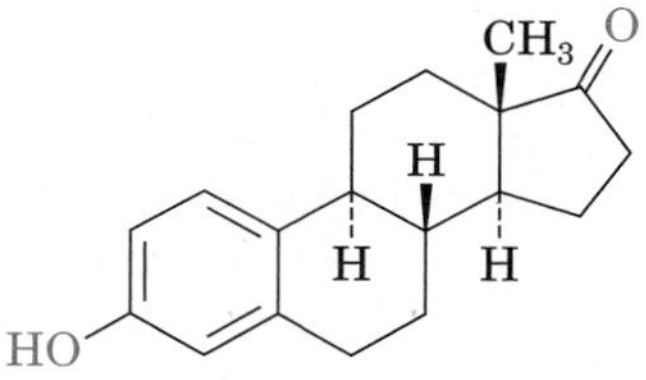

Estrone 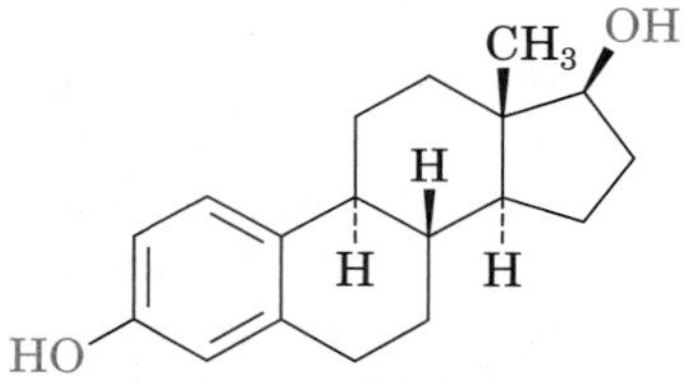 **Estradiol** **Progesterone (a progestin)**

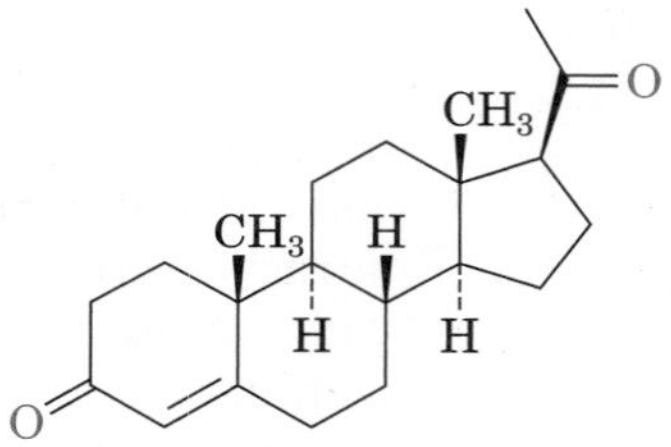

(Estrogens)

Adrenocortical Hormones

Adrenocortical steroids are secreted by the adrenal glands, small organs located near the upper end of each kidney. There are two types of adrenocortical steroids, called *mineralocorticoids* and *glucocorticoids.* Mineralocorticoids, such as aldosterone, control tissue swelling by regulating cellular salt balance between Na^+ and K^+. Glucocorticoids, such as hydrocortisone, are involved in the regulation of glucose metabolism and in the control of inflammation. Glucocorticoid ointments are widely used to bring down the swelling from exposure to poison oak or poison ivy.

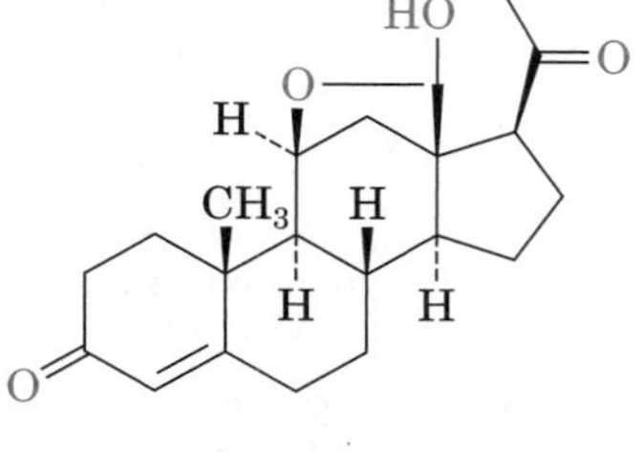

Aldosterone (a mineralocorticoid) **Hydrocortisone (a glucocorticoid)**

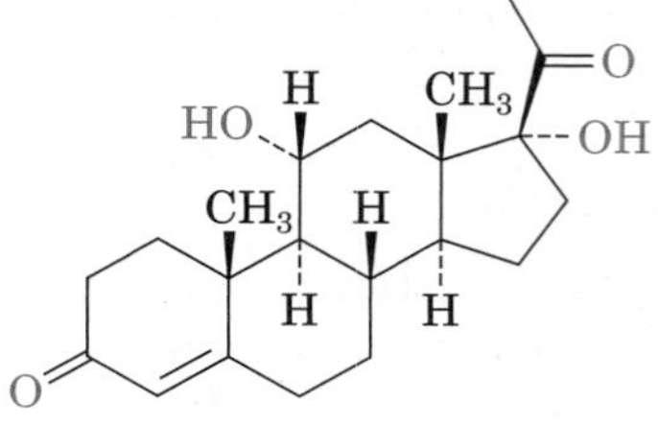

Synthetic Steroids

In addition to the many hundreds of steroids isolated from plants and animals, thousands more have been synthesized in pharmaceutical laboratories in a search for new drugs. The idea is to start with a natural hormone, carry out a chemical modification of the structure, and then see what biological properties the modified steroid has.

Among the best-known synthetic steroids are the oral contraceptives and the anabolic agents. Most birth-control pills are a mixture of two compounds, a synthetic estrogen, such as ethynylestradiol, and a synthetic progestin, such as norethindrone. Anabolic steroids, such as stanozolol (detected in several athletes during the 1988 Olympic Games), are synthetic androgens that mimic the tissue-building effects of natural testosterone.

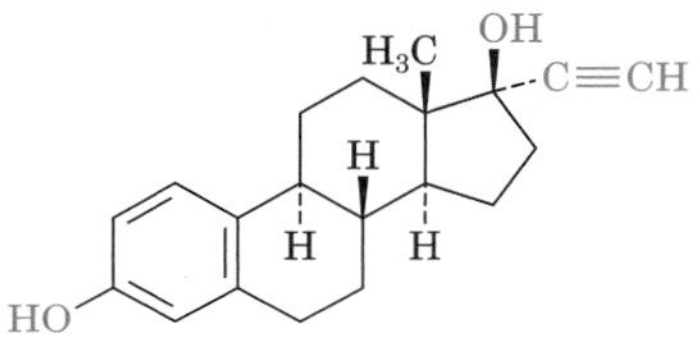

Ethynylestradiol (a synthetic estrogen)

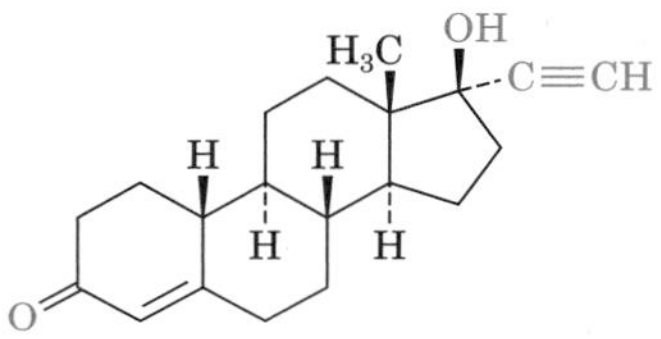

Norethindrone (a synthetic progestin)

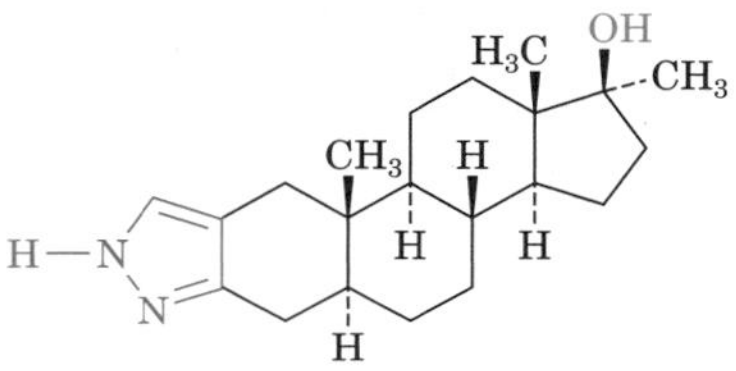

Stanozolol (an anabolic agent)

PROBLEM

16.6 Look at the structure of cholesterol shown at the beginning of this chapter, and tell whether the hydroxyl group is axial or equatorial.

PROBLEM

16.7 Look at the structure of progesterone, and identify all the functional groups in the molecule.

PROBLEM

16.8 Look at the structures of estradiol and ethynylestradiol, and point out the differences. What common structural feature do they share that makes both estrogens?

16.6 Nucleic Acids and Nucleotides

The **nucleic acids, deoxyribonucleic acid (DNA)** and **ribonucleic acid (RNA),** are the chemical carriers of a cell's genetic information. Coded in a cell's DNA is all the information that determines the nature of the cell, controls cell growth and division, and directs biosynthesis of the enzymes and other proteins required for all cellular functions.

Just as proteins are polymers of amino acid units, nucleic acids are polymers of individual building blocks called **nucleotides** linked together to form a long chain. Each nucleotide is composed of a **nucleoside** bonded to a phosphate group, and each nucleoside is composed of an aldopentose sugar joined to a heterocyclic amine base (Section 12.6).

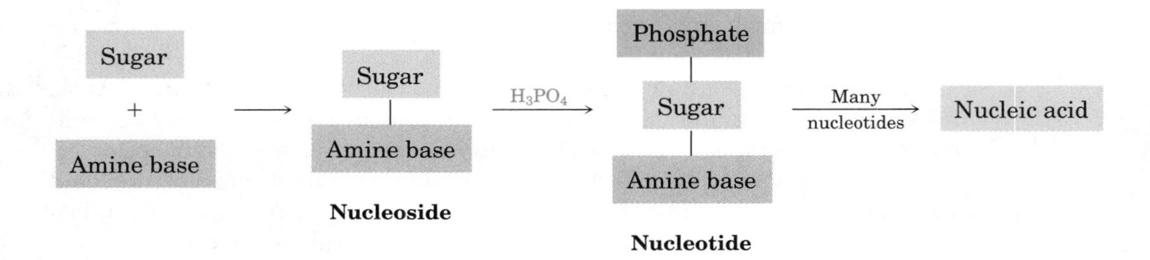

The sugar component in RNA is ribose, and the sugar in DNA is 2-deoxyribose. (The prefix *2-deoxy* means that oxygen is missing from C2 of ribose.)

Ribose **2-Deoxyribose**

There are four different heterocyclic amine bases in DNA. Two are substituted *purines* (adenine and guanine), and two are substituted *pyrimidines* (cytosine and thymine). Adenine, guanine, and cytosine also occur in RNA, but thymine is replaced in RNA by a different pyrimidine base called uracil.

Pyrimidine **Cytosine** **Uracil (RNA)** **Thymine (DNA)**

Purine **Adenine** **Guanine**

In both DNA and RNA, the heterocyclic amine base is bonded to C1′ of the sugar, and the phosphoric acid is bonded by a phosphate ester linkage to the C5′ sugar position. Thus, nucleosides and nucleotides have the general structure shown in Figure 16.3. (In referring to nucleic acids, numbers with a prime superscript refer to positions on the sugar, and numbers without a prime refer to positions on the heterocyclic amine base.) The complete structures of all four deoxyribonucleotides and all four ribonucleotides are shown in Figure 16.4 (p. 520).

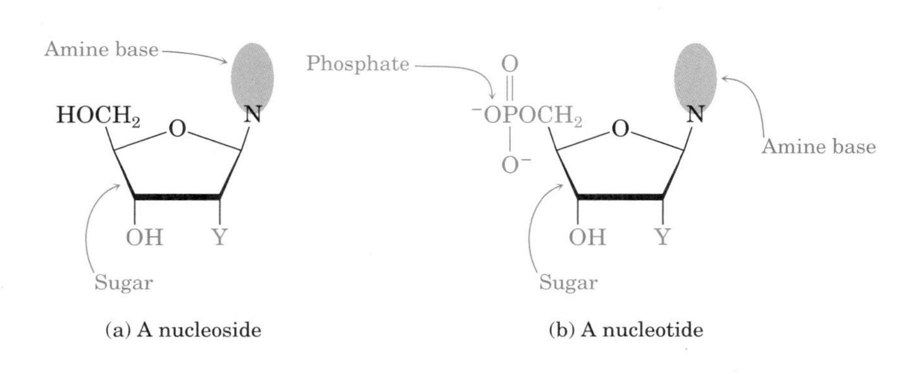

Figure 16.3 General structures of (a) a nucleoside and (b) a nucleotide. When Y = H, the sugar is deoxyribose; when Y = OH, the sugar is ribose.

Though chemically similar, DNA and RNA differ in size and have different roles within the cell. Molecules of DNA are enormous. With molecular weights of up to 150 billion and lengths of up to 12 cm, they are found mostly in the nucleus of the cell. Molecules of RNA, by contrast, are much smaller (as low as 35,000 mol wt) and are found mostly outside the cell nucleus. We'll consider the two kinds of nucleic acids separately, beginning with DNA.

16.7 Structure of DNA

Nucleotides join together in DNA by forming a phosphate ester bond between the 5′-phosphate group on one nucleotide and the 3′-hydroxyl group on the sugar of another nucleotide (Figure 16.5, p. 521). One end of the nucleic acid polymer thus has a free hydroxyl at C3′ (the **3′ end**), and the other end has a phosphate at C5′ (the **5′ end**).

Just as the structure of a protein depends on the sequence in which individual amino acids are connected, the structure of a nucleic acid depends on the sequence of individual nucleotides. To carry the analogy further, just as a protein has a polyamide backbone with different side chains attached

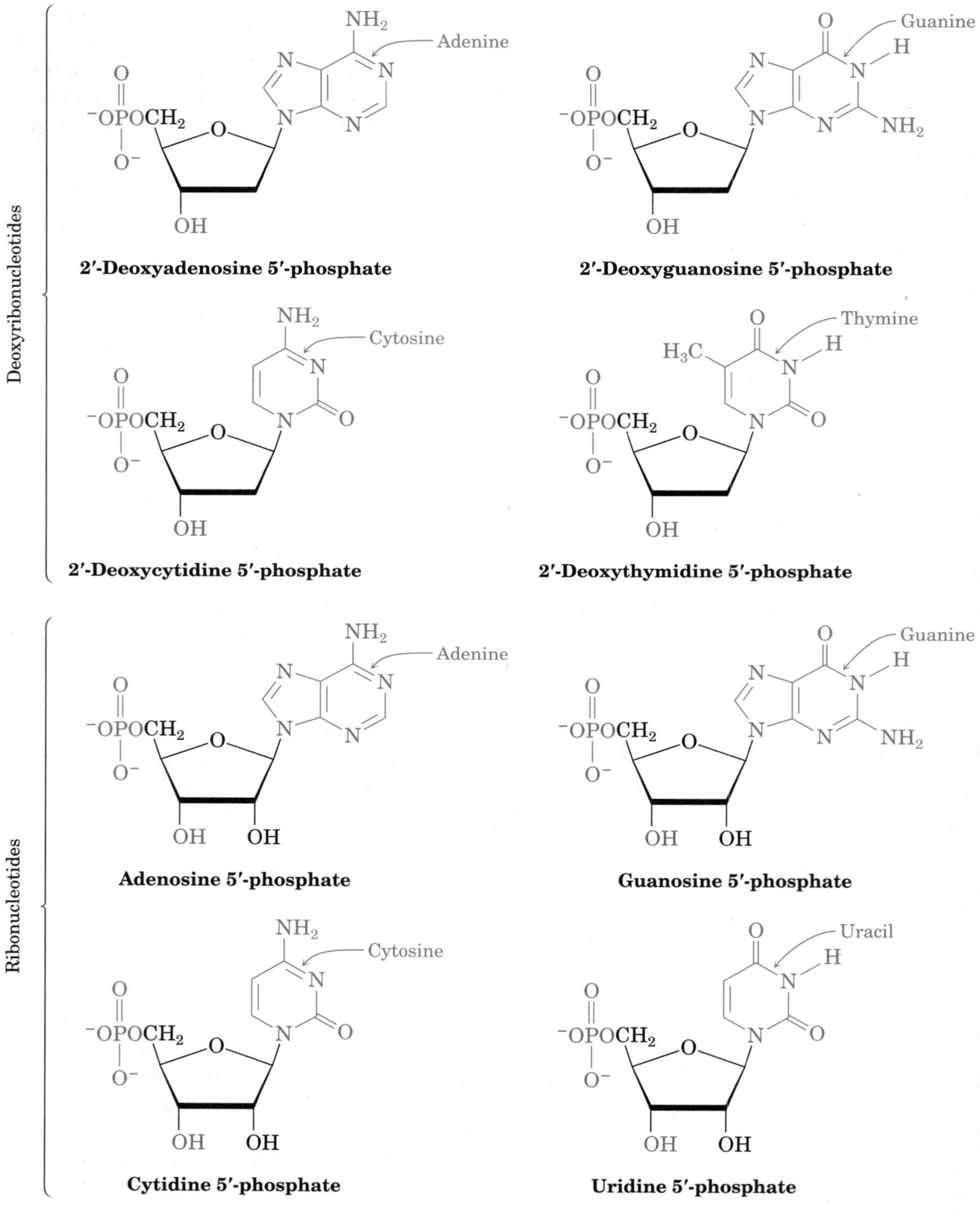

Deoxyribonucleotides

Adenine

2′-Deoxyadenosine 5′-phosphate

Guanine

2′-Deoxyguanosine 5′-phosphate

Cytosine

2′-Deoxycytidine 5′-phosphate

Thymine

2′-Deoxythymidine 5′-phosphate

Ribonucleotides

Adenine

Adenosine 5′-phosphate

Guanine

Guanosine 5′-phosphate

Cytosine

Cytidine 5′-phosphate

Uracil

Uridine 5′-phosphate

Figure 16.4 Structures of the four deoxyribonucleotides and the four ribonucleotides.

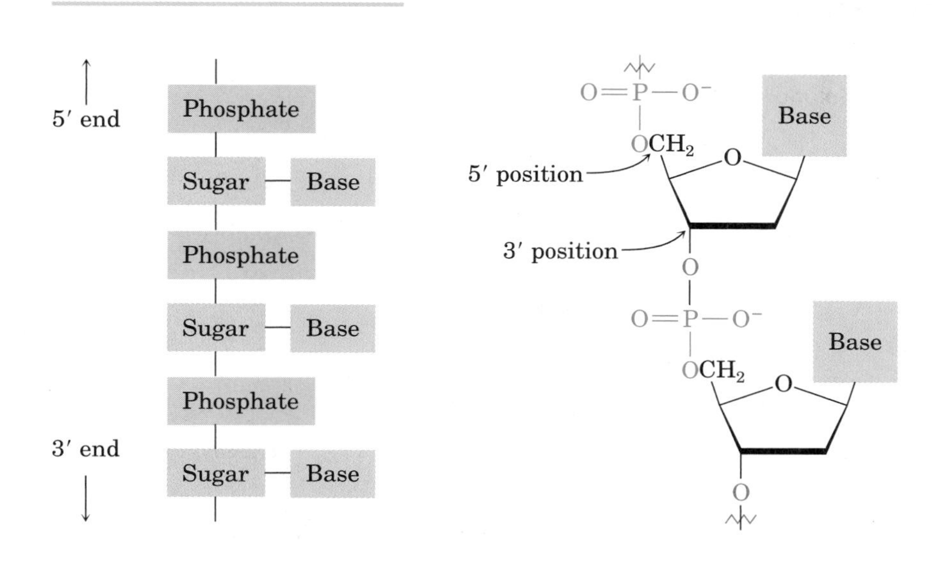

Figure 16.5 *Generalized structure of DNA.*

to it, a nucleic acid has an alternating sugar–phosphate backbone with different amine bases attached.

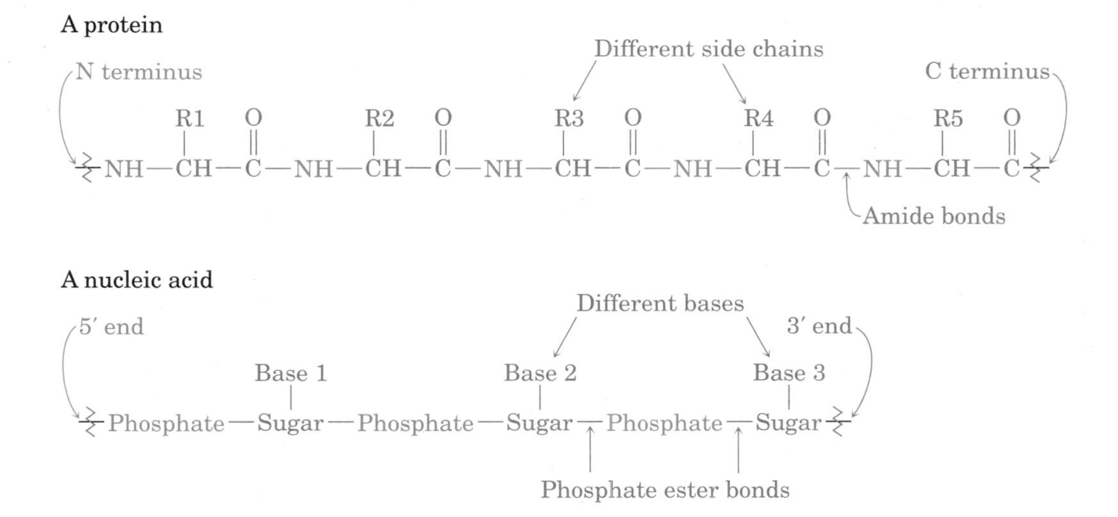

The sequence of nucleotides in a chain is described by starting at the 5′ end and identifying the bases in order of occurrence. Rather than write the full name of each nucleotide, though, it's easier to use abbreviations: A for adenosine, T for thymine, G for guanosine, and C for cytidine. Thus, a typical sequence might be written as T-A-G-G-C-T.

PRACTICE PROBLEM 16.2

Draw the full structure of the DNA dinucleotide C-T.

Solution

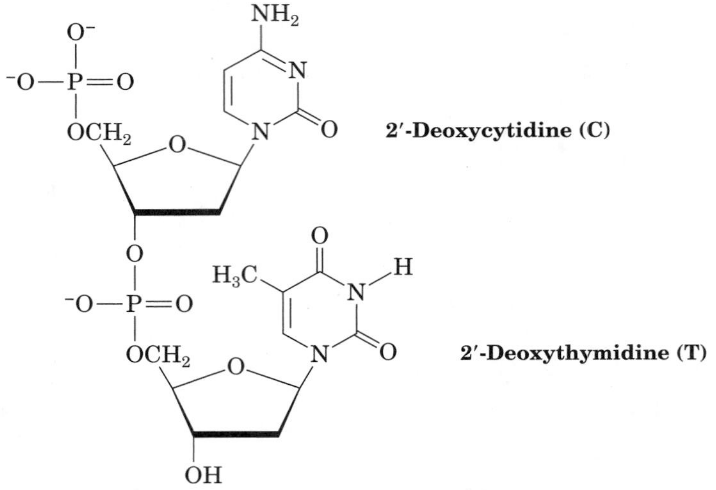
NH2
N
O−
−O—P=O
OCH2
O
N
O
2′-Deoxycytidine (C)
O
H3C
N
H
−O—P=O
OCH2
O
N
O
2′-Deoxythymidine (T)
OH

PROBLEM

16.9 Draw the full structure of the DNA dinucleotide A-G.

PROBLEM

16.10 Draw the full structure of the RNA dinucleotide U-A.

16.8 Base Pairing in DNA: The Watson–Crick Model

Samples of DNA isolated from different tissues of the same species have the same proportions of heterocyclic bases, but samples from different species can have greatly different proportions of bases. Human DNA, for example, contains about 30% each of A and T and about 20% each of G and C. The bacterium *Clostridium perfringens,* however, contains about 37% each of A and T and only 13% each of G and C. Note that in both examples, the bases occur in pairs; A and T are usually present in equal amounts, as are G and C. Why should this be?

In 1953, James Watson and Francis Crick made their classic proposal for the secondary structure of DNA. According to the Watson–Crick model, DNA consists of two polynucleotide strands coiled around each other in a **double helix.** The two strands run in opposite directions and are held together by hydrogen bonds between specific pairs of bases. Guanine (G) and cytosine (C) hydrogen bond to each other but not to A or T. Similarly, A and T hydrogen bond to each other but not to G or C.

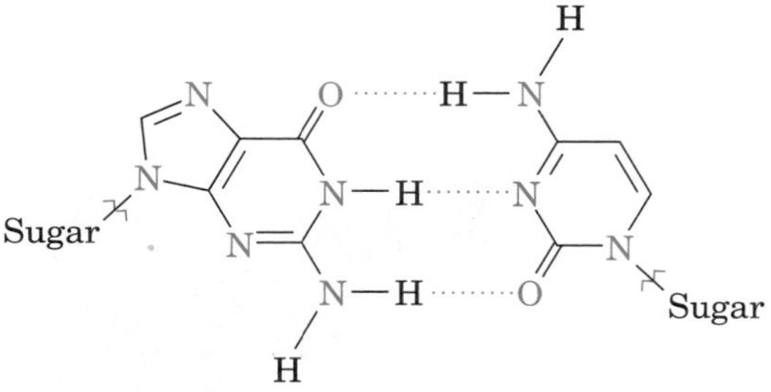

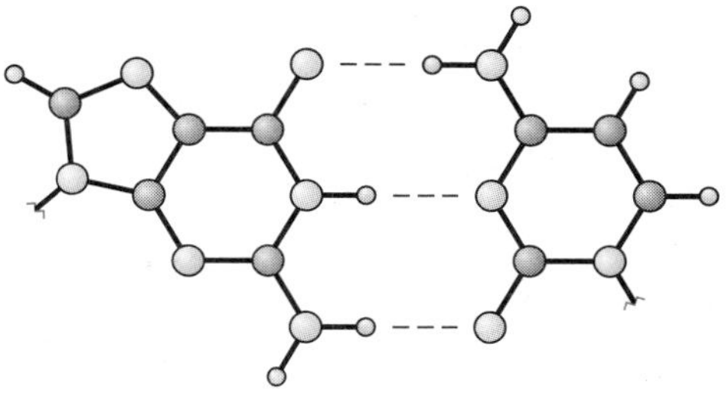

(Guanine) G :::::: C (Cytosine)

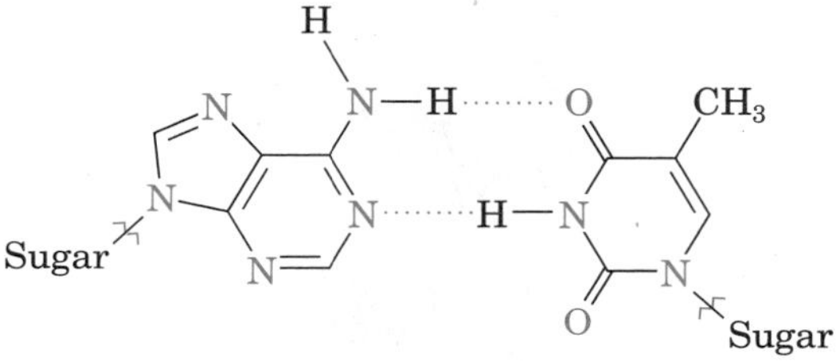

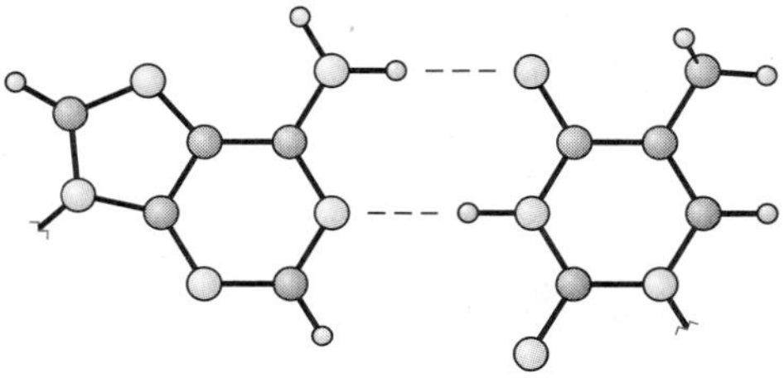

(Adenine) A :::::: T (Thymine)

The two strands of the DNA double helix are not identical; rather, they're complementary. Whenever a G occurs in one strand, a C occurs opposite it in the other strand. When an A occurs in one strand, a T occurs in the other strand. This complementary pairing of bases explains why A and T are always found in equal amounts, as are G and C. Figure 16.6 (p. 524) illustrates this base pairing and shows how the two complementary strands coil into the double helix. X-ray measurements reveal that the DNA double helix is 20 Å wide, that there are 10 base pairs in each full turn, and that each turn is 34 Å in height.

A helpful mnemonic device to remember the pairing of the four DNA bases is the phrase "pure silver taxi":

Pure	Silver	Taxi
Pur	Ag	TC
The purine bases,	A and G,	pair with T and C.

The two strands of the double helix coil in such a way that two kinds of "grooves" result, a *major groove* 12 Å wide and a *minor groove* 6 Å wide. The major groove is slightly deeper than the minor groove, and both are lined by potential hydrogen bond donors and acceptors. As a result, a variety of flat, polycyclic aromatic molecules are able to *intercalate,* or fit sideways between the strands. Many cancer-causing and cancer-preventing agents are thought to function by interacting with DNA in this way.

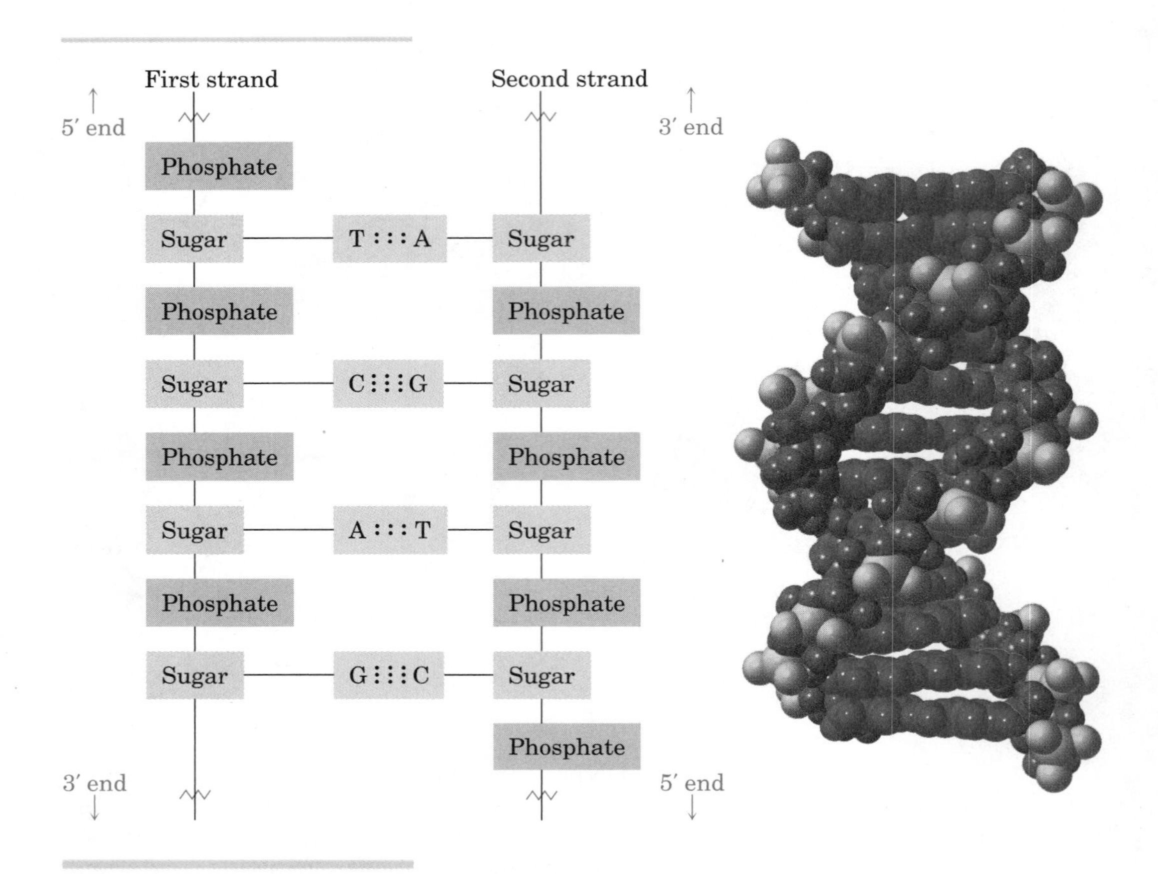

Figure 16.6 Complementarity of base pairing in the DNA double helix as shown by this computer-generated structure. The sugar–phosphate backbone runs along the outside of the helix, while the amine bases form hydrogen bonds to one another on the inside.

PRACTICE PROBLEM 16.3

What sequence of bases on one strand of DNA is complementary to the sequence T-A-T-G-C-A-T on another strand?

Solution Since A and G form complementary pairs with T and C, respectively, go through the given sequence replacing A by T, G by C, T by A, and C by G:

Original: T-A-T-G-C-A-T

Complement: A-T-A-C-G-T-A

PROBLEM

16.11 What sequence of bases on one strand of DNA is complementary to the following sequence on another strand?

G-G-C-T-A-A-T-C-C-G-T

16.9 Nucleic Acids and Heredity

DNA is the chemical repository of an organism's genetic information, which is stored as a sequence of deoxyribonucleotides strung together in the DNA chain. For the information to be preserved and passed on to future generations, a mechanism must exist for copying DNA. For the information to be used, a mechanism must exist for decoding the DNA message and implementing the instructions it contains.

What Crick called the "central dogma of molecular genetics" says that the function of DNA is to store information and pass it on to RNA. The function of RNA, in turn, is to read, decode, and use the information received from DNA to make proteins. By decoding the right bit of DNA at the right time, an organism uses genetic information to synthesize the thousands of proteins necessary for functioning.

Replication ⟲ **DNA** —Transcription→ **RNA** —Translation→ **Proteins**

Three processes take place in the transfer of genetic information:

1. **Replication** is the process by which identical copies of DNA are made so that genetic information can be preserved and handed down to offspring.
2. **Transcription** is the process by which the genetic messages are read and carried out of the cell nucleus.
3. **Translation** is the process by which the genetic messages are decoded and used to build proteins.

16.10 Replication of DNA

DNA **replication** is an enzyme-catalyzed process that begins by a partial unwinding of the double helix. As the DNA strands separate and bases are exposed, new nucleotides line up on each strand in an exactly complementary manner, A to T and C to G, and two new strands begin to grow. Each new strand is complementary to its old template strand, and two new DNA double helices are produced (Figure 16.7, p. 526).

Crick probably described the process best when he used the analogy of the two DNA strands fitting together like a hand in a glove. The hand and glove separate, a new hand forms inside the glove, and a new glove forms around the hand. Two identical copies now exist where only one existed before.

The process by which the nucleotides are joined to create new DNA strands involves many steps and many different enzymes. Addition of new nucleotide units to the growing chain is catalyzed by the enzyme *DNA poly-*

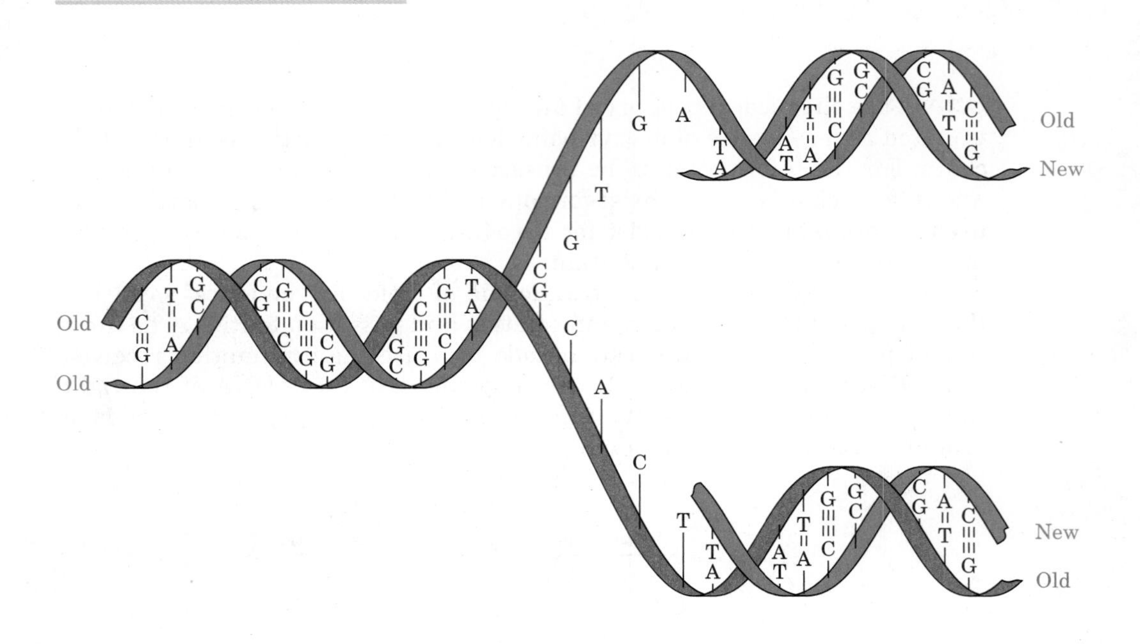

Figure 16.7 Schematic representation of DNA replication.

merase and occurs by addition of a 5′-mononucleotide triphosphate to the free 3′-hydroxyl group of the growing chain.

5′ end

3′ end

Template strand

New strand

5′ end

The magnitude of the replication process is staggering. The nucleus of a human cell contains 46 chromosomes (23 pairs), each of which consists of one very large DNA molecule. Each chromosome, in turn, is made up of several thousand DNA segments called *genes,* and the sum of all genes in a human cell (the *genome*) is estimated to be approximately 3 billion base pairs. Despite the size of these massive molecules, the base sequence is faithfully copied during replication, with an error occurring only about once each 10–100 billion bases.

16.11 Structure and Synthesis of RNA: Transcription

RNA is structurally similar to DNA. Both are sugar–phosphate polymers, and both have heterocyclic bases attached. The only differences are that RNA contains ribose rather than 2-deoxyribose and uracil rather than thymine. Uracil in RNA forms strong hydrogen bonds to its complementary base, adenine, just as thymine does in DNA.

There are three major kinds of ribonucleic acid, each of which serves a specific function. All three are much smaller molecules than DNA, and all remain single-stranded rather than double-stranded.

Messenger RNA (mRNA) carries genetic messages from DNA to *ribosomes,* small granular particles in the cytoplasm of a cell where protein synthesis occurs.

Ribosomal RNA (rRNA) provides the physical makeup of ribosomes.

Transfer RNA (tRNA) transports specific amino acids to the ribosomes, where they are joined together to make proteins.

The conversion of the information in DNA into proteins begins in the nucleus of cells with the synthesis of mRNA by the process of **transcription.** Several turns of the DNA double helix unwind, forming a "bubble" and exposing the bases of the two strands. Ribonucleotides line up in the proper order by hydrogen bonding to their complementary bases on DNA, bond formation occurs in the 5′ → 3′ direction, and the growing RNA molecule unwinds from DNA (Figure 16.8).

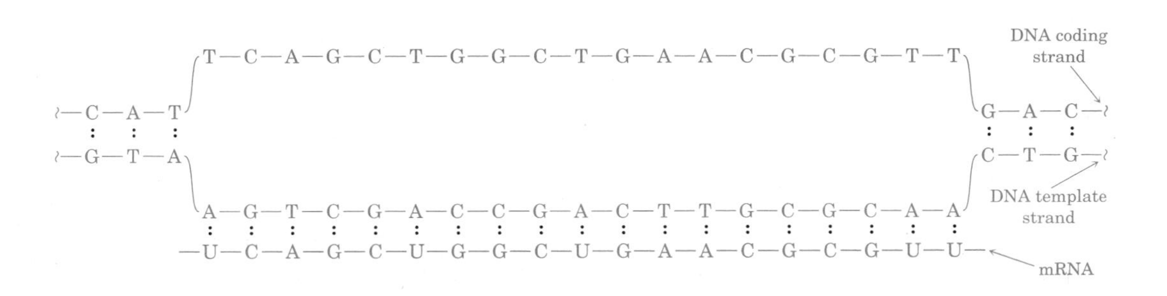

Figure 16.8 Synthesis of RNA using a DNA segment as template.

Unlike what happens in DNA replication, where both strands are copied, only one of the two DNA strands is transcribed into mRNA. The strand that contains the gene is called the **coding strand,** or **sense strand,** and the strand that gets transcribed is called the **template strand,** or **antisense strand.**

Transcription of DNA by the process just discussed raises many questions. How does the DNA know where to unwind? Where along the chain does one gene stop and the next one start? How do the ribonucleotides know the right place along the template strand to begin lining up and the right place to stop? The picture that has emerged in the last decade is that a DNA chain contains specific base sequences called *promoter sites* that lie at positions 10 base pairs and 35 base pairs upstream from the beginning of the coding region and signal the beginning of a gene. Similarly, there are other base sequences near the end of the gene that signal a stop.

PRACTICE PROBLEM 16.4

What RNA base sequence is complementary to the following DNA base sequence?

T-A-A-G-C-C-G-T-G

Solution Go through the sequence replacing A by U, G by C, T by A, and C by G:

Original DNA: T-A-A-G-C-C-G-T-G

Complementary RNA: A-U-U-C-G-G-C-A-C

PROBLEM

16.12 Show how uracil can form strong hydrogen bonds to adenine, just as thymine can.

PROBLEM

16.13 What RNA base sequence is complementary to the following DNA base sequence?

G-A-T-T-A-C-C-G-T-A

PROBLEM

16.14 From what DNA base sequence was the following RNA sequence transcribed?

U-U-C-G-C-A-G-A-G-U

16.12 RNA and Protein Biosynthesis: Translation

The primary cellular function of RNA is to direct biosynthesis of the thousands of diverse peptides and proteins required by an organism. The mechanics of protein biosynthesis are directed by mRNA and take place on *ribosomes,* small granular particles in the cytoplasm of a cell that consist of about 60% rRNA and 40% protein. The specific ribonucleotide sequence in mRNA acts like a long coded sentence to specify the order in which different amino acid residues are to be joined.

Each "word," or **codon,** along the mRNA chain consists of a sequence of three ribonucleotides that is specific for a given amino acid. For example, the series UUC on mRNA is a codon directing incorporation of the amino acid phenylalanine into the growing protein. Of the $4^3 = 64$ possible triplets of the four bases in RNA, 61 code for specific amino acids (most amino acids are specified by more than one codon) and 3 code for chain termination. Table 16.3 shows the meaning of each codon.

Table 16.3 Codon Assignments of Base Triplets

		Third base (3′ end)			
First base (5′ end)	*Second base*	*U*	*C*	*A*	*G*
U	U	Phe	Phe	Leu	Leu
	C	Ser	Ser	Ser	Ser
	A	Tyr	Tyr	Stop	Stop
	G	Cys	Cys	Stop	Trp
C	U	Leu	Leu	Leu	Leu
	C	Pro	Pro	Pro	Pro
	A	His	His	Gln	Gln
	G	Arg	Arg	Arg	Arg
A	U	Ile	Ile	Ile	Met
	C	Thr	Thr	Thr	Thr
	A	Asn	Asn	Lys	Lys
	G	Ser	Ser	Arg	Arg
G	U	Val	Val	Val	Val
	C	Ala	Ala	Ala	Ala
	A	Asp	Asp	Glu	Glu
	G	Gly	Gly	Gly	Gly

The message carried by mRNA is read by tRNA in a process called **translation.** There are 61 different tRNA's, one for each of the 61 codons in Table 16.3 that specifies an amino acid. A typical tRNA is roughly the shape of a cloverleaf, as shown in Figure 16.9 (p. 530). It consists of about 70–100 ribonucleotides and is bonded to a specific amino acid by an ester linkage through the 3′-hydroxyl on ribose at the end of the tRNA. Each tRNA also contains in its chain a segment called an **anticodon,** a sequence of three ribonucleotides that is complementary to the codon sequence. For example, the codon sequence UUC present on mRNA is read by a phenylalanine-bearing tRNA having the complementary anticodon sequence AAG.

As each successive codon on mRNA is read, appropriate tRNA's bring the correct amino acids into position for enzyme-mediated transfer to the growing peptide. When synthesis of the proper protein is completed, a "stop" codon signals the end, and the protein is released from the ribosome. The process of protein biosynthesis is illustrated schematically in Figure 16.10 (p. 531).

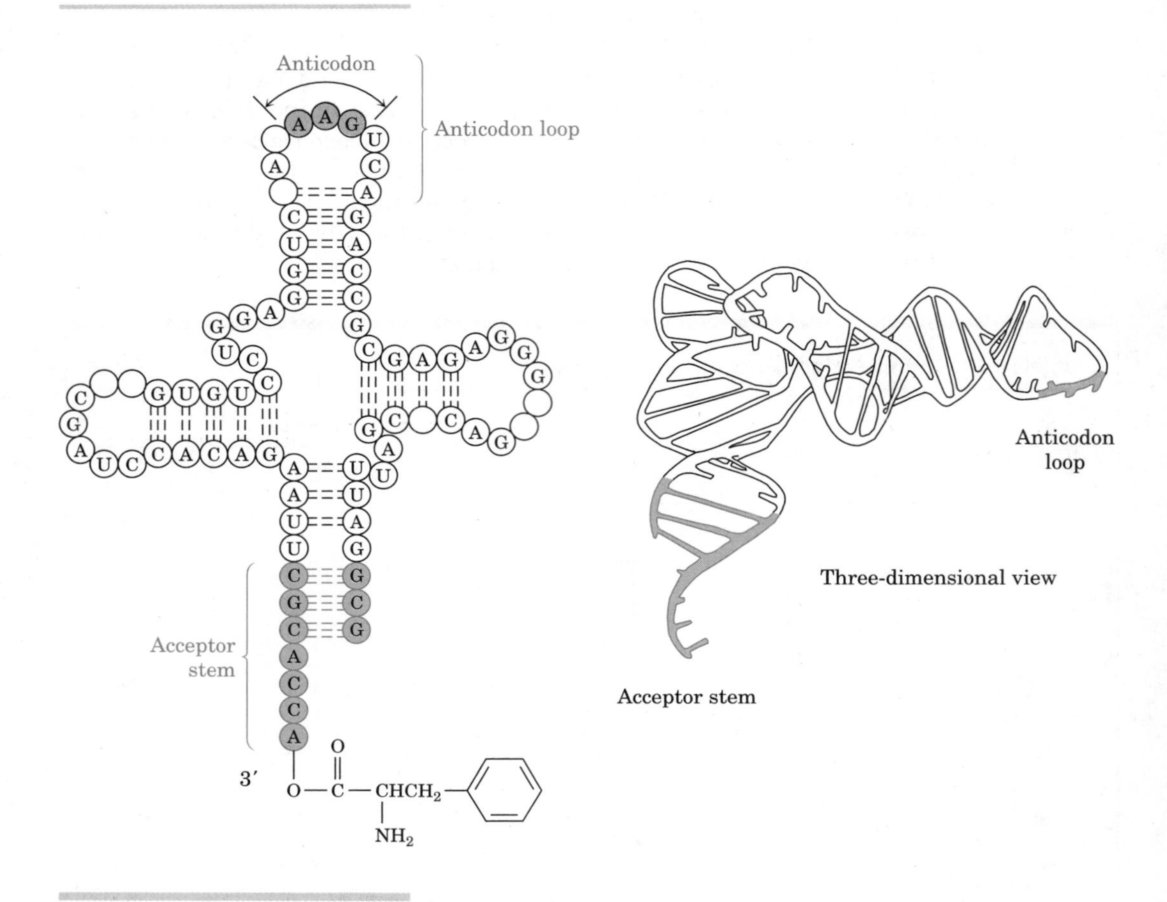

Figure 16.9 Structure of a tRNA molecule. The tRNA is a roughly cloverleaf-shaped molecule containing an anticodon triplet on one "leaf" and a covalently attached amino acid residue at its 3′ end. The example shown is a yeast tRNA that codes for phenylalanine. The nucleotides not specifically identified are chemically modified analogs of the four usual nucleotides.

PRACTICE PROBLEM 16.5

Give a codon sequence for valine.

Solution According to Table 16.3, there are four codons for valine: GUU, GUC, GUA, GUG.

PRACTICE PROBLEM 16.6

What amino acid sequence is coded by the mRNA base sequence AUC-GGU?

Solution Table 16.3 indicates that AUC codes for isoleucine and GGU codes for glycine. Thus, AUC-GGU codes for Ile-Gly.

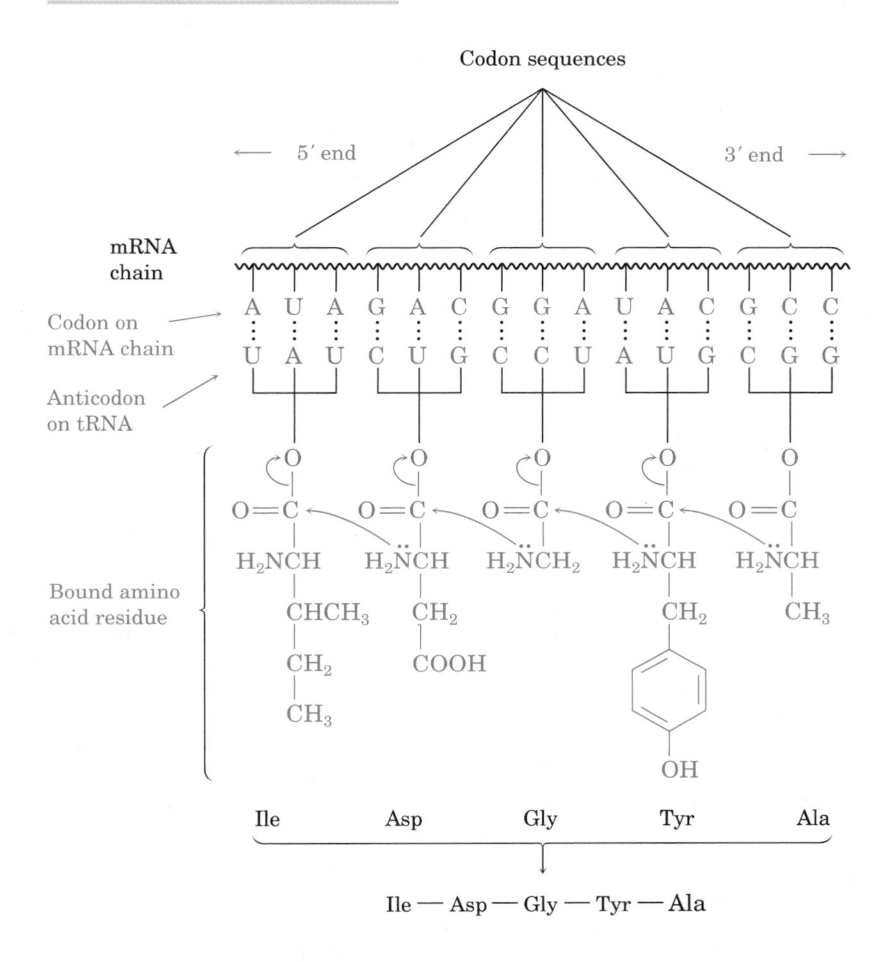

Figure 16.10 A schematic representation of protein biosynthesis. The mRNA containing codon base sequences is read by tRNA containing complementary anticodon base sequences. Transfer RNA assembles proper amino acids into position for incorporation into the peptide.

PROBLEM

16.15 List codon sequences for the following amino acids:
(a) Ala
(b) Phe
(c) Leu
(d) Tyr

PROBLEM

16.16 What amino acid sequence is coded by the following mRNA base sequence?

CUU-AUG-GCU-UGG-CCC-UAA

PROBLEM

16.17 What anticodon sequences of tRNA's are coded by the mRNA in Problem 16.16?

PROBLEM

16.18 What was the base sequence in the original DNA strand on which the mRNA sequence in Problem 16.16 was made?

16.13 Sequencing DNA

One of the greatest scientific revolutions in history is now underway in molecular biology as scientists are learning how to manipulate and harness the genetic machinery of organisms. None of the extraordinary advances of the past decade would have been possible, however, were it not for the discovery in 1977 of methods for sequencing immense DNA chains to find their messages.

There are two methods of DNA sequencing in general use. Both operate along similar lines, but the *Sanger method,* often called the *dideoxy method,* uses enzymatic reactions while the *Maxam–Gilbert method* uses chemical techniques. There are five steps to the Maxam–Gilbert method:

Step 1 The first problem in DNA sequencing is to cleave the enormous DNA chain at specific points to produce smaller, more manageable pieces. This problem has been solved by the use of enzymes called **restriction endonucleases.** Each different restriction enzyme, of which more than 200 are available, cleaves a DNA molecule at well-defined points along the chain where a specific base sequence occurs. For example, the restriction enzyme *Alu I* cleaves the linkage between G and C in the four-base sequence AG-CT. If the original DNA molecule is cut with another restriction enzyme, still other segments are produced whose sequences partially overlap those produced by the first enzyme. Sequencing of all the segments, followed by identification of the overlapping sections, then allows complete DNA sequencing.

Step 2 After cleavage of the DNA into smaller pieces, called *restriction fragments,* each fragment is radioactively tagged by enzymatically incorporating a labeled ^{32}P phosphate group onto the 5′-hydroxyl of the terminal nucleotide. The double-stranded fragments are then separated into single strands by heating, and the strands are isolated. Imagine, for example, that we now have a single-stranded DNA fragment with the following partial structure:

(5′ end) ^{32}P-G-A-T-C-A-G-C-G-A-T- (3′ end)

Step 3 The labeled DNA sample is divided into four subsamples and subjected to four parallel sets of chemical reactions under conditions that cause:

(a) Splitting of the DNA chain next to A
(b) Splitting of the DNA chain next to G
(c) Splitting of the DNA chain next to C
(d) Splitting of the DNA chain next to *both* T and C

Mild reaction conditions are used so that *only a few of the many possible splittings occur in each reaction.* Thus, the pieces shown in Table 16.4 would be produced:

Table 16.4 Splitting of a DNA Fragment Under Four Sets of Conditions

Cleavage conditions	*Labeled DNA pieces produced*
Original DNA fragment	^{32}P-G-A-T-C-A-G-C-G-A-T-
Next to A	^{32}P-G ^{32}P-G-A-T-C ^{32}P-G-A-T-C-A-G-C-G + Larger pieces
Next to G	^{32}P-G-A-T-C-A ^{32}P-G-A-T-C-A-G-C + Larger pieces
Next to C	^{32}P-G-A-T ^{32}P-G-A-T-C-A-G + Larger pieces
Next to C + T	^{32}P-G-A ^{32}P-G-A-T ^{32}P-G-A-T-C-A-G ^{32}P-G-A-T-C-A-G-C-G-A + Larger pieces

Cleavages next to A and G are accomplished by treating the restriction fragment with dimethyl sulfate, $(CH_3O)_2SO_2$. Deoxyadenosine (A) is methylated at N3 (S_N2 reaction), and deoxyguanosine (G) is methylated at N7, but T and C aren't affected. Treatment of the methylated DNA with an aqueous solution of the secondary amine piperidine then brings about destruction of the methylated nucleotides and opening of the DNA chain at both the 3′ and 5′ positions next to the methylated bases. By working carefully, it's possible to find reaction conditions that are selective for cleavage either at A or at G.

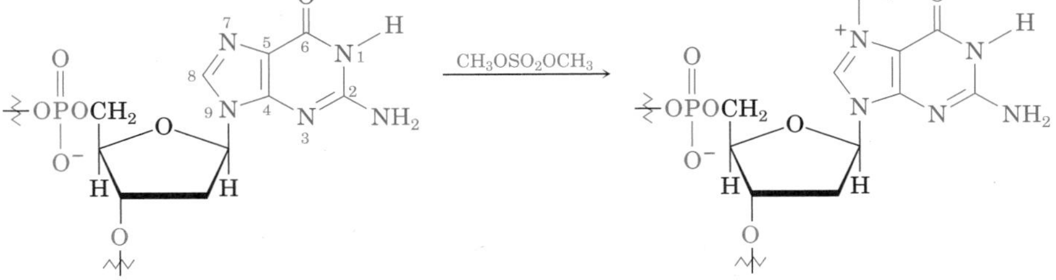

$\xrightarrow{CH_3OSO_2OCH_3}$

Deoxyguanosine

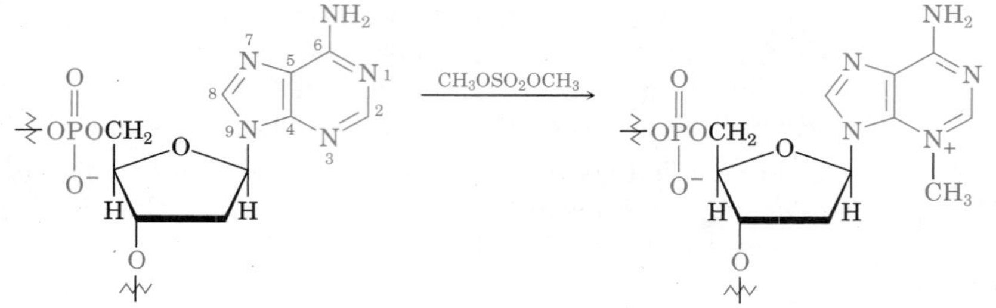

$\xrightarrow{CH_3OSO_2OCH_3}$

Deoxyadenosine

Breaking the DNA chain next to C and T is accomplished by treatment of DNA with hydrazine, H_2NNH_2, followed by heating with aqueous piperidine. Although conditions that are selective for cleavage next to T have not been found, selective cleavage next to C is accomplished by carrying out the hydrazine reaction in 5 M NaCl solution.

Step 4 Product mixtures from each of the four cleavage reactions are separated by electrophoresis (Section 15.3). When a mixture is placed at one end of a strip of buffered polyacrylamide gel and a voltage is applied to the two ends of the strip, each DNA piece moves along the gel at a rate that depends on the number of negatively charged phosphate groups it contains (that is, on the number of nucleotides it has). Smaller pieces move rapidly, and larger pieces move more slowly. The technique is so sensitive that up to 600 DNA pieces, differing in size by only one nucleotide, can be separated.

Once separated, the locations of the DNA cleavage products are detected by exposing the gel to a photographic plate. Each radioactive end piece containing a ^{32}P label appears as a dark band on the photographic plate, but nonradioactive pieces from the middle of the chain aren't seen. The gel electrophoresis pattern shown in Figure 16.11 would be obtained in our hypothetical example.

Step 5 The DNA sequence is read directly from the gel. The band that appears farthest from the origin is the terminal mononucleotide (the smallest piece) and can't be identified. Because the terminal mononucleotide appears in the A column, though, it must have been produced by splitting *next to* an A. Thus, the *second* nucleotide in the DNA sequence is an A. The second farthest band from the origin is a dinucleotide that appears only in the T + C column and is produced by splitting next to the third nucleotide. But because this piece does *not* appear in the C column, the third nucleotide is not a C and must therefore be a T. The third farthest band appears in both C and T + C columns, so the fourth nucleotide is a C.

Continuing in this manner, the entire sequence of the DNA fragment is read from the gel simply by noting in what column the successively larger labeled polynucleotide pieces appear. The sequence can be checked by determining the sequence of the complementary strand.

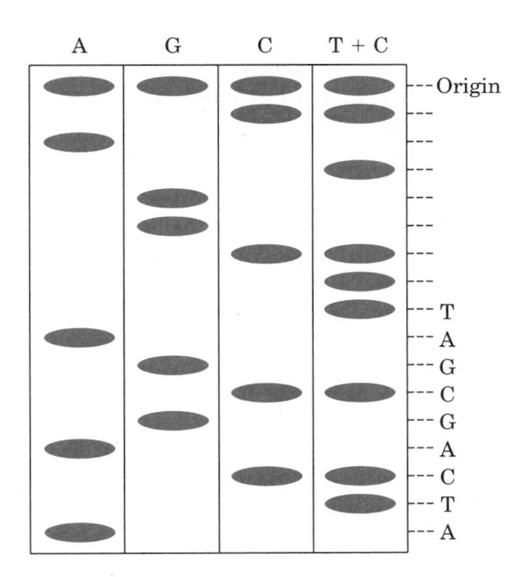

Figure 16.11 Representation of a gel electrophoresis pattern. The products of the four cleavage experiments are placed at the top of the gel, and a voltage is applied between top and bottom. Smaller products migrate along the gel at a faster rate and thus appear at the bottom. The DNA sequence is read from the position of the radioactive spots.

The Maxam–Gilbert method of DNA sequencing is so efficient that a single worker can sequence up to 2000 base pairs per day. The genome of the yeast *Saccharomyces cerevisiae,* containing more than 12 million base pairs and 6000 genes, has been sequenced, and work is well under way to sequence the 3 billion base pairs of the entire human genome. As long as 10 years and as much as several billion dollars will be needed to complete the job.

PROBLEM

16.19 Show the labeled products you would obtain if the following DNA fragment were subjected to each of the four cleavage reactions:

^{32}P-A-A-C-A-T-G-G-C-G-C-T-T-A-T-G-A-C-G-A

PROBLEM

16.20 Sketch what you would expect the gel electrophoresis pattern to look like if the DNA segment in Problem 16.19 were sequenced.

PROBLEM

16.21 Finish assigning the sequence to the gel electrophoresis pattern shown in Figure 16.11.

16.14 The Polymerase Chain Reaction

The invention of the **polymerase chain reaction (PCR)** by Kary Mullis in 1986 has been described as being to genes what Gutenberg's invention of the printing press was to the written word. Just as the printing press produces multiple copies of a book, PCR produces multiple copies of a given DNA sequence. Starting from less than 1 *picogram* of DNA with a chain length of 10,000 nucleotides (1 pg = 10^{-12} g; about 100,000 molecules), PCR makes it possible to obtain several micrograms (1 μg = 10^{-6} g; about 10^{11} molecules) in just a few hours.

The key to the polymerase chain reaction is *Taq* DNA polymerase, a heat-stable enzyme isolated from the thermophilic bacterium *Thermus aquaticus* found in a hot spring in Yellowstone National Park.[1] *Taq* polymerase is able to take a single strand of DNA and, starting from a short "primer" piece that is complementary to one end of the chain, finish constructing the entire complementary strand. The overall process takes three steps, as shown schematically in Figure 16.12.

Step 1 The double-stranded DNA to be amplified is heated in the presence of *Taq* polymerase, Mg^{2+} ion, the four deoxyribonucleotide triphosphate monomers (dNTP's), and a large excess of two short DNA primer pieces of about 20 bases each. Each primer is complementary to the sequence at the end of one of the target DNA segments. At a temperature of 95°C, double-stranded DNA spontaneously breaks apart into two single strands.

Step 2 The temperature is lowered to between 37°C and 50°C, allowing the primers, because of their relatively high concentration, to anneal to a complementary sequence at the end of each target strand.

Step 3 The temperature is then raised to 72°C, and *Taq* polymerase catalyzes the addition of further nucleotides to the two primed DNA strands. When replication of each strand is finished, *two* copies of the original DNA now exist. Repeating the denature–anneal–synthesize cycle a second time yields four DNA copies, repeating a third time yields eight copies, and so on, in an exponential series.

PCR has been automated, and 30 or so cycles can be carried out in an hour, resulting in a theoretical amplification factor of 2^{30} ($\sim 10^9$). In practice, however, the efficiency of each cycle is less than 100%, and an experimental amplification of about 10^6–10^8 is routinely achieved for 30 cycles.

---

[1]Two new heat-stable DNA polymerase enzymes have recently become available, Vent polymerase and *Pfu* polymerase, both isolated from bacteria growing near geothermal vents in the ocean floor. The error rate of both enzymes is substantially less than that of *Taq*.

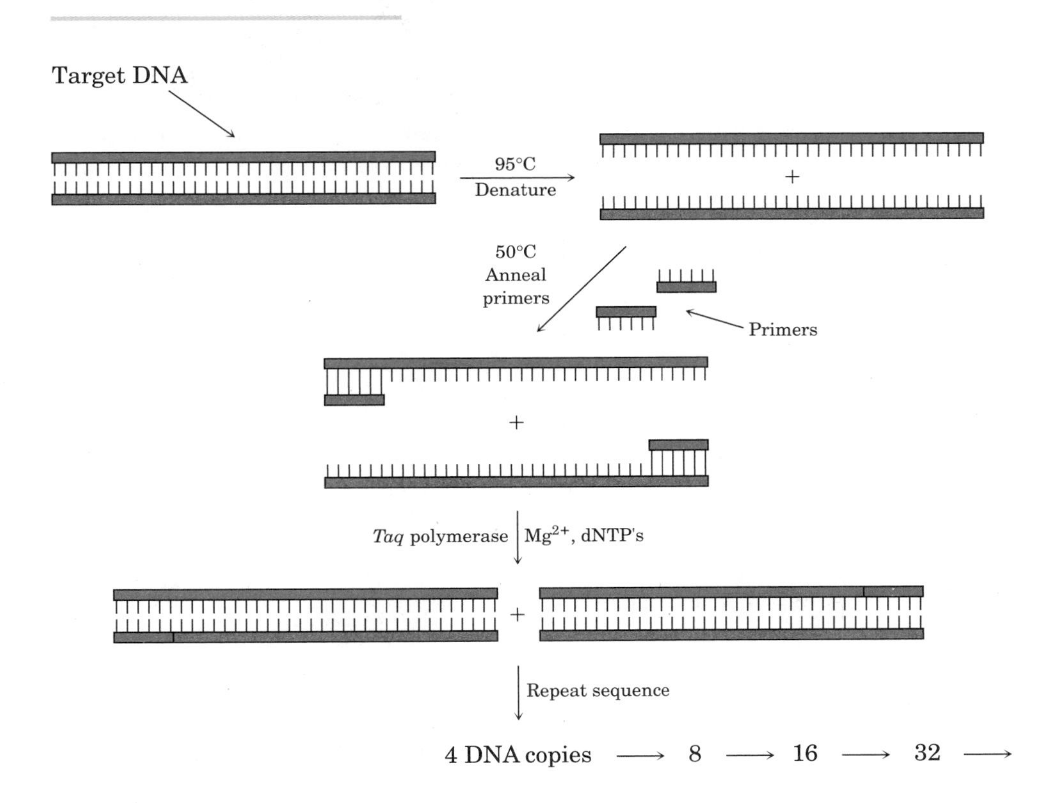

Figure 16.12 The polymerase chain reaction. Double-stranded DNA is heated to 95°C in the presence of two short primer sequences, each of which is complementary to the end of one of the strands. After the DNA denatures, the temperature is lowered and the primer sequences anneal to the strand ends. Raising the temperature in the presence of *Taq* polymerase, Mg^{2+}, and a mixture of the four deoxynucleotide triphosphates (dNTP's) effects strand replication, producing two DNA copies. Each further repetition of the sequence again doubles the number of copies.

INTERLUDE

Cholesterol and Heart Disease

This coronary artery is partially blocked by deposits of cholesterol.

We read a lot about the relationship between cholesterol and heart disease. What are the facts? It's well established that a diet rich in saturated animal fats often leads to an increase in blood serum cholesterol, at least in sedentary, overweight people. Conversely, a diet lower in saturated fats and higher in polyunsaturated fats (PUFA's) leads to a lower serum cholesterol level. Studies have shown that a serum cholesterol level greater than 300 mg/dL (a normal value is 150–200 mg/dL) is weakly correlated with an increased incidence of *atherosclerosis,* a form of heart disease in which cholesterol deposits build up on the inner walls of coronary arteries, blocking the flow of blood to the heart muscles.

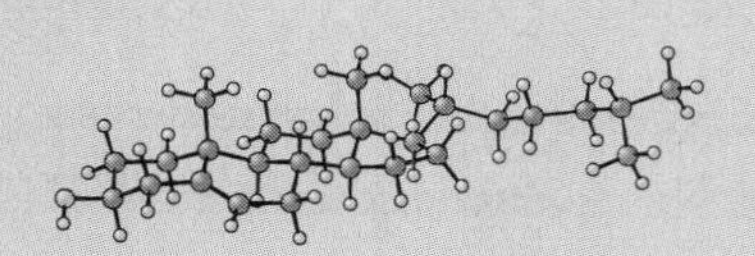

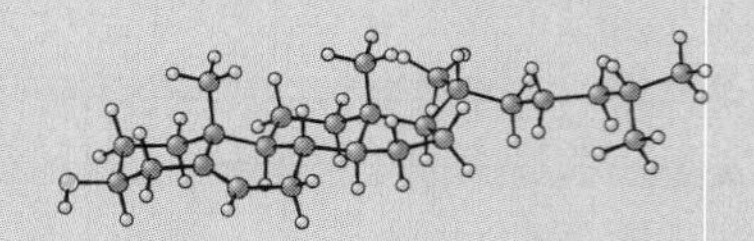

Cholesterol Stereo View

A better indication of a person's risk of heart disease comes from a measurement of blood lipoprotein levels. *Lipoproteins* are complex molecules with both lipid and protein parts that transport lipids through the body. They can be divided into four types according to density, as shown in Table 16.5. People with a high level of high-density lipoproteins (HDL's)

Table 16.5 Serum Lipoproteins

Name	*Density (g/mL)*	*% Lipid*	*% Protein*
Chylomicrons	<0.94	98	2
VLDL's (very-low-density lipoproteins)	0.940–1.006	90	10
LDL's (low-density lipoproteins)	1.006–1.063	75	25
HDL's (high-density lipoproteins)	1.063–1.210	60	40

(*continued*)➤

seem to have a decreased risk of heart disease. As a rule of thumb, a person's risk drops about 25% for each increase of 5 mg/dL in HDL concentration. Normal values are about 45 mg/dL for men and 55 mg/dL for women, perhaps explaining why women are generally less susceptible than men to heart disease.

Chylomicrons and very-low-density lipoproteins (VLDL's) act primarily as carriers of triglycerides from the intestines to peripheral tissues, whereas LDL's and HDL's act as carriers of cholesterol to and from the liver. Present evidence suggests that LDL's transport cholesterol as its fatty acid ester *to* peripheral tissues, whereas HDL's remove cholesterol as its stearate ester *from* dying cells and transport it back to the liver. If LDL's deliver more cholesterol than is needed, and if insufficient HDL's are present to remove it, the excess is deposited in arteries. The higher the HDL level, the less the likelihood of deposits and the lower the risk of heart disease.

Not surprisingly, the most important factor in gaining high HDL levels is a generally healthy lifestyle. Obesity, smoking, and lack of exercise lead to low HDL levels; regular exercise and a sensible diet lead to high HDL levels. Distance runners and other endurance athletes have HDL levels nearly 50% higher than the general population.

Summary and Key Words

adrenocortical hormone, 516
androgen, 515
anticodon, 529
antisense strand, 528
coding strand, 528
codon, 529
deoxyribonucleic acid (DNA), 517
double helix, 522
3′ end, 519
5′ end, 519
estrogen, 516
fatty acid, 508
hormone, 515
lipid, 507
lipid bilayer, 514
messenger RNA (mRNA), 527
micelle, 512
nucleic acid, 517

Lipids are the naturally occurring substances isolated from plants and animals by extraction with organic solvents. Animal fats and vegetable oils are the most widely occurring lipids. Both fats and oils are **triacylglycerols,** triesters of glycerol with long-chain **fatty acids. Phosphoglycerides** such as lecithin and cephalin are closely related to fats. The glycerol backbone in these molecules is esterified to two fatty acids and one phosphate ester. **Sphingolipids,** another major class of **phospholipids,** have an amino alcohol such as sphingosine for their backbone.

Steroids are plant and animal lipids with a characteristic four-ring carbon skeleton. Steroids occur widely in body tissue and have many different kinds of physiological activity. Among the more important kinds of steroids are the sex hormones (**androgens** and **estrogens**) and the **adrenocortical** hormones.

The **nucleic acids, DNA (deoxyribonucleic acid)** and **RNA (ribonucleic acid),** are biological polymers that act as chemical carriers of an organism's genetic information. Enzyme-catalyzed hydrolysis of a nucleic acid yields **nucleotides,** which consist of a purine or pyrimi-

nucleoside, 518
nucleotide, 517
phosphoglyceride, 513
phospholipid, 513
polymerase chain reaction (PCR), 536
polyunsaturated fatty acid (PUFA), 510
replication, 525
restriction endonuclease, 532
ribonucleic acid (RNA), 517
ribosomal RNA (rRNA), 527
sense strand, 528
sphingolipid, 514
steroid, 515
template strand, 528
transcription, 527
transfer RNA (tRNA), 527
translation, 529
triacylglycerol, 508

dine heterocyclic amine base linked to C1′ of a pentose sugar (ribose in RNA and 2-deoxyribose in DNA), with the sugar in turn linked through its C5′ hydroxyl to a phosphate group.

Molecules of DNA consist of two polynucleotide strands held together by hydrogen bonds between heterocyclic bases on the different strands and coiled into a **double-helix** conformation. Adenine (A) and thymine (T) form hydrogen bonds to each other, as do cytosine (C) and guanine (G). The two strands of DNA are complementary rather than identical.

Three main processes take place in deciphering the genetic information in DNA: **Replication** of DNA is the process by which identical DNA copies are made. This occurs when the DNA double helix unwinds, complementary deoxyribonucleotides line up in order, and two new DNA molecules are produced. **Transcription** is the process by which RNA is produced. This occurs when a segment of the DNA double helix unwinds and complementary ribonucleotides line up to produce **messenger RNA (mRNA)**. **Translation** is the process by which mRNA directs protein synthesis. Each mRNA has a three-base segment called a **codon** along its chain. Codons are recognized by small molecules of **transfer RNA (tRNA)**, which carry and then deliver the appropriate amino acids needed for protein synthesis.

Sequencing of DNA fragments is done by the Maxam–Gilbert method, in which chemical reactions are carried out to cause specific cleavages of the DNA chain, followed by separation of the pieces by electrophoresis. The DNA sequence is read directly from the electrophoresis pattern.

ADDITIONAL PROBLEMS

16.22 Write representative structures for the following:
(a) A fat (b) A vegetable oil (c) A steroid

16.23 Write the structures of the following molecules:
(a) Sodium stearate (b) Ethyl linoleate (c) Glyceryl palmitodioleate

16.24 Show the products you would expect to obtain from the reaction of glyceryl trioleate with the following:
(a) Excess Br_2 in CCl_4 (b) H_2/Pd (c) NaOH, H_2O (d) $KMnO_4$, H_3O^+
(e) $LiAlH_4$, then H_3O^+

16.25 How would you convert oleic acid into the following substances?
(a) Methyl oleate (b) Methyl stearate (c) Nonanedioic acid

16.26 Eleostearic acid, $C_{18}H_{30}O_2$, is a rare fatty acid found in tung oil. On oxidation with $KMnO_4$, eleostearic acid yields 1 part pentanoic acid, 2 parts oxalic acid ($HO_2C–CO_2H$), and 1 part nonanedioic acid. Propose a structure for eleostearic acid.

16.27 Stearolic acid, $C_{18}H_{32}O_2$, yields oleic acid on catalytic hydrogenation over the Lindlar catalyst. Propose a structure for stearolic acid.

16.28 Draw the products you would obtain from treatment of cholesterol with the following reagents:
(a) Br_2 (b) H_2, Pd catalyst (c) CH_3COCl, pyridine

16.29 If the average molecular weight of soybean oil is 1500, how many grams of NaOH are needed to saponify 5.00 g of the oil?

16.30 Define the following terms:
(a) Steroid (b) DNA (c) Base pair
(d) Codon (e) Lipid (f) Transcription

16.31 The DNA from sea urchins contains about 32% A and about 18% G. What percentages of T and C would you expect in sea urchin DNA? Explain.

16.32 What DNA sequence is complementary to the following sequence?

G-A-A-G-T-T-C-A-T-G-C

16.33 Give codons for the following amino acids:
(a) Ile (b) Asp (c) Thr

16.34 Draw the complete structure of the ribonucleotide codon UAC. For what amino acid does this sequence code?

16.35 Draw the complete structure of the deoxyribonucleotide sequence from which the mRNA codon in Problem 16.34 was transcribed.

16.36 What amino acids do the following ribonucleotide codons code for?
(a) AAU (b) GAG (c) UCC (d) CAU (e) ACC

16.37 From what DNA sequences were each of the mRNA codons in Problem 16.36 transcribed?

16.38 What anticodon sequences of tRNA's are coded by each of the codons in Problem 16.36?

16.39 If the gene sequence T-A-A-C-C-G-G-A-T on DNA were miscopied during replication and became T-G-A-C-C-G-G-A-T, what effect would the mutation have on the sequence of the protein produced?

16.40 Give an mRNA sequence that codes for synthesis of metenkephalin, a small peptide with morphine-like properties:

Tyr-Gly-Gly-Phe-Met

16.41 Give a DNA gene sequence that will code for metenkephalin (see Problem 16.40).

16.42 Human and horse insulin both have two polypeptide chains with one chain containing 21 amino acids and the other containing 30 amino acids. How many nitrogen bases are present in the DNA to code for each chain?

16.43 Human and horse insulin (see Problem 16.42) differ in primary structure at two amino acids: at the 9th position in one chain (human has Ser and horse has Gly) and at the 30th position in the other chain (human has Thr and horse has Ala). How must the DNA differ?

16.44 What amino acid sequence is coded by the following mRNA sequence?

CUA-GAC-CGU-UCC-AAG-UGA

16.45 What anticodon sequences of tRNA's are coded by the mRNA in Problem 16.44? What was the base sequence in the original DNA strand on which this mRNA was made? What was the base sequence in the DNA strand *complementary* to that from which this mRNA was made?

16.46 Look up the structure of angiotensin II (Figure 15.2), and give an mRNA sequence that codes for its synthesis.

16.47 Diethylstilbestrol (DES) exhibits estradiol-like activity even though it is structurally unrelated to steroids. Once used widely as an additive in animal feed, DES has been implicated as a causative agent in several types of cancers. Look up the structure of estradiol (Section 16.5), and show how DES can be drawn so that it is sterically similar to estradiol.

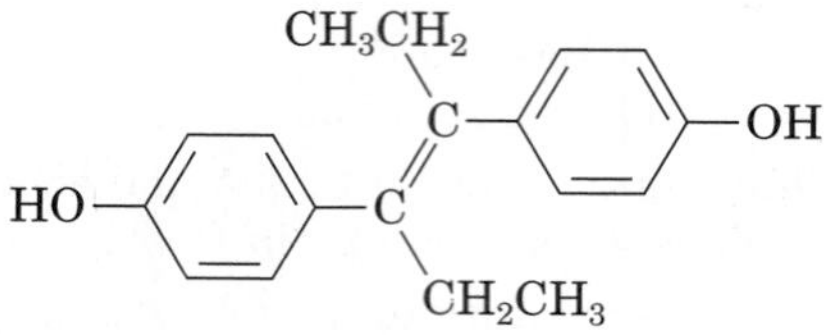

Diethylstilbestrol

16.48 How many stereocenters are present in estradiol (see Problem 16.47)? Label them.

16.49 What products would you obtain from reaction of estradiol (Problem 16.47) with the following reagents?
(a) NaOH, then CH_3I (b) CH_3COCl, pyridine (c) Br_2 (1 equiv)

16.50 *Nandrolone* is an anabolic steroid sometimes taken by athletes to build muscle mass. Compare the structures of nandrolone and testosterone, and point out their structural similarities.

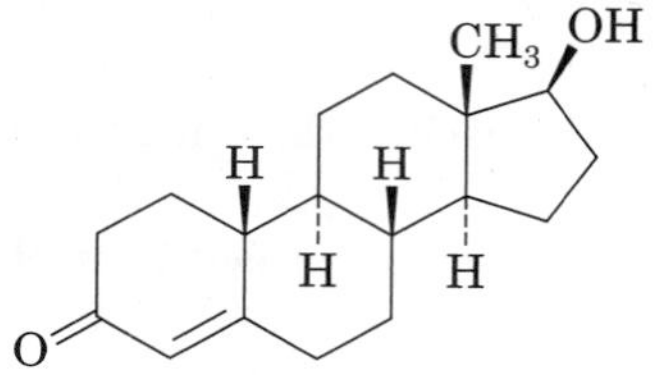

Nandrolone (an anabolic steroid)

Visualizing Chemistry

16.51 Cholic acid, a major steroidal constituent of human bile, has the following structure. Label the three –OH groups in the molecule as axial or equatorial (gray = C, red = O, light green = H).

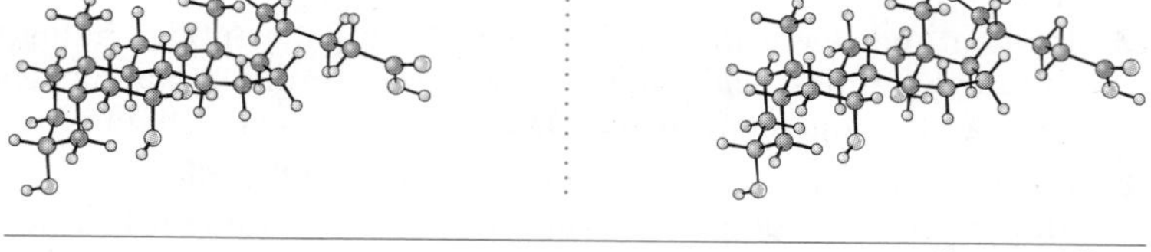

Stereo View

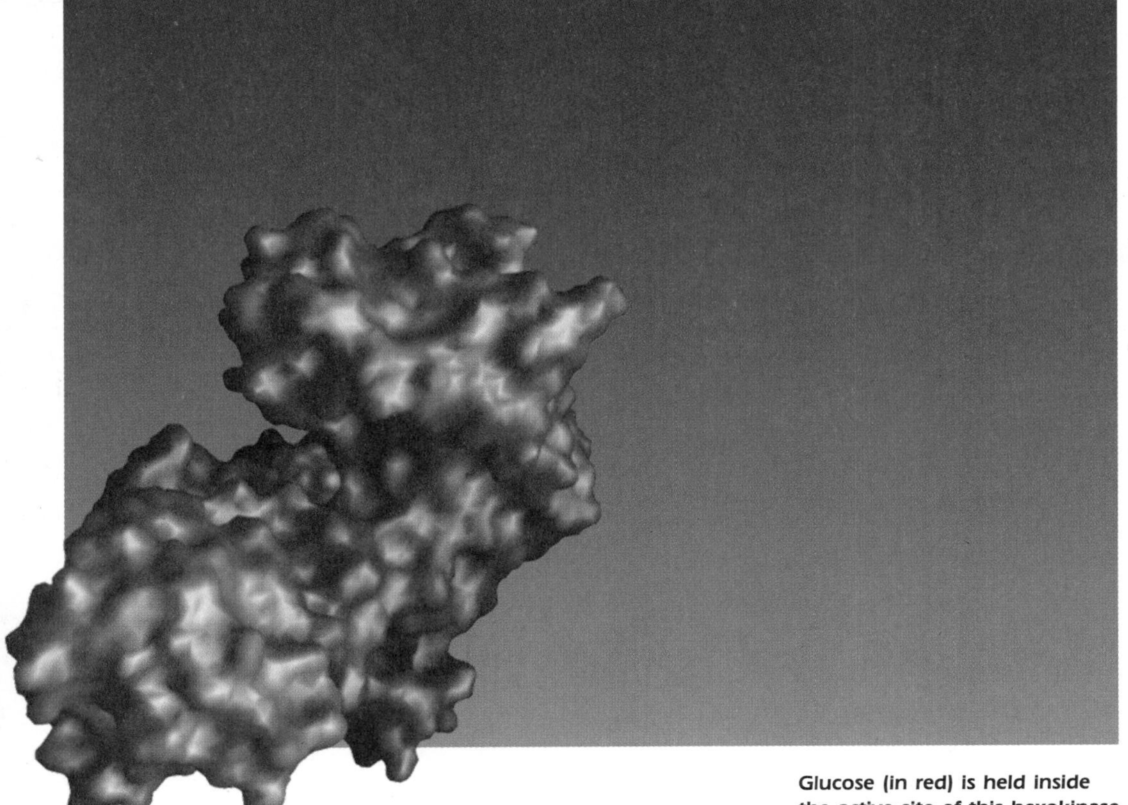

Glucose (in red) is held inside the active site of this hexokinase enzyme during glycolysis.

17 The Organic Chemistry of Metabolic Pathways

The organic chemical reactions that take place in even the smallest and simplest living organism are more complex than those carried out in any laboratory. Yet the reactions in living organisms, regardless of their complexity, follow the same rules of reactivity we've developed in the preceding chapters.

In this chapter, we'll look at some of the pathways by which organisms carry out their chemistry, focusing primarily on how they break down fats and carbohydrates. Our emphasis will be on the organic chemistry of the various pathways and on recognizing the similarities between mechanisms of biological reactions and mechanisms of analogous laboratory reactions.

17.1 An Overview of Metabolism and Biochemical Energy

The many reactions that go on in the cells of living organisms are collectively called **metabolism.** The pathways that break down larger molecules into smaller ones are known as **catabolism;** the pathways that put smaller molecules together to synthesize larger biomolecules are **anabolism.** Catabolism can be divided into the four stages shown in Figure 17.1.

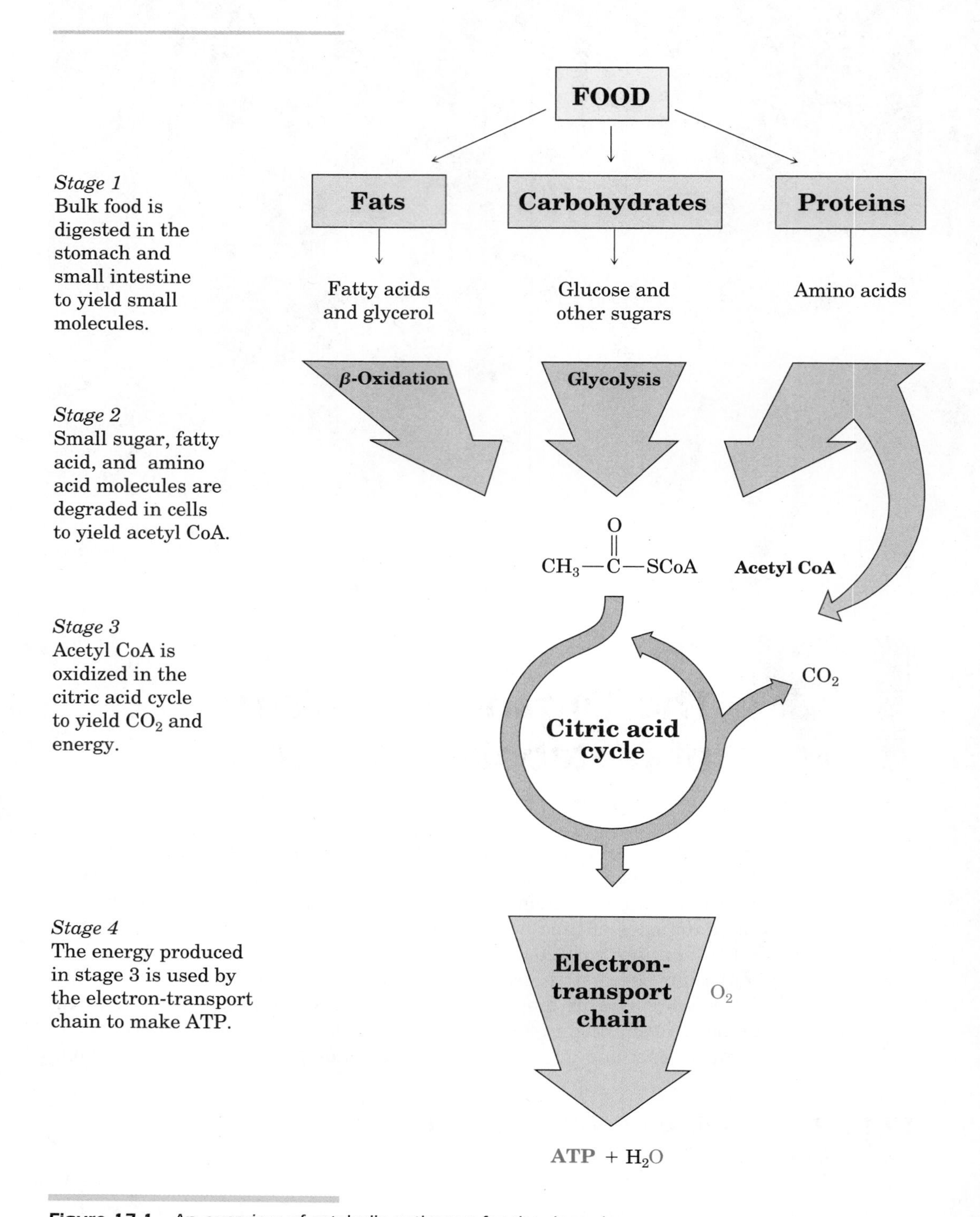

Figure 17.1 An overview of catabolic pathways for the degradation of food and the production of biochemical energy. The ultimate products of food catabolism are CO_2, H_2O, and adenosine triphosphate, ATP.

In the first catabolic stage, **digestion,** food is broken down by hydrolysis of ester, glycoside (acetal), and peptide (amide) bonds to yield fatty acids, simple sugars, and amino acids. These small molecules are further degraded in the second stage of catabolism to yield two-carbon acetyl groups attached by a *thiol ester* bond (RCOSR′, the sulfur analog of an ester) to the large carrier molecule *coenzyme A*. The resultant compound, *acetyl coenzyme A* (*acetyl CoA*), is an intermediate in the breakdown of all main classes of food molecules.

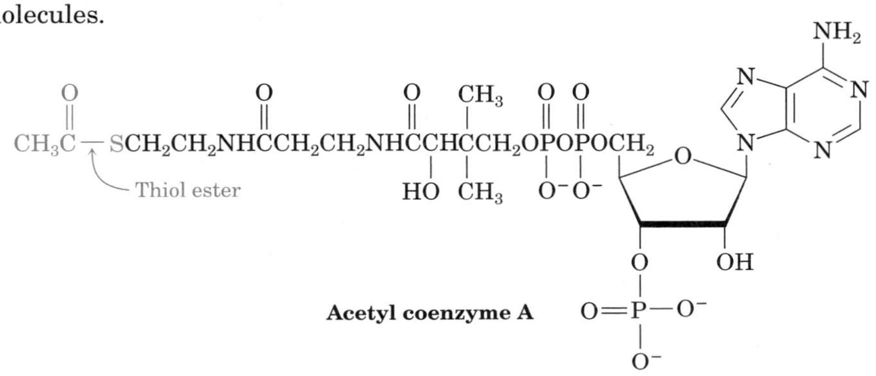

$CH_3C(=O)-SCH_2CH_2NHC(=O)CH_2CH_2NHC(=O)CH(OH)C(CH_3)_2CH_2OP(=O)(O^-)OP(=O)(O^-)OCH_2$–(ribose 3′-phosphate)–adenine

Thiol ester

Acetyl coenzyme A

Acetyl groups are oxidized in the third stage of catabolism, the **citric acid cycle,** to yield CO_2. This stage also releases a large amount of energy that is used in the fourth stage, the **electron-transport chain,** to make molecules of the nucleotide *adenosine triphosphate* (*ATP*). As the final result of food catabolism, ATP has been called the "energy currency" of the cell. Catabolic reactions "pay off" in ATP by synthesizing it from *adenosine diphosphate* (*ADP*) plus hydrogen phosphate ion, HPO_4^{2-}. Anabolic reactions "spend" ATP by transferring a phosphate group to another molecule, thereby regenerating ADP. The entire process of energy production thus revolves around the ATP $\rightleftharpoons$ ADP interconversion:

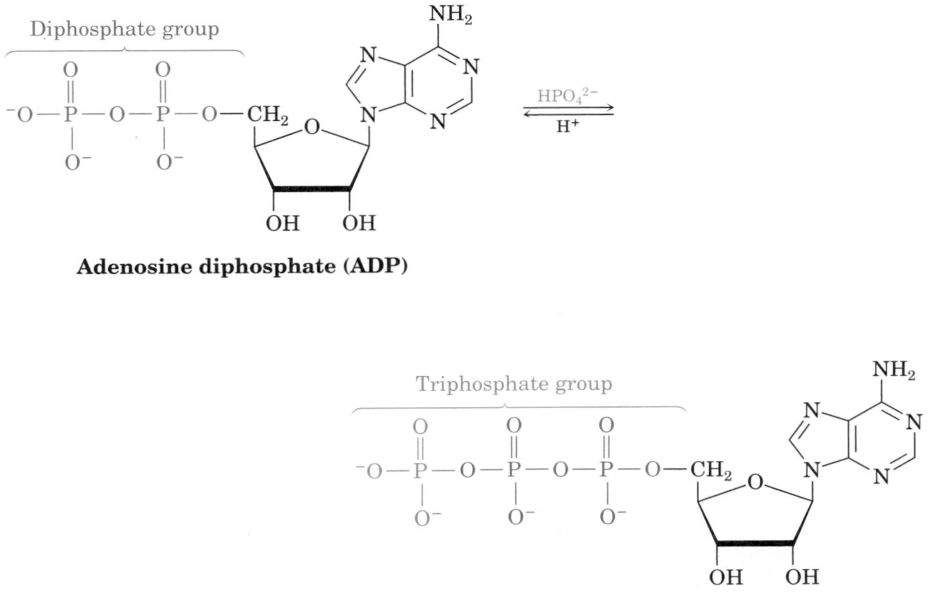

Diphosphate group

$^-O-P(=O)(O^-)-O-P(=O)(O^-)-O-CH_2$–ribose–adenine

Adenosine diphosphate (ADP)

$$\underset{H^+}{\overset{HPO_4^{2-}}{\rightleftharpoons}}$$

Triphosphate group

$^-O-P(=O)(O^-)-O-P(=O)(O^-)-O-P(=O)(O^-)-O-CH_2$–ribose–adenine

Adenosine triphosphate (ATP)

Note that both ADP and ATP are **phosphoric acid anhydrides,** which contain $-\overset{O}{\overset{\|}{P}}-O-\overset{O}{\overset{\|}{P}}-$ linkages analogous to that of the $-\overset{O}{\overset{\|}{C}}-O-\overset{O}{\overset{\|}{C}}-$ linkages in carboxylic acid anhydrides (Section 10.8). Just as carboxylic anhydrides react with alcohols by breaking a C–O bond and forming a carboxylic acid ester, phosphoric anhydrides react with alcohols by breaking a P–O bond and forming a phosphate ester, $ROPO_3^{2-}$.

How does the body use ATP? Recall from Section 3.9 that energy (actually, *free energy, G*) must be released for a chemical reaction to have a favorable equilibrium constant and occur spontaneously. This means that the free-energy change for the reaction (ΔG) must be negative. If ΔG is positive, then the reaction is unfavorable, and the process can't occur spontaneously. What normally happens in order for an energetically unfavorable reaction to occur is that it is "coupled" to an energetically favorable reaction so that the *overall* free-energy change for the two reactions together is favorable. Take, for example, the *phosphorylation* reaction of glucose to yield glucose 6-phosphate, an important step in the breakdown of dietary carbohydrates. The reaction of glucose with HPO_4^{2-} ion is energetically unfavorable by 13.8 kJ (3.3 kcal):

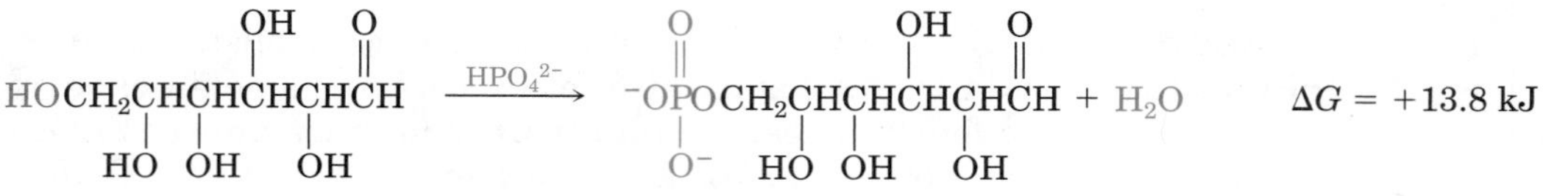

$$HOCH_2CH(OH)CH(OH)CH(OH)CH(OH)CHO \xrightarrow{HPO_4^{2-}} {}^{-}O\overset{O}{\overset{\|}{P}}(O^-)OCH_2CH(OH)CH(OH)CH(OH)CH(OH)CHO + H_2O \qquad \Delta G = +13.8\text{ kJ}$$

Glucose → **Glucose 6-phosphate**

With ATP, however, glucose undergoes an energetically favorable reaction to yield glucose 6-phosphate plus ADP. The effect is as if the ATP reacted with the water produced in the unfavorable HPO_4^{2-} reaction, making the overall process favorable by about 16.7 kJ (4.0 kcal). We therefore say that ATP "drives" the phosphorylation reaction of glucose:

$$\begin{array}{lrcll} & \text{Glucose} + HPO_4^{2-} & \longrightarrow & \text{Glucose 6-phosphate} + H_2O & \Delta G = +13.8\text{ kJ} \\ & \text{ATP} + H_2O & \longrightarrow & \text{ADP} + HPO_4^{2-} & \Delta G = -30.5\text{ kJ} \\ \hline \text{Net:} & \text{Glucose} + \text{ATP} & \longrightarrow & \text{Glucose 6-phosphate} + \text{ADP} & \Delta G = -16.7\text{ kJ} \end{array}$$

It's this ability to drive otherwise unfavorable phosphorylation reactions that makes ATP so useful. The resultant phosphates are much more reactive substances than the compounds they are derived from.

PROBLEM

17.1 One of the steps in fat metabolism is the reaction of glycerol (1,2,3-propanetriol) with ATP to yield glycerol 1-phosphate. Write the reaction, and draw the structure of glycerol 1-phosphate.

PROBLEM

17.2 The reaction of ATP with an alcohol to yield ADP and an alkyl phosphate is analogous to that of a carboxylic acid anhydride with an alcohol to yield a carboxylate ion and an ester (Section 10.8). Show the mechanism of the reaction of ATP with methanol to yield methyl phosphate, $CH_3OPO_3^{2-}$.

17.2 Catabolism of Fats: β-Oxidation Pathway

The catabolism of fats and oils (triacylglycerols) begins with their hydrolysis in the stomach and small intestine to yield glycerol plus fatty acids. Glycerol is then phosphorylated by reaction with ATP and oxidized to yield glyceraldehyde 3-phosphate, which enters the carbohydrate catabolic pathway. (We'll discuss this in more detail in Section 17.3.)

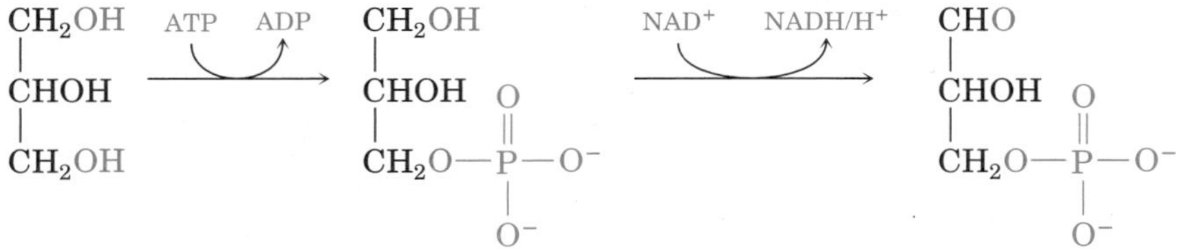

$$\underset{\textbf{Glycerol}}{CH_2OH{-}CHOH{-}CH_2OH} \xrightarrow{ATP \;\to\; ADP} \underset{\textbf{Glycerol 1-phosphate}}{CH_2OH{-}CHOH{-}CH_2O{-}PO_3^{2-}} \xrightarrow{NAD^+ \;\to\; NADH/H^+} \underset{\textbf{Glyceraldehyde 3-phosphate}}{CHO{-}CHOH{-}CH_2O{-}PO_3^{2-}}$$

Note how the above reactions are written. It's common practice when writing biochemical transformations to show only the structure of the substrate, while abbreviating the structures of coenzymes (Section 15.12) and other reactants. The curved arrow intersecting the usual straight reaction arrow in the first step shows that ATP is also a reactant and that ADP is a product. The coenzyme *nicotinamide adenine dinucleotide* (NAD^+) is required in the second step, and *reduced nicotinamide adenine dinucleotide* (*NADH*) plus a proton are products. We'll see shortly that NAD^+ is often involved as a biochemical oxidizing agent for converting alcohols to ketones or aldehydes.

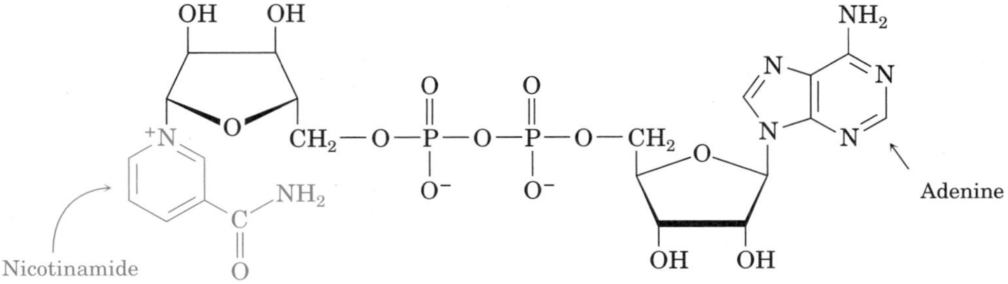

Nicotinamide — Adenine

Nicotinamide adenine dinucleotide (NAD^+)

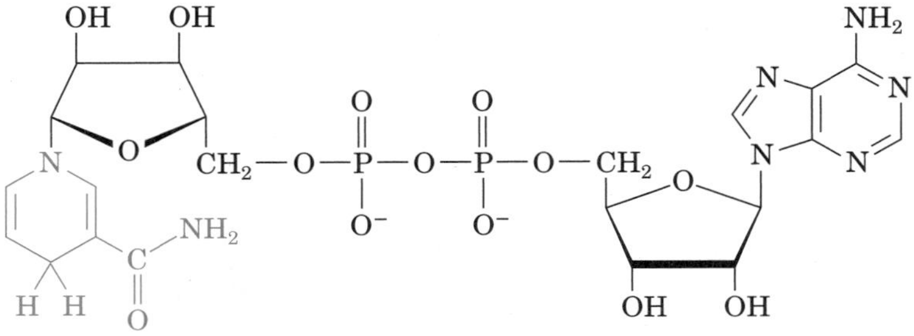

Reduced nicotinamide adenine dinucleotide (NADH)

Note also that glyceraldehyde 3-phosphate is written with its phosphate group dissociated; that is, as $-OPO_3^{2-}$ rather than $-OPO_3H_2$. It's standard practice in writing biochemical structures to show carboxylic acids and phosphoric acids as their anions, because they exist in this form at the physiological pH found in the cells of organisms.

Fatty acids are catabolized by a repetitive four-step sequence of enzyme-catalyzed reactions called the *fatty acid spiral,* or ***β*-oxidation pathway,** shown in Figure 17.2. Each passage through the pathway results in the cleavage of a two-carbon acetyl group from the end of the fatty acid chain, until the entire molecule is ultimately degraded. As each acetyl group is produced, it enters the citric acid cycle and is further catabolized (see Section 17.4).

Step 1 Introduction of a double bond The β-oxidation pathway begins when a fatty acid forms a thiol ester with coenzyme A to give a fatty acyl CoA. Two hydrogen atoms from C2 and C3 are then removed by an acyl CoA dehydrogenase enzyme to yield an unsaturated acyl CoA. This kind of oxidation—the introduction of a conjugated double bond into a molecule—occurs frequently in biochemical pathways and is usually carried out by the coenzyme *flavin adenine dinucleotide (FAD).* Reduced $FADH_2$ is the by-product.

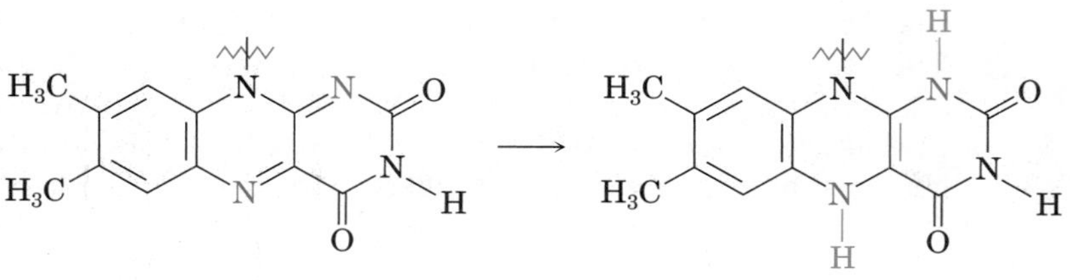

FAD → **$FADH_2$**

Step 2 Addition of water The unsaturated acyl CoA produced in step 1 reacts with water to yield a β-hydroxyacyl CoA in a process catalyzed by the enzyme enoyl CoA hydratase. Though we haven't specifically studied such processes in previous chapters, the reaction is closely analogous to the nucleophilic addition of water to a ketone to yield a hydrate (Section 9.7). In ketone hydration, water as nucleophile adds *directly* to a C=O group, yielding an alkoxide ion. In step 2 of the β-oxidation pathway, water as nucleophile adds to a double bond *conjugated with* a C=O group, yielding an enolate ion. In both reactions, the carbonyl oxygen atom withdraws electrons from the nearby carbon atom, making that carbon atom electrophilic.

Direct addition

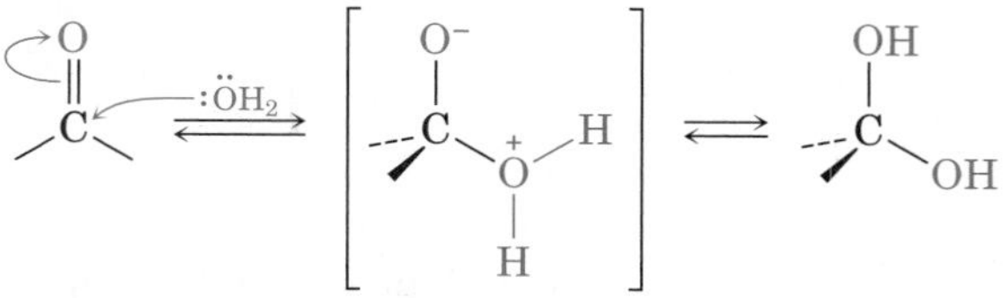

$$\mathrm{RCH_2CH_2CH_2CH_2\overset{\overset{\large O}{\|}}{C}SCoA}$$

Step 1
A double bond is introduced by enzyme-catalyzed removal of hydrogens from C2 and C3.

$$\xrightarrow{\text{FAD} \;\to\; \text{FADH}_2}$$

$$\mathrm{RCH_2CH_2CH{=}CH\overset{\overset{\large O}{\|}}{C}SCoA}$$

Step 2
Water adds to the double bond to yield an alcohol.

$$\xrightarrow{H_2O}$$

$$\mathrm{RCH_2CH_2\overset{\overset{\large OH}{|}}{C}HCH_2\overset{\overset{\large O}{\|}}{C}SCoA}$$

Step 3
The alcohol is oxidized by NAD^+ to give a ketone.

$$\xrightarrow{\text{NAD}^+ \;\to\; \text{NADH/H}^+}$$

$$\mathrm{RCH_2CH_2\overset{\overset{\large O}{\|}}{C}CH_2\overset{\overset{\large O}{\|}}{C}SCoA}$$

Step 4
The bond between C2 and C3 is broken by nucleophilic attack of coenzyme A on the C3 carbonyl group in a retro-Claisen reaction to yield acetyl CoA and a chain-shortened fatty acid.

$$\xrightarrow{\text{HSCoA}}$$

$$\mathrm{RCH_2CH_2\overset{\overset{\large O}{\|}}{C}SCoA + CH_3\overset{\overset{\large O}{\|}}{C}SCoA}$$

Figure 17.2 The four steps of the β-oxidation pathway, resulting in the cleavage of an acetyl group from the end of the fatty acid chain. The chain-shortening step is a retro-Claisen reaction of a β-keto ester.

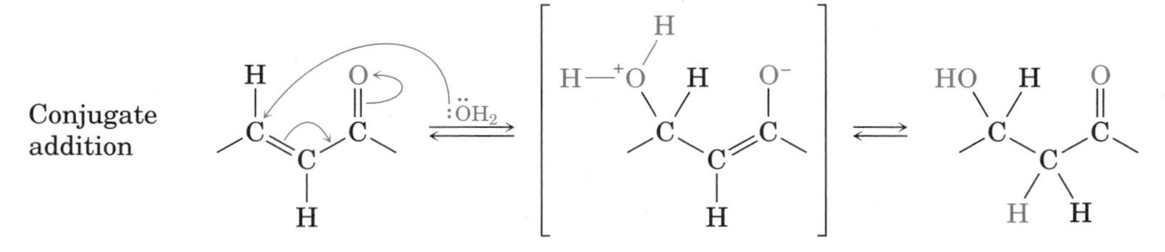

Step 3 Alcohol oxidation The β-hydroxyacyl CoA from step 2 is oxidized to a β-ketoacyl CoA in a reaction catalyzed by the enzyme L-3-hydroxyacyl CoA dehydrogenase. As in the oxidation of glycerol

1-phosphate to glyceraldehyde 3-phosphate mentioned earlier, alcohol oxidation in the β-oxidation pathway requires NAD$^+$ as a coenzyme and yields reduced NADH/H$^+$ as by-product.

It's useful when thinking about enzyme-catalyzed oxidation and reduction reactions to recognize that a hydrogen *atom* is equivalent to a hydrogen *ion,* H$^+$, plus an electron, e$^-$. Thus, for the two hydrogen atoms removed in the oxidation of an alcohol, 2 H atoms = 2 H$^+$ + 2 e$^-$. When NAD$^+$ is involved, both electrons accompany one H$^+$, in effect adding a hydride ion, H:$^-$, to NAD$^+$ to give NADH. The second hydrogen removed from the oxidized substrate enters the solution as H$^+$.

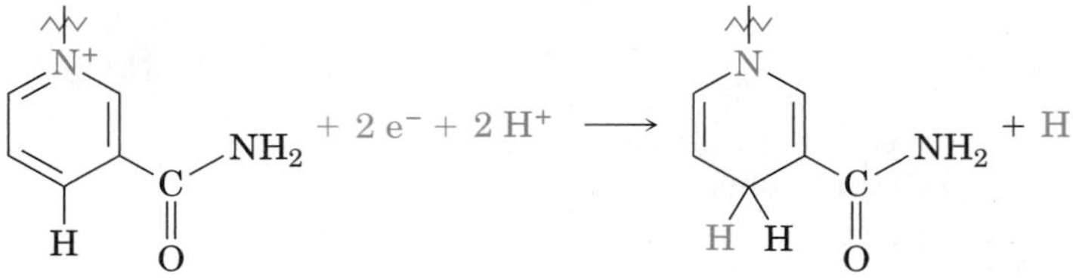

$+ 2\,e^- + 2\,H^+ \longrightarrow$ $+ H^+$

NAD$^+$ **NADH/H$^+$**

The mechanism of oxidation of glycerol 1-phosphate has many analogies in the laboratory, and is similar in some respects to that of the hydration reaction in step 2. Thus, a hydride ion expelled from the alcohol acts as a nucleophile and adds to the C=C–C=N$^+$ part of NAD$^+$ in much the same way that water acts as a nucleophile and adds to the C=C–C=O part of the unsaturated acyl CoA.

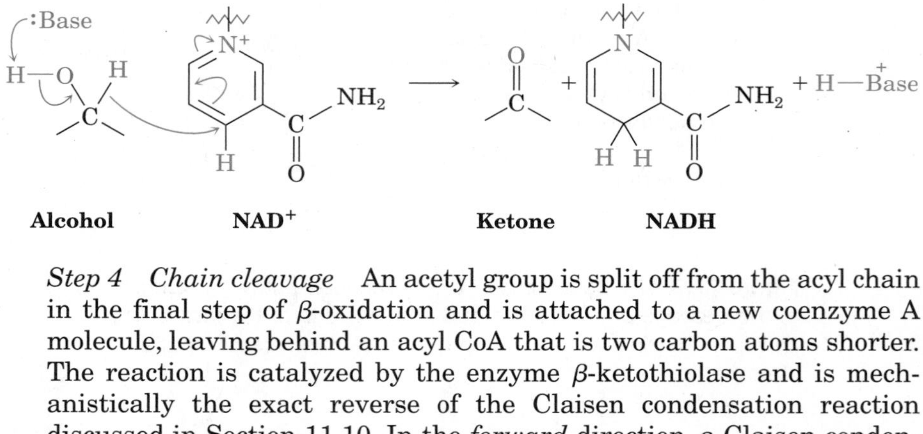

Alcohol **NAD$^+$** **Ketone** **NADH**

Step 4 *Chain cleavage* An acetyl group is split off from the acyl chain in the final step of β-oxidation and is attached to a new coenzyme A molecule, leaving behind an acyl CoA that is two carbon atoms shorter. The reaction is catalyzed by the enzyme β-ketothiolase and is mechanistically the exact reverse of the Claisen condensation reaction discussed in Section 11.10. In the *forward* direction, a Claisen condensation joins two esters together to form a β-keto ester product. In the *reverse* direction, a retro-Claisen reaction splits a β-keto ester (or β-keto thiol ester) apart to form two esters (or two thiol esters).

The reaction occurs by nucleophilic addition of coenzyme A to the keto group of the β-keto acyl CoA to yield an alkoxide ion intermediate, followed by cleavage of the C2–C3 bond with expulsion of an acetyl CoA enolate ion. Protonation of the enolate ion gives acetyl CoA, and the chain-shortened acyl CoA enters another round of the β-oxidation pathway for further degradation.

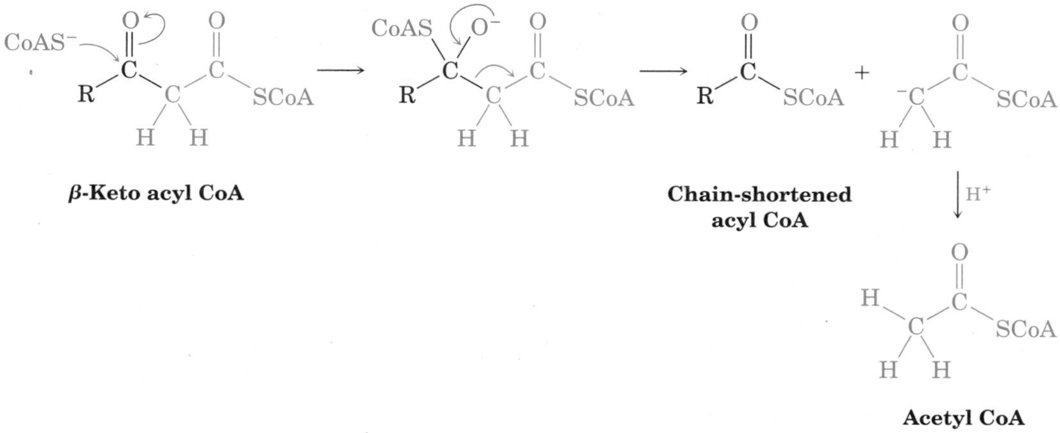

$$CoAS^- + R\overset{O}{\overset{\|}{C}}CH_2\overset{O}{\overset{\|}{C}}SCoA \longrightarrow R\overset{CoAS\ \ O^-}{C}CH_2\overset{O}{\overset{\|}{C}}SCoA \longrightarrow R\overset{O}{\overset{\|}{C}}SCoA + {}^{-}CH_2\overset{O}{\overset{\|}{C}}SCoA$$

β-Keto acyl CoA **Chain-shortened acyl CoA**

$$\downarrow H^+$$

$$H\text{—}CH_2\overset{O}{\overset{\|}{C}}SCoA$$

Acetyl CoA

Look at the catabolism of myristic acid shown in Figure 17.3 to see the overall results of the β-oxidation pathway. The first passage converts the C_{14} myristyl CoA into the C_{12} lauryl CoA plus acetyl CoA; the second passage converts lauryl CoA into the C_{10} capryl CoA plus acetyl CoA; the third passage converts capryl CoA into the C_8 caprylyl CoA; and so on. Note that the last passage produces *two* molecules of acetyl CoA because the precursor has four carbons.

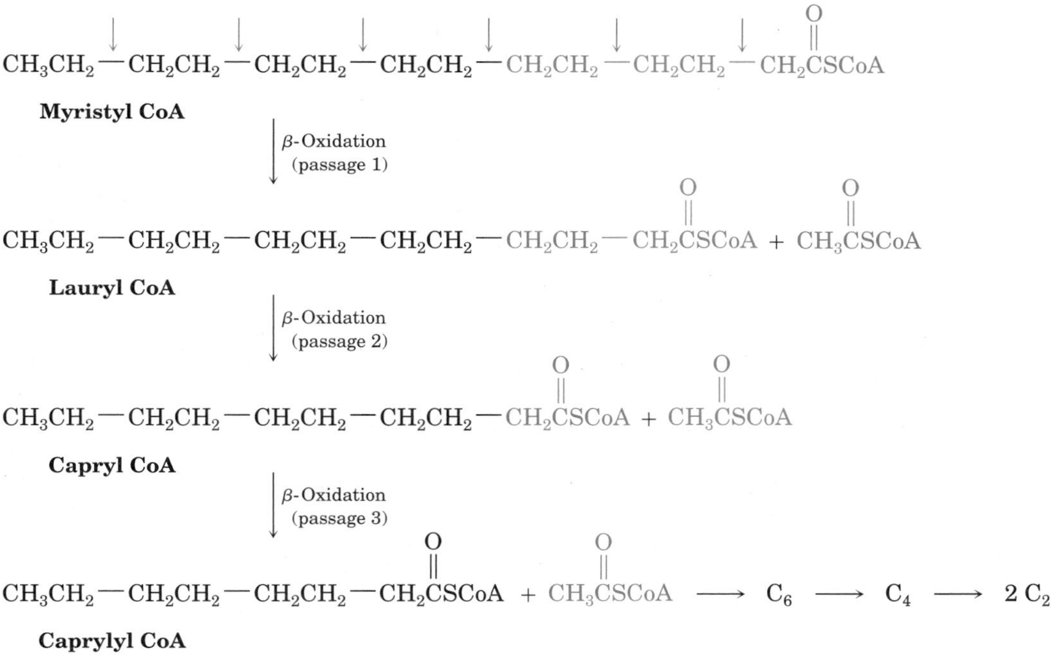

$$CH_3CH_2—CH_2CH_2—CH_2CH_2—CH_2CH_2—CH_2CH_2—CH_2CH_2—CH_2\overset{O}{\overset{\|}{C}}SCoA$$

Myristyl CoA

$\downarrow$ β-Oxidation (passage 1)

$$CH_3CH_2—CH_2CH_2—CH_2CH_2—CH_2CH_2—CH_2CH_2—CH_2\overset{O}{\overset{\|}{C}}SCoA + CH_3\overset{O}{\overset{\|}{C}}SCoA$$

Lauryl CoA

$\downarrow$ β-Oxidation (passage 2)

$$CH_3CH_2—CH_2CH_2—CH_2CH_2—CH_2CH_2—CH_2\overset{O}{\overset{\|}{C}}SCoA + CH_3\overset{O}{\overset{\|}{C}}SCoA$$

Capryl CoA

$\downarrow$ β-Oxidation (passage 3)

$$CH_3CH_2—CH_2CH_2—CH_2CH_2—CH_2\overset{O}{\overset{\|}{C}}SCoA + CH_3\overset{O}{\overset{\|}{C}}SCoA \longrightarrow C_6 \longrightarrow C_4 \longrightarrow 2\ C_2$$

Caprylyl CoA

Figure 17.3 Catabolism of the C_{14} myristic acid in the β-oxidation pathway yields seven molecules of acetyl CoA after six passages.

You can predict how many molecules of acetyl CoA will be obtained from a given fatty acid simply by counting the number of carbon atoms and dividing by 2. For example, the C_{14} myristic acid yields seven molecules of acetyl CoA after six passages along the β-oxidation pathway. The number of passages is always 1 less than the number of acetyl CoA molecules produced because the last passage cleaves a C_4 chain into two acetyl CoA's.

Most fatty acids have an even number of carbon atoms, so that none are left over after β-oxidation. Those fatty acids with an odd number of carbon atoms or with double bonds require additional steps for degradation, but all carbon atoms are ultimately released for further oxidation in the citric acid cycle (see Section 17.4).

PROBLEM

17.3 Write the equations for the remaining passages of the β-oxidation pathway following those shown in Figure 17.3.

PROBLEM

17.4 How many molecules of acetyl CoA are produced by catabolism of the following fatty acids, and how many passages of the β-oxidation pathway are needed?
(a) Palmitic acid, $CH_3(CH_2)_{14}COOH$ (b) Arachidic acid, $CH_3(CH_2)_{18}COOH$

17.3 Catabolism of Carbohydrates: Glycolysis

Glycolysis is a series of ten enzyme-catalyzed reactions that break down glucose molecules into 2 equivalents of pyruvate, $CH_3COCO_2^-$. The steps of glycolysis, also called the *Embden–Meyerhoff pathway* after its discoverers, are summarized in Figure 17.4 (p. 554).

Steps 1–3 Phosphorylation and isomerization Glucose, produced by the digestion of dietary carbohydrates, is first phosphorylated by reaction with ATP in a reaction catalyzed by the enzyme hexokinase. The glucose 6-phosphate that results is then isomerized by phosphoglucose isomerase to fructose 6-phosphate. As the open-chain structures in Figure 17.4 show, this isomerization reaction takes place by keto–enol tautomerism (Section 11.1), since both glucose and fructose share a common enol:

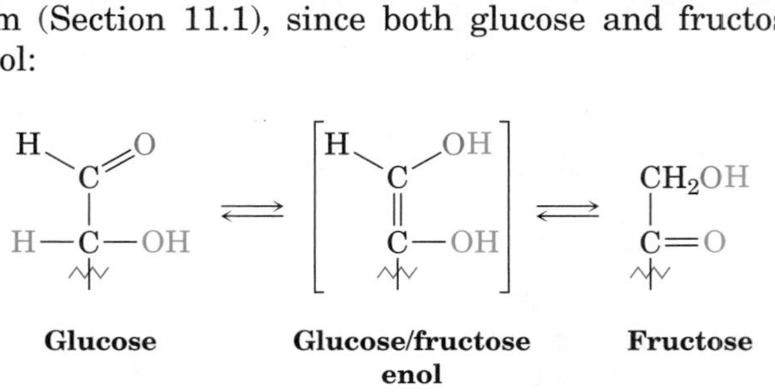

H–C(=O)–(H–C–OH) ⇌ [H–C(OH)=C–OH] ⇌ CH_2OH–C=O

Glucose **Glucose/fructose enol** **Fructose**

Fructose 6-phosphate is then converted to fructose 1,6-bisphosphate (the "bis" prefix means two) by phosphofructokinase-catalyzed reaction with ATP. The result is a molecule ready to be split into the two three-carbon intermediates that will ultimately become two molecules of pyruvate.

Steps 4–5 Cleavage and isomerization Fructose 1,6-bisphosphate is cleaved in step 4 into two, three-carbon monophosphates, one an aldose and one a ketose. The bond between C3 and C4 in fructose 1,6-bisphosphate breaks, and a C=O group is formed. Mechanistically, this cleavage is the reverse of an aldol reaction (Section 11.8) and is carried out by an aldolase enzyme. (A *forward* aldol reaction joins two ketones/aldehydes to give a β-hydroxy ketone or aldehyde; a *retro* aldol reaction cleaves a β-hydroxy ketone or aldehyde into two ketones/aldehydes):

$CH_2OPO_3^{2-}$ | C=O | HO—C—H | H—C—O—H (:Base) | H—C—OH | $CH_2OPO_3^{2-}$ ⟶ [$CH_2OPO_3^{2-}$ | C—O^- ‖ HO—C—H] + H—C(=O) | H—C—OH | $CH_2OPO_3^{2-}$ $\xrightarrow{H^+}$ $CH_2OPO_3^{2-}$ | C=O | CH_2OH

Fructose 1,6-bisphosphate (a β-hydroxy ketone)

Glyceraldehyde 3-phosphate

Dihydroxyacetone phosphate

Glyceraldehyde 3-phosphate continues on in the glycolysis pathway, but dihydroxyacetone phosphate must first be isomerized by the enzyme triose phosphate isomerase. As in the glucose-to-fructose conversion of step 2, the isomerization of dihydroxyacetone phosphate to glyceraldehyde 3-phosphate takes place by keto–enol tautomerization through a common enol. The net result of steps 4 and 5 is the production of two glyceraldehyde 3-phosphate molecules, both of which pass down the rest of the pathway. Thus, each of the remaining five steps of glycolysis takes place twice for every glucose molecule that enters at step 1.

Steps 6–8 Oxidation and phosphorylation Glyceraldehyde 3-phosphate is oxidized and phosphorylated by the coenzyme NAD^+ in the presence of the enzyme glyceraldehyde 3-phosphate dehydrogenase and phosphate ion, HPO_4^{2-}, (abbreviated P_i). Transfer of a phosphate group from the carboxyl of 1,3-bisphosphoglycerate to ADP then yields 3-phosphoglycerate, which is isomerized to 2-phosphoglycerate. The phosphorylation is catalyzed by phosphoglycerate kinase, and the isomerization is catalyzed by phosphoglyceromutase.

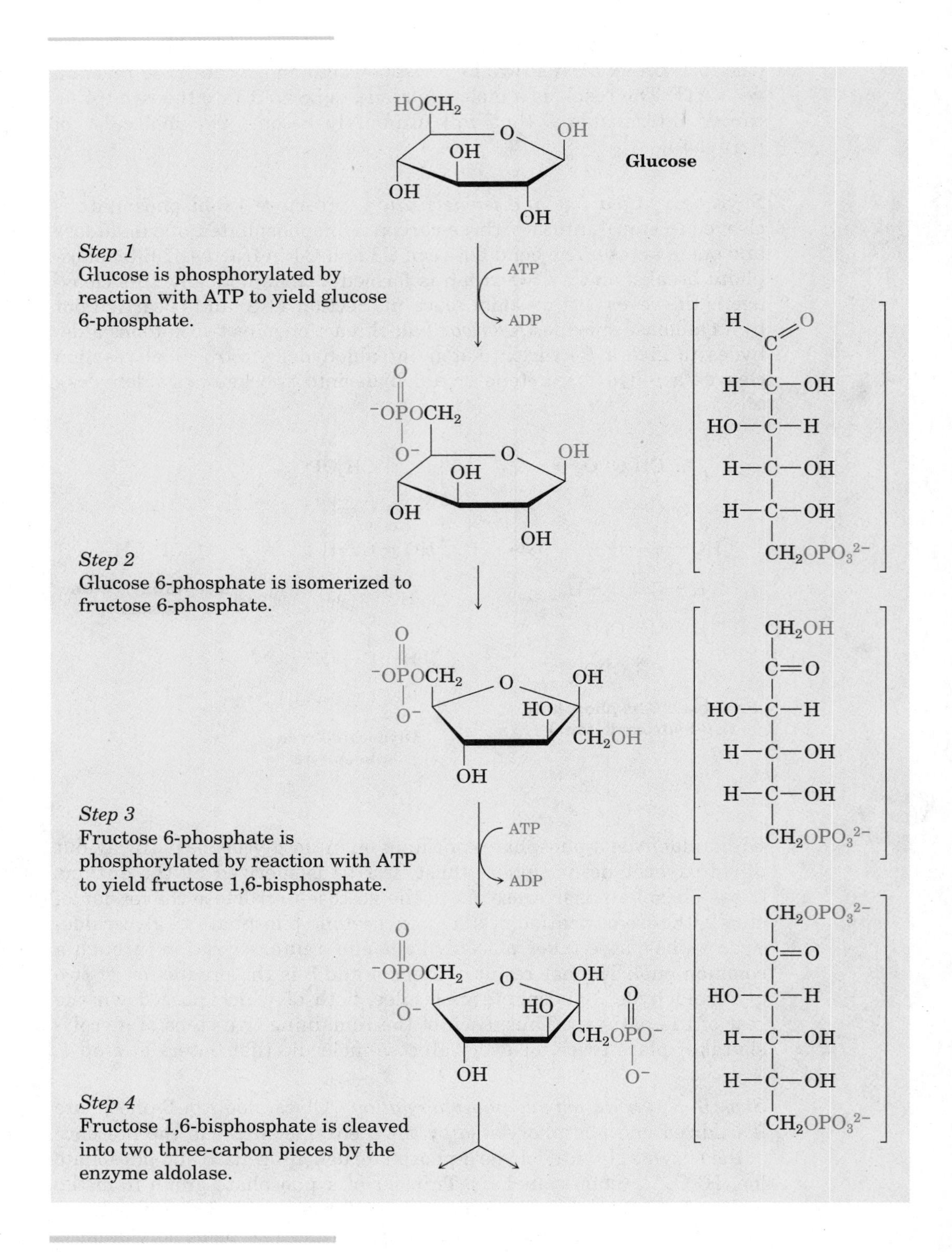

Figure 17.4 The ten-step glycolysis pathway for catabolizing glucose to pyruvate.

Step 5
Dihydroxyacetone phosphate, one of the products of step 4, is isomerized to glyceraldehyde 3-phosphate, the other product of step 4.

$$^{2-}O_3POCH_2CH(OH)CHO \rightleftharpoons {}^{2-}O_3POCH_2C(=O)CH_2OH$$

Step 6
Glyceraldehyde 3-phosphate is oxidized and phosphorylated to yield 3-phosphoglyceroyl phosphate.

NAD^+, P_i → $NADH/H^+$

$$^{2-}O_3POCH_2CH(OH)C(=O)OPO_3^{2-}$$

[$O=C(OPO_3^{2-})$—H—C—OH—$CH_2OPO_3^{2-}$]

Step 7
A phosphate is transferred from the carboxyl group to ADP, resulting in synthesis of an ATP and yielding 3-phosphoglycerate.

ADP → ATP

$$^{2-}O_3POCH_2CH(OH)C(=O)O^-$$

[$O=C(O^-)$—H—C—OH—$CH_2OPO_3^{2-}$]

Step 8
A phosphate group is transferred from the C3 hydroxyl to the C2 hydroxyl, giving 2-phosphoglycerate.

$$HOCH_2CH(OPO_3^{2-})C(=O)O^-$$

[$O=C(O^-)$—H—C—OPO_3^{2-}—CH_2OH]

Step 9
Dehydration occurs to yield phosphoenolpyruvate (PEP).

→ H_2O

$$H_2C=C(OPO_3^{2-})—C(=O)O^-$$

[$O=C(O^-)$—C—OPO_3^{2-} =CH_2]

Step 10
A phosphate is transferred from PEP to ADP, yielding pyruvate and ATP.

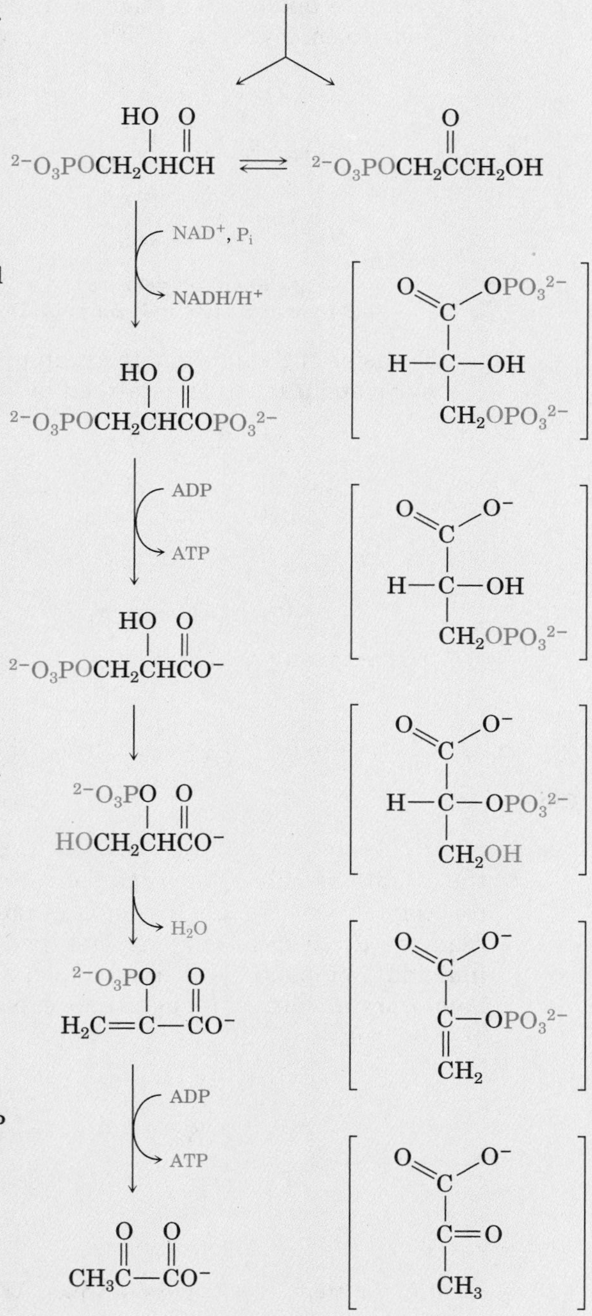

ADP → ATP

$$CH_3C(=O)—C(=O)O^-$$

[$O=C(O^-)$—C=O—CH_3]

Steps 9–10 Dehydration and dephosphorylation Like the β-hydroxy carbonyl compounds produced in aldol reactions (Section 11.9), 2-phosphoglycerate undergoes a ready dehydration by an E2 mechanism, yielding phosphoenolpyruvate (PEP). The process is catalyzed by enolase.

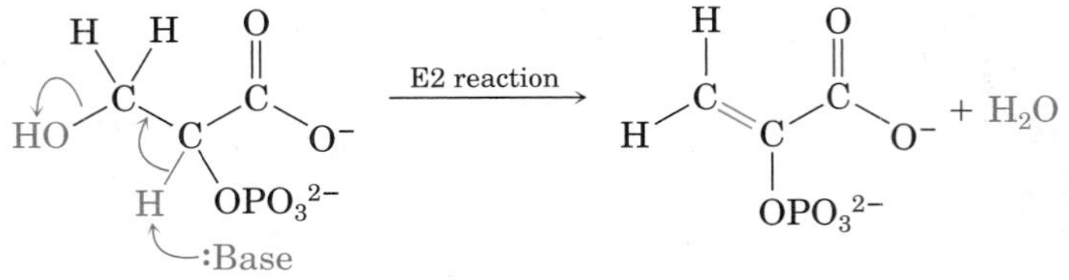

$$\text{HO–CH}_2\text{–CH(OPO}_3^{2-}\text{)–CO}_2^- \xrightarrow{\text{E2 reaction}} \text{H}_2\text{C=C(OPO}_3^{2-}\text{)–CO}_2^- + H_2O$$

2-Phosphoglycerate (a β-hydroxy carbonyl compound) **Phosphoenolpyruvate (PEP)**

Transfer of the phosphate group to ADP then generates ATP and gives pyruvate, a reaction catalyzed by pyruvate kinase.

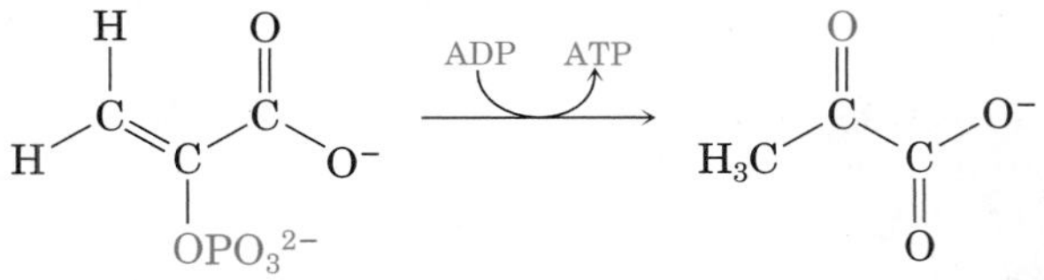

$$\text{H}_2\text{C=C(OPO}_3^{2-}\text{)–CO}_2^- \xrightarrow{\text{ADP} \;\; \text{ATP}} \text{H}_3\text{C–CO–CO}_2^-$$

Phosphoenolpyruvate **Pyruvate**

The net result of glycolysis can be summarized by the following equation:

$$C_6H_{12}O_6 + 2\,NAD^+ + 2\,P_i + 2\,ADP \longrightarrow 2\,CH_3\overset{O}{\overset{\|}{C}}-\overset{O}{\overset{\|}{C}}O^- + 2\,NADH + 2\,ATP + 2\,H_2O + 2\,H^+$$

Glucose **Pyruvate**

Pyruvate can undergo several further transformations, depending on the conditions and on the organism. Most commonly, pyruvate is converted to acetyl CoA through a complex, multistep sequence of reactions that requires three different enzymes and four different coenzymes. All the individual steps are well understood and are well precedented by simple laboratory analogies, though their explanations are a bit outside the scope of this book.

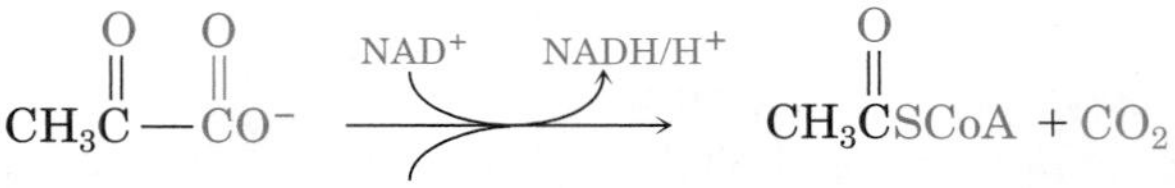

$$CH_3\overset{O}{\overset{\|}{C}}-\overset{O}{\overset{\|}{C}}O^- \xrightarrow[\text{HSCoA}]{\text{NAD}^+ \;\; \text{NADH/H}^+} CH_3\overset{O}{\overset{\|}{C}}SCoA + CO_2$$

Pyruvate **Acetyl CoA**

PROBLEM

17.5 Identify the steps in glycolysis in which ATP is produced.

PROBLEM

17.6 Propose a mechanism for the isomerization of dihydroxyacetone phosphate to glyceraldehyde 3-phosphate in step 5 of glycolysis.

17.4 The Citric Acid Cycle

The first two stages of catabolism result in the conversion of fats and carbohydrates into acetyl groups that are bonded through a thiol ester link to coenzyme A. These acetyl groups then enter the third stage of catabolism—the **citric acid cycle,** also called the *tricarboxylic acid (TCA) cycle* or *Krebs cycle* after Hans Krebs, who unraveled its complexities in 1937. The eight steps of the citric acid cycle, along with brief descriptions, are given in Figure 17.5 (p. 558).

As its name implies, the citric acid *cycle* is a closed loop of reactions in which the product of the last step is a reactant in the first step. The intermediates are constantly regenerated and flow continuously through the cycle, which operates as long as the oxidizing coenzymes NAD^+ and FAD are available. To meet this condition, the reduced coenzymes NADH and $FADH_2$ must be reoxidized via the electron-transport chain, which in turn relies on oxygen as the final electron acceptor. Thus, the cycle is also dependent on the availability of oxygen and on the operation of the electron-transport chain.

Steps 1–2 Addition to oxaloacetate Acetyl groups from acetyl CoA enter the citric acid cycle by nucleophilic addition to the ketone carbonyl group of oxaloacetate to give citrate. The addition is an aldol reaction (Section 11.8) of an enolate ion from acetyl CoA, and is catalyzed by the enzyme citrate synthetase.

$$\underset{\textbf{Oxaloacetate}}{O{=}C(COO^-)(CH_2COO^-)} + {}^-{:}CH_2\overset{O}{\overset{\|}{C}}SCoA \longrightarrow \left[{}^-O{-}C(CH_2COSCoA)(COO^-)(CH_2COO^-) \right] \xrightarrow{H_2O \quad HSCoA} \underset{\textbf{Citrate}}{HO{-}C(CH_2COO^-)(COO^-)(CH_2COO^-)}$$

Citrate, a tertiary alcohol, is next converted into its isomer, isocitrate, a secondary alcohol. The isomerization occurs in two steps, both of which are catalyzed by the same aconitase enzyme. The initial step is an E2 dehydration of the same sort that occurs in step 9 of glycolysis (Figure 17.4). The second step is a nucleophilic addition of water of the same sort that occurs in step 2 of the β-oxidation pathway (Figure 17.2). Note that the dehydration of citrate takes place specifically *away* from the carbon atoms of the acetyl group that added to oxaloacetate in step 1.

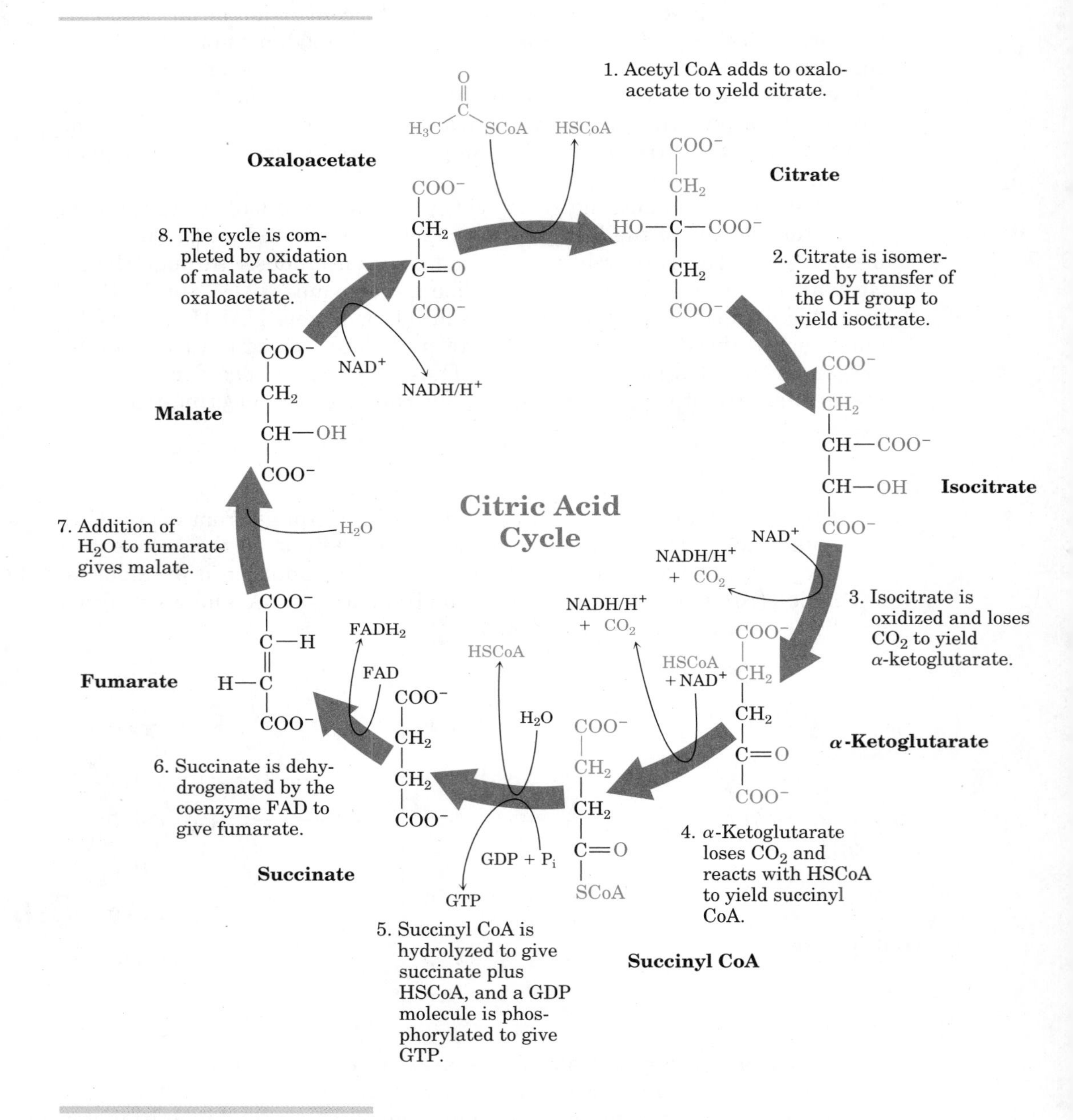

Figure 17.5 The citric acid cycle, an eight-step series of reactions that results in the conversion of an acetyl group into two molecules of CO_2 plus reduced coenzymes.

$$\underset{\textbf{Citrate}}{HO{-}C(CH_2COO^-)(CH_2COO^-){-}COO^-} \xrightarrow[\text{Aconitase}]{-H_2O} \underset{\textbf{Aconitate}}{^-OOC{-}CH_2{-}C(COO^-){=}CH{-}COO^-} \xrightarrow[\text{Aconitase}]{+H_2O} \underset{\textbf{Isocitrate}}{^-OOC{-}CH_2{-}CH(COO^-){-}CH(OH){-}COO^-}$$

Steps 3–4 Oxidative decarboxylations Isocitrate, a secondary alcohol, is oxidized by NAD^+ in step 3 to give a ketone, which loses CO_2 to give α-ketoglutarate. Catalyzed by the enzyme isocitrate dehydrogenase, the decarboxylation is a typical reaction of a carboxylic acid that has a second carbonyl group two atoms away (a β-keto acid). A similar kind of decarboxylation reaction occurs in the malonic ester synthesis (Section 11.6).

$$^-OOC{-}C(=O){-}CH(CH_2COO^-){-}C(=O)O^- \longrightarrow \left[\,^-OOC{-}C(O^-){=}CH{-}CH_2COO^-\,\right] \longrightarrow \underset{\alpha\textbf{-Ketoglutarate}}{^-OOC{-}C(=O){-}CH_2{-}CH_2COO^-} + CO_2$$

The transformation of α-ketoglutarate to succinyl CoA in step 4 is a multistep process analogous to the transformation of pyruvate to acetyl CoA that we saw in the previous section. Like the pyruvate conversion, the α-ketoglutarate conversion requires a number of different enzymes and coenzymes.

Steps 5–6 Hydrolysis and dehydrogenation of succinyl CoA Succinyl CoA is hydrolyzed to succinate in step 5. The reaction is catalyzed by succinyl CoA synthetase and is coupled with phosphorylation of guanosine diphosphate (GDP) to give guanosine triphosphate (GTP). Succinate is then dehydrogenated by FAD and succinate dehydrogenase to give fumarate—a process analogous to that of the first step in the β-oxidation pathway.

Steps 7–8 Regeneration of oxaloacetate Catalyzed by the enzyme fumarase, nucleophilic addition of water to fumarate yields malate in a reaction similar to that of step 2 in the β-oxidation pathway. Oxidation with NAD^+ then gives oxaloacetate in a step catalyzed by malate dehydrogenase, and the cycle has returned to its starting point, ready to revolve again.

The net result of the citric acid cycle can be summarized as:

$$\text{Acetyl CoA} + 3\ NAD^+ + FAD + ADP + P_i + 2\ H_2O \longrightarrow \text{HSCoA} + 3\ NADH + 3\ H^+ + FADH_2 + ATP + 2\ CO_2$$

PROBLEM

17.7 Which of the substances in the citric acid cycle are tricarboxylic acids (thus giving the cycle its alternate name)?

PROBLEM

17.8 Write mechanisms for the reactions in step 2 of the citric acid cycle, the dehydration of citrate and the addition of water to aconitate.

17.5 Catabolism of Proteins: Transamination

The catabolism of proteins is more complex than that of fats and carbohydrates because each of the 20 amino acids is degraded through its own unique pathway. The general idea, however, is that the amino nitrogen atoms are removed and the substances that remain are converted into compounds that enter the citric acid cycle.

Most amino acids lose their nitrogen atom by a **transamination** reaction in which the $-NH_2$ group of the amino acid changes places with the keto group of α-ketoglutarate. The products are a new α-keto acid and glutamate:

$$\underset{\textbf{An amino acid}}{\overset{NH_3^+}{\overset{|}{RCHCOO^-}}} + \underset{\textbf{α-Ketoglutarate}}{^-OOCCH_2CH_2\overset{O}{\overset{||}{C}}COO^-} \rightleftharpoons \underset{\textbf{An α-keto acid}}{R\overset{O}{\overset{||}{C}}COO^-} + \underset{\textbf{Glutamate}}{^-OOCCH_2CH_2\overset{NH_3^+}{\overset{|}{C}}HCOO^-}$$

Transaminations use pyridoxal phosphate, a derivative of vitamin B_6, as cofactor. As shown in Figure 17.6 for the reaction of alanine, the key step in transamination is nucleophilic addition of the amino acid $-NH_2$ group to the pyridoxal aldehyde group to yield an imine (Section 9.9). Loss of a proton from the α position then results in a bond rearrangement to give a different imine, which is hydrolyzed (the exact reverse of imine formation) to yield pyruvate and a nitrogen-containing derivative of pyridoxal phosphate. Pyruvate is converted into acetyl CoA (Section 17.3), which enters the citric acid cycle for further catabolism. The pyridoxal phosphate derivative transfers its nitrogen atom to α-ketoglutarate by the reverse of the steps in Figure 17.6, thereby forming glutamate and regenerating pyridoxal phosphate for further use.

Glutamate, which now contains the nitrogen atom of the former amino acid, next undergoes an *oxidative deamination* to yield ammonium ion and regenerated α-ketoglutarate. The oxidation of the amine to an imine is similar to the oxidation of a secondary alcohol to a ketone, and is carried out by NAD^+. The imine is then hydrolyzed (Section 9.9).

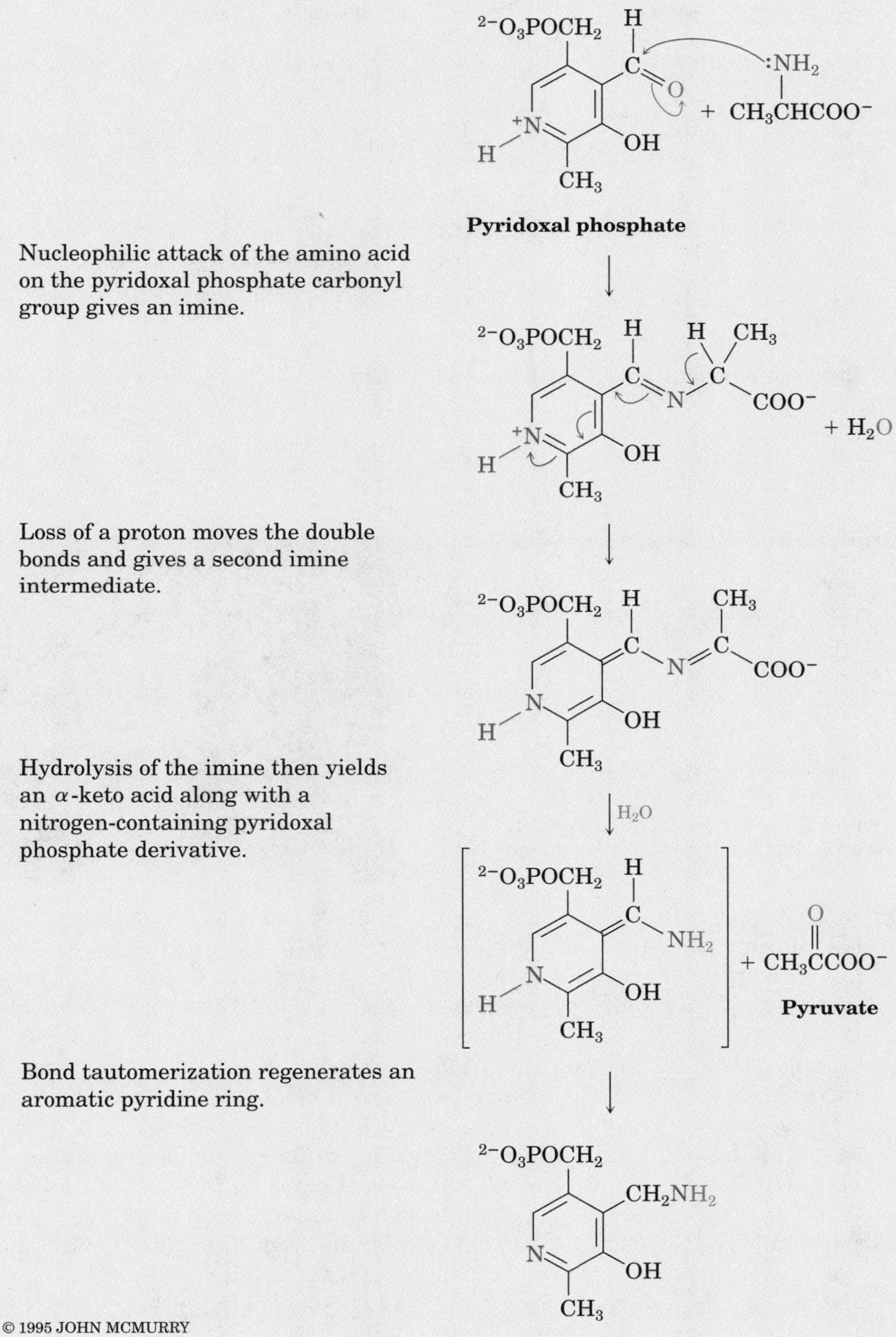

$^{2-}O_3POCH_2$ H C O :NH_2 + CH_3CHCOO^- ^+N H OH CH_3

Pyridoxal phosphate

Nucleophilic attack of the amino acid on the pyridoxal phosphate carbonyl group gives an imine.

$^{2-}O_3POCH_2$ H C N H CH_3 C COO^- + H_2O ^+N H OH CH_3

Loss of a proton moves the double bonds and gives a second imine intermediate.

$^{2-}O_3POCH_2$ H C N C CH_3 COO^- N H OH CH_3

Hydrolysis of the imine then yields an α-keto acid along with a nitrogen-containing pyridoxal phosphate derivative.

H_2O

$^{2-}O_3POCH_2$ H C NH_2 N H OH CH_3 + CH_3CCOO^- (O)

Pyruvate

Bond tautomerization regenerates an aromatic pyridine ring.

$^{2-}O_3POCH_2$ CH_2NH_2 N OH CH_3

Figure 17.6 Oxidative deamination of alanine requires the cofactor pyridoxal phosphate and yields pyruvate as product.

$$^{-}O_2CCH_2CH_2CH(NH_3^+)CO_2^- \xrightarrow{NAD^+ \quad NADH/H^+} \left[^{-}O_2CCH_2CH_2C(=NH)CO_2^- \right] \xrightarrow{H_2O,\ H^+} {}^{-}O_2CCH_2CH_2C(=O)CO_2^- + NH_4^+$$

Glutamate → **α-Ketoglutarate**

PROBLEM

17.9 Show the product from transamination of leucine.

INTERLUDE

Basal Metabolism

Riders in the Tour de France need up to 6000 kcal per day to fuel their prodigious energy needs.

The minimum amount of energy per unit time an organism expends to stay alive is called the organism's *basal metabolic rate (BMR).* This rate is measured by monitoring respiration and finding the rate of oxygen consumption, which is proportional to the amount of energy used. Assuming an average dietary mix of fats, carbohydrates, and proteins, approximately 4.82 kcal are required for each liter of oxygen consumed.

The average basal metabolic rate for humans is about 65 kcal/h, or 1600 kcal/day. Obviously, the rate varies for different people, depending on sex, age, weight, and physical condition. As a rule, the BMR is lower for older people than for younger people, is lower for females than for males, and is lower for people in good physical condition than for those who are out of shape and overweight. A BMR substantially above the expected value indicates an unusually rapid metabolism, perhaps caused by a fever or some biochemical abnormality.

(*continued*)➤

The total number of calories a person needs each day is the sum of the basal requirement plus the energy used for physical activities, as shown in Table 17.1. A relatively inactive person needs about 30% above basal requirements per day, a lightly active person needs about 50% above basal, and a very active person such as an athlete or construction worker may need 100% above basal requirements. Some endurance runners in ultradistance events can use as many as 10,000 calories per day above the basal level. Each day that your caloric intake is above what you use, fat is stored in your body and your weight rises. Each day that your caloric intake is below what you use, fat in your body is metabolized and your weight drops.

Table 17.1 Energy Cost of Various Activities[a]

Activity	*kcal/min*	*Activity*	*kcal/min*
Sleeping	1.2	Tennis	7–9
Sitting, reading	1.6	Basketball	9–10
Standing still	1.8	Walking up stairs	10–18
Walking	3–6	Running	9–22

[a]For a 70 kg man.

Summary and Key Words

Metabolism is the sum of all chemical reactions in the body. Reactions that break down larger molecules into smaller ones are called **catabolism;** reactions that build up larger molecules from smaller ones are called **anabolism.** Although the details of specific biochemical pathways are sometimes complex, all the reactions that occur follow the normal rules of organic chemical reactivity.

The catabolism of fats begins with **digestion,** in which ester bonds are hydrolyzed to give glycerol and fatty acids. The fatty acids are degraded in the four-step ***β*-oxidation pathway** by removal of two carbons at a time, yielding acetyl CoA. Catabolism of carbohydrates begins with the hydrolysis of glycoside bonds to give glucose, which is degraded in the ten-step **glycolysis** pathway. Pyruvate, the initial product of glycolysis, is then converted into acetyl CoA. The acetyl groups produced by degradation of fats and carbohydrates next enter the eight-step **citric acid cycle,** where they are further degraded into CO_2.

Protein catabolism is more complex than that of fats or carbohydrates because each of the 20 different amino acids is degraded by its own unique pathway. In general, the amino nitrogen atoms are removed and the substances that remain are converted into compounds that enter the citric acid cycle. Most amino acids lose their nitrogen atom by **transamination,** a reaction in which the $-NH_2$ group of the amino acid changes places with the keto group of an α-keto acid such as α-ketoglutarate. The products are a new α-keto acid and glutamate.

The energy released in all catabolic pathways is used in the **electron-transport chain** to make molecules of *adenosine triphosphate* (ATP), the final result of food catabolism. ATP couples with and drives many otherwise unfavorable reactions.

ADDITIONAL PROBLEMS

17.10 What chemical events occur during the digestion of food?

17.11 What is the difference between digestion and metabolism?

17.12 What is the difference between anabolism and catabolism?

17.13 Draw the structure of adenosine *mono*phosphate (AMP), an intermediate in some biochemical pathways.

17.14 What general kind of reaction does ATP carry out?

17.15 What general kind of reaction does NAD^+ carry out?

17.16 What general kind of reaction does FAD carry out?

17.17 What substance is the starting point of the citric acid cycle, reacting with acetyl CoA in the first step and being regenerated in the last step?

17.18 Lactate, a product of glucose catabolism in oxygen-starved muscles, can be converted into pyruvate by oxidation. What coenzyme do you think is needed? Write the equation in the normal biochemical format using a curved arrow.

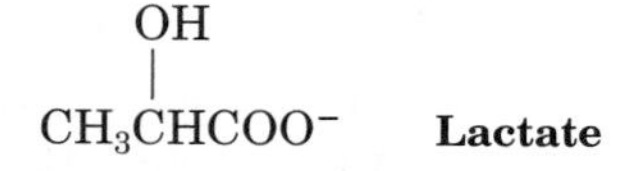

17.19 How many moles of acetyl CoA are produced by catabolism of the following substances?
(a) 1.0 mol glucose
(b) 1.0 mol palmitic acid ($C_{15}H_{31}CO_2H$)
(c) 1.0 mol maltose

17.20 How many grams of acetyl CoA (mol wt = 809.6 amu) are produced by catabolism of the following substances?
(a) 100.0 g glucose (b) 100.0 g palmitic acid (c) 100.0 g maltose

17.21 Which of the substances listed in Problem 17.20 is the most efficient precursor of acetyl CoA on a weight basis?

17.22 List the sequence of intermediates involved in the catabolism of glycerol from hydrolyzed fats to yield acetyl CoA.

17.23 Write the equation for the final step in the β-oxidation pathway of any fatty acid with an even number of carbon atoms.

17.24 Show the products of each of the following reactions:

$$\text{(a)}\ CH_3CH_2CH_2CH_2CH_2\overset{O}{\overset{\|}{C}}SCoA \xrightarrow[\text{Acetyl SCoA dehydrogenase}]{\text{FAD} \rightarrow \text{FADH}_2} ?$$

$$\text{(b) Product of (a)} + H_2O \xrightarrow[\text{hydratase}]{\text{Enoyl SCoA}} ?$$

$$\text{(c) Product of (b)} \xrightarrow[\beta\text{-Hydroxyacyl SCoA dehydrogenase}]{\text{NAD}^+ \rightarrow \text{NADH/H}^+} ?$$

17.25 What is the structure of the α-keto acid formed by transamination of each of the following amino acids?
(a) Valine (b) Phenylalanine (c) Methionine

17.26 What enzyme cofactor is associated with transamination?

17.27 Fatty acids are synthesized in the body by the *lipogenesis* cycle, which begins with acetyl CoA. The first step in lipogenesis is the condensation of two acetyl CoA molecules to yield acetoacetyl CoA, which undergoes three further enzyme-catalyzed steps, yielding butyryl CoA. Based on the kinds of reactions that occur in the β-oxidation pathway, what do you think are the three further steps of lipogenesis?

$$\underset{\textbf{Acetoacetyl CoA}}{CH_3\overset{O}{\overset{\|}{C}}CH_2\overset{O}{\overset{\|}{C}}SCoA} \xrightarrow{\text{3 steps}} \underset{\textbf{Butyryl CoA}}{CH_3CH_2CH_2\overset{O}{\overset{\|}{C}}SCoA}$$

17.28 In the *pentose phosphate* pathway for degrading sugars, ribulose 5-phosphate is converted to ribose 5-phosphate. Propose a mechanism for the isomerization.

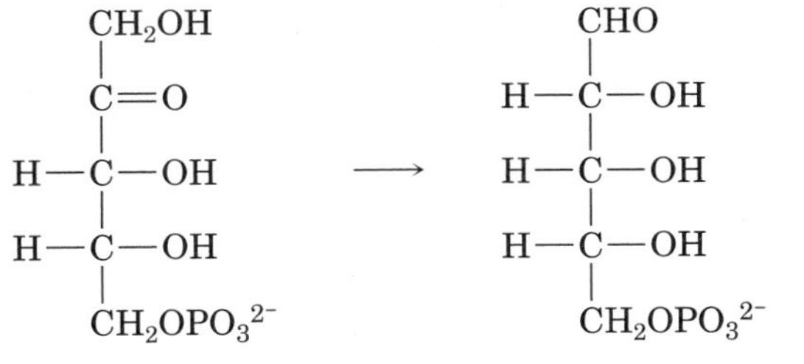

$$\begin{array}{c} CH_2OH \\ | \\ C{=}O \\ | \\ H{-}C{-}OH \\ | \\ H{-}C{-}OH \\ | \\ CH_2OPO_3^{2-} \end{array} \longrightarrow \begin{array}{c} CHO \\ | \\ H{-}C{-}OH \\ | \\ H{-}C{-}OH \\ | \\ H{-}C{-}OH \\ | \\ CH_2OPO_3^{2-} \end{array}$$

Ribulose 5-phosphate **Ribose 5-phosphate**

17.29 Another step in the pentose phosphate pathway for degrading sugars (see Problem 17.28) is the conversion of ribose 5-phosphate to glyceraldehyde 3-phosphate. What kind of organic process is occurring? Propose a mechanism for the conversion.

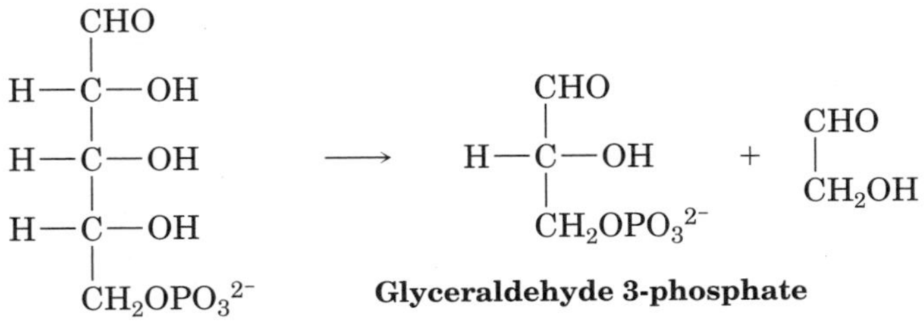

$$\begin{array}{c} CHO \\ | \\ H{-}C{-}OH \\ | \\ H{-}C{-}OH \\ | \\ H{-}C{-}OH \\ | \\ CH_2OPO_3^{2-} \end{array} \longrightarrow \begin{array}{c} CHO \\ | \\ H{-}C{-}OH \\ | \\ CH_2OPO_3^{2-} \end{array} + \begin{array}{c} CHO \\ | \\ CH_2OH \end{array}$$

Ribose 5-phosphate **Glyceraldehyde 3-phosphate**

17.30 One of the steps in the *gluconeogenesis* pathway for synthesizing glucose in the body is the reaction of pyruvate with CO_2 to yield oxaloacetate. Tell what kind of reaction is occurring, and suggest a mechanism.

$$CO_2 + CH_3\overset{O}{\overset{\|}{C}}-\overset{O}{\overset{\|}{C}}O^- \longrightarrow {}^-O\overset{O}{\overset{\|}{C}}CH_2\overset{O}{\overset{\|}{C}}-\overset{O}{\overset{\|}{C}}O^-$$

Pyruvate **Oxaloacetate**

17.31 Another step in gluconeogenesis (see Problem 17.30) is the conversion of oxaloacetate to phosphoenolpyruvate by decarboxylation and phosphorylation. Propose a mechanism.

$${}^-O\overset{O}{\overset{\|}{C}}CH_2\overset{O}{\overset{\|}{C}}-\overset{O}{\overset{\|}{C}}O^- \xrightarrow{\text{ATP} \to \text{ADP}} H_2C=\overset{{}^{2-}O_3PO}{\overset{|}{C}}-\overset{O}{\overset{\|}{C}}O^- + CO_2$$

Oxaloacetate **Phosphoenolpyruvate**

17.32 The primary fate of acetyl CoA under normal metabolic conditions is degradation in the citric acid cycle to yield CO_2. When the body is stressed by prolonged starvation, however, acetyl CoA is converted into compounds called *ketone bodies,* which can be used by the brain as a temporary fuel. The biochemical pathway for the synthesis of ketone bodies from acetyl CoA is shown. Fill in the missing information represented by the four question marks.

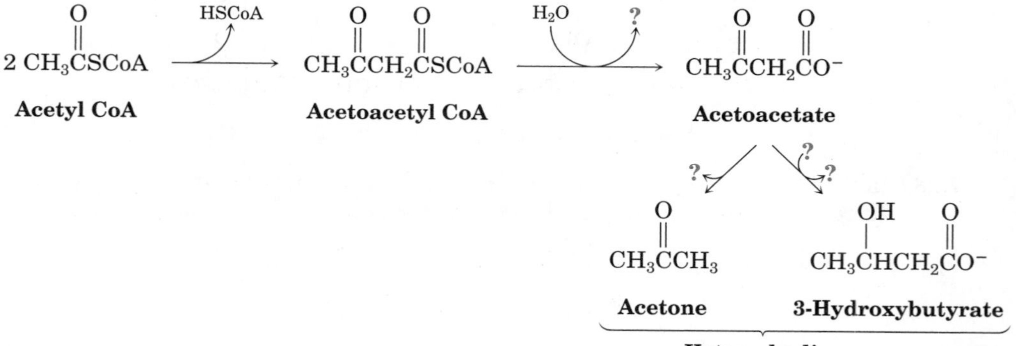

17.33 The initial reaction in Problem 17.32, conversion of two molecules of acetyl CoA to one molecule of acetoacetyl CoA, is a Claisen reaction. Assuming that there is a base present, show the mechanism of the reaction.

17.34 The amino acid tyrosine is metabolized by a series of steps that include the following transformations. Propose a mechanism for the conversion of fumaroylacetoacetate into fumarate plus acetoacetate.

$$\text{Tyrosine} \Longrightarrow {}^-O_2C\text{–CH=CH–CO–CH}_2\text{–CO–CH}_2\text{–}CO_2^- \longrightarrow CH_3COCH_2CO_2^- + {}^-O_2C\text{–CH=CH–}CO_2^-$$

$$CH_3COCH_2CO_2^- \longrightarrow CH_3\overset{O}{\overset{\|}{C}}SCoA$$

Fumaroylacetoacetate **Acetoacetate** **Fumarate**

17.35 Propose a mechanism for the conversion of acetoacetate into acetyl CoA (Problem 17.34).

Appendix A: Nomenclature of Polyfunctional Organic Compounds

Judging from the number of incorrect names that appear in the chemical literature, it's probably safe to say that relatively few practicing organic chemists are fully conversant with the rules of organic nomenclature. Simple hydrocarbons and monofunctional compounds present few difficulties, because the basic rules for naming such compounds are logical and easy to understand. Problems are often encountered with polyfunctional compounds, however. Whereas most chemists could correctly identify hydrocarbon **1** as 3-ethyl-2,5-dimethylheptane, few could correctly identify polyfunctional compound **2.** Should we consider **2** as an ether? As an ethyl ester? As a ketone? As an alkene? It is, of course, all four, but it has only one correct name: ethyl 3-(4-methoxy-2-oxo-3-cyclohexenyl)propanoate.

CH_3CHCH_3

$CH_3CH_2CHCH_2CHCH_2CH_3$

CH_3

1. 3-Ethyl-2,5-dimethylheptane

O O $COCH_2CH_3$ CH_3O

2. Ethyl 3-(4-methoxy-2-oxo-3-cyclohexenyl)propanoate

Naming polyfunctional organic compounds isn't really much harder than naming monofunctional ones. All that's required is a knowledge of nomenclature for monofunctional compounds and a set of additional rules. In the following discussion, it's assumed that you have a good command of the rules of nomenclature for monofunctional compounds that were given throughout the text as each new functional group was introduced. A list of where these rules can be found is given in Table A.1.

Table A.1 Where to Find Nomenclature Rules for Simple Functional Groups

Functional group	*Text section*	*Functional group*	*Text section*
Acid anhydrides	10.1	Amines	12.1
Acid halides	10.1	Aromatic compounds	5.4
Alcohols	8.1	Carboxylic acids	10.1
Aldehydes	9.3	Cycloalkanes	2.7
Alkanes	2.3	Esters	10.1
Alkenes	3.1	Ethers	8.1
Alkyl halides	7.1	Ketones	9.3
Alkynes	4.13	Nitriles	10.1
Amides	10.1	Phenols	8.1

The name of a polyfunctional organic molecule has four parts:

1. **Suffix**—the part that identifies the principal functional-group class to which the molecule belongs
2. **Parent**—the part that identifies the size of the main chain or ring
3. **Substituent prefixes**—parts that identify what substituents are located on the main chain or ring
4. **Locants**—numbers that tell where substituents are located on the main chain or ring

To arrive at the correct name for a complex molecule, you must identify the four name parts and then express them in the proper order and format. Let's look at the four parts.

The Suffix—Functional-Group Precedence

A polyfunctional organic molecule can contain many different kinds of functional groups, but for nomenclature purposes, we must choose just one suffix. It's not correct to use two suffixes. Thus, keto ester **3** must be named either as a ketone with an *-one* suffix or as an ester with an *-oate* suffix, but it can't be named as an *-onoate*. Similarly, amino alcohol **4** must be named either as an alcohol (*-ol*) or as an amine (*-amine*), but it can't properly be named as an *-olamine*. The only exception to this rule is in naming compounds that have double or triple bonds. For example, the unsaturated acid $H_2C{=}CHCH_2COOH$ is 3-butenoic acid, and the acetylenic alcohol $HC{\equiv}CCH_2CH_2CH_2CH_2OH$ is 5-hexyn-1-ol.

$CH_3C(=O)CH_2CH_2C(=O)OCH_3$

3. Named as an ester with a keto (oxo) substituent: **methyl 4-oxopentanoate**

4. Named as an alcohol with an amino substituent: **5-amino-2-pentanol**

How do we choose which suffix to use? Functional groups are divided into two classes, **principal groups** and **subordinate groups,** as shown in Table A.2. Principal groups may be cited either as prefixes or as suffixes, while subordinate groups may be cited only as prefixes. Within the principal groups, an order of precedence has been established. The proper suffix for a given compound is determined by identifying all the functional groups present and then choosing the principal group of highest priority. For example, Table A.2 indicates that keto ester **3** must be named as an ester rather than as a ketone, since an ester functional group is higher in priority than a ketone. Similarly, amino alcohol **4** must be named as an alcohol rather than as an amine. The correct name of **3** is methyl 4-oxopentanoate, and the correct name of **4** is 5-amino-2-pentanol. Further examples follow:

5. Named as a cyclohexanecarboxylic acid with an oxo substituent: **4-oxocyclohexanecarboxylic acid**

Table A.2 Classification of Functional Groups for Purposes of Nomenclature[a]

Functional-group class	*Structure*	*Name when used as suffix*	*Name when used as prefix*
Principal groups			
Carboxylic acids	—COOH	-oic acid -carboxylic acid	carboxy
Carboxylic anhydrides	—C(=O)—O—C(=O)—	-oic anhydride -carboxylic anhydride	
Carboxylic esters	—COOR	-oate -carboxylate	alkoxycarbonyl
Acid halides	—COCl	-oyl halide carbonyl halide	halocarbonyl
Amides	—$CONH_2$	-amide -carboxamide	amido
Nitriles	—C≡N	-nitrile -carbonitrile	cyano
Aldehydes	—CHO	-al -carbaldehyde	formyl
	=O		oxo
Ketones	=O	-one	oxo
Alcohols	—OH	-ol	hydroxy
Phenols	—OH	-ol	hydroxy
Thiols	—SH	-thiol	mercapto
Amines	—NH_2	-amine	amino
Imines	=NH	-imine	imino
Alkenes	C=C	-ene	alkenyl
Alkynes	C≡C	-yne	alkynyl
Alkanes	C—C	-ane	alkyl
Subordinate groups			
Ethers	—OR		alkoxy
Sulfides	—SR		alkylthio
Halides	—F, —Cl, —Br, —I		halo
Nitro	—NO_2		nitro
Azides	N=N=N		azido
Diazo	=N=N		diazo

[a]Principal functional groups are listed in order of decreasing priority, but the subordinate functional groups have no established priority order. Principal functional groups may be cited either as prefixes or as suffixes; subordinate functional groups may be cited only as prefixes.

$$HO\overset{O}{\overset{\|}{C}}-\underset{CH_3}{\underset{|}{\overset{CH_3}{\overset{|}{C}}}}-CH_2CH_2CH_2\overset{O}{\overset{\|}{C}}Cl$$

6. Named as a carboxylic acid with a chlorocarbonyl substituent:
5-chlorocarbonyl-2,2-dimethylpentanoic acid

$$CH_3\overset{CHO}{\overset{|}{C}}HCH_2CH_2CH_2\overset{O}{\overset{\|}{C}}OCH_3$$

7. Named as an ester with an oxo substituent:
methyl 5-methyl-6-oxohexanoate

The Parent—Selecting the Main Chain or Ring

The parent, or base, name of a polyfunctional organic compound is usually easy to identify. If the group of highest priority is part of an open chain, we simply select the longest chain that contains the largest number of principal functional groups. If the highest-priority group is attached to a ring, we use the name of that ring system as the parent. For example, compounds **8** and **9** are isomeric aldehydo acids, and both must be named as acids rather than as aldehydes according to Table A.2. The longest chain in compound **8** has seven carbons, and the substance is therefore named 6-methyl-7-oxoheptanoic acid. Compound **9** also has a chain of seven carbons, but the longest chain that contains both of the principal functional groups has only three carbons. The correct name of **9** is 3-oxo-2-pentylpropanoic acid.

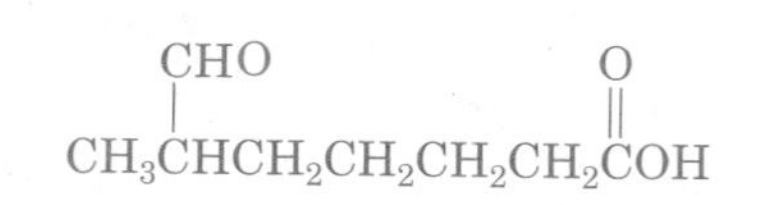

8. Named as a substituted heptanoic acid:
6-methyl-7-oxoheptanoic acid

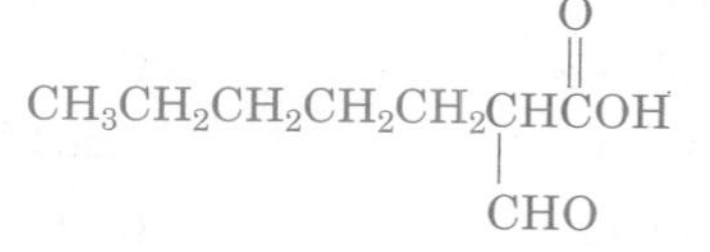

9. Named as a substituted propanoic acid:
3-oxo-2-pentylpropanoic acid

Similar rules apply for compounds **10–13,** which contain rings. Compounds **10** and **11** are isomeric keto nitriles, and both must be named as nitriles according to Table A.2. Substance **10** is named as a benzonitrile since the –CN functional group is a substituent on the aromatic ring, but substance **11** is named as an acetonitrile since the –CN functional group is on an open chain. The correct names are 2-acetyl-4-methylbenzonitrile **(10)** and (2-acetyl-4-methylphenyl)acetonitrile **(11).** Compounds **12** and **13** are both keto acids and must be named as acids. The correct names are 3-(2-oxocyclohexyl)propanoic acid **(12)** and 2-(3-oxopropyl)cyclohexanecarboxylic acid **(13).**

10. Named as a substituted benzonitrile:
2-acetyl-4-methylbenzonitrile

11. Named as a substituted acetonitrile:
(2-acetyl-4-methylphenyl)acetonitrile

12. Named as a carboxylic acid:
3-(2-oxocyclohexyl)propanoic acid

13. Named as a carboxylic acid:
2-(3-oxopropyl)cyclohexanecarboxylic acid

The Prefixes and Locants

With the suffix and parent name established, the next step is to identify and number all substituents on the parent chain or ring. These substituents include all alkyl groups and all functional groups other than the one cited in the suffix. For example, compound **14** contains three different functional groups (carboxyl, keto, and double bond). Because the carboxyl group is highest in priority, and because the longest chain containing the functional groups is seven carbons long, **14** is a heptenoic acid. In addition, the main chain has an oxo (keto) substituent and three methyl groups. Numbering from the end nearer the highest-priority functional group, we find that **14** is 2,5,5-trimethyl-4-oxo-2-heptenoic acid. Note that the final *-e* of heptene is deleted in the word *heptenoic.* This deletion occurs when the name would have two adjacent vowels (thus, *heptenoic* has the final "e" deleted, but *heptenenitrile* retains the "e"). Look back at some of the other compounds we've named to see other examples of how prefixes and locants are assigned.

$$CH_3CH_2-C(CH_3)_2-C(=O)CH=C(CH_3)C(=O)OH$$

14. Named as a heptenoic acid:
2,5,5-trimethyl-4-oxo-2-heptenoic acid

Writing the Name

Once the name parts have been established, the entire name is written out. Several additional rules apply:

1. *Order of prefixes* When the substituents have been identified, the main chain has been numbered, and the proper multipliers such as *di-* and *tri-* have been assigned, the name is written with the

substituents listed in alphabetical, rather than numerical, order. Multipliers such as *di-* and *tri-* are not used for alphabetization purposes, but the prefix *iso-* is used.

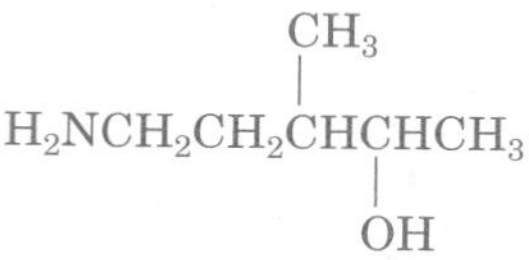

$H_2NCH_2CH_2CH(CH_3)CH(OH)CH_3$

15. 5-Amino-3-methyl-2-pentanol
(*NOT* 3-methyl-5-amino-2-pentanol)

2. *Use of hyphens; single- and multiple-word names* The general rule is to determine whether the principal functional group is itself an element or compound. If it is, then the name is written as a single word; if it isn't, then the name is written as multiple words. For example, methylbenzene (one word) is correct because the parent—benzene—is itself a compound. Diethyl ether, however, is written as two words because the parent—ether—is a class name rather than a compound name. Some further examples are shown below:

$H_3C—Mg—CH_3$

16. Dimethylmagnesium
(one word, since magnesium is an element)

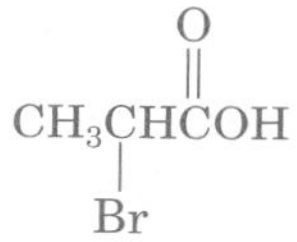

$CH_3CH(Br)COOH$

17. 2-Bromopropanoic acid
(two words, since "acid" is not a compound)

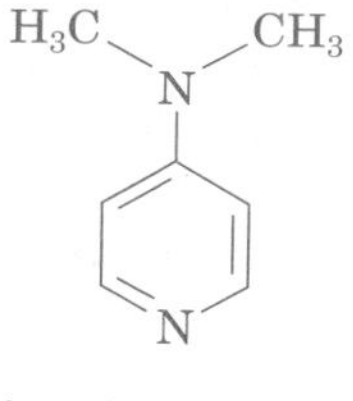

18. 4-(Dimethylamino)pyridine
(one word, since pyridine is a compound)

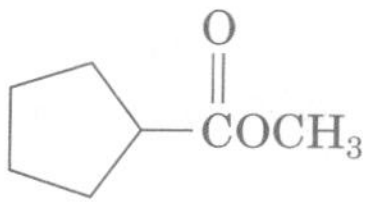

19. Methyl cyclopentanecarboxylate

3. *Parentheses* Parentheses are used to denote complex substituents when ambiguity would otherwise arise. For example, chloromethylbenzene has two substituents on a benzene ring, but (chloromethyl)benzene has only one complex substituent. Note that the expression in parentheses is not set off by hyphens from the rest of the name.

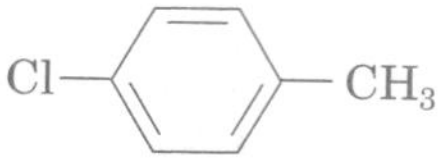

$Cl—C_6H_4—CH_3$

20. *p*-Chloromethylbenzene
(two substituents)

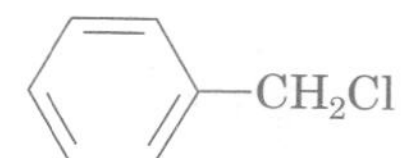

21. (Chloromethyl)benzene
(one complex substituent)

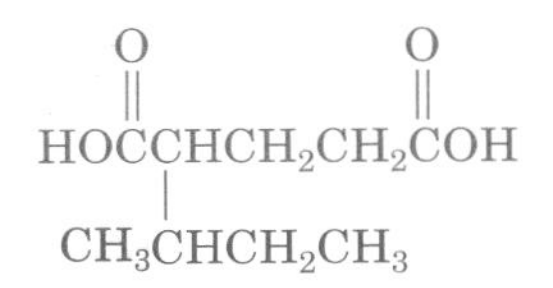

22. **2-(1-Methylpropyl)pentanedioic acid**
(The 1-methylpropyl group is a complex substituent on C2 of the main chain.)

Additional Reading

Further explanations of the rules of organic nomenclature can be found in the following references:

1. "A Guide to IUPAC Nomenclature of Organic Compounds," CRC Press, Boca Raton, FL, 1993.
2. J. G. Traynham, "Organic Nomenclature: A Programmed Introduction," Prentice Hall, Englewood Cliffs, NJ, 1985.
3. "Nomenclature of Organic Chemistry, Sections A, B, C, D, E, F, and H," International Union of Pure and Applied Chemistry, Pergamon Press, Oxford, 1979.

Appendix B: Glossary

Absorption spectrum (Section 13.1): A plot of wavelength of incident light versus amount of light absorbed. Organic molecules show absorption spectra in both the infrared and ultraviolet regions of the electromagnetic spectrum.

Acetal (Section 9.8): A functional group consisting of two ether-type oxygen atoms bonded to the same carbon, $R_2C(OR')_2$.

Achiral (Section 6.2): Lacking handedness. A molecule is achiral if it has a plane of symmetry and is thus superimposable on its mirror image.

Acid chloride (Section 9.1): A substance with the general formula RCOCl.

Acidity constant, K_a (Section 1.12): A value that expresses the strength of an acid in water solution. The larger the K_a, the stronger the acid.

Activating group (Section 5.9): An electron-donating group such as hydroxyl (–OH) or amino ($-NH_2$) that increases the reactivity of an aromatic ring toward electrophilic aromatic substitution.

Activation energy, E_{act} (Section 3.10): The difference in energy between ground state and transition state. The amount of activation energy required by a reaction determines the rate at which the reaction proceeds.

Acyl group (Section 5.8): A name for the $-\overset{\overset{\displaystyle O}{\|}}{C}-R$ group.

Acylation (Section 5.8): The introduction of an acyl group, –COR, onto a molecule. For example, acylation of an aromatic ring yields a ketone ($ArH \rightarrow ArCOR$).

1,4-Addition (Section 4.10): The addition of an electrophile to carbons 1 and 4 of a conjugated diene.

Addition reaction (Section 3.5): The reaction that occurs when two reactants combine to form a single new product with no atoms left over.

Alcohol (Section 8.1): A compound with an –OH group bonded to a saturated, sp^3-hybridized carbon atom.

Aldaric acid (Section 14.8): The dicarboxylic acid that results from oxidation of an aldose.

Aldehyde (Section 9.1): A substance with the general formula RCHO.

Alditol (Section 14.8): The polyalcohol that results from reduction of the carbonyl group of a monosaccharide.

Aldol reaction (Section 11.8): A carbonyl condensation reaction between two ketones or aldehydes leading to a β-hydroxy ketone or aldehyde product.

Aldonic acid (Section 14.8): The monocarboxylic acid that results from mild oxidation of an aldose.

Aldose (Section 14.1): A simple sugar with an aldehyde carbonyl group.

Alicyclic (Section 2.7): Referring to an aliphatic cyclic hydrocarbon such as a cycloalkane or cycloalkene.

Aliphatic (Section 2.2): Referring to a nonaromatic hydrocarbon such as a simple alkane, alkene, or alkyne.

Alkaloid (Chapter 12 Interlude): A naturally occurring compound that contains a basic amine functional group.

Alkane (Section 2.2): A compound that contains only carbon and hydrogen and has only single bonds.

Alkene (Section 3.1): A hydrocarbon with one or more carbon–carbon double bonds.

Alkoxide ion (Section 8.3): The anion formed by loss of H^+ from an alcohol.

Alkyl group (Section 2.2): A part structure, formed by removing a hydrogen from an alkane.

Alkylation (Sections 5.8, 11.6): The introduction of an alkyl group onto a molecule. For example, aromatic rings can be alkylated to yield arenes (ArH $\rightarrow$ ArR).

Alkyne (Section 4.13): A hydrocarbon that has a carbon–carbon triple bond.

Allylic (Section 4.10): The position next to a double bond.

Alpha position (Section 11.1): The position next to a carbonyl carbon.

Alpha-substitution reaction (Section 11.1): A reaction that results in substitution of a hydrogen on the α carbon of a carbonyl compound.

Amide (Section 9.1): A substance with the general formula $RCONH_2$.

Amine (Section 12.1): An organic derivative of ammonia, RNH_2, R_2NH, or R_3N.

α-Amino acid (Section 15.1): A compound with an amino group attached to the carbon atom next to the carboxyl group, $RCH(NH_2)COOH$.

Amino sugar (Section 14.11): A sugar with an $-NH_2$ group in place of an $-OH$ group on one carbon.

Amplitude (Section 13.1): The height of a wave from midpoint to peak.

Anabolism (Section 17.1): Metabolic reactions that synthesize larger molecules from smaller precursors.

Androgen (Section 16.5): A steroidal male sex hormone such as testosterone.

Angle strain (Section 2.9): The strain introduced into a molecule when a bond angle is deformed from its ideal value.

Anomeric center (Section 14.6): The hemiacetal carbon in a pyranose or furanose sugar.

Anomers (Section 14.6): Cyclic stereoisomers of sugars that differ only in their configurations at the hemiacetal (anomeric) carbon.

Anti periplanar geometry (Section 7.9): Reaction geometry in which all reacting atoms lie in a plane, with one group on top and another on the bottom of the molecule.

Anti stereochemistry (Section 4.5): Referring to opposite sides of a double bond or molecule. For example, an anti addition reaction is one in which the two carbon atoms of the double bond react on different faces.

Anticodon (Section 16.12): A sequence of three bases on tRNA that read the codons on mRNA and bring the correct amino acids into position for protein synthesis.

Antisense strand (Section 16.11): An alternate name for the template strand of DNA.

Apoenzyme (Section 15.12): The protein part of an enzyme that needs a cofactor for biological activity.

Aromatic (Section 5.1): The class of compounds that contain a benzene-like six-membered ring with three double bonds.

Arylamine (Section 12.1): An amine that has its nitrogen atom bonded to an aromatic ring, $Ar{-}NH_2$.

Atomic number, *Z* (Section 1.1): The number of protons in an atom's nucleus.

Atomic weight (Section 1.1): The average mass, in atomic mass units (amu), of a large number of atoms of an element.

Axial bond (Section 2.10): A bond to chair cyclohexane that lies along the ring axis perpendicular to the rough plane of the ring.

Azo compound (Section 12.5): A compound containing the –N=N– functional group.

Basicity constant, K_b (Section 12.3): A value that expresses the strength of a base in water solution. The larger the K_b, the stronger the base.

Benzyl (Section 5.4): The $C_6H_5CH_2$– group.

Benzylic (Section 5.11): The position next to an aromatic ring.

β-oxidation pathway (Section 17.2): A series of four enzyme-catalyzed reactions that cleave two carbon atoms at a time from the end of a fatty acid chain.

β-pleated sheet (Section 15.10): A secondary structure in which a protein chain folds back on itself so that two sections of the chain run parallel.

Bimolecular reaction (Section 7.7): A reaction that occurs between two molecules.

Bond angle (Section 1.7): The angle formed between two adjacent bonds.

Bond dissociation energy (Section 1.6): The energy needed to break a covalent bond to produce two radical fragments.

Bond length (Section 1.6): The equilibrium distance between the nuclei of two atoms that are bonded to each other.

Bond strength (Section 1.6): The amount of energy needed to break a bond to produce two radical fragments.

Branched-chain alkane (Section 2.2): An alkane that contains a branching arrangement of carbon atoms in its chain.

Bromonium ion (Section 4.5): A species with a positively charged, divalent bromine atom, R_2Br^+.

Brønsted–Lowry acid (Section 1.12): A substance that donates a hydrogen ion (proton, H^+) to a base.

Brønsted–Lowry base (Section 1.12): A substance that accepts a hydrogen ion, H^+, from an acid.

C-terminal amino acid (Section 15.4): The amino acid with a free –COOH group at one end of a protein chain.

Carbanion (Section 9.6): A carbon anion, or substance that contains a trivalent, negatively charged carbon atom ($R_3C:^-$).

Carbocation (Section 3.8): A carbon cation, or substance that contains a trivalent, positively charged carbon atom having six electrons in its outer shell (R_3C^+).

Carbocycle (Section 12.6): A cyclic molecule that has only carbon atoms in the ring.

Carbohydrate (Section 14.1): A polyhydroxy aldehyde or polyhydroxy ketone. Carbohydrates can be either simple sugars such as glucose or complex sugars such as cellulose.

Carbonyl condensation reaction (Section 11.1): A reaction between two carbonyl compounds in which the α carbon of one partner bonds to the carbonyl carbon of the other.

Carbonyl group (Section 9.1): The carbon–oxygen double bond functional group, C=O.

Carboxyl group (Section 10.1): The –COOH group.

Carboxylation (Section 10.4): The addition of CO_2 to a molecule.

Carboxylic acid (Section 9.1): A substance with the general formula RCOOH.

Catabolism (Section 17.1): Metabolic reactions that break down large molecules.

Chain-growth polymer (Section 10.12): A polymer produced by chain reaction of a monofunctional monomer.

Chain reaction (Section 7.2): A reaction that, once initiated, sustains itself in an endlessly repeating cycle of propagation steps. The radical chlorination of alkanes is an example.

Chair cyclohexane (Section 2.9): A three-dimensional conformation of cyclohexane that resembles the rough shape of a chair. The chair form of cyclohexane has neither angle strain nor eclipsing strain.

Chemical shift (Section 13.7): The position on the NMR chart where a nucleus absorbs. By convention, the chemical shift of tetramethylsilane (TMS) is set at zero, and all other absorptions usually occur downfield (to the left on the chart).

Chiral (Section 6.2): Handed. A chiral molecule does not have a plane of symmetry, is not superimposable on its mirror image, and thus exists in right- and left-handed forms.

Cis–trans isomers (Section 2.8): Stereoisomers that differ in their stereochemistry about a double bond or a ring.

Citric acid cycle (Section 17.4): The third stage of catabolism, in which acetyl groups are degraded to CO_2.

Claisen condensation reaction (Section 11.10): A carbonyl condensation reaction between two esters leading to formation of a β-keto ester product.

Coding strand (Section 16.11): The strand of the DNA double helix that contains genes.

Codon (Section 16.12): A three-base sequence on the mRNA chain that encodes the genetic information necessary to cause specific amino acids to be incorporated into proteins.

Coenzyme (Section 15.12): A small organic molecule that acts as an enzyme cofactor.

Cofactor (Section 15.12): A small nonprotein part of an enzyme necessary for biological activity.

Complex carbohydrate (Section 14.1): A carbohydrate composed of two or more simple sugars linked together by acetal bonds.

Condensed structure (Section 2.2): A shorthand way of drawing structures in which bonds are understood rather than shown.

Configuration (Section 6.6): The three-dimensional arrangement of atoms bonded to a stereocenter.

Conformation (Section 2.5): The exact three-dimensional shape of a molecule at any given instant, assuming that rotation around single bonds is frozen.

Conformers (Section 2.5): Conformational isomers that differ only in rotation around a single bond.

Conjugate acid (Section 1.12): The product that results when a base accepts H^+.

Conjugate base (Section 1.12): The anion that results from dissociation of an acid.

Conjugated protein (Section 15.9): A protein composed of both an amino acid part and a non-amino acid part.

Conjugation (Section 4.10): A series of alternating single and multiple bonds with overlapping *p* orbitals.

Constitutional isomers (Section 2.2): Isomers such as butane and 2-methylpropane, which have their atoms connected in a different order.

Coupling (Section 13.10): The interaction of neighboring nuclear spins that results in spin–spin splitting.

Coupling constant, *J* (Section 13.10): The magnitude of the spin–spin splitting interaction between nuclei whose spins are coupled.

Covalent bond (Section 1.5): A bond formed by sharing electrons between two nuclei.

Cycloalkane (Section 2.7): An alkane with a ring of carbon atoms.

D sugar (Section 14.3): A sugar whose hydroxyl group at the stereocenter farthest from the carbonyl group points to the right when the molecule is drawn in Fischer projection.

Deactivating group (Section 5.9): An electron-withdrawing substituent that decreases the reactivity of an aromatic ring towards electrophilic aromatic substitution.

Decarboxylation (Section 11.6): A reaction that involves loss of CO_2. β-Keto acids decarboxylate readily on heating.

Dehydration (Section 4.9): Elimination of water from an alcohol to yield an alkene.

Dehydrohalogenation (Section 4.9): Elimination of HX from an alkyl halide to yield an alkene on treatment with a strong base.

Delocalization (Section 4.11): A spreading out of electron density over a conjugated π electron system.

Delta (δ) scale (Section 13.7): The arbitrary scale used for defining the position of NMR absorptions; $1\ \delta = 1$ ppm of spectrometer frequency.

Deoxy sugar (Section 14.11): A sugar with an –OH group missing from one carbon.

Deoxyribonucleic acid, DNA (Section 16.6): A biopolymer of deoxyribonucleotide units.

Deshielding (Section 13.6): An effect observed in NMR that causes a nucleus to absorb downfield because of a withdrawal of electron density from the nucleus.

Dextrorotatory (Section 6.3): An optically active substance that rotates the plane of polarization of plane-polarized light in a right-handed (clockwise) direction.

Diastereomer (Section 6.7): Stereoisomers that have a non-mirror-image relationship.

1,3-Diaxial interaction (Section 2.11): A spatial interaction between two axial substituents separated by three carbons in a substituted chair cyclohexane.

Diazotization (Section 12.5): The conversion of a primary amine, RNH_2, into a diazonium salt, RN_2^+, by treatment with nitrous acid.

Digestion (Section 17.1): The first stage of catabolism, in which food molecules are hydrolyzed to yield fatty acids, amino acids, and monosaccharides.

Disaccharide (Section 14.1): A complex carbohydrate having two simple sugars bonded together.

Disulfide link (Section 15.5): A sulfur–sulfur link between two cysteine residues in a peptide.

DNA (Section 16.6): See Deoxyribonucleic acid.

Double helix (Section 16.8): The conformation into which double-stranded DNA coils.

Downfield (Section 13.7): The left-hand portion of the NMR chart.

E1 reaction (Section 7.10): An elimination reaction that takes place in two steps through a unimolecular mechanism.

E2 reaction (Section 7.9): An elimination reaction that takes place in a single step through a bimolecular mechanism.

Eclipsed conformation (Section 2.5): The geometric arrangement around a carbon–carbon single bond in which the bonds on one carbon are parallel to the bonds on the neighboring carbon as viewed in a Newman projection.

Edman degradation (Section 15.7): A method for selectively cleaving the N-terminal amino acid from a peptide.

Electromagnetic spectrum (Section 13.1): The range of electromagnetic energy, including infrared, ultraviolet, and visible radiation.

Electron shell (Section 1.1): An imaginary layer around the nucleus occupied by electrons.

Electron subshell (Section 1.1): The *s, p, d,* or *f* part of an electron shell.

Electron-transport chain (Section 17.1): The fourth stage of catabolism, in which ATP is synthesized.

Electronegativity (Section 1.11): The ability of an atom to attract electrons and thereby polarize a covalent bond. Electronegativity generally increases from right to left and from bottom to top of the periodic table.

Electrophile (Section 3.6): An "electron-lover," or substance that accepts an electron pair from a nucleophile in a polar bond-forming reaction.

Electrophilic aromatic substitution reaction (Section 5.5): The substitution of an electrophile for a hydrogen atom on an aromatic ring.

Electrophoresis (Section 15.3): A technique for separating charged organic molecules, particularly proteins and amino acids, by placing them in an electric field.

Elimination reaction (Section 3.5): The reaction that occurs when a single reactant splits apart into two products.

Enantiomers (Section 6.1): Stereoisomers that have a mirror-image relationship, with opposite configurations at all stereocenters.

3′ End (Section 16.7): The end of a nucleic acid chain that has a free sugar hydroxyl group.

5′ End (Section 16.7): The end of a nucleic acid chain that has a phosphoric acid unit.

Endothermic (Section 3.9): A reaction that absorbs heat and therefore has a positive ΔH.

Enol (Section 11.1): A vinylic alcohol, C=C–OH.

Enolate ion (Section 11.1): The anion of an enol; a resonance-stabilized α-keto carbanion.

Enone (Section 11.9): An unsaturated ketone.

Entgegen, *E* (Section 3.4): A term used to describe the stereochemistry of a carbon–carbon double bond in which high-priority groups on each carbon are on opposite sides of the double bond.

Enzyme (Section 15.11): A biological catalyst. Enzymes are large proteins that catalyze specific biochemical reactions.

Epoxide (Section 8.10): A three-membered-ring ether functional group.

Equatorial bond (Section 2.10): A bond to cyclohexane that lies along the rough equator of the ring. (See Axial bond.)

Equilibrium constant, K_{eq} (Section 3.9): A value that expresses the extent to which a given reaction takes place at equilibrium.

Essential amino acid (Section 15.1): An amino acid that must be obtained in the human diet.

Ester (Section 9.1): A substance with the general formula RCO_2R'.

Estrogen (Section 16.5): A female steroid sex hormone.

Ether (Section 8.1): A compound with two organic groups bonded to the same oxygen atom, R–O–R′.

Exothermic (Section 3.9): A reaction that releases heat and therefore has a negative ΔH.

Fat (Section 16.2): A solid triacylglycerol derived from animal sources.

Fatty acid (Section 16.2): A long straight-chain carboxylic acid found in fats and oils.

Fibrous protein (Section 15.9): A protein that consists of polypeptide chains arranged side by side in long threads.

Fingerprint region (Section 13.2): The complex region of the infrared spectrum from 1500 cm^{-1} to 400 cm^{-1}.

Fischer esterification reaction (Section 10.6): The conversion of a carboxylic acid into an ester by acid-catalyzed reaction with an alcohol.

Fischer projection (Section 14.2): A method for depicting the configuration of a stereocenter using crossed lines. Horizontal bonds come out of the plane of the page, and vertical bonds go back into the plane of the page.

Frequency (Section 13.1): The number of electromagnetic wave cycles that travel past a fixed point in a given unit of time, usually expressed in reciprocal seconds, s^{-1}, or Hertz.

Friedel–Crafts reaction (Section 5.8): The introduction of an alkyl or acyl group onto an aromatic ring by an electrophilic substitution reaction.

Functional group (Section 2.1): An atom or group of atoms that is part of a larger molecule and has a characteristic chemical reactivity.

Furanose (Section 14.5): The five-membered ring structure of a simple sugar.

Geminal (Section 9.7): Referring to two groups attached to the same carbon atom.

Globular protein (Section 15.9): A protein that is coiled into a compact, nearly spherical shape.

Glycol (Section 8.11): A 1,2-diol such as ethylene glycol, $HOCH_2CH_2OH$.

Glycolysis (Section 17.3): A series of ten enzyme-catalyzed reactions that break down a glucose molecule into two pyruvate molecules.

Glycoside (Section 14.8): A cyclic acetal formed by reaction of a sugar with another alcohol.

Grignard reagent (Section 7.4): An organomagnesium halide, RMgX.

Haworth projection (Section 14.5): A view of a furanose or pyranose sugar in which the ring is flat and viewed edge-on with the hemiacetal oxygen atom at the upper right.

Heat of reaction, ΔH (Section 3.9): The amount of heat released or absorbed in a reaction.

α-Helix (Section 15.10): A common secondary structure in which a protein chain coils into a spiral.

Hemiacetal (Section 9.8): A compound that has one –OR group and one –OH group bonded to the same carbon atom.

Heterocycle (Section 12.6): A cyclic molecule whose ring contains more than one kind of atom.

Heterogenic (Section 3.6): Electronically unsymmetrical formation of a covalent bond by combination of an anion and a cation.

Heterolytic (Section 3.6): Electronically unsymmetrical breaking of a covalent bond to yield an anion and a cation.

Holoenzyme (Section 15.12): The combination of enzyme and cofactor.

Homogenic (Section 3.6): Electronically symmetrical formation of a covalent bond by combination of two radicals.

Homolytic (Section 3.6): Electronically symmetrical breaking of a covalent bond to yield two radicals.

Hormone (Section 16.5): A chemical messenger secreted by a specific gland and carried through the bloodstream to affect a target tissue.

Hybrid orbital (Section 1.7): An orbital derived from a combination of ground-state atomic orbitals. Hybrid orbitals, such as the sp^3, sp^2, and *sp* hybrids of carbon, are strongly directed and form strong bonds.

Hydration (Section 4.4): Addition of water to a molecule, such as occurs when alkenes are treated with strong aqueous acid.

Hydrocarbon (Section 2.2): A compound made up of only carbon and hydrogen.

Hydrogen bond (Section 8.2): An attraction between a hydrogen atom bonded to an electronegative element and an electron lone pair on another atom.

Hydrogenation (Section 4.6): Addition of hydrogen to a double or triple bond to yield a saturated product.

Hydroquinone (Section 8.8): A compound that contains a *p*-dihydroxybenzene group.

Hydroxylation (Section 4.7): The addition of one or more –OH groups to a molecule.

Imine (Section 9.9): A compound with a C=N functional group.

Inductive effect (Section 1.11): The electron-attracting or electron-withdrawing effect that is transmitted through σ bonds.

Infrared spectroscopy, IR (Section 13.1): A kind of optical spectroscopy that uses infrared energy. IR spectroscopy is particularly useful in organic chemistry for determining the kinds of functional groups in molecules.

Initiator (Section 7.2): A substance with an easily broken bond that is used to initiate a radical chain reaction.

Integration (Section 13.9): A means of electronically measuring the ratios of the number of nuclei responsible for each peak in an NMR spectrum.

Intermediate (Section 3.11): A species that is formed during the course of a multistep reaction but is not the final product.

Intramolecular, intermolecular (Section 8.11): Reactions that occur within the same molecule are *intra*molecular; reactions that occur between two molecules are *inter*molecular.

Ionic bond (Section 1.4): A bond between two ions due to the electrical attraction of unlike charges.

Isoelectric point (Section 15.3): The pH at which the number of positive charges and the number of negative charges on a protein or amino acid are exactly balanced.

Isomers (Section 2.2): Compounds with the same molecular formula but different structures.

Kekulé structure (Section 1.5): A representation of a molecule in which a line between atoms represents a covalent bond.

Ketone (Section 9.1): A substance with the general formula $R_2C=O$.

Ketose (Section 14.1): A simple sugar with a ketone carbonyl group.

L sugar (Section 14.3): A sugar whose hydroxyl group at the stereocenter farthest from the carbonyl group points to the left when the molecule is drawn in Fischer projection.

Leaving group (Section 7.6): The group that is replaced in a substitution reaction.

Levorotatory (Section 6.3): An optically active substance that rotates the plane of polarization of plane-polarized light in a left-handed (counterclockwise) direction.

Lewis acid (Section 1.13): A substance with a vacant low-energy orbital that can accept an electron pair from a base.

Lewis base (Section 1.13): A substance that donates an electron lone pair to an acid.

Lewis structure (Section 1.5): A representation of a molecule showing covalent bonds as pairs of electron dots between atoms.

Line-bond structure (Section 1.5): A representation of a molecule showing covalent bonds as lines between atoms.

1,4′ Link (Section 14.9): A glycosidic link between the C1 carbonyl group of one sugar and the C4 hydroxyl group of another sugar.

Lipid (Section 16.1): A naturally occurring substance that can be isolated from plants or animals by extraction with a nonpolar organic solvent.

Lipid bilayer (Section 16.4): The double layer of phospholipids that makes up cell walls.

Lipophilic (Section 16.3): Fat-loving. Long, nonpolar hydrocarbon chains tend to cluster together in polar solvents because of their lipophilic properties.

Lone-pair electrons (Section 1.5): A nonbonding electron pair that occupies a valence orbital.

Major groove (Section 16.8): The larger of two grooves in double helical DNA.

Malonic ester synthesis (Section 11.6): A method for forming α-substituted acetic acids by reaction of diethyl malonate with an alkyl halide, followed by decarboxylation.

Markovnikov's rule (Section 4.2): A guide for determining the regiochemistry (orientation) of electrophilic addition reactions. In the addition of HX to an alkene, the hydrogen atom becomes bonded to the alkene carbon that has fewer alkyl substituents.

Mass number, *A* (Section 1.1): The total number of protons and neutrons in an atom's nucleus.

Mechanism (Section 3.6): A complete description of how a reaction occurs. A mechanism accounts for all reactants and all products, and describes the details of each individual step in the overall reaction process.

Mercapto group (Section 8.12): An alternative name for the thiol group, –SH.

Meso (Section 6.8): A compound that contains one or more stereocenters but is nevertheless achiral because it has a symmetry plane.

Metabolism (Section 17.1): The total of all reactions in living organisms.

Micelle (Section 16.3): A spherical cluster of soaplike molecules that aggregate in aqueous solution. The ionic heads of the molecules lie on the outside where they are solvated by water, and the organic tails bunch together on the inside of the micelle.

Minor groove (Section 16.8): The smaller of two grooves in double helical DNA.

Molecule (Section 1.5): A group of atoms joined by covalent bonds.

Monomer (Sections 4.8, 10.12): The starting unit from which a polymer is made.

Monosaccharide (Section 14.1): A simple sugar.

Mutarotation (Section 14.6): The spontaneous change in optical rotation observed when a pure anomer of a sugar is dissolved in water and equilibrates to an equilibrium mixture of anomers.

$n + 1$ rule (Section 13.10): The signal of a proton with n neighboring protons splits into $n + 1$ peaks in the NMR spectrum.

N-terminal amino acid (Section 15.4): The amino acid with a free $-NH_2$ group at one end of a protein chain.

Newman projection (Section 2.5): A way of viewing a molecule's spatial arrangement by looking end-on at a carbon–carbon bond.

Nitrile (Section 10.1): A compound with a –C≡N functional group.

Nonbonding electron (Section 1.5): A valence electron not used for bonding.

Normal alkane (Section 2.2): A straight-chain alkane.

Nuclear magnetic resonance, NMR (Section 13.5): A spectroscopic technique that provides information about the carbon–hydrogen framework of a molecule.

Nucleophile (Section 3.6): An electron-rich species that can donate an electron pair to an electrophile in a polar reaction.

Nucleophilic acyl substitution reaction (Section 10.5): A substitution reaction that replaces one nucleophile bonded to a carbonyl group by another.

Nucleophilic addition reaction (Section 9.6): A reaction that involves the addition of a nucleophile to a carbonyl group.

Nucleophilic substitution reaction (Section 7.5): A substitution reaction in which one nucleophile replaces another.

Nucleoside (Section 16.6): A nucleic acid constituent, consisting of a sugar residue bonded to a heterocyclic purine or pyrimidine base.

Nucleotide (Section 16.6): A nucleic acid constituent, consisting of a sugar residue bonded both to a heterocyclic purine or pyrimidine base and to phosphoric acid.

Nylon (Section 10.12): A polyamide prepared by reaction between a diacid and a diamine.

Optical activity (Section 6.3): The ability of a chiral molecule in solution to rotate plane-polarized light.

Orbital (Section 1.1): A region of space occupied by a given electron or pair of electrons.

Oxidation (Section 4.7): The addition of oxygen to a molecule or removal of hydrogen from it.

Oxirane (Section 8.10): An alternative name for an epoxide.

Paraffin (Section 2.4): A common name for an alkane.

Peptide (Section 15.1): A small amino acid polymer in which the individual amino acid residues are linked by amide bonds. (See Protein.)

Periplanar (Section 7.9): A conformation in which two bonds to neighboring atoms lie in the same plane.

Phenol (Section 8.1): A compound with an –OH group bonded to an aromatic ring.

Phenoxide ion (Section 8.3): The anion formed by loss of H^+ from a phenol.

Phenyl (Section 5.4): The $-C_6H_5$ group.

Phosphoglyceride (Section 16.4): A phospholipid in which glycerol has ester links to two fatty acids and to phosphoric acid.

Phospholipid (Section 16.4): A lipid that contains a phosphate residue.

Phosphoric acid anhydride (Section 17.1): A functional group containing the P–O–P linkage.

Phosphorylation (Section 17.1): A reaction that transfers a phosphate group from a phosphoric anhydride to an alcohol.

Pi (π) bond (Section 1.9): A covalent bond formed by sideways overlap of two *p* orbitals.

Plane of symmetry (Section 6.2): An imaginary plane that bisects a molecule such that one half of the molecule is the mirror image of the other half.

Plane-polarized light (Section 6.3): Ordinary light that has its electric vectors in a single plane rather than in random planes.

Polar covalent bond (Section 1.11): A covalent bond in which the electrons are shared unequally between the atoms.

Polar reaction (Section 3.6): A reaction in which bonds are made when a nucleophile donates two electrons to an electrophile, and bonds are broken when one fragment leaves with both electrons from the bond.

Polarity (Section 1.11): The unsymmetrical distribution of electrons in a molecule that results when one atom attracts electrons more strongly than another.

Polycyclic aromatic hydrocarbon (Section 5.12): A molecule that has two or more benzene rings fused together.

Polyester (Section 10.12): A polymer prepared by reaction between a diacid and a dialcohol.

Polymer (Sections 4.8, 10.12): A large molecule made up of repeating smaller units.

Polymerase chain reaction, PCR (Section 16.14): A method for amplifying small amounts of DNA to prepare large amounts.

Polysaccharide (Section 14.1): A complex carbohydrate having many simple sugars bonded together by acetal links.

Polyunsaturated fatty acid, PUFA (Section 16.2): A fatty acid with more than one double bond in its chain.

Primary amine (Section 12.1): An amine with one organic substituent on nitrogen, RNH_2.

Primary structure (Section 15.10): The amino acid sequence of a protein.

Propagation step (Section 7.2): A step in a radical chain reaction that carries on the chain.

Protecting group (Section 9.8): A group that is temporarily introduced into a molecule to protect a functional group from reaction elsewhere in the molecule.

Protein (Section 15.1): A large biological polymer containing 50 or more amino acid residues.

Pyranose (Section 14.5): The six-membered ring structure of a simple sugar.

Quaternary ammonium salt (Section 12.1): A compound with four organic substituents attached to a positively charged nitrogen, $R_4N^+ X^-$.

Quaternary structure (Section 15.10): The way in which several protein molecules aggregate together to yield a larger structure.

Quinone (Section 8.8): A compound that contains the cyclohexadienedione functional group.

R group (Section 2.2): A general symbol used for an organic partial structure.

***R,S* convention** (Section 6.6): A method for defining the absolute configuration around a stereocenter.

Racemic mixture (Section 6.10): A 50:50 mixture of the two enantiomers of a chiral substance.

Radical (Section 3.6): A species that has an odd number of electrons, such as the chlorine radical, $Cl\cdot$.

Radical reaction (Section 3.6): A reaction in which bonds are made by donation of one electron from each of two reagents, and bonds are broken when each fragment leaves with one electron.

Radical substitution reaction (Section 7.2): A substitution reaction that takes place by a radical mechanism.

Reaction energy diagram (Section 3.10): A graph depicting the energy changes that occur during a reaction.

Reaction rate (Section 7.7): The exact speed of a reaction under defined conditions.

Rearrangement reaction (Section 3.5): The reaction that occurs when a single reactant undergoes a reorganization of bonds and atoms to give an isomeric product.

Reducing sugar (Section 14.8): A sugar that reduces Ag^+ in the Tollens test or Cu^{2+} in the Fehling's or Benedict's tests.

Reduction (Section 4.6): The addition of hydrogen to a molecule or the removal of oxygen from it.

Regiochemistry (Section 4.9): A term describing the orientation of a reaction that occurs on an unsymmetrical substrate.

Regiospecific (Section 4.2): Describing the orientation of an addition reaction that occurs on an unsymmetrical substrate and leads to a single product.

Replication (Section 16.10): The process by which double-stranded DNA uncoils and is replicated to produce two new copies.

Resolution (Section 6.10): Separation of a racemic mixture into its pure component enantiomers.

Resonance forms (Section 4.11): Representations of a molecule that differ only in where the bonding electrons are placed.

Resonance hybrid (Section 4.11): The composite structure of a molecule described by different resonance forms.

Restriction endonuclease (Section 16.13): An enzyme that is able to cut a DNA strand at a specific base sequence in the chain.

Ribonucleic acid, RNA (Section 16.6): A biopolymer of ribonucleotide units.

Ring-flip (Section 2.11): The molecular motion that converts one chair conformation of cyclohexane into another chair conformation, thereby interconverting axial and equatorial bonds.

RNA (Section 16.6): See Ribonucleic acid.

Saccharide (Section 14.1): A sugar.

Salt bridge (Section 15.10): The ionic attraction between charged amino acid side chains that helps stabilize a protein's tertiary structure.

Sandmeyer reaction (Section 12.5): The conversion of an aryl diazonium salt into an aryl halide by reaction with a cuprous halide.

Saponification (Section 10.9): An old term for the base-induced hydrolysis of an ester to yield a carboxylic acid salt.

Saturated (Section 2.2): A compound that has only single bonds.

Sawhorse structure (Section 2.5): A perspective view of the conformation around single bonds.

Secondary amine (Section 12.1): An amine with two organic substituents on nitrogen, R_2NH.

Secondary structure (Section 15.10): The specific way in which segments of a protein chain are oriented into a regular pattern.

Sense strand (Section 16.11): An alternate name for the coding strand of DNA.

Sequence rules (Sections 3.4, 6.6): A series of rules for assigning relative priorities to substituent groups on a double-bond carbon atom or on a stereocenter.

Shielding (Section 13.6): An effect observed in NMR that causes a nucleus to absorb toward the right (upfield) side of the chart. Shielding is caused by donation of electron density to the nucleus. (See Deshielding.)

Sigma (σ) bond (Section 1.6): A covalent bond formed by head-on overlap of atomic orbitals.

Simple protein (Section 15.9): A protein composed entirely of amino acids.

Simple sugar (Section 14.1): A carbohydrate like glucose that can't be hydrolyzed to smaller molecules.

Skeletal structure (Section 2.6): A shorthand way of drawing structures that shows only bonds, not atoms.

S_N1 reaction (Section 7.8): A nucleophilic substitution reaction that takes place in two steps through a carbocation intermediate.

S_N2 reaction (Section 7.7): A nucleophilic substitution reaction that takes place in a single step by back-side displacement of the leaving group.

***sp* hybrid orbital** (Section 1.10): An atomic orbital formed by combination of one *s* and one *p* atomic orbital.

sp^2 hybrid orbital (Section 1.9): An atomic orbital formed by combination of one *s* and two *p* atomic orbitals.

sp^3 hybrid orbital (Section 1.7): An atomic orbital formed by combination of one *s* and three *p* atomic orbitals.

Specific rotation, $[\alpha]_D$ (Section 6.4): The amount by which an optically active compound rotates plane-polarized light under standard conditions.

Sphingolipid (Section 16.4): A phospholipid based on the sphingosine backbone rather than on glycerol.

Spin–spin splitting (Section 13.10): The splitting of an NMR signal into a multiplet caused by an interaction between nearby magnetic nuclei whose spins are coupled.

Staggered conformation (Section 2.5): The three-dimensional arrangement of atoms around a carbon–carbon single bond in which the bonds on one carbon bisect the bond angles on the second carbon as viewed end-on.

Step-growth polymer (Section 10.12): A polymer produced by a series of polar reactions between two difunctional monomers.

Stereocenter (Section 6.2): An atom in a molecule that is a cause of chirality.

Stereochemistry (Section 6.1): The branch of chemistry concerned with the three-dimensional arrangement of atoms in molecules.

Stereoisomers (Section 2.8): Isomers that have their atoms connected in the same order but with a different three-dimensional arrangement.

Steric strain (Section 2.11): The strain imposed on a molecule when two groups are too close together and try to occupy the same space.

Steroid (Section 16.5): A lipid whose structure is based on a characteristic tetracyclic carbon skeleton.

Straight-chain alkane (Section 2.2): An alkane whose carbon atoms are connected in a row.

Substitution reaction (Section 3.5): The reaction that occurs when two reactants exchange parts to give two products.

Sulfide (Section 8.12): A compound that has two organic groups bonded to a sulfur atom, R–S–R′.

Syn stereochemistry (Section 4.6): A syn addition reaction is one in which the two ends of the double bond are attacked from the same face.

Tautomers (Section 11.1): Isomers that are rapidly interconverted.

Template strand (Section 16.11): The strand of the DNA double helix that is used for transcription.

Tertiary amine (Section 12.1): An amine with three organic substituents on nitrogen, R_3N.

Tertiary structure (Section 15.10): The way in which a protein molecule is oriented into an overall three-dimensional shape.

Thiol (Section 8.12): A compound with the –SH functional group.

Tollens' reagent (Section 9.5): A solution of Ag^+ in aqueous NH_3; useful for oxidizing aldehydes to carboxylic acids.

Transamination (Section 17.5): A reaction in which the $-NH_2$ group of an amine changes places with the keto group of an α-keto acid.

Transcription (Section 16.11): The process by which RNA is synthesized from DNA.

Transition state (Section 3.10): An activated complex between reactants, representing the highest energy point on a reaction curve.

Translation (Section 16.12): The process by which the genetic information transcribed from DNA onto mRNA is read by tRNA and used to direct protein synthesis.

Triacylglycerol (Section 16.2): A lipid such as animal fat and vegetable oil; a triester of glycerol with long-chain fatty acids.

Ultraviolet (UV) spectroscopy (Section 13.3): An optical spectroscopy employing ultraviolet irradiation. UV spectroscopy provides structural information about the extent of electron conjugation in organic molecules.

Unimolecular (Section 7.8): A reaction step that involves only one molecule.

Unsaturated (Section 3.1): A molecule that has one or more double or triple bonds and thus has fewer hydrogens than the corresponding alkane.

Upfield (Section 13.7): The right-hand portion of the NMR chart.

Valence shell (Section 1.4): The outermost electron shell of an atom.

Vinylic (Section 4.13): A term that refers to a substituent attached to a double-bond carbon atom.

Vitamin (Section 15.11): A small organic molecule that must be obtained in the diet and is required for proper growth and functioning.

Wavelength, λ (Section 13.1): The length of a wave from peak to peak.

Wavenumber (Section 13.2): A unit of frequency measurement equal to the reciprocal of the wavelength in centimeters, cm^{-1}.

Williamson ether synthesis (Section 8.6): The reaction of an alkoxide ion with an alkyl halide to yield an ether.

Zaitsev's rule (Section 4.9): A rule stating that E2 elimination reactions normally yield the more highly substituted alkene as major product.

Zusammen, *Z* (Section 3.4): A term used to describe the stereochemistry of a carbon–carbon double bond in which the two high-priority groups on each carbon are on the same side of the double bond.

Zwitterion (Section 15.2): A neutral dipolar molecule whose positive and negative charges are not adjacent.

Appendix C: Answers to Selected In-Text Problems

The following answers are meant only as a quick check. Full answers and explanations for all problems are provided in the accompanying *Study Guide and Solutions Manual.*

Chapter 1

1.1 (a) 1 (b) 2 (c) 3

1.2 (a) Boron: $1s^2 2s^2 2p$ (b) Phosphorus: $1s^2 2s^2 2p^6 3s^2 3p^3$ (c) Oxygen: $1s^2 2s^2 2p^4$ (d) Argon: $1s^2 2s^2 2p^6 3s^2 3p^6$

1.3 Cl, C, H, H, H

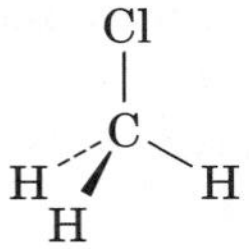

1.4 (a) 2 (b) 6 (c) 7 **1.5** (a) Oxygen (b) Bromine

1.6 (a) CCl_4 (b) AlH_3 (c) CH_2Cl_2 (d) SiF_4

1.7 (a) H:C:Cl: with :Cl: above and :Cl: below; H—C—Cl with Cl above and Cl below (b) :S:H with H above; S—H with H above (c) H:C:N:H with H above and H H below; H—C—N—H with H above and H H below

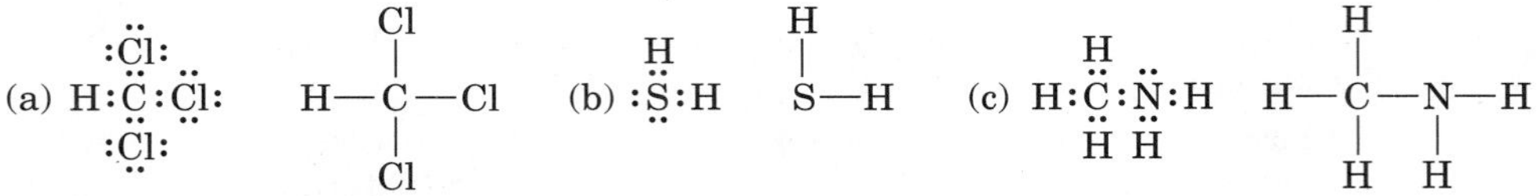

1.8 Ionic: LiI, KBr, $MgCl_2$ Covalent: CH_4, CH_2Cl_2, Cl_2

1.9 H:C:C:H with H H above and H H below; H—C—C—H with H H above and H H below

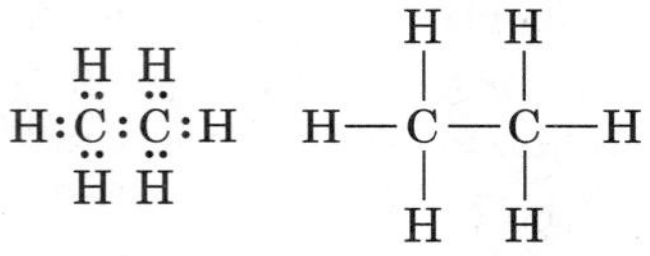

1.10 Cl, C, Cl, Cl, Cl

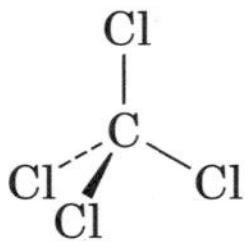

1.11 Carbon uses its second shell for bonding; hydrogen uses its first shell.

1.12 All bond angles are near 109°.

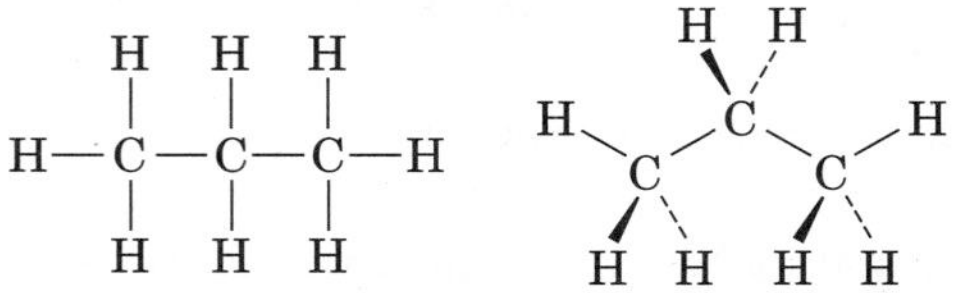

H—C—C—C—H with H H H above and H H H below; H, C, H, H, C, C, H, H, H, H, H

1.13 C_2H_7 has too many hydrogens for a compound with two carbons.

1.14 H—C—C—H with H above the first C, O double-bonded above the second C, and H below the first C; H:C:C:H with H above the first C, :O: above the second C, and H below the first C

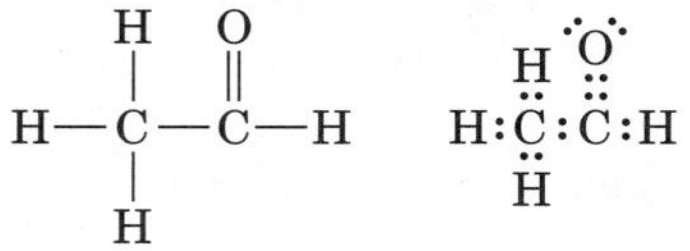

1.15 The CH_3 carbon is sp^3, the double-bond carbons are sp^2, and the C=C–C bond angle is approximately 120°.

1.16 All carbons are sp^2, and all bond angles are near 120°.

1.17 The CH_3 carbon is sp^3, the triple-bond carbons are *sp*, and the C≡C–C bond angle is approximately 180°.

1.18 (a) H (b) Br (c) Cl

1.19 (a) C is $\delta+$, Br is $\delta-$ (b) C is $\delta+$, N is $\delta-$ (c) Li is $\delta+$, C is $\delta-$ (d) H is $\delta+$, N is $\delta-$ (e) C is $\delta+$, O is $\delta-$ (f) Mg is $\delta+$, C is $\delta-$ (g) C is $\delta+$, F is $\delta-$

1.20 CCl_4 and $ClO_2 < TiCl_3 < MgCl_2$

1.21 Formic acid: $K_a = 1.8 \times 10^{-4}$; picric acid: $K_a = 0.42$

1.22 Picric acid is stronger.

1.23 Water is stronger.

1.24 (a) No (b) No

1.25 Lewis acids: $MgBr_2$, $B(CH_3)_3$, $^+CH_3$; Lewis bases: $(CH_3)_2NH$, $(CH_3)_3P$; both: CH_3CH_2OH

1.26 (a) $CH_3CH_2OH + HCl \rightarrow CH_3CH_2OH_2^+\ Cl^-$; $(CH_3)_2NH + HCl \rightarrow (CH_3)_2NH_2^+\ Cl^-$; $(CH_3)_3P + HCl \rightarrow (CH_3)_3PH^+\ Cl^-$

(b) $HO^- + CH_3^+ \rightarrow HO{-}CH_3$; $HO^- + B(CH_3)_3 \rightarrow HO{-}B(CH_3)_3^-$; $HO^- + MgBr_2 \rightarrow HO{-}MgBr_2^-$

Chapter 2

2.1 (a) Carboxylic acid, double bond (b) Carboxylic acid, aromatic ring, ester (c) Aldehyde, alcohol

2.2 (a) CH_3OH (b) benzene ring (c) $CH_3\overset{O}{\overset{\|}{C}}OH$ (d) CH_3NH_2

(e) $CH_3\overset{O}{\overset{\|}{C}}CH_2CH_2NH_2$ (f) $H_2C{=}CHCH{=}CH_2$

2.3

$CH_3CH_2CH_2CH_2CH_2CH_3$ $\quad CH_3\overset{CH_3}{\overset{|}{C}}HCH_2CH_2CH_3$ $\quad CH_3CH_2\overset{CH_3}{\overset{|}{C}}HCH_2CH_3$

$CH_3\underset{CH_3}{\underset{|}{\overset{CH_3}{\overset{|}{C}}}}CH_2CH_3$ $\quad CH_3\overset{CH_3}{\overset{|}{C}}H\underset{CH_3}{\underset{|}{C}}HCH_3$

2.5

$CH_3CH_2CH_2CH_2CH_2{-}$ $\quad CH_3CH_2CH_2\underset{CH_3}{\underset{|}{C}}H{-}$ $\quad CH_3CH_2\underset{CH_2CH_3}{\underset{|}{C}}H{-}$ $\quad CH_3\overset{CH_3}{\overset{|}{C}}HCH_2CH_2{-}$

$CH_3CH_2\overset{CH_3}{\overset{|}{C}}HCH_2{-}$ $\quad CH_3CH_2\underset{CH_3}{\underset{|}{\overset{CH_3}{\overset{|}{C}}}}{-}$ $\quad CH_3\overset{CH_3}{\overset{|}{C}}H\underset{CH_3}{\underset{|}{C}}H{-}$ $\quad CH_3\underset{CH_3}{\underset{|}{\overset{CH_3}{\overset{|}{C}}}}CH_2{-}$

2.6 (a) $CH_3\overset{CH_3}{\overset{|}{C}}H\underset{CH_3}{\underset{|}{C}}HCH_3$ (b) $CH_3CH_2\overset{CH_3CHCH_3}{\overset{|}{C}}HCH_2CH_3$ (c) $CH_3\underset{CH_3}{\underset{|}{\overset{CH_3}{\overset{|}{C}}}}CH_2CH_3$

2.7 (a)

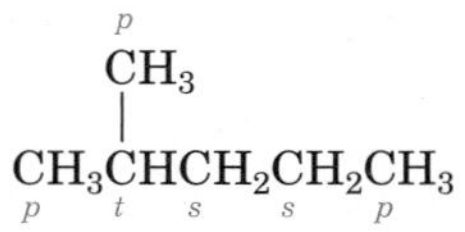

$$\begin{array}{c} \overset{p}{CH_3} \\ | \\ \underset{p}{CH_3}\underset{t}{CH}\underset{s}{CH_2}\underset{s}{CH_2}\underset{p}{CH_3} \end{array}$$

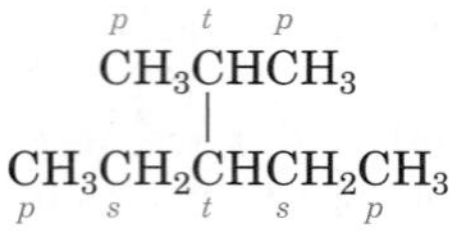

(b)

$$\begin{array}{c} \overset{p}{CH_3}\overset{t}{CH}\overset{p}{CH_3} \\ | \\ \underset{p}{CH_3}\underset{s}{CH_2}\underset{t}{CH}\underset{s}{CH_2}\underset{p}{CH_3} \end{array}$$

(c)

$$\begin{array}{c} \overset{p}{CH_3} \quad\quad\quad CH_3 \\ | \quad\quad\quad\quad | \\ \underset{p}{CH_3}\underset{t}{CH}\underset{s}{CH_2}-\underset{q}{C}-\underset{p}{CH_3} \\ | \\ \underset{p}{CH_3} \end{array}$$

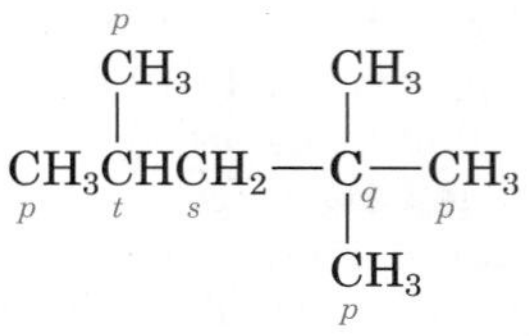

2.8 (a) Pentane, 2-methylbutane, 2,2-dimethylpropane (b) 3,4-Dimethylhexane
(c) 2,4-Dimethylpentane (d) 2,2,5-Trimethylheptane

2.9 (a)

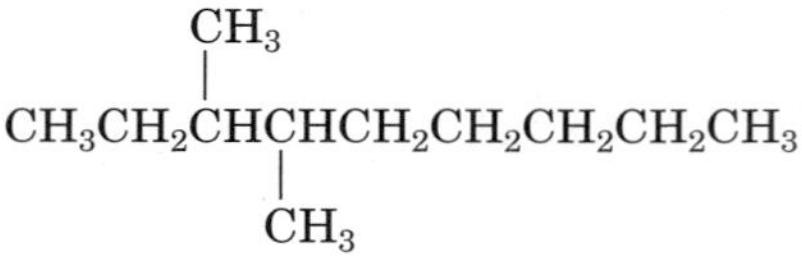

$$\begin{array}{c} CH_3 \\ | \\ CH_3CH_2CHCHCH_2CH_2CH_2CH_2CH_3 \\ | \\ CH_3 \end{array}$$

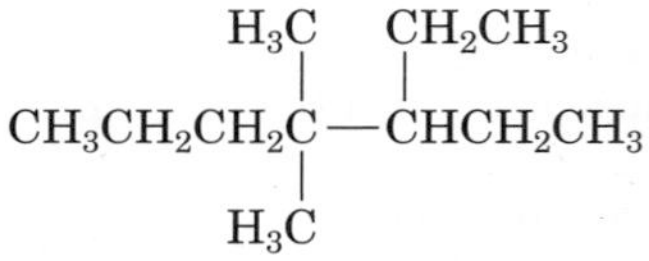

(b)

$$\begin{array}{c} H_3C \quad CH_2CH_3 \\ | \quad\quad | \\ CH_3CH_2CH_2C-CHCH_2CH_3 \\ | \\ H_3C \end{array}$$

(c)

$$\begin{array}{c} CH_3 \quad CH_2CH_2CH_3 \\ | \quad\quad | \\ CH_3CCH_2CHCH_2CH_2CH_2CH_3 \\ | \\ CH_3 \end{array}$$

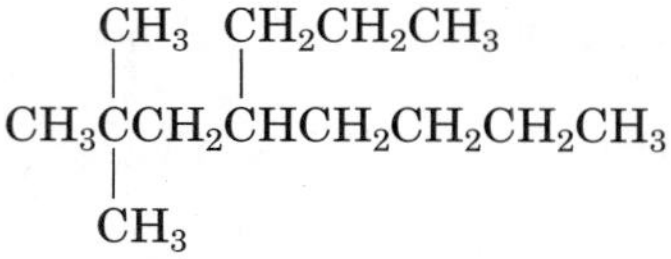

(d)

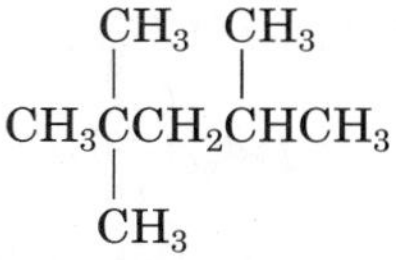

$$\begin{array}{c} CH_3 \quad CH_3 \\ | \quad\quad | \\ CH_3CCH_2CHCH_3 \\ | \\ CH_3 \end{array}$$

2.10

CH_3; H, H, H, H, H

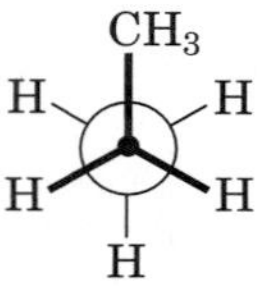

Most stable conformation (staggered)

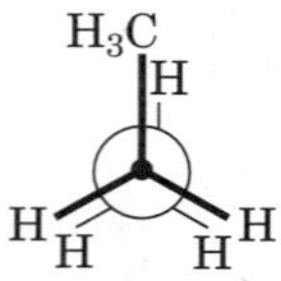

H_3C; H, H, H, H, H

Least stable conformation (eclipsed)

2.12

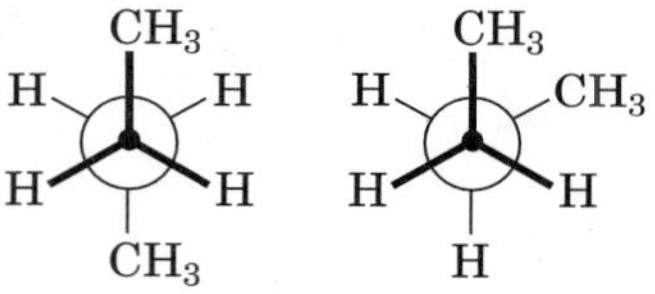

CH_3; H, H, H, H; CH_3 $\quad$ CH_3; H, CH_3, H, H; H

Staggered butane

H_3C; CH_3; H, H, H, H $\quad$ H_3C; H; H, H, CH_3, H

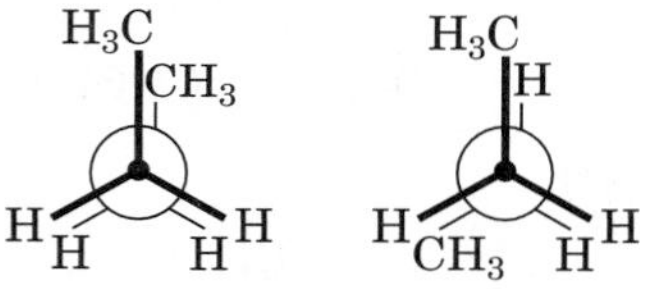

Eclipsed butane

2.13 The first staggered conformation of butane is the most stable.

2.14 (a) C_5H_5N (b) $C_6H_{10}O$ (c) C_8H_7N

2.16 (a) 1,4-Dimethylcyclohexane (b) 1-Ethyl-3-methylcyclopentane
(c) Isopropylcyclobutane

2.17 (a) 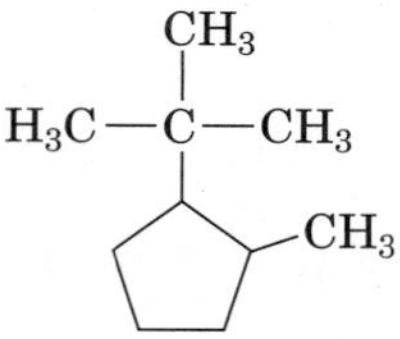 $H_3C-C(CH_3)-CH_3$ attached to cyclopentane bearing CH_3 (b) cyclobutane with CH_3, CH_3 (c) CH_3CH_2– cyclohexane –$CHCH_3$ with CH_3

2.18 H_3C, H, Cl, H

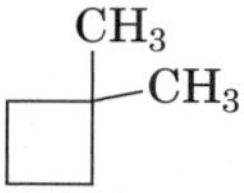

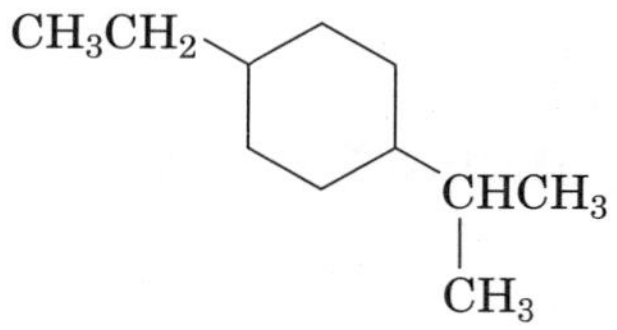

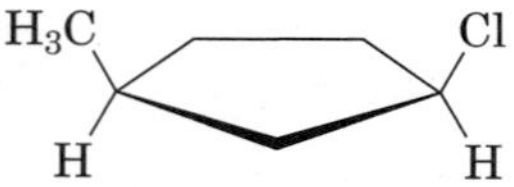

2.19 Br, H, Br, H $\quad$ Br, H, H, Br

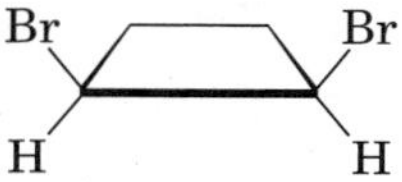

Cis 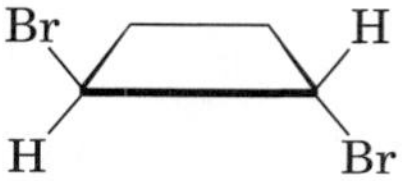 $\quad$ Trans

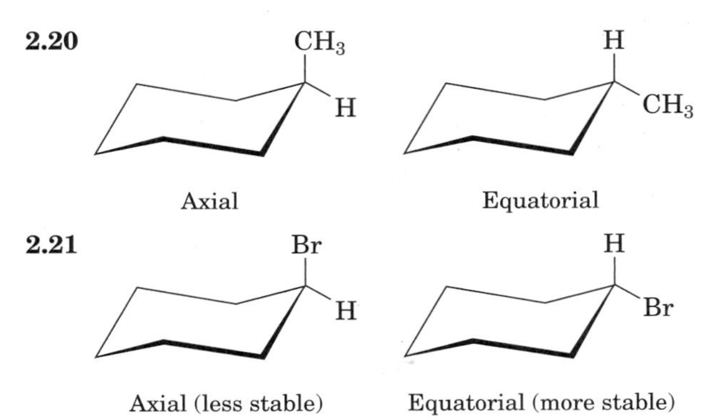

2.22–2.23 Axial and equatorial positions alternate on each side of a ring.

Chapter 3

3.1 (a) 4-Methyl-1-pentene (b) 3-Heptene (c) 1,5-Heptadiene (d) 2-Methyl-3-hexene

3.2 (a) 1,2-Dimethylcyclohexene (b) 4,4-Dimethylcycloheptene (c) 3-Isopropylcyclopentene

3.3 (a) $CH_3CH_2CH_2CH_2C(CH_3){=}CH_2$ (b) $(CH_3)_3CCH{=}CHCH_3$

(c) $H_2C{=}CHCH_2CH_2C(CH_3){=}CH_2$ (d) $CH_3CH_2CH_2CH{=}C(CH_2CH_3)C(CH_3)_3$

3.4 Compounds (c), (d), (e), and (f) can exist as pairs of isomers.

3.5 Trans is more stable.

3.6 A trans double bond cannot exist in a six-membered ring.

3.7 (a) Br (b) Br (c) CH_2CH_3 (d) OH (e) CH_2OH (f) CH=O

3.8 CO_2CH_3 is higher. **3.9** Isopropyl is higher. **3.10** (a) *Z* (b) *E*

3.11 (a) Substitution (b) Elimination (c) Addition

3.12 (a) $\overset{\delta+}{C}{=}\overset{\delta-}{O}$ Ketone (b) $-\overset{\delta+}{C}-\overset{\delta-}{Cl}$ Alkyl chloride (c) $-\overset{\delta+}{C}-\overset{\delta-}{OH}$ Alcohol (d) $-\overset{\delta-}{C}-\overset{\delta+}{Li}$ Alkyllithium

3.13 (a) $CH_3\underset{\delta+}{C}(\overset{\delta-}{=O})CH_3$ Ketone (b) $CH_3\underset{\delta+}{CH_2}-\underset{\delta-}{Cl}$ Alkyl halide (c) $\underset{\delta+}{CH_3}-\underset{\delta-}{SH}$ Thiol (d) $(CH_3\underset{\delta-}{CH_2})_4\overset{\delta+}{Pb}$ [each CH_2CH_3 $\delta-$] Organometallic

3.14 Electrophile: (a), (c); nucleophile: (b), (d), (e) **3.15** $(CH_3)_3C^+$

3.16 2-Chloropentane and 3-chloropentane

3.17 $\Delta H = -10$ kJ/mol is more exothermic.

3.18 $K_{eq} = 1000$ is more exothermic.

3.19 $E_{act} = 60$ kJ/mol is faster; K_{eq} is not predictable.

Chapter 4

4.1 (a) 2-Chlorobutane (b) 2-Iodo-2-methylpentane (c) Chlorocyclohexane
4.2 (a) Cyclopentene (b) 3-Hexene (c) 1-Isopropylcyclohexene
(d) Cyclohexylethylene

4.3 (a) 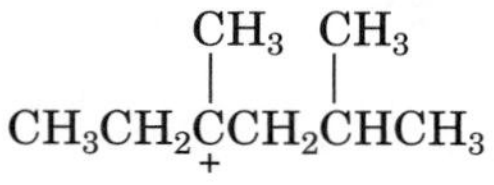 $CH_3CH_2\overset{+}{C}(CH_3)CH_2CH(CH_3)CH_3$ (b) 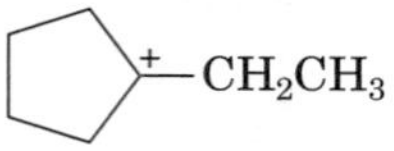 cyclopentyl cation $-CH_2CH_3$

4.4 (a) 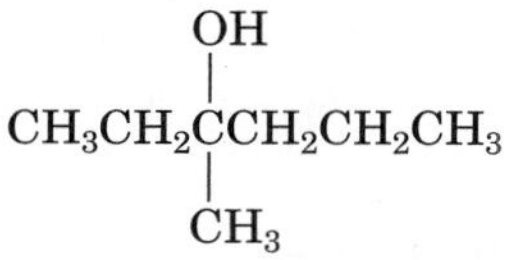 $CH_3CH_2C(OH)(CH_3)CH_2CH_2CH_3$ (b) 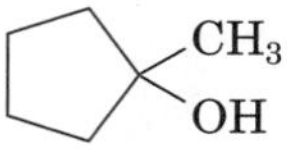 1-methylcyclopentanol (CH_3, OH) (c) 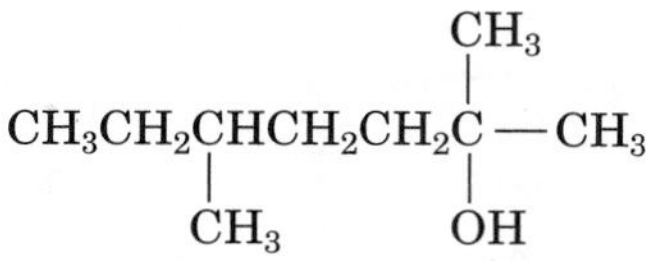 $CH_3CH_2CH(CH_3)CH_2CH_2C(CH_3)(OH)-CH_3$

4.5 (a) 1-Butene or 2-butene (b) 3-Methyl-2-pentene or 2-ethyl-1-butene
(c) 1,2-Dimethylcyclohexene or 2,3-dimethylcyclohexene
4.6 *trans*-1,2-Dibromo-1,2-dimethylcyclohexane
4.8 (a) 2-Methylpentane (b) 1,1-Dimethylcyclopentane

4.9 (a) 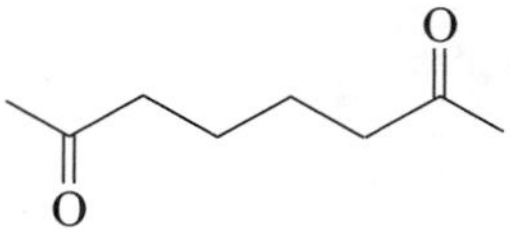 (diketone skeletal structure, O, O) (b) 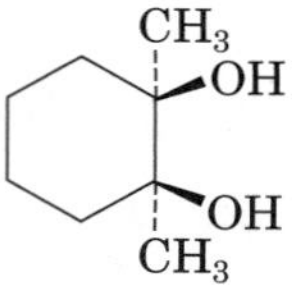 cyclohexane with CH_3, OH, OH, CH_3

4.10 (a) 2-Methylpropene (b) 3-Hexene

4.11 (polytetrafluoroethylene chain: C, F F F F F F F F / F F F F F F F F)

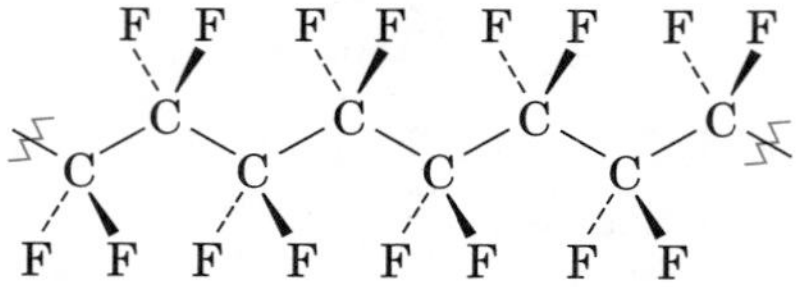

4.12 2-Methyl-2-butene (major) and 2-methyl-1-butene

4.13 (a) 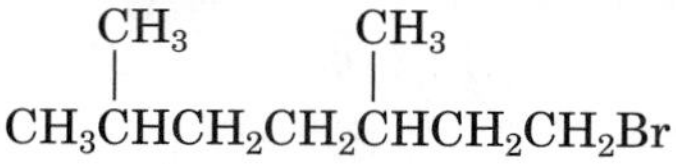 $CH_3CH(CH_3)CH_2CH_2CH(CH_3)CH_2CH_2Br$ (b) 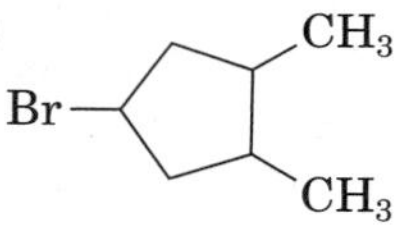 Br-cyclopentane with CH_3, CH_3

4.14 (a) 2-Methyl-2-pentene (major) and 2-methyl-1-pentene
(b) 2,3-Dimethyl-2-pentene (major) plus three others

4.15 (a) 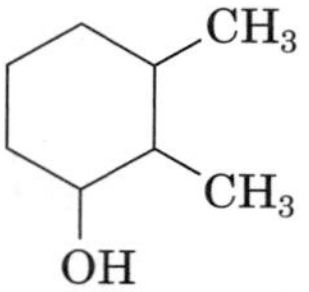 cyclohexane with CH_3, CH_3, OH (b) 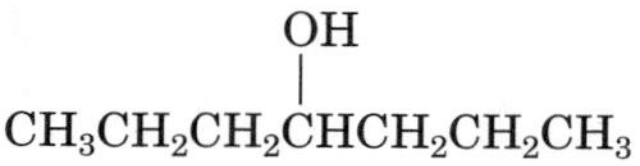 $CH_3CH_2CH_2CH(OH)CH_2CH_2CH_3$

4.16 1,4-Dibromo-2-butene and 3,4-dibromo-1-butene **4.17** Four products

4.18 $CH_3CH{=}CH\overset{+}{C}H{-}CH_3$

4.19 (a) 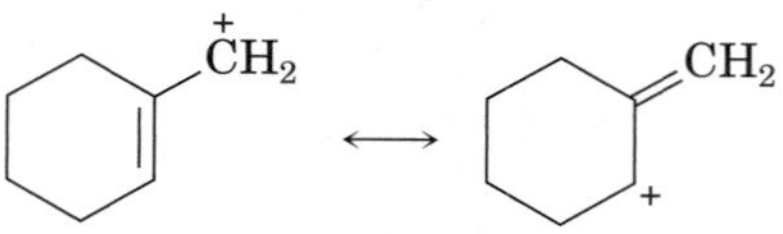 cyclohexenyl $\overset{+}{C}H_2$ $\longleftrightarrow$ $=CH_2$ cyclohexyl cation (+) (b) $CH_3-C(=\ddot{O}:)-\ddot{C}H_2^{-}$ $\longleftrightarrow$ $CH_3-C(-\ddot{O}:^{-})=CH_2$

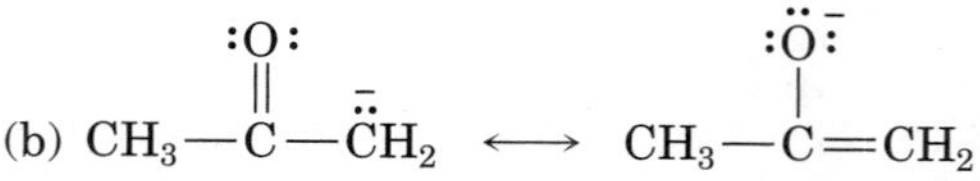

(c) + $\longleftrightarrow$ + $\longleftrightarrow$ +

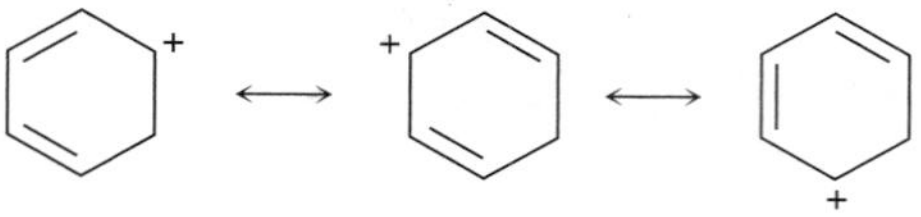

4.20 (a) 6-Methyl-3-heptyne (b) 3,3-Dimethyl-1-butyne (c) 5-Methyl-2-hexyne
(d) 2-Hepten-5-yne
4.21 (a) 1,2-Dichloro-1-pentene (b) 4-Bromo-3-heptene and 3-bromo-3-heptene
(c) *cis*-6-Methyl-3-heptene
4.22 4-Octanone **4.23** (a) 1-Pentyne (b) 3-Hexyne
4.24 (a) 1-Bromo-3-methylbutane + acetylene
(b) 1-Bromopropane + 1-propyne or bromomethane + 1-pentyne
(c) Bromomethane + 3-methyl-1-butyne

Chapter 5

5.1 Br Br Br Br Br Br Br Br

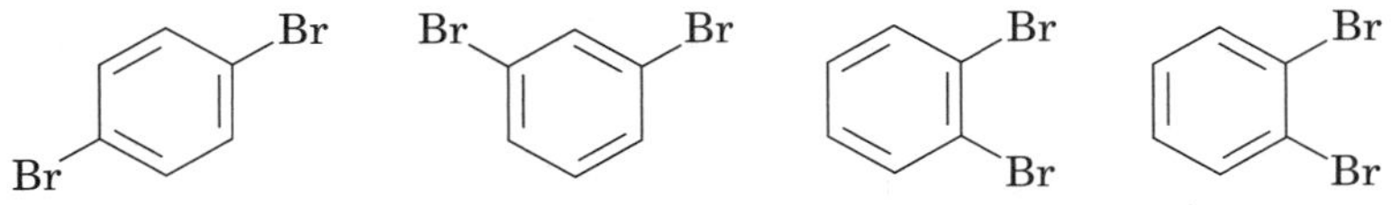

5.2 Two Kekulé structures are resonance forms. **5.3** (a) meta (b) para (c) ortho
5.4 (a) *m*-Bromochlorobenzene (b) Isobutylbenzene (c) *p*-Bromoaniline
5.6 *o*-, *m*-, and *p*-bromotoluene
5.7 H $\overset{+}{N}O_2$ ⟶ [NO_2 H +] ⟶ NO_2 + H^+

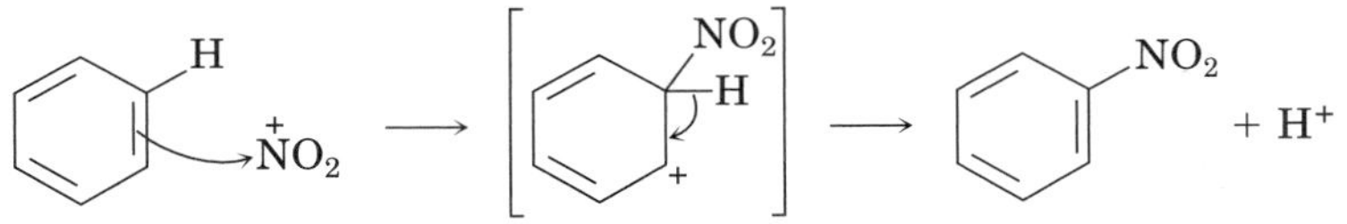

Carbocation intermediate

5.8 *p*-Xylene has one kind of ring position; *o*-xylene has two. **5.9** Three
5.10 Electrophilic substitution of D^+ for H^+
5.11 (a) Ethylbenzene (b) 1-Ethyl-2,5-dimethylbenzene
5.12 (a) *tert*-Butylbenzene (b) Propanoylbenzene, $C_5H_5COCH_2CH_3$
5.13 (a) Nitrobenzene < Toluene < Phenol
(b) Benzoic acid < Chlorobenzene < Benzene < Phenol
(c) Benzaldehyde < Bromobenzene < Benzene < Aniline
5.14 (a) *m*-Dinitrobenzene (b) *o*- and *p*-Bromonitrobenzene (c) *o*- and *p*-Nitrotoluene
(d) *m*-Nitrobenzoic acid (e) 1,4-Dimethyl-2-nitrobenzene
5.17 (a) *m*-Chlorobenzoic acid (b) *o*-Benzenedicarboxylic acid

5.18 ⟷ ⟷

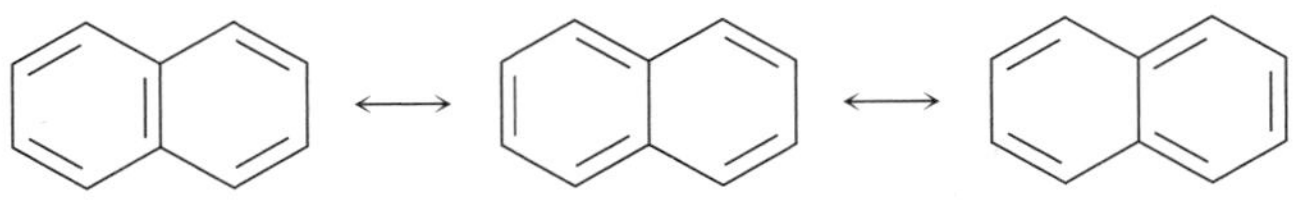

5.19 (a) **1.** CH_3Cl, $AlCl_3$; **2.** CH_3COCl, $AlCl_3$ (b) **1.** Cl_2, $FeCl_3$; **2.** HNO_3, H_2SO_4
5.20 (a) **1.** Br_2, $FeBr_3$; **2.** CH_3Cl, $AlCl_3$ (b) **1.** 2 CH_3Cl, $AlCl_3$; **2.** Br_2, $FeBr_3$
5.21 **1.** CH_3Cl, $AlCl_3$; **2.** $KMnO_4$, H_2O; **3.** Cl_2, $FeCl_3$

Chapter 6

6.1 Chiral: screw, beanstalk, shoe **6.2** Chiral: (b), (c) **6.3** Chiral: (b)
6.4 (a) N CH_3 N * (b) O CH_3 * (c) H_3C CH_3 * CH_3 O *

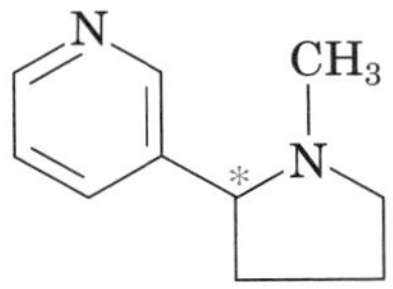

Nicotine **Muscone (musk oil)** **Camphor**

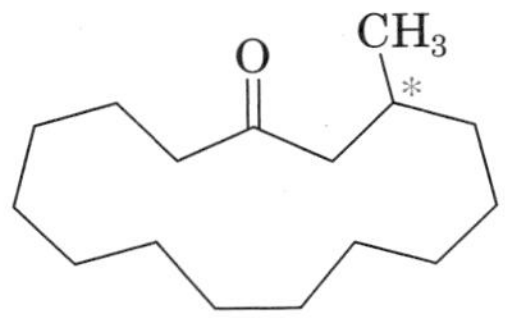

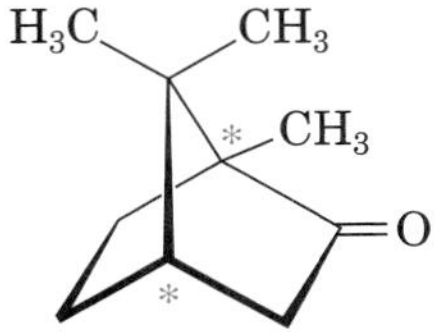

6.5 COOH / C / H, CH$_3$, NH$_2$ and COOH / C / H$_3$C, H, NH$_2$

6.6 Levorotatory **6.7** +16.1°

6.8 (a) –Br, –CH_2CH_2OH, –CH_2CH_3, –H (b) –OH, –CO_2CH_3, –CO_2H, –CH_2OH
(c) –Br, –Cl, –CH_2Br, –CH_2Cl

6.9 (a) *S* (b) *S* (c) *R*

6.10 H / C / HO, $CH_2CH_2CH_3$, CH_3

6.11 (a) *R, R* (b) *S, R* (c) *R, S*

6.12 Molecules (b) and (c) are enantiomers (mirror images). Molecule (a) is the diastereomer of (b) and (c).

6.13 *R, R* **6.14** Meso: (a) and (c) **6.15** Meso: (a) and (c)

6.16 Six stereocenters; 64 stereoisomers **6.17** The product is the pure *S* ester.

6.18 (a) Constitutional isomers (b) Diastereomers

Chapter 7

7.1 (a) 2-Bromobutane (b) 3-Chloro-2-methylpentane (c) 1-Chloro-3-methylbutane
(d) 1,3-Dichloro-3-methylbutane (e) 1-Bromo-4-chlorobutane
(f) 4-Bromo-1-chloropentane

7.2 (a) $CH_3CH_2CH_2C(CH_3)_2CH(Cl)CH_3$ (b) $CH_3CH_2CH_2C(Cl)_2CH(CH_3)_2$
(c) $CH_3CH_2C(Br)(CH_2CH_3)_2$ (d) $CH_3CH(Cl)CH_2CH(CH_3)CH(Br)CH_3$

7.3 1-Chloro-3-methylpentane, 2-chloro-3-methylpentane, 3-chloro-3-methylpentane, 3-(chloromethyl)pentane. The first two are chiral.

7.4 All hydrogens in 2,2-dimethylpropane are equivalent, so only one monochloro product can form.

7.5 (a) 2-Methyl-2-propanol + HCl (b) 4-Methyl-2-pentanol + PBr_3
(c) 5-Methyl-1-hexanol + PBr_3 (d) 2,4-Dimethyl-2-hexanol + HCl

7.6 (a) 4-Bromo-2-methylhexane (b) 1-Chloro-1-methylcyclohexane
(c) 1-Chloro-3,3-dimethylcyclopentane

7.7 React the halide with Mg, and then treat the Grignard reagent with D_2O.

7.8 **1.** PBr_3; **2.** Mg, ether; **3.** H_2O

7.9 (a) 2-Iodobutane (b) $(CH_3)_2CHCH_2SH$ (c) $C_6H_5CH_2CN$

7.10 (a) 1-Bromobutane + NaOH (b) 1-Bromo-3-methylbutane + NaN_3

7.11 (a) Rate is tripled. (b) Rate is quadrupled.

7.12 (*R*)-1-Methylpentyl acetate, $CH_3CO_2CH(CH_3)CH_2CH_2CH_2CH_3$

7.13 The product is racemic when 50% of the starting material has reacted.

7.14 (a) Reaction with $CH_3CH_2CH_2Br$ is faster.
(b) Reaction with $(CH_3)_2CHCH_2Cl$ is faster.

7.15 $CH_3I > CH_3Br > CH_3F$

7.16 (a) Rate is unchanged. (b) Rate is doubled.

7.17 Racemic 3-bromo-3-methyloctane

7.18 (a) 2-Methyl-2-pentene (b) 2,3,5-Trimethyl-2-hexene (c) [cyclohexane ring]$=CHCH_3$

7.19 (*Z*)-1-Bromo-1,2-diphenylethylene

7.20 Rate is tripled.

7.21 (a) S_N2 (b) E2 (c) S_N1

Chapter 8

8.1 (a) 5-Methyl-2,4-hexanediol (b) 2-Methyl-4-phenyl-2-butanol
(c) 4,4-Dimethylcyclohexanol (d) *trans*-2-Bromocyclopentanol
8.2 Secondary: (a), (c), (d); tertiary: (b)
8.4 (a) Diisopropyl ether (b) Cyclopentyl propyl ether
(c) *p*-Bromoanisole or 4-bromo-1-methoxybenzene (d) Ethyl isobutyl ether
8.5 (a) Methanol < *p*-Methylphenol < Phenol < *p*-Nitrophenol
(b) Benzyl alcohol < *p*-Methoxyphenol < *p*-Bromophenol < 2,4-Dibromophenol
8.7 (a) $NaBH_4$ (b) $LiAlH_4$
8.8 (a) C_6H_5CHO, C_6H_5COOH, $C_6H_5CO_2R$ (b) $C_6H_5COCH_3$ (c) Cyclohexanone
8.10 (a) $CH_3CH_2CH_2O^- + CH_3Br$ (b) $C_6H_5O^- + CH_3Br$
(c) $(CH_3)_2CHO^- + C_6H_5CH_2Br$
8.11 (a) Bromoethane > Chloroethane > 2-Bromopropane > 2-Chloro-2-methylpropane
8.12 (a) 1-Phenylethanol (b) 2-Methyl-1-propanol (c) Cyclopentanol
8.13 (a) Cyclohexanone (b) Hexanoic acid (c) 2-Hexanone
8.14 (a) Cyclohexanone (b) Hexanal (c) 2-Hexanone
8.15 (a) **1.** CH_3Cl, $AlCl_3$; **2.** SO_3, H_2SO_4; **3.** NaOH, heat
8.16 (a) Ethanol + iodoethane (b) Cyclohexanol + iodoethane
(c) 2-Iodo-2-methylpropane + ethanol
8.17 *cis*-2,3-Epoxybutane **8.18** *trans*-2,3-Epoxybutane
8.19 The product is a racemic mixture of *R,R* and *S,S* diols.
8.20 The product is a meso diol.
8.21 (a) 2-Butanethiol (b) 2,2,6-Trimethyl-4-heptanethiol (c) 3-Cyclopentene-1-thiol
8.22 (a) Ethyl methyl sulfide (b) *tert*-Butyl ethyl sulfide (c) *o*-Di(methylthio)benzene
8.23 (a) **1.** HBr; **2.** $Na^+ \ ^-SH$ (b) **1.** $LiAlH_4$; **2.** PBr_3; **3.** $Na^+ \ ^-SH$

Chapter 9

9.1 (a) 2-Pentanone (b) $CH_3CH_2CH_2CH{=}CHCHO$ (c) $CH_3CH_2COCH_2CH_2CHO$
(d) Cyclopentanone
9.2 (a) 2-Methyl-3-pentanone (b) 3-Phenylpropanal (c) 2,6-Octanedione
(d) *trans*-2-Methylcyclohexanecarbaldehyde (e) Pentanedial
(f) *cis*-2,5-Dimethylcyclohexanone
9.4 (a) PCC (b) **1.** $LiAlH_4$; **2.** PCC (c) **1.** $KMnO_4$; **2.** $LiAlH_4$; **3.** PCC
9.5 (a) PCC (b) H_3O^+, $HgSO_4$ (c) $KMnO_4$, H_3O^+
9.6 (a) **1.** H_3O^+; **2.** PCC (b) **1.** CH_3COCl, $AlCl_3$; **2.** $NaBH_4$
9.7 (a) Pentanoic acid (b) 2,2-Dimethylhexanoic acid (c) No reaction
9.8 $(CH_3)_2C(OH)CN$ **9.10** Propanal is more reactive because it is less hindered.
9.11 $CCl_3CH(OH)_2$ **9.12** Labeled water adds reversibly to the carbonyl group.
9.13

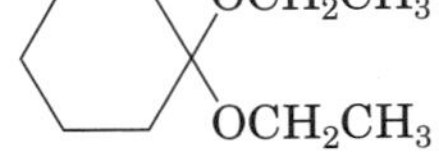

9.15 **1.** $HOCH_2CH_2OH$, H^+ catalyst; **2.** $LiAlH_4$; **3.** H_3O^+
9.16 (a) Cyclohexanone oxime (b) Cyclohexanone 2,4-DNP (c) Cyclohexanol
9.17 (a) 1-Methylcyclopentanol (b) 1,1-Diphenylethanol (c) 3-Methyl-3-hexanol
9.18 (a) Acetone + CH_3MgBr (b) Cyclohexanone + CH_3MgBr
(c) 3-Pentanone + CH_3MgBr or 2-butanone + CH_3CH_2MgBr

Chapter 10

10.1 (a) 3-Methylbutanoic acid (b) 4-Bromopentanoic acid (c) 4-Hexenoic acid
(d) 2-Ethylpentanoic acid (e) *trans*-2-Methylcyclohexanecarboxylic acid

10.3 (a) 4-Methylpentanoyl chloride (b) 2-Methylbutanenitrile (c) 4-Pentenamide (d) 2-Ethylbutanenitrile (e) Cyclopentyl 2,2-dimethylpropanoate (f) 2,3-Dimethyl 2-butenoyl chloride (g) Benzoic anhydride (h) Isopropyl cyclopentanecarboxylate

10.5 (a) $C_6H_5CO_2^-\ ^+Na$ (b) $(CH_3)_3CCO_2P^-\ ^+K$

10.6 Methanol < Phenol < *p*-Nitrophenol < Acetic acid < Sulfuric acid

10.7 (a) $CH_3CH_2COOH < BrCH_2COOH < BrCH_2CH_2COOH$ (b) Ethanol < Benzoic acid < *p*-Cyanobenzoic acid

10.8 (a) C_6H_5COOH (b) $(CH_3)_2CHCH_2CH_2COOH$

10.9 **1.** NaCN; **2.** NaOH, H_2O. Iodobenzene cannot be converted to benzoic acid by this method.

10.10 **1.** Mg, ether; **2.** CO_2; **3.** H_3O^+. Iodomethane can be converted to acetic acid by this method.

10.11 (a) CH_3COCl (b) $CH_3CH_2CO_2CH_3$ (c) $CH_3CO_2COCH_3$ (d) $CH_3CO_2CH_3$

10.12 The electron-withdrawing trifluoromethyl group polarizes the carbonyl carbon.

10.13 (a) C_6H_5COCl (b) $C_6H_5CO_2CH_3$ (c) $C_6H_5CH_2OH$ (d) $C_6H_5CO_2^-\ ^+Na$

10.14 (a) Acetic acid + 1-butanol (b) Butanoic acid + methanol

10.15 [structure: six-membered lactone ring, C=O and ring O]
O
O

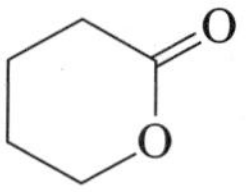

10.16 (a) Propanoyl chloride + methanol (b) Acetyl chloride + ethanol (c) Acetyl chloride + cyclohexanol

10.18 (a) Propanoyl chloride + NH_3 (b) 3-Methylbutanoyl chloride + CH_3NH_2 (c) Propanoyl chloride + $(CH_3)_2NH$ (d) Benzoyl chloride + diethylamine

10.20 [structure: benzene ring with ortho substituents]
O ‖ $COCH_3$
COH ‖ O

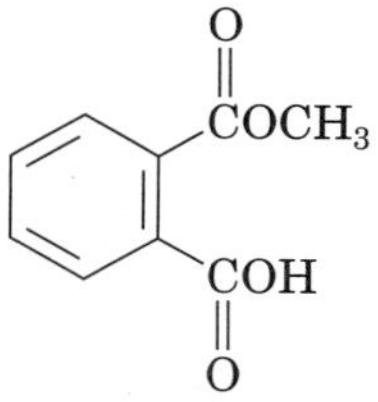

10.21 (a) Isopropyl alcohol + acetic acid (b) Methanol + cyclohexanecarboxylic acid

10.22 Reaction of an acid with an alkoxide ion gives the unreactive carboxylate ion.

10.23 (a) $CH_3CH_2CH_2CH(CH_3)CH_2OH$ (b) $C_6H_5OH + C_6H_5CH_2OH$

10.24 $HOCH_2CH_2CH_2CH_2OH$

10.25 (a) Ethyl benzoate + 2 CH_3MgBr (b) Ethyl acetate + 2 C_6H_5MgBr (c) Ethyl pentanoate + 2 CH_3CH_2MgBr

10.26 (a) H_2O, NaOH (b) **1.** H_2O, NaOH; **2.** $LiAlH_4$ (c) $LiAlH_4$

10.27 [structure: pyrrolidine ring]
H_3C
H_3C N
H

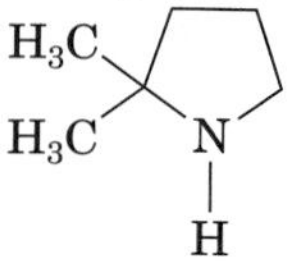

10.28 (a) $CH_3CH_2CN + CH_3CH_2MgBr$
(b) $CH_3CH_2CN + (CH_3)_2CHMgBr$ or $(CH_3)_2CHCN + CH_3CH_2MgBr$
(c) $C_6H_5CN + CH_3MgBr$ or $CH_3CN + C_6H_5MgBr$
(d) $C_6H_{11}CN + C_6H_{11}MgBr$

10.29 (a) **1.** $Na^+\ ^-CN$; **2.** $LiAlH_4$ (b) **1.** $Na^+\ ^-CN$; **2.** CH_3CH_2MgBr

10.30 [polymer repeat unit]
$\left(-\overset{O}{\overset{\|}{C}}-C_6H_4-\overset{O}{\overset{\|}{C}}-NH-C_6H_4-NH- \right)_n$

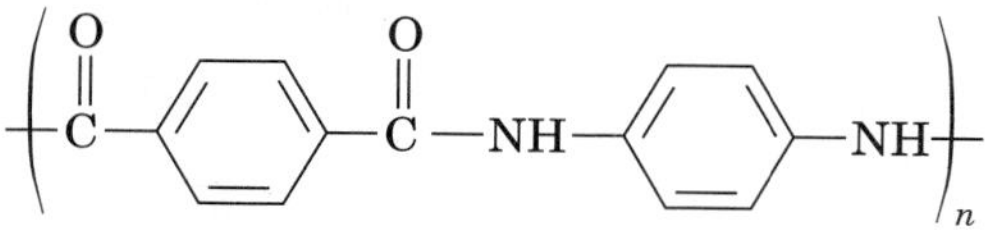

Chapter 11

11.2 (a) 4 (b) 3 (c) 3 (d) 4 (e) 3

11.3 Enolization can occur in either direction from the carbonyl group.

11.4 (a)

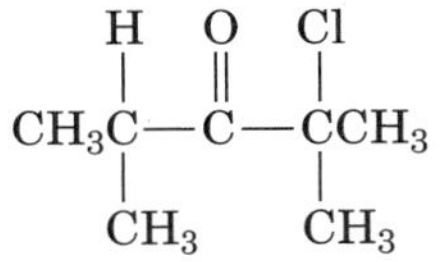

H O Cl

$CH_3C-C-CCH_3$

CH_3 CH_3

(b) 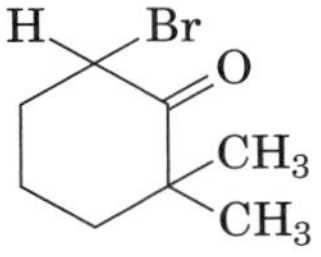 H, Br, O, CH_3, CH_3 (cyclohexanone ring)

11.5 **1.** Br_2; **2.** Pyridine, heat **11.6** The enol intermediate is achiral.

11.7 (a) $CH_3\mathbf{CH_2}CHO$ (b) $(CH_3)_3CCOC\mathbf{H_3}$ (c) $\mathbf{CH_3}COO\mathbf{H}$ (d) $CH_3CH_2\mathbf{CH_2}C{\equiv}N$

11.8 (a) $CH_3CH_2CH{=}CH(OH)$ (b) $CH_3CH{=}C(OH)CH_3$ and $CH_3CH_2C(OH){=}CH_2$

11.10 (a) CH_3CH_2Br (b) $C_6H_5CH_2Br$ (c) $(CH_3)_2CHCH_2CH_2Br$

11.11 (a) Alkylate with $(CH_3)_2CHCH_2Br$.

(b) Alkylate first with $CH_3CH_2CH_2Br$ and then with CH_3Br.

11.12 Only (a) undergoes an aldol reaction.

11.13 (a)

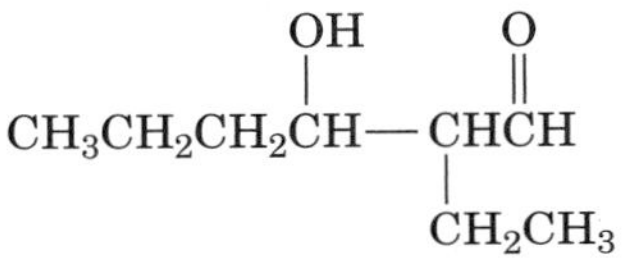

OH O

$CH_3CH_2CH_2CH-CHCH$

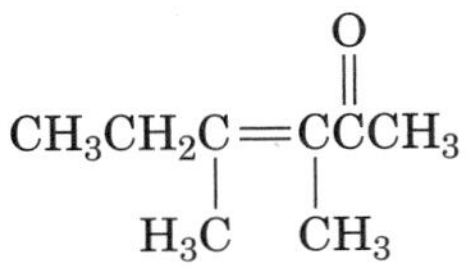

CH_2CH_3

(b) 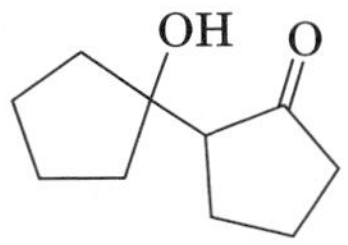 OH O (cyclopentane rings)

(c)

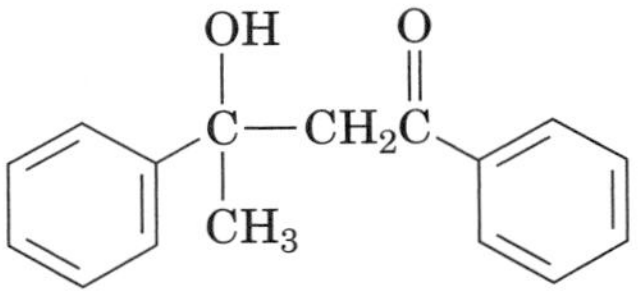

OH O

$C-CH_2C$

CH_3

(with phenyl rings)

O O

11.15 $CH_3CH_2C{=}CCCH_3$ and $CH_3CH_2C{=}CHCCH_2CH_3$

H_3C CH_3 CH_3

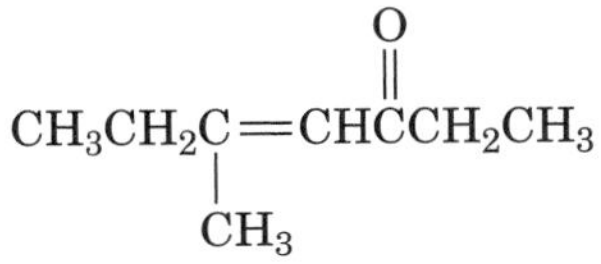

11.16 Only (c) undergoes a Claisen reaction.

Chapter 12

12.1 (a) Primary (b) Secondary (c) Tertiary (d) Quaternary

12.3 (a) *N*-Methylethylamine (b) Tricyclohexylamine (c) *N*-Methylpyrrole

(d) *N*-Methyl-*N*-propylcyclohexylamine (e) 1,3-Butanediamine

12.5 *p*-Methylaniline is stronger.

12.6 *N*-Methylcyclopentylammonium bromide

12.7 (a) $CH_3CH_2NH_2$ (b) NaOH (c) CH_3NHCH_3 (d) $(CH_3)_3N$

12.8 (a) 3 $CH_3CH_2Br + NH_3$ (b) 4 $CH_3Br + NH_3$

12.9 (a) Propanamide (b) *N*-Propylpropanamide (c) Benzamide

12.10 (a) 3-Methylbutanenitrile (b) Benzonitrile

12.11 (a) **1.** CH_3Cl, $AlCl_3$; **2.** $KMnO_4$, H_2O; **3.** HNO_3, H_2SO_4; **4.** H_2, Pt catalyst

(b) **1.** HNO_3, H_2SO_4; **2.** H_2/Pt catalyst; **3.** 3 Br_2

12.12 $CH_3COCl + (CH_3CH_2)_2NH \rightarrow CH_3CON(CH_2CH_3)_2$

12.13 **1.** Br_2, $FeBr_3$; **2.** HNO_3, H_2SO_4; **3.** Fe, H_3O^+; **4.** $NaNO_2$, H_2SO_4; **5.** CuCN

12.14 (a) **1.** CH_3Cl, $AlCl_3$; **2.** $KMnO_4$, H_2O; **3.** Br_2, $FeBr_3$

(b) **1.** HNO_3, H_2SO_4; **2.** Br_2, $FeBr_3$; **3.** Fe, H_3O^+; **4.** $NaNO_2$, H_2SO_4; **5.** CuCl

12.17 (a) *N*-Methyl-2-bromopyrrole (b) *N*-Methyl-2-methylpyrrole

(c) *N*-Methyl-2-acetylpyrrole

12.19

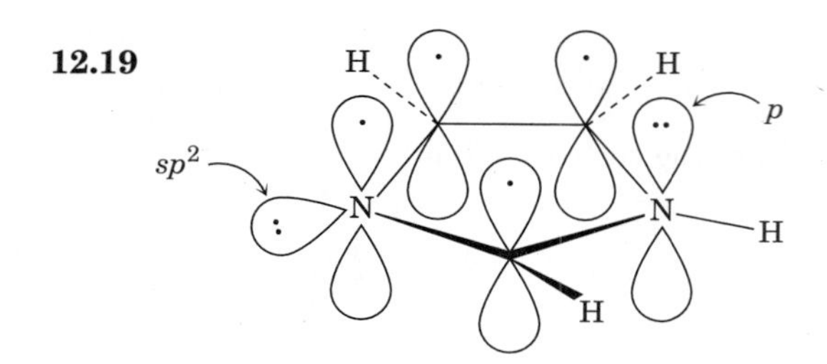

Chapter 13

13.1 IR: $\varepsilon = 2 \times 10^{-19}$ J; X ray: $\varepsilon = 7 \times 10^{-17}$ J
13.2 $\lambda = 9.0 \times 10^{-4}$ cm is higher in energy.
13.3 (a) Ketone or aldehyde (b) Nitro (c) Nitrile or alkyne (d) Carboxylic acid (e) Alcohol and ester
13.4 (a) CH_3CH_2OH has an –OH absorption.
(b) 1-Hexene has a double-bond absorption.
(c) Propanoic acid has a very broad –OH absorption.
13.5 (a), (c), (d), and (f) have UV absorptions.
13.6 1,3,5-Hexatriene absorbs at a longer wavelength.
13.7 The energy used by NMR spectroscopy is less than that used by IR spectroscopy.
13.8 (a) ^{1}H, 1; ^{13}C, 1 (b) ^{1}H, 1; ^{13}C, 1 (c) ^{1}H, 2; ^{13}C, 2 (d) ^{1}H, 1; ^{13}C, 1 (e) ^{1}H, 1; ^{13}C, 1 (f) ^{1}H, 1; ^{13}C, 1 (g) ^{1}H, 2; ^{13}C, 2 (h) ^{1}H, 2; ^{13}C, 2 (i) ^{1}H, 1; ^{13}C, 2
13.9 The vinylic C–H protons are nonequivalent.
13.10 (a) 126 Hz (b) 2.1 δ (c) 210 Hz
13.11 (a) 7.27 δ (b) 3.05 δ (c) 3.47 δ (d) 5.30 δ
13.12 (a) 0.88 δ (b) 2.17 δ (c) 7.17 δ (d) 2.22 δ
13.13 Two peaks; 3:2 ratio
13.14 (a) Doublet and 10-line multiplet (b) Doublet and quartet (c) Singlet and two triplets (d) Singlet, triplet, and quartet (e) Triplet and quintet (f) Singlet, doublet, and septet
13.15 (a) CH_3OCH_3 (b) $CH_3CO_2CH_3$ (c) $(CH_3)_2CHCl$
13.16 (a) 1 (b) 5 (c) 4 (d) 7
13.17 (a) 1-Heptene (b) 2-Methylpentane (c) 1-Chloro-2-methylpropane

Chapter 14

14.1 (a) Aldotetrose (b) Ketopentose (c) Ketohexose (d) Aldopentose

14.2

CHO
HO—+—H
CH_2OH

14.3

H
H_3C—+—Cl
CH_2CH_3
R

H
Cl—+—CH_3
CH_2CH_3
S

14.4 (a) *S* (b) *R* (c) *S* **14.5** (a) L (b) D (c) D
14.8 There are 16 D- and 16 L-aldoheptoses.

14.10 CH_2OH, OH, O, OH, OH, OH

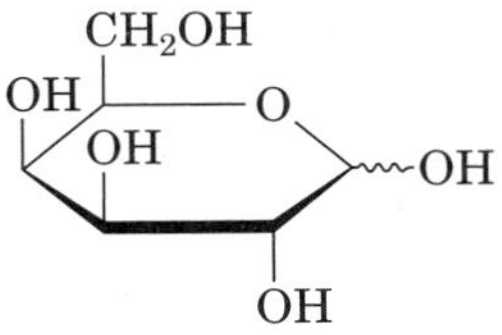

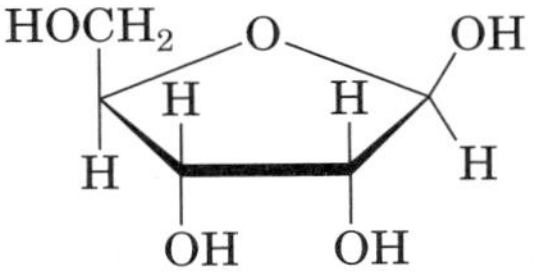

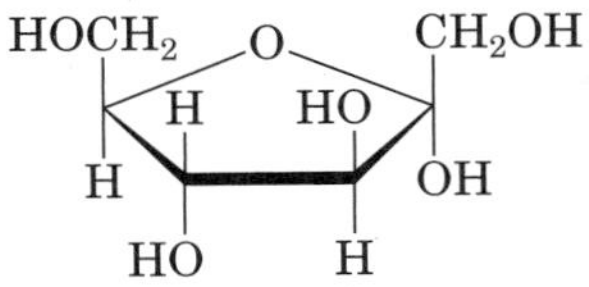

14.11 $HOCH_2$, O, OH, H, H, H, H, OH, OH

14.12 $HOCH_2$, O, CH_2OH, H, HO, H, OH, HO, H

α-D-Fructofuranose

$HOCH_2$, O, OH, H, HO, H, CH_2OH, HO, H

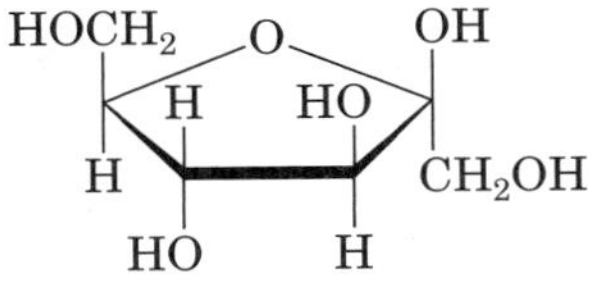

β-D-Fructofuranose

14.13 36% α anomer; 64% β anomer
14.14 28% α anomer; 72% β anomer

14.15 HO, CH_2OH, O, HO, OH, OH

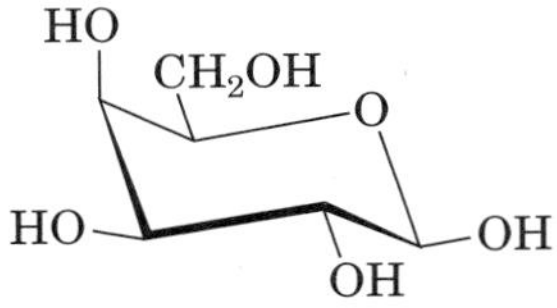

14.16

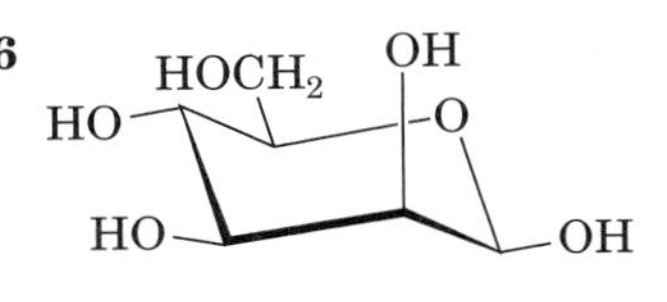

14.19 D-Galactitol is a meso compound.
14.20 An alditol has a $-CH_2OH$ group at both ends; either could have been a $-CHO$ group in the parent sugar.
14.21 D-Allaric acid is a meso compound; D-glucaric acid is not.
14.22 D-Allose and D-galactose yield meso aldaric acids; the other six D-aldohexoses yield optically active aldaric acids.

Chapter 15

15.1 Aromatic: Phe, Tyr, Trp, His; sulfur-containing: Cys, Met; alcohols: Ser, Trp; hydrocarbon side chains: Ala, Ile, Leu, Val
15.2 The sulfur atom in the $-CH_2SH$ group of cysteine makes the side chain higher in priority than the $-COOH$ group.

15.4 $C_6H_5-CH_2CH(^+NH_3)C(=O)-O^-$

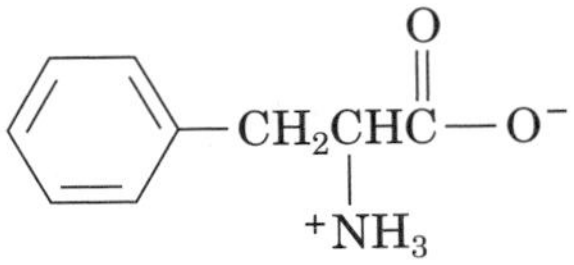

15.6 (a) $H_3\overset{+}{N}CH_2CH_2CH_2CH_2CH(^+NH_3)COOH$ (b) $^-OOCCH_2CH(^+NH_3)COO^-$

(c) $H_2NCH_2CH_2CH_2CH_2CH(NH_2)COO^-$ (d) $CH_3CH(^+NH_3)COO^-$

15.7 (a) Toward (+): Glu > Val; toward (−): none (b) Toward (+): Phe; toward (−): Gly
(c) Toward (+): Phe > Ser; toward (−): none
15.9 Val-Tyr-Gly, Tyr-Gly-Val, Gly-Val-Tyr, Val-Gly-Tyr, Tyr-Val-Gly, Gly-Tyr-Val
15.12 Trypsin: Asp-Arg + Val-Tyr-Ile-His-Pro-Phe;
Chymotrypsin: Asp-Arg-Val-Tyr + Ile-His-Pro-Phe
15.13 Arg-Pro-Leu-Gly-Ile-Val **15.14** The tripeptide is cyclic.
15.17 (a) Lyase (b) Hydrolase (c) Oxidoreductase

Chapter 16

16.1 $CH_3(CH_2)_{20}COOH + HO(CH_2)_{27}CH_3$
16.2 Glyceryl monooleate distearate is higher melting.
16.3 The fat molecule with stearic acid esterified to the central –OH group of glycerol is optically inactive.
16.4 $[CH_3(CH_2)_7CH{=}CH(CH_2)_7COO^-]_2\ Mg^{2+}$
16.6 Equatorial hydroxyl **16.7** Two ketones, double bond
16.8 Both have an aromatic ring. **16.11** C-C-G-A-T-T-A-G-G-C-A
16.12

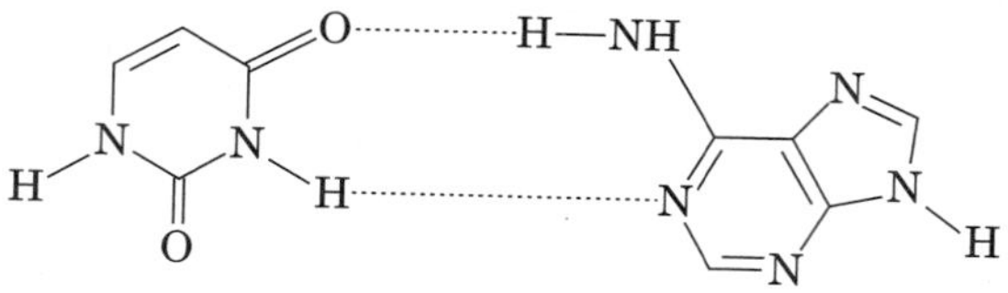

Uracil **Adenine**

16.13 C-U-A-A-U-G-G-C-A-U **16.14** A-A-G-C-G-T-C-T-C-A
16.15 (a) GCU, GCC, GCA, GCG (b) UUU, UUC
(c) UUA, UUG, CUU, CUC, CUA, CUG (d) UAU, UAC
16.16 Leu-Met-Ala-Trp-Pro-Stop **16.17** GAA-UAC-CGA-ACC-GGG-AUU
16.18 GAA-TAC-CGA-ACC-GGG-ATT
16.19 A cleavage: ^{32}P-A, ^{32}P-A-A-C, ^{32}P-A-A-C-A-T-G-G-C-G-C-T-T, ^{32}P-A-A-C-A-T-G-G-C-G-C-T-T-A-T-G, ^{32}P-A-A-C-A-T-G-G-C-G-C-T-T-A-T-G-A-C-G
G cleavage: ^{32}P-A-A-C-A-T, ^{32}P-A-A-C-A-T-G, ^{32}P-A-A-C-A-T-G-G-C
^{32}P-A-A-C-A-T-G-G-C-G-C-T-T-A-T, ^{32}P-A-A-C-A-T-G-G-C-G-C-T-T-A-T-G-A-C
C cleavage: ^{32}P-A-A, ^{32}P-A-A-C-A-T-G-G, ^{32}P-A-A-C-A-T-G-G-C-G
^{32}P-A-A-C-A-T-G-G-C-G-C-T-T-A-T-G-A
C + T cleavage: ^{32}P-A-A, ^{32}P-A-A-C-A, ^{32}P-A-A-C-A-T-G-G, ^{32}P-A-A-C-A-T-G-G-C-G, ^{32}P-A-A-C-A-T-G-G-C-G-C, ^{32}P-A-A-C-A-T-G-G-C-G-C-T, ^{32}P-A-A-C-A-T-G-G-C-G-C-T-T-A, ^{32}P-A-A-C-A-T-G-G-C-G-C-T-T-A-T-G-A
16.21 T-C-G-G-T-A-C

Chapter 17

17.1 $HOCH_2CH(OH)CH_2OH + ATP \rightarrow HOCH_2CH(OH)CH_2OPO_3^{2-} + ADP$
17.2 The reaction is a nucleophilic acyl substitution reaction at the phosphorus atom.
17.4 (a) 8 acetyl CoA; 7 passages (b) 10 acetyl CoA; 9 passages
17.5 Steps 7 and 10
17.6 Keto–enol tautomerization through a common enol.
17.7 Citrate and isocitrate
17.8 E2 elimination of water, followed by conjugate addition
17.9 $(CH_3)_2CHCH_2COCO_2^-$

Index

References in boldface type refer to pages where terms are defined.

Credits

Project Development Editor: Beth Wilbur
Developmental Editor: Keith Dodson
Editorial Assistants: Leigh Hamilton, Georgia Jurickovich
Marketing Team: Christine Davis, Caroline Croley, Michele Mootz
Production Service: Phyllis Niklas
Production Editor: Jamie Sue Brooks
Manuscript Editor: Phyllis Niklas
Art Coordination/Technical Illustration: Dovetail Publishing Services
Chapter opening illustrations: Kenneth Eward, BioGrafx
Interior Design: Nancy Benedict
Cover Design: Vernon T. Boes
Cover Photo: IFA/West Stock
Photo Researcher: Stuart Kenter Associates
Type Composition: York Graphic Services
Cover Printing: Phoenix Color Corp.
Printing and Binding: R. R. Donnelley & Sons, Willard

Photo Credits

Page 29: Tom McCarthy/SKA
Page 68: Tenneco, Inc.
Page 100: François Gohier/Photo Researchers, Inc.
Page 140: McGlynn/The Image Works
Page 175: David Madison
Page 209: Shelby Thorner
Page 243: Louisa Preston
Page 272: Leslie O'Shaughnessy Studios
Page 300: Mary M. Thacher/Photo Researchers, Inc.
Page 342: Andrew McClenaghan/Photo Researchers, Inc.
Page 372: "L'Ortolano" by Giuseppe Arcimboldo/Cremona, Museo Civico/Art Resource
Page 398: Timothy Ross/The Image Works
Page 432: Photo Researchers, Inc.
Page 468: Tom McCarthy/SKA
Page 500: B. Daemmrich/The Image Works
Page 538: Visuals Unlimited
Page 562: P. Pavani/AFP

Periodic Table of the Elements

Atomic number	Name	Symbol	Atomic weight
22	Titanium	Ti	47.90

[a]Mass number of most stable or best-known isotope

[b]Mass of the isotope of longest half-life

Atomic weights are based on carbon-12. Atomic weights in parentheses indicate the most stable or best-known isotope.

Period	1A	2A	3B	4B	5B	6B	7B	8B	8B	8B	1B	2B	3A	4A	5A	6A	7A	8A
1	1 Hydrogen **H** 1.0079																	2 Helium **He** 4.00260
2	3 Lithium **Li** 6.941	4 Beryllium **Be** 9.01218											5 Boron **B** 10.81	6 Carbon **C** 12.011	7 Nitrogen **N** 14.0067	8 Oxygen **O** 15.9994	9 Fluorine **F** 18.99840	10 Neon **Ne** 20.179
3	11 Sodium **Na** 22.98977	12 Magnesium **Mg** 24.305											13 Aluminum **Al** 26.98154	14 Silicon **Si** 28.086	15 Phosphorus **P** 30.97376	16 Sulfur **S** 32.06	17 Chlorine **Cl** 35.453	18 Argon **Ar** 39.948
4	19 Potassium **K** 39.098	20 Calcium **Ca** 40.08	21 Scandium **Sc** 44.9559	22 Titanium **Ti** 47.90	23 Vanadium **V** 50.9414	24 Chromium **Cr** 51.996	25 Manganese **Mn** 54.9380	26 Iron **Fe** 55.847	27 Cobalt **Co** 58.9332	28 Nickel **Ni** 58.69	29 Copper **Cu** 63.546	30 Zinc **Zn** 65.38	31 Gallium **Ga** 69.72	32 Germanium **Ge** 72.61	33 Arsenic **As** 74.9216	34 Selenium **Se** 78.96	35 Bromine **Br** 79.904	36 Krypton **Kr** 83.80
5	37 Rubidium **Rb** 85.4678	38 Strontium **Sr** 87.62	39 Yttrium **Y** 88.9059	40 Zirconium **Zr** 91.22	41 Niobium **Nb** 92.9064	42 Molybdenum **Mo** 95.94	43 Technetium **Tc** 98.9062[b]	44 Ruthenium **Ru** 101.07	45 Rhodium **Rh** 102.9055	46 Palladium **Pd** 106.4	47 Silver **Ag** 107.868	48 Cadmium **Cd** 112.40	49 Indium **In** 114.82	50 Tin **Sn** 118.71	51 Antimony **Sb** 121.75	52 Tellurium **Te** 127.60	53 Iodine **I** 126.9045	54 Xenon **Xe** 131.30
6	55 Cesium **Cs** 132.9054	56 Barium **Ba** 137.34	* 57 Lanthanum **La** 138.9055	72 Hafnium **Hf** 178.49	73 Tantalum **Ta** 180.9479	74 Wolfram (Tungsten) **W** 183.85	75 Rhenium **Re** 186.2	76 Osmium **Os** 190.2	77 Iridium **Ir** 192.22	78 Platinum **Pt** 195.09	79 Gold **Au** 196.9665	80 Mercury **Hg** 200.59	81 Thallium **Tl** 204.37	82 Lead **Pb** 207.2	83 Bismuth **Bi** 208.9804	84 Polonium **Po** (210)[a]	85 Astatine **At** (210)[a]	86 Radon **Rn** (222)[a]
7	87 Francium **Fr** (223)[a]	88 Radium **Ra** 226.0254[b]	** 89 Actinium **Ac** (227)[a]	104 Rutherfordium **Rf** (261)[a]	105 Dubnium **Db** (262)[a]	106 Seaborgium **Sg** (263)[a]	107 Bohrium **Bh** (262)	108 Hassium **Hs** (265)	109 Meitnerium **Mt** (268)	110 – (269)	111 – (272)	112 – (277)						

Transition elements: groups 3B–2B

Inner transition elements

Series	Period														
* Lanthanide series	6	58 Cerium **Ce** 140.12	59 Praseodymium **Pr** 140.9077	60 Neodymium **Nd** 144.24	61 Promethium **Pm** (145)[a]	62 Samarium **Sm** 150.4	63 Europium **Eu** 151.96	64 Gadolinium **Gd** 157.25	65 Terbium **Tb** 158.9254	66 Dysprosium **Dy** 162.50	67 Holmium **Ho** 164.9304	68 Erbium **Er** 167.26	69 Thulium **Tm** 168.9342	70 Ytterbium **Yb** 173.04	71 Lutetium **Lu** 174.97
** Actinide series	7	90 Thorium **Th** 232.0381[b]	91 Protactinium **Pa** 231.0359[b]	92 Uranium **U** 238.029	93 Neptunium **Np** 237.0482	94 Plutonium **Pu** (242)[a]	95 Americium **Am** (243)[a]	96 Curium **Cm** (247)[a]	97 Berkelium **Bk** (249)[a]	98 Californium **Cf** (251)[a]	99 Einsteinium **Es** (254)[a]	100 Fermium **Fm** (253)[a]	101 Mendelevium **Md** (256)[a]	102 Nobelium **No** (254)[a]	103 Lawrencium **Lr** (257)[a]